THEORY OF STRUCTURE AND MECHANICS OF YARNS

THEORY OF STRUCTURE AND MECHANICS OF YARNS

Bohuslav Neckář
and
Dipayan Das

WOODHEAD PUBLISHING INDIA PVT LTD

New Delhi

Published by Woodhead Publishing India Pvt. Ltd.
Woodhead Publishing India Pvt. Ltd.,
303, Vardaan House, 7/28, Ansari Road,
Daryaganj, New Delhi - 110002, India
www.woodheadpublishingindia.com

First published 2018, Woodhead Publishing India Pvt. Ltd.
© Woodhead Publishing India Pvt. Ltd., 2018
Reprint, 2020

Woodhead Publishing India Pvt. Ltd. ISBN: 978-93-85059-40-7
Woodhead Publishing India Pvt. Ltd. e-ISBN: 978-93-85059-88-9

Typeset by Allen Smalley, Chennai

Printed and bound in India by Replika Press Pvt. Ltd.

Contents

Preface

THE IMAGINATION IS MORE IMPORTANT
THAN THE KNOWLEDGE

Albert Einstein

There exist many books in the world that deal with manufacturing and properties of different kinds of yarns. Nevertheless, this book is not intended towards such type of publication. It is also not a compilation of several hundred references citing the work of other authors. Of course, the authors' own work is cited in this book, but there is no attempt made to summarize the information and knowledge available in the literature.

Our goal was – above all – to prepare an original scientific book on a very specific but traditional fibrous assembly – single and mostly twisted yarn. Although yarn is perhaps a 27,000-year-old product[1] of human civilization, the technological activities of yarn production are not much explored scientifically. Why is it so? Since their existence, yarns were developed and used purely on an empirical basis. As a result, a high volume of experience was obtained over many years without much scientific research. (The first scientific knowledge on yarn was published as late as sometime between 18th and 19th centuries, but the major scientific studies began from the second half of 20th century.) In the last few decades, many new avenues were opened up for deeper scientific understanding of yarns in the light of newer necessities and possibilities – recent technological innovations, novel end-uses (often called 'technical textiles'), advanced mathematical tools connecting computers, latest laboratory equipments, etc.

This book does not provide a complete recipe on how to carry out different activities in relation to yarn – it is not a 'handbook' of yarn production, properties, and applications. The authors of this book made attempts to scientifically explain why specific behaviours are observed in yarns with a goal to develop a better understanding of our yarns. This is considered to be a typical character of the so-called basic research work. According to the Frascati

---

1   See, for example, the results of archeological research in Czechia stated in the book by Adovasio, J.M., Hyland, D.C. and Soffer, O. Textiles and Cordage: A Preliminary assessment, In: Svoboda, J. (ed.), Pavlov I – Northwest. The Dolní Věstonice Studies, Volume 4, Brno, 1997, pp. 403–424.

Manual 2015[2], 'Basic research is experimental or theoretical work under-taken primarily to acquire new knowledge of the underlying foundations of phenomena and observable facts, without any particular application or use in view… Oriented basic research is carried out with the expectation that it will produce a broad base of knowledge likely to form the basis of the solution to recognized or expected current or future problems or possibilities'.

As in many fields, mathematics creates the basic tool for expressing our scientific understanding on yarn. (Feynman[3] considered mathematics as a language and simultaneously a method of cognition, that is, language and logic together, as an instrument of thinking.) However, the textile specialists are not always having enough experience with mathematical operations. Therefore, in this book, the derivations of mathematical expressions are provided step by step so that the reader, if having less experience in formulation and manipulation of mathematical expressions, can easily follow the text. (In our first book[4], we wrote 'The authors do not like the idiom "The reader can himself easily derive…", the so-called "easy derivation" may represent a work of one month!'.) This results in relatively large number of equations which might cause a repulsive view. Nevertheless, to keep the logical continuity of the text, some special mathematical formulations are given separately as appendixes. Let us note that the derived equations, except equations with physical dimensions stated in the subscripts, are valid in any coherent unit system (for example, international system of units – SI).

Theory of structure and mechanics of yarns represents a relatively large and non-trivial complex of relations, despite the fact that it accounts for a small segment of theory of fibrous assemblies. The simpler parts described in this book can be useful for teaching of undergraduate students in colleges or universities as well as for the technologists in textile industries. The relatively difficult parts are made for the postgraduate as well as doctoral students and also for the researchers working in the area of structure–property relationship in yarns. The most complicated theories described in this book can inspire our academic colleagues who are professionally oriented towards basic research.

---

2   Frascati Manual 2015, Guidelines for Collecting and Reporting Data on Research and Experimental Development, OECD (Organization for Economic Co-operation and Development), Chapter 2, Section 2.5, 2015.

3   Feynman R.P. (Nobel award winner in physics in 1965), Character of Physical Laws, Chapter 2, Penguin Books, London, 1992.

4   Nečkář, B. and Das, D. Theory of Structure and Mechanics of Fibrous Assemblies. Woodhead Publishing India Pvt. Ltd., New Delhi, 2012.

   We are thankful to both of our universities, Technical University of Liberec (TUL) and Indian Institute of Technology Delhi (IITD), for their support to our research work and publications. We also want to extend our thanks to the students and colleagues of our departments – Department of Textile Technologies and Structures at the Faculty of Textiles in TUL and Department of Textile Technology in IITD – for their help and support while writing this book.

September 2018   Bohuslav Neckář
                                                        Dipayan Das

Introduction

This book consists of 12 chapters describing different problems and solutions on the structure and mechanics of yarns. Chapters 1 and 2 have a basic character, Chapters 3–7 present highly fascinating themes on yarn structure and Chapters 8–12 contain extremely interesting topics on structural mechanics[1] of yarns.

Chapter 1 introduces the basic characteristics of fibers - the building blocks of yarns and their mutual relations. This chapter also presents one of the oldest concepts on yarns, regarding the relationship among yarn count, twist and diameter proposed by Koechlin.

Chapter 2 is relatively small and rather atypical one. It gives a short overview of the spinning processes in a little unusual manner.

Chapter 3 is the first of the three chapters (3, 8 and 11) that has a purely theoretical character. It presents a general view of yarn structure, deals with the general paths of fibers, classifies them structurally and introduces relatively universal characteristics of yarn structure. (The well-known helical and migration models are hereby considered as the simple cases of more general equations.) Maybe this chapter can inspire theoretical scientists for future development of superior models that will be more complex but more realistic.

Chapter 4 describes the helical model of fibers in yarns. The corresponding relations and structural consequences of traditional helical model – number of fibers in yarn cross section, models of yarn retraction and limit of yarn twist – are covered in this chapter.

Chapter 5 introduces the classical Treloar's model of radial fiber migration in yarns in a new style. This chapter also includes Hearle's characteristics of radial fiber migration. In the middle, it shows that this model is not in good agreement with the experimental results. It then describes the formulation of equidistant migration model which shows much better correspondence with the experimental results. The modified characteristics of fiber migration in accordance with this model are also included in this chapter.

---

1   J.W.S. Hearle, P. Grosberg and S. Backer used the term 'structural mechanics' probably for the first time in their book entitled 'Structural Mechanics of Fibers, Yarns, and Fabrics – Volume 1' published by John Wiley & Sons, New York in the year of 1969. This term covers the study of mechanical problems in the light of specific structures of fibrous assemblies. It is also found to be used in other branches of study, for example, mechanics of rocks.

Chapter 6 discusses on mass unevenness of staple fiber yarns in the light of binomial and Poisson slivers. It includes the influences of fiber direction and fiber length on the mass irregularity of yarns. Further, it describes a model of mass irregularity, based on random aggregates of fibers, which explains the disproportion between limit irregularity and actual mass irregularity of yarns.

Chapter 7 deliberates on the phenomenon of yarn hairiness. It documents the formulation of a completely new probabilistic model of yarn hairiness. It discusses on how this model can be applied in practice to evaluate the hairiness of yarns. Further, it displays a comparison between this model and the experimental results of yarn hairiness.

Chapter 8 is the second of the triplet of purely theoretical chapters. It introduces internal yarn mechanics in the light of continuum mechanics. The differential equation of radial equilibrium of forces in yarns as well as the principles, conditions and procedures of solving this equation are discussed. It is here shown that many inputs are required to solve this equation; however, a few of them are unfortunately not enough known till date. (For example, the stress and strain tensors in relation to fibrous assembly are not known as much as required.) Needless to say, if such inputs will be precisely known in future, a more accurate output can be obtained. We, therefore, feel that this chapter can also be an inspiration for future scientific activities in this direction.

Chapter 9 covers the tensile behaviour of staple fiber yarns. It documents the formulation of an expression for fiber stress utilization in yarns according to Gegauff's concept. It then demonstrates an original way to generalize the classical helical model to a model of randomly oriented fibers. Finally, it puts on record a satisfactory correspondence between the generalized model and the experimental results.

Chapter 10 discusses on yarn strength in relation to gauge length. It at first presents two models (Peirce and Weibull) of 'independent' strengths and then documents our original model of 'dependent' strengths [Stationary, Ergodic, Markovian, Gaussian (SEMG) stochastic process] along with the comparison between theoretical and experimental results. It is shown that the dependent strength model can explain the yarn strength at different gauge lengths in a better manner than the independent strength model.

Chapter 11, the last purely theoretical chapter, presents our original concept of solving the problem of fiber-to-fiber slippage in yarns. Unfortunately, the practical application of the proposed constitutive theory of fiber-to-fiber slippage requires a set of complex inputs such as the regulations of fiber-to-fiber friction and pressure distribution in yarn, which are not fully known till date. Maybe, in future, if these inputs will be precisely known, this theory can find suitable applications. This chapter can therefore be the starting point of future scientific activities.

Chapter 12 describes a method to predict 'optimum' yarn twist. It also proposes a semi-empirical model (Solovev's concept) for determination of yarn strength.

It is now quite clear that this book gives different views to the problems of yarns. Therefore, our readers may wonder, how to read or rather study this book? Evidently, this book cannot be studied as fiction. It is necessary to ask yourself, 'What do I need to know about yarns?' Whatever your answer is, we recommend you to recapitulate the basic set of elementary knowledge and the symbols used in this book, that is, Chapter 1, in each case. (Maybe, the briefly and simply written Chapter 2 can broaden your mind a little.) You may then ask yourself, 'Why do I need to broaden my knowledge?' If you are a young student or a young technologist then you can study the first sections of the relevant chapters. These sections are usually written in a simpler manner as compared to the other sections. If you are a PhD student or a specialist in applied research or a creative technologist, then you may select the chapter corresponding to the themes of your interest and read it very attentively. The authors tried to formulate the text of this book systematically, logically, precisely and completely as far as possible. The mathematical derivations were carried out in accordance with the level of mathematical education in technical universities. Finally, we come to an end saying that the triplet of purely theoretical chapters (3, 8 and 11) can be considered as an 'invitation' to our colleagues, who are working in the area of basic research, to carry out their scholarly research work on these themes.

Basic properties of single fibers and yarns

The basic structural element of the yarns[1] considered here is fiber. A fiber is sufficiently long and thin and featured by many characteristics such as length, diameter, aspect ratio, fineness, surface area, specific surface area, cross-sectional area, cross-sectional shape, strength, breaking elongation, etc. As known, numerous fibers constitute the yarn. Like fibers, the yarn is also featured by its characteristics like total length of fibers in a defined portion of yarn, total surface area occupied by fibers in that portion of yarn, etc. In this chapter, the basic characteristics of individual fiber are described and their relations to those of the yarn are derived. Here, the yarn is considered to be made up of homogeneous or heterogeneous fibers.

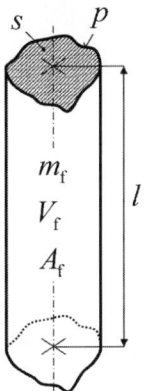

Figure 1.1 Scheme of a fiber

1.1 Fiber characteristics: definitions and relations

Starting parameters

Figure 1.1 illustrates a staple fiber of length l, mass m_f, volume V_f and surface area A_f. Let us assume a homogenous portion of the yarn that contains

---

1. Preliminarily, we intuitively imagine a yarn as a very 'long' and 'thin' twisted bundle of fibers – natural fibers or manufactured fibers.

N number of such identical fibers. If L, m, V, A represent total length, mass, volume and surface area of all fibers in that portion of the yarn, respectively, then

$$L = Nl, \quad m = Nm_f, \quad V = NV_f, \quad A = NA_f \tag{1.1}$$

Fiber density ρ

By using Equation (1.1), the fiber density can be expressed as

$$\rho = m_f / V_f = m/V. \tag{1.2}$$

Table 1.1 reports on density of some commonly used fibers.

Table 1.1 Fiber density values according to Goswami et al. [1]

Fiber	$\rho(\mathrm{kg\,m^{-3}})$
Cotton	1520
Linen, jute	1520
Wool	1310
Natural silk	1340
Viscose	1500
Acetate	1320
Polyester	1360
Polyamide	1140
Polyacrylonitrile	1300
Polypropylene	910

Fiber fineness t

In practice, it is often necessary to specify the fineness of fibers. The fiber fineness is usually defined by fiber mass per unit length. In other words, it is sometimes called 'linear density' or 'titre'. By using Equation (1.1), the fiber fineness can be expressed as

$$t = \frac{m_f}{l} = \frac{m}{L}. \tag{1.3}$$

In industry, fibers such as cotton, wool, manufactured fibers and microfibers are used. The fineness of some commonly used fibers is given in Table 1.2.

Table 1.2 Fineness of different types of fibers

Fibrous material	Fineness (dtex)
Microfibers	<1
Cotton and compatible manufactured fibers	About 1.6
Wool and compatible manufactured fibers	About 3.5
Carpet fibers, industrial fibers	>7

By applying Equations (1.2) and (1.3), we get

$$t = \frac{V_f}{l}\rho = \frac{V}{L}\rho, \ V_f = \frac{tl}{\rho}, \ V = \frac{tL}{\rho}, \ \frac{V_f}{l} = \frac{V}{L} = \frac{t}{\rho}. \tag{1.4}$$

Equation (1.4) points out a limitation for the use of t as a measure of fiber fineness. In the fiber-based-product industry, we think about 'fineness' in terms of fiber geometry, particularly 'area of cross section' or 'diameter'. As such, the use of fiber volume per unit length (the ratio V_f/l or V/L) would be more logical to express the fineness of fibers. Then, the standard value of fiber fineness t must be divided by the numerical value of fiber density ρ to obtain the numerical value of fiber volume per unit length. Otherwise, if we compare the numerical values of fineness t of two fibers having different densities, we may find that the 'heavier' fiber (higher value of ρ) is thinner than the 'lighter' one (smaller value of ρ).

It is thus more logical to use V/L than t to express fiber fineness. However, in industrial practice, the latter is preferred to the former, because the laboratory methods for measurement of t are easier than those for measurement of V/L.

Fiber cross-sectional area s

The shaded area shown in Figure 1.1 represents fiber cross-sectional area, which is formed by intersecting a plane perpendicular to the axis of the fiber. Assuming that fiber cross-sectional area s is same throughout its length (or generally speaking we consider the expression s as the mean fiber cross-sectional area), the volume of individual fiber is expressed as $V_f = sl$, and the volume of all fibers in the yarn is expressed by $V = sL$. By substituting this expression into Equation (1.4), we obtain the expression for fiber cross-sectional area as follows:

$$t = s\rho, s = \frac{V_f}{l} = \frac{V}{L} = \frac{t}{\rho}. \tag{1.5}$$

The expression for fiber cross-sectional area is identical to the expression for fiber volume per unit length. Hence we realize that fiber cross-sectional area is an important measure of fiber geometry, i.e., the 'size' of a fiber.

Equivalent fiber diameter d

Let us consider a cylindrical fiber as shown in Figure 1.2a. The fiber cross-sectional shape is circular and the fiber cross-sectional area is given by $s = \pi d^2/4$, where d stands for fiber diameter. By applying Equation (1.5), we find the following expression for fiber diameter:

$$d = \sqrt{\frac{4s}{\pi}} = \sqrt{\frac{4t}{\pi\rho}}, t = \frac{\pi d^2}{4}\rho. \tag{1.6}$$

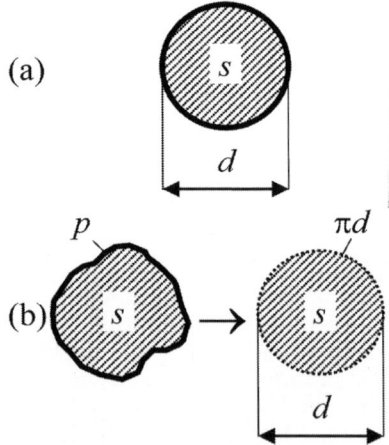

Figure 1.2 Fiber cross section

Let us now consider a fiber with non-circular cross-sectional shape; such a fiber is presented in a classical geometrical way without any defined diameter in Figure 1.2b. The variable d, calculated from Equation (1.6), expresses the diameter of an (imaginary) equivalent circle of same cross-sectional area (Figure 1.2b) and this diameter is known as equivalent fiber diameter. The correct value of fiber cross-sectional area can be calculated without considering the real shape of the fiber. The area of a circular cross section is

$$s = \frac{\pi d^2}{4}.$$ (1.7)

By applying Equations (1.5) and (1.7), the fiber volume can be expressed as

$$V_f = sl = \frac{\pi d^2}{4} l.$$ (1.8)

Analogously, the volume of a given portion of yarn can be stated as

$$V = sL = \frac{\pi d^2}{4} L.$$ (1.9)

After calculating the value of d, Equation (1.6) can be used to estimate fiber fineness.

Fiber aspect ratio Λ

Fiber length and fiber diameter are frequently used to characterize fiber geometry. It is then reasonable to introduce an expression for fiber aspect ratio, which is defined by the ratio of fiber length l to fiber diameter d. This is expressed in the following equation:

$$\Lambda = l/d.$$ (1.10)

Typical values of fiber aspect ratio are given in Table 1.3.

The numerical values of fiber aspect ratio are in the order of thousands. In this connection, let us here introduce one interesting example. If we enlarge our model of staple fiber, we may get a pipe of 1 cm diameter (such a pipe is frequently used for supplying gas in chemical laboratories), and if we consider the aspect ratio of that pipe same as that of the fiber, for example, $\Lambda = 2000$; then the length of that pipe will be around $20\,\text{m}$. We may recognize from this example that the application of mechanical force at one end of a fiber may not affect the other end of that fiber.

Table 1.3 Typical values of fiber aspect ratio

Fiber	Aspect ratio Λ
Cotton	1500
Wool	3000
Flax*	1250
Ramie	3000

* Single fiber.

Perimeter p and shape factor q of fiber cross section

The real perimeter p encloses the real cross section of a fiber as shown in Figure 1.2b. The perimeter of an imaginary equivalent circle of cross-sectional area s is πd. It is well known from geometry that a circle is the shortest possible curve enclosing a given area, therefore, $p \geq \pi d$. Then, $p/(\pi d) \geq 1$, and

$$q = \frac{p}{\pi d} - 1 \geq 0, \quad p = \pi d(1+q). \tag{1.11}$$

The numerical values of q, given by Malinowska [2], depend on fiber cross-sectional shape, in other words, on shape factor ($q = 0$ for cylindrical fibers). The numerical value of fiber shape factor q becomes higher when the shape of fiber cross section is irregular, i.e., far away from circular shape. Some typical values of fiber shape factor are given in Table 1.4. Note that Morton and Hearle [3] used another definition of fiber shape factor, which is equal to zero for a fiber with circular cross section.

A lot of useful information can be obtained by enlarging the images of fiber cross section obtained by using microscopy technique. Nowadays computers are used for evaluation of fiber perimeter and fiber cross section (image processing technique). By applying the value of q from Table 1.4 into Equation (1.11), the perimeter p of fiber cross section can be calculated.

Table 1.4 Cross-sectional shape and shape factor

Shape of fiber cross section	q [1]
Circle – ideal (◯)	0
Circle – real fiber	0–0.07
Triangle – ideal (△)	0.29
Triangle – real fiber	0.09–0.12
Mature cotton	0.20–0.35
Irregular saw	>0.60

Fiber-specific surface area a

The fiber surface area is expressed by $A_f = pl$ (Figure 1.1). The exact fiber surface area should include the areas of cross sections of the two ends. Usually, these two areas are negligibly small. In fact, they are much smaller than that of the cylindrical surface. Hence these two areas can be ignored safely so that the resulting error will be negligibly small. Thus, by rearranging Equation (1.11), we find the following expression for fiber surface area

$$A_f = pl = \pi d (1+q) l. \tag{1.12}$$

Analogously, by using Equations (1.1) and (1.12), the surface area of fibers in a fibrous assembly A can be expressed as follows:

$$A = NA_f = \pi d (1+q) Nl = \pi d (1+q) L. \tag{1.13}$$

According to the definition of fiber density expressed in Equation (1.2) and by applying Equation (1.8) or (1.9), we obtain the following expressions for the mass of a single fiber and also for the mass of a given portion of yarn, respectively:

$$m_f = V_f \rho = \left(\pi d^2 / 4 \right) l \rho, \tag{1.14}$$

$$m = V \rho = \left(\pi d^2 / 4 \right) L \rho. \tag{1.15}$$

Fiber-specific surface area is expressed by surface area per unit mass of fiber. By applying Equations (1.12) and (1.14) or Equations (1.13) and (1.15), we get the following expression for fiber-specific surface area:

$$a = \frac{A_f}{m_f} = \frac{A}{m} = \frac{\pi d (1+q) l}{\left(\pi d^2 / 4 \right) l \rho} = \frac{\pi d (1+q) L}{\left(\pi d^2 / 4 \right) L \rho} = \frac{4 (1+q)}{\rho d}. \tag{1.16}$$

An alternative expression for a is obtained by substituting Equation (1.6) into (1.16) as follows:

$$a = \frac{4 (1+q)}{\rho} \sqrt{\frac{\pi \rho}{4t}} = 2 \sqrt{\pi} \frac{1+q}{\sqrt{\rho t}}. \tag{1.17}$$

The fiber surface characteristics strongly affect the end-use properties (sorption, hand, etc.) of fibrous assemblies, which are very important from the consumer point of view. In particular, they strongly influence the physiological and comfort properties of end products.

Fiber surface area per unit volume γ

The ratio of fiber surface area to fiber volume is a useful measure of the geometrical structure of fibers. This is expressed by $\gamma = A_f / V_f = A / V$. By applying Equations (1.12) and (1.8) or Equations (1.13) and (1.9), we get the following expression for fiber surface area per unit volume:

$$\gamma = \frac{A_f}{V_f} = \frac{A}{V} = \frac{\pi d\left(1+q\right)l}{\left(\pi d^2/4\right)l} = \frac{\pi d\left(1+q\right)L}{\left(\pi d^2/4\right)L} = \frac{4\left(1+q\right)}{d}. \tag{1.18}$$

By substituting Equation (1.6) into (1.18), we get an alternative expression for γ as

$$\gamma = 4\left(1+q\right)\sqrt{\frac{\pi\rho}{4t}} = 2\sqrt{\pi}\left(1+q\right)\sqrt{\frac{\rho}{t}}. \tag{1.19}$$

By rearranging Equations (1.16) and (1.18), we can also find that

$$\gamma = a\rho. \tag{1.20}$$

The fiber surface area per unit volume γ is generally a geometrical variable, it does not depend on fiber density, and it is more useful than the fiber-specific surface area a.

According to Equation (1.18), the inverse of fiber surface area per unit volume is directly proportional to the equivalent fiber diameter. The numerical value of $1/\gamma$ has a dimension of fiber length and it is, to some extent, a measure of fiber 'thickness'. This is a very useful variable, e.g., for calculating the size of pores between fibers – see Reference [4].

Tensile stress

In engineering mechanics, the ratio of the applied force F to the cross-sectional area defines mechanical (or engineering) stress σ'. (In SI system, it is measured in the unit of $1\,\mathrm{Nm^{-2}} = 1\,\mathrm{Pa}$.) In fiber/textile technology, it has become a tradition to use the term specific stress σ, which is expressed by the ratio of the applied force to the linear density of the fiber (or yarn). (In SI system, its unit is $1\,\mathrm{N\,Mtex^{-1}}$.) By using Equation (1.5), the relationship between specific stress and mechanical stress can be obtained as

$$\sigma = \frac{F}{t} = \frac{F}{s\rho} = \frac{\sigma'}{\rho}. \tag{1.21}$$

The specific stress at which the fiber (or yarn) breaks is called tenacity.

Earlier, the so-called 'breaking length' $L = R$ was defined by the length required to break the fiber under its own weight. According to Equation (1.3), the mass of a fiber of length R is Rt. By using the acceleration due to gravity $g = 9.81\,\mathrm{ms^{-2}}$, the weight (gravitational force) is $F = Rtg$. According to Equation (1.21), $\sigma = F/t = Rg$, where $\sigma_{\left[\mathrm{cN\,tex^{-1}}\right]} = 0.981\,R_{\left[\mathrm{km}\right]}$. Thus, the tenacity in cN tex^{-1} is approximately equal to the breaking length in kilometre. Also, the unit of force F was earlier expressed by the so-called 'pond' $[\text{p}]$, which was known as 'gram force'. It is then valid to write that $F_{[\text{p}]} = F_{[\text{N}]}/0.00981 = F_{[\text{cN}]}/0.981$. 'Denier' $T_{[\text{den}]} = 9t_{[\text{tex}]} = 0.9t_{[\text{dtex}]}$ is frequently used to express fiber fineness. Accordingly, fiber tenacity in 'pond per denier' is expressed by $\sigma_{[\text{p/den}]} = F_{[\text{p}]}/T_{[\text{den}]} = \sigma_{\left[\text{cN dtex}^{-1}\right]}/0.8829$.

The specific tensile stress is not a reasonable expression for fibers possessing different densities, because fiber fineness depends on fiber density. In such case, it is recommended to consider the stress as mechanical (engineering) stress $\sigma' = \sigma\rho$. The reason for standardizing specific stress remains the same as for standardizing fineness.

Example. We consider a polyester fiber whose tenacity is $\sigma = 0.43\,\text{N tex}^{-1}$ and density is $\rho = 1360\,\text{kg m}^{-3}$. The calculated value of mechanical (engineering) strength, according to Equation (1.21), is $\sigma' = 585\,\text{MPa}$. Similarly, for a cotton fiber with tenacity $\sigma = 0.32\,\text{N tex}^{-1}$ and density $\rho = 1520\,\text{kg m}^{-3}$, the mechanical (engineering) strength is $\sigma' = 487\,\text{MPa}$. If we compare the tenacities (N tex^{-1}) of the fibers then we find that polyester fiber exhibits 33% higher tenacity than cotton fiber. But, if we compare the mechanical (engineering) strength (MPa) of the fibers, then we observe that polyester fiber offers only 20% higher specific (engineering) strength than cotton fiber. Here we may remind that ordinary steel has mechanical (engineering) strength of $\sigma' = 500\,\text{MPa}$. This means that the mechanical (engineering) strengths of both fibers are almost comparable with that of ordinary steel.

1.2 Characteristics of general fibrous assemblies

Packing density μ – definition

The cotton wool is a very fine and soft product that it is being used for surgical dressings, while during the middle age, the wooden stakes were used as an execution tool. Interestingly, both of these materials are composed of cellulose. This peculiar example is given to show that the behaviour of an ultimate material does not only depend on the constituent material, but also on the compactness of the final product.

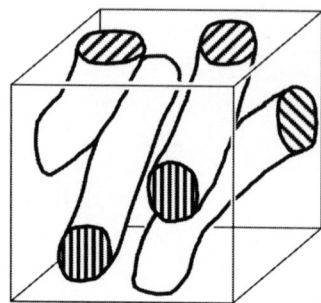

Figure 1.3 Section of a fibrous assembly in a three-dimensional plane

Figure 1.3 illustrates a three-dimensional section of a fibrous assembly (e.g., yarn) of total volume V_c. The volume of fibers in this section is V, thus $V \leq V_c$. The difference between the volumes $V_c - V$ expresses the volume of air present in the three-dimensional plane, i.e., among the empty spaces between the fibers.

The fiber compactness is measured by the ratio of the volume occupied by the fibers to the total volume of the fibrous assembly. This is expressed in the following equation:

$$\mu = \frac{V}{V_c}, \mu \in \langle 0,1 \rangle. \tag{1.22}$$

In textile literature, the variable μ is called packing density. (In chemical technology, it is termed as 'volume fraction'.) Some typical values of packing density of different fibrous assemblies are summarized in Table 1.5.

Table 1.5 Typical packing density values

Fibrous assemblies	μ
Linear textiles	
Monofilament	1
Limit structure*	0.907
Hard twisted silk	0.75–0.85
Wet spun linen yarn	About 0.65
Combed cotton yarn	0.5–0.6
Carded cotton yarn	0.38–0.55
Worsted yarn	0.38–0.50
Woolen yarn	0.35–0.45

Cotton roving	0.10–0.20
Sliver	About 0.03
Other textiles	
Woven fabric	0.15–0.30
Knitted fabric	0.10–0.20
Cotton wool**	0.02–0.04
Leather (textiles)**	0.005–0.02
Other materials	
Earthenware**	0.20–0.23
Wood**	0.3–0.7
Animal leather**	0.33–0.66

* Theoretical value [4].** Piller and Trávníček [5] and Marschik [6].

It is important to observe that the textile materials contain relatively high volume of air. This imparts softness, porosity, pleasant hand, good drapability, etc. (If we buy a package of ordinary cotton wool from a shop, we pay actually 97% of our money for air and only 3% for cotton wool.) At the same time, textiles are strong and relatively stress resistant (thanks to the good mechanical properties of textile fibers). The presence of both behaviours together determines the typical end-use of textiles.

Packing density μ – areal interpretation

Figure 1.4a illustrates an infinitely thin section of a fibrous assembly. The total volume of this section is

$$dV_c = ab \, dh = S_c \, dh, \tag{1.23}$$

where $ab = S_c$ denotes the total area of the 'upper wall' of the section. In this section, there are N number of fiber segments. (We can ignore the curvatures of the fiber segments because they are infinitely small.) A typical j-th fiber segment ($j = 1,2,\cdots,N$) is shown in Figure 1.4b. The volume of this elementary section is expressed as a product of the projected area and the perpendicular height, i.e., $s_j^* \, dh$. The total volume of all fiber segments is

$$dV = \sum_{j=1}^{N}\left(s_j^* \, dh\right) = dh \sum_{j=1}^{N} s_j^* = S \, dh, \tag{1.24}$$

where $\sum_{j=1}^{N} s_j^* = S$ is the total sectional area of all fibers present in the upper wall section. Now we can express the sectional area packing density using the general definition given in Equation (1.22). This is done as follows:

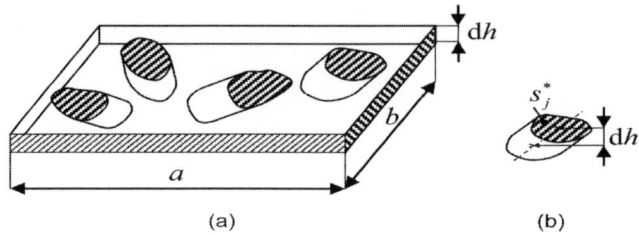

(a)   (b)

Figure 1.4 Fiber segments in an elementary section of a fibrous assembly

$$\mu = \frac{dV}{dV_c} = \frac{S\,dh}{S_c\,dh} = \frac{S}{S_c}. \tag{1.25}$$

It is obvious from the above expression that the packing density can be expressed as a ratio of the sectional area of all fibers to the total area of the fibrous assembly including 'the empty spaces' and fibers. The expression stated in Equation (1.25) can be considered as the areal interpretation of packing density.

Such expression is called local packing density and may be used as a measure of compactness of small areas around the sectional plane (the upper wall section) of the fibrous assembly. If we assume that the observed fibrous assembly has the same packing density in all sections, then we can use Equation (1.25) as the packing density of the whole assembly. This assumption is very useful to express the packing density of yarns and other linear textiles.

Packing density μ – density interpretation

The three-dimensional section of a fibrous assembly (e.g., yarn), shown in Figure 1.3, has mass m and total volume V_c. The density ρ^* of such fibrous assembly is then given by the fraction m/V_c as shown in the following equation:

$$\rho^* = m/V_c, \left(V_c = m/\rho^*\right). \tag{1.26}$$

The mass m refers to the mass of fibers only. (The mass of air and the mass of adhesives imparted during the finishing process are not considered.) The volume of these fibers is V. According to Equation (1.2), the fiber density is $\rho = m/V, V = m/\rho$. By applying Equations (1.26) and (1.2) into (1.22), we find

$$\mu = \frac{m/\rho}{m/\rho^*} = \frac{\rho^*}{\rho}. \tag{1.27}$$

Equation (1.27) gives another expression for packing density, where it is defined as a ratio of density of the fibrous assembly to the density of the constituent fiber. This expression is known as the density interpretation of packing density.

This interpretation of packing density is applicable for formulating the media-continuum models that utilize the idea of mass elements. The density ρ^* of such an element can be divided by an 'arbitrary' constant ρ to get the packing density factor μ of this element. Actually, the packing density factor at any arbitrary point of a three-dimensional space takes either 1 if only fibrous material occupies the space or 0 if there is no fibrous material in the space.

The density interpretation is also useful for practical determination of packing density. It is difficult to use Equation (1.22) directly for the calculation of packing density, because fiber volume cannot be measured directly in ordinary textile laboratories. The mass of the assembly can be simply found by weighing the assembly and its total volume can be also easily calculated from its macrodimensions. Then, the density of the fibrous assembly can be estimated by using Equation (1.26). The fiber density can be obtained from Table 1.1. By applying these two density values in Equation (1.27), we can estimate the packing density. (Earlier Marschik [6] directly used ρ^* in lieu of packing density as a measure of fiber compactness.)

Porosity (permeability) ψ

The measure of fiber compactness in a fibrous assembly can also be characterized by the presence of relative amount of air in the fibrous assembly. For example, the fibrous assembly shown in Figure 1.3 has a total volume V_c including fiber volume V. Then, the volume of air (volume of pores) among the fibers in the assembly is

$$V_p = V_c - V. \tag{1.28}$$

The relative volume of air is expressed by porosity $\psi = V_p/V_c$. By applying Equations (1.22) and (1.28) into the expression of porosity, we find the following expression:

$$\psi = \frac{V_p}{V_c} = \frac{V_c - V}{V_c} = 1 - \frac{V}{V_c} = 1 - \mu. \tag{1.29}$$

Note: Yarns, alike other fibrous assemblies, can be constituted from one type of fibers or different types of fibers. It is required to determine the mean parameters of fiber blends in this case. The way for this determination as well as the mutual relations among mean parameters of fiber blend are derived in Reference [4].

1.3   Yarn characteristics: definitions and relations

Yarn, within the scope of this book, is considered as a twisted bundle of fibers. The easiest picture of yarn is shown in Figure 1.5. Note that this yarn is created from fibers of same properties and it does not have any hairiness and unevenness.

Initial quantities

Let us consider that (1) the yarn consists of fibers possessing same density ρ and same fineness t and (2) the twist, diameter, fineness and number of fibers present in the cross section of the yarn are Z, D, T and n, respectively.

A portion of this yarn having length l and mass m is shown in Figure 1.5.

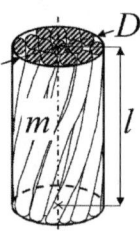

Figure 1.5 Scheme of yarn

Yarn fineness

Based on international standards, the yarn fineness (yarn 'count') is defined as follows:

$$T = \frac{m}{l}. \tag{1.30}$$

Note: The unit of yarn fineness, according to the international system of units (SI system), is $1\,\text{Mtex} = 1\,\text{kg}/1\,\text{m}$. However, the unit used in practice is $1\,\text{tex} = 1\,\text{g}/1\,\text{km} = 10^{-6}\,\text{Mtex}$. (These units are same for fibers too.) Nevertheless, we derive all mathematical expressions under a coherent system of units (e.g., SI system) in this book. This allows us to work without constants of dimensions. We recommend the readers to adapt himself/herself to our equations with suitable units for numerical calculations as well as practical applications. In the past, a lot of other units were used for characterization of yarn fineness and some of them are still being used in praxis. A few older quantities of yarn fineness are given in Table 1.6.

Table 1.6 An older type of yarn 'numbering' system [7]

Symbol	Name	Unit of		Conversion to T [tex]
		Length, l*	**Mass, m**	
		'Length' type, m/l		
Nm	Number metric**	1 km	1 kg	$T = 1000/Nm$
Ne_c	Number English (cotton)**	840 yards	1 English pound	$T = 590/Ne_c$
Ne_{w1}	Number English (wool)	560 yards	1 English pound	$T = 886/Ne_{w1}$
Ne_{w2}	Number English (wool)	1120 yards	1 English pound	$T = 443/Ne_{w2}$
Ne_f	Number English (flax)	300 yards	1 English pound	$T = 1654/Ne_f$
Nf_c	Number French (cotton)	1 km	0.5 kg	$T = 500/Nf_c$
Nf_w	Number French (wool)	714 yards	0.5 kg	$T = 766/Nf_w$
Nd	Old Dutch number	840 yards	0.5 kg	$T = 651/Nd$
Na	Old Austrian number	1487.5 Wiener cubits	1 Austrian pound	$T = 483/Na$
		'Weight' type, l/m		
Td	Legal titre**	450 m	1 denier	$T = 0.111 Td$
Tm	Titre metric	500 m	1 denier	$T = 0.1 Tm$
Ts	Scottish number of jute	14,400 yards	1 English pound	$T = 34.4 Ts$

*The lengths used for numbering system follow the terms like 'skein', 'hank', 'spindle', etc.
**At present, the non-standard 'numbering' system is more frequently used.

Let us denote the volume of fibers present in the given portion of the yarn by V. Then, the mass of this portion of the yarn (equal to fiber mass) is $m = V\rho$, so that the following expression can be written from Equation (1.30):

$$T = \frac{V}{l}\rho. \tag{1.31}$$

It infers that a suitable characteristic of geometrically interpreted "yarn fineness" can be the ratio V/l (fiber volume per unit length of yarn), but the standardized quantity of yarn fineness is obtained after multiplication of V/l by ρ – compare this with Equation (1.4).

Note: It is because of the fact that a given length of yarn can be easily weighed in a spinning mill, whereas the fiber volume in a yarn is relatively complicated to determine and it requires a special laboratory.

So, a yarn made up of 'lighter fibers' (e.g., polypropylene, $\rho = 910\,\mathrm{kg\,m^{-3}}$) may seem to be more 'large' (higher V/l ratio) than a yarn prepared from 'heavy fibers' (e.g., viscose, $\rho = 1500\,\mathrm{kg\,m^{-3}}$), though both of the yarns may have same fineness T. We therefore need to use the ratio $T/\rho = V/l$ for comparing the 'size' of yarns.

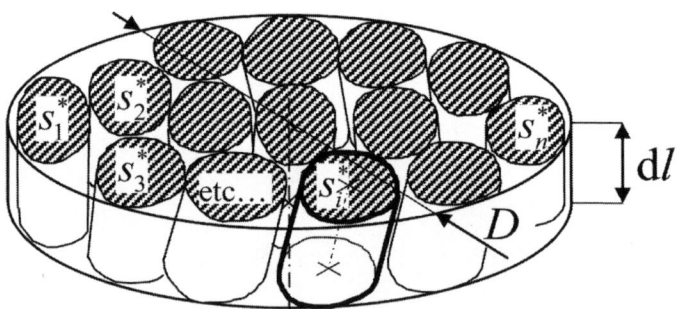

Figure 1.6 Scheme of an elementary yarn section

Substance cross section

An elementary cross section of a yarn is displayed in Figure 1.6. The shaded areas $s_1^*, s_2^*, \ldots, s_i^*, \ldots, s_n^*$ represent the sectional areas of n fibers present in the cross section of the yarn. Johansen [8] called the sum of sectional areas of all fibers present in yarn cross section as the substance cross-sectional area of yarn. This is stated in the following equation:

$$S = \sum_{i=1}^{n} s_i^* .$$
(1.32)

The volume of an i-th (general) elementary fiber segment (thick lines shown in Figure 1.6) is $s_i^* \, dl$, so that the total volume of all elementary fiber segments is $dV = \sum_{i=1}^{n} \left(s_i^* \, dl \right) = dl \left(\sum_{i=1}^{n} s_i^* \right) = dl \, S$. The mass of such segments is $dm = dV \rho = dl \, S \rho$, and the (local) fineness, according to Equations (1.30) and (1.31), is

$$T = \frac{dl \, S \rho}{dl} = S \rho, \quad S = \frac{T}{\rho} = \frac{V}{l}.$$
(1.33)

Note: It requires to state that the substance cross-sectional area of yarn is a right quantity for characterization of fineness as a pure geometrical interpretation. [Compare the last expression with that mentioned in Equation (1.5).]

Mean sectional area of fiber

It is evident that the mean sectional area of fibers in yarn cross section, according to Equation (1.32), is

$$\bar{s}^* = \frac{1}{n} \sum_{i=1}^{n} s_i^* = \frac{S}{n}.$$
(1.34)

Substance diameter

Let us imagine a yarn made up of synthetic fibers as shown in Figure 1.7a. The diameter of the yarn is D. Further, let us (imaginatively) compress this yarn from all sides (thick arrows shown in Figure 1.7b) so that the air is completely removed from the yarn. As a result, the shape of the yarn is changed to that of a compact ring, shown in Figure 1.7b. The cross-sectional area of the ring denotes the substance cross-sectional area S of the yarn. The diameter of the ring indicates the substance diameter D_S of the yarn. By using Equation (1.33), it is valid to write that

$$S = \frac{\pi D_\mathrm{S}^2}{4}, \quad D_\mathrm{S} = \sqrt{\frac{4S}{\pi}} = \sqrt{\frac{4T}{\pi \rho}}.$$
(1.35)

Note: The substance diameter D_S of yarn is smaller than the real diameter D of yarn in all practical cases.

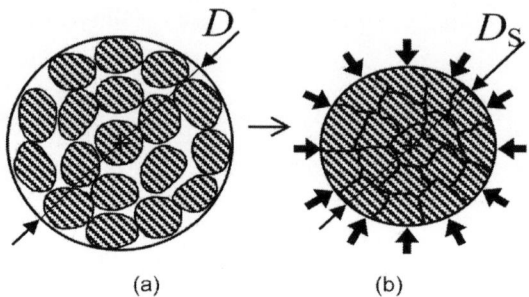

(a)   (b)

Figure 1.7 Scheme of yarn 'compression'

Relative fineness

The ratio of yarn fineness T to fiber fineness t is called relative fineness of yarn τ. This is stated in the following equation:

$$\tau = \frac{T}{t}.$$ (1.36)

Note: In a lot of handbooks, this ratio is called 'number n of fibers' present in yarn cross section. We will show later on that this interpretation is not correct.

By using Equations (1.5) and (1.33) in (1.36), we get

$$\tau = \frac{S\rho}{s\rho} = \frac{S}{s}.$$ (1.37)

By substituting Equations (1.7) and (1.35) in the last expression, we obtain

$$\tau = \frac{\pi D_S^2/4}{\pi d^2/4} = \left(\frac{D_S}{d}\right)^2.$$ (1.38)

Coefficient k_n

We define coefficient k_n as the ratio fiber cross-sectional area to the mean sectional area of fiber in yarn cross section. This is stated as

$$k_n = \frac{s}{\overline{s}^*}.$$ (1.39)

The i-th (general) fiber element, shown earlier in Figure 1.6, is now shown in Figure 1.8. Let us consider that the height perpendicular to the sectional plane is dh, the sectional area of the fiber element is s_i^*, the angle of inclination of the fiber element from the yarn axis is ϑ_i (evidently, $\cos\vartheta_i = dl/dx_i$), the length of the fiber element is dx_i and the cross-sectional area (section A-A) of the fiber element is s. The volume of this fiber element can be expressed as $s_i^* dl$ and/or sdx_i. Then, from the equivalency of both the expressions, we obtain

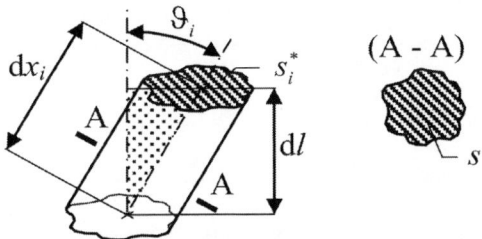

Figure 1.8 One fiber element

$$s_i^* dl = sdx_i, \quad s_i^* = s\frac{dx_i}{dl} = \frac{s}{\cos\vartheta_i}. \tag{1.40}$$

So, the substance cross-sectional area of yarn, according to Equation (1.32), can be rearranged as

$$S = \sum_{i=1}^{n}\frac{s}{\cos\vartheta_i} = s\sum_{i=1}^{n}\frac{1}{\cos\vartheta_i}. \tag{1.41}$$

The mean sectional area of fiber in yarn cross section can be obtained by using Equations (1.34) and (1.41) as

$$\overline{s}^* = \frac{s}{n}\sum_{i=1}^{n}\frac{1}{\cos\vartheta_i}, \tag{1.42}$$

and, by using Equations (1.39) and (1.42), we obtain the following expression for the coefficient k_n:

$$k_n = \frac{s}{\dfrac{s}{n}\displaystyle\sum_{i=1}^{n}\dfrac{1}{\cos\vartheta_i}} = \frac{1}{\dfrac{1}{n}\displaystyle\sum_{i=1}^{n}\dfrac{1}{\cos\vartheta_i}}, \quad \left(\frac{1}{k_n} = \frac{1}{n}\sum_{i=1}^{n}\frac{1}{\cos\vartheta_i}\right). \tag{1.43}$$

It can be seen that

1. The quantity k_n is expressed as a harmonic mean of fiber inclination angles ϑ_i.

2. If $\vartheta_i = 0$ for all fibers, i.e., when the yarn is assumed to be a bundle of perfectly parallel fibers, then $k_n = 1$; for all other cases, $k_n < 1$.

3. k_n can be used as a measure of inclination of fibers in yarn.

Table 1.7 Some values of k_n

Yarn type	Approximate value of k_n
Cotton type, ring spun	0.95
Cotton type, rotor (OE)	0.80
Wool yarns, combed	0.94

The typical values of k_n, based on authors' practical experience, are illustrated in Table 1.7. (The method of determination of k_n is mentioned later on.)

Number of fibers

The expression for the number n of fibers in yarn cross section is given in Equation (1.34). It is valid that $n = S/\overline{s}^*$. By rearranging this expression and by using Equations (1.37) and (1.39), we obtain

$$n = \frac{S}{\overline{s}^*} = \left(\frac{S}{s}\right)\left(\frac{s}{\overline{s}^*}\right) = \tau k_n .$$

(1.44)

It is shown that the equation $n = \tau$ is valid only for $k_n = 1$, i.e., for a parallel bundle of fibers. In other cases, $n < \tau$.

Note: The last expression is also used for experimental evaluation of coefficient k_n. For this, we need to measure the real yarn fineness T and the real fiber fineness t. Then we can calculate the real value τ by using Equation (1.36). Further, we must analyze a set of experimentally prepared cross sections of yarn (microscopic technique in conjunction with image analysis) and determine the real (average) number of fibers n. Then, we will be able to calculate the numerical value of coefficient k_n.

Packing density

The fiber volume in the given portion of the yarn, shown in Figure 1.5, can be expressed from Equation (1.33) as $V = Tl/\rho = Sl$. The total volume of that portion of the yarn is evidently $V_c = (\pi D^2/4)l$. Based on the definition of packing density as expressed in Equation (1.22), the packing density of the yarn can be expressed as

$$\mu = \frac{V}{V_c} = \frac{Tl/\rho}{(\pi D^2/4)l} = \frac{Sl}{(\pi D^2/4)l}, \qquad \mu = \frac{4T}{\pi D^2 \rho} = \frac{4S}{\pi D^2}. \qquad (1.45)$$

By using the substance cross sectional area as expressed in Equation (1.35), we can also write

$$\mu = \frac{4\dfrac{\pi D_S^2}{4}}{\pi D^2} = \left(\frac{D_S}{D}\right)^2. \qquad (1.46)$$

Yarn diameter

The yarn body, as shown in Figure 1.5, evokes an idea of a cylinder with diameter[2] D. By using Equations (1.45) and (1.46), the diameter of the yarn can be expressed as

$$D = \sqrt{\frac{4T}{\pi \mu \rho}} = \sqrt{\frac{4S}{\pi \mu}} = \frac{D_S}{\sqrt{\mu}}. \qquad (1.47)$$

Often, the following coefficients are used:

$$K = \sqrt{\frac{4}{\pi \mu \rho}} = \frac{2}{\sqrt{\pi \mu \rho}}, \quad K_S = \sqrt{\frac{4}{\pi \mu}} = \frac{2}{\sqrt{\pi \mu}}. \qquad (1.48)$$

Then, we can formally write the equation of yarn diameter as

$$D = K\sqrt{T} = K_S \sqrt{S}. \qquad (1.49)$$

---

2. Yarn diameter is easy to imagine by seeing an idealized picture of a yarn, shown in Figure 1.5. However, the determination of diameter of a real yarn body, including yarn hairiness, yarn unevenness, etc., is much more complicated. This problem will be studied later on in this book.

K is called coefficient of yarn diameter. It depends on fiber mass density ρ, i.e., type of fibers, and moreover on packing density μ of the yarn. It follows that K does not need to be a constant, even not for the yarns prepared from same type of fibers. Nevertheless, different handbooks recommended the use of coefficient of yarn diameter K as a constant parameter[3] for yarns prepared from same type of fibers. It is because the variability of packing density is not considered to be too high among the most common types of yarns.

The coefficient K_S is, however, not used in industrial practice. It corresponds to substance cross section of yarn; therefore, we call it substance coefficient of yarn diameter. This coefficient depends only on the packing density of yarn. Being a dimensionless quantity, it is very much suitable for theoretical work.

Twist

The traditional staple yarns are strengthened by means of twisting the bundle of fibers. The number of coils inserted per unit length of yarn is called twist Z. This is expressed as

$$Z = \frac{N_C}{l},$$
(1.50)

where N_C denotes the number of coils inserted in a length l of yarn. Its dimension is usually expressed in m^{-1} or $inch^{-1}$.

Twist intensity

The quantity DZ is therefore dimensionless. Nevertheless, we usually use the following dimensionless quantity:

$$\kappa = \pi DZ,$$
(1.51)

where κ is called twist intensity. The geometrical sense of twist intensity is depicted in Figure 1.9.

---

3.   See Section 1.4 for more details.

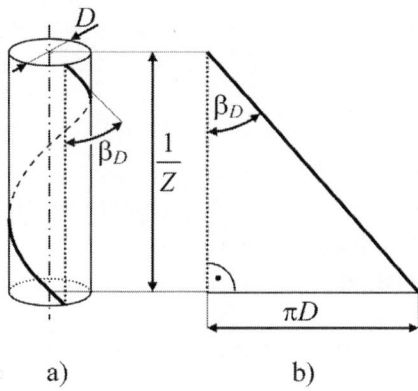

Figure 1.9 Ideal fiber on yarn surface

The cylinder, shown in Figure 1.9a, symbolizes a portion of a yarn of length $1/Z$. It contains only one coil of ideal fibers. The peripheral fibers are lying at the cylindrical surface of the yarn, i.e., at yarn diameter D. We assume that these fibers follow the helical shape. The angle β_D is defined as an angle between the tangent to the fiber helix and the direction of yarn axis. Unrolling of the cylindrical surface would result in a triangle, shown in Figure 1.9b. It is then valid to write that

$$\tan \beta_D = \pi D / (1/Z) = \pi D Z = \kappa. \tag{1.52}$$

So, the twist intensity represents the tangent of angle β_D of the peripheral fibers in a yarn. This variable is evidently dimensionless.

Twist coefficients – definitions

Twist coefficients are often used to characterize the 'level' of twisting. Some of them are introduced in Table 1.8.

Table 1.8 Some twist coefficients

Twist coefficient		Type		Twist exponent
		Common (industrially used)	**Substance (used in theory)**	
General		$\alpha = ZT^q$	$\alpha_S = ZS^q$	q
Special	Koechlin	$\alpha = Z\sqrt{T}$	$\alpha_S = Z\sqrt{S}$	$1/2$
	Phrix	$a = ZT^{2/3}$	$a_S = ZS^{2/3}$	$2/3$

Note: Internationally, Koechlin's common twist coefficient is mostly used in practice. However, Phrix common twist coefficient is used in Czech Republic [9, 10] – and sometimes, in relation to rotor spinning technology, this is used in abroad too.

Koechlin's twist coefficients

The definition of the most-widespread Koechlin's coefficients is given by rearranging Equations (1.35), (1.46) and (1.51) as

$$\alpha_S = Z\sqrt{S} = \frac{\kappa}{\pi D}\sqrt{\frac{\pi D_S^2}{4}} = \frac{\kappa}{2\sqrt{\pi}}\left(\frac{D_S}{D}\right) = \frac{\kappa\sqrt{\mu}}{2\sqrt{\pi}}. \tag{1.53}$$

Then, by using Equations (1.33) and (1.53), we obtain

$$\alpha = Z\sqrt{T} = Z\sqrt{S\rho} = \alpha_S\sqrt{\rho} = \frac{\kappa\sqrt{\mu\rho}}{2\sqrt{\pi}}. \tag{1.54}$$

Let us note that Koechlin's substance twist coefficient α_S is a dimensionless quantity.

Overview of dimensionless quantities

The following dimensionless quantities are introduced earlier: relative fineness τ, coefficient k_n, packing density μ, substance coefficient of yarn diameter K_S, twist intensity κ and Koechlin's substance twist coefficient α_S.

Note: It is possible to use four dimensionless quantities τ, κ, μ and k_n, in place of a set of initial parameters T, Z, D and n. All other quantities can be calculated by using the equations derived so far in this chapter.

1.4   Yarn fineness, twist and diameter in accordance with Koechlin's concept

The earlier-stated quantities and their mutual relations are the results of definitions and systematic derivations. They are valid independent to the type of material used and the technology employed for creation of yarns. On the contrary, the relation among yarn fineness T, twist Z and diameter D is a result of special behaviours of fibrous material in the yarns.

By twisting, the fibrous material of a yarn gets compressed in consequence of working of action forces. Nevertheless, this fibrous material resists against deformation by its reaction forces. The equilibrium of action and reaction forces determines the final diameter of the yarn. So, the relations among T, Z and D cannot be solved on the basis of definitions and relations only. They should be rather determined by taking into account of the mechanical behaviour of fibrous material.

Probably, the oldest semi-empirical solution to the above-mentioned problem was given by Koechlin[4] in the first quarter of the 19th century.

General assumptions of Koechlin's concept

Koechlin thought about yarns that are

1. prepared from same fibrous material,
2. spun by employing same type of spinning technology and
3. designed for same (and/or analogical) purpose of end-use.

For such yarns, he determined a relation between yarn twist, fineness and diameter based on two special assumptions.

Koechlin's first special assumption

This assumption can be stated as follows: 'The packing density of yarn is an increasing function of yarn twist intensity only'. It can be mathematically written as

$$\mu = f(\kappa). \tag{1.55}$$

We, experimentally, observe that the packing density increases with an increase of twist, i.e., also with twist intensity. Nevertheless, this is not the only parameter that decides packing density. According to our recent knowledge, the packing density depends – besides twist intensity – also on other variables, for example, yarn fineness.

Note: A suitable substitution of the first assumption demands the knowledge of mechanics of fibrous assemblies. Nevertheless, this is not fully known till date. However, the models of compression of fibrous assembly, as described in our book [4], can be used.

By accepting the assumption as stated in Equation (1.55) and by using Equations (1.53), (1.54) and (1.48), we can write the following expressions

---

4. Kolundzic [11] quoted that 'Koechlin presented his known equation, through which twist and number (fineness) of yarn are regulated'. Nevertheless, Budnikov [12] reported that Koechlin's formula was empirical and already known.

$$\alpha_S = \frac{\kappa\sqrt{f(\kappa)}}{2\sqrt{\pi}} ,$$ (1.56)

$$\alpha = \frac{\kappa\sqrt{f(\kappa)\rho}}{2\sqrt{\pi}} ,$$ (1.57)

$$K_S = \frac{2}{\sqrt{\pi f(\kappa)}} ,$$ (1.58)

$$K = \frac{2}{\sqrt{\pi f(\kappa)\rho}} .$$ (1.59)

Let us note that all four quantities, i.e., α_S, α, K_S, K, are functions of twist intensity κ only.[5]

Koechlin's second special assumption

This assumption can be stated as follows: 'The yarns of different finenesses should have the same numerical value of twist intensity'.[6] Mathematically, it can be expressed as

$$\kappa = \text{constant} .$$ (1.60)

This assumption (recommendation) needs a logical discussion. The yarns having the same (and/or analogical) purpose of end-use should be logically same (or similar) in terms of all properties. However, it is not possible in an 'absolute' sense. (The yarns with different finenesses cannot have all properties same – such a requirement is often called as 'contradictio in adjecto'.) We, therefore, think about geometrical properties only. If all yarns shall be geometrically similar then, based on the regulations of geometrical similarity, the corresponding angles must be same. It means that the angle β_D must be same, and then $\tan\beta_D = \kappa$ must be same too.

By applying Equation (1.60) in equations from (1.55) to (1.59), we can write

---

5.   The density ρ, as mentioned in Equations (1.57) and (1.59), is a constant when we think about same fibrous material in accordance with the first assumption.

6.   It indicates to same (and/or analogical) purpose of end-use, according to the third general assumption.

$$\mu = \text{constant}, \tag{1.61}$$

$$\alpha_S = \text{constant}, \tag{1.62}$$

$$\alpha = \text{constant}, \tag{1.63}$$

$$K_S = \text{constant}, \tag{1.64}$$

$$K = \text{constant}. \tag{1.65}$$

The previous results are usually used in two following directions:

1. For determination of suitable twist for yarns with different finenesses. By using the definition of α from Table 1.8 and Equation (1.63), it is valid to write that

$$Z = \frac{\alpha}{\sqrt{T}}, \quad \text{e.g.} \quad Z_{\left[m^{-1}\right]} = \frac{\alpha_{\left[m^{-1}ktex^{1/2}\right]}}{\sqrt{T_{[ktex]}}} = \frac{\alpha_{\left[m^{-1}ktex^{1/2}\right]}}{\sqrt{T_{[tex]}/1000}} = \frac{31.623\,\alpha_{\left[m^{-1}ktex^{1/2}\right]}}{\sqrt{T_{[tex]}}}. \tag{1.66}$$

A suitable constant value of α is known from our practical experience. (For example, we can use $\alpha = 120\,m^{-1}ktex^{1/2}$ for cotton carded ring-spun yarns.)

Note: For twist coefficient α, the dimension that is frequently used is $m^{-1}ktex^{1/2}$. It follows namely the older characteristic of yarn fineness – 'number metric', $T = 1000/Nm$ – see Table 1.6. (Earlier, the following expression was used $Z_{\left[m^{-1}\right]} = \alpha_{\left[m^{-1}ktex^{1/2}\right]}\sqrt{Nm_{\left[ktex^{-1}\right]}}$.)

Besides this, the English twist coefficient α_e (cotton) is also used in some countries. Its formula[7] is $Z_{\left[inch^{-1}\right]} = \alpha_e\sqrt{Ne_c}$; Ne_c – see Table 1.6. (If, for example, $\alpha = 120\,m^{-1}ktex^{1/2}$, then the corresponding English twist coefficient (cotton) is $\alpha_e = 3.97$.)

2. For determination of diameter of yarns with different finenesses. By using Equations (1.49) and (1.65), it is valid to write that

---

7. The physical dimension of α_e is quite complicated. It is expressed in $inch^{-1}(840\,yards)^{-1/2}\,pound^{1/2}$, because Ne_c, according to Table 1.6, is expressed in $(840\,yards)\,pound^{-1}$. Nevertheless, the dimensions are not used in industrial practice.

$$D = K\sqrt{T}, \quad \text{e.g.} \quad D_{[\text{mm}]} = K_{\left[\text{mm tex}^{-1/2}\right]}\sqrt{T_{[\text{tex}]}} \ . \tag{1.67}$$

A suitable constant value of K is also known from our practical experience. (For example, we can recommend $K = 0.0395\,\text{mm tex}^{-1/2}$ for cotton carded ring-spun yarns.)

1.5   Empirical corrections to Koechlin's concept

The traditional equations of Koechlin's theoretical concept are very easy so that they are used in different practical applications worldwide. Nevertheless, they are not too precise. This is mainly because of the first special assumption of Koechlin which is not found to be in good accordance with the experimental results.

Table 1.9 Some proposals of twist exponents

Twist exponent q	Authors	Year	Remark
0.5	Koechlin	1828	1, 2
0.6	Staub	1900	1, 2
0.644	Johansen	1902	Warp, 1, 2
0.785	Laetch	1905	Warp, 1, 2
0.720			Weft, 1, 2
0.62–0.75		1941	1, 2
0.666			3
0.666	Phrix	1942	2, 4, 5
0.6	Neckář	1971	6
0.577			6
0.570	Salaba	1975	7
0.551			7

Note: 1: Kolundzic [11]; 2: Banke [13]; 3: Koch and Wagner [14]; 4: Simon [15]; 5: Czech branch standard [9]; 6: Neckář [16]; 7: Salaba [17].

Intervals of twist coefficients

The authors of different handbooks of spinning solved the above-mentioned disproportion by dividing the whole interval of possible yarn finenesses (same type of yarns) into a few groups – for example, a group of fine yarns, a group of coarse yarns, etc. Then, they recommended a little different values of Koechlin's twist coefficient for different yarn groups.

Modification of twist exponent

In practice, the finer yarns are twisted a little more than the coarser yarns to obtain good properties. Therefore, a lot of different authors proposed an empirical generalization of Equation (1.66). In opposite to the twist exponent resulting from Koechlin's concept ($q = 0.5$ – compare Table 1.8), authors recommended different modified values for this exponent. Table 1.9 illustrates some proposed twist exponents.

It is interesting to note that all authors recommended twist exponents greater than 0.5 ($q > 0.5$). They are mostly not too far from 0.6. The quantity $q = 0.666$ recommended by Phrix (and Laetch) is used in Czech Republic[8] for single yarns. Then, Equation (1.66) can be modified as follows:

$$Z = \frac{a}{T^{2/3}}, \quad \text{e.g.} \quad Z_{[m^{-1}]} = \frac{a_{[m^{-1}\text{ktex}^{2/3}]}}{T_{[\text{ktex}]}^{2/3}} = \frac{a_{[m^{-1}\text{ktex}^{2/3}]}}{\left(T_{[\text{tex}]}/1000\right)^{2/3}} = \frac{100\, a_{[m^{-1}\text{ktex}^{2/3}]}}{T_{[\text{tex}]}^{2/3}}. \quad (1.68)$$

(Note that this type of twist exponent is expressed with traditional symbol a in place of α.)

Modification of equation for diameter

The traditional expression stated in Equation (1.67) has two problems in relation to practical application.

1. The finer yarns are a little thinner than the coarser yarns.

---

8   This practice was originated in Czech Republic just after the World War II. At that time, only the mechanical calculator was available in industry. And it was necessary to use a logarithmic table for exponentiation ($Z = \alpha/T^q$) with a general exponent q. But it was too difficult to use it for routine calculations. Nevertheless, a simple mechanical instrument called 'sliding rule' was popular during those days. Using this instrument, the square and cubic roots were possible to calculate very easily and quickly.

As it was quickly possible to calculate $T^q = T^{2/3} = \sqrt[3]{T^2}$, the Czech branch-standard elected Phrix twist exponent as $q = 2/3 = 0.666$. By the way, the oft-used name 'Phrix' denotes the name of a German company, not the name of any author.

2. It is not true that a high twisted yarn is thinner (more compressed) than a low twisted yarn.

It is possible, for example, to suggest an empirically generalized equation as follows

$$D = Q_\alpha T^w \alpha^v, \quad \text{e.g.} \quad D_{[\text{mm}]} = Q_\alpha T^w_{[\text{tex}]} \alpha^v_{[\text{m}^{-1}\text{ktex}^{1/2}]} \tag{1.69}$$

where Q_α, w, v are suitable parameters.

Note: The numerical value of parameter Q_α along with its physical dimensions can be used in the last equation.

For example, it is possible to use the following expression for cotton carded ring-spun yarns:

$$D_{[\text{mm}]} = 0.1018 T^{0.553}_{[\text{tex}]} \alpha^{-0.220}_{[\text{m}^{-1}\text{ktex}^{1/2}]} . \tag{1.70}$$

By expressing μ from Equation (1.47), then substituting D from Equation (1.69), and applying α according to Equation (1.66), we can obtain the following expression for yarn packing density:

$$\mu = \frac{4T_{[\text{tex}]}}{\pi D^2_{[\text{mm}]} \rho_{[\text{kg m}^{-3}]}} = \frac{4T_{[\text{tex}]}}{\pi \ Q^2_\alpha T^{2w}_{[\text{tex}]} \alpha^{2v}_{[\text{m}^{-1}\text{ktex}^{1/2}]} \ \rho_{[\text{kg m}^{-3}]}}$$

$$= \frac{4T_{[\text{tex}]}}{\pi \rho_{[\text{kg m}^{-3}]} Q^2_\alpha T^{2w}_{[\text{tex}]} \left(\dfrac{Z_{[\text{m}^{-1}]}\sqrt{T_{[\text{tex}]}}}{\sqrt{1000}} \right)^{2v}} = \frac{4 \cdot 1000^v}{\pi \rho_{[\text{kg m}^{-3}]} Q^2_\alpha} Z^{-2v}_{[\text{m}^{-1}]} T^{1-2w-v}_{[\text{tex}]}. \tag{1.71}$$

By substituting $Q_\alpha = 0.1018$, $w = 0.553$ and $v = -0.220$ from Equations (1.70) to (1.71), we obtain the following expression for packing density of yarn:

$$\mu = 0.01768 Z^{0.440}_{[\text{m}^{-1}]} T^{0.114}_{[\text{tex}]} . \tag{1.72}$$

Based on Equations (1.66) and (1.68), it is possible to write that $a_{[\text{m}^{-1}\text{ktex}^{2/3}]} = 0.3162 T^{1/6}_{[\text{tex}]} \alpha_{[\text{m}^{-1}\text{ktex}^{1/2}]}$, so that the following expression can also be obtained from Equation (1.70):

$$D_{[\text{mm}]} = 0.079 T^{0.59}_{[\text{tex}]} a^{-0.22}_{[\text{m}^{-1}\text{ktex}^{2/3}]} . \tag{1.73}$$

1.6   References

[1] Goswami, B. C., Martindale, J. G., and Scardino, F. L., Textile Yarns: Technology, Structure, and Applications, John Wiley & Sons, New York, 1977.

[2] Malinowska, K., Prace Inst. Wlok. (Research Report), 29, 1979 (In Polish).

[3] Morton, W. E. and Hearle, J. W. S., Physical Properties of Textile Fibers, The Textile Institute and Butterworth and Co. (Publishers) Ltd., London, 1962.

[4] Neckář, B. and Das, D., Theory of Structure and Mechanics of Fibrous Assemblies, Woodhead Publishing India Pvt. Ltd., New Delhi, 2012.

[5] Piller, B. and Trávniček, Z., Synthetická vlákna díl. 1 (Synthetic Fibers, Part I), SNTL Publisher, Prague, 1956 (In Czech).

[6] Marschik, S., Physicalisch-technische Untersuchungen von Gespinsten und Geweben (Physical and Technical Investigation of Yarns and Fabrics), Carl Gerolds Sohn, Wien, 1904 (In German).

[7] Neckář, B., Příze: tvorba, struktura, vlastnosti (Yarns: Creation, Structure, Properties), SNTL Publisher, Prague, 1990 (In Czech).

[8] Johansen, O, Handbuch der Baumwollspinnerei Band I (Handbook of Cotton Spinning, Volume I), Verlag von B. F. Voigt, Leipzig, 1930 (In German).

[9] Czech Branch-Standard, Režné jednoduché bezvřetenové bavlnářské příze (Grey single OE yarns), No. ON 802120, MP ČSR and BP, Hradec Králové, 1976 (In Czech).

[10] Czech Association of Cotton industry, Pravidla technické exploatace v bavlnářských přádelnách (Regulations of Technical Exploatation for Cotton Spinning Mills), Hradec Králové, 1961 (In Czech).

[11] Kolundzic, B., Koechlinov obrazac i osnova jednaczina za prediva (Koechlin's formula and fundamental equation for yarns), Tekstilna industrija, 13, 251–255, 1965 (In Serbian).

[12] Budnikov, V. I., Osnovy prjadenija Czast 2 (Principles of Spinning, Part 2), Gizlegprom Publisher, Moscow, 1945 (In Russian).

[13] Banke, K. H., Übersicht zur Problematik der Drehungsberechnung von Gespinsten (Summary to problems of twist calculation of yarn), Faserforschung und Textiltechnik, 8, 280–285, 1957 (In German).

[14] Koch, P.-A. and Wagner, E., Textile Prüfungen (Textile Testing), 8th Edition, Wuppertal-Elbersfeld, Spohr, 1966 (In German).

[15] Simon, J., Přádelnictví (Spinning), Teaching Book, Technical University of Liberec, 1964 (In Czech).

[16] Neckář, B., Struktura a vlastnosti bavlněné příze mykané (Structure and properties of cotton carded yarns), Research Report S 72-IX, State Textile Research Institute, Liberec, 1971 (In Czech).

[17] Salaba, J., Geometrie a vlastnosti staplových přízí (Geometry and Properties of Staple Yarns), PhD Thesis, Technical University of Liberec, 1974 (In Czech).

Creation of yarns

2.1   General structure of technological process

Yarn manufacturing technology[1], where the input fiber material is changed to the output fibrous product – yarn, is relatively complicated. Besides yarn, a variety of other fibrous assemblies are also being produced by this technology. They are semi-finished products, commonly known as lap, sliver, roving, etc.

Technological operations

The technological operations consist of many activities by which the input fiber material is changed step by step to the output fibrous product. In principle, two modes of changes take place during the spinning operation – one in terms of aggregation and the other in terms of modification (Table 2.1).

---

1   According to a very old edition of Mayer's Lexicon [1], J. Beckman introduced the term 'technology' and its detachment as a separate scientific branch in 1770. The conversion of initial material to a useful object creates a common attribute which is defined differently in literature. The change of material substance ('chemical' technologies) or change of form and shape ('mechanical' technologies) represents different traditional interpretations of technology. Lately, these conversions are comprehended by a larger sense of natural sciences.

Table 2.1 Types of changes occur in spinning technology [2]

Mode of change	Variant	Activity
In terms of aggregation	Divergence	Opening
		Separation
	Convergence	Concentration
		Joining
In terms of modification	Reconstruction	Change of fiber shapes
		Change of fiber configuration
	Changes of contacts	Modification of character of fiber-to-fiber contacts
		Change of number and configuration of fiber-to-fiber contacts
	Modification of mass	Modification of fiber surfaces
		Modification of whole fiber substance

The changes in terms of aggregation always pertain to groups of fibers and indicate to their mutual divergence or convergence. By opening, the fiber material remains completely intact in terms of its mass. But, by separation, many new groups of fibers can be generated. Similarly, by concentration, only one formation is changed. But, by joining, many partial formations are united to form a higher complex.

The changes in terms of modification pertain to individual fibers as well as to fibrous assemblies. Reconstruction means a geometrical change either in terms of geometry of individual fibers or in the light of arrangement of fibers. The changes of fiber-to-fiber contacts can be understood as a modification of character of individual contacts or in terms of changes of number and/or configuration of contacts in the fibrous assembly. The modification of mass includes possibilities of physical, chemical, and other different modifications of fiber surfaces and/or modification of mass of fibers of the whole assembly.

This relatively general classification characterizes the changes observed in the fibers and/or in the fibrous assemblies. Some of these changes are in principle necessary for creation of yarns, while some other are necessary for smooth running of the technological processes. However, sometimes 'undesirable' changes can also occur in the spinning process.

Table 2.2 Types of activities [2]

Classification of activities		Main types of transformations (Table 2.1)	Traditional terms
Group	**Activity**		
Preparation of fiber material	Selection of raw material for yarns	Opening, separation	Cleaning, removal of short fibers
	Modification of fiber mass	Modification of fiber surface and whole fiber substance	Change of fiber properties*
Arrangement	Fiber blending	Joining, change of mutual fiber configuration	Blending
	Dividing and joining of fiber flow	Separation, joining	Creation of fiber flow
	Change of fineness and mass unevenness	Separation, joining, change of mutual fiber configuration	Attenuation, improvement of evenness
	Change of density of fiber flow	Opening Concentration	Opening, fiber concentration
	Directional ranking operations	Change of fiber shapes, change of mutual fiber configuration	Parallelization
Joining	Intensive mutual contacting of fibers	Change of number and configuration of contacts	Twisting, Strengthening
Transfer	Manipulation, adjustment, transport		Winding

*Primarily, shortening of fibers takes place due to carding and drafting.

In practice, there exist different types of activities in the spinning process. They are differentiated by the fiber materials used, technical equipments employed, and technological procedures adapted. However, these activities are often similar in principle, if not they are same. According to the traditional literature, the activities include fiber opening, fiber cleaning, removal of short fibers, fiber blending, doubling, attenuation, improvement of evenness, condensation, twisting, winding, etc. Nevertheless, we can consider such activities in the light of transformation of fibrous assemblies. There exist four basic groups of activities such as preparation of fiber material, arrangement, joining, and transfer (see Table 2.2).

Spinning technologies

Traditionally, it is possible to characterize the operations of existing spinning technologies into two groups: (a) preparatory to spinning process and (b) spinning process.

Opening and blending are considered to be preparatory processes. In case of cotton, the impurities are mechanically removed by opening and cleaning. But, in case of wool, cleaning is done chemically by scouring, followed by drying. The relatively clean fibers then follow carding process, where the fiber clumps are reduced to individual fibers and/or small fiber bundles, finer impurities are removed, and undesirable (short, etc.) fibers are separated as card strips. Simultaneously, a sliver with partly oriented fibers is created. (More special processes are used for preparation of flax fiber bundles.)

The slivers are then doubled and drafted whereby the mixing of fiber materials takes place, the longitudinal orientation of fibers enhances, and the unevenness of fiber flow reduces. The combing operation removes short fibers and also increases fiber parallelization. The final operation usually represents formation of roving, which takes place by attenuating the slivers and partial strengthening of the fiber strand. The above-mentioned traditional principles of operations of preparatory processes are generally preserved, in spite of the development of new machines or improvement of machine elements. (Probably, we find more intrinsic differences only in case of preparation of broken tow by using crush-cutting or crush-breaking convertors.)

A more heterogeneous development can be observed in spinning process. Traditionally, the final preparation of material during spinning creates a very fine and thin strand by drafting of roving on a ring spinning frame. Nevertheless, such a thin strand can also be prepared from a 'thick' (drawn) sliver by using combing roller on an open end (OE)-rotor-spinning machine.

Note: In a limited sense, the term 'spinning' means creation of single twisted yarn from a thin strand of more or less parallel fibers. Then, the traditional term 'OE spinning' is not fully correct. There, the interruption of fiber flow, i.e., 'open end', is originated before twisting, already by creation of a thin strand of fibers.

Finally, strengthening is realized by the process of twisting of single staple fiber yarn. Different spindles are used for traditional (monotonous) twisting – mule, ring, flyer, centrifugal, two-for-one, etc. Some ring spinning frames can condensate fiber strand and then twist the fibers from the hairiness region so as to incorporate them into the yarn body; this is known as the so-called compact spinning[2].

In some cases, the fiber strands, issuing from the drafting system of ring spinning frame, are divided into very fine strands (solo-spinning) or two exceedingly thin fiber strands are put together (Siro-spinning) and then such structures are twisted together. The resulting yarns look like plied yarns; however, they are single twisted yarns with a better organization of their internal structure. (See Chapter 2.3.5 of Reference [2] for more details.)

Yarns, twisted alternately by false-twist, i.e., twisted in clock-wise and anti-clock-wise directions, create another group. Such yarns do not exist independently, but a couple of them, plied immediately after twisting, create a relatively stable formation – the so-called self-twist yarns. (For example, Repco machine inserts false-twist by rolling two yarns in-between moving cylinders. In an alternative to this, the recurrent 'whirlwind' is used.)

An alternative type of twisting includes rotation of free end (OE) of yarn. It is mostly realized by rotors of different constructions, by rotation of frictional cylinders, or by force generated by swirling airflow. Different machines of the so-called rotor spinning represent the first case. Friction spinning machines, often called as DREF spinning, represent the second case. There exist also machines which are spinning in 'whirlwind' and machines which work with electrostatic field (electro spinning).

Except 'proper' twisting, 'partial' twisting is also used for strengthening of staple fiber yarns. Only fibers and fiber ends from yarn surface are intensively twisted around the non-twisted yarn core by the so-called air-jet spinning (type Murata, etc.)[3].

---

2   Such yarns are smoother; therefore, the resulting fabric gives better appearance. Nevertheless, the compact yarns are not always advantageous. This is because of the following reasons. (1) The handle of the fabrics made from compact yarns can be partially hard. (The hairs function as a soft and flexible layer on yarn surface.) (2) The surface of compact yarns can absorb only a small volume of viscose liquids, for example, adhesive, which can impose certain problems during technical usage. (3) As compared to compact yarns, the non-compact yarns impart partially higher cover effect in a woven or a knitted structure as the yarns with significant hairs have a little higher 'cover diameter'; therefore, we can sometimes use a little smaller quantity of such yarns for production of the same type of textile structures.

3   This technology is more advantageous for finer yarns, because the fine yarns provide larger surface in relation to their volume, so, relatively more fiber material can be twisted around the yarn core.

In principle, there exist methods of creation of twist-less staple fiber yarns, e.g., adhesive yarns, however, such technologies (e.g., Bobtex system) have not been very popular worldwide.

2.2   Outline of principles of fiber twisting in yarn

We here again mention that the dominant principle of yarn creation consists of traditional (monotonous) twisting. One (or more) continuous thin strand(s) is (are) provided as input to the twisting process. The approximately helical organization of fibers, which is originated by turning of the strand(s), implicates strengthening of the yarn by this way.

Principles of fiber twisting

The complete action of yarn creation can take place in a very small space, containing only very short length of a thin strand (or many strands). This is called 'twisting around a point'. In other case, a relatively long length of a thin strand (or many strands) is turned by gradual insertion of twists. This is called 'twisting around a length'.

'Coaxial' twisting assumes a common axis of a thin strand (or many strands) and of the resulting yarn, while 'non-coaxial twisting' considers different directions of the axes of the thin strand (or strands) and the resulting yarn. Further, we classify the coaxial twisting in two variants: 'symmetrical' and 'asymmetrical'; and the non-coaxial twisting also in two variants: 'one strand' or 'more strands'. So, there exist eight principles of twisting altogether [2] (Table 2.3).

Table 2.3 Principles of fiber twisting in yarn [2]

In Variant 1, a thin cylindrical strand, consisting of fibers, is considered. The fibers are concentrated and twisted in a small and slightly conical space of the yarn. In case of common method of spinning, the idea of cylindrical form of a thin strand is not usually realized. Nevertheless, it is possible to find this variant in many places – as a certain simplification of reality.

Variant 2 assumes an asymmetrical thin strand, for example, in a form of flat ribbon. In this case, at first, a helical body with more or less rectangular cross section is originated according to either option (a) or option (b). In option (a), a flat and ribbon-shaped strand is finally deformed itself to the cylindrical shape of a twisted yarn. In option (b), the flat and ribbon-shaped strand takes the yarn shape quickly. This way of twisting corresponds to the traditional view of yarn creation with a spinning triangle being the output of the drafting process on a ring spinning frame.

Variant 3 considers twisting of two or more thin strands together. This is generally known from the traditional production of plied yarns; however, this is also seen during the creation of siro-spun yarn. The traditional twisting of fibers in the spinning triangle, which is sometimes idealized as winding of a couple of flat ribbons, also corresponds to this variant.

Variant 4 assumes winding of one flat and thin strand – usually ribbon – around the axis of the yarn. The generated 'tube' is then (continuously) compressed to 'compact' cylindrical body by the subsequent turns. Balls [3] and later Hearle et al. [4] thought over this type of twisting in case of ring spinning. Rohlena et al. [5] demonstrated the reality of this twisting in case of rotor spinning. A mathematical model of ribbon winding is given in Reference [2].

Note: The twist of second order, derived in Section 4.6 and illustrated in Figure 4.12 of this book, follows such mechanism too.

Variant 5 represents gradual twisting around a long length of a cylindrical strand. It approximately corresponds to creation of yarns by mule spinning. Probably, this is the oldest form of twisting that resulted in Koechlin's assumptions for twist coefficient and other basic quantities of yarns.

Variant 6 depicts gradual twisting around a long length of a non-cylindrical strand. It partly gives a more accurate description of twisting of a thin strand in mule spinning. This variant can be viewed as an alternative to Variant 2(a).

Variant 7 shows gradual twisting around a long length of a couple of strands or more strands. This type is not usually seen during yarn spinning. Nevertheless, this can be seen in case of hand-made production of ropes, lines, etc.

Variant 8 expresses twisting of a thin strand in analogy to winding of a ribbon in Variant 4 but around a relatively long length. It can be locally realized by twisting in mule spinning under certain geometrical and mechanical (tension) conditions. It had been used in some models as a simplification of twisting according to Variant 4.

The above description of eight variants characterizes typical situations of yarn creation by twisting. Evidently, there exist possibilities for transition from one variant to another.

2.3   Notes to methodology of studying yarns

Difficulty during observations

There exist different transient states of yarn creation in relation to instantaneous conditions. One has to be careful about it while studying the creation of yarns. The other difficulty lies in spatio-temporal problem of yarn creation. According to the variants of 'twisting around a point', which are most common, a thin strand forms a yarn in a very short time – usually in several milliseconds and the most important action takes place in 10 µs only. Further, the dimension of space, in which the conversion from fibers to yarn takes place, is commensurable with yarn diameter, i.e., in tenth of a millimetre, and in some of the processes, this is done in even smaller regions. Further, the optical opacity makes the study of yarn creation very difficult. A lot of partial processes that are taking place in the yarn body cannot be observed by traditional optical methods and instruments[4]. Nevertheless, only a limiting sphere of knowledge is possible to obtain by carrying out analysis of samples from different places and at different moments of spinning activities. The yarn creation is a markedly dynamic process and there exist different formations that are not stable in terms of static equilibrium. Therefore, a direct observation is not ideal for enough deeper understanding of yarn and its creation[5]. Alternatively, it is better to use indirect methods of cognition and understanding, adherent to the creation of suitable models.

---

4   Of all the microscopic methods, the Morton's 'tracer fiber technique' [6–8] is still very popular [4]. Based on its principle, Stejskal and Kasparek [9,10] developed a special instrument, called 'Omest', which they and later Neckář [2] used for experimental analysis of yarn structure. At present, there exist many other techniques – for example, 'micro-CT' (computed tomography) – which might be used in future for studying yarn structure.

5   Such situation also exists in many other branches of natural sciences. For example, it is often not possible to quickly observe and interpret nuclear and sub-nuclear particles in nuclear physics on one hand and some processes and objects of universe in astrophysics on the other hand.

Modelling

The term 'model' indicates an object of different nature which corresponds to the reality to a certain degree and in a quantitative sense. Except material models (dummy copies), there exist different ideal (non-material) models – structural (isomorphs) models, and functional (hetero- or homo-morphs) models. The structural models are equivalent to the original object in terms of their internal structures and properties, but the functional models are equivalent to the original object only in terms of their properties (not in terms of their internal structure). Based on the degree of cognition, it is possible to distinguish between the models of observation and experiments, which generate knowledge for a suitable configuration and generalization, and the theoretical models, which provide knowledge and understanding based on verified theories and hypothesis.

In this book, we prefer ideal, structural, and theoretical models, which are formulated above all mathematically; so we speak about mathematical models[6] in short.

The models of yarn 'oscillate' in-between two concepts – continuous object and discontinuous object. If the object to be modelled is interpreted as having a very high number of particles such that the particles completely fill the space that the object occupies then the idea of continuum can be applied. In such case, mathematical analysis can be used as a modelling tool. On the contrary, if the object to be modelled is interpreted as having a very less number of particles such that the particles partially fill the space that the object occupies then the object can be considered to be discontinuous. In such case, the concept of 'particle to particle' can be used for modelling; however, a special calculus does not exist for this in reality. In yarn, different hybrid variants of both approaches are often applied, mostly under the concept of 'fiber – bundle – yarn'. We at first create a representative discontinuous model for several fibers as an 'average' building block of the whole formation. Then, we interpret their behaviours as those of differential elements and use them for solving in case of whole yarn by using continuum techniques such as integration.

---

6   Mathematical models can be used for numerical calculations too. But, it is not the main reason for creating mathematical models. As mathematics is a 'language' of our understanding, it does make sense to model the objects and processes by using mathematics.

2.4   References

[1]   Mayers Konversations Lexikon, 5th Edition, Leipzig, Wien, 1897 (In German).

[2]   Nečkář, B., Příze: Tvorba, struktura, vlastnosti (Yarns: Creation, structure, properties), SNTL Publisher, Prague, 1990 (In Czech).

[3]   Balls, W. L., Studies of Quality in Cotton, Macmillan Publisher, London, 1928.

[4]   Hearle, J. W. S., Grosberg, P., and Backer, S., Structural Mechanics of Fibers, Yarns, and Fabrics, Wiley-Interscience, New York, 1969.

[5]   Rohlena, V., Open-end Spinning, SNTL Publisher, Prague, 1974.

[6]   Morton, W. E. and Summers, R. J., Fiber arrangement in card slivers, *Journal of Textile Institute*, 40, P106–P116, 1949.

[7]   Morton, W. E. and Yen, R. J., The arrangement of fibers in fibro yarns, *Journal of Textile Institute*, 43, T60-T66, 1952.

[8]   Morton, W. E., The arrangement of fibers in single yarns, *Textile Research Journal*, 26, 325–331, 1956.

[9]   Stejskal, A. and Kasparek, J., Přístroj pro zjišťování polohy jednotlivých vláken v přízi (Device for detecting the position of the individual fibers in the yarn), Czech Patent No. 117179, 1966.

[10] Stejskal, A., Přístroj OMEST pro optické zkoumání vnitřní struktury příze (OMEST-instrument for optical investigation of internal yarn structure), *Jemna Mechanika a Optika*, 13, 377–382, 1968 (In Czech).

General description of yarn structure

3.1 Description of fiber paths and fiber elementary vectors

Introduction

Yarn is known to be a very old product of human civilization. It has been considered as a thin and twisted bundle of staple fibers for the last 30,000 years[1]. Nevertheless, the yarns became more or less reproducible from the time of development of industrial spinning machines, i.e., approximately 200 years ago. This has allowed us to think about specific regulations that are determining the geometrical configuration – i.e., structure – of different yarns, and the consequences of yarn structure to different physical properties, at most the mechanical properties. Such ideas led to mathematical formulation of structural models of yarn.

It is quite surprising to see that there exist innumerable mathematical models of yarn structure in literature. Why there exists such a 'jungle' of models, when we all think about one category of objects – twisted yarns – being more or less similar mutually? This is probably because the internal structure of yarns is usually very complicated and each model assumes a few aspects of fiber paths and their mutual configuration. The other aspects, which are really existing and significantly influencing yarn structure, are either fully neglected or trivially interpreted. Let us illustrate this in terms of two well-known models of yarn structure.

Let us first take the case of 'helical model of fibers in yarns'. This model assumes the real helical trend of fibers in yarns, but neglects the effect of entangled fibers in yarns, namely fiber loops. Further, the transverse distribution of fiber paths is interpreted such that the packing density is assumed to be same at all places of the yarn, which is too far from reality. Also, this model does not say anything about the mass unevenness of yarns, at most about the longitudinal unevenness.

Let us then consider another (stochastic) model that takes into account of real probabilistic mechanism of transformation of fibers into yarn, e.g.,

---

1   See, e.g., Adovasio et al. [1].

Martindale's model of mass unevenness of yarns. Though this model considers the dominant influences causing the mass unevenness of yarns, but the other geometrical relations determining the yarn structure are either fully neglected or trivially considered.

We like to mention here that this book is not going to offer a general and complex model of yarn structure. Nevertheless, we wish to formulate the equations of fiber paths as generally as possible in this chapter. Numerous models presented in this book will be considered to be the special cases of relatively general equations mentioned in this chapter. This way perhaps allows us to create a unique system for classification of different models of yarn structure.

General fiber path

In Figure 3.1, the dot-and-dash straight line represents the ζ-axis of yarn. The two mutually perpendicular axes x, y determine the transverse plane to ζ-axis (yarn axis) so that the triplet x, y, ζ creates a Cartesian system of yarn coordinates. The cylinder shown in Figure 3.1 explains 'yarn body'[2], the thick curved line represents one general fiber inside the yarn.

Note: We often think about the axis of fiber when we say 'fiber path'.

We usually prefer the system of cylindrical coordinates to describe the path of fibers inside the yarns. As shown in Figure 3.1, the radius $r \geq 0$ is measured from the yarn axis ζ, the angle φ is measured from the x-axis and the axial coordinate ζ refers to point C.

Let us define the starting point $A \equiv r_A, \varphi_A, \zeta_A$ and the end point $B \equiv r_B, \varphi_B, \zeta_B$ of fiber in such a way that the inequality $\zeta_A \leq \zeta_B$ is valid. Further, let us define the positive direction of angle φ in accordance with the direction of (technological) twisting of the yarn.

Furthermore, let us define the quantity l (fiber length coordinate) as a distance between the starting point A and a general point on the fiber (e.g., point C), measured along the path of the fiber. So, $l = l_A = 0$ at the starting point A, and $l = l_B = l_f$ at the end point B. (The length l_f denotes the entire length of the fiber.) Let us note that the length l is always a positive quantity, which is increasing from the starting point A to the end point B.

---

2   This intuitive term will be described more exactly later on.

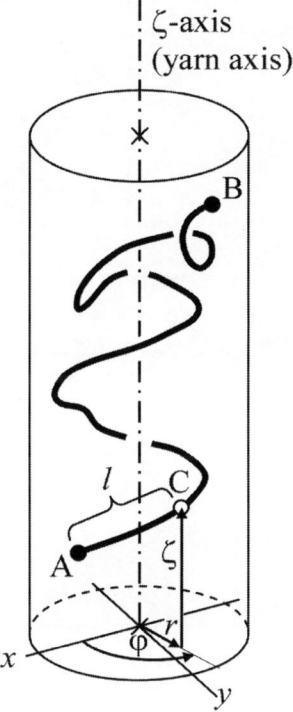

Figure 3.1 Scheme of a general fiber in yarn

The full description of a definite fiber path is given by the following triplet of smooth parametric functions, which are continuous and their first derivatives are also continuous:

$$r = r(l), \quad r \ge 0, \tag{3.1}$$

$$\varphi = \varphi(l), \tag{3.2}$$

$$\zeta = \zeta(l), \tag{3.3}$$

where the parameter $l \in \langle 0, l_f \rangle$. As a matter of fact, it is valid to write that

$$\left. \begin{array}{l} r_A = r(0), \quad \varphi_A = \varphi(0), \quad \zeta_A = \zeta(0), \\ r_B = r(l_f), \quad \varphi_B = \varphi(l_f) \quad \zeta_B = \zeta(l_f). \end{array} \right\} \tag{3.4}$$

The derivatives of the parametric functions[3] shown in Equations (3.1), (3.2) and (3.3) are as follows:

$$dr/dl = r'(l),$$ (3.5)

$$d\varphi/dl = \varphi'(l),$$ (3.6)

$$d\zeta/dl = \zeta'(l).$$ (3.7)

Equations (3.5), (3.6) and (3.7) denote three differential equations which can also determine the path of a fiber when the boundary condition is known, that is, the coordinates of the starting point A are known. The fiber path is then given by the triplet of (differential) equations as follows:

$$dr = r'(l)dl, \quad r_{l=0} = r_A,$$ (3.8)

$$d\varphi = \varphi'(l)dl, \quad \varphi_{l=0} = \varphi_A,$$ (3.9)

$$d\zeta = \zeta'(l)dl, \quad \zeta_{l=0} = \zeta_A.$$ (3.10)

Fiber element

As shown in Figure 3.2a, an elementary fiber length $dl > 0$ creates a diagonal vector (from starting point U to final point V) in the elementary body. The increment of height of this body is $d\zeta$, angular increment is $d\varphi$ so that the increment of width of this body is $rd\varphi$ and the increment of radius is dr.

Note: The differential increments as shown in Figure 3.2a are all positive. Nevertheless, we distinguish between the elementary increment and the elementary 'distance' in this chapter. An elementary increment can be positive (real increment) or negative (decrement in reality) in relation to the 'direction' of change. They are, e.g., dr, $d\varphi$ and/or $rd\varphi$, $d\zeta$. (Only the elementary increment of fiber length dl is always introduced as a positive quantity by means of definition.) On the contrary, the 'lengths' (absolute values) of elementary distances are always thought to be positive. They are

---

3   The derivatives are defined for all values of l, because we assume smooth parametric equations. Often, we will use the symbols r', φ', ζ' for the derivatives in this chapter; however, in other chapters, we prefer to use the ratio of differentials, e.g., $dr(l)/dl$, for the derivatives.

designated by symbols of absolute value in this chapter; e.g., $|dr|$, $|d\varphi|$ and $r|d\varphi|$, $|d\zeta|$ [4].

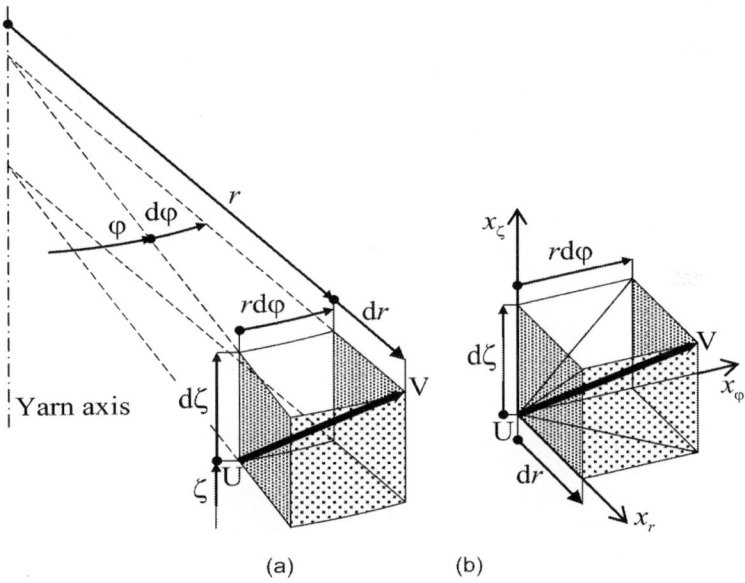

(a)   (b)

Figure 3.2 Fiber elementary vector and corresponding elementary prism

Each of them is differentially so small that it can be interpreted as an elementary prism, as shown in Figure 3.2b. Here, we define the local Cartesian system of axes x_r, x_φ, x_ζ that begins with the starting point U and the positive orientation in the direction of increase of r, φ, and ζ (independent of the orientation of the elementary vector UV).

Directional angles and cosines

The direction of elementary vector UV can be determined by three directional angles ϑ_r, ϑ_φ and ϑ_ζ. There exist eight different directions of the elementary vector UV. They are graphically illustrated in Figure 3.3a–h.

---

4   We understand the absolute values $|dr|$, $|d\varphi|$ and $|d\zeta|$ geometrically. We see them as infinitely short elementary distances, i.e., positive distances, in a three-dimensional space.

Note: Each angle is measured from the corresponding axis to the vector UV and takes a positive value in the interval $(0, \pi)$. If the 'arch' of an angle intersects a 'wall' of elementary prism then such intersection is marked by symbol • in Figure 3.3.

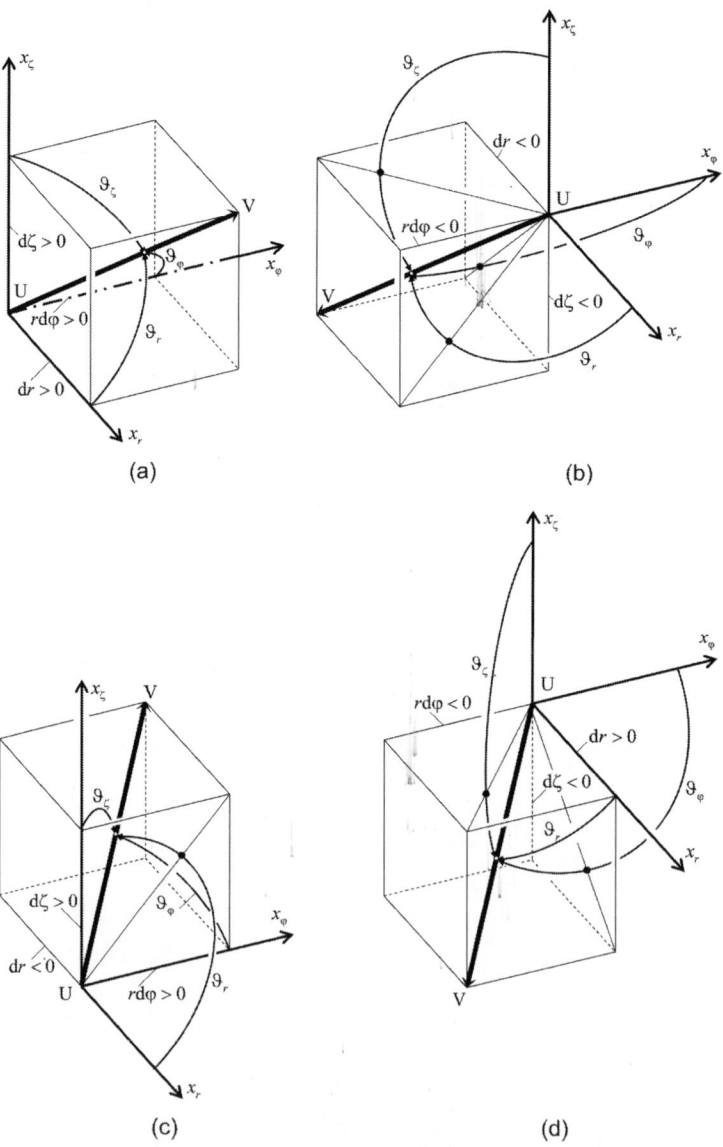

Figure 3.3. Directional angles – schemes from (a) to (d)

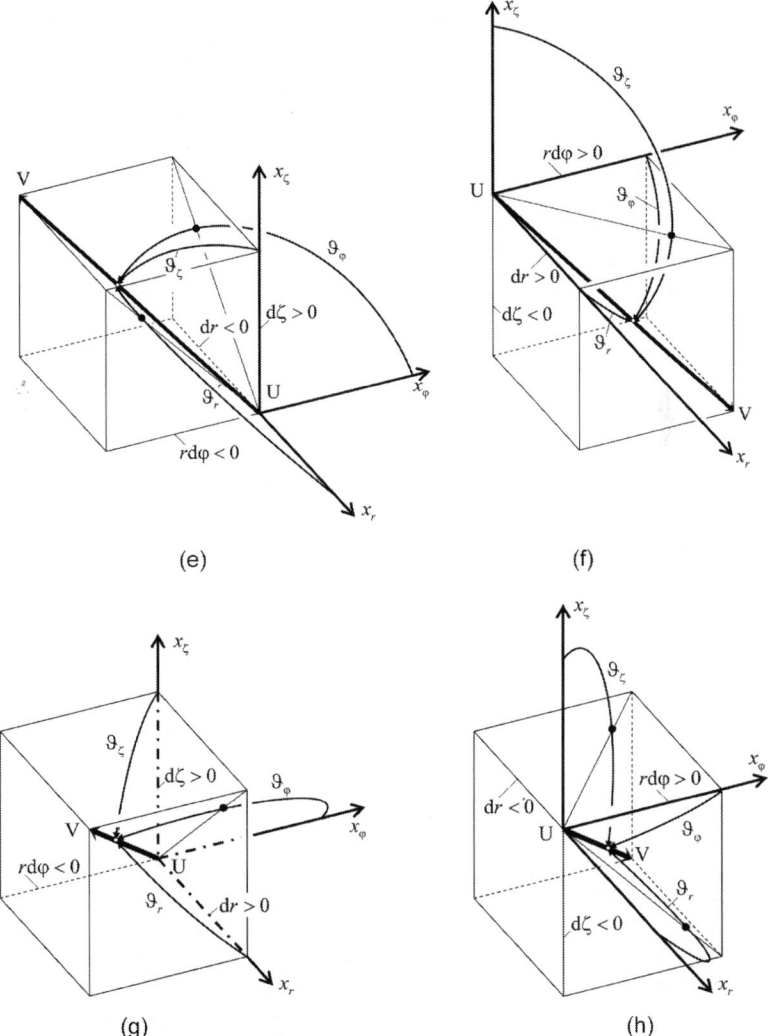

Figure 3.3 Directional angles – schemes from (e) to (h)

It is evident that

$$\cos \vartheta_r = \frac{dr}{dl} , \tag{3.11}$$

$$\cos \vartheta_\varphi = \frac{rd\varphi}{dl} , \tag{3.12}$$

Table 3.1 Signs of increments at edges, directional cosines and directional angles

Figure 3.3	(a)	(b)	(c)	(d)	(e)	(f)	(g)	(h)
dr	>0	<0	<0	>0	<0	>0	>0	<0
$r\,d\varphi$	>0	<0	>0	<0	<0	>0	<0	>0
$d\zeta$	>0	<0	>0	>0	>0	<0	>0	<0
$\cos\vartheta_r = dr/dl$	>0	<0	<0	>0	<0	>0	>0	<0
$\cos\vartheta_\varphi = r\,d\varphi/dl$	>0	<0	>0	<0	<0	>0	<0	>0
$\cos\vartheta_\zeta = d\zeta/dl$	>0	<0	>0	>0	>0	<0	>0	<0
$\vartheta_r \in$	$\left(0,\dfrac{\pi}{2}\right)$	$\left(\dfrac{\pi}{2},\pi\right)$	$\left(\dfrac{\pi}{2},\pi\right)$	$\left(0,\dfrac{\pi}{2}\right)$	$\left(\dfrac{\pi}{2},\pi\right)$	$\left(0,\dfrac{\pi}{2}\right)$	$\left(0,\dfrac{\pi}{2}\right)$	$\left(\dfrac{\pi}{2},\pi\right)$
$\vartheta_\varphi \in$	$\left(0,\dfrac{\pi}{2}\right)$	$\left(\dfrac{\pi}{2},\pi\right)$	$\left(0,\dfrac{\pi}{2}\right)$	$\left(\dfrac{\pi}{2},\pi\right)$	$\left(\dfrac{\pi}{2},\pi\right)$	$\left(0,\dfrac{\pi}{2}\right)$	$\left(\dfrac{\pi}{2},\pi\right)$	$\left(0,\dfrac{\pi}{2}\right)$
$\vartheta_\zeta \in$	$\left(0,\dfrac{\pi}{2}\right)$	$\left(\dfrac{\pi}{2},\pi\right)$	$\left(0,\dfrac{\pi}{2}\right)$	$\left(0,\dfrac{\pi}{2}\right)$	$\left(0,\dfrac{\pi}{2}\right)$	$\left(\dfrac{\pi}{2},\pi\right)$	$\left(0,\dfrac{\pi}{2}\right)$	$\left(\dfrac{\pi}{2},\pi\right)$

$$\cos \vartheta_\zeta = \frac{d\zeta}{dl} .$$

(3.13)

The signs of the increments at the edges of the prism and the signs of cosines and the intervals of angles are shown in Figure 3.4. They are also reported in Table 3.1.

Note: The directional cosines according to Equations (3.11), (3.12) and (3.13) are closely related to the derivatives of the parametric equations in accordance with Equations (3.5), (3.6) and (3.7). They are as follows:

$$r'(l) = \cos \vartheta_r, \quad \varphi'(l) = \cos \vartheta_\varphi / r, \quad \zeta'(l) = \cos \vartheta_\zeta .$$

(3.14)

Each diagonal dl of the elementary prism, shown in Figure 3.2b (or prisms shown by eight schemes in Figure 3.3), follows the Pythagorean theorem:

$$d^2 l = d^2 r + (r d\varphi)^2 + d^2 \zeta .$$

(3.15)

By applying Equations (3.11), (3.12) and (3.13) in Equation (3.15), we obtain

$$1 = \frac{d^2 r}{d^2 l} + \frac{(r d\varphi)^2}{d^2 l} + \frac{d^2 \zeta}{d^2 l} = \cos^2 \vartheta_r + \cos^2 \vartheta_\varphi + \cos^2 \vartheta_\zeta .$$

(3.16)

This is a very well-known general rule of directional cosines.

The expression for the increment of fiber length can also be obtained from Equation (3.15). This is stated as follows:

$$dl = \sqrt{d^2 r + (r d\varphi)^2 + d^2 \zeta} .$$

(3.17)

The directional cosines according to Equations (3.11) to (3.13) can either take a positive value or a negative value. The absolute values of the directional cosines are determined as follows:

$$|\cos \vartheta_r| = \frac{|dr|}{dl} ,$$

(3.18)

$$|\cos \vartheta_\varphi| = \frac{r |d\varphi|}{dl} ,$$

(3.19)

$$|\cos \vartheta_\zeta| = \frac{|d\zeta|}{dl} .$$

(3.20)

Angles of projections and their tangents

The projections of elementary vector UV of length d*l* onto the walls of the elementary prism[5] (see, e.g., Figure 3.2b, dotted lines) allow us to determine the direction of fiber element by another method, based on three angles among the local axes x_r, x_φ, x_ζ and the projections of fiber element.

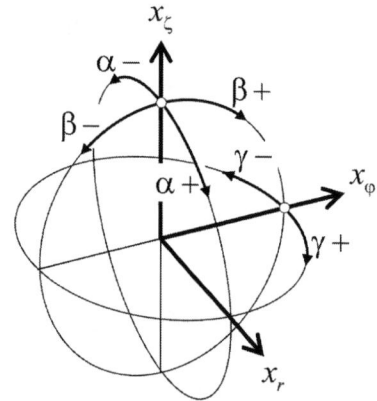

Figure 3.4 Determination of angles α, β, γ

We determine the angles α, β and γ according to Figure 3.4. Each of these angles is lying in the interval $(-\pi/2, \pi/2)$. The angle α lies in the plane created by the axes x_ζ and x_r, the angle β lies in the plane created by the axes x_ζ and x_φ and the angle γ lies in the plane created by the axes x_φ and x_r. The positive directions are marked by α+,β+,γ+, while the negative directions are marked α−,β−,γ−. This is shown in Figure 3.4.

Eight different types of directions of elementary vector UV are graphically illustrated in Figure 3.5a–h. Three projections of elementary vector UV (length d*l*) onto the walls of the elementary prism are represented by the thick dashed lines in Figure 3.5.

---

5   The walls which contain the starting point U should always contain two local axes.

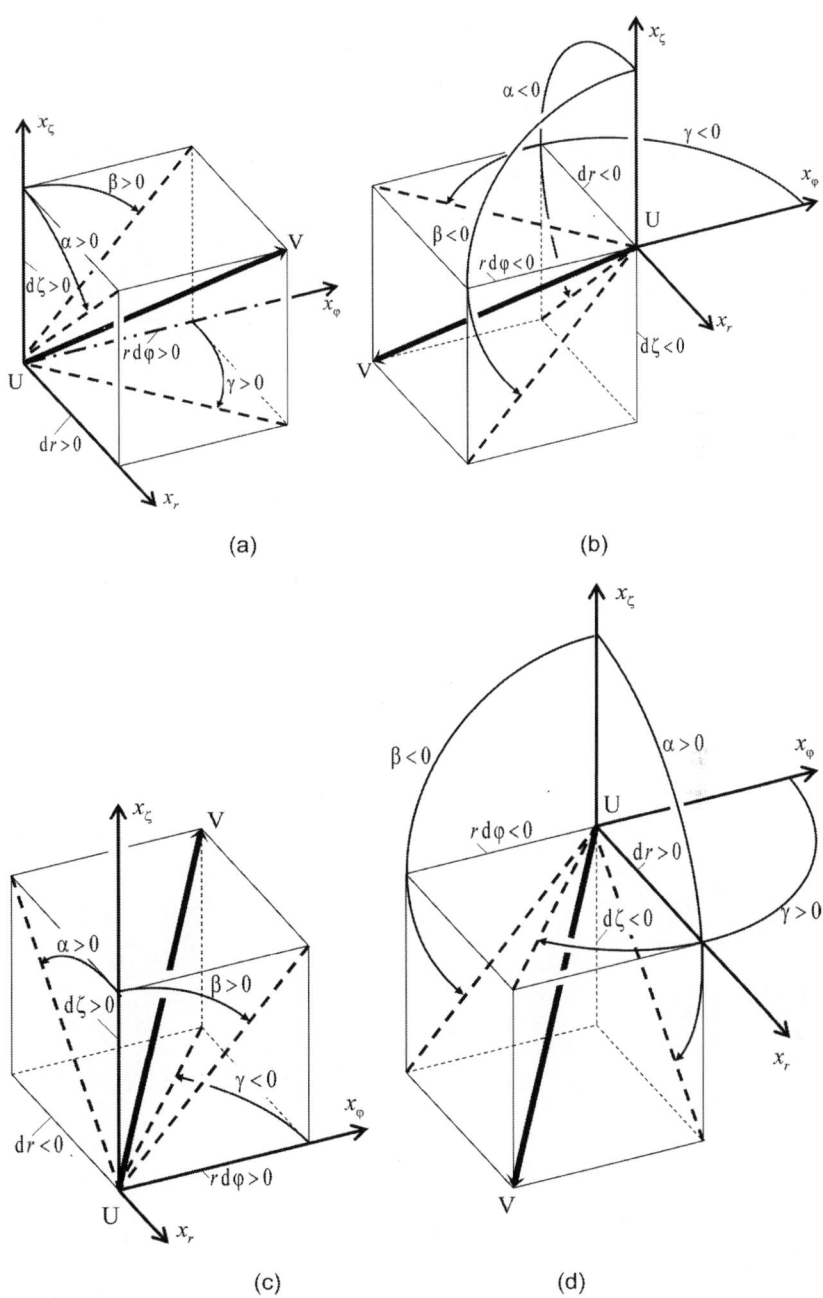

Figure 3.5 Angles of projections – schemes from (a) to (d)

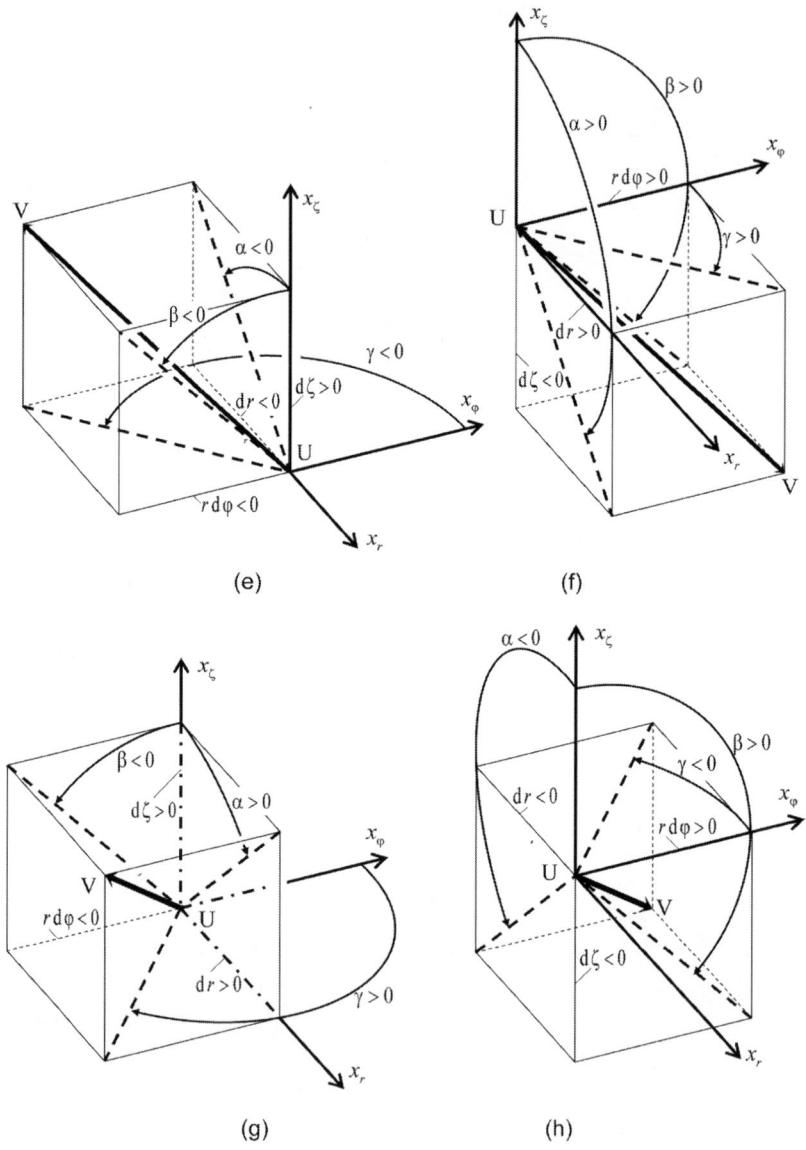

Figure 3.5 Angles of projections – schemes from (e) to (h)

It is evident that

$$\tan \alpha = \frac{dr}{d\zeta},$$

$$(3.21)$$

Table 3.2 Signs of edge increments, projection tangents and projection angles

Figure	(a)	(b)	(c)	(d)	(e)	(f)	(g)	(h)
dr	>0	<0	<0	>0	<0	>0	>0	<0
$r\,d\varphi$	>0	<0	>0	<0	<0	>0	<0	>0
$d\zeta$	>0	<0	>0	<0	>0	<0	>0	<0
$\tan\alpha = dr/d\zeta$	>0		<0		<0		>0	
$\tan\beta = r\,d\varphi/d\zeta$	>0		>0		<0		<0	
$\tan\gamma = dr/(r\,d\varphi)$	>0		<0		>0		<0	
$\alpha \in$	$\left(0, \dfrac{\pi}{2}\right)$	$\left(-\pi, -\dfrac{\pi}{2}\right)$	$\left(-\dfrac{\pi}{2}, 0\right)$	$\left(\dfrac{\pi}{2}, \pi\right)$	$\left(-\dfrac{\pi}{2}, 0\right)$	$\left(\dfrac{\pi}{2}, \pi\right)$	$\left(0, \dfrac{\pi}{2}\right)$	$\left(-\pi, -\dfrac{\pi}{2}\right)$
$\beta \in$	$\left(0, \dfrac{\pi}{2}\right)$	$\left(-\pi, -\dfrac{\pi}{2}\right)$	$\left(0, \dfrac{\pi}{2}\right)$	$\left(-\pi, -\dfrac{\pi}{2}\right)$	$\left(-\dfrac{\pi}{2}, 0\right)$	$\left(\dfrac{\pi}{2}, \pi\right)$	$\left(-\dfrac{\pi}{2}, 0\right)$	$\left(\dfrac{\pi}{2}, \pi\right)$
$\gamma \in$	$\left(0, \dfrac{\pi}{2}\right)$	$\left(-\pi, -\dfrac{\pi}{2}\right)$	$\left(-\dfrac{\pi}{2}, 0\right)$	$\left(\dfrac{\pi}{2}, \pi\right)$	$\left(-\pi, -\dfrac{\pi}{2}\right)$	$\left(0, \dfrac{\pi}{2}\right)$	$\left(\dfrac{\pi}{2}, \pi\right)$	$\left(-\dfrac{\pi}{2}, 0\right)$

$$\tan \beta = \frac{r\,d\varphi}{d\zeta},\tag{3.22}$$

$$\tan \gamma = \frac{dr}{r\,d\varphi}.\tag{3.23}$$

The signs of the increments at the edges of the prism, tangents and intervals of angles are shown in Figure 3.5. They are also reported in Table 3.2.

The following relationship can be obtained by mutually comparing Equations (3.21), (3.22) and (3.23):

$$\tan \alpha = \tan \beta \tan \gamma.\tag{3.24}$$

(It can be seen that only two angles are mutually independent.) Further, by comparing Equations (3.11), (3.12) and (3.13) with Equations (3.21), (3.22) and (3.23), respectively, we get

$$\tan \alpha = \cos \vartheta_r / \cos \vartheta_\zeta,\tag{3.25}$$

$$\tan \beta = \cos \vartheta_\varphi / \cos \vartheta_\zeta,\tag{3.26}$$

$$\tan \gamma = \cos \vartheta_r / \cos \vartheta_\varphi.\tag{3.27}$$

By using Equations (3.15), (3.13), (3.25), (3.26) and (3.27), the following expressions are valid to write

$$\frac{d^2 l}{d^2 \zeta} = \frac{1}{\cos^2 \vartheta_\zeta} = \frac{d^2 r}{d^2 \zeta} + \frac{(r\,d\varphi)^2}{d^2 \zeta} + \frac{d^2 \zeta}{d^2 \zeta} = \tan^2 \alpha + \tan^2 \beta + 1,$$

$$\cos^2 \vartheta_\zeta = \frac{1}{\tan^2 \alpha + \tan^2 \beta + 1}.\tag{3.28}$$

It is also valid that

$$\left| \cos \vartheta_\zeta \right| = \frac{1}{\sqrt{\tan^2 \alpha + \tan^2 \beta + 1}}.\tag{3.29}$$

The derivatives of the parametric equations, according to Equation (3.14), can be rearranged by using Equations (3.25) and (3.26) as follows:

$$r'(l) = \cos \vartheta_r = \tan \alpha \cos \vartheta_\zeta.\tag{3.30}$$

$$\varphi'(l) = \cos \vartheta_\varphi / r = \frac{1}{r} \tan \beta \cos \vartheta_\zeta .$$ (3.31)

$$\zeta'(l) = \cos \vartheta_\zeta .$$ (3.32)

Note: The knowledge of three directional cosines appears to be sufficient for determination of the tangents according to Equations (3.25) to (3.27), but this is not fully valid. If we know the values of $\tan \alpha$ and $\tan \beta$ [$\tan \gamma$ is given by Equation (3.24)], then we can evaluate $\cos^2 \vartheta_\zeta$ according to Equation (3.28), but we are not able to determine the sign of $\cos \vartheta_\zeta$.

The quantities $\tan \alpha$, $\tan \beta$ and $\tan \gamma$ can be positive or negative. Nevertheless, we define their absolute values as follows:

$$|\tan \alpha| = \frac{|dr|}{|d\zeta|} ,$$ (3.33)

$$|\tan \beta| = \frac{r|d\varphi|}{|d\zeta|} ,$$ (3.34)

$$|\tan \gamma| = \frac{|dr|}{r|d\varphi|} .$$ (3.35)

In analogy to Equations (3.24) to (3.27), it is valid that

$$|\tan \alpha| = |\tan \beta||\tan \gamma| ,$$ (3.36)

$$|\tan \alpha| = |\cos \vartheta_r| / |\cos \vartheta_\zeta| ,$$ (3.37)

$$|\tan \beta| = |\cos \vartheta_\varphi| / |\cos \vartheta_\zeta| ,$$ (3.38)

$$|\tan \gamma| = |\cos \vartheta_r| / |\cos \vartheta_\varphi| .$$ (3.39)

Radial migration

If the values of $\tan \alpha$ for different fiber elements are mostly different from zero, then we speak about radially migrating fibers and the yarns made up of such fibers are called as yarns with radial migration of fibers. If it is valid that $\tan \alpha = 0$ for all fiber elements, then each fiber is lying on a cylindrical surface and the yarns prepared from such fibers are called yarns with cylindrical structure.

Simply organized structure

The real yarn is often near to an idea of a fiber organization, where the increment $d\zeta$ is positive for all elementary vectors UV and for all fibers. In other words, the path of each fiber is always increasing along the yarn axis. Then only, the cases (a), (c), (e) and (g) as shown in Figures 3.3 and 3.5 and reported in Tables 3.1 and 3.2 are valid for all fiber elementary vectors. We will term such a structure by simply organized structure of fibers in yarn.

The following expressions obtained from Equations (3.7), (3.13) and Tables 3.1 (a), (c), (e) and (g) are valid for simply organized structure:

$$d\zeta > 0, \quad \zeta'(l) = \cos \vartheta_\zeta = \left|\cos \vartheta_\zeta\right| > 0, \quad \vartheta_\zeta \in (0, \pi/2). \tag{3.40}$$

Note: The parametric equation $\zeta = \zeta(l)$ remains always as an increasing function.

In this case, we can write the following expressions which are obtained by using Equation (3.40) in (3.29), and further Equations (3.25) and (3.26):

$$\cos \vartheta_\zeta = \frac{1}{\sqrt{\tan^2 \alpha + \tan^2 \beta + 1}}, \tag{3.41}$$

$$\cos \vartheta_r = \tan \alpha \cos \vartheta_\zeta = \frac{\tan \alpha}{\sqrt{\tan^2 \alpha + \tan^2 \beta + 1}}, \tag{3.42}$$

$$\cos \vartheta_\varphi = \tan \beta \cos \vartheta_\zeta = \frac{\tan \beta}{\sqrt{\tan^2 \alpha + \tan^2 \beta + 1}}. \tag{3.43}$$

The expressions mentioned in Equations (3.30), (3.31) and (3.32) for derivatives of the parametric equations can be rearranged by using Equations (3.41), (3.42) and (3.43) as follows:

$$r'(l) = \cos \vartheta_r = \frac{\tan \alpha}{\sqrt{\tan^2 \alpha + \tan^2 \beta + 1}}, \tag{3.44}$$

$$\varphi'(l) = \cos \vartheta_\varphi / r = \frac{1}{r} \frac{\tan \beta}{\sqrt{\tan^2 \alpha + \tan^2 \beta + 1}}, \tag{3.45}$$

$$\zeta'(l) = \cos\vartheta_\zeta = \frac{1}{\sqrt{\tan^2\alpha + \tan^2\beta + 1}}. \tag{3.46}$$

Twist of element

Traditionally, the twist represents the number of coils (number of 'turns' at 2π radians, i.e., at 360°) in relation to yarn length. Especially, the 'number of coils' of a fiber element is $d\varphi/(2\pi)$ and its corresponding length along yarn axis is $d\zeta$. So, the twist z of the element is determined by the following equation:

$$z = \frac{d\varphi/(2\pi)}{d\zeta}, \quad \left(\frac{d\varphi}{d\zeta} = 2\pi z\right). \tag{3.47}$$

Note: The twist of fiber element is positive if the 'sense of rotation' corresponds to the (technological) direction of yarn twist surrounding the element. Nevertheless, it is negative when the fiber element has an opposite sense of rotation.

By substituting the last relation in Equations (3.22) and (3.31), we get

$$\tan\beta = 2\pi rz, \tag{3.48}$$

$$\varphi'(l) = \cos\vartheta_\varphi/r = 2\pi z \cos\vartheta_\zeta. \tag{3.49}$$

Note: Equation (3.48) can be used in all previous equations.

Fiber migration due to variable twist

If the twists z of different fiber elements vary, then we speak about fiber migration due to variable twist, and the yarn made up of such fibers is called yarn with migration of fibers due to variable twist. If the twists of all fiber elements are equal to one common value z, then each fiber lies on a helical surface[6] and the yarn prepared from such fibers are called yarn with helical-surface structure.

---

6   The helical surface of a fiber is given by expression $\varphi = \varphi_A + 2\pi z\,\zeta$, which follows Equation (3.47).

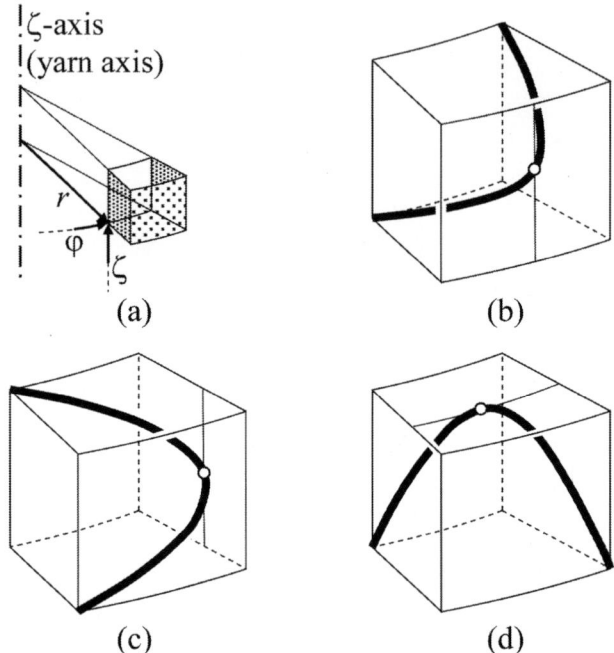

Figure 3.6 Turning points O – examples, (a) orientation of following schemes, (b) radial turning point, (c) turning point of twist, (d) axial turning point

Fiber turning points

Till now, we derived the relations for fiber elementary vectors by assuming that all three elementary vectors of coordinates ($dr, d\varphi, d\zeta$) are different from zero. (See, e.g., Tables 3.1 and 3.2.) Nevertheless, these elementary increments can be equal to zero at some definite number of points on the fiber.

The radius r can increase and then decrease along the fiber and the radial turning point ($dr = 0$, i.e., $\cos \vartheta_r = 0$) lies between these two parts – see Figure 3.6b.

Similarly, the angular coordinate φ can increase and then decrease along the fiber so that the point where $d\varphi = 0$, i.e., $\cos \vartheta_\varphi = 0$, lies between these two parts – Figure 3.6c. It is also valid that the twist of element at this point, according to Equation (3.47), is zero $(z = 0)$. Therefore, we will label this point as a turning point of twist.

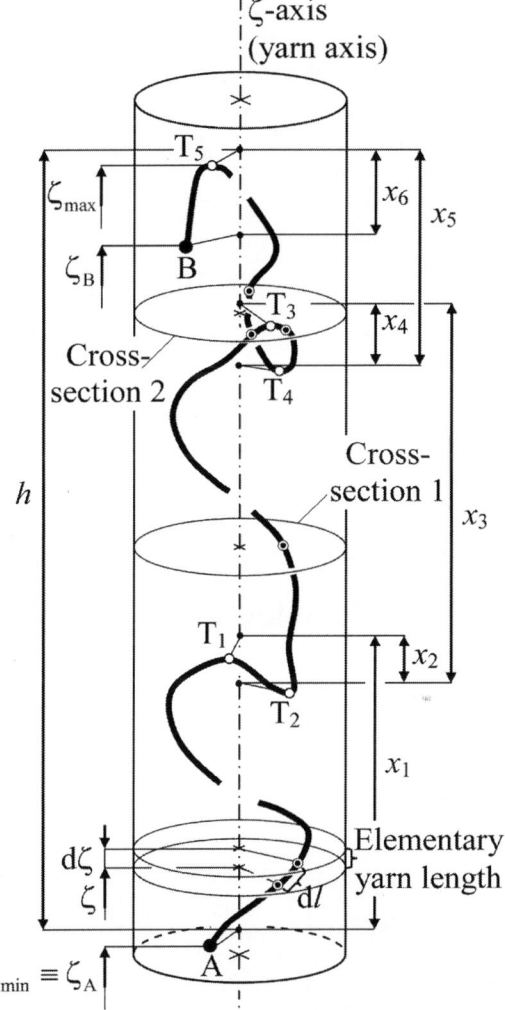

Figure 3.7 Axial turning points on a general fiber in yarn

Finally, the axial coordinate ζ can increase and then decrease along the fiber so that the axial turning point ($d\zeta = 0$, i.e., $\cos\vartheta_\zeta = 0$) lies between these two parts – Figure 3.6d. [In a special case of simply organized structure, according to Equation (3.40), such axial points do not exist.]

Note: Besides the cases described earlier and illustrated in Figure 3.6 (increase and then decrease of coordinates), the turning points may also result in an opposite event, i.e., first decrease and then increase of coordinates.

Axial migration

If the axial turning points exist on the fiber, then we speak about an axially migrating fiber and the yarns prepared from such fibers are called yarns with axial migration of fibers. If the axial turning points do not exist on the fiber, then such fiber is not axially migrating and the yarns made from such fibers evidently follow the simply organized structure of fibers in yarn.

Note: The fibers having a turning point are always considered to be migrating fibers, owing to radial migration or migration due to variable twist or axial migration.

Functions of l

Let us remind that the previous quantities – the cylindrical coordinates r, φ, ζ, angles ϑ_r, ϑ_φ, ϑ_ζ, α, β, γ, the goniometrical functions $\cos\vartheta_r$, $\cos\vartheta_\varphi$, $\cos\vartheta_\zeta$, $\tan\alpha$, $\tan\beta$, $\tan\gamma$, and their absolute values as well as the twist of element z – are determined at each point of fiber path, i.e., for each of fiber length parameter l.

3.2   A structural characteristic with respect to axial arrangement of fiber paths

Lengths of fiber in yarn

Let us imagine a general fiber path (fiber length l_f) in the yarn that is going from its starting point A to the end point B. The fiber length coordinate along the fiber at the starting point A is $l_A = 0$ and the same at the end point B is $l_B = l_f$. Further, let us imagine that the axial turning points $T_1, T_2, \ldots, T_K$ $(K \geq 0)$ are present on the fiber. These points have the length coordinates along the fiber path as follows: $0 < l_{T_1} < l_{T_2} < \cdots < l_{T_K} < l_f$. The ζ-coordinates of the above-mentioned points are $\zeta_A = \zeta(l_A) = \zeta(0)$, $\zeta(l_{T_1})$, $\zeta(l_{T_2})$, ..., $\zeta(l_{T_K})$, $\zeta_B = \zeta(l_B) = \zeta(l_f)$. An example of such fiber is illustrated in Figure 3.7. (In this example, the number of axial turning points is $K = 5$.)

The minimum and maximum ζ-coordinates are determined as follows:

$$\zeta_{min} = \min\left\{\zeta(0), \zeta(l_{T_1}), \ldots, \zeta(l_{T_K}), \zeta(l_f)\right\}, \tag{3.50}$$

$$\zeta_{max} = \max\left\{\zeta(0), \zeta(l_{T_1}), \ldots, \zeta(l_{T_K}), \zeta(l_f)\right\}. \tag{3.51}$$

The minimum ζ-coordinate represents a point where the length coordinate is $l = l_{min}$ and the maximum ζ-coordinate represents a point where the length coordinate is $l = l_{max}$. (It is valid that $\zeta_{min} = \zeta(0)$ and $\zeta_{max} = \zeta(l_{T_5})$ in the example illustrated in Figure 3.7 so that the corresponding length coordinates along the fiber path are $l_{min} = l_A = 0$ and $l_{max} = l_{T_5}$.)

Fiber elevation

The axial fiber elevation is stated as follows:

$$h = \zeta_{max} - \zeta_{min} = \zeta(l_{max}) - \zeta(l_{min}), \quad h > 0. \tag{3.52}$$

It is also possible to express this by using Equation (3.13) as follows:

$$h = \int_{l=l_{min}}^{l=l_{max}} d\zeta = \int_{l_{min}}^{l_{max}} \left(\frac{d\zeta}{dl}\right) dl = \int_{l_{min}}^{l_{max}} \cos\vartheta_\zeta \, dl. \tag{3.53}$$

(Especially, $h = \int_0^{l_{T_5}} \cos\vartheta_\zeta \, dl$, as shown in Figure 3.7.)

Note: In the example illustrated in Figure 3.7, the following relations are valid: $\int_0^{l_{T_1}} \cos\vartheta_\zeta \, dl > 0$, $\int_{l_{T_1}}^{l_{T_2}} \cos\vartheta_\zeta \, dl < 0$, $\int_{l_{T_2}}^{l_{T_3}} \cos\vartheta_\zeta \, dl > 0$, $\int_{l_{T_3}}^{l_{T_4}} \cos\vartheta_\zeta \, dl < 0$ and $\int_{l_{T_4}}^{l_{T_5}} \cos\vartheta_\zeta \, dl > 0$. Let us note that the fiber portion from T_1 to T_2 and the fiber portion from T_3 to T_4 bring negative values of the corresponding components of integral h, because ζ-coordinate is decreasing with increasing length l. (There, $d\zeta < 0$, i.e., $\cos\vartheta_\zeta < 0$ – compare among cases (b), (d), (f), (h) in Figure 3.3 and Table 3.1.)

Fiber projected length

In another concept, the fiber portions AT_1, T_1T_2, ..., T_KB create the axially projected lengths $x_1, x_2, \ldots, x_K, x_{K+1}$, which we understand all now as positive distances. Then, the whole axially projected distance (projected length) is $H = \sum_{i=1}^{K+1} x_i$. This can be expressed by using Equation (3.20) as follows:

$$H = \int_{l=0}^{l_f} |d\zeta| = \int_0^{l_f} \left(\frac{|d\zeta|}{dl}\right) dl = \int_0^{l_f} |\cos\vartheta_\zeta| \, dl. \tag{3.54}$$

(Especially, H is the summation of six positive values $x_1 + x_2 + \cdots + x_6$, as shown in Figure 3.7.) It is clear that the following inequalities are valid among fiber length l_f, fiber projected length H and fiber elevation h such that

$$l_f \geq H \geq h.$$
(3.55)

Fiber intersections with yarn cross section

If fiber elevation h is smaller than the fiber projected length H, then the fiber intersects with some of yarn cross sections more than once. (For example, the fiber, shown in Figure 3.7, intersects yarn cross-section number 1 only once (⦿), whereas the same fiber intersects yarn cross-section number 2 thrice.)

Then, the mean number of fiber intersections with a given number of yarn cross sections (on the axial fiber elevation h) is given by the following ratio:

$$\upsilon = \frac{H}{h} = \frac{\displaystyle\int_0^{l_f} \left|\cos \vartheta_\zeta\right| dl}{\displaystyle\int_{l_{min}}^{l_{max}} \cos \vartheta_\zeta \, dl}.$$
(3.56)

In Figure 3.8, a general fiber element in yarn is shown to be oblique in relation to the vertical direction of yarn axis. The cross-sectional area s of the fiber is evidently smaller than the sectional area s^* of the fiber. The volume of such fiber element can be formulated by two equivalent expressions: $s\,dl$ and $s^* \left|d\zeta\right|$. It is thus possible to write that

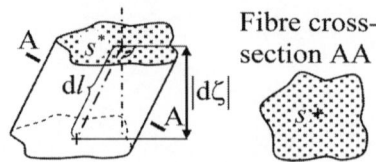

Fibre cross-section AA

Figure 3.8 Element of a fiber

$$s^* = s \frac{dl}{\left|d\zeta\right|} = \frac{s}{\left|\cos \vartheta_\zeta\right|}.$$
(3.57)

[Equation (3.20) was used for rearrangement.]

Fiber volume and sections

The volume of the whole fiber is expressed by

$$v = \int_0^{l_f} s \, \mathrm{d}l \,,$$ (3.58)

where, in general, the fiber cross-sectional area s is also a function of l. The mean value of fiber cross-sectional area is written as

$$s_f = \frac{v}{l_f} = \frac{1}{l_f} \int_0^{l_f} s \, \mathrm{d}l \,.$$ (3.59)

The mean value of fiber sectional area is $s_f^* = v/H$. This can be expressed by applying Equations (3.54) and (3.58) as follows:

$$s_f^* = \frac{v}{H} = \frac{\displaystyle\int_0^{l_f} s \, \mathrm{d}l}{\displaystyle\int_0^{l_f} \left|\cos \vartheta_\zeta\right| \mathrm{d}l} \,.$$ (3.60)

Coefficient k_n of fiber

Coefficient k_n is generally defined as (mean) fiber cross-sectional area divided by (mean) fiber sectional area. We must distinguish the coefficient related to fiber element – $s/s^* = \left|\cos \vartheta_\zeta\right|$, according to Equation (3.57), from the coefficient k_n related to the whole fiber, as follows:

$$k_n = \frac{s_f}{s_f^*} = \frac{H}{l_f} \,,$$ (3.61)

[Equations (3.59) and (3.60) were used to obtain the last expression.] Further, this must be distinguished from the coefficient related to the whole yarn; which will be introduced later on.

Spinning-in coefficient K_F of fiber

Let us define the spinning-in coefficient[7] of fiber as follows:

---

7   Here we used the same symbols as used by Kasparek.

$$K_F = \frac{h}{l_f} \qquad (3.62)$$

Note: The term 'spinning-in coefficient' was first introduced by Kasparek [2] for analyzing the internal structure of yarns. Nevertheless, there are two differences between the definition of spinning-in coefficient given by Kasparek and our definition as mentioned in Equation (3.62). (1) Kasparek did not include the distances of fiber loops and fiber ends protruding from the yarn 'body' (yarn cylinder in Figure 3.7[8]) to the axial yarn elevation h. (2) Kasparek did not use the full fiber length l_f, but used only the length of fiber path projected to a plane, which was parallel to yarn axis[9].

The mean number υ of fiber intersections with a given number of yarn cross sections, as determined by Equation (3.56), can also be expressed by using the coefficients k_n and K_F stated in Equations (3.61) and (3.62). This is shown as follows:

$$\upsilon = \frac{H}{h} = \frac{k_n}{K_F}. \qquad (3.63)$$

Fiber mass and fineness

The fiber elementary volume $s\,dl$ in an elementary yarn length as shown in Figure 3.7 has a local fiber density ρ so that its mass can be expressed as $\rho s\,dl$. Then, the mass of the whole fiber is

$$m = \int_0^{l_f} \rho s\,dl, \qquad (3.64)$$

where, in general, the fiber local density ρ is a function of l. According-ing to Equations (3.58) and (3.64), the (mean) value of fiber density is expressed as

---

8   The determination of 'yarn diameter' is relatively difficult. This issue will be discussed later on.

9   The cotton yarns were of main interest, so that each fiber length would have to be measured individually inside the yarn and this could be very difficult. On the contrary, Kasparek obtained the measurable projections of fiber path to a plane, which was parallel to yarn axis. (He used 'tracer fiber technique' in his Omest instrument.)

$$\rho_f = \frac{m}{v} = \frac{\int_0^{l_f} \rho s\, dl}{\int_0^{l_f} s\, dl} \,, \tag{3.65}$$

and the fineness of the fiber is given by

$$t = \frac{m}{l_f} = \frac{1}{l_f} \int_0^{l_f} \rho s\, dl \,. \tag{3.66}$$

Equivalent fiber

Let us introduce an imaginary fiber – called equivalent fiber – substituting our real fiber in yarn. This is shown in Figure 3.9. We then assume that

1. the equivalent fiber is lying parallel to yarn axis,
2. the elevation of the equivalent fiber is equal to that h of a real fiber,
3. the volume and mass of the equivalent fiber are equal to the volume v and mass m of a real fiber (so that the mean fiber density ρ_f remains the same), respectively,
4. the equivalent fiber has cross-sectional area s_{eq}, which is a constant.

The volume of the equivalent fiber, as shown in Figure 3.9b, is $v = s_{eq} h$. Nevertheless, it is also valid to write from Equations (3.60) and (3.56) that $v = s_f^* H = s_f^* \upsilon h$. Then, the following expression can be obtained from the equivalency of the right-hand sides of the last two expressions:

$$s_{eq} = s_f^* \upsilon \,. \tag{3.67}$$

Consequently, the cross-sectional area of the equivalent fiber s_{eq} is the product of mean sectional area of the real fiber s_f^* and the mean number of fiber sections υ in a given cross section of the yarn.

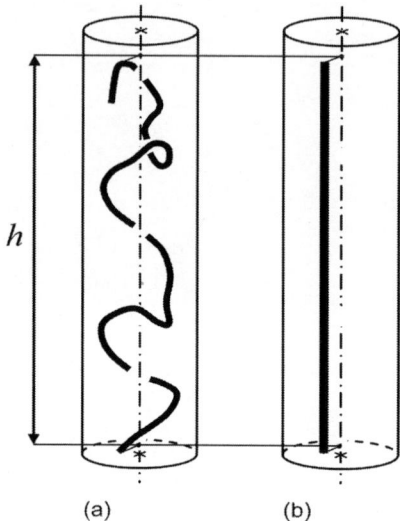

h

(a)　　　　　(b)

Figure 3.9 Introduction to equivalent fiber:
(a) original fiber and (b) equivalent fiber

The equivalent fiber has fineness $t_{eq} = m/h$. According to Equations (3.66) and (3.62), the following expression is valid to write:

$$t_{eq} = \frac{m}{h} = \frac{t\, l_f}{K_F\, l_f} = \frac{t}{K_F}.$$ (3.68)

Note: The concept of substituting a real fiber by an equivalent fiber can be sometimes useful for creation of structural models of yarn.

Fiber in simply organized structure

The simply organized structure satisfies the relations given by Equation (3.40). The axial turning points do not exist in this case. Then, Equations (3.50) and (3.51) can be expressed as follows:

$$\zeta_{min} = \zeta_A = \zeta(l_A) = \zeta(0),$$ (3.69)

$$\zeta_{max} = \zeta_B = \zeta(l_B) = \zeta(l_f),$$ (3.70)

i.e., $l_{min} = l_A = 0$ and $l_{max} = l_B = l_f$.

By applying these results and Equation (3.40) in (3.52), (3.53) and (3.54), we obtain the axial fiber elevation h and the axially projected length H as follows:

$$h = H = \zeta(l_{\rm f}) - \zeta(0) = \int_0^{l_{\rm f}} \cos\vartheta_\zeta \, {\rm d}l , \qquad (3.71)$$

and then we can write the following expression:

$$l_{\rm f} \geq H = h . \qquad (3.72)$$

Let us note that the axial fiber elevation and the axially projected length are same in this case.

By using Equations (3.56) and (3.72), the mean number of fiber intersections with a given number of yarn cross section can be expressed in the following way:

$$\upsilon = \frac{H}{h} = 1 . \qquad (3.73)$$

So, a fiber in a simply organized structure intersects the yarn cross section (lying in-between $\zeta_{\rm A}$ and $\zeta_{\rm B}$) only once.

Yarn

Let us imagine a yarn of a very long length L and fineness T and containing a very large number N of fibers. (Fibers are denoted by subscripts $i = 1, 2, \ldots, N$.) Each fiber has its own set of parameters ($l_{\min}, l_{\max}, \ l_{\rm f}, H, h,$ $\upsilon, v, s_{\rm f}, s_{\rm f}^*, k_n, K_F, m, \rho_{\rm f}, t, s_{\rm eq}, t_{\rm eq}$) as mentioned above and also its own functions (above all $\cos\vartheta_\zeta, s, \rho$) of longitudinal fiber coordinate l. Now, we formulate a set of mean parameters related to such yarn.

Mean fiber length

The length of all fibers in the yarn is $\sum_{i=1}^{N}[l_{\rm f}]_i$ [10] so that the mean length of fibers in the yarn is

$$\overline{l}_{\rm f} = \frac{1}{N} \sum_{i=1}^{N}[l_{\rm f}]_i . \qquad (3.74)$$

---

10 By using the symbol $\sum_{i=1}^{N}[\ ]_i$ of summation, we have in mind that all symbols and expressions related to i-th fiber are given inside the square brackets. The mean fiber parameters in yarn are marked by horizontal bar line.

Mean fiber axially projected length

The mean fiber axially projected length in the yarn can be expressed in accordance with Equation (3.54) as follows:

$$\bar{H} = \frac{1}{N}\sum_{i=1}^{N}[H]_i = \frac{1}{N}\sum_{i=1}^{N}\left[\int_0^{l_f}|\cos\vartheta_\zeta|\,dl\right]_i. \tag{3.75}$$

Mean fiber axial elevation

The mean fiber axial elevation in the yarn is expressed according to Equation (3.53) as follows:

$$\bar{h} = \frac{1}{N}\sum_{i=1}^{N}[h]_i = \frac{1}{N}\sum_{i=1}^{N}\left[\int_{l_{min}}^{l_{max}}\cos\vartheta_\zeta\,dl\right]_i. \tag{3.76}$$

Mean number of intersections per fiber

The mean number of fiber intersections with yarn cross section per fiber can be a suitable characteristic of axial migration. It is expressed in accordance with Equation (3.56) as follows:

$$\bar{\upsilon} = \frac{\bar{H}}{\bar{h}} = \frac{\sum_{i=1}^{N}\left[\int_0^{l_f}|\cos\vartheta_\zeta|\,dl\right]_i}{\sum_{i=1}^{N}\left[\int_{l_{min}}^{l_{max}}\cos\vartheta_\zeta\,dl\right]_i}. \tag{3.77}$$

Mean volume of fiber

The total volume of fibers in the yarn can be expressed in accordance with Equation (3.58) as follows:

$$V = \sum_{i=1}^{N}[v]_i = \sum_{i=1}^{N}\left[\int_0^{l_f}s\,dl\right]_i, \tag{3.78}$$

and then the mean volume of fiber is expressed as follows:

$$\bar{v} = \frac{V}{N} = \frac{1}{N}\sum_{i=1}^{N}[v]_i = \frac{1}{N}\sum_{i=1}^{N}\left[\int_0^{l_f}s\,dl\right]_i. \tag{3.79}$$

Mean fiber cross-sectional area

The mean cross-sectional area of fiber is expressed by the ratio of the total fiber volume to the total fiber length. In analogy to Equation (3.59), the following expression can be obtained by applying Equations (3.74), (3.78) and/or (3.79)

$$\overline{s}_f = \frac{V}{N\overline{l}_f} = \frac{\overline{v}}{\overline{l}_f} = \frac{\sum_{i=1}^{N}\left[\int_0^{l_f} s\,dl\right]_i}{\sum_{i=1}^{N}[l_f]_i}.$$ (3.80)

Mean sectional area of fiber

The mean sectional area of fiber in yarn cross section can be expressed by the ratio of total fiber volume to the summation of axially projected lengths of all fibers. In analogy to Equation (3.60), the following expression can be obtained by applying Equations (3.75) and (3.79):

$$\overline{s}_f^* = \frac{V}{N\overline{H}} = \frac{\overline{v}}{\overline{H}} = \frac{\sum_{i=1}^{N}\left[\int_0^{l_f} s\,dl\right]_i}{\sum_{i=1}^{N}\left[\int_0^{l_f} |\cos\vartheta_\zeta|\,dl\right]_i}.$$ (3.81)

Coefficient $\overline{k}_n$

By definition, the coefficient $\overline{k}_n$ related to the whole yarn can be expressed as $\overline{k}_n = \overline{s}_f/\overline{s}_f^*$. By using Equations (3.80), (3.81), (3.74) and (3.75), we can write

$$\overline{k}_n = \frac{\overline{s}_f}{\overline{s}_f^*} = \frac{\overline{H}}{\overline{l}_f} = \frac{\sum_{i=1}^{N}\left[\int_0^{l_f} |\cos\vartheta_\zeta|\,dl\right]_i}{\sum_{i=1}^{N}[l_f]_i}.$$ (3.82)

This expression is analogous to Equation (3.61).

Spinning-in coefficient

As per the definition given by Equation (3.62), the spinning-in coefficient $\overline{K}_F$ related to the whole yarn can be expressed by the ratio of mean fiber axial elevation to mean fiber length. By using Equations (3.74) and (3.76), we obtain the following expression:

$$\bar{K}_F = \frac{\bar{h}}{\bar{l}_f} = \frac{\sum_{i=1}^{N}\left[\int_0^{l_f}\cos\vartheta_\zeta\,dl\right]_i}{\sum_{i=1}^{N}[l_f]_i}. \tag{3.83}$$

The last expression is analogous to Equation (3.62).

Alternative formula of υ

The mean number of intersections made by one fiber with the yarn cross section, expressed by Equation (3.77), can be rearranged by using Equations (3.82) and (3.83) as follows:

$$\bar{\upsilon} = \frac{\bar{H}}{\bar{h}} = \frac{\bar{k}_n}{\bar{K}_F}. \tag{3.84}$$

This rearrangement resembles the same procedure that was carried out to obtain Equation (3.63).

Mean mass of fiber

The total mass of fibers in the yarn is obviously

$$M = \sum_{i=1}^{N}\left[\int_0^{l_f}\rho s\,dl\right]_i, \tag{3.85}$$

and the mean mass of one fiber in the yarn is

$$\bar{m} = \frac{M}{N} = \frac{1}{N}\sum_{i=1}^{N}\left[\int_0^{l_f}\rho s\,dl\right]_i. \tag{3.86}$$

Mean fiber fineness

The fineness of the yarn of (very long) length L is

$$T = \frac{M}{L} = \frac{N\bar{m}}{L} = \frac{1}{L}\sum_{i=1}^{N}\left[\int_0^{l_f}\rho s\,dl\right]_i, \tag{3.87}$$

and the mean fiber fineness, in accordance with Equations (3.74) and (3.86), is expressed by

$$\bar{t} = \frac{\bar{m}}{\bar{l}_f} = \frac{\sum\limits_{i=1}^{N}\left[\int\limits_{0}^{l_f}\rho s\, dl\right]_i}{\sum\limits_{i=1}^{N}\left[l_f\right]_i}.$$

(3.88)

Mean fiber density

By using Equations (3.79) and (3.86), the mean fiber density can be expressed as follows:

$$\bar{\rho}_f = \frac{M}{V} = \frac{\bar{m}}{\bar{v}} = \frac{\sum\limits_{i=1}^{N}\left[\int\limits_{0}^{l_f}\rho s\, dl\right]_i}{\sum\limits_{i=1}^{N}\left[\int\limits_{0}^{l_f} s\, dl\right]_i}.$$

(3.89)

Mean equivalent fiber

Let us now think about the (fictive) mean equivalent fiber. According to the introduction of equivalent fiber as given in Figure 3.9, the mean equivalent fiber (1) is lying parallel to yarn axis, (2) has the mean fiber axial elevation $\bar{h}$, (3) has the mean fiber volume $\bar{v}$ and mean fiber mass $\bar{m}$ (so that the mean fiber mass density $\bar{\rho}_f = \bar{m}/\bar{v}$), (4) has fiber cross-sectional area $\bar{s}_{eq}$, which is a constant. So, the following expression must be valid:

$$\bar{v} = \bar{s}_{eq}\bar{h}.$$

(3.90)

Nevertheless, we can also express the mean fiber volume from Equation (3.81) and together with Equation (3.77) as follows:

$$\bar{v} = \overline{s_f^*\, \bar{H}} = \overline{s_f^*\, \bar{\upsilon}\bar{h}}.$$

(3.91)

By comparing Equations (3.90) and (3.91), we obtain the expression for cross-sectional area of mean equivalent fiber as follows:

$$\bar{s}_{eq} = \overline{s_f^*\, \bar{\upsilon}}.$$

(3.92)

The last equation is analogically reported in Equation (3.67).

Equivalent yarn

Let us now define an equivalent yarn as shown in Figure 3.10. We imagine the equivalent yarn

(a) as a parallel bundle of mean equivalent fibers,

(b) where the fibers are lying immediately one after another in the longitudinal direction,

(c) where the length of the equivalent yarn is equal to the length L of the real yarn,

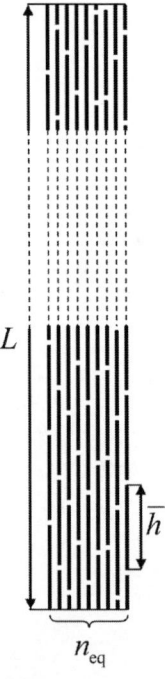

Figure 3.10 Equivalent yarn

(d) where the number n_{eq} of mean equivalent fibers is same in each cross section and

(e) where the cumulative volume of mean equivalent fibers is equal to the fiber volume in the real yarn.

The total volume of fibers in the equivalent yarn can be expressed as follows:

$$V = L \bar{s}_{eq} n_{eq} = L \bar{s}_f^* \bar{\upsilon} n_{eq}.$$

(3.93)

[Equation (3.92) was used for rearrangement.]

Mean number of fibers

The quantity υ denotes the mean number of intersections made by one (real) fiber with the yarn cross section. Thus, the product $\bar{\upsilon}n_{eq}$ expresses the mean number $\bar{n}$ of intersections (mean number of sectional areas $\overline{s_f^*}$) made by all fibers present in the yarn cross section. So, we can rewrite Equation (3.93) expressing the total fiber volume as $V = L\,\overline{s_f^*}\,\bar{n}$. The same volume can be expressed from Equation (3.79) as $V = N\bar{v}$. By comparing the right-hand sides of the last two equations and applying Equations (3.80), (3.82), (3.87) and (3.88) step by step, the following expression can be derived:

$$L\,\overline{s_f^*}\,\bar{n} = N\bar{v},$$

$$\bar{n} = \frac{N\bar{v}}{\overline{s_f^*}\,L} = \frac{N\,\overline{s_f}\,\overline{l_f}}{\overline{s_f^*}\,L} = \left(\frac{\overline{s_f}}{\overline{s_f^*}}\right)\left(\frac{N\bar{m}}{L}\right)\left(\frac{\overline{l_f}}{\bar{m}}\right) = \bar{k}_n\frac{T}{t}. \tag{3.94}$$

Mean substance cross-sectional area

In a (local) yarn cross section, there are n number of fiber (oblique) sections $s_1^*, s_2^*, \ldots, s_n^*$. They are illustrated by a set of shaded areas in Figure 3.11. The summation of all fiber sectional areas is the so-called substance cross-sectional area[11] S of yarn. This is expressed as follows: $S = s_1^* + s_2^* + \cdots + s_n^*$. The mean value of substance cross-sectional area of yarn $\bar{S}$ is then a product of mean number of fiber sections $\bar{n}$ and mean area of fiber section $\overline{s_f^*}$ in the yarn. This is expressed by $\bar{S} = \bar{n}\,\overline{s_f^*}$. By applying Equations (3.94), (3.81), (3.82), (3.89) and (3.88) step by step, we obtain the following relations:

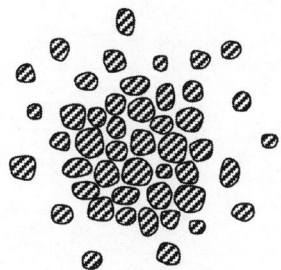

Figure 3.11 Scheme of yarn cross section

---

11   This term was coined by Johansen [3].

$$\overline{S} = \overline{n}\,\overline{s_f^*} = \overline{k}_n \frac{T}{t}\frac{\overline{v}}{\overline{H}} = \overline{k}_n \frac{T}{t}\frac{\overline{v}}{\overline{k_n l_f}} = \frac{T}{t}\frac{\overline{v}}{\overline{l_f}} = \frac{T}{t}\frac{\overline{v}\,\overline{\rho}_f}{\overline{l}_f}\frac{1}{\overline{\rho}_f} = \frac{T}{t}\frac{\overline{m}}{\overline{l}_f}\frac{1}{\overline{\rho}_f} = \frac{T}{t}\,\overline{t}\,\frac{1}{\overline{\rho}_f} = \frac{T}{\overline{\rho}_f}.$$

$$(3.95)$$

Local fineness of yarn

Let us choose a value of axial coordinate ζ of the yarn. The elementary yarn length ('plate' length $\mathrm{d}\zeta$) is lying on this coordinate – see Figure 3.7. This length $\mathrm{d}\zeta$ is thought as a positive increment of axial coordinate. If a fiber passes through such elementary plate, then it 'leaves' an (always positive) elemental length $\mathrm{d}l$ inside it, so that, according to Equation (3.20), it is valid to write that $\mathrm{d}l = |\mathrm{d}\zeta|/|\cos\vartheta_\zeta|$. However, because we understand an elementary yarn length (differential "plate") $\mathrm{d}\zeta$ as a positive quantity, we can write it as follows:

$$\mathrm{d}l = \frac{\mathrm{d}\zeta}{|\cos\vartheta_\zeta|}. \qquad (3.96)$$

Note: The elementary (positive) quantity $\mathrm{d}\zeta$ (differential length of yarn) is common for all elementary vectors passing through the differential length. On the contrary, the lengths $\mathrm{d}l$ of several such elementary vectors are generally diverse inside this "plate". In Equation (3.96), each value of $|\cos\vartheta_\zeta|$ is given by the corresponding length coordinate l of the fiber element.

The fibers passing through this plate follow $\zeta_{\min} < \zeta < \zeta_{\max}$. (The rest of the fibers in the yarn are lying completely out of the above-mentioned elementary yarn length at the chosen coordinate ζ.) Each fiber passes through the given coordinate value ζ once or more. We identify the fiber elementary vectors passing through the coordinate ζ in terms of fiber length coordinates l, as stated in Equation (3.3). So, all fibers offer together n number of fiber elementary vectors (ordered numbers $j = 1, 2, \ldots, n$) in the above-mentioned elementary yarn length.

By applying Equation (3.96), the mass of fiber element is $\rho s\,\mathrm{d}l = \rho s\,\mathrm{d}\zeta/|\cos\vartheta_\zeta|$. Then, the mass of elementary yarn length is

$$dm = d\zeta \sum_{j=1}^{n} \left(\frac{\rho s}{|\cos \vartheta_\zeta|} \right)_j {}^{12}. \tag{3.97}$$

and the fineness of elementary yarn length is

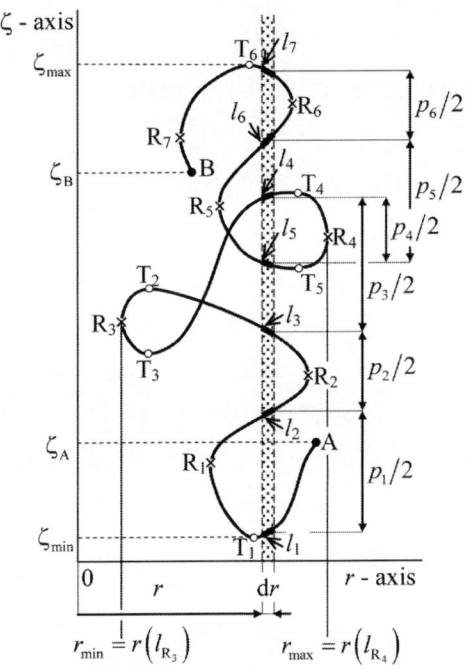

Figure 3.12 r–ζ relation of fiber

$$T_{local} = \frac{dm}{d\zeta} = \sum_{j=1}^{n} \left(\frac{\rho s}{|\cos \vartheta_\zeta|} \right)_j. \tag{3.98}$$

The values T_{local} are generally different at different cross sections (plates) of the yarn. The variability of T_{local} along the yarn is known as 'yarn uneven-

---

12   The values of $\rho, s, |\cos \vartheta_\zeta|$ are different for different elemental vectors on fibers – different subscript j. For each fiber, they depend on l, where the relevant fiber length coordinates l are determined as the roots of Equation (3.3) for a chosen common value ζ.

ness' and this is experimentally measured and evaluated (by, e.g., Uster evenness tester).

    Note: Evidently, it is not technically possible to measure the local yarn finenesses on infinitely short yarn portions $d\zeta$. The commercially available instruments evaluate yarn unevenness on a short but definite length (a few millimetre).

3.3   A structural characteristic with respect to radial arrangement of fiber paths

Radial arrangement of fiber

The couple of parametric functions $r(l), \zeta(l), l \in \langle 0, l_f \rangle$, according to Equations (3.1) and (3.3), determine the r–ζ relation of a fiber as a (smooth) curve; see Figure 3.12.

    Note: Such curve is generally not a simple function and/or a simple curve.

    The length coordinate along the fiber at the starting point A is $l = l_A = 0$ and the same at the end point B is $l = l_B = l_f$. We can also see all axial turning points $T_1, T_2, \ldots, T_K$ and all radial turning points $R_1, R_2, \ldots, R_J$ in the scheme of Figure 3.12. (In this figure, as an example, $K = 6$ and $J = 7$.) The radial turning points have the length coordinates along the fiber path $0 < l_{R_1} < l_{R_2} < \cdots < l_{R_J} < l_f$ and the radial coordinates of the mentioned points are $r_A = r(l_A) = r(0)$, $r(l_{R_1})$, $r(l_{R_2})$, $\ldots$, $r(l_{R_J})$, $r_B = r(l_B) = r(l_f)$.

    The minimum and maximum of the introduced r-coordinates are as follows:

$$r_{\min} = \min\left\{ r(0), r(l_{R_1}), \ldots, r(l_{R_J}), r(l_f) \right\}, \tag{3.99}$$

$$r_{\max} = \max\left\{ r(0), r(l_{R_1}), \ldots, r(l_{R_J}), r(l_f) \right\}. \tag{3.100}$$

[It is valid that $r_{\min} = r(l_{R_3})$ and $r_{\max} = r(l_{R_4})$ in the example illustrated in Figure 3.12.] The fiber is passing through all radial positions in the interval $r \in (r_{\min}, r_{\max})$ once or more. On the contrary, it does not pass through the radial positions lying outside this interval.

Differential layer and length of fiber element

Let us now think about a 'differential layer' in the yarn. The two cylinders, having radii r and $r + dr$ and the common axis – yarn axis, define a differential layer[13] as shown in Figure 3.13. The elementary thickness dr of this layer is thought as a positive increment of radius. If a fiber passes through such layer, then it leaves an (always positive) elementary length dl inside it and it is valid to write, according to Equation (3.18), that $dl = |dr|/|\cos \vartheta_r|$. However, because we understand an elementary thickness of differential layer dr as a positive quantity, we can write the aforesaid expression in the following manner:

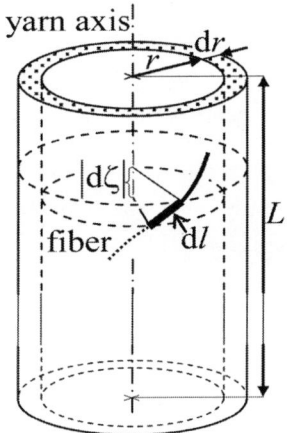

Figure 3.13 Fiber element in differential layer

$$dl = \frac{dr}{|\cos \vartheta_r|} = \frac{dr}{|\tan \alpha||\cos \vartheta_\varsigma|}. \tag{3.101}$$

[Equation (3.37) was used for rearrangement.]

Note: The elementary (positive) quantity dr (thickness of differential layer) is common for all elementary vectors passing through the differential layer. On the contrary, the lengths dl of several elementary vectors passing through the differential layer are generally different inside this layer.

---

13   This term was used, for the first time, by Braschler [4].

Note: In Equation (3.101), the values of $|\cos\vartheta_r|$ as well as $|\tan\alpha|$ and $|\cos\vartheta_\varsigma|$ are given by the corresponding length coordinate l of the selected fiber element.

Fiber elementary vectors at radius r

Let us imagine the differential layer of thickness dr which is situated at radius r (dotted strip in Figure 3.12). If a fiber meets the relation $r \notin (r_{min}, r_{max})$, then it does not pass through such differential layer; therefore, the number of fiber elementary vectors passing through the above-stated differential layer is $Q = 0$.

On the contrary, if it is valid that $r \in (r_{min}, r_{max})$, then the fiber passes through the above-mentioned differential layer once or more. The length coordinates l of all elementary vectors passing through the above-stated differential layer at radius r are $l_1, l_2, \ldots, l_Q \equiv \{l_j\}_{j=1}^{Q}$, $Q \ge 1$; they are the roots of Equation (3.1) at a given value of r. (There are $l_1, l_2, \ldots, l_7$, i.e., $Q = 7$, in the differential layer shown in Figure 3.12.)

The fiber leaves the total length $\sum_{j=1}^{Q}[dl]_j$ [14] in the differential layer. Then the corresponding fiber volume is

$$dv_r = \sum_{j=1}^{Q}[s\,dl]_j = dr\sum_{j=1}^{Q}\left[\frac{s}{|\cos\vartheta_r|}\right]_j = dr\sum_{j=1}^{Q}\left[\frac{s}{|\tan\alpha||\cos\vartheta_\varsigma|}\right]_j = dr\sum_{j=1}^{Q}\left[\frac{s^*}{|\tan\alpha|}\right]_j .$$

(3.102)

[Equations (3.101) and (3.57) were used for rearrangement.]

Note: The quantities mentioned inside the square brackets, i.e., s, s^*, $|\cos\vartheta_r|$, $|\tan\alpha|$ and $|\cos\vartheta_\varsigma|$, generally take different values for different

---

14   We use the general convention that a sum is equal to zero if a superscript is smaller than a subscript. In this case, $\sum_{j=1}^{Q}[dl]_j = 0$ if $Q = 0$ and $\sum_{j=1}^{Q}[dl]_j > 0$ if $Q \ge 1$.

subscript j in relation to the fiber length coordinates $l_1, l_2, \ldots, l_Q$ of the fiber elementary vectors[15].

If we want to 'add up' (integrate) fiber volumes $\mathrm{d}v_r$ over all radii, then the 'selection' of fiber elementary vectors according to different radii does not make a sense. We add up (integrate) elementary fiber volumes $s\,\mathrm{d}l$ over all fiber elementary vectors, and we must evidently obtain the whole fiber volume v according to Equation (3.58). Thus,

$$
\begin{aligned}
v = \int_{r=0}^{\infty} \mathrm{d}v_r &= \int_{r=0}^{\infty} \left\{ \sum_{j=1}^{Q} \left[\frac{s}{|\cos \vartheta_r|} \right]_j \right\} \mathrm{d}r \\
&= \int_{r=0}^{\infty} \left\{ \sum_{j=1}^{Q} \left[\frac{s}{|\tan \alpha||\cos \vartheta_\zeta|} \right]_j \right\} \mathrm{d}r = \int_{r=0}^{\infty} \left\{ \sum_{j=1}^{Q} \left[\frac{s^*}{|\tan \alpha|} \right]_j \right\} \mathrm{d}r.
\end{aligned}
\tag{3.103}
$$

Note: Generally, we think about the limits of integration from 0 to ∞, because if $r < r_{\min}$ or $r > r_{\max}$ then $Q = 0$ so that the contribution to the sum is equal to zero.

Cumulative axially projected length at a given radius

As shown in Figure 3.13, the axially projected length of one element of fiber is $|\mathrm{d}\zeta|$. Then, according to Equations (3.20) and (3.33), the following expressions are valid to write

$$
|\mathrm{d}\zeta| = |\cos \vartheta_\zeta| \mathrm{d}l = \frac{\mathrm{d}r}{|\tan \alpha|}.^{16}
\tag{3.104}
$$

The cumulative axially projected length of all elementary vectors of the fiber, which are lying inside the differential layer at radius r, is

$$
\mathrm{d}H_r = \sum_{j=1}^{Q} \left[|\mathrm{d}\zeta| \right]_j = \sum_{j=1}^{Q} \left[|\cos \vartheta_\zeta| \mathrm{d}l \right]_j = \mathrm{d}r \sum_{j=1}^{Q} \left[\frac{1}{|\tan \alpha|} \right]_j.
\tag{3.105}
$$

---

15 We use the general convention that the variables inside the square or curly brackets, following symbol Σ, have generally different values in relation to the subscript at the right-hand side of the brackets.

16 Because   always stands for positive thickness of the differential layer, therefore, the symbol of absolute value is not necessary for it.

Note: The quantities $\left|\cos\vartheta_\zeta\right|$ and $\left|\tan\alpha\right|$ also generally take different values for different subscript j in relation to the length coordinates $l_1,l_2,...,l_Q$ of the fiber elementary vectors.

If we want to add up the axially projected lengths dH_r over all radii, then the selection of fiber elementary vectors, passing at individual radii ($j=1,2,...,Q$), does not make a sense. We add up (integrate) elementary axially projected lengths over all fiber elementary vectors and we must obtain the whole axially projected lengths of fiber H as defined by Equation (3.54). By applying Equation (3.105), we obtain

$$H = \int_{r=0}^{\infty} dH_r = \int_0^{\infty}\left\{\sum_{j=1}^{Q}\left[\frac{1}{\left|\tan\alpha\right|}\right]_j\right\}dr. \tag{3.106}$$

Mean packing density at a given radius

Let us think again about the yarn of a very long length L, having fineness T (mass of yarn $M = TL$) and containing very high number N of fibers.

Further, let us determine a differential layer at the chosen radius r inside the yarn. The cross section of this differential layer, i.e., differential annulus in Figure 3.13, has the (dotted) area $2\pi r\,dr$ [17]. The volume of the whole differential layer is then

$$dV_{c,r} = 2\pi r\,dr\,L. \tag{3.107}$$

The volume dV_r of all fiber elementary vectors passing through the differential layer (of all N fibers) is the sum of the volumes dv of all individual fibers. By using Equation (3.102), we can write

$$dV_r = \sum_{i=1}^{N}\{dv_r\}_i = dr\sum_{i=1}^{N}\left\{\sum_{j=1}^{Q}\left[\frac{s}{\left|\cos\vartheta_r\right|}\right]_j\right\}_i$$

$$= dr\sum_{i=1}^{N}\left\{\sum_{j=1}^{Q}\left[\frac{s}{\left|\tan\alpha\right|\left|\cos\vartheta_\zeta\right|}\right]_j\right\}_i = dr\sum_{i=1}^{N}\left\{\sum_{j=1}^{Q}\left[\frac{s^*}{\left|\tan\alpha\right|}\right]_j\right\}_i. \tag{3.108}$$

---

17   This volume is $\pi\left(r+dr\right)^2 - \pi r^2 = 2\pi r\,dr + \pi\,d^2 r = 2\pi r\,dr\left[1 + dr/(2r)\right]$ and the second term in the square brackets can be neglected.

(The values of the quantities s, s^*, $|\cos \vartheta_r|$, $|\tan \alpha|$ and $|\cos \vartheta_\zeta|$, vary among the fibers – subscript i – and also among length coordinates $l_1, l_2, \ldots, l_Q$ of the same fiber – subscript j. The number Q varies only among fibers.)

Finally, the (mean) packing density at radius r is equal to the ratio of the last two expressions:

$$
\mu_r = \frac{dV_r}{dV_{c,r}} = \frac{1}{2\pi rL} \sum_{i=1}^{N} \left\{ \sum_{j=1}^{Q} \left[\frac{s}{|\cos \vartheta_r|} \right]_j \right\}_i
$$

$$
= \frac{1}{2\pi rL} \sum_{i=1}^{N} \left\{ \sum_{j=1}^{Q} \left[\frac{s}{|\tan \alpha||\cos \vartheta_\zeta|} \right]_j \right\}_i = \frac{1}{2\pi rL} \sum_{i=1}^{N} \left\{ \sum_{j=1}^{Q} \left[\frac{s^*}{|\tan \alpha|} \right]_j \right\}_i .
$$

$$(3.109)$$

Yarn diameter

The values of packing density μ_r are different at different radii. They are relatively high in the central part of the yarn and they decrease with the increase of radius to very small values in the region of hairiness sphere. Therefore, a suitable value of packing density μ_r can determine a borderline radius $r_D = D/2$ which separates a 'compact' part of the yarn ($r < r_D$) from the hairiness sphere ($r > r_D$). Then, yarn diameter is meant by twice such radius.

Total axially projected length at radius r

The sum of the values dH_r from Equation (3.105) for all N fibers in yarn represents the sum of axially projected length of all fiber elementary vectors lying in the differential layer at radius r. This is called dX_r and it is valid to write that

$$
dX_r = \sum_{i=1}^{N} \left\{ dH_r \right\}_i = \sum_{i=1}^{N} \left\{ \sum_{j=1}^{Q} \left[|\cos \vartheta_\zeta| dl \right]_j \right\}_i = dr \sum_{i=1}^{N} \left\{ \sum_{j=1}^{Q} \left[\frac{1}{|\tan \alpha|} \right]_j \right\}_i . \quad (3.110)
$$

Mean axially projected length of fiber – an alternative

If we want to add up (integrate) the total axially projected length dX_r over all radii, then the selection of fiber elementary vectors according to different radii does not make a sense. We add up (integrate) elementary projected lengths over all elementary vectors of all fibers and we must evidently obtain the cumulative fiber axially projected length $\int_{r=0}^{\infty} dX_r = \sum_{i=1}^{N}[H]_i = N\bar{H}$ – see Equation (3.75). By applying Equation (3.110), we obtain

$$\bar{H} = \frac{1}{N}\int_{r=0}^{\infty} dX_r = \frac{1}{N}\sum_{i=1}^{N}\left(\int_{0}^{\infty}\left\{\sum_{j=1}^{Q}\left[\frac{1}{|\tan\alpha|}\right]_j\right\}dr\right)_i. \tag{3.111}$$

Mean sectional area of fiber at radius r

The mean sectional area of fiber in the whole yarn cross section, according to Equation (3.81), is $\bar{s_f^*} = V/(N\bar{H})$, where V is the volume of the fibers in the whole yarn, $N\bar{H}$ is the sum of projected lengths of all these fibers in the yarn. Now – with an imagination of only one differential layer in the place of earlier complete yarn – we can use an analogical equation for mean sectional area $\bar{s_r^*}$ of fiber in the differential layer at radius r. However, we shall use the volume dV_r in the place of earlier volume V and the sum of axially projected length dX_r in the place of earlier quantity $N\bar{H}$. Thus, we can write

$$\bar{s_r^*} = \frac{dV_r}{dX_r} = \frac{\sum_{i=1}^{N}\left\{\sum_{j=1}^{Q}\left[\frac{s}{|\cos\vartheta_r|}\right]_j\right\}_i}{\sum_{i=1}^{N}\left\{\sum_{j=1}^{Q}\left[\frac{1}{|\tan\alpha|}\right]_j\right\}_i} = \frac{\sum_{i=1}^{N}\left\{\sum_{j=1}^{Q}\left[\frac{s}{|\tan\alpha||\cos\vartheta_\zeta|}\right]_j\right\}_i}{\sum_{i=1}^{N}\left\{\sum_{j=1}^{Q}\left[\frac{1}{|\tan\alpha|}\right]_j\right\}_i}$$

$$= \frac{\sum_{i=1}^{N}\left\{\sum_{j=1}^{Q}\left[\frac{s^*}{|\tan\alpha|}\right]_j\right\}_i}{\sum_{i=1}^{N}\left\{\sum_{j=1}^{Q}\left[\frac{1}{|\tan\alpha|}\right]_j\right\}_i}. \tag{3.112}$$

Mean number of fibers in differential annulus

The total area of (dotted) differential annulus shown in Figure 3.13 is $2\pi r dr$. The (mean) sectional area of fibers inside the differential annulus is then $2\pi r dr\,\mu_r$. The (mean) number of fibers in differential annulus (i.e., the mean number of fiber sections in the cross section of the differential layer) is then

$$dn = \frac{2\pi r dr\, \mu_r}{s_r^*} = \frac{2\pi r dr\, \dfrac{1}{2\pi r L}\displaystyle\sum_{i=1}^{N}\left\{\sum_{j=1}^{Q}\left[\dfrac{s^*}{|\tan\alpha|}\right]_j\right\}_i}{\displaystyle\sum_{i=1}^{N}\left\{\sum_{j=1}^{Q}\left[\dfrac{s^*}{|\tan\alpha|}\right]_j\right\}_i \Bigg/ \displaystyle\sum_{i=1}^{N}\left\{\sum_{j=1}^{Q}\left[\dfrac{1}{|\tan\alpha|}\right]_j\right\}_i}$$

$$= \frac{1}{L}\sum_{i=1}^{N}\left\{\sum_{j=1}^{Q}\left[\frac{1}{|\tan\alpha|}\right]_j\right\}_i dr \qquad (3.113)$$

Note: The 'sum' (integral) of mean number of fibers from all differential annuli (i.e., over all radii) must give the mean number of fibers $\bar{n}$ in yarn cross section. By integrating Equation (3.113) and using Equations (3.111) and (3.91) for rearrangement, we obtain the following expression:

$$\bar{n} = \int_{r=0}^{\infty} dn = \frac{1}{L}\sum_{i=1}^{N}\left(\int_{0}^{\infty}\left\{\sum_{j=1}^{Q}\left[\frac{1}{|\tan\alpha|}\right]_j\right\} dr\right)_i = \frac{N}{L}\bar{H} = \frac{N\bar{v}}{s_f^* L}. \qquad (3.114)$$

This gives us the same value as obtained by the earlier derived expression in Equation (3.94).

Balanced radial migration

If the radial increment dr is positive, then the fiber extends its radial coordinate to a given place at radius r (measured along the fiber path, i.e., with increase of length coordinate l.) We can then say that the fiber is going 'from inside to outside' of cylinder at radius r. [See the elementary vectors in Figure 3.5(a), (d), (f), (g).] Then, the quantity $\cos\vartheta_r = dr/dl$, according to Equation (3.11), must be positive too. In an opposite case, i.e., if the fiber is going 'from outside to inside' cylinder at radius r, the radial increment dr is negative so that the quantity $\cos\vartheta_r = dr/dl$ is also negative. So, the quantity $\cos\vartheta_r/|\cos\vartheta_r| = 1$ when the fiber is going 'from inside to outside' and $\cos\vartheta_r/|\cos\vartheta_r| = -1$ when the fiber is going 'from outside to inside'.

Note: If the number of elementary vectors of a fiber at radius r is more than one, then the elementary vectors alternately move from inside to outside and from outside to inside. See, e.g., Figure 3.12 – the elementary vectors l_2, l_4, l_6 go 'from inside to outside' and the elementary vectors l_1, l_3, l_5, l_7 go 'from outside to inside'.

Thinking about all fiber elementary vectors being in the differential layer at radius r (i.e., N fibers in the yarn, each passing the differential layer with its own Q elemental vectors), we can define the characteristic σ_r of balanced radial migration at radius r as follows:

$$\sigma_r = \frac{\sum\limits_{i=1}^{N}\left\{\sum\limits_{j=1}^{Q}\left[\frac{\cos\vartheta_r}{|\cos\vartheta_r|}\right]_j\right\}_i}{\sum\limits_{i=1}^{N}\{Q\}_i} = \frac{\sum\limits_{i=1}^{N}\left\{\sum\limits_{j=1}^{Q}\left[\frac{\tan\alpha\cos\vartheta_\zeta}{|\tan\alpha||\cos\vartheta_\zeta|}\right]_j\right\}_i}{\sum\limits_{i=1}^{N}\{Q\}_i}. \tag{3.115}$$

[Equation (3.30) was used for rearrangement.] It is evident that $\sigma_r \in \langle -1, 1 \rangle$.

If $\sigma_r = 0$ for all radii, then we speak about balanced radial migration in yarn. (If σ_r is not significantly different from zero, then we speak about roughly balanced radial migration.)

Note: We can anticipate that the fibers in a traditional yarn have a roughly balanced radial migration. On the contrary, we can expect that the value σ_r is very significantly different from zero in case of, e.g., friction-spun ("DREF") yarns.

In the case of simply organized structure of fibers in the yarn, Equation (3.40) is valid and the characteristic σ_r, expressed in Equation (3.115) of balanced radial migration, obtains the following form:

$$\sigma_r = \frac{\sum\limits_{i=1}^{N}\left\{\sum\limits_{j=1}^{Q}\left[\frac{\tan\alpha}{|\tan\alpha|}\right]_j\right\}_i}{\sum\limits_{i=1}^{N}\{Q\}_i}. \tag{3.116}$$

Symmetrical radial migration

Let us think about yarn with balanced radial migration. Furthermore, if, for each element with angles α, β, there exists a 'mirror' element $-\alpha, \beta$ at the same radius r, then we speak about symmetrical radial migration.

Half-period of radial migration at radius r

In the text prior to Equation (3.102), it was mentioned that if it is valid that $r \in (r_{min}, r_{max})$ then the fiber passes the surface of the cylinder (differential layer) at radius r with fiber length coordinates $l_1, l_2, \ldots, l_Q \equiv \{l_j\}_{j=1}^Q$ and $Q \geq 1$ – roots of Equation (3.1). (In Figure 3.12, $Q = 7$.). We say:

- If the number of points passing the radius r is $Q = 0$, i.e., $r \notin (r_{min}, r_{max})$, then the fiber does not pass the radius r. The number of such fibers in the yarn is N_0.

- If the number of points passing the radius r is $Q = 1$ then the fiber passes radius r only one time, i.e., non-periodically. The number of fibers of that kind in the yarn is N_1.

- If the number of points passing the radius r is $Q \geq 2$ then such fiber passes radius r periodically. The number of such fibers in the yarn is N_p.

Evidently, the earlier introduced number of all fibers in the yarn (of length L) is

$$N = N_0 + N_1 + N_p. \tag{3.117}$$

In the last case, the fiber passes the radius r from inside to outside and from outside to inside alternately (see Figure 3.12). The ζ-distance between the neighbouring points on the fiber path create the so-called local half-periods $p/2$ of radial migration ('half-periods' in short); e.g., the distances $p_1/2, p_2/2, \ldots, p_6/2$ in Figure 3.12[18].

The set $\{l_j\}_{j=1}^Q$ of length coordinates of points define the corresponding ζ-coordinates $\zeta(l_1), \zeta(l_2), \ldots, \zeta(l_Q)$ according to Equation (3.3). It is then valid to write that

---

18 In a periodic function, the distance between two neighbouring intersection points type 'up–down' or 'down–up' ('outside–inside' or 'inside–outside' here) is known as a half-period. When this value is doubled, we obtain what is called local period.

$$\frac{p_j}{2} = \left|\zeta\left(l_{j+1}\right) - \zeta\left(l_j\right)\right| = \left|\int_{l=l_j}^{l=l_{j+1}} d\zeta\right|$$

$$= \left|\int_{l_j}^{l_{j+1}} \cos \vartheta_\zeta dl\right|, \, j = 1, 2, \cdots Q-1, Q \geq 2 \tag{3.118}$$

[Equation (3.13) was used for rearrangement.] It is evident that the number of local half-periods is $Q-1$ ($Q \geq 2$).

Note: The absolute value expressed in Equation (3.115) is necessary, because each half-period is defined as a distance, i.e., a positive quantity. It is shown that, e.g., the difference $\zeta(l_5) - \zeta(l_4)$ is negative in Figure 3.12, while the half-period $p_4/2$ is understood also as a positive quantity. Nevertheless, all existing differences $\zeta\left(l_{j+1}\right) - \zeta\left(l_j\right)$ are positive in a simply organized structure, as defined by Equation (3.40).

The mean value of half-period of radial migration at a given radius r is then

$$\frac{p}{2} = \frac{\displaystyle\sum_{i=1}^{N_p}\left\{\sum_{j=1}^{Q-1}\left[\frac{p_j}{2}\right]_j\right\}_i}{\displaystyle\sum_{i=1}^{N_p}\left\{Q-1\right\}_i} = \frac{\displaystyle\sum_{i=1}^{N_p}\left\{\sum_{j=1}^{Q-1}\left[\left|\zeta\left(l_{j+1}\right) - \zeta\left(l_j\right)\right|\right]_j\right\}_i}{\displaystyle\sum_{i=1}^{N_p}\left\{Q-1\right\}_i}$$

$$= \frac{\displaystyle\sum_{i=1}^{N_p}\left\{\sum_{j=1}^{Q-1}\left[\left|\int_{l_j}^{l_{j+1}} \cos \vartheta_\zeta dl\right|\right]_j\right\}_i}{\displaystyle\sum_{i=1}^{N_p}\left\{Q-1\right\}_i}. \tag{3.119}$$

Note: The summations expressed in Equation (3.119) are concerned with periodical fibers only. The lengths l_j are generally different on the same fiber and also among the periodical fibers. The numbers Q are generally different among the periodical fibers.

3.4   A structural characteristic with respect to twisted fiber paths

Angular twist of fiber

The couple of parametric functions $\varphi(l)$ [19] and $\zeta(l)$, $l \in \langle 0, l_f \rangle$, according to Equations (3.2) and (3.3), determine the φ–ζ relation of a fiber as a (smooth) curve; see, e.g., Figure 3.14.

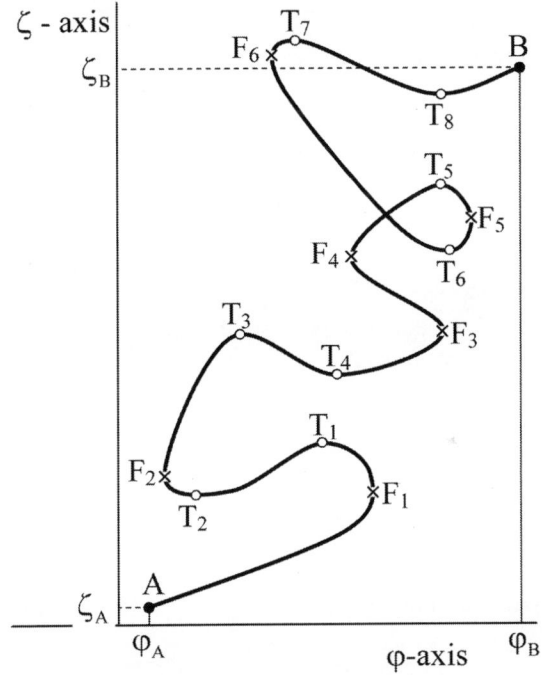

Figure 3.14 φ–ζ relation of fiber

Note: In general, such curve is not a simple function and/or a simple curve.

As earlier, the length coordinate along the fiber at starting point A is $l = l_A = 0$, and the same at end point B is $l = l_B = l_f$. We can also see all axial

---

19   We chose the positive direction of rotation, i.e., the direction of increase of angle φ, in accordance with the direction of (technological) twisting of the yarn. (See also the introductory text of Chapter 3.)

turning points $T_1, T_2, \ldots, T_K$ and all twisted turning points $F_1, F_2, \ldots, F_W$ in Figure 3.14. (Here, as an example, $K = 8$ and $W = 6$.)

Note: The axial turning points are lying on the length coordinates $l_{T_1} < l_{T_2} < \cdots < l_{T_K}$ (all roots of equation $\cos \vartheta_\zeta = 0$), and the twisted turning points are lying on the length coordinates $l_{F_1} < l_{F_2} < \cdots < l_{F_W}$ (all roots of equation $\cos \vartheta_\varphi = 0$).

Fiber portions

Let us imagine all fiber portions between the neighbouring turning points (axial turning points and twisted turning points together) including end points A and B. Then, the signs of the elementary quantities $d\varphi$ and $d\zeta$ are always same at all places on one yarn portion. There exist four different types of such portions:

1. Portions where each $d\varphi > 0$ and $d\zeta > 0$ [type of elementary vectors (a) and (c) in Figure 3.5] – see, e.g., the portions AF_1, F_2T_3, T_4F_3, F_4T_5, F_6T_7, T_8B in Figure 3.14.

2. Portions where each $d\varphi < 0$ and $d\zeta > 0$ [type of elementary vectors (e) and (g) in Figure 3.5] – see, e.g., the portions F_1T_1, T_2F_2, F_3F_4, T_6F_6 in Figure 3.14.

3. Portions where each $d\varphi > 0$ and $d\zeta < 0$ [type of elementary vectors (f) and (g) in Figure 3.5] – see, e.g., the portions T_3T_4, T_5F_5, T_7T_8 in Figure 3.14.

4. Portions where each $d\varphi < 0$ and $d\zeta < 0$ [type of elementary vectors (b) and (d) in Figure 3.5] – see, e.g., the portions T_1T_2, F_5T_6 in Figure 3.14.

Evidently, the sign of $\tan\beta = r d\varphi / d\zeta$ – see Equation (3.22) – is always positive on each place of each portion from the previous groups 1 and 4. On the contrary, the sign of $\tan\beta = r d\varphi / d\zeta$ is always negative on each place of each portion from groups 2 and 3. So, the fiber portions from groups 1 and 4 follow 'the rotation of technological twist', whereas the fiber portions from the groups 2 and 3 follow 'the rotation in the opposite direction'.

Elementary angular quantity $d\varphi^*$

Let us introduce the following elementary angular quantity

$$d\varphi^* = d\varphi \frac{|d\zeta|}{d\zeta}. \tag{3.120}$$

It is valid that

(a) $\left|d\varphi^*\right| = \left|d\varphi\right|$,

(b) $d\varphi^* = d\varphi$ when $d\zeta > 0$ [20] and

(c) $d\varphi^* = -d\varphi$ when $d\zeta < 0$.

In other words, the angular increment $d\varphi^*$ is positive in case when the elementary place on the fiber follows the sense of rotation of technological twist and it is negative in the opposite case.

By using Equations (3.22) and (3.20) in (3.120), we obtain the expression as follows:

$$d\varphi^* = d\varphi \frac{\left|d\zeta\right|}{d\zeta} = \frac{1}{r}\left(\frac{r\,d\varphi}{d\zeta}\right)\left(\frac{\left|d\zeta\right|}{dl}\right)dl = \frac{1}{r}\tan\beta\left|\cos\vartheta_\zeta\right|dl. \tag{3.121}$$

(The quantities r, $\tan\beta$ and $\cos\vartheta_\zeta$ are functions of fiber length coordinate l.)

Twist of element

The projected length of fiber element is evidently $\left|d\zeta\right|$. So, the twist of the element must be $z = \left[d\varphi^*/(2\pi)\right]/\left|d\zeta\right|$. By using Equation (3.121), we obtain the following expression:

$$z = \frac{d\varphi^*/(2\pi)}{\left|d\zeta\right|} = \left[d\varphi\frac{\left|d\zeta\right|}{d\zeta}\right]\frac{1}{\left|d\zeta\right|2\pi} = \frac{d\varphi/(2\pi)}{d\zeta}, \quad \left(\frac{d\varphi}{d\zeta} = 2\pi z\right), \tag{3.122}$$

which is equal to the definition given by Equation (3.47), as expected.

Twist of fiber

The total sum (integral) of all incremental contributions of $d\varphi^*$ on the fiber is

$$\varphi^* = \int_{l=0}^{l=l_f} d\varphi^* = \int_0^{l_f} \frac{1}{r}\tan\beta\left|\cos\vartheta_\zeta\right|dl. \tag{3.123}$$

20　This relation is valid for whole fiber in the case of simply organized structure – see Equation (3.40).

The total sum (integral) of all incremental contributions of $|\mathrm{d}\zeta|$ on the fiber is $H = \int_{l=0}^{l_f} |\mathrm{d}\zeta| = \int_0^{l_f} |\cos\vartheta_\zeta|\,\mathrm{d}l$, which is the fiber projected length determined by Equation (3.54).

The ratio between last two quantities determines the twist z_f of fiber as follows:

$$z_f = \frac{\varphi^*/(2\pi)}{H} = \frac{\displaystyle\int_0^{l_f} \frac{1}{r}\tan\beta\,|\cos\vartheta_\zeta|\,\mathrm{d}l}{2\pi\displaystyle\int_0^{l_f} |\cos\vartheta_\zeta|\,\mathrm{d}l}, \quad \left(\frac{\varphi^*}{H} = 2\pi z_f\right). \tag{3.124}$$

Structural twist of yarn

As shown, a very long length L of yarn is formed from a very high number of fibers N ($i = 1, 2, \ldots, N$). Each fiber has its own (total) angle φ^* and projected length H. By using Equations (3.123) and (3.54), the structural twist of this yarn can be determined by

$$Z = \frac{\displaystyle\sum_{i=1}^{N}[\varphi^*]_i \Big/ (2\pi)}{\displaystyle\sum_{i=1}^{N}[H]_i} = \frac{\displaystyle\sum_{i=1}^{N}\left[\int_0^{l_f} \frac{1}{r}\tan\beta\,|\cos\vartheta_\zeta|\,\mathrm{d}l\right]_i}{2\pi\displaystyle\sum_{i=1}^{N}\left[\int_0^{l_f} |\cos\vartheta_\zeta|\,\mathrm{d}l\right]_i}, \quad \left(\frac{\displaystyle\sum_{i=1}^{N}[\varphi^*]_i}{\displaystyle\sum_{i=1}^{N}[H]_i} = 2\pi Z\right).$$

$$\tag{3.125}$$

Note: Besides the structural twist of yarn Z, we know the 'technological' twist of yarn Z_{TECH}, which is commonly used while spinning (e.g., number of spindle revolution per unit length of yarn). These two quantities can be same sometimes but need not to be same always. If they are (significantly) different, then we must think about the causes of such effect (e.g., initial structure of the 'untwisted' fiber bundle).

Twist at given radius

In Section 3.3, the differential layer was introduced at a general radius r (e.g., see Figure 3.13), where some fibers passed such layer Q times. The number of elementary vector passing at radius r can be $Q = 0$ (The fiber does not pass through the above-mentioned differential layer.) or $Q \geq 1$ (The fiber passes through the above-stated differential layer once or more on its length coordinates $l_1, l_2, \ldots, l_Q$.).

The elementary angular quantity $d\varphi^*$, related to a fiber element, is determined by Equation (3.120). This elementary quantity was formerly rearranged in Equation (3.121); however, it is possible to use an alternative rearrangement as follows:

$$d\varphi^* = d\varphi \frac{|d\zeta|}{d\zeta} = \frac{1}{r}\left(\frac{r\,d\varphi}{d\zeta}\right)\left(\frac{|d\zeta|}{dr}\right)dr = \frac{1}{r}\frac{\tan\beta}{|\tan\alpha|}dr \ . \tag{3.126}$$

Note: The elementary (positive) quantity dr (thickness of differential layer) is common for all elementary vectors passing through the differential layer. Therefore, the ratio $dr/|d\zeta| = |dr|/|d\zeta|$, according to Equation (3.33), represents the quantity $|\tan\alpha|$.

The total sum of elementary contributions of $d\varphi^*$ from all passing elements of one fiber is $\sum_{j=1}^{Q}\left[d\varphi^*\right]_j$ [21]. The cumulative axially projected length of all such fiber elements (lying inside the differential layer at radius r) is analogically $dH_r = \sum_{j=1}^{Q}\left[|d\zeta|\right]_j$ – see also Equation (3.105).

Let us think again about the yarn of a very long length L, containing a very high number N of fibers. Each fiber passes or can pass the differential layer so that the total sum of elementary contributions of angular quantities $d\varphi^*$ from all passing elements of all fibers is

$$d\varphi_r^* = \sum_{i=1}^{N}\left\{\sum_{j=1}^{Q}\left[d\varphi^*\right]_j\right\}_i = \sum_{i=1}^{N}\left\{\sum_{j=1}^{Q}\left[\frac{1}{r}\frac{\tan\beta}{|\tan\alpha|}dr\right]_j\right\}_i = \frac{dr}{r}\sum_{i=1}^{N}\left\{\sum_{j=1}^{Q}\left[\frac{\tan\beta}{|\tan\alpha|}\right]_j\right\}_i \ .$$

$$\tag{3.127 [22]}$$

The total axially projected length dX_r of all elements at radius r was derived in Equation (3.110). So, by knowing the quantities $d\varphi_r^*$ according to

---

21 We use the general convention again that a sum is equal to zero if a superscript is smaller than a subscript, i.e., if $Q = 0$. In this case, it can be written from definition that $\sum_{j=1}^{Q}\left[d\varphi^*\right]_j = 0$.

22 The values of the quantities vary among fibers – subscript i – and also among length coordinates $l_1, l_2, \ldots, l_Q$ of the same fiber – subscript j. The number Q varies only among fibers. (See the 'regulations' of square and/or curly brackets mentioned before.)

Equation (3.127) and $\mathrm{d}X_r$ according to Equation (3.110), the twist at radius r can be determined as follows:

$$Z_r = \frac{\mathrm{d}\varphi_r^*/(2\pi)}{\mathrm{d}X_r} = \frac{\dfrac{\mathrm{d}r}{r}\sum_{i=1}^{N}\left\{\sum_{j=1}^{Q}\left[\dfrac{\tan\beta}{|\tan\alpha|}\right]_j\right\}_i}{2\pi\,\mathrm{d}r\sum_{i=1}^{N}\left\{\sum_{j=1}^{Q}\left[\dfrac{1}{|\tan\alpha|}\right]_j\right\}_i} = \frac{\sum_{i=1}^{N}\left\{\sum_{j=1}^{Q}\left[\dfrac{\tan\beta}{|\tan\alpha|}\right]_j\right\}_i}{2\pi r\sum_{i=1}^{N}\left\{\sum_{j=1}^{Q}\left[\dfrac{1}{|\tan\alpha|}\right]_j\right\}_i}.$$

(3.128)

3.5   Idea of regular yarn

Four assumptions of regular yarn

The general relations mentioned in the previous sections of Chapter 3 are relatively complicated and bring – in its generality – only limited possibilities of their utilization in mathematical modelling of yarns[23]. Therefore, let us formulate a simpler idea for 'building' of a yarn which is nevertheless 'sufficiently near' to filament yarns and also to many staple yarns. Such an idea creates the concept of 'regular yarn'.

We determine regular yarn using five assumptions. Let us now introduce the first four assumptions.

1. The regular yarn is a kind of simply organized yarn, determined according to Equation (3.40). Therefore, Equations (3.41) to (3.46), (3.69) to (3.73) and (3.116) are valid.

2. We can consider the model fibers as 'infinitely long' in the regular yarn. We are then imagining a filament yarn and/or a staple yarn where each (real) staple fiber is 'starting' at the 'end point' of the previous staple fiber. (The real staple fibers follow 'one after another' in the yarn.)

3. The regular yarn has a very long length L and it is consisting of N number of such infinitely long model fibers. Without loosing generality, we can think that – for us – each starting point A of fiber is lying at the starting place of yarn (ζ-coordinate is equal to 0) and the end point B of fiber is lying at the end place of the yarn (ζ-coordinate is

---

23 Nevertheless, if we substitute the small but some finite quantities instead of elementary increments in the equations then a lot of derived relations can also be utilized by various ways of experimental evaluation of real structure of yarns.

equal to L). Then, the model fiber (originally having infinite length) has the finite length l_f. It is valid that

$$l_{min} = l_A = 0, \quad l_{max} = l_B = l_f, \tag{3.129}$$

and according to Equations (3.69) and (3.70)

$$\zeta_{min} = \zeta_A = \zeta(l_A) = \zeta(0) = 0,$$
$$\zeta_{max} = \zeta_B = \zeta(l_B) = \zeta(l_f) = L. \tag{3.130}$$

We also obtain the following expression for axial fiber elevation and axially projected length for each fiber by using Equations (3.71) and (3.72) as follows:

$$h = H = \int_0^{l_f} \cos \vartheta_\zeta \, dl = L \leq l_f. \tag{3.131}$$

So, the number of intersections of the fiber in each yarn cross section, according to Equation (3.73), is $\upsilon = H/h = 1$.

Evidently, the number of fibers in each yarn cross section is same and it is equal to the number of fibers in the yarn

$$n = N. \tag{3.132}$$

4. The local fiber cross-sectional area s and the local mass density ρ are same at each point on each fiber:

$$s = \text{const.}, \quad \rho = \text{const.} \tag{3.133}$$

By applying these four assumptions, the following equations are valid for the fibers in a regular yarn.

Fiber volume

According to Equations (3.58) and (3.133), the fiber volume is

$$v = \int_0^{l_f} s \, dl = s l_f. \tag{3.134}$$

Fiber cross-sectional area

According to Equations (3.59) and (3.134), the (mean) fiber cross-sectional area is

$$s_{\mathrm{f}} = v/l_{\mathrm{f}} = s .$$ (3.135)

Fiber intersectional area

According to Equations (3.131) and (3.134), the mean value of fiber intersectional area in yarn cross sections is

$$s_{\mathrm{f}}^* = v/H = sl_{\mathrm{f}}/L .$$ (3.136)

Coefficient k_n of fiber

According to Equations (3.61), (3.135) and (3.136), the coefficient k_n related to the whole fiber is

$$k_n = s_{\mathrm{f}}/s_{\mathrm{f}}^* = L/l_{\mathrm{f}} .$$ (3.137)

Spinning-in coefficient

According to Equations (3.62) and (3.131), the spinning-in coefficient is

$$K_F = h/l_{\mathrm{f}} = L/l_{\mathrm{f}} .$$ (3.138)

[By using Equation (3.63) and the last two expressions, we obtain the relation $\upsilon = k_n/K_F = 1$ which corresponds to Equation (3.73), which was derived for simple organized structure.]

Fiber mass

According to Equations (3.64) and (3.133), the fiber mass is

$$m = \int_0^{l_{\mathrm{f}}} \rho s \, \mathrm{d}l = \rho s l_{\mathrm{f}} .$$ (3.139)

Fiber density

According to Equations (3.65), (3.134) and (3.139), the (mean) value of fiber density is

$$\rho_{\mathrm{f}} = m/v = \rho .$$ (3.140)

Fiber fineness

According to Equations (3.66) and (3.139), the fiber fineness is

$$t = m/l_f = \rho s . \tag{3.141}$$

Fifth assumption of regular yarn

Let us introduce the following (fifth) assumption of 'ergodicity':

It is further assumed that the structural characteristics of all N fibers in the regular yarn are same (e.g., fiber length, coefficient k_n, period of radial migration, etc.). It allows us to obtain the information about the whole yarn based on studying of only one enough long fiber. (Do not forget that yarn length L is very high and $l_f \geq L$ is very high too. So, we can find all possible situations according to one fiber only.)

Note: If the model fiber paths in a regular yarn are interpreted as realizations of a random process, then such process is called ergodic[24] random process. Generally, fiber paths need not to have a random character (they can be deterministic curves); however, this idea of ergodicity remains the same also in this case.

Note: Based on this assumption, the earlier expression $\sum_{i=1}^{N}\{\ \}_i$ is reduced to a simpler multiplication $N \cdot \{\ \}$.

By applying the fifth assumption, the following equations are valid for a whole regular yarn.

Fiber length

According to Equation (3.74), the (mean) fiber length in the yarn is

$$\bar{l}_f = Nl_f/N = l_f . \tag{3.142}$$

Axially projected length of fibers

According to Equations (3.75) and (3.131), the (mean) axially projected length of fibers in the yarn is

$$\bar{H} = NH/N = H = L . \tag{3.143}$$

---

24 See a text book on theory of probability and random processes to know about the mathematical introduction of the term ergodicity.

Axial elevation of fibers

According to Equations (3.76) and (3.131), the (mean) axial elevation of fibers in the yarn is

$$\bar{h} = Nh/N = h = L .$$

(3.144)

Number of fiber intersections with yarn cross section

According to Equations (3.77), (3.143) and (3.144), the (mean) number of fiber intersections with yarn cross section per fiber is

$$\bar{\upsilon} = \bar{H}/\bar{h} = 1 .$$

(3.145)

Total volume of fibers

According to Equations (3.78) and (3.134), the total volume of fibers in the yarn is

$$V = Nv = Nsl_{\mathrm{f}} .$$

(3.146)

Volume of fiber

According to Equations (3.79), (3.134) and (3.146), the (mean) volume of fiber in the yarn is

$$\bar{v} = V/N = sl_{\mathrm{f}} = v .$$

(3.147)

Fiber cross-sectional area

According to Equations (3.80), (3.142) and (3.147), the (mean) fiber cross-sectional area in the yarn is

$$\bar{s}_{\mathrm{f}} = \bar{v}/\bar{l}_{\mathrm{f}} = s .$$

(3.148)

Sectional area of fiber in yarn cross sections

According to Equations (3.81), (3.143) and (3.147), the (mean) sectional area of fiber in yarn cross section is

$$\bar{s_{\mathrm{f}}^*} = \bar{v}/\bar{H} = sl_{\mathrm{f}}/L .$$

(3.149)

Coefficient $\bar{k}_n$ related to whole yarn

According to Equations (3.82), (3.148) and (3.149), the coefficient $\bar{k}_n$ related to the whole yarn is

$$\overline{k}_n = \overline{s}_f \big/ \overline{s}_f^* = L/l_f \; .$$

(3.150)

Spinning-in coefficient related to whole yarn

According to Equations (3.83), (3.142) and (3.144), the spinning-in coefficient related to the whole yarn is

$$\overline{K}_F = \overline{h}\big/\overline{l}_f = L/l_f \; .$$

(3.151)

Number of intersections of yarn cross section

According to Equations (3.84), (3.150) and (3.151), the (mean) number of intersections in yarn cross section per fiber is

$$\overline{\upsilon} = \overline{k}_n \big/ \overline{K}_F = 1 \; .$$

(3.152)

[The last expression represents an alternative derivation of Equation (3.145).]

Total mass of fibers and mass of one fiber

According to Equations (3.85) and (3.133), the total mass of fibers in the yarn is

$$M = N \int_0^{l_f} \rho s \; \mathrm{d}l = N\rho s l_f \; .$$

(3.153)

So, according to Equations (3.86), (3.139) and (3.153), the (mean) mass of one fiber is

$$\overline{m} = M/N = \rho s l_f = m \; .$$

(3.154)

Yarn fineness

According to Equations (3.87) and (3.153), the fineness of the yarn is

$$T = M/L = N\rho s l_f / L \; .$$

(3.155)

Fiber fineness

According to Equations (3.88), (3.141), (3.142) and (3.154), the (mean) fiber fineness is

$$\overline{t} = m/l_f = \rho s = t \; .$$

(3.156)

Fiber density

According to Equations (3.89), (3.140), (3.146) and (3.153), the (mean) value of fiber density is

$$\bar{\rho}_f = M/V = \rho_f = \rho .$$
(3.157)

Note: This is evident from assumption (4).

Cross-sectional area of equivalent fiber

According to Equations (3.92), (3.149) and (3.152), the cross-sectional area of (mean) equivalent fiber is

$$\bar{s}_{eq} = \overline{s_f^*\, \upsilon} = sl_f/L .$$
(3.158)

Number of fibers in yarn cross section

According to Equations (3.94), (3.150), (3.155) and (3.156), the (mean) number of fibers in yarn cross section is

$$\bar{n} = \bar{k}_n\, T/\bar{t} = \left(L/l_f\right)\left(N\rho sl_f/L\right)/\left(\rho s\right) = N = n .$$
(3.159)

Note: This is evident from assumption (3), Equation (3.132).

Substance cross-sectional area of yarn

According to Equations (3.95), (3.150), (3.155) and (3.157), the mean value of substance cross-sectional area of yarn is

$$\bar{S} = T/\bar{\rho}_f = \left(N\rho sl_f/L\right)/\rho = Nsl_f/L = Ns/k_n .$$
(3.160)

Local fineness of yarn

According to Equations (3.98), (3.40), (3.132) and (3.133), the local fineness of yarn (elementary length of yarn) is

$$T_{local} = \rho s \sum_{j=1}^{N}\left(\frac{1}{\cos\vartheta_\varsigma}\right)_j .$$
(3.161)

Note: The local finenesses are generally different at different cross sections ('plates') of the yarn. It is because the fifth assumption of ergodicity does not imply that the sum in the previous equation should be same in each yarn cross section. Nevertheless, we usually find that the variability of T_{local} is not too significant in case of model regular yarn. (One can consider a filament yarn as a practical example.)

Fibers passing through differential layer

In Sections 3.3, we studied the fiber elements in a so-called differential layer at radius r. Because of assumption (5) (assumption of ergodicity) we assume that each fiber from our regular yarn is passing through the given differential layer Q times, where Q is a function of r, which is common for all fibers.

The following rearrangements are valid in this case.

Volume of one fiber in the differential layer

According to Equations (3.102) and (3.133), the volume of one fiber in the differential layer is

$$dv_r = dr \, s \sum_{j=1}^{Q} \left[\frac{1}{|\cos \vartheta_r|} \right]_j .$$

(3.162)

Cumulative axially projected length in the differential layer

Equation (3.105), i.e., $dH_r = dr \sum_{j=1}^{Q} \left[1/|\tan \alpha| \right]_j$, is valid for cumulative axially projected length of all elements for one fiber in the differential layer.

Volume of all elements

According to Equations (3.108), (3.133) and (3.162), the volume dV_r of all the fiber elementary vectors passing through the differential layer (for all N fibers) is

$$dV_r = N \, dv_r = dr \, Ns \sum_{j=1}^{Q} \left[\frac{1}{|\cos \vartheta_r|} \right]_j .$$

(3.163)

Packing density at radius r

According to Equations (3.109) and (3.133), the (mean) packing density at radius r in the yarn is

$$\mu_r = \frac{1}{2\pi r L} N \sum_{j=1}^{Q} \left[\frac{s}{|\cos \vartheta_r|} \right]_j = \frac{s}{2\pi r} \left(\frac{NQ}{L} \right) \left(\frac{1}{Q} \sum_{j=1}^{Q} \left[\frac{1}{|\cos \vartheta_r|} \right]_j \right).$$

(3.164)

Note: Note that the quantity NQ/L represents the mean number of all protruding elements per unit length of the differential layer. Also, note that the quantity $\sum_{j=1}^{Q}\left[1/\left|\cos\vartheta_r\right|\right]_j / Q$ is the mean of reciprocal values of $\left|\cos\vartheta_r\right|$.

Sum of axially projected length in the differential layer

According to Equations (3.110) and (3.105), the sum of axially projected length of all fiber elements lying in the differential layer in the yarn is

$$dX_r = N\, dH_r = N\, dr \sum_{j=1}^{Q}\left[\frac{1}{\left|\tan\alpha\right|}\right]_j . \tag{3.165}$$

Mean sectional area of fiber

According to Equations (3.112), (3.163) and (3.165), the mean sectional area of fiber (in yarn cross section) in the differential layer at radius r is

$$\overline{s_r^*} = \frac{dV_r}{dX_r} = \frac{s\sum_{j=1}^{Q}\left[\dfrac{1}{\left|\cos\vartheta_r\right|}\right]_j}{\sum_{j=1}^{Q}\left[\dfrac{1}{\left|\tan\alpha\right|}\right]_j} . \tag{3.166}$$

Mean number of fibers in differential annulus

According to Equations (3.113), the mean number of fibers in the differential annulus (i.e., mean number of fiber sections in the cross section of the differential layer) is

$$dn = \frac{N}{L}\sum_{j=1}^{Q}\left[\frac{1}{\left|\tan\alpha\right|}\right]_j dr = \left(\frac{NQ}{L}\right)\left(\frac{1}{Q}\sum_{j=1}^{Q}\left[\frac{1}{\left|\tan\alpha\right|}\right]_j\right)dr . \tag{3.167}$$

Characteristic of radial balanced migration

According to Equations (3.115) and (3.42), the characteristic σ_r of the balanced radial migration at radius r is

$$\sigma_r = \frac{1}{Q}\sum_{j=1}^{Q}\left[\frac{\cos\vartheta_r}{\left|\cos\vartheta_r\right|}\right]_j = \frac{1}{Q}\sum_{j=1}^{Q}\left[\frac{\tan\alpha}{\left|\tan\alpha\right|}\right]_j . \tag{3.168}$$

Number of periodical fibers

Based on assumptions (2) and (3), each fiber is very ('infinitely') long so that it passes each radius $r \leq r_{max}$ minimum two times ($Q \geq 2$), but at most Q is very high. Then, each fiber path is periodic so that the following equation is valid:

$$N_p = N, \quad (N_0 = 0, N_1 = 0). \tag{3.169}$$

[Compare it with Equation (3.117).]

The fiber passes the surface of the cylinder at radius r with its length coordinates $l_1, l_2, \ldots, l_Q \equiv \{l_j\}_{j=1}^{Q}$ – roots of Equation (3.1)[25] (e.g., $Q = 7$ in Figure 3.12). The expression

$$\int_{l_1}^{l_2} \cos \vartheta_\zeta dl + \int_{l_2}^{l_3} \cos \vartheta_\zeta dl + \cdots + \int_{l_{Q-1}}^{l_Q} \cos \vartheta_\zeta dl = \sum_{j=1}^{Q-1} \left[\int_{l_j}^{l_{j+1}} \cos \vartheta_\zeta dl \right]_j = \int_{l_1}^{l_Q} \cos \vartheta_\zeta dl$$

$$\tag{3.170}[26]$$

represents the axial ζ-distance between the first and last elements passing through the fiber length coordinates l_1 and l_Q. This length is a little smaller than the length $\int_{l_A}^{l_B} \cos \vartheta_\zeta dl = \int_0^{l_f} \cos \vartheta_\zeta dl$, because Equation (3.170) does not incorporate the 'ends' of fiber – from l_A to l_1 and from l_Q to l_A. Nevertheless, the fiber is very long so that the influence of the above-mentioned ends is negligible. Therefore, it is valid to write that

$$\sum_{j=1}^{Q-1} \left[\int_{l_j}^{l_{j+1}} \cos \vartheta_\zeta dl \right]_j \doteq \int_0^{l_f} \cos \vartheta_\zeta dl = L. \tag{3.171}$$

[Equation (3.131) was used.] We can also approximately write that

$$Q - 1 \doteq Q, \tag{3.172}$$

because we assume that Q is a very high number.

---

25 Compare it with the discussion made prior to Equation (3.102).

26 Do not forget that all values of $\cos \vartheta_\zeta$ are positive in the case of regular yarn.

Half-period and period of radial migration

According to Equations (3.119), (3.40), (3.169), (3.170), (3.171) and (3.172), the half-period of radial migration at a given radius is

$$
\frac{p}{2} = \frac{N \sum_{j=1}^{Q-1} \left[\int_{l_j}^{l_{j+1}} \cos \vartheta_\zeta \, dl \right]_j}{N(Q-1)} = \frac{NL}{N(Q-1)} \doteq \left(\frac{L}{NQ} \right) N = \frac{L}{Q}. \tag{3.173}
$$

The period of radial migration p is twice of half-period $p/2$. After formulation of the quantity QN/L from Equation (3.164) and by using this in Equation (3.173), we obtain the following expression:

$$
p = 2\frac{p}{2} = 2\left(\frac{L}{QN} \right) N = 2N \frac{s}{2\pi r \mu_r} \left(\frac{1}{Q} \sum_{j=1}^{Q} \left[\frac{1}{|\cos \vartheta_r|} \right]_j \right)
$$

$$
= \frac{Ns}{\pi r \mu_r} \left(\frac{1}{Q} \sum_{j=1}^{Q} \left[\frac{1}{|\cos \vartheta_r|} \right]_j \right). \tag{3.174}
$$

Note: The last expression was already derived by Neckář et al. [5].

Elementary angular quantity $d\varphi^*$

In Section 3.3, the elementary angular quantity $d\varphi^*$ was introduced according to Equation (3.120). According to the first assumption of the regular yarn, we can use Equations (3.40) and also (3.22) there, so that the following relations are obtained:

$$
d\varphi^* = d\varphi = \frac{1}{r} \left(\frac{r \, d\varphi}{d\zeta} \right) d\zeta = \frac{1}{r} \tan \beta \, d\zeta, \tag{3.175}
$$

[Compare it with Equation (3.121).]

Twist of element

According to Equations (3.122) and (3.175), the twist of element is

$$
z = \frac{d\varphi/(2\pi)}{d\zeta} = \frac{\tan \beta}{2\pi r}, \quad \left(d\varphi = \frac{\tan \beta}{r} d\zeta \right). \tag{3.176}
$$

Twist of fiber

According to Equations (3.123), (3.131), (3.175) and (3.176), the sum (integral) of all incremental contributions of $d\varphi^* = d\varphi$ is

$$\varphi^* = \int\limits_{l=0}^{l=l_f} d\varphi = \int\limits_{\zeta=0}^{\zeta=H=L} d\varphi = \int\limits_0^L \frac{\tan\beta}{r} d\zeta = 2\pi \int\limits_0^L z\, d\zeta \ . \tag{3.177}$$

Note: We introduced the quantities $\tan\beta$, r and also z primarily as a suitable function of fiber length coordinate l. Nevertheless, the ζ coordinate is always an increasing function of l [simple organized structure, Equation (3.40)], so that we can consider our quantities as a suitable functions of ζ too.

According to Equations (3.124), (3.131), (3.176) and (3.177), the twist of fiber can be expressed as

$$z_f = \frac{\varphi^*/(2\pi)}{L} = \frac{2\pi\int\limits_0^L z\, d\zeta}{2\pi L} = \frac{1}{L}\int\limits_0^L z\, d\zeta \ . \tag{3.178}$$

Structural twist of yarn

According to Equations (3.125), (3.131) and (3.177), the structural twist of yarn is expressed as

$$Z = \frac{N\,\varphi^*/(2\pi)}{NL} = \frac{1}{2\pi L}\int\limits_0^L \frac{\tan\beta}{r} d\zeta = \frac{1}{L}\int\limits_0^L z\, d\zeta = z_f \ . \tag{3.179}$$

Twist at radius r

According to Equations (3.128), the twist at radius r can be defined by

$$Z_r = \frac{N\sum\limits_{j=1}^{Q}\frac{\tan\beta}{|\tan\alpha|}}{2\pi r\,N\sum\limits_{j=1}^{Q}\frac{1}{|\tan\alpha|}} = \frac{\sum\limits_{j=1}^{Q}\frac{\tan\beta}{|\tan\alpha|}}{2\pi r\sum\limits_{j=1}^{Q}\frac{1}{|\tan\alpha|}}, \quad \left(2\pi r Z_r = \frac{\sum\limits_{j=1}^{Q}\frac{\tan\beta}{|\tan\alpha|}}{\sum\limits_{j=1}^{Q}\frac{1}{|\tan\alpha|}}\right). \tag{3.1}$$

3.6   Classification of structural models

Overview of models

It is known that the structure of a real yarn is generally very complicated. Therefore, there is no single structural model that is valid universally, but each of them is only more or less approximation of reality.

Table 3.3 gives an overview of some existing models. There are two groups of concepts – deterministic and stochastic. The deterministic models ignore the random aspects of fiber paths and interpret them as regular paths, described by deterministic mathematical relations. On the contrary, the stochastic models interpret fiber paths as a set of random curves ('random process'). Besides that both groups of models simplify the real yarn structure by a set of other simplifications, often connected with our assumptions of simple structure or even regular structure, introduced in the previous sections.

Let us now briefly comment on the key approaches and models mentioned in Table 3.3.

Table 3.3 Some model ideas of yarn structure

Deterministic phenomena and models					
$\tan\beta = \dfrac{r\,\mathrm{d}\varphi}{\mathrm{d}\zeta} = 2\pi rz$	$\tan\alpha = \dfrac{\mathrm{d}r}{\mathrm{d}\zeta}$	General	Simple		Regular
$z = 0$	$\tan\alpha = 0$	–	**Ideal parallel fibers**		
	$\tan\alpha \neq 0$	(radial migration only)			
$z = \text{const.}$	$\tan\alpha = 0$	–	**Helical models**		
	$\tan\alpha \neq 0$	(other types of rad. migration)	**'Ideal' radial migration** **Equidistant migration**		
$z \neq \text{const.}$	$\tan\alpha = 0$	(other cylindrical structures)			
	$\tan\alpha \neq 0$	**Yarn diameter** **'Parallel' trend of fibers** ($Z \doteq 0$) *Fiber length compensation* *Regular bring-in migration* *Migration from spinning triangle*			
Stochastic phenomena and models					
		General	Simple		Regular
		Hairiness			–
		(more general models)	Fibers follow Markovian and other random processes		
		Yarn irregularity			–
		Yarn imperfections			
		Random bring-in migration			
		Fiber overlapping			

Notes:
1. The empty rectangles are designed as logically invalid.
2. The ideas expressed in bold will be fully presented later on in this book. (Concepts of parallel fiber bundle were presented in our earlier book [6].)
3. The ideas shown in italic represent mostly experimental findings and sometimes as their first qualitative interpretations. They will be mentioned in this book too.
4. The ideas stated within parentheses are yet to be solved.

Ideal parallel fibers

The bundle of perfectly parallel fibers (infinity long and/or staple) is a special case of simple and/or regular yarn structure. This was described in our earlier book [6].

Helical models

They are the oldest models of yarns based on simple and/or regular yarn structure. All the fibers follow helical path, i.e., each fiber path is lying on its own radius (without radial migration) and the twist of each element of each fiber is a common constant z. The helical models are discussed by many authors and they will be described in Chapter 4.

Ideal radial migration

This model is also based on simple and/or regular yarn structure. The twist of each element of each fiber is a common constant as in case of a helical model; however, the fiber paths are changed moreover according to special regulations inside the yarn. The model of ideal radial migration, based on Trelor's idea [7] and Hearle and his coworkers' work [8], will be described in Chapter 5.

Equidistant migration

This variant is an alternative to previous model, based on alternative regulations inside the yarn. Such model, based on Neckář's concept [9], will be formulated in Chapter 5.

Yarn diameter

A model for calculation of yarn diameter can but need not assume special structural assumptions. However, such models must work also with mechanical regulations and assumptions in different cases. A certain versions are presented in Chapters 8.

Parallel trend of fibers ($Z \doteq 0$)

In opposite to the ideal bundle of perfectly parallel fibers, this group of models considers differently crimped fibers. Nevertheless, their main 'trend' is (roughly) parallel so that the twist of each whole fiber is $Z \doteq 0$. (Such an idea is often close to the real 'parallel fiber bundle'.) Such bundles (infinitely long fibers and/or staple fibers) were partly described in our earlier book [6].

Note: These models are not fully deterministic. They also have probabilistic components (probability density functions, etc.) describing the variability in crimp of short fiber portions.

Fiber length compensation

Morton [10] interpreted the radial migration of fibers, observed experimentally, as a result of geometrical necessity of different fiber lengths at different radial positions in a yarn. The fiber portions lying at a higher radius are 'lacking' of lengths in contrary to the fiber portions lying at smaller radius which have relatively 'surplus' of lengths. Therefore, the fiber portions from higher radii have a tendency to press themselves nearer to yarn axis, and the fiber portions from smaller radii have a tendency to be 'pushed' to higher radial positions. Neckář [9] showed that such mechanism brings migration due to variable twist. This idea will be mentioned in Chapter 5.

Regular bring-in migration

The twisted fiber bundles often take the flat ribbon shape. If the starting fiber bundle has a twist (e.g., regular protecting twist of filament yarns) then – due to ribbon type of twisting – it will be transformed to radial migration, as shown by Hearle and Bose [11, 12]. This idea will be mentioned in Chapter 5.

Random bring-in migration

The fiber portions in a starting fiber bundle (or fiber ribbon) have variable directions. (See our earlier book [6] for fiber orientation in a bundle.) This directional variability is transformed in a way to the directional variability of fibers in the yarn, i.e., to radial migration and migration due to variable twist.

Migration from spinning triangle

The complicated movement of fiber portions in the spinning triangle (ring spinning), observed experimentally, can create some kind of small fiber aggregates ('before-twisting') which come to yarn body as a result of migration. This idea of Neckář [9] will be mentioned in Chapter 5.

Hairiness

The parts of fibers lying on the yarn periphery create a randomly arranged hairiness sphere. Yarn hairiness is often measured experimentally but there exist only a few theoretical models. One stochastic model of the above-mentioned phenomenon, based on Neckář's concept [9], is presented in Chapter 7.

Fibers following Markovian and other random processes

In some articles, the fiber paths are interpreted as different realizations of a stationary and ergodic random process [10, 13, 14]. Evidently, the idea of simple and/or regular structure must be used in this case.

Yarn irregularity

Traditionally, yarn irregularity is characterized by different quantities (e.g., yarn 'CV') and/or functions (e.g., 'spectrogram') of mass variability along yarn axis. Sometimes, a geometrical variability (e.g., yarn diameter) is analyzed similarly. Yarn irregularity is a result of natural random processes and specific technological situations while creation of yarns in the spinning mills. (Irregularity of filament yarns is usually not practically significant.) Stochastic models of yarn irregularity are presented by Martindale [15] and other authors. The concept of 'bundle theory' of yarn unevenness according to Neckář [16] will be presented in Chapter 6.

Yarn imperfections

The yarn imperfections (thick places, thin places, neps, etc.) are usually measured together with yarn irregularity. There exists quite rich experimental experience, but only a few probabilistic models are available in literature [17].

3.7   References

[1]   Adovasio, J. M., Hyland, D. C., and Soffer, O., Textiles and Cordage: A Preliminary Assessment, In: Svoboda, J. (Ed.), Pavlov I – Northwest, The Dolnì Věstonìce Studies, Volume 4, Brno, 403–424, 1997.

[2]   Kasparek, J., Geometric and Mechanical Properties of Open-end Yarn in "Open-end Spinning, Rohlena, V. (Ed.), Elsevier Scientific Publishing Company, Amsterdam, 214, 1975.

[3]   Johansen, O., Handbuch der Baumwollspinnerei, Rohweberei und Fabrikanlagen, Volume I, Verlag von B. F. Voigt, Leipzig, 1935 (In German).

[4]   Braschler, E., Die Festigkeit von Baumwollgespinsten, Leeman and Company, Zurich and Leipzig, 1935 (In German).

[5]   Neckář, B., Soni, M. K., and Das, D., Modeling of radial fiber migration in yarns, *Textile Research Journal*, 76, 486–491, 2006.

[6]   Neckář, B. and Das, D., Theory of Structure and Mechanics of Fibrous Assemblies, Woodhead Publishing India Pvt. Ltd., New Delhi, 2012.

[7]   Treloar, L. R. G., A migration filament theory of yarn properties, *Journal of Textile Institute*, 56, T359–T380, 1965.

[8]   Hearle, J. W. S., Gupta, B. S., Merchant, V. B., Migration of fibers in yarns: Part I: Characterization and idealization of migration behaviour, *Textile Research Journal*, 35, 329–334, 1965.

[9]   Neckář, B., Příze: Tvorba, struktura, vlastnosti, (Yarns: Creation, structure, properties), SNTL Publisher, Prague, 1990 (In Czech).

[10] Morton, W. E., The arrangement of fibers in single yarns, *Textile Research Journal*, 26, 325–331, 1956.

[11] Hearle, J. W. S. and Bose, O. N., The form of yarn twisting: Part I: Ideal cylindrical and ribbon twisted forms, *Journal of Textile Institute*, 57, T294–T307, 1966.

[12] Hearle, J. W. S. and Bose, O. N., The form of yarn twisting, Part II: Experimental studies, *Journal of Textile Institute*, 57, T308–T320, 1966.

[13] Hearle, J. W. S. and Gupta, B. S., Migration of Fibers in Yarns, Part III: A study of migration in staple fiber rayon yarns, *Textile Research Journal*, 35, 788–795, 1965.

[14] Hearle, J. W. S. and Gupta, B. S., Migration of Fibers in Yarns, Part IV: A study of migration in a continuous filament yarn, *Textile Research Journal*, 35, 885–889, 1965.

[15] Martindale, J. G., A new method of measuring the irregularity of yarns with some observations on the origin of irregularities in worsted slivers and yarns, *Journal of Textile Institute*, 36, T39–T47, 1945.

[16] Neckář, B., Neuere Erkenntisse zur garnungleichmässigkeit, *Melliand Textilberichte*, 70, 480–486, 1989 (In German).

[17] Ursiny, P., Theory of Spinning, Technical University of Liberec, Liberec, 1992 (In Czech).

Helical model of fibers in yarns

4.1   Introduction to helical model

Axiomatic introduction

The internal structure of staple fiber yarns is very complex. Nevertheless, a vast majority of these yarns intuitively remind us to be consisting of a set of more or less helically twisted fibers. This inspired many scientists and researchers to idealize the internal structure of yarns mathematically under the term helical model of fibers in yarns. In fact, the helical model and its applications are considered to be one of the oldest concepts of yarn modelling which are published by many authors worldwide.

The helical model of fibers in yarns is based on the following assumptions (axioms):

1. All fibers have same cross sections. [1]

2. All fibers are 'infinitely long'.[2]

3. The axes of all fibers are creating 3D curves – helixes – having same direction of rotation.

4. All helixes have one common axis which is basically the yarn axis.

5. The height of each coil of fiber is same.

    Often, the idealized helical model is discussed by considering one more assumption as follows:

6. The local packing density is same at all places inside the yarn.

---

1   This assumption may not be necessary always, for example, in case of blended yarns. But a large number of problems are solved by considering this assumption.

2   In the case of staple fiber yarn, we imagine that each staple fiber is 'starting' at the 'end-point' of the previous fiber. But, in reality, the staple fibers follow 'one after another' in such yarns.

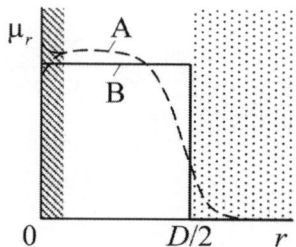

Figure 4.1 Radial packing density μ_r against yarn radius r
A: real yarn, B: idealized yarn, ▨: sphere near yarn axis,
▨ : sphere of hairiness

Note: Figure 4.1 introduces a scheme of radial packing density μ_r in relation to yarn radius r in case of real and idealized yarns.

One coil of a helical fiber is displayed in Figure 4.2. The dashed cylinder represents a portion of a yarn with diameter D. It displays the length of yarn corresponding to one coil of fiber. If the twist in such yarn is Z (number of coils per unit length), then the length of the portion of the yarn is evidently $1/Z$. The thick line represents a fiber (fiber axis), which is lying at a general radius $r \leq D/2$ on the surface of the imaginary cylinder, shown by continuous lines. The slope of the fiber is characterized by the angle β between the direction of yarn axis and the tangent to the fiber path.

The triangle shown in Figure 4.2b is originated by unrolling the surface of the aforesaid imaginary cylinder. (The coil of fiber creates the hypotenuse of the triangle.) Then, the angle β can be expressed as follows:

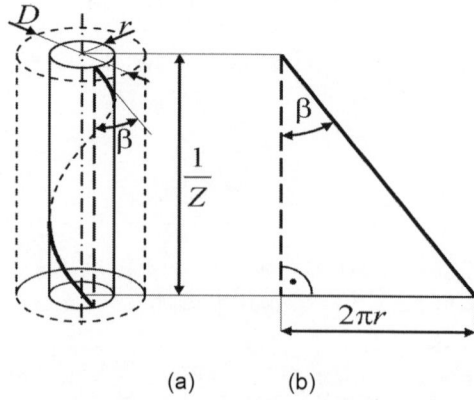

(a)   (b)

Figure 4.2 One coil of helical fiber

$$\tan \beta = \frac{2\pi r}{1/Z} = 2\pi r Z \,, \tag{4.1}$$

so that

$$\cos \beta = \frac{1}{\sqrt{1 + \tan^2 \beta}} = \frac{1}{\sqrt{1 + (2\pi r Z)^2}} \,. \tag{4.2}$$

Note: If especially $r = D/2$ (fiber lies on the surface of the yarn), then $\tan \beta = \tan \beta_D = \pi D Z = \kappa$. This is identical with Equation (1.52). (It can be noticed that Figure 4.2 is similar to Figure 1.9.)

Differential layer and annulus

Figure 4.3 shows the cross section of a yarn with diameter D. Here, the hatched areas represent fiber sections.

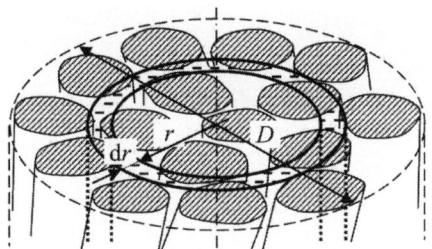

Figure 4.3 Differential layer in yarn cross section

Let us imagine that there are two concentric cylinders around the yarn axis and the difference between their radii r and $r + dr$ is infinitesimally small. Let us call the space between the cylinders as differential layer, as suggested by Braschler [1] and the section of the differential layer in yarn cross section as differential annulus.

The area occupied by the differential annulus is $2\pi r \, dr$. This consists of the area of fiber sections (hatched areas) and the area occupied by the air (dashed areas). If the fiber packing density in this differential layer is μ_r, which is generally a function of radius r, then the total sectional area of fibers inside the differential annulus is

$$dS = 2\pi r \, dr \, \mu_r,{}^3 \tag{4.3}$$

---

3   This is considered to be the elementary substance cross-sectional area S of yarn.

This is in agreement with the area-based interpretation of packing density in accordance with Equation (1.25).

According to Equations (1.40) and (4.2), the sectional area s^* of one fiber lying at radius r [4] is

$$s_r^* = \frac{s}{\cos\beta} = s\sqrt{1+\left(2\pi rZ\right)^2} \,.$$
(4.4)

Then the 'number of fibers' dn present in the differential layer can be expressed from Equations (4.3) and (4.4) as follows:

$$dn = \frac{dS}{s_r^*} = \frac{2\pi r\, dr\, \mu_r}{s\sqrt{1+\left(2\pi rZ\right)^2}} = \frac{2\pi}{s}\frac{\mu_r r}{\sqrt{1+\left(2\pi rZ\right)^2}} dr \,.$$
(4.5)

Substance cross-sectional area – helical model[5]

The following expression can be written by using Equation (4.3):

$$S = \int_{r=0}^{r=D/2} dS = 2\pi \int_0^{D/2} \mu_r r\, dr \,.$$
(4.6)

(Mean) yarn packing density – helical model

The following expression can be obtained by substituting Equation (4.6) to Equation (1.45):

$$\mu = \frac{4S}{\pi D^2} = \frac{8}{D^2} \int_0^{D/2} \mu_r r\, dr \,.$$
(4.7)

Number of fibers in yarn cross section – helical model

The following expression is valid to write by using Equation (4.5):

$$n = \int_{r=0}^{r=D/2} dn = \frac{2\pi}{s} \int_0^{D/2} \frac{\mu_r r}{\sqrt{1+\left(2\pi rZ\right)^2}} dr \,.$$
(4.8)

---

4   It means that the fiber axis is lying at radius r.

5   Here, the first to fifth assumptions of helical model are valid, but the sixth assumption pertaining to ideal helical model is not valid.

Coefficient k_n – helical model

It is valid to write that $k_n = n/\tau$ from Equation (1.44) and $\tau = S/s$ from Equation (1.37). By using these expressions and Equations (4.6) and (4.8), the following expression can be obtained:

$$k_n = \frac{n}{\tau} = \frac{ns}{S} = \frac{2\pi}{S} \int_0^{D/2} \frac{\mu_r r}{\sqrt{1 + (2\pi r Z)^2}} \, dr \tag{4.9}$$

or

$$k_n = \frac{2\pi \int_0^{D/2} \dfrac{\mu_r r}{\sqrt{1 + (2\pi r Z)^2}} \, dr}{2\pi \int_0^{D/2} \mu_r r \, dr} = \frac{\int_0^{D/2} \dfrac{\mu_r r}{\sqrt{1 + (2\pi r Z)^2}} \, dr}{\int_0^{D/2} \mu_r r \, dr}. \tag{4.10}$$

Note: The upper limit of the integrals expressed in Equations (4.6) to (4.10) is used as a suitable yarn radius $D/2$. It is therefore implied that all the fibrous material is present inside the cylindrical yarn.

The knowledge of μ_r as a function of radius is necessary for solving Equations (4.6) to (4.10). This can be determined either experimentally or on the basis of a special model of internal yarn mechanics. Both ways are however very difficult. On the contrary, a set of much easier relations can be obtained for ideal helical model considering all the six assumptions mentioned earlier.

(Mean) yarn packing density – ideal helical model

By using Equation (4.7) and considering $\mu_r = $ constant according to the sixth assumption, the following expression can be obtained:

$$\mu = \frac{4S}{\pi D^2} = \frac{8}{D^2} \mu_r \int_0^{D/2} r \, dr = \frac{8}{D^2} \mu_r \frac{D^2}{8} = \mu_r. \tag{4.11}$$

Substance cross-sectional area – ideal helical model

By using Equations (4.6) and (4.11), it must be valid to write that

$$S = 2\pi\mu \int_0^{D/2} r \, dr = 2\pi\mu \frac{D^2}{8} = \frac{\pi D^2}{4} \mu. \tag{4.12}$$

[This expression is identical to Equation (1.45).]

Number of fibers in yarn cross section – ideal helical model

The following expression can be written by using Equations (4.11) and (4.8):

$$n = \frac{2\pi\mu}{s} \int\limits_0^{D/2} \frac{r}{\sqrt{1+(2\pi r Z)^2}}\, dr .\qquad(4.13)$$

The aforesaid integral can be solved by the following way:

$$\int\limits_0^{D/2} \frac{r\, dr}{\sqrt{1+(2\pi r Z)^2}} = \int\limits_1^{\sqrt{1+(\pi D Z)^2}} \frac{x\, dx}{x(2\pi Z)^2} = \frac{1}{(2\pi Z)^2}\left[\sqrt{1+(\pi D Z)^2}-1\right],$$

Substitution: $x^2 = 1+(2\pi r Z)^2$, (4.14)

$2x\, dx = (2\pi Z)^2 2r\, dr, \quad r\, dr = x\, dx/(2\pi Z)^2,$

$$n = \frac{2\pi\mu}{s} \int\limits_0^{D/2} \frac{r}{\sqrt{1+(2\pi r Z)^2}}\, dr = \frac{2\pi\mu}{s}\left\{\frac{1}{(2\pi Z)^2}\left[\sqrt{1+(\pi D Z)^2}-1\right]\right\}$$

$$= \frac{\mu}{2\pi Z^2 s}\left[\sqrt{1+(\pi D Z)^2}-1\right] = \frac{2}{(\pi D Z)^2}\left[\frac{\dfrac{\pi D^2}{4}\mu}{s}\right]\left[\sqrt{1+(\pi D Z)^2}-1\right]$$

$$= \frac{2}{(\pi D Z)^2}\left[\frac{S}{s}\right]\left[\sqrt{1+(\pi D Z)^2}-1\right] = \frac{2\tau}{(\pi D Z)^2}\left[\sqrt{1+(\pi D Z)^2}-1\right].\qquad(4.15)$$

[Equations (1.45) and (1.37) were also obtained by rearrangement.]

Coefficient k_n – ideal helical model

By using the first relation expressed in Equations (4.9) and (4.15), the following expression can be obtained:

$$k_n = \frac{n}{\tau} = \frac{1}{\tau}\frac{2\tau}{(\pi D Z)^2}\left[\sqrt{1+(\pi D Z)^2}-1\right] = \frac{2}{(\pi D Z)^2}\left[\sqrt{1+(\pi D Z)^2}-1\right].\qquad(4.16)$$

The last equation is possible to rearrange by using the expression for twist intensity in accordance with Equation (1.52). This is shown below:

$$k_n = \frac{2}{\tan^2 \beta_D} \left[\sqrt{1 + \tan^2 \beta_D} - 1 \right] = 2 \frac{\cos^2 \beta_D}{\sin^2 \beta_D} \left[\frac{1}{\cos \beta} - 1 \right] = 2 \frac{\cos^2 \beta_D}{\sin^2 \beta_D} \frac{1 - \cos \beta_D}{\cos \beta_D}$$

$$= 2 \frac{\cos^2 \beta_D}{\sin^2 \beta_D} \frac{(1 - \cos \beta_D)(1 + \cos \beta_D)}{\cos \beta_D (1 + \cos \beta_D)} = 2 \frac{\cos^2 \beta_D}{\sin^2 \beta_D} \frac{\sin^2 \beta_D}{\cos \beta_D (1 + \cos \beta_D)}$$

$$= \frac{2 \cos \beta_D}{1 + \cos \beta_D}. \tag{4.17}$$

The graphical interpretation of Equation (4.17) is shown in Figure 4.4. It can be observed that

- k_n decreases with an increase of β_D (increase of twist).
- $k_n = 0$ for $\beta_D = 0$ (parallel fiber bundle).
- In case of common ring yarns (cotton type), the angle β_D often lies between 20° and 30°. This corresponds to k_n ranging from 0.93 to 0.97.
- The typical value of the slope of peripheral fiber is $\beta_D = 25°$; it corresponds to $k_n = 0.95$.

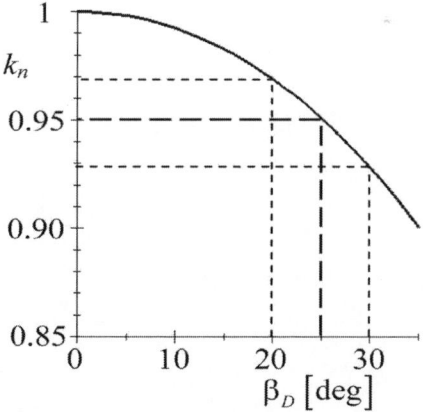

Figure 4.4 Dependence of k_n on slope of peripheral fiber β_D in ideal helical model

Note: It is interesting to note that – independent to a lot of simplifying assumptions – the value of coefficient $k_n = 0.95$ is often experimentally observed in case of ring yarns (cotton type).

4.2   An alternative introduction to helical model

In this section, the helical model of fibers in yarns is going to be introduced in a different way than it was done in the earlier section. Here, the equations of helical model are derived based on the most general relations mentioned in Chapter 3. Nevertheless, it does not bring any new result in relation to the earlier one. Our readers, who are mostly oriented to applications, may like to skip this section and continue with Section 4.3 without any problem.

Assumptions

The helical model of fibers in yarn is based on the following assumptions:

1. Helical model is a special type of regular yarn, according to Section 3.5. [Besides other things, each of N number of infinity long fibers has 'same characteristics' according to the fifth assumption of 'ergodicity' – see the text written after Equation (3.141).]

2. The twist of each fiber element z is a constant, which is common to all fiber elements, that is,
$$z = \text{constant}. \tag{4.18}$$

3. The value of $|\tan \alpha|$ is same for all the fiber elements lying at a given radius, and this quantity is limited to zero. This is shown as follows.
$$|\tan \alpha| ...\text{parameter of } r, \ |\tan \alpha| \to 0 \ [6]. \tag{4.19}$$

Note: This triplet of assumptions correspond to the preliminary introduction of helical model as given in Table 3.3.

Twists

The (structural) twist of yarn can be evaluated by substituting Equation (4.18) into (3.179) as follows:
$$Z = \frac{1}{L} z \int_0^L d\zeta = z. \tag{4.20}$$

So, the twist of each fiber element is same as the twist of the yarn. Moreover, by applying Equation (4.20) in (3.176), the following expression can be obtained:
$$\tan \beta = 2\pi r Z. \tag{4.21}$$

[This is identical to Equation (4.1).]

---

6   The symbol '→' means that the quantity is limited to a given value.

Angles $\vartheta_\zeta, \vartheta_r$

The angles $\vartheta_r, \vartheta_\varphi, \vartheta_\zeta$ and the angles α, β, γ were introduced in Chapter 3. Among others, Equation (3.28) is generally valid. However, based on the third assumption and by using Equations (4.19) and (4.21), the following expression can be obtained:

$$\left|\cos \vartheta_\zeta\right| = \frac{1}{\sqrt{\tan^2 \alpha + \tan^2 \beta + 1}} \rightarrow \frac{1}{\sqrt{\tan^2 \beta + 1}} = \frac{1}{\sqrt{1 + (2\pi r Z)^2}}. \quad (4.22)$$

Equation (3.37) can be expressed in the following way:

$$\left|\cos \vartheta_r\right| = \left|\cos \vartheta_\zeta\right|\left|\tan \alpha\right| \rightarrow \frac{\left|\tan \alpha\right|}{\sqrt{1 + (2\pi r Z)^2}} \rightarrow 0. \quad (4.23)$$

Note: The expressions mentioned in Equations (4.22) and (4.23) represent common parameters for all the fiber elements, which are lying at radius r.

Period of radial migration at a radius r

According to Equations (3.174) and (4.23), the period p of radial migration at radius r is expressed as follows:

$$p = \frac{Ns}{\pi r \mu_r}\left(\frac{1}{Q}\sum_{j=1}^{Q}\left[\frac{1}{\left|\cos \vartheta_r\right|}\right]_j\right) = \frac{Ns}{\pi r \mu_r}\left(\frac{1}{Q}\sum_{j=1}^{Q}\left[\frac{\sqrt{1 + (2\pi r Z)^2}}{\left|\tan \alpha\right|}\right]_j\right)$$

$$= \frac{Ns}{\pi r \mu_r}\frac{\sqrt{1 + (2\pi r Z)^2}}{\left|\tan \alpha\right|}\left(\frac{1}{Q}\sum_{j=1}^{Q}[1]_j\right) = \frac{Ns}{\pi r \mu_r}\frac{\sqrt{1 + (2\pi r Z)^2}}{\left|\tan \alpha\right|} \rightarrow \infty. \quad (4.24)$$

Note: Let us remind the meaning of the symbols introduced in Chapter 3: L stands for a very long length of yarn, $N = n$ refers to the number of fibers present in yarn cross section according to Equation (3.132) [Each fiber has 'same characteristics' according to the fifth assumption of 'ergodicity' – see the text under Equation (3.141).] and Q denotes the number of fiber elements passing the differential layer situated at radius r per one fiber.

An alternative expression for the period p of radial migration at radius r as presented in Equation (3.173) can be written by using Equation (4.24) as follows:

$$p = 2\frac{L}{Q} \rightarrow \infty, \quad \frac{Q}{L} \rightarrow 0. \quad (4.25)$$

Note: The ratio Q/L represents the number of fiber elements per one fiber per unit length of yarn. This is limited to zero. This indicates that the fiber elements are very sparsely distributed at radius r.

Packing density

The packing density at radius r was determined according to Equation (3.164). Applying Equation (4.23), the following expression can be obtained:

$$\mu_r = \frac{s}{2\pi r}\left(\frac{NQ}{L}\right)\left(\frac{1}{Q}\sum_{j=1}^{Q}\left[\frac{1}{|\cos\vartheta_r|}\right]_j\right) = \frac{sN}{2\pi r}\frac{Q}{L}\left(\frac{1}{Q}\sum_{j=1}^{Q}\left[\frac{\sqrt{1+(2\pi rZ)^2}}{|\tan\alpha|}\right]_j\right)$$

$$= \frac{sN}{2\pi r}\frac{Q}{L}\frac{\sqrt{1+(2\pi rZ)^2}}{|\tan\alpha|}\left(\frac{1}{Q}\sum_{j=1}^{Q}[1]_j\right) = \frac{sN}{2\pi r}\frac{Q}{L}\frac{\sqrt{1+(2\pi rZ)^2}}{|\tan\alpha|},$$

$$\mu_r = \frac{sN}{2\pi r}\cdot 0\cdot\infty\ldots\text{indefinite expression} \tag{4.26}$$

Number of fibers in differential annulus

The number dn of fibers present in the differential annulus can be expressed from Equation (3.167) as follows:

$$\frac{dn}{dr} = \left(\frac{NQ}{L}\right)\left(\frac{1}{Q}\sum_{j=1}^{Q}\left[\frac{1}{|\tan\alpha|}\right]_j\right) = N\frac{Q}{L}\frac{1}{|\tan\alpha|}\left(\frac{1}{Q}\sum_{j=1}^{Q}[1]_j\right) = N\frac{Q}{L}\frac{1}{|\tan\alpha|},$$

$$\frac{dn}{dr} = N\cdot 0\cdot\infty\ldots\text{indefinite expression}, \tag{4.27}$$

Nevertheless, by substituting Equation (4.27) into (4.26), the following expression can be obtained:

$$\mu_r = \frac{s}{2\pi r}\sqrt{1+(2\pi rZ)^2}\left(N\frac{Q}{L}\frac{1}{|\tan\alpha|}\right) = \frac{s}{2\pi r}\sqrt{1+(2\pi rZ)^2}\left(\frac{dn}{dr}\right),$$

i.e., $$dn = \frac{2\pi}{s}\frac{\mu_r r}{\sqrt{1+(2\pi rZ)^2}}dr,$$

$$dn = \frac{2\pi\mu_r r}{s_r^*}dr,\text{ where } s_r^* = s\sqrt{1+(2\pi rZ)^2}\ . \tag{4.28}$$

Note: The expression for the sectional area of fiber s_r^* lying at radius r corresponds to Equation (4.4), but it also corresponds to Equation (3.166) after substituting Equation (4.23) into it; that is

$$\overline{s_r^*} = \frac{s\sum_{j=1}^{Q}\left[\dfrac{1}{|\cos\vartheta_r|}\right]_j}{\sum_{j=1}^{Q}\left[\dfrac{1}{|\tan\alpha|}\right]_j} = \frac{s\sum_{j=1}^{Q}\left[\dfrac{\sqrt{1+(2\pi rZ)^2}}{|\tan\alpha|}\right]_j}{\sum_{j=1}^{Q}\left[\dfrac{1}{|\tan\alpha|}\right]_j}$$

$$= \frac{s\dfrac{\sqrt{1+(2\pi rZ)^2}}{|\tan\alpha|}\sum_{j=1}^{Q}[1]_j}{\dfrac{1}{|\tan\alpha|}\sum_{j=1}^{Q}[1]_j} = s\sqrt{1+(2\pi rZ)^2}.$$

Equation (4.28), derived now, is identical to Equation (4.5), derived in Section 4.1. Therefore, the expressions for the number n of fibers present in yarn cross-section according to Equation (4.8) and/or Equation (4.15), and the coefficient k_n according to Equation (4.10) and/or Equations (4.16) and (4.17) are now valid.

Image of helical model in the light of derived equations

The helical model is perhaps the easiest among all deterministic models of yarn structure. Usually, this group of models assumes an image of regular yarns where the twist of fiber elements z and the values $|\tan\alpha|$ are common constants for all such elements present at same radius r of yarn. The helical model is only a limiting case of approximation of real (staple) yarns. Strictly thinking, such yarn should be spontaneously disintegrated because the staple fibers are not actually 'entangled'. Nevertheless, many equations of 'fully helical' model are valid also for the following 'semi-helical' model. Let us imagine that the value $|\tan\alpha|$ is very small, then $|\tan\alpha| \ll 1+(2\pi rZ)^2$, but not precisely equal to zero; hence the limit expressed in Equation (4.19) is not valid. Then, $|\cos\vartheta_\zeta| = 1\big/\sqrt{1+(2\pi rZ)^2}$ according to Equation (4.22) is useful as an enough good approximation, as well as $|\cos\vartheta_r| \rightarrow |\tan\alpha|\big/\sqrt{1+(2\pi rZ)^2}$ according to Equation (4.23). Then, the period p of radial migration at radius r takes a very high value, but not infinity, so that the ratio Q/L according to Equation (4.25) is very small, but not equal to zero. The quantities μ_r and $\mathrm{d}n/\mathrm{d}r$ according to Equations (4.26) and (4.27) are definite quantities at radius r, now. However, Equation (4.28) – identical to Equation (4.5) – remains valid without any change, so that Equations (4.8), (4.15), (4.10), (4.16), (4.17) are also valid. It infers that the expressions of helical model can be used for 'very slow' radial migration of fibers.

4.3   Retraction and limit of twisting – hypothesis of neutral radius

Quantities

Yarns are shortened due to the process of twisting. An idealized scheme of such process is shown in Figure 4.5. Let us assume that the starting (non-twisted) bundle contains n number of infinitely long parallel fibers (Figure 4.5a). The starting length of the bundle is ζ_0. By twisting the length of bundle is shortened so that after twisting the yarn length becomes ζ (Figure 4.5b). The difference between lengths is $\zeta_0 - \zeta$.

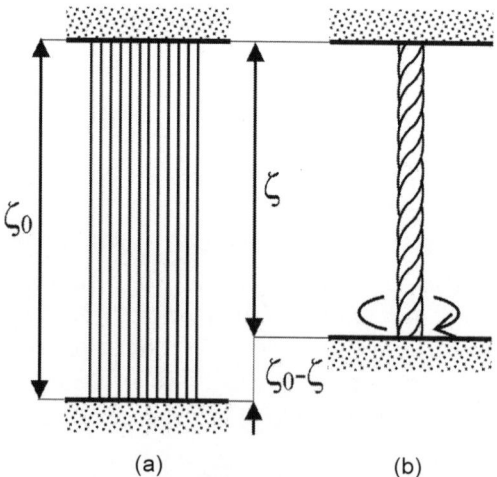

(a)   (b)

Figure 4.5 Twisting of fiber bundle: (a) before twisting, (b) after twisting

The relative shortening due to twisting is usually characterized by yarn contraction as follows:

$$\gamma = \frac{\zeta_0 - \zeta}{\zeta} = \frac{\zeta_0}{\zeta} - 1, \tag{4.29}$$

or by yarn retraction as follows:

$$\delta = \frac{\zeta_0 - \zeta}{\zeta_0} = 1 - \frac{\zeta}{\zeta_0}.$$ (4.30)

The following relation is valid between these two quantities.

$$\gamma = \frac{\delta}{1 - \delta}, \quad \delta = \frac{\gamma}{1 + \gamma}.$$ (4.31)

Note: We, however, prefer to use term yarn retraction in this book.

Let us introduce a set of quantities, which are necessary for solving the problem of yarn retraction. They are shown in Table 4.1.

Table 4.1 Quantities and their symbols

Name of quantity	Before twisting (starting value), Figure 4.5a	After twisting (final or yarn value), Figure 4.5b	Note
Length of bundle/yarn	ζ_0	ζ	Figure 4.5
Yarn retraction	0	$\delta = 1 - \zeta/\zeta_0$	Equation (4.30)
Number of fibers		n	*
Mass of fibers		m	*
Volume of fibers	V_0	V	**
Number of coils	0	N_C	
Fineness (count)	$T_0 = m/\zeta_0$	$T = m/\zeta$	Equation (1.30)
Twist	0	$Z = N_C/\zeta$	Equation (1.50)
Imaginary twist	$Z_i = N_C/\zeta_0$		***
Twist coefficient (Koechlin)	0	$\alpha = Z\sqrt{T}$	Equation (1.54)
Imaginary twist coefficient	$\alpha_i = Z_i\sqrt{T_0}$		***

*Independent of twisting.
**Generally, volume of fibers is changed due to twisting.
***Imaginary ('mixed') quantities.

Meaning of imaginary quantities

According to Table 4.1, the imaginary twist is $Z_i = N_C/\zeta_0$. We find the number of coils N_C in the finally twisted yarn (Figure 4.5b), but the length ζ_0 is measured on the untwisted bundle (Figure 4.5a). Therefore, Z_i is not a real quantity. It says how many coils will be put to the bundle at its starting length[7]. Similarly, the imaginary twist coefficient $\alpha_i = Z_i\sqrt{T_0} = N_C\sqrt{T_0}/\zeta_0$ is not also a real quantity, because the number of coils N_C is obtained from the finally twisted yarn (Figure 4b), whereas the starting fineness T_0 and the starting length ζ_0 are obtained from the starting untwisted bundle. So, the imaginary twist coefficient α_i is a relative characteristic of twist associated with the starting values of the bundle.

The following relations can be derived from the expressions mentioned in Table 4.1:

$$\frac{Z_i}{Z} = \frac{\zeta}{\zeta_0} = 1 - \delta, \quad Z = \frac{Z_i}{1 - \delta}, \tag{4.32}$$

$$\frac{T_0}{T} = \frac{\zeta}{\zeta_0} = 1 - \delta, \quad T = \frac{T_0}{1 - \delta}, \tag{4.33}$$

$$\alpha = Z\sqrt{T} = \frac{Z_i}{1 - \delta}\sqrt{\frac{T_0}{1 - \delta}} = \frac{Z_i\sqrt{T_0}}{(1 - \delta)^{3/2}} = \frac{\alpha_i}{(1 - \delta)^{3/2}}. \tag{4.34}$$

Neutral radius and neutral position

Because of twisting, the central fibers lying at small yarn radius should be shortened and the surface fibers lying at a radius near to $D/2$ should be elongated. The fibers lying at a special radius – neutral radius r_n – will have their final length same as their starting length. This is illustrated in Figure 4.6a. Figure 4.6b is obtained by unrolling the cylinder of radius r_n. The twisted yarn has length ζ, but the fiber at neutral radius r_n has its original length ζ_0

---

7   However, the starting length ζ_0 'disappears' and a new shorter length ζ 'appears' by the process of inserting coils.

– see also Figure 4.5. So, it is valid to write the following expression by using Equation (4.1)

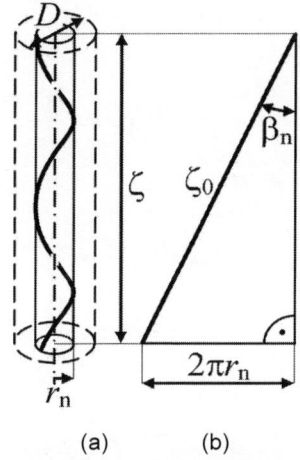

(a)   (b)

Figure 4.6 Fiber on neutral radius

$$\frac{\zeta}{\zeta_0} = \cos\beta_n = \frac{1}{\sqrt{1 + \tan^2 \beta_n}} = \frac{1}{\sqrt{1 + (2\pi r_n Z)^2}} = \frac{1}{\sqrt{1 + \left(\frac{2r_n}{D}\right)^2 (\pi D Z)^2}}$$

$$= \frac{1}{\sqrt{1 + x_n^2 (\pi D Z)^2}}. \tag{4.35}$$

The quantity x_n denotes the so-called neutral position as follows:

$$x_n = 2r_n / D. \tag{4.36}$$

(What is the suitable value of x_n? This will be discussed later on.)

Yarn retraction

The following expression for yarn retraction can be obtained by using Equations (4.30), (4.35) and (1.51):

$$\delta = 1 - \frac{\zeta}{\zeta_0} = 1 - \frac{1}{\sqrt{1 + x_n^2 (\pi D Z)^2}} = 1 - \frac{1}{\sqrt{1 + x_n^2 \tan^2 \beta_D}} = 1 - \frac{1}{\sqrt{1 + x_n^2 \kappa^2}}.$$

$$\tag{4.37}$$

The last equation expresses yarn retraction as a function of yarn twist intensity.

Nevertheless, twist intensity κ can be expressed from Equation (1.54), i.e., $\kappa = 2\sqrt{\pi}\,\alpha/\sqrt{\mu\rho}$, and by applying it in Equation (4.37), the following expression can be obtained:

$$\delta = 1 - \frac{1}{\sqrt{1 + x_n^2 \left(\dfrac{2\sqrt{\pi}\,\alpha}{\sqrt{\mu\rho}}\right)^2}} = 1 - \frac{1}{\sqrt{1 + x_n^2 \dfrac{4\pi\,\alpha^2}{\mu\rho}}} \,. \tag{4.38}$$

As shown, yarn retraction is now a function of yarn twist coefficient α.

Finally, yarn retraction can be expressed also as a function of the imaginary twist coefficient α_i. By using Equation (4.34) in (4.38), the following expression can be obtained:

$$\delta = 1 - \frac{1}{\sqrt{1 + x_n^2 \dfrac{4\pi}{\mu\rho}\left[\dfrac{\alpha_i}{(1-\delta)^{3/2}}\right]^2}} = 1 - \frac{1}{\sqrt{1 + x_n^2 \dfrac{4\pi}{\mu\rho}\dfrac{\alpha_i^2}{(1-\delta)^3}}} \,. \tag{4.39}$$

Because the quantity δ is present at both sides of the last expression, it must be rearranged explicitly in terms of δ. This leads to the solution of a cubic equation as shown in Appendix 1 and the resulting expression is given in Equation (A1.7) as follows:

$$\delta = 1 - \frac{2}{\sqrt{3}}\cos\left[\frac{\pi}{3} \pm \frac{1}{3}\arccos\left(\frac{6\sqrt{3}\,\pi x_n^2}{\mu}\frac{\alpha_i^2}{\rho}\right)\right]. \tag{4.40}$$

Yarn retraction must be $\delta = 0$ in case of untwisted bundle, i.e., at $\alpha_i = 0$. Then

$$0 = 1 - \frac{2}{\sqrt{3}}\cos\left[\frac{\pi}{3} \pm \frac{1}{3}\arccos 0\right] = 1 - \frac{2}{\sqrt{3}}\cos\left[\frac{\pi}{3} \pm \frac{\pi}{6}\right] = 1 - \frac{2}{\sqrt{3}}\cos\frac{2\pi \pm \pi}{6}.$$

Sign '+': $0 = 1 - \dfrac{2}{\sqrt{3}}\cos\dfrac{3\pi}{6} = 1 - \dfrac{2}{\sqrt{3}}\cos\dfrac{\pi}{2} = 1 - 0 = 1 \ldots$ is not valid.

Sign '−': $0 = 1 - \dfrac{2}{\sqrt{3}}\cos\dfrac{\pi}{6} = 1 - \dfrac{2}{\sqrt{3}}\dfrac{\sqrt{3}}{2} = 1 - 1 = 0 \ldots$ is valid.

The minus sign in Equation (4.40) is physically real.

Limit of twisting

The quantity $6\sqrt{3}\,\pi x_n^2\alpha_i^2/(\mu\rho)$ mentioned in Equation (4.40) must not be higher than one[8]. It is therefore valid to write that

$$\frac{6\sqrt{3}\,\pi x_n^2}{\mu}\frac{\alpha_i^2}{\rho}\leq 1, \quad \alpha_i \leq \frac{1}{\sqrt{6\sqrt{3}\,\pi}}\frac{\sqrt{\mu\rho}}{x_n}=0.17501\frac{\sqrt{\mu\rho}}{x_n}. \tag{4.41}$$

It is thus shown that the imaginary twist coefficient is limited.

What does this mean? Earlier it was shown that $\alpha_i = Z_i\sqrt{T_0} = N_C\sqrt{T_0}/\zeta_0$, so that the maximum number of coils N_C embedded in the starting length ζ_0 has an upper limit. It is therefore not possible to insert more than the maximum number of coils in the starting length of the yarn.

In the limiting case (maximum twisting, i.e., saturated twist), the following expressions can be obtained from Equations (4.40) and (4.41):

$$\frac{6\sqrt{3}\,\pi x_n^2}{\mu}\frac{\alpha_i^2}{\rho}=1, \quad \alpha_i = \frac{1}{\sqrt{6\sqrt{3}\,\pi}}\frac{\sqrt{\mu\rho}}{x_n}=0.17501\frac{\sqrt{\mu\rho}}{x_n}. \tag{4.42}$$

$$\delta = 1 - \frac{2}{\sqrt{3}}\cos\left[\frac{\pi}{3}\pm\frac{1}{3}\arccos(1)\right]=1-\frac{2}{\sqrt{3}}\cos\left[\frac{\pi}{3}\right]=1-\frac{1}{\sqrt{3}}=0.42265. \tag{4.43}$$

By using Equation (4.43) in (4.37), the value of angle β_D can be obtained for the limiting case as follows:

$$\delta = 1-\frac{1}{\sqrt{3}}=1-\frac{1}{\sqrt{1+x_n^2\tan^2\beta_D}}, \quad \frac{1}{3}=\frac{1}{1+x_n^2\tan^2\beta_D}, \quad 2=x_n^2\tan^2\beta_D,$$

$$\tan\beta_D = \sqrt{2}/x_n, \quad \beta_D = \arctan\left(\sqrt{2}/x_n\right). \tag{4.44}$$

It is shown from Equations (4.43) and (4.44) that the yarn retraction δ, twist intensity $\kappa = \tan\beta_D$, and the angle β_D are also limited.

---

8   Generally, the quantity x in the function of $\arccos x$ has a sense of cosine. Therefore, x must not be higher than 1.

Neutral position – ideal helical model

The expressions that will follow in this section are related to the neutral position $x_n = 2r_n/D$ – see Equation (4.36). But, there is a problem regarding the numerical value to be used for this quantity. Obuch [2] and Besset [3] considered the image of the ideal helical model where the local packing density is same at all places of yarn cross section with diameter D (assumption 6 in Section 4.1). According to Equation (4.11), it is then valid to write that $\mu = \mu_r$.

The cross section of such an idealized yarn is illustrated in Figure 4.7. Obuch [2] and Besset [3] assumed that a constant part of all fibrous material in a yarn is generally lying inside the cylinder with neutral radius r_n, and this part, as a first approximation, is one half, that is, one half is lying inside and the other half is lying outside of radius r_n. The following expressions are then valid to write:

$$\pi r_n^2 \mu = \frac{1}{2}\frac{\pi D^2}{4}\mu, \quad \frac{r_n^2}{D^2} = \frac{1}{8}, \quad \frac{r_n}{D} = \frac{1}{2\sqrt{2}},$$

$$x_n = \frac{2r_n}{D} = \frac{2}{2\sqrt{2}} = \frac{1}{\sqrt{2}} = 0.70711. \tag{4.45}$$

Figure 4.7 Fiber areas

We can use this value of x_n in Equations (4.35) to (4.44), especially for twisted filament yarns.

Note: Zurek [4] cited another recommendation of Besset [3] regarding the neutral position: for cotton yarns $x_n = 0.50$, for combed woollen yarns $x_n = 0.53$, and for woollen yarns $x_n = 0.39$. A comparison of the theoretical relations with experimental results and a specific problem of the staple yarns in this regard are shown in Section 4.6.

Limit of twisting – ideal helical model

By applying the value $x_n = 1/\sqrt{2}$, according to Equation (4.45) in the previous expressions, the following expressions are obtained for the limit of twisting, i.e., saturated twist.

The maximum of imaginary twist coefficient is resulted from Equation (4.42) as follows:

$$\alpha_i = \frac{\sqrt{2}\sqrt{\mu\rho}}{\sqrt{6\sqrt{3}\,\pi}} = 0.24750\sqrt{\mu\rho}\,. \tag{4.46}$$

The maximum yarn retraction is independent of x_n and according Equation (4.43), it is still valid to write that $\delta = 1 - 1/\sqrt{3} = 0.42265$.

The limit value of twist intensity and the value of angle β_D can be obtained from Equation (4.44) as follows:

$$\tan\beta_D = \sqrt{2}\sqrt{2} = 2, \quad \beta_D = \arctan(2) = 1.1071\,\mathrm{rad}\,(63.435\,\mathrm{deg})\,. \tag{4.47}$$

4.4 Retraction and limit of twisting – hypothesis of constant fiber volume

Here, the general expressions, as mentioned in Equations (4.29) to (4.34), are valid along with Figure 4.5 and Table 4.1.

Basic formula of yarn retraction

The starting fiber bundle as shown in Figure 4.5a contains n number of fibers and each fiber has length ζ_0 and cross-sectional area s. The starting fiber volume in this bundle can therefore be expressed as follows:

$$V_0 = ns\zeta_0, \quad \zeta_0 = \frac{V_0}{ns}\,. \tag{4.48}$$

After twisting, the twisted yarn, as shown in Figure 4.5b, has a substance cross-sectional area S – see Equation (1.32) – and length ζ. The fiber volume in this yarn is

$$V = S\zeta, \quad \zeta = \frac{V}{S}\,. \tag{4.49}$$

By using Equations (4.30), (4.48), (4.49), (1.34) and (1.39), the expression for yarn retraction can be written as follows:

$$\delta = 1 - \frac{\zeta}{\zeta_0} = 1 - \frac{V/S}{V_0/ns} = 1 - \frac{V}{V_0}\frac{s}{S/n} = 1 - \frac{V}{V_0}\frac{s}{s^*} = 1 - \frac{V}{V_0}k_n. \tag{4.50}$$

It is shown that the yarn retraction depends on yarn geometry in terms of coefficient k_n and further on the ratio V/V_0 of final and starting volume of fibers in the yarn.

Braschler's assumption

In general, the volume of fibers changes as a result of twisting. The peripheral fibers have a tendency to be elongated and the fibers near the yarn axis can be shortened due to twisting; such a process is usually accompanied with changes in volume. Braschler [1], however, considered that such changes in volume at yarn centre and yarn periphery are very small and can be mutually compensated. The following expressions are then valid from Equation (4.50):

$$V = V_0, \tag{4.51}$$

$$\delta = 1 - k_n. \tag{4.52}$$

Yarn retraction

By applying Equation (4.10) – which is valid for the helical model of yarn – in Equation (4.52), yarn retraction can be expressed as follows:

$$\delta = 1 - \frac{\displaystyle\int_0^{D/2} \frac{\mu_r r}{\sqrt{1 + (2\pi r Z)^2}}\, dr}{\displaystyle\int_0^{D/2} \mu_r r\, dr} = 1 - \frac{\displaystyle\int_0^{D/2} \frac{\mu_r r}{\sqrt{1 + (2r/D)^2 (\pi D Z)^2}}\, dr}{\displaystyle\int_0^{D/2} \mu_r r\, dr}. \tag{4.53}$$

The twist intensity $\kappa = \pi D Z$ can be expressed from Equation (1.54) as $\tan\beta_D = \kappa = 2\sqrt{\pi}\,\alpha/\sqrt{\mu\rho}$, and by applying this in Equation (4.53) yarn retraction can be expressed as a function of twist coefficient α in the following manner:

$$\delta = 1 - \frac{\displaystyle\int_0^{D/2} \frac{\mu_r r}{\sqrt{1 + (2r/D)^2 \frac{4\pi \alpha^2}{\mu\rho}}}\, dr}{\displaystyle\int_0^{D/2} \mu_r r\, dr}. \tag{4.54}$$

Finally, yarn retraction can also be expressed as a function of imaginary twist coefficient α_i. By using Equation (4.34) in (4.54), the following expression is obtained:

$$\delta = \frac{\displaystyle\int_0^{D/2} \frac{\mu_r r}{\sqrt{1+(2r/D)^2\,\dfrac{4\pi\,\alpha_i^2}{\mu\rho(1-\delta)^3}}}\,dr}{\displaystyle\int_0^{D/2} \mu_r r\,dr}. \tag{4.55}$$

One can see that the term δ (yarn retraction) appears on both sides of the expression. It is however not possible to rearrange such expression explicitly in terms of δ in this general shape. (A pertinent rearrangement requires the expression of local packing density μ_r as a function of r.)

Yarn retraction – ideal helical model

It is easier to solve for yarn retraction under the consideration of ideal helical model. By applying Equation (4.16) – which is valid for ideal helical model – in Equation (4.52), the following expression is obtained for yarn retraction:

$$\delta = 1 - \frac{2}{(\pi DZ)^2}\left[\sqrt{1+(\pi DZ)^2}-1\right]$$

$$= \frac{\left(\sqrt{1+(\pi DZ)^2}+1\right) - \dfrac{2}{(\pi DZ)^2}\left[\sqrt{1+(\pi DZ)^2}-1\right]\left(\sqrt{1+(\pi DZ)^2}+1\right)}{\sqrt{1+(\pi DZ)^2}+1}$$

$$= \frac{\sqrt{1+(\pi DZ)^2}+1 - \dfrac{2}{(\pi DZ)^2}\left\{1+(\pi DZ)^2-1\right\}}{\sqrt{1+(\pi DZ)^2}+1},$$

$$\delta = \frac{\sqrt{1+(\pi DZ)^2}-1}{\sqrt{1+(\pi DZ)^2}+1} = \frac{\sqrt{1+\tan^2\beta_D}-1}{\sqrt{1+\tan^2\beta_D}+1} = \frac{\sqrt{1+\kappa^2}-1}{\sqrt{1+\kappa^2}+1}. \tag{4.56}$$

[Equation (1.52) was used for rearrangement.] By using Equation (4.17), the following expression can be obtained:

$$\delta = 1 - \frac{2\cos\beta_D}{1+\cos\beta_D} = \frac{(1+\cos\beta_D)-2\cos\beta_D}{1+\cos\beta_D} = \frac{1-\cos\beta_D}{1+\cos\beta_D}. \tag{4.57}$$

$$\delta = \frac{1-\cos\beta_D}{1+\cos\beta_D} = \frac{\left(\cos^2\frac{\beta_D}{2}+\sin^2\frac{\beta_D}{2}\right)-\left(\cos^2\frac{\beta_D}{2}-\sin^2\frac{\beta_D}{2}\right)}{\left(\cos^2\frac{\beta_D}{2}+\sin^2\frac{\beta_D}{2}\right)+\left(\cos^2\frac{\beta_D}{2}-\sin^2\frac{\beta_D}{2}\right)}$$

$$= \frac{2\sin^2\frac{\beta_D}{2}}{2\cos^2\frac{\beta_D}{2}} = \tan^2\frac{\beta_D}{2} \qquad\qquad . \text{ (4.58)}^9$$

Note: Braschler [1], for the first time, derived Equation (4.58).

The expression for twist intensity can be obtained from Equation (1.54) as $\kappa = 2\sqrt{\pi}\,\alpha/\sqrt{\mu\rho}$. By using this formula in Equation (4.56), yarn retraction can be expressed as a function of twist coefficient α in the following manner:

$$\delta = \frac{\sqrt{1+\left(\frac{2\sqrt{\pi}\,\alpha}{\sqrt{\mu\rho}}\right)^2}-1}{\sqrt{1+\left(\frac{2\sqrt{\pi}\,\alpha}{\sqrt{\mu\rho}}\right)^2}+1} = \frac{\sqrt{1+\frac{4\pi\alpha^2}{\mu\rho}}-1}{\sqrt{1+\frac{4\pi\alpha^2}{\mu\rho}}+1}. \qquad (4.59)$$

Finally, by using Equation (4.34) in (4.59), yarn retraction can be found to be related to the imaginary twist coefficient α_i as follows

$$\delta = \frac{\sqrt{1+\frac{4\pi}{\mu\rho}\frac{\alpha_i^2}{(1-\delta)^3}}-1}{\sqrt{1+\frac{4\pi}{\mu\rho}\frac{\alpha_i^2}{(1-\delta)^3}}+1}. \qquad (4.60)$$

Because the term δ (yarn retraction) appears on both sides of the last expression, this is required to be rearranged explicitly in terms of δ. This is done as follows:

$$A = \frac{4\pi\alpha_i^2}{\mu\rho(1-\delta)^3} \quad\ldots\text{helping expression.}$$

---

9   The following goniometrical formulas were used: $\cos^2\frac{\beta_D}{2}+\sin^2\frac{\beta_D}{2}=1$
and $\cos^2\frac{\beta_D}{2}-\sin^2\frac{\beta_D}{2}=\cos\left(2\frac{\beta_D}{2}\right)=\cos\beta_D$.

$$\delta = \frac{\sqrt{1+A}-1}{\sqrt{1+A}+1}, \quad \delta\sqrt{1+A}+\delta = \sqrt{1+A}-1, \quad 1+\delta = \sqrt{1+A}(1-\delta),$$

$$(1+\delta)^2 = \left[\sqrt{1+A}(1-\delta)\right]^2, \quad (1+\delta)^2 = (1+A)(1-\delta)^2,$$

$$(1+\delta)^2(1-\delta) = (1+A)(1-\delta)^3,$$

$$1+2\delta+\delta^2-\delta-2\delta^2-\delta^3 = 1-3\delta+3\delta^2-\delta^3+\frac{4\pi\alpha_i^2}{\mu\rho},$$

$$0 = 4\delta^2-4\delta+\frac{4\pi\alpha_i^2}{\mu\rho}, \quad 0 = \delta^2-\delta+\frac{\pi\alpha_i^2}{\mu\rho}. \tag{4.61}$$

The last expression is a quadratic equation whose two roots are mentioned as follows:

$$\delta = \frac{1\pm\sqrt{1-\dfrac{4\pi\,\alpha_i^2}{\mu\,\rho}}}{2} = \frac{1}{2}\pm\frac{1}{2}\sqrt{1-\frac{4\pi\,\alpha_i^2}{\mu\,\rho}}. \tag{4.62}$$

Because in reality $\delta = 0$ when $\alpha_i = 0$, only the minus sign stands physically real.

Limit of twisting – ideal helical model

Evidently, it must be valid that

$$1-\frac{4\pi\,\alpha_i^2}{\mu\,\rho}\geq 0, \quad \alpha_i \leq \sqrt{\frac{\mu\rho}{4\pi}} = 0.28209\sqrt{\mu\rho}. \tag{4.63}$$

It is shown that the imaginary twist coefficient is limited, which is similar to the case of hypothesis of neutral radius as shown in Equation (4.41).

Note: According to Equation (4.46), it was obtained $\alpha_i \leq 0.24750\sqrt{\mu\rho}$ in case of hypothesis of neutral radius. It is shown that this limiting twist is not too far from the limiting twist derived in Equation (4.63).

In the limiting case (maximum or saturated twist), the following expressions are obtained from Equations (4.62) and (4.63):

$$1-\frac{4\pi\,\alpha_i^2}{\mu\,\rho} = 0, \quad \delta = \frac{1}{2}\pm\frac{1}{2}\sqrt{0} = \frac{1}{2}. \tag{4.64}$$

Equation (4.56) is rearranged as follows:

$$\delta\sqrt{1+\tan^2\beta_D}+\delta=\sqrt{1+\tan^2\beta_D}-1, \quad 1+\delta=\sqrt{1+\tan^2\beta_D}\,(1-\delta),$$

$$\left(\frac{1+\delta}{1-\delta}\right)^2=1+\tan^2\beta_D, \quad \tan\beta_D=\sqrt{\left(\frac{1+\delta}{1-\delta}\right)^2}-1. \tag{4.65}$$

When Equation (4.64) is substituted into (4.65), the following values for limiting or saturated twist are obtained:

$$\left.\begin{array}{l} \tan\beta_D=\sqrt{\left(\dfrac{1+1/2}{1-1/2}\right)^2-1}=\sqrt{8}=2\sqrt{2}=2.8284, \\[3mm] \beta_D=\arctan\left(2\sqrt{2}\right)=1.2310\,\text{rad}\;\left(70.529\,\text{deg}\right). \end{array}\right\} \tag{4.66}$$

4.5   Retraction and limit of twisting – hypothesis of zero axial force

Fiber strain

A general fiber lying at a radius r of the helical yarn is shown in Figure 4.8a. Figure 4.8b is originated due to unrolling of cylindrical surface. The starting length of each fiber is ζ_0 (Figure 4.5), but the actual length of this fiber – lying in a portion of yarn of length ζ – is l. Then, the following expression is valid for angle β:

$$\cos\beta=\zeta/l. \tag{4.67}$$

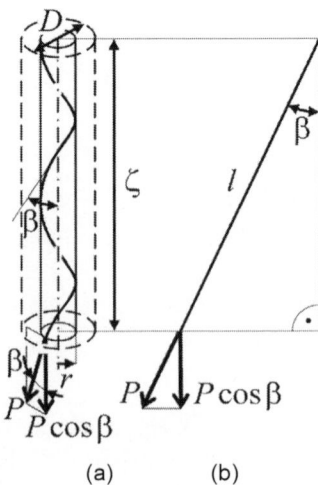

(a)   (b)

Figure 4.8 Tensioned fiber in helical yarn

Then, fiber strain can be expressed as follows:

$$\varepsilon = \frac{l - \zeta_0}{\zeta_0} = \frac{l}{\zeta_0} - 1 = \frac{\zeta/\zeta_0}{\cos\beta} - 1 = \frac{1 - \delta}{\cos\beta} - 1 . \tag{4.68}$$

[Equations (4.30) and (4.67) are used.]

Forces

Because of fiber strain, an axial force P is acted on the fiber. In this connection, Budnikov [5] considered the following relation between the axial stress σ and axial strain ε of fiber under Hook's law:

$$\sigma = E\varepsilon, \quad E \ldots \text{parameter of fiber (Young's modulus)} . \tag{4.69}$$

As the fiber cross-sectional area is s, the axial force P in fiber is $P = \sigma s = E\varepsilon s$ and the axial component of force P, i.e., $P\cos\beta$ (Figure 4.8b), is

$$P\cos\beta = \sigma s \cos\beta = E\varepsilon s \cos\beta = Es\left[\frac{1 - \delta}{\cos\beta} - 1\right]\cos\beta = Es\left[(1 - \delta) - \cos\beta\right] . \tag{4.70}$$

The number of fibers dn present inside the differential annulus in yarn cross section (Figure 4.3) can be determined according to Equation (4.5). By using Equation (4.2), the following expression can be written:

$$dn = \frac{2\pi}{s}\cos\beta \, \mu_r r dr . \tag{4.71}$$

Then, the following expression is valid for the component of axial force:

$$\begin{aligned}
dF &= P\cos\beta \, dn = Es\left[(1 - \delta) - \cos\beta\right]\frac{2\pi}{s}\cos\beta \, \mu_r r dr \\
&= 2\pi E\left[(1 - \delta)\cos\beta - \cos^2\beta\right]\mu_r r dr .
\end{aligned} \tag{4.72}$$

The total axial force in yarn is then takes the following expression:

$$\begin{aligned}
F &= \int_{r=0}^{r=D/2} dF = \int_{0}^{D/2} 2\pi E\left[(1 - \delta)\cos\beta - \cos^2\beta\right]\mu_r r dr \\
&= 2\pi E(1 - \delta)\int_{0}^{D/2}\cos\beta \, \mu_r r dr - 2\pi E\int_{0}^{D/2}\cos^2\beta \, \mu_r r dr
\end{aligned}$$

$$= 2\pi E\left(1-\delta\right) \int\limits_{0}^{D/2} \frac{\mu_r r dr}{\sqrt{1+\left(2\pi r Z\right)^2}} \quad - 2\pi E \int\limits_{0}^{D/2} \frac{\mu_r r dr}{1+\left(2\pi r Z\right)^2}. \tag{4.73}$$

[Equations (4.2) and (4.72) were used for rearrangement.]

Yarn retraction

Let us express the quantity δ from Equation (4.73) in the following manner

$$\delta = 1 - \frac{\dfrac{F}{2\pi E} + \displaystyle\int\limits_{0}^{D/2} \dfrac{\mu_r r dr}{1+\left(2\pi r Z\right)^2}}{\displaystyle\int\limits_{0}^{D/2} \dfrac{\mu_r r dr}{\sqrt{1+\left(2\pi r Z\right)^2}}}. \tag{4.74}$$

Nevertheless, the axial force must be equal to zero $\left(F=0\right)$ in case of yarn retraction. Then, the following formula is valid for yarn retraction:

$$\delta = 1 - \frac{\displaystyle\int\limits_{0}^{D/2} \dfrac{\mu_r r dr}{1+\left(2\pi r Z\right)^2}}{\displaystyle\int\limits_{0}^{D/2} \dfrac{\mu_r r dr}{\sqrt{1+\left(2\pi r Z\right)^2}}}. \tag{4.75}$$

The twist intensity can be expressed from Equations (1.52) and (1.54) as $\pi D Z = \kappa = 2\sqrt{\pi}\,\alpha/\sqrt{\mu\rho}$. By using this formula in Equation (4.75), the yarn retraction can be expressed as a function of twist coefficient α as follows:

$$\delta = 1 - \frac{\displaystyle\int\limits_{0}^{D/2} \dfrac{\mu_r r dr}{1+\left(2r/D\right)^2\left(\pi D Z\right)^2}}{\displaystyle\int\limits_{0}^{D/2} \dfrac{\mu_r r dr}{\sqrt{1+\left(2r/D\right)^2\left(\pi D Z\right)^2}}} = 1 - \frac{\displaystyle\int\limits_{0}^{D/2} \dfrac{\mu_r r dr}{1+\left(\dfrac{2r}{D}\right)^2 \dfrac{4\pi\alpha^2}{\mu\rho}}}{\displaystyle\int\limits_{0}^{D/2} \dfrac{\mu_r r dr}{\sqrt{1+\left(\dfrac{2r}{D}\right)^2 \dfrac{4\pi\alpha^2}{\mu\rho}}}}. \tag{4.76}$$

Finally, the yarn retraction can be expressed as a function of imaginary twist coefficient α_i. By using Equation (4.34) in (4.76), the following expression can be obtained:

$$\delta = 1 - \dfrac{\displaystyle\int_0^{D/2} \dfrac{\mu_r r\,dr}{1 + \left(\dfrac{2r}{D}\right)^2 \dfrac{4\pi\alpha_i^2}{\mu\rho(1-\delta)^3}}}{\displaystyle\int_0^{D/2} \dfrac{\mu_r r\,dr}{\sqrt{1 + \left(\dfrac{2r}{D}\right)^2 \dfrac{4\pi\alpha_i^2}{\mu\rho(1-\delta)^3}}}}. \tag{4.77}$$

Here again the term δ (yarn retraction) appears on both sides of the last equation. It is however not possible to rearrange such expression explicitly in terms of δ. (To obtain a pertinent rearrangement, it is necessary to know the expression for local packing density μ_r as a function of radius r.)

Yarn retraction – ideal helical model

The packing density $\mu_r = \mu$ is a constant in the ideal helical model. Then, the yarn retraction can be expressed according to Equation (4.75) as follows:

$$\delta = 1 - \dfrac{\mu \displaystyle\int_0^{D/2} \dfrac{r\,dr}{1 + (2\pi r Z)^2}}{\mu \displaystyle\int_0^{D/2} \dfrac{r\,dr}{\sqrt{1 + (2\pi r Z)^2}}} = 1 - \dfrac{\displaystyle\int_0^{D/2} \dfrac{r\,dr}{1 + (2\pi r Z)^2}}{\displaystyle\int_0^{D/2} \dfrac{r\,dr}{\sqrt{1 + (2\pi r Z)^2}}}. \tag{4.78}$$

The definite integral expressed in the denominator is already solved in Equation (4.14), and the definite integral shown in the numerator can be solved as follows:

$$\int_0^{D/2} \dfrac{r\,dr}{1 + (2\pi r Z)^2} = \int_1^{\sqrt{1+(\pi D Z)^2}} \dfrac{x\,dx}{x^2\,(2\pi Z)^2} = \dfrac{1}{(2\pi Z)^2}$$

$$\cdot \int_1^{\sqrt{1+(\pi D Z)^2}} \dfrac{dx}{x} = \dfrac{1}{(2\pi Z)^2}\Big[\ln|x|\Big]_1^{\sqrt{1+(\pi D Z)^2}}$$

Substitution: $x^2 = 1 + (2\pi r Z)^2$, $2x\,dx = (2\pi Z)^2\,2r\,dr$, $r\,dr = x\,dx/(2\pi Z)^2$

$$= \dfrac{1}{(2\pi Z)^2}\ln\sqrt{1 + (\pi D Z)^2}. \tag{4.79}$$

By applying Equations (4.14) and (4.79) in (4.78), the following expression is obtained:

$$\delta = 1 - \frac{\dfrac{1}{(2\pi Z)^2} \ln \sqrt{1 + (\pi DZ)^2}}{\dfrac{1}{(2\pi Z)^2}\left[\sqrt{1 + (\pi DZ)^2} - 1\right]}$$

$$= 1 - \frac{\ln \sqrt{1 + (\pi DZ)^2}}{\sqrt{1 + (\pi DZ)^2} - 1} = 1 - \frac{\ln \sqrt{1 + \tan^2 \beta_D}}{\sqrt{1 + \tan^2 \beta_D} - 1} = 1 - \frac{\ln \sqrt{1 + \kappa^2}}{\sqrt{1 + \kappa^2} - 1}. \qquad (4.80)$$

[Equation (1.52) was used for rearrangement.]

The twist intensity can be expressed from Equation (1.54) as $\kappa = 2\sqrt{\pi}\, \alpha / \sqrt{\mu\rho}$. By using this formula in Equation (4.80), the following expression can be obtained:

$$\delta = 1 - \frac{\ln \sqrt{1 + \dfrac{4\pi\alpha^2}{\mu\rho}}}{\sqrt{1 + \dfrac{4\pi\alpha^2}{\mu\rho}} - 1}. \qquad (4.81)$$

Further, the yarn retraction can be expressed as a function of imaginary twist coefficient α_i by using Equation (4.34) in (4.81) in the following manner:

$$\delta = 1 - \frac{\ln \sqrt{1 + \dfrac{4\pi\alpha_i^2}{\mu\rho(1-\delta)^3}}}{\sqrt{1 + \dfrac{4\pi\alpha_i^2}{\mu\rho(1-\delta)^3}} - 1}. \qquad (4.82)$$

Here again the term δ (yarn retraction) appears on both sides of the last equation. It is however not possible to rearrange such expression explicitly in terms of δ.

Parametric formulation of relation $\alpha_i - \delta$ - ideal helical model

Nevertheless, the yarn retraction δ can be solved parametrically by assigning imaginary twist coefficient α_i. This is because the twist intensity can be formulated from Equation (1.54) as $\kappa = 2\sqrt{\pi}\,\alpha/\sqrt{\mu\rho}$. By substituting Equations (4.34) and (4.80) to this expression, the following expression can be written:

$$\kappa = \frac{2\sqrt{\pi}\,\alpha}{\sqrt{\mu\rho}} = \frac{2\sqrt{\pi}\,\alpha_i}{\sqrt{\mu\rho}\,(1-\delta)^{3/2}} = \frac{2\sqrt{\pi}\,\alpha_i}{\sqrt{\mu\rho}\left(\dfrac{\ln\sqrt{1+\kappa^2}}{\sqrt{1+\kappa^2}-1}\right)^{3/2}},$$

$$\alpha_i = \frac{\sqrt{\mu\rho}}{2\sqrt{\pi}}\kappa\left(\frac{\ln\sqrt{1+\kappa^2}}{\sqrt{1+\kappa^2}-1}\right)^{3/2}. \tag{4.83}$$

Equations (4.80) and (4.83) create a pair of parametrical equations of function $\alpha_i - \delta$ by means of parameter κ. The imaginary twist coefficient α_i can be determined from Equation (4.83) and the corresponding yarn retraction δ can be determined from Equation (4.80) for each value[10] of the twist intensity κ.

Limit of twisting – ideal helical model

The character of the function $\delta - \alpha_i$ in case of limiting case of twisting (or saturated twisting) is displayed in Appendix 2. According to Equation (A2.17), the following expression can be derived:

$$\sqrt{1 + \frac{4\pi\alpha_i^2}{\mu\rho(1-\delta)^3}} = 9.5802. \tag{4.84}$$

By using this value in Equation (4.82), the following expression is obtained for yarn retraction

$$\delta = 1 - \frac{\ln 9.5802}{9.5802 - 1} = 0.73664. \tag{4.85}$$

---

10   But the numerical value of κ must not be 'too high'.

From Equations (4.84) and (4.85), the imaginary twist coefficient α_i is calculated as follows:

$$\frac{4\pi\alpha_i^2}{\mu\rho(1-0.73664)^3} = 9.5802^2 - 1, \quad \alpha_i^2 = \left(9.5802^2 - 1\right)\frac{\mu\rho}{4\pi}\left(1-0.73664\right)^3,$$

$$\alpha_i = \sqrt{\mu\rho}\,\frac{\sqrt{(9.5802^2-1)(1-0.73664)^3}}{2\sqrt{\pi}} = \sqrt{\mu\rho}\cdot 0.36326. \tag{4.86}$$

Let us use Equation (1.52) and Equation (A2.1) from Appendix 2 as follows:

$$\sqrt{1+\tan^2\beta_D} = \sqrt{1+\kappa^2} = \sqrt{1+\frac{4\pi\alpha^2}{\mu\rho}} = \sqrt{1+\frac{4\pi\alpha_i^2}{\mu\rho(1-\delta)^3}}. \tag{4.87}$$

By applying the values of yarn retraction and imaginary twist coefficient obtained from Equations (4.84) to (4.87), the following expression is obtained:

$$\sqrt{1+\tan^2\beta_D} = \sqrt{1+\kappa^2} = \sqrt{1+\frac{4\pi\alpha^2}{\mu\rho}} = 9.5802. \tag{4.88}$$

It is now easy to evaluate that

$$\tan\beta_D = \sqrt{9.5802^2-1} = 9.5279. \tag{4.89}$$

$$\beta_D = \arctan 9.5279 = 1.4662\,\text{rad}\,(84.008\,\text{deg}). \tag{4.90}$$

4.6   Yarn retraction in ideal helical model – comparison of hypotheses

In the earlier sections, three types of theoretical concepts were introduced for solving the problem of yarn retraction. The basic results derived from the three types of concepts under ideal helical model are summarized in Table 4.2. Here, symbol ① relates to the hypothesis of neutral radius (Section 4.3), symbol ② represents the hypothesis of constant fiber volume (Section 4.4) and symbol ③ refers to the hypothesis of yarn axial force (Section 4.5).

Table 4.2 $\delta - \alpha_i$ relation and parameters of limit (saturated) twisting by ideal helical model

Type	$\delta - \alpha_i$ relation	Parameters for limiting case of twisting (state of saturated twist)			
		$\dfrac{\alpha_i}{\sqrt{\mu\rho}}$ *	δ	$\tan\beta_D$	$\beta_D\,[\deg]$
①	$\delta = 1 - \dfrac{2}{\sqrt{3}}\cos\left[\dfrac{\pi}{3} \pm \dfrac{1}{3}\arccos\left(\dfrac{3\sqrt{3}}{\mu}\pi\dfrac{\alpha_i^2}{\rho}\right)\right]^{**}$	0.248	0.423	2	63.4
②	$\delta = \dfrac{1}{2} \pm \dfrac{1}{2}\sqrt{1 - \dfrac{4\pi}{\mu}\dfrac{\alpha_i^2}{\rho}}$	0.282	0.5	2.83	70.5
③	$\delta = 1 - \dfrac{\ln\sqrt{1+\kappa^2}}{\sqrt{1+\kappa^2}-1},$	0.363	0.737	9.53	84.0
	$\alpha_i = \dfrac{\sqrt{\mu\rho}}{2\sqrt{\pi}}\kappa\left(\dfrac{\ln\sqrt{1+\kappa^2}}{\sqrt{1+\kappa^2}-1}\right)^{3/2}, \kappa\ldots\text{parameter}$				

*Dimensionless quantity.

**Equation (4.40) considering $r_n = 1/\sqrt{2}$ according to Equation (4.45).

Real and hypothetical behaviours of functions

The graphical interpretation of the derived relations is shown in Figure 4.9. Here the thick lines represent the real parts of the function $\delta - \alpha_i$, and the thin lines indicate the hypothetical (non-real) parts of the aforesaid function.

Note: It is theoretically possible to create the structures also from the hypothetical (non-real) parts 'artificially', but not by 'natural' twisting process. However, such 'artificial' structure would be spontaneously transpose – 'jump over' (when axial force is limiting to zero) – with the real one (corresponding to the thick part)[11].

As shown in Figure 4.9, the three theoretical concepts result in three different curves, the first part of the curves is very much similar and close to one another (Figure 4.10). In practice, the real fiber bundles correspond to the region shown by the hatched lines. It is therefore not too important to know which of the three theoretical concepts is preferred to in practice.

---

11   Mr. M. Konopasek drew attention to this fact by discussion with one of the authors of this book in 1967.

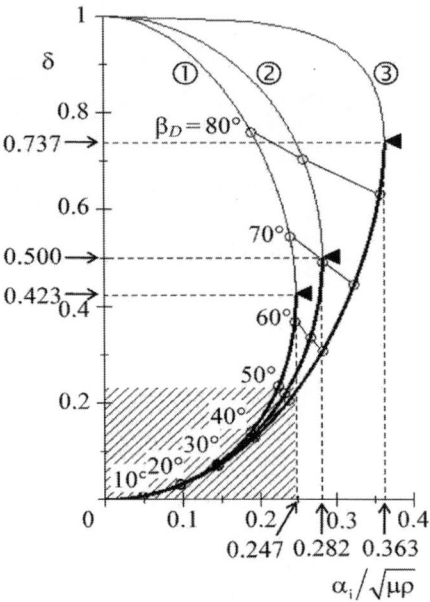

Figure 4.9 Thick parts – real relation, thin parts – hypothetical (unreal) relation.
① : Equation (4.40) with Equation (4.45); ②: Equation (4.62); ③: Equation (4.80) and
Equation (4.83); : limit twisting (saturated twist); hatched area: experimentally verified

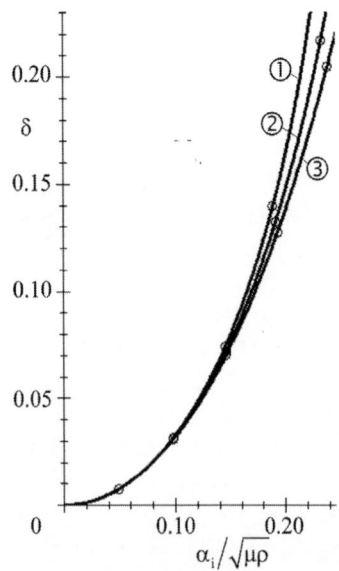

Figure 4.10 Yarn retraction δ as the functions of α_i.
Magnified hatched area from Figure 4.9

Experimental results

Marko and Neckář [6] evaluated retraction of flat polyamide filament yarns (type Perlon) prepared from fibers of 5.55 dtex $(5\,\text{den})$ fineness. To obtain yarn fineness (count) similar to the commercially produced yarns, a few fibers were taken out from the original filament yarns. Accordingly, the starting values of yarn fineness T_0 were obtained as shown in Figure 4.11.

The values of twisted length of fiber bundles ζ and the different values of number of entered coils N_C were measured on a modified twist tester by keeping the starting length as $\zeta_0 = 0.5\,\text{m}$ and a small pre-tension as $0.176\,\text{cN}\,\text{dtex}^{-1}$ $(0.2\,\text{p}\,\text{den}^{-1})$. The yarn retraction was calculated from Equation (4.30), i.e., $\delta = 1 - \zeta/\zeta_0$, the imaginary twist was determined from $Z_\text{i} = N_\text{C}/\zeta_0$, and the imaginary twist coefficient was calculated from $\alpha_\text{i} = Z_\text{i}\sqrt{T_0}$ in accordance with Table 4.1.

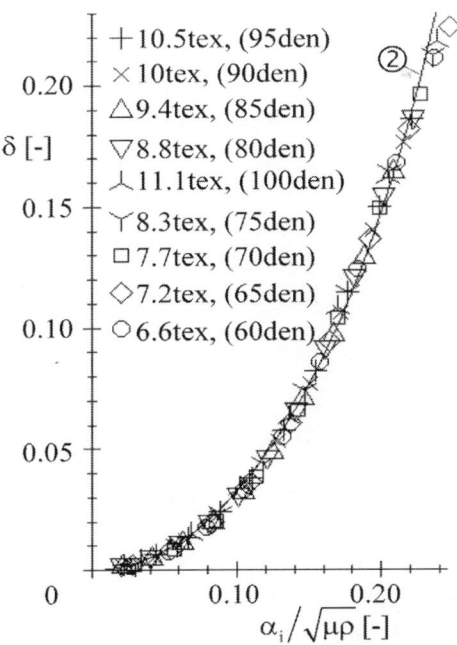

Figure 4.11 Experimental results of filament yarn retraction (Perlon). Parameters of theoretical curve (2): $\mu = 0.75$, $\rho = 1150\,\text{kg}\,\text{m}^{-3}$

The experimentally measured values represent the points as shown in Figure 4.11. The values $\alpha_i / \sqrt{\mu \rho}$ are evaluated from α_i using $\mu = 0.75$ and $\rho = 1150 \, \text{kg m}^{-3}$ (see Table 1.1).

Note: While twisting the flat filament yarn, it was observed that the fibers 'sat down' on one another when the first few coils were inserted, and then they stayed in their position. So, the packing density was practically a constant (here 0.75), independent of the level of twisting in case of flat filament yarns.

It is shown from Figure 4.11 that

1. The behaviour is independent of yarn fineness – all points create one common function. It means that the yarn retraction is related to α_i only.

2. The experimental results correspond to the theoretically derived relation very well. [The smooth curve corresponds to Equation (4.62).]

3. The maximum of yarn retraction, which was possible to achieve practically, lies near to 0.23. It was not possible to insert more coils inside the yarn body[12]. It can be stated that:

   (a) The theoretically predicted existence of limiting case of twisting, i.e., saturated twist [Equations (4.40), (4.62) and (4.84) in Sections 4.3–4.5] was verified experimentally.

   (b) However, this phenomenon appears to come earlier than it was predicted by theoretical equations. (Experimental value around $\delta = 0.23$ in relation to theoretical values 0.42, 0.5, 0.74, experimental value around $\alpha_i / \sqrt{\mu \rho} = 0.23$ in relation to theoretical values 0.25, 0.28, 0.36 – see Table 4.2.)

Why does the saturated twist come earlier than that predicted theoretically? It is definitely because all assumptions are not fully valid in practice. Probably, the most important among them is related to the axial asymmetry of real twisted yarns. Here, the geometrical influences in axially quite symmetric twisted structure are thought, but the real filament yarn is not however so perfectly symmetrical. So, the forces acting on the fibers of the yarn (mainly near the yarn surface) create the resultant 'axially compressing' force which does not work precisely along the yarn axis but at a distance from it. This eccentricity brings a bending moment to the twisted yarn. Therefore, if the

---

12   About the so-called 'twist of second order', see later on.

twisting is near to limiting (saturated) case then the so-called 'twist of second order' appears.

Twist of second order

What happens to a yarn when we insert more coils than that corresponds to the limiting case of (saturated) twisting? Such a process is illustrated in Figure 4.12.

Figure 4.12a shows a filament yarn before reaching the saturated twist. In Figure 4.12b, the filament yarn is not able to 'absorb' more coils in the form of regular twist; therefore the yarn deviates from its straight shape and starts taking helical form. Then, according to Figure 4.12c, the helical shape is transformed to the first coil of second order. Further twisting leads to more coils of second order (Figure 4.12d) to the yarn and the parts of the yarn are now called to have the twist of second order.

a)
b)
c)
d)

Figure 4.12 Creation of twist of second order [6]

Note: The origin of twist of second order often brings critical problems. For example, this can cause yarn breakage during false-twist texturing, where a high yarn twist is used.

4.7   Notes to retraction of staple yarns

The derived equations, as stated in Sections 4.3–4.5, describe the process of yarn retraction due to twist in case of filament yarns very well – see one example displayed in Figure 4.11[13]. However, it is very difficult to find an

---

13   Figure 4.11 represents the highest correlation between theory and experiment during the entire professional history of both the authors.

expression for yarn retraction in case of staple yarns. The staple fibers or fiber portions usually slip and/or are straightened from their original crimped shape. Such processes create something as 'false draft' in the yarn, where a part of the fibers (at most from the outer layers) is 'passive', and – as a result of this – the retraction of yarn is usually found to be smaller.

Further, an increase in compression of fiber materials due to twisting implies that more fibers, especially from the outer layers of yarn, come together – after straightening – come for creation of yarn retraction by means of their mechanical activities. So, the volume of 'passive' fibers decreases continually by twisting. The entire process is very complicated. Even an adequate structural model of yarn retraction is still not fully known.

Experimental observations

It is relatively difficult to evaluate yarn retraction for staple yarns experimentally. This is because the retraction values are usually very small, i.e., a few percentages only, which is often expressed as an experimental error in measurements of length. Therefore, there exist a very few experimental results of retraction of staple yarns in literature.

Nevertheless, a set of experimental results δ_{EXP} on yarn retraction for cotton yarns were published in relation to yarn finenesses T and (Koechlin's) twist coefficient α in the book of Sokolov [7][14] as follows. The representation of the data is shown in Table 4.3 and Figure 4.13.

Table 4.3 Experimental data of yarn retraction for cotton yarns [7]

$T\,[\text{tex}]\,(Nm)$	$\alpha\,[\text{m}^{-1}\text{ktex}^{1/2}]$						
	82	91	100	109	118	127	136
100 (10)	0.0467	0.0480	0.0537	0.0600	0.0672	0.0742	0.0822
62.5 (16)	0.0332	0.0339	0.0391	0.0446	0.0507	0.0566	0.0638
50 (20)	0.0250	0.0293	0.0339	0.0391	0.0447	0.0508	0.0576
41.7 (24)	0.0236	0.0261	0.0306	0.0355	0.0410	0.0468	0.0534
35.7 (28)	0.0216	0.0239	0.0281	0.0329	0.0382	0.0438	0.0503
29.4 (34)	0.0179	0.0215	0.0254	0.0331*	0.0351	0.0408	0.0470
25 (40)	0.0163	0.0200	0.0237	0.0281	0.0386*	0.0386	0.0448
20.8 (48)	0.0149	0.0182	0.0219	0.0263	0.0314	0.0366	0.0426
18.5(54)	0.0140	0.0173	0.0202	0.0244	0.0292	0.0345	0.0405
16.6 (60)	0.0116	0.0147	0.0182	0.0222	0.0270	0.0321	0.0379
11.8 (85)	0.0116	0.0147	0.0182	0.0215	0.0329*	–	–

* A typical value (probably misprint).

---

14   In Reference [7], the values of $1 - \delta_{EXP}$ were originally published.

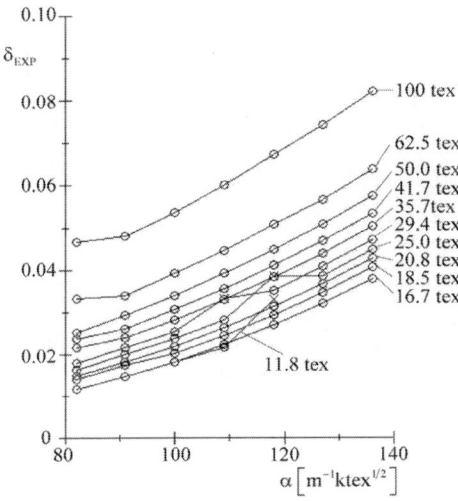

Figure 4.13 Afoncikov's retraction of cotton yarns according [7] and
Table 4.3. O: right values; ×: atypical values

Empirical equations

Let us modify (Braschler's) Equation (4.56) for calculation of yarn retraction
δ as follows:

$$\delta = \left[\sqrt{1 + \left(\pi D_{\delta[\text{mm}]} Z_{[\text{mm}^{-1}]} \right)^2} - 1 \right] \Bigg/ \left[\sqrt{1 + \left(\pi D_{\delta[\text{mm}]} Z_{[\text{mm}^{-1}]} \right)^2} + 1 \right]. \qquad (4.91)$$

Note that this equation uses a 'reduced' (smaller) diameter D_δ in
place of the original yarn diameter D. The twist in yarn, according to
Equation (1.52), is expressed by $Z_{[\text{m}^{-1}]} = \sqrt{1000}\, \alpha_{[\text{m}^{-1}\text{ktex}^{1/2}]} \Big/ \sqrt{T_{[\text{tex}]}}$ or

$Z_{[\text{mm}^{-1}]} = \alpha_{[\text{m}^{-1}\text{ktex}^{1/2}]} \Big/ \sqrt{1000\, T_{[\text{tex}]}}$.

Further, let us take the following empirical relation for the experimental
data of yarn retraction as reported in Table 4.3

$$D_\delta = D \left[1 - \frac{1.17 \left(\mu_m - \mu \right)}{0.00126\, T_{[\text{tex}]}^{1.88} + 1} \right] \qquad (4.92)$$

where μ denotes actual yarn packing density and μ_m represents theoretically maximum yarn packing density. Equation (1.47), in the form of

$D_{[\text{mm}]} = \sqrt{4 T_{[\text{tex}]} \Big/ \left(\pi \mu \rho_{[\text{kg\,m}^{-3}]} \right)}$, is now valid for yarn diameter D. Finally, let

us use the following semi-empirical equation for yarn packing density μ:

$$\frac{\mu^{2,5}}{\left[1-\left(\mu/\mu_m\right)^3\right]^3} = Q_{[m^2\,tex^{-1/2}]} \left(Z_{[m^{-1}]} T_{[tex]}^{1/4}\right)^2.$$

(4.93)[15]

(Q is a typical parameter for the given type of yarns.)

By using the numerical values $\mu_m = 0.8$, fiber density $\rho = 1500\,\mathrm{kg\,m}^{-3}$ and parameter $Q = 9.15 \cdot 10^{-8}\,\mathrm{m}^2\,\mathrm{tex}^{-1/2}$ for cotton yarns, let us calculate yarn retraction step by step as follows.

1. At first, we need to calculate the packing density μ of yarn, when the values T and Z are given. This can be done by using a numerical method for finding of root from Equation (4.93).

2. Then we calculate yarn diameter D.

3. Further, we calculate the 'reduced' diameter D_δ by using Equation (4.92).

4. Finally, we calculate the yarn retraction by using Equation (4.91).

Figure 4.14 illustrates the comparison between the experimental results of yarn retraction $\delta = \delta_{EXP}$ (Table 4.3) and the values $\delta = \delta_{COMP}$ obtained by computation, barring the coarsest yarn ($T = 100$ tex). Clearly, a very good correlation is reported between experimental and calculated values.

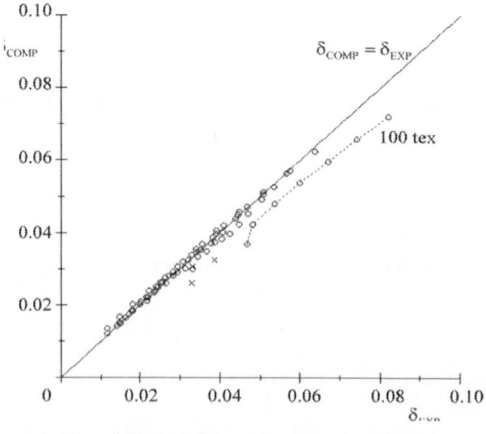

Figure 4.14 Comparison between experimental and computed values of yarn retraction. O: right values; ×: atypical values

---

15   This equation will be discussed in detail in another chapter of this book.

Note: Probably, the fibers in the 'peripheral' layers of the coarsest yarn of 100 tex count do not play a significant role on the reduction of yarn retraction.

Non-constant packing density

The ideal helical model that assumes constant packing density at each and every place inside the yarn was discussed in Sections 4.3–4.6. However, the real staple yarns have significantly non-constant function of local packing density μ_r against yarn radius r. This is illustrated in Figure 4.1.

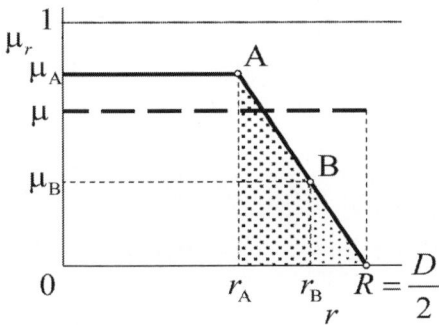

Figure 4.15 Trapezoidal shape of function μ_r against radius r

As a simplification, the function of local packing density μ_r in the staple yarns can be thought to follow a 'trapezoidal model' against yarn radius r, as shown in Figure 4.15. It is assumed that the packing density $\mu_r = \mu_A$ is a constant in the internal part of yarn body, i.e., till the point A lying at radius r_A. In the external part, i.e., at radii $r \in (r_A, R)$, μ_r decreases linearly from μ_A (at radius r_A) to zero (at radius R).

For yarn radius, i.e., for one half of yarn diameter, we use here the following symbol:

$$R = D/2 . \tag{4.94}$$

Then, the twist intensity can be expressed by means of Equations (1.51) and (4.94) in the following form:

$$\kappa = \pi D Z = 2\pi R Z . \tag{4.95}$$

Let us take the following relative quantities

$$\rho_A = r_A/R , \tag{4.96}$$

$$\rho_B = r_B/R,$$ (4.97)

$$\xi = \mu/\mu_A.$$ (4.98)

The behaviour of function μ_r describes the following expressions that are resulting from Figure 4.15:

$$\mu_r = \mu_A \text{ for } r \in \langle 0, r_A \rangle, \quad \mu_r = -\mu_A \frac{r}{R - r_A} + \mu_A \frac{R}{R - r_A} \text{ for } r \in (r_A, R).$$

(4.99)

Let us further take that the packing density at radius r_B (point B lying on the decreasing line as shown in Figure 4.15) is μ_B. Then, the following relations are valid from the geometrical similarity of dotted triangles as shown in Figure 4.15:

$$\frac{\mu_B}{\mu_A} = \frac{R - r_B}{R - r_A}, \quad r_B = R - \frac{\mu_B}{\mu_A}(R - r_A).$$ (4.100)

Note: If the packing density is a constant, i.e., $\mu_r = \mu$ for all radii, then the behaviour of the local packing density corresponds to the horizontal dashed line as shown in Figure 4.15. Then, it is valid to write that $r_A = r_B = R$.

The (mean) yarn packing density μ is given by Equation (4.7). By using Equation (4.94), Equation (A3.18) from Appendix 3 and Equation (4.98), we can write

$$\mu = \frac{8}{D^2} \int_0^{D/2} \mu_r r \, dr = \frac{2}{R^2} \mu_A R^2 \frac{1 + \rho_A + \rho_A^2}{6}, \quad \xi = \frac{\mu}{\mu_A} = \frac{1 + \rho_A + \rho_A^2}{3}.$$

(4.101)

Thus also

$$\rho_A^2 + \rho_A + (1 - 3\xi) = 0, \quad \rho_A = \frac{-1 \pm \sqrt{1 - 4(1 - 3\xi)}}{2} = -\frac{1}{2} + \sqrt{3}\sqrt{\xi - \frac{1}{4}}.$$

(4.102)[16]

Evidently, it must be valid that $\xi \geq 1/4$. Nevertheless, it must be also valid that $\rho_A = r_A/R \geq 0$ and using Equation (4.102), we can write

---

16 The sign + bears the physical sense only.

$$-\frac{1}{2}+\sqrt{3}\sqrt{\xi-\frac{1}{4}}\geq 0,\quad \sqrt{3}\sqrt{\xi-\frac{1}{4}}\geq\frac{1}{2},\quad 3\left(\xi-\frac{1}{4}\right)\geq\frac{1}{4},\quad 12\xi-3\geq 1,\quad \xi\geq\frac{1}{3}.$$

$$(4.103)$$

Note: If $\xi=1/3$, then $\rho_A=r_A=0$, and the trapezoidal function μ_r obtains a triangular shape.

Specific sources of retraction of staple yarns

Equation (4.53) introduces yarn retraction δ as a function of local packing density μ_r according to Braschler's idea of constant fiber volume. By applying our model of trapezoidal function μ_r, it is possible to substitute the integrals expressed in Equation (4.53) in accordance with Equations (A3.18) and (A3.20) from Appendix 3. This results in the following expression:

$$\delta=1-\frac{\displaystyle\int_0^{D/2}\frac{\mu_r r}{\sqrt{1+(2\pi rZ)^2}}\,dr}{\displaystyle\int_0^{D/2}\mu_r r\,dr}=1-\frac{\displaystyle\int_0^{R}\frac{\mu_r r}{\sqrt{1+(2\pi rZ)^2}}\,dr}{\displaystyle\int_0^{R}\mu_r r\,dr}$$

$$=1-\frac{\dfrac{\mu_A R^2}{\kappa^2(1-\rho_A)}\left\{\dfrac{1}{2}\sqrt{1+\kappa^2}-\dfrac{\rho_A}{2}\sqrt{1+\rho_A^2\kappa^2}-1+\rho_A+\dfrac{1}{2\kappa}\ln\dfrac{\sqrt{1+\kappa^2}+\kappa}{\sqrt{1+\rho_A^2\kappa^2}+\rho_A\kappa}\right\}}{\mu_A R^2\dfrac{1+\rho_A+\rho_A^2}{6}},$$

$$\delta=1-\frac{\dfrac{3}{\kappa^2(1-\rho_A)}\left\{\sqrt{1+\kappa^2}-\rho_A\sqrt{1+\rho_A^2\kappa^2}-2+2\rho_A+\dfrac{1}{\kappa}\ln\dfrac{\sqrt{1+\kappa^2}+\kappa}{\sqrt{1+\rho_A^2\kappa^2}+\rho_A\kappa}\right\}}{1+\rho_A+\rho_A^2}.$$

$$(4.104)$$

In the last equation, yarn retraction δ is expressed as a function of two quantities only: (1) twist intensity κ and (2) relative radius $\rho_A=r_A/R$. The relative radius can be naturally determined by means of $\xi=\mu/\mu_A$ according to Equation (4.102). The yarn retraction is shown as a function of twist intensity in Figure 4.16 for different values of $\xi\in\langle 1/3,1\rangle$. (Equations (4.102) and (4.104) were used.)

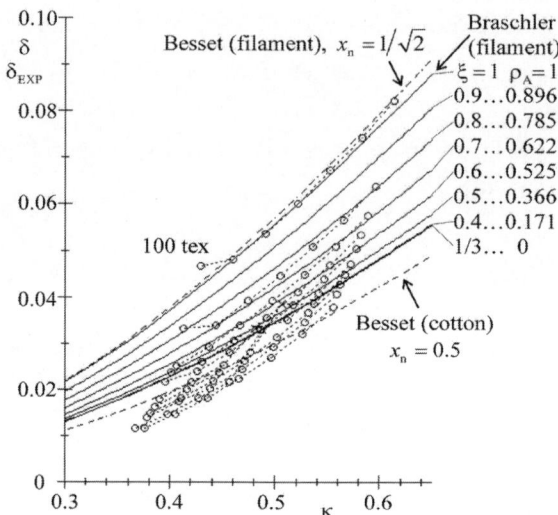

Figure 4.16 Yarn retraction as the function of twist intensity.
—: Equations (4.104) with (4.102); - - -: Besset's Equation (4.37);
O: experimental data according to Table 4.3

Note: If $\xi = 1$, i.e., $\rho_A = 1$, then Equation (4.104) is limited to Equation (4.56) which was derived earlier for constant packing density and it is valid for filament yarns. Except this, there are also two dashed curves representing Besset's concept (see Section 4.3) in Figure 4.16.

The small circles shown in Figure 4.16 represent the experimental values of yarn retraction δ_{EXP} as mentioned in Table 4.3 in relation to twist intensity κ. Here, Equation (1.47) is used for yarn diameter, where the yarn packing density μ is evaluated as a root of Equation (4.93).

Figure 4.16 shows that

1. The trapezoidal shape of function μ_r leads to smaller values of yarn retraction; the increasing value of μ_r (smaller value of ξ) results in the smaller value of δ for a given yarn.

2. The retraction of very coarse yarns (100 tex) is practically same as the retraction of filament yarns (idea of constant packing density inside the yarn).

3. The experimental values of yarn retraction follow a steep slope as the trends of calculated values, i.e., ξ and ρ_A are shown to increase with the increase in yarn twist intensity κ. It means that the contribution of peripheral part of yarn (following the decreasing trend of local

packing density μ_r) decreases with an increase in twist[17]. (Probably, the initial 'free' fibers from yarn periphery gradually take part in the process of yarn compression and also yarn retraction.)

4. Many experimental values of yarn retraction (especially from fine yarns) are lying under the set of permissible curves. It means that the trapezoidal shape of function μ_r cannot fully explain the decreasing trend of yarn retraction in staple yarns.

Non-functionable fibers

In the previous section, the fourth point indicates that all the fibers are not playing significant roles in determining the yarn retraction. Let us simplify our image of such process by introducing a radius $r_B \in \langle r_A, R \rangle$ – see Figure 4.15. The fibers from outside this radius do not influence on yarn retraction; as if they 'do not exist' in the yarn and also as if r_B is the 'highest' radius, i.e., a new 'yarn radius' for determination of yarn retraction. So, the original expression mentioned in Equation (4.53) obtains a new form as stated below

$$\delta = 1 - \frac{\int\limits_0^{r_B} \dfrac{\mu_r r}{\sqrt{1 + (2\pi r Z)^2}}\, dr}{\int\limits_0^{r_B} \mu_r r\, dr} . \tag{4.105}$$

[Equation (4.99) is still valid for local packing density μ_r.]

The aforesaid expression can be solved by using the integrals stated in Equations (A3.13) and (A3.15) in Appendix 3. Then, the following expression can be obtained:

$$\delta = 1 - \frac{\int\limits_0^{r_B} \dfrac{\mu_r r}{\sqrt{1 + (2\pi r Z)^2}}\, dr}{\int\limits_0^{r_B} \mu_r r\, dr}$$

---

17 We partly observed a similar trend by experimental analysis of local packing density in yarn cross sections.

$$= 1 - \frac{\frac{\mu_A R^2}{\kappa^2 (1-\rho_A)} \left\{ \left(1-\frac{\rho_B}{2}\right)\sqrt{1+\rho_B^2 \kappa^2} - \frac{\rho_A}{2}\sqrt{1+\rho_A^2 \kappa^2} - 1 + \rho_A + \frac{1}{2\kappa}\ln\frac{\sqrt{1+\rho_B^2\kappa^2}+\rho_B\kappa}{\sqrt{1+\rho_A^2\kappa^2}+\rho_A\kappa}\right\}}{\mu_A R^2 \dfrac{-\rho_A^3 + 3\rho_B^2 - 2\rho_B^3}{6(1-\rho_A)}},$$

$$\delta = 1 - \frac{\frac{3}{\kappa^2}\left\{(2-\rho_B)\sqrt{1+\rho_B^2\kappa^2} - \rho_A\sqrt{1+\rho_A^2\kappa^2} - 2 + 2\rho_A + \frac{1}{\kappa}\ln\frac{\sqrt{1+\rho_B^2\kappa^2}+\rho_B\kappa}{\sqrt{1+\rho_A^2\kappa^2}+\rho_A\kappa}\right\}}{-\rho_A^3 + 3\rho_B^2 - 2\rho_B^3}. \quad (4.106)$$

In the last expression, the yarn retraction δ is expressed as a function of three quantities: (1) twist intensity κ, (2) relative radius $\rho_A = r_A/R$ and (3) relative radius $\rho_B = r_B/R$. We determine the quantity ρ_B by means of Equations (4.96), (4.97) and (4.100) as follows:

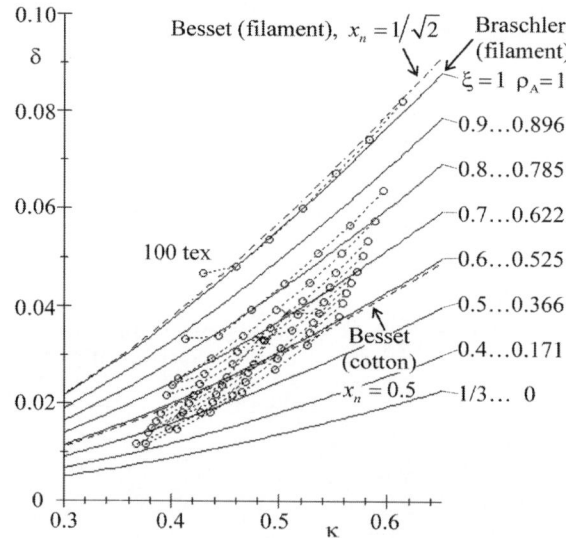

Figure 4.17 Yarn retraction as the function of twist intensity.
—: Equations (4.102), (4.106) and (4.107) by $\mu_B/\mu_A = 0.5$;
- - -: Besset's Equation (4.37); O: experimental data according to Table 4.3

$$\frac{r_B}{R} = \frac{R}{R} - \frac{\mu_B}{\mu_A}\left(\frac{R}{R} - \frac{r_A}{R}\right), \quad \rho_B = 1 - \frac{\mu_B}{\mu_A}(1-\rho_A). \quad (4.107)$$

($\mu_B \in \langle 0, \mu_A \rangle$, so that $\rho_B \in \langle \rho_A, 1 \rangle$ – see Figure 4.15.)

An example by using $\mu_B/\mu_A = 0.5$, i.e., $\rho_B = (1 + \rho_A)/2$, is shown in Figure 4.17. (The method for obtaining this graph is same as that for obtaining Figure 4.16.)

Except the points from 1 to 3 – found already in Figure 1.16 – the example displayed in Figure 4.17 shows that the experimental values, reported in Table 4.3, can lie inside the set of theoretical curves following Equation (4.106) for suitable values of μ_B/μ_A. It means that there exists always a suitable pair of values ρ_A & ρ_B , for which Equation (4.106) results in yarn retraction identical to the experimental one. Moreover, the whole set of suitable pairs ρ_A & ρ_B as stated in Equation (4.106) leads to the same experimentally determined value of yarn retraction. It shows that Equation (4.106) can be the basis for structural model of retraction of staple yarns. Unfortunately, the physically well-founded pair ρ_A & ρ_B (and/or μ_A/μ_B but also μ) is a result of complicated mechanical processes in the yarn, which are still not enough known.

Note: If the exact relationship between ρ_A & ρ_B (and/or μ_A & μ_B) is not known, it would be probably possible to create empirical relations for ρ_A & ρ_B (and/or μ_A & μ_B) as a function of yarn fineness and yarn twist coefficient[18]. Such an expression would have an empirical character entirely. In this case, it is easier to use fully empirical expressions as stated in Equations (4.91), (4.92) and (4.93). Nevertheless, Equation (4.106) can be considered as a starting point for future research work.

Staple yarn retraction – a resume

We believe that the cause of (mostly) smaller values of staple yarn retraction (in relation to filament yarns) is resulting from the synergistic effect of the following influences:

1. The fiber packing density is more at the inner part of the yarn as compared to the outer part of the yarn. (This is shown in a simplified way by the trapezoidal model in Figure 4.15.) Of course, the fibers lying at smaller radii create smaller yarn retraction as they are deviating more from the helical geometry.

---

18 Our very preliminary results indicate that the relative radius ρ_A is higher in case of coarser yarns as compared to finer yarns and in case of high twisted yarns as compared to low twisted yarns. The relative radius $\rho_B \geq \rho_A$ is increasing (approaching to 1) by increasing yarn twist.

2. The fibers from the outer part of the yarn lie in a region surrounding the region of smaller compression. They are more free, they can easily slip and/or become straighten (false draft), without any significant contribution to creation of yarn retraction. (In a simplified way, the fibers actually do not exist at higher radii, i.e., $r > r_B$, for creation of yarn retraction – Figure 4.15.)

3. The characteristic parameters (ρ_A & ρ_B and/or μ_A & μ_B), as stated in Equation (4.106), depend on yarn fineness and twist coefficient by means of many complicated processes of internal yarn mechanics, which are still not enough known.

4.8   References

[1]   Braschler, E., Die Festigkeit von Baumwollgespinsten (The tenacity of cotton yarns), Zürich, 1935 (In German).

[2]   Obuch, I.G., O chrapovikie i plotnosti poczatka i usadke prjazi na vaterach (Ratchet-wheel, cop density and yarn retraction on ring spun machine), *Chlopcatobumaznaja promyslennost*, 1–2, 1936 (In Russian).

[3]   Besset, M.C., Influence de la torsion sur la longueur d'un files (Influence of twist to the yarn length), *L'Industrie Textile*, 737, April 1948 (In French).

[4]   Zurek, W., Struktura przedzy (Yarn structure) Wydawnictwa Naukovo-Techniczne, Warszawa, 1971 (In Polish).

[5]   Budnikov, V.I., Osnovy prjadenija, Cast 2 (Principles of Spinning, Part 2) Moscow-Leningrad, 1945 (In Russian).

[6]   Marko, J. and Neckář, B., Sur la teorie du tordage des files artificiales et synthétiques (A contribution to the theory of twisting man-made filament yarns), *Věda a výzkum v textilním průmyslu*, 11, 1970 (Czech journal, in French).

[7]   Sokolov, G.V., Voprosy teorii krucenija voloknistych materialov (Problems of theory of twisting of fiber materials), Gizlegprom, Moscow, 1957 (In Russian).

Models of fiber migration in yarns

5.1 Fundamental equation of radial fiber migration

It is generally known that the real fibers follow more or less different trajectories than the ideal helical fiber in a twisted yarn. Above all, each individual fiber changes its radial position inside the yarns – more in case of staple fiber yarns but less in case of filament yarns. This phenomenon is termed as radial migration of fibers in yarns.

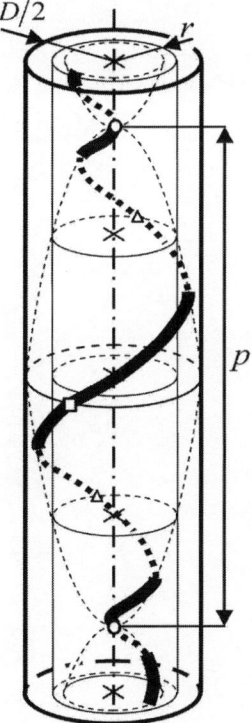

Figure 5.1 Deterministic model of a migrating fiber

Treloar's idea of fibers path

However, although the fiber path inside a yarn is a random quantity, this character is ignored in numerous deterministic models that interpret fiber trajectory as a regular path and describe it by means of deterministic mathematical equations. Probably, the very first model of radial migration of fibers in yarns was created by Treloar [1]. It was imagined that each fiber starts from the yarn axis, i.e., $r = 0$ (point ○ in Figure 5.1), and rotate around yarn axis such that the radius of fiber path is continually increasing from a general radius r (point △) to yarn surface, i.e., $r = D/2$, where D stands for yarn diameter. Then, the fiber path breaks on yarn surface (point □) and continues to return to yarn axis, reflecting a mirror image (Figure 5.1). This fiber path is further repeated along its complete length (infinitely long). In reality, the fibers follow different trajectories inside a yarn in terms of axial movement and angular rotation around yarn axis.

Nevertheless, the image of fiber path is considered to be too strict for a variety of models like Treloar's model of radial migration of fibers in yarns [1]. It is therefore considered more generally that a fiber 'starts' from a smaller radius $r \geq 0$, then the radius is continually increasing to a higher one $r \leq D/2$, afterwards the path breaks and the radius is decreasing continuously before the fiber starts its movement from a new smaller radius.

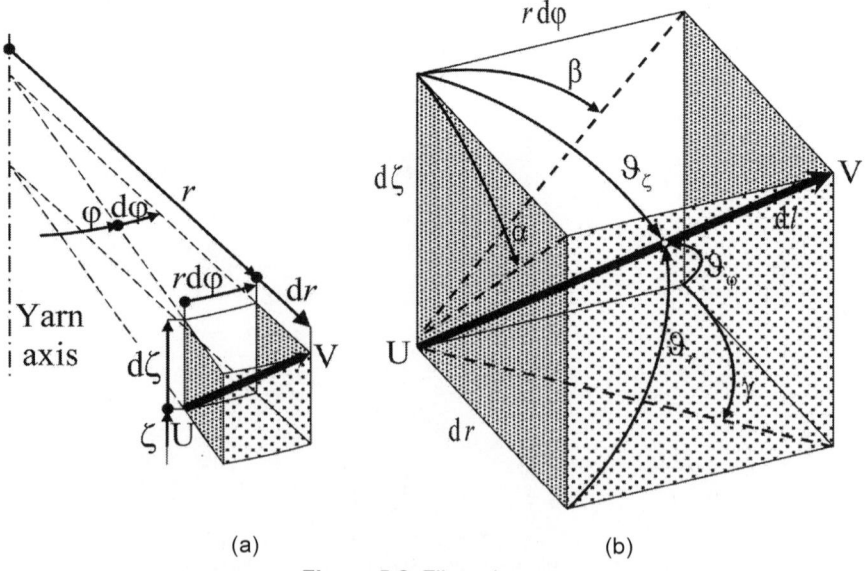

(a)   (b)

Figure 5.2 Fiber element

Fiber element

Figure 5.2a displays a general fiber element UV together with a corresponding elementary parallelepiped. We usually prefer the system of cylindrical coordinates in the yarn, i.e., radius $r \geq 0$ measured from the yarn axis, angle φ measured around the yarn axis and axial coordinate ζ[1]. A general location of fiber element UV of length $\mathrm{d}l$ is shown in Figure 5.2a while it is magnified in Figure 5.2b.

We introduce the angles $\vartheta_r, \vartheta_\varphi, \vartheta_\zeta$ (to the fiber element UV) and also the angles α, β, γ ('on the walls' of the elementary parallelepiped) for characterization of this element[2].

The following relations can be deduced from Figure 5.2b

$$\tan \alpha = \mathrm{d}r/\mathrm{d}\zeta, \quad \tan \beta = r\mathrm{d}\varphi/\mathrm{d}\zeta, \quad \tan \gamma = \mathrm{d}r/(r\,\mathrm{d}\varphi), \tag{5.1}$$

$$\mathrm{d}^2 l = \mathrm{d}^2 r + (r\mathrm{d}\varphi)^2 + \mathrm{d}^2 \zeta = \left[1 + (r\mathrm{d}\varphi/\mathrm{d}r)^2 + (\mathrm{d}\zeta/\mathrm{d}r)^2\right]\mathrm{d}^2 r$$

$$= \frac{\left[(\mathrm{d}r/\mathrm{d}\zeta)^2 + (r\mathrm{d}\varphi/\mathrm{d}\zeta)^2 + 1\right]}{(\mathrm{d}r/\mathrm{d}\zeta)^2}\mathrm{d}^2 r = \frac{\tan^2 \alpha + \tan^2 \beta + 1}{\tan^2 \alpha}\mathrm{d}^2 r, \tag{5.2}$$

$$\cos^2 \vartheta_r = \frac{\mathrm{d}^2 r}{\mathrm{d}^2 l} = \frac{\tan^2 \alpha}{\tan^2 \alpha + \tan^2 \beta + 1}. \tag{5.3}$$

If we think about 'purely' radial migration of fibers, then we usually accept the following set of assumptions:

1. The twist of each fiber or fiber element is a common positive constant Z, which is equal to the value of (technological) twist of yarn. Then, the following expression, already derived for helical model, is valid to write

$$\tan \beta = 2\pi r Z \quad \left(\text{also } \tan \beta_D = 2\pi \frac{D}{2} Z = \pi D Z\right). \tag{4.1}$$

---

1   See Figure 3.1 in Chapter 3 for a general introduction of cylindrical coordinates.

2   See Figures 3.3 and 3.5 in Chapter 3 for a more general introduction of angles.

2. The value of $|\tan \alpha|$ is a common constant for all fiber elements lying at same radius r. Then, $|\tan \alpha|$ is a function of radius r only. (Let us note that if the fiber is moving 'from inside to outside', then the value of $\tan \alpha$ is positive, but if the fiber is moving 'from outside to inside', then the value of $\tan \alpha$ is negative.)

   Note: At a given radius, each fiber element possesses the same values of $\tan \beta$ and $|\tan \alpha|$ so that, according to Equation (5.3), all fiber elements possess the same value of $\cos^2 \vartheta_r$.

3. Each fiber is infinitely long in a yarn so that it passes (longitudinally) through the complete (very long) length L of yarn. The number N of fibers in each cross section of the yarn is a constant.

   Note: We are imagining a filament yarn and/or a staple yarn, where each (real) staple fiber starts from the 'end-point' of the previous staple fiber. (The real staple fibers follow 'one after another' in the yarn.)

4. Each infinitely long fiber passes through all possible positions in the (very long) yarn with the same frequency. Each fiber passes a given radius $\upsilon(r)$ times, where $\upsilon(r)$ is a function of r only.

   Note: The frequency with which a fiber passes through a given radius r was denoted by symbol Q in Chapter 3. Now, we use a new symbol $\upsilon(r)$ here. (The earlier symbol Q will be used in another sense – as a parameter – in the present chapter.) It can be noted that this change of symbols needs to be automatically assumed whenever the expressions from Chapter 3 are referred to in this chapter.

   Note: If the fiber paths can be interpreted as realizations of a random process, then such process can be thought of an 'ergodic'[3] random process. Generally, the fiber paths do not need to have a random character (they can also follow deterministic curves); however, the idea of 'ergodicity' remains the same also in this case. (See also assumption 5 in Chapter 3, Section 3.5.)

5. The fiber cross-sectional area s and the fiber density ρ are common constants at all places for all fibers.

---

3   See a book on theory of probability and random processes for a mathematical introduction of the term ergodicity.

Packing density at a given radius

Based on the aforesaid assumptions, we can now formulate the expression for packing density μ_r at a given radius r. Let us imagine a differential layer as shown in Figure 4.3, i.e., the elementary space between two cylinders having radii r and $r + dr$. An elementary fiber length dl, inside the differential layer, has volume $s\,dl$. Each fiber passes through the above-mentioned differential layer $\upsilon(r)$ times in a (very long) length L of yarn. All N fibers pass through the differential layer $N\upsilon(r)$ times and occupy fiber volume $N\upsilon(r)s\,dl$.

The area occupied by the differential annulus as shown in Figure 4.3 is $2\pi r\,dr$ so that the total volume of the differential layer in the length L of yarn becomes $2\pi r\,dr\,L$.

Then, the packing density μ_r of the differential layer at radius r is $\mu_r = \left[N\upsilon(r)s\,dl\right]/(2\pi r\,dr\,L)$. By applying Equations (4.1) and (5.3), we obtain the following expression for μ_r:

$$\mu_r = \frac{N\upsilon(r)s\,dl}{2\pi r\,dr\,L} = \frac{\upsilon(r)}{L}\frac{Ns}{2\pi r}\frac{dl}{dr},$$

$$\mu_r^2 = \left(\frac{\upsilon(r)}{L}\right)^2\left(\frac{Ns}{2\pi r}\right)^2\frac{d^2l}{d^2r} = \left(\frac{\upsilon(r)}{L}\right)^2\frac{(NsZ)^2}{(2\pi rZ)^2}\frac{\tan^2\alpha + \tan^2\beta + 1}{\tan^2\alpha},$$

$$\mu_r^2 = \left[\frac{2NsZ}{p(r)}\right]^2\frac{\tan^2\alpha + \tan^2\beta + 1}{\tan^2\alpha\tan^2\beta}, \tag{5.4}$$

where

$$p(r) = 2L/\upsilon(r). \tag{5.5}$$

Period of migration $p(r)$ – definition

Each fiber passes $\upsilon(r)$ times through the differential layer (at radius r) in a very long length L of yarn. It passes many times from inside to outside, then from outside to inside. So, it is evident that the mean axial distance between the neighbouring 'passes' on one fiber is $L/\upsilon(r)$. The mean axial distance between the neighbouring places where the fiber passes the differential in the same direction is then $2L/\upsilon(r)$. (Here, the movement of fibers either from inside to outside or from outside to inside is considered.) Such distance is called (mean) period of radial migration and it is denoted by $p(r)$ here – see Equation (5.5). Generally, the (mean) period of migration can be a function of radius r^4 by means of $\upsilon(r)$.

Angle α of migration

The angle α characterizes the changes of radius, i.e., radial migration. We can express this by rearrangement of Equation (5.4) as follows:

$$\left[\frac{\mu_r p(r)}{2NsZ}\right]^2 = \frac{\tan^2\alpha + \tan^2\beta + 1}{\tan^2\alpha\,\tan^2\beta}, \left[\frac{\mu_r p(r)}{2NsZ}\right]^2 \tan^2\alpha\,\tan^2\beta = \tan^2\alpha + \tan^2\beta + 1,$$

$$\tan^2\alpha = \frac{1 + \tan^2\beta}{\left[\dfrac{\mu_r p(r)}{2NsZ}\right]^2 \tan^2\beta - 1}.$$

$$(5.6)$$

The angle α changes its value in relation to radius r, because (a) $\tan\beta = 2\pi r Z$ according to Equation (4.1), (b) μ_r is (generally) a function of radius and (c) $p(r)$ is (generally) a function of radius. We call Equation (5.6) as the fundamental equation of radial fiber migration.

Period of migration $p(r)$ – evaluation

Rearranging Equation (5.4) we obtain:

$$\left[\frac{\mu_r p(r)}{2NsZ}\right]^2 = \frac{\tan^2\alpha + \tan^2\beta + 1}{\tan^2\alpha\,\tan^2\beta}, \quad p^2(r) = \left[\frac{2NsZ}{\mu_r}\right]^2 \frac{\tan^2\alpha + \tan^2\beta + 1}{\tan^2\alpha\,\tan^2\beta},$$

$$p(r) = \frac{2NsZ}{\mu_r} \frac{\sqrt{\tan^2\alpha + \tan^2\beta + 1}}{|\tan\alpha|\tan\beta}. \qquad (5.7)$$

---

4   We will show later on that if each fiber is starting precisely from the yarn axis and directly going to yarn surface, then – according to our assumptions – $p(r)$ must be constant. Nevertheless, it is not required to be a constant in other models.

5.2   An alternative way to determine the fundamental equation of radial migration

In this section, the model of radial fiber migration is introduced in a different way than it was done earlier. Here, the equations of radial fiber migration are derived based on the most general relations presented in Chapter 3. Nevertheless, it does not bring any new relation as compared to the previous one. We suggest that our readers, who are mostly oriented to applications, may like to skip this section and continue with Section 5.3.

Idea of regular yarn

The idea of a so-called regular yarn is presented in Section 3.5, based on five assumptions: (1) it is a type of simply organized yarn, (2) all fibers are infinitely long, (3) yarn is composed of N such fibers, (4) fiber cross-sectional areas s and densities ρ are same at each and every point on each fiber and (5) fiber paths are ergodic. (See Section 3.5 for more details.)

These assumptions fulfil the assumptions 3–5 stated in the previous section. However, the assumptions 1 and 2, i.e., constant value Z and constant value $|\tan \alpha|$, are even more strict than the idea of (1) simple organized yarn in Section 3.5. Summarily, the expressions for a regular yarn, presented in Section 3.5, are also valid in the present case, but moreover we must use also the constant value of Z and the constant value of $|\tan \alpha|$ in these equations.

Packing density at a given radius

The following expression was derived for a given radius r in Equations (3.164)[5] and (3.42):

$$
\begin{aligned}
\mu_r &= \frac{s}{2\pi r}\left(\frac{N\upsilon(r)}{L}\right)\left(\frac{1}{\upsilon(r)}\sum_{j=1}^{\upsilon(r)}\left[\frac{1}{|\cos\vartheta_r|}\right]_j\right) \\
&= \frac{s}{2\pi r}\left(\frac{N\upsilon(r)}{L}\right)\left(\frac{1}{\upsilon(r)}\sum_{j=1}^{\upsilon(r)}\left[\frac{\sqrt{\tan^2\alpha + \tan^2\beta + 1}}{|\tan\alpha|}\right]_j\right).
\end{aligned}
$$

(5.8)

The expression stated above within the square brackets represents the average value from all $\upsilon(r)$ elements of each fiber, lying in the differential

---

5   Do not forget that the symbol Q appeared in Chapter 3 was renamed to the symbol $\upsilon(r)$ in this chapter – see the note stated for the fourth assumption.

layer at a general radius r. However, we think about constant values of $\tan\beta = 2\pi rZ$ and $|\tan\alpha|$ at a given radius r. So, it is valid to write that

$$\frac{1}{\upsilon(r)}\sum_{j=1}^{\upsilon(r)}\left[\frac{\sqrt{\tan^2\alpha+\tan^2\beta+1}}{|\tan\alpha|}\right]_j = \frac{\sqrt{\tan^2\alpha+\tan^2\beta+1}}{|\tan\alpha|}, \qquad (5.9)$$

and then

$$\mu_r = \frac{s}{2\pi r}\left(\frac{N\upsilon(r)}{L}\right)\left(\frac{\sqrt{\tan^2\alpha+\tan^2\beta+1}}{|\tan\alpha|}\right),$$

$$\mu_r^2 = \left(\frac{\upsilon(r)}{L}\right)^2\frac{(NsZ)^2}{(2\pi rZ)^2}\frac{\tan^2\alpha+\tan^2\beta+1}{\tan^2\alpha}$$

$$= \left(\frac{\upsilon(r)}{2L}\right)^2(2NsZ)^2\frac{\tan^2\alpha+\tan^2\beta+1}{\tan^2\alpha\tan^2\beta},$$

$$\mu_r^2 = \left[\frac{2NsZ}{p(r)}\right]^2\frac{\tan^2\alpha+\tan^2\beta+1}{\tan^2\alpha\tan^2\beta}, \text{ where } p(r)=2\frac{L}{\upsilon(r)}. \qquad (5.10)$$

The last expression fully corresponds to Equations (5.4) and (5.5) stated in the previous section. Therefore, Equation (5.6) derived for $\tan^2\alpha$ (the fundamental equation of radial migration) and Equation (5.7) derived for $p(r)$ (mean period of migration) must be valid too.

5.3   Treloar's ideal radial fiber migration model

Assumptions

Besides assumptions 1–5 stated in Section 5.1, Treloar [1] assumed that

6. each fiber follows a fully regular path, i.e., each fiber starts from yarn axis, i.e., $r = 0$ (point O in Figure 5.1), and rotate around the yarn axis such that the radius of fiber path is continually increasing from a general radius r (point $\triangle$) to yarn surface, i.e., $r = D/2$ (point $\square$), where D stands for yarn diameter. Then, the fiber path breaks on yarn surface and continues to return to yarn axis (see Figure 5.1[6]). So the fiber path is periodically repeated. This path of a fiber is further

---

6   We will show later on that this idea is not always fully valid.

repeated along its complete length (infinitely long). The trajectories of the fibers in the yarn differ in terms of the position of points of yarn axis (point O) and the starting value of angle φ of rotation around the yarn axis.

Further, it was considered that

7. the local packing density μ_r is a constant and this is equal to yarn packing density μ.

Basic equation of ideal radial fiber migration

It is evident from assumption 6 that the period of migration $p(r)$ takes a constant value $p(r) = p$ for each radius. This length p is shown in Figure 5.1. Simultaneously, the packing density $\mu_r = \mu$ is a constant (assumption 7), the number of fibers in yarn cross section N is a constant (assumption 3), fiber cross-sectional area s is a constant (assumption 5) and yarn twist Z is also a constant (assumption 1). So, we can write

$$K = \frac{\mu_r p(r)}{2NsZ} = \frac{\mu p}{2NsZ} \ldots \text{a constant} - \text{parameter of yarn.} \qquad (5.11)$$

By applying this dimensionless parameter K in Equation (5.6), i.e., the fundamental equation of radial fiber migration, we obtain the basic equation of ideal radial fiber migration as follows:

$$\tan^2 \alpha = \frac{1 + \tan^2 \beta}{K^2 \tan^2 \beta - 1}. \qquad (5.12)$$

Meaning of parameter K

Let the yarn be originated from a very thin sliver, say of fineness T_0. If we think that a set of N parallel fibers creates such a sliver, then its substance cross-sectional area is $S_0 = Ns$ and also $S_0 = T_0/\rho$ according to Equation (1.33) mentioned in Section 1.3. The initial fineness T_0 changes its value to a new and higher value T due to yarn retraction δ originating from yarn twist Z. Then, Equation (4.33), i.e., $T = T_0/(1-\delta)$, is valid[7] (see also Figure 4.5).

---

7   This relation was derived in Chapter 4. Nevertheless, the above-mentioned relation is generally valid, as follows from the derivation in the context of Figure 4.5.

According to Equation (1.33), it is also valid that $S = T/\rho$ for substance cross-sectional area of a twisted yarn. Finally, we can also write that $S = \mu \pi D^2/4$ from Equation (1.45). By using all these relations, it is valid to write that

$$Ns = S_0 = T_0/\rho = T(1-\delta)/\rho = (1-\delta)S = (1-\delta)\mu \pi D^2/4. \qquad (5.13)$$

We use the last expression stated in Equation (5.11) together with Equation (4.1):

$$K = \frac{\mu p}{2\left[(1-\delta)\mu \pi D^2/4\right]Z} = \frac{2p}{D}\frac{1}{1-\delta}\frac{1}{\tan\beta_D} = \frac{2p}{D}\frac{1}{1-\delta}\frac{1}{\pi DZ}. \qquad (5.14)$$

This expression indicates a logical sense of parameter K.

Differential equations

Each fiber path is described by a pair of differential equations. The first one can be written from Equations (4.1) and (5.1) as follows:

$$\tan\beta = 2\pi rZ = rd\varphi/d\zeta, \quad d\varphi = 2\pi Z\, d\zeta. \qquad (5.15)$$

The second one can be derived from Equations (4.1), (5.1) and (5.12) as follows:

$$\tan\alpha = \frac{dr}{d\zeta} = \pm\sqrt{\frac{1+\tan^2\beta}{K^2\tan^2\beta-1}}$$

$$= \pm\sqrt{\frac{1+(2\pi rZ)^2}{K^2(2\pi rZ)^2-1}}, \quad d\zeta = \pm\sqrt{\frac{K^2(2\pi rZ)^2-1}{1+(2\pi rZ)^2}}\, dr, \qquad (5.16)$$

where the plus sign relates to the part of the fiber going from inside to outside ($dr/d\zeta > 0$) and the minus sign relates to the part of the fiber going from outside to inside ($dr/d\zeta < 0$).

Domain of definition

Equation (5.16) is defined if the relation $K^2(2\pi rZ)^2 \geq 1$ is valid. So, we can write the following inequality:

$$K \geq \frac{1}{2\pi rZ}, \quad r \geq \frac{1}{2\pi Z K}. \tag{5.17}$$

Alternatively, we obtain the following expression by using Equations (5.14) and (4.1) in the inequality stated in Equation (5.17):

$$r \geq \frac{1}{2\pi Z K} = \frac{1}{2\pi Z\left(\dfrac{2p}{D}\dfrac{1}{1-\delta}\dfrac{1}{\pi DZ}\right)}, \quad r \geq \frac{(D/2)^2}{p}(1-\delta). \tag{5.18}$$

The last inequality shows that the fiber path is not defined for too small values of radius. It means that assumption 6 cannot be fulfilled.

Note: The border value of r – Equation (5.18) – is often very small, because mostly the period p is much higher than the square of yarn radius $D/2$ and $(1-\delta)$ is smaller than one.

Fiber path

Let us now introduce the starting point of the first fiber path in terms of the cylindrical coordinates (first limiting conditions) for differential Equations (5.15) and (5.16) as follows:

$$r = r_0 = \frac{1}{2\pi Z K}, \quad \varphi = \varphi_0 = 0, \quad \zeta = \zeta_0 = 0. \tag{5.19}$$

The process of solving Equation (5.15) is trivial. By considering the borderline condition as mentioned in Equation (5.19), we obtain

$$\varphi = 2\pi Z \zeta. \tag{5.20}$$

But the process of solving Equation (5.16) is more difficult. At first, let us solve this relation with sign + (from inside to outside). By integrating Equation (5.16), considering the borderline conditions according to Equation (5.19), we obtain the following expression:

$$\int_0^\zeta d\zeta' = \int_{\frac{1}{2\pi ZK}}^r \sqrt{\frac{K^2(2\pi r'Z)^2 - 1}{1+(2\pi r'Z)^2}}\, dr', \quad \zeta = \frac{1}{2\pi Z}\int_{1/K}^{2\pi rZ}\sqrt{\frac{K^2 x^2 - 1}{1+x^2}}\, dx,$$

Substitution: $2\pi r'Z = x$, $dr' = \dfrac{dx}{2\pi Z}$, $2\pi Z\zeta = \displaystyle\int_{1/K}^{2\pi rZ}\sqrt{\frac{K^2 x^2 - 1}{1+x^2}}\, dx.$ \qquad (5.21)

(The integration variables were renamed to ζ' and r' to avoid any confusion with the upper limits of the integrals.)

Especially, if radius $r = D/2$, then the ζ-coordinate must be equal to one half of period p, i.e., $\zeta = p/2$. So it is valid to write that

$$2\pi Z(p/2) = \pi Zp = \int_{1/K}^{\pi DZ} \sqrt{\frac{K^2 x^2 - 1}{1 + x^2}} \, dx .$$ (5.22)

Note: Note that the quantities $2\pi Z\zeta, 2\pi rZ, \pi Zp, \pi DZ$ as mentioned in Equations (5.21) and (5.22) are dimensionless. They are arrived at by multiplying the quantities $\zeta, r, p, D/2$ to a common parameter $2\pi Z$. Therefore, the relation between $2\pi Z\zeta$ and $2\pi rZ$ characterizes the relation between ζ and r very well. Similarly, the relation between $\pi Zp, \pi DZ$ characterizes the relation between p and D very well.

Dividing Equation (5.22) by twist intensity $\kappa = \tan\beta_D = \pi DZ$, we rearrange Equation (5.22) as follows:

$$\frac{p}{D} = \frac{1}{\pi DZ} \int_{1/K}^{\pi DZ} \sqrt{\frac{K^2 x^2 - 1}{1 + x^2}} \, dx .$$ (5.23)

The integral expressed in Equations (5.21), (5.22) and (5.23) does not have any analytical solution. (It leads to elliptical integral.) This must be solved by using a suitable numerical integration technique. By using this technique, we found the first part (+) of the curves as shown in Figure 5.3. (The stated dimensionless values p/D are resulting from Equation (5.23).) The first part of fiber path breaks on yarn surface ($r = D/2$ and $\zeta = p/2$, i.e., $2\pi rZ = \pi DZ$ and $2\pi Z\zeta = 2\pi Z(p/2)$) and then, according to assumptions 1 and 2 as mentioned in Section 5.1 and assumption 6 in this Section, the fiber continues to return (from outside to inside) to a borderline value $r = r_0 = 2\pi ZK$ as a mirror image shown in Figure 5.3. So, at the border value of r ($2\pi rZ = 1/K$), the value of ζ obtains the value of p ($2\pi Z\zeta = 2\pi Zp$). The second part of fiber path, shown in Figure 5.3, represents the solution of Equation (5.16) by using the above-mentioned borderline conditions and sign (minus).

Further, the fiber path continues with new borderline conditions $- 2\pi rZ = 1/K$, $2\pi Z\zeta = 2\pi Zp$ and sign + or − in the same style as illustrated in Figure 5.3.

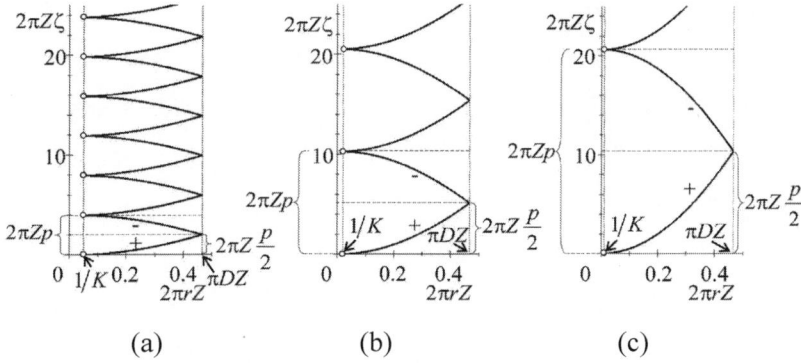

(a)   (b)   (c)

Figure 5.3 Characteristics of fiber paths in $\zeta - r$ ($2\pi Z\,\zeta$-$2\pi Z\,r$) graphs.

Parameters used: $\pi DZ = \tan\beta_D = \tan 25° = 0.46631$,

Central cylinder

The description of fiber paths, given earlier, is still not completed, because the fiber trajectories are not yet determined inside a small cylinder, where according to Equation (5.18), $r < 1/(2\pi ZK)$. Note that each fiber is not required to possess the same trajectory inside this cylinder.

Note: Let us think about the differential area surroundings the yarn axis where $r \to 0$. It is valid that $\lim\limits_{r\to0}\tan\beta = \lim\limits_{r\to0}(2\pi rZ) = 0$ according to Equation (4.1). By using this relation with the limit of Equation (5.12), we can

write that $\lim\limits_{r\to0}\left(\tan^2\alpha\right) = \dfrac{1 + \lim\limits_{r\to0}\left(\tan^2\beta\right)}{\lim\limits_{r\to0}\left(K^2\right)\lim\limits_{r\to0}\left(\tan^2\beta\right) - 1} = \dfrac{1}{\lim\limits_{r\to0}\left(K^2\right)\cdot 0 - 1}$. If $\tan^2\alpha$

possesses a value in the stated differential area, then $\lim\limits_{r\to0}\left(\tan^2\alpha\right) \in \langle 0,\infty)$, so

that $\left[\lim\limits_{r\to0}\left(K^2\right)\cdot 0\right] \in \langle 1,\infty)$ and thus the necessary assumption is

$\lim\limits_{r\to0}\left(K^2\right) = \infty$. By using Equation (5.11), it must be valid to write that

$\lim\limits_{r\to0}\left(\dfrac{\mu p}{2NsZ}\right)^2 = \dfrac{1}{\left(2NsZ\right)^2}\lim\limits_{r\to0}\left(\mu p\right) = \infty$. This expression is fulfilled when

$\lim\limits_{r\to0}\mu = \infty$ (infinitely high density of fiber material around the yarn axis) and/

or $\lim\limits_{r\to0}p = \infty$ (no radial fiber migration). Both are not possible to happen in

practice.

After accepting the assumptions 3 (number N of fibers is a constant) and 5 (fiber cross-sectional area s is a constant) from Section 5.1 and the assumption 7 (local packing density $\mu_r = \mu$ is a constant) from this section, we can express two quantities: (1) the total volume V_c° of section lengths p of central cylinder and (2) fiber volume V° inside it. By using Equations (5.19) and (1.22), it is valid to write that

$$V_c^{\circ} = \pi \left(\frac{1}{2\pi Z\,K} \right)^2 p = \frac{\pi p}{\left(2\pi Z\,K \right)^2}, \quad V^{\circ} = V_c^{\circ}\mu = \frac{\pi p \mu}{\left(2\pi Z\,K \right)^2}. \quad (5.24)$$

Fiber length in central cylinder

Each fiber touched the borderline radius r_0 stated in Equation (5.19) just once per period p of migration. (See points O as shown in Figure 5.3.) In these points, the fiber passes (can pass) through the borderline radius and creates (can create) something as a flat loop[8] in the central cylinder. The fiber creates the average flat loop length l_1° on the length p. All N fibers create total fiber length Nl_1° and the fiber volume $V^{\circ} = Nl_1^{\circ}s$. By using Equations (5.24) and (5.11), we can express the average length per fiber in the central cylinder on length p as follows:

$$V^{\circ} = \frac{\pi p \mu}{\left(2\pi Z\,K \right)^2} = Nl_1^{\circ}s, \quad l_1^{\circ} = \frac{V^{\circ}}{Ns} = \frac{\pi p \mu}{Ns\left(2\pi Z\,K \right)^2} = \frac{\pi p \mu}{Ns4\pi^2 Z^2}\frac{1}{K^2}$$

$$= \frac{p\mu}{2NsZ}\frac{1}{2\pi Z}\frac{1}{K^2} = K\frac{1}{2\pi Z}\frac{1}{K^2} = \frac{1}{2\pi Z K}. \quad (5.25)$$

Fiber length in main part of yarn

Let us think about the first part (sign +) of fiber path – see graphs in Figure 5.3. This fiber portion starts at point $r = 1/(2\pi ZK)$, $\zeta = 0$ (i.e., $2\pi rZ = 1/K$, $2\pi Z\zeta = 0$) and ends at point $r = D/2$, $\zeta = p/2$ (i.e., $2\pi rZ = \pi DZ$, $2\pi Z\zeta = 2\pi Z(p/2)$). Let us take the length of the above-mentioned fiber portion as $l_1^*/2$. So, the double of length $l_1^*/2$, i.e., l_1^* corresponds to the length of one fiber on the yarn height of one period p.

The following expression can be obtained from the first expression of Equations (5.6) and (5.11):

---

8   Treloar [1] imagined that the fiber parts inside the central cylinder as something like the felloes of a wheel.

$$\frac{\tan^2 \alpha + \tan^2 \beta + 1}{\tan^2 \alpha} = \left[\frac{\mu p}{2NsZ}\right]^2 \tan^2 \beta = K^2 \tan^2 \beta. \tag{5.26}$$

By using this result and Equation (4.1) in Equation (5.2), we can write

$$\mathrm{d}^2 l = \frac{\tan^2 \alpha + \tan^2 \beta + 1}{\tan^2 \alpha} \mathrm{d}^2 r = K^2 \tan^2 \beta \mathrm{d}^2 r = K^2 \left(2\pi rZ\right)^2 \mathrm{d}^2 r, \tag{5.27}$$

$$\mathrm{d}l = K \, 2\pi rZ \, \mathrm{d}r.$$

By integrating this differential equation, we obtain the following expressions:

$$\int_0^{l_1^*/2} \mathrm{d}l = \frac{l_1^*}{2} = K \int_{1/(2\pi ZK)}^{D/2} 2\pi rZ \, \mathrm{d}r = \frac{K}{2\pi Z} \int_{1/K}^{\pi DZ} x \, \mathrm{d}x = \frac{K}{4\pi Z}\left[\left(\pi DZ\right)^2 - \frac{1}{K^2}\right]$$

Substitution: $2\pi rZ = x$, $\mathrm{d}r = \mathrm{d}x/(2\pi Z)$, $l_1^* = \frac{K}{2\pi Z}\left[\left(\pi DZ\right)^2 - \frac{1}{K^2}\right]$

$$= \frac{K\left(2\pi Z\right)^2 \left(D/2\right)^2}{2\pi Z} - \frac{1}{2\pi ZK} = K\left(2\pi Z\right)\left(\frac{D}{2}\right)^2 - \frac{1}{2\pi ZK}. \tag{5.28}$$

Total fiber length on one period

The total fiber length l_1 on one period is the sum of the fiber length of main part l_1^* according to Equation (5.28) and the fiber length in central cylinder l_1° according to Equation (5.25). Thus

$$l_1 = l_1^* + l_1^\circ = \left[K\left(2\pi Z\right)\left(\frac{D}{2}\right)^2 - \frac{1}{2\pi ZK}\right] + \left[\frac{1}{2\pi ZK}\right] = K\left(2\pi Z\right)\left(\frac{D}{2}\right)^2. \tag{5.29}$$

Total fiber length – alternative derivation

The total fiber length l_1 can be independently determined more easily. Let us consider that the yarn portion having length equal to the period of migration p. The total volume of the yarn portion is $V_c = p\pi D^2/4$, and the fiber volume inside this is $V = \mu V_c = \mu p\pi D^2/4$. The same fiber volume can also be expressed by the expression $V = Nl_1 s$. By comparing both expressions and applying Equation (5.11), we obtain

$$V = Nl_1 s = \mu p\frac{\pi D^2}{4}, \qquad l_1 = \frac{\pi\mu p}{Ns}\left(\frac{D}{2}\right)^2 = \left(\frac{\mu p}{2NsZ}\right)2\pi Z\left(\frac{D}{2}\right)^2 = K\,2\pi Z\left(\frac{D}{2}\right)^2.$$

This result is identical to Equation (5.29).

Introduction to yarn retraction

Yarn retraction δ is defined in Section 4.3 by Equation (4.30) and Figure 4.5.[9] Moreover, each fiber has the same length in the current migration model[10]. If the starting structure – a parallel fiber bundle – contains N fibers of length l_1, then such a bundle is transformed through spinning process to a yarn portion of length p. So, in place of earlier symbol (starting length) ζ_0, we now use l_1 and in place of earlier symbol (final length) ζ, we now use p in Equation (4.30). By applying Equation (5.29), we obtain yarn retraction δ as follows:

$$\delta = 1 - \frac{p}{l_1} = 1 - \frac{p}{K\left(2\pi Z\right)\left(D/2\right)^2} = 1 - \frac{2\left(p/D\right)}{K\left(\pi DZ\right)}. \tag{5.30}$$

Note: We can also obtain this relation easily by rearranging Equation (5.14).

The yarn retraction depends, besides twist intensity, on K and p/D. Generally, each yarn has – besides twist intensity $\kappa = \tan\beta_D = \pi DZ$ [Equation (1.51)] – its own parameters, which are not required to be constant. Equation (5.23) mutually links these quantities. We can write

$$\frac{p}{D} = \frac{1}{\pi DZ} \int_{1/K}^{\pi DZ} \sqrt{\frac{K^2 x^2 - 1}{1 + x^2}}\, dx$$

$$= \frac{\pi DZ - 1/K}{\pi DZ} \int_0^1 \sqrt{\frac{K^2\left[y\left(\pi DZ - 1/K\right) + 1/K\right]^2 - 1}{1 + \left[y\left(\pi DZ - 1/K\right) + 1/K\right]^2}}\, dy,$$

---

9   Chapter 4 is devoted to helical models. Nevertheless, Equation (4.29) and Figure 4.5 determine yarn retraction quit generally in this chapter. Let us remind that we consider the starting structure as a bundle of parallel fibers.

10   This statement is not quite precise. The length l_1^* in the main part of the yarn is really the same for all fibers in this model. Nevertheless, the length l_1° is only the average value of fiber portions inside the small central cylinder. Therefore, the lengths $l_1 = l_1^* + l_1^{\circ}$ of fibers on a yarn portion of length p can have small differences. However, these fiber lengths become equal on a long length of yarn.

Substitution:

$$y = \frac{x - 1/K}{\pi DZ - 1/K}, \quad x = y(\pi DZ - 1/K) + 1/K, \quad dx = dy(\pi DZ - 1/K),$$

$$\frac{p}{D} = \frac{K\pi DZ - 1}{K\pi DZ} \int_0^1 \sqrt{\frac{\left[y(K\pi DZ - 1) + 1\right]^2 - 1}{1 + \left[y(K\pi DZ - 1) + 1\right]^2 / K^2}} \, dy. \tag{5.31}$$

The relationship among quantities K, p/D, and $\kappa = \tan\beta_D = \pi DZ$ according to Equation (5.31) is illustrated in Figure 5.4. [The integral expressed in Equation (5.31) leads to an elliptical integral; it can be solved only numerically.]

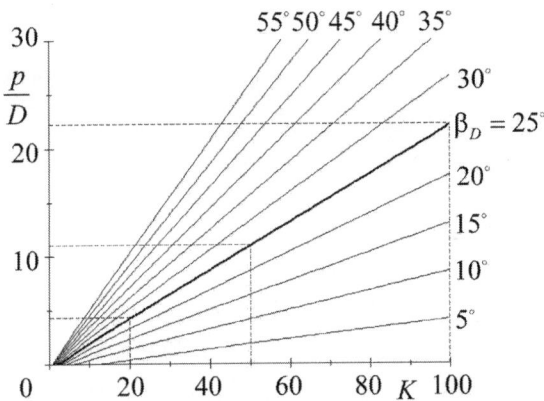

Figure 5.4 Relationship among K, p/D,

and β_D ($\tan\beta_D = \pi DZ$) according to Equation (5.31)

Note: The calculated curves are very flat, but they are not precisely straight lines. The thick line corresponds to the value $\pi DZ = \tan\beta_D = \tan 25° = 0.46631$, which is used in examples shown in Figure 5.3. Also, the values of p/D for $K = 20, 50, 100$ correspond to the values shown in Figure 5.3 – see dashed lines in Figure 5.4.

Here, a question can be raised. Which of the relation between K and p/D is really valid for the yarns with different levels of twist intensity? We can derive only two 'borderline' concepts of yarn retraction: (1) idea of constant value of K and (2) idea of constant value of p/D.

Yarn retraction using common constant K

Let us assume that all yarns have a common constant value of parameter K. It means that the relation between angles α and β is same [according to Equation (5.12)] for all yarns with different twists.

Then, we obtain the following expression after substitution of p from Equation (5.22) to (5.30):

$$\delta = 1 - \frac{2\dfrac{1}{\pi Z}\displaystyle\int_{1/K}^{\pi DZ}\sqrt{\dfrac{K^2x^2-1}{1+x^2}}\,dx}{KD(\pi DZ)} = 1 - \frac{2}{K(\pi DZ)^2}\int_{1/K}^{\pi DZ}\sqrt{\frac{K^2x^2-1}{1+x^2}}\,dx. \quad (5.32)$$

The resulting general expression shown in Equation (5.32) needs discussion. At first, let us study the limit curve of yarn retraction for very high values of K, i.e., when $K \to \infty$. Then, the following expression is valid to write by using Equation (A4.18) from Appendix 4:

$$\lim_{K\to\infty}\delta = 1 - \lim_{K\to\infty}\left[\frac{2}{K(\pi DZ)^2}\int_{1/K}^{\pi DZ}\sqrt{\frac{K^2x^2-1}{1+x^2}}\,dx\right]$$

$$= 1 - \frac{2}{(\pi DZ)^2}\lim_{K\to\infty}\int_{1/K}^{\pi DZ}\sqrt{\frac{x^2-1/K}{1+x^2}}\,dx$$

$$= 1 - \frac{2}{(\pi DZ)^2}\int_{0}^{\pi DZ}\frac{x\,dx}{\sqrt{1+x^2}} = 1 - \frac{2}{(\pi DZ)^2}\left[\sqrt{1+(\pi DZ)^2}-1\right]. \quad (5.33)$$

[This derivation is similar to that of Equation (4.14).] Further, by rearrangement according to Equation (4.56), we obtain the following final expression:

$$\lim_{K\to\infty}\delta = \frac{\sqrt{1+(\pi DZ)^2}-1}{\sqrt{1+(\pi DZ)^2}+1} = \frac{\sqrt{1+\tan^2\beta_D}-1}{\sqrt{1+\tan^2\beta_D}+1} = \frac{\sqrt{1+\kappa^2}-1}{\sqrt{1+\kappa^2}+1}, \quad (5.34)$$

which is identical to yarn retraction derived in Braschler's ideal helical model, Equation (4.56).

Furthermore, let us study a case with a very small value of K. Let us imagine that the borderline value of radius is $r_0 \ge D/2$, so that the whole yarn is lying inside the central cylinder as mentioned earlier. If especially $r_0 = D/2$, then according to Equations (5.19), (5.22), (5.25), (5.29), we obtain the following relations:

$$r_0 = \frac{1}{2\pi Z\,K} = \frac{D}{2}, \quad K = \frac{1}{\pi D Z}, \tag{5.35}$$

$$\pi Z p = \int_{\pi D Z}^{\pi D Z} \sqrt{\frac{K^2 x^2 - 1}{1 + x^2}}\; dx = 0, \quad p = 0, \tag{5.36}$$

$$l_1^{\square} = \frac{1}{2\pi Z}\pi D Z = \frac{D}{2}, \tag{5.37}$$

$$l_1 = \frac{1}{\pi D Z}(2\pi Z)\left(\frac{D}{2}\right)^2 = \frac{D}{2}. \tag{5.38}$$

Finally, we obtain the following value of yarn retraction by using Equations (5.35) and (5.36) in (5.30):

$$\lim_{K \to \frac{1}{\pi D Z}} \delta = 1 - \frac{0}{\dfrac{1}{\pi D Z}(2\pi Z)(D/2)^2} = 1 - \frac{0}{D/2} = 1. \tag{5.39}$$

Note: In this theoretical result, the fibers create flat loops in yarn cross section only, without any change of axial ζ-coordinate. (The quantity δ is not defined for values $r_0 > D/2$.)

If $K > 1/(\pi D Z)$ then the yarn retraction depends on twist intensity $\pi D Z$, according to Equation (5.32). The graphical interpretation of this statement is shown in Figure 5.5. (The numerical integration was used.) Each curve is defined at the right-hand side from the borderline value $K = 1/(\pi D Z)$, $\delta = 1$ (vertical thin dotted lines). It is shown that the higher is the value of K, the nearer is the behaviour of function $\pi D Z - \delta$ to the ideal helical model according to Equation (5.34) or (4.56).

Evidently, the decreasing nature of the retraction curves shown in Figure 5.5 does not accord to the reality; however, the increasing part of the curves is more or less acceptable and it is valid for higher values of twist intensity $\pi D Z$. It is also visible that the non-real region is smaller and less important for the higher values of parameter K. (The radius $1/K$ of central cylinder is smaller for higher values of K as shown in Figure 5.3.)

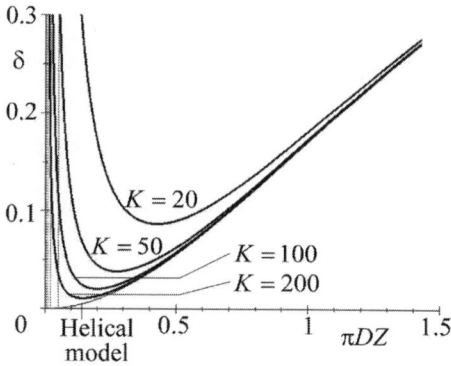

Figure 5.5 Yarn retraction for constant K. Thick lines – Equation (5.32). Thin line – Equations (4.56) and (5.34)

The experimental results showed that the period p is relatively high and in consequence of it the parameter K is also high. Therefore, Treloar [1] proposed to use an approximation of the previous equations which will be described in the next section.

Yarn retraction using common constant p/D

Let us assume that all yarns have a common constant value of parameter p/D. Then, for each value of twist intensity πDZ, we can find a suitable value of K corresponding to Equation (5.31).

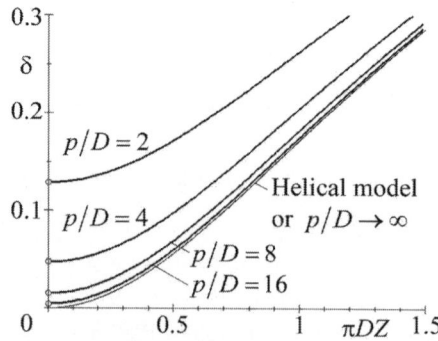

Figure 5.6 Yarn retraction by constant value of p/D.
Thin line – Equation (4.56)

Note: Let us consider that we know the value of p/D. Then, for a given value of twist intensity πDZ and the chosen value of K, we are able to quantify (by using numerical method) the integral expressed in Equation (5.31). Further, using a numerical method for finding the root of K, we can gradually find the suitable value of K, because Equation (5.31) is valid.

Finally, we use the same value of p/D, the determined value K and the given twist intensity πDZ in Equation (5.30) to evaluate yarn retraction δ.

The examples of the curves, determined by the described algorithm, are illustrated in Figure 5.6. Nevertheless, the described algorithm cannot be used for $\tan\beta_D = \pi DZ \to 0$, because at a given value p/D and $\pi DZ \to 0$, the value $K \to \infty$ (Compare it with Figure 5.4.), and the product $K \cdot \pi DZ \to \infty \cdot 0$ is an indefinite expression. Therefore, let us introduce the following quantity C; this product can be a real quantity as follows:

$$C = K\pi DZ .$$

(5.40)

We use Equation (5.40) in (5.31),

$$\frac{p}{D} = \frac{C-1}{C}\int_0^1 \sqrt{\frac{\left[y(C-1)+1\right]^2 - 1}{1+\left[y(C-1)+1\right]^2 / K^2}}\, dy ,$$

(5.41)

so that p/D is a function of C and K, now. Because it is valid that $\tan\beta_D = \pi DZ \to 0$ for $K \to \infty$, we can write

$$\lim_{\pi DZ \to 0}\frac{p}{D} = \lim_{K \to \infty}\left\{\frac{C-1}{C}\int_0^1 \sqrt{\frac{\left[y(C-1)+1\right]^2 - 1}{1+\left[y(C-1)+1\right]^2 / K^2}}\, dy\right\} .$$

(5.42)

Let us solve the last equation by using Equation (A4.12) from Appendix 4:

$$\lim_{\pi DZ \to 0}\frac{p}{D} = \frac{C-1}{C}\int_0^1 \sqrt{\left[y(C-1)+1\right]^2 - 1}\, dy = \frac{1}{C}\int_1^C \sqrt{t^2 - 1}\, dt$$

$$= \frac{1}{2C}\left[C\sqrt{C^2 - 1} - \ln\left(C + \sqrt{C^2 - 1}\right)\right]$$

$$= \frac{1}{2}\sqrt{C^2 - 1} - \frac{1}{2C}\ln\left(C + \sqrt{C^2 - 1}\right) .$$

(5.43)

Because we know the same value p/D, we can find the root C of Equation (5.43) by using a numerical method.

We then rearrange the yarn retraction according to Equation (5.30) using (5.40) as follows:

$$\lim_{\pi DZ \to 0} \delta = 1 - \frac{2(p/D)}{K(\pi DZ)} = 1 - \frac{2(p/D)}{C}. \tag{5.44}$$

The coordinates for the small rings shown along the δ-axis in Figure 5.6 were calculated in this way.

It is evident that a positive value of yarn retraction exists also in case of untwisted yarn (points $\bigcirc$ in Figure 5.6 at $\pi DZ = 0$). How could this be explained? The fibers in such a yarn structure have only angles α, while angles β are all equal to zero[11]. It corresponds to an ideal model of entangled filament yarn. So it is visible that Treloar's original model can also be a starting point for the modelling of entangled yarns.

5.4   Treloar's approximation

Assumptions

Equation (5.12) of ideal radial fiber migration can be rearranged as follows:

$$\tan^2 \alpha = \frac{1 + \tan^2 \beta}{K^2 \tan^2 \beta - 1} = \frac{1}{\cos^2 \beta \left(K^2 \tan^2 \beta - 1 \right)} = \frac{1}{K^2 \sin^2 \beta - \cos^2 \beta}$$

$$= \frac{1}{K^2 - K^2 \cos^2 \beta - \cos^2 \beta} = \frac{1}{K^2 - \cos^2 \beta \left(K^2 + 1 \right)}. \tag{5.45}$$

Let us assume that K is high. Then, it is approximately valid that $K^2 + 1 \cong K^2$ so that

$$\tan^2 \alpha \cong \frac{1}{K^2 - K^2 \cos^2 \beta} = \frac{\sin^2 \beta + \cos^2 \beta}{K^2 \sin^2 \beta} = \frac{1}{K^2} \left(1 + \frac{1}{\tan^2 \beta} \right). \tag{5.46}$$

Let us also assume that $\tan^2 \beta \ll 1$ for major parts of fibers in common yarns. So, it is approximately valid that $1/\tan^2 \beta \gg 1$, $1 + 1/\tan^2 \beta \cong 1/\tan^2 \beta$, and we can write Equation (5.46) in the following form:

---

11   Let us note that we think always about the term 'yarn retraction' in relation to a parallel fiber bundle.

$$\tan^2 \alpha = \frac{1}{K^2 \tan^2 \beta}, \quad \tan \alpha = \frac{\pm 1}{K \tan \beta},$$

$$\tan \alpha = \frac{dr}{d\zeta} = \frac{\pm 1}{K(2\pi r Z)}, \quad d\zeta = \pm K \, 2\pi r Z \, dr. \tag{5.47}$$

[Do not forget that $\tan \alpha = dr/d\zeta$ according to Equation (5.1) and $\tan \beta = 2\pi r Z$ according to Equation (4.1).] Let us take a note that these expressions are valid for all radii $r \in \langle 0, D/2 \rangle$, in contrary to the original model discussed in the previous section.

Without loosing generality, let us use the following starting point for the first part of fiber path (sign +, from inside to outside) such that

$$r = r_0 = 0, \quad \varphi = \varphi_0 = 0, \quad \zeta = \zeta_0 = 0. \tag{5.48}[12]$$

By using Equation (5.47) with sign + and Equation (5.14), the axial length ζ_D of the above-mentioned part of fiber path is

$$\int_0^{\zeta_D} d\zeta = K \, 2\pi Z \int_0^{D/2} r \, dr, \quad \zeta_D = K \, 2\pi Z \left(\frac{D}{2} \right)^2$$

$$= K D \frac{\pi D Z}{4} = \frac{2p}{D} \frac{1}{1-\delta} \frac{1}{\pi D Z} D \frac{\pi D Z}{4} = \frac{p/2}{1-\delta}. \tag{5.49}$$

Though it seems that the axial length ζ_D should be equal to one-half of the period, but it is actually not so, as stated in Equation (5.49).

Approximated equation

Because to obtain a right period of migration (same as in original model), let us use a modified parameter $K' = K(\ -\delta)$ in place of former parameter K from Equation (5.14). Thus

$$K' = K(1-\delta) = \frac{2p}{D} \frac{1}{1-\delta} \frac{1}{\pi D Z}(1-\delta) = \frac{2p}{D} \frac{1}{\pi D Z}, \quad \left(\frac{p}{D} = \frac{K'}{2}(\pi D Z) \right). \tag{5.50}$$

Note: By using K' in place of K in earlier derived Equation (5.49), we obtain the relation $\zeta_{D/2} = p/2$, which fully corresponds to our idea.

---

12   Unlike Equation (5.19), let us remind that the radius is defined from zero, now.

By using K' in place of K in earlier derived Equation (5.47), we obtain the approximated equation as follows:

$$\tan^2 \alpha = \frac{1}{K'^2 \tan^2 \beta}.$$ (5.51)

Differential equation

We utilize Equations (5.1) and (4.1) in (5.51) so that we obtain the following differential equation for the fiber path:

$$\tan \alpha = \frac{\pm 1}{K' \tan \beta}, \quad \frac{dr}{d\zeta} = \frac{\pm 1}{K'(2\pi r Z)}, \quad d\zeta = \pm K' 2\pi r Z \, dr.$$ (5.52)

Note: The second differential Equation (5.15), i.e., $d\varphi = 2\pi Z \, d\zeta$, is still valid.

Fiber path

For the first part of the fiber path, we use $+$ sign and the starting point according to Equation (5.48) in the previous differential equation. Thus,

$$\int_0^\zeta d\zeta^* = K' 2\pi Z \int_0^r r^* \, dr^*, \zeta = K' \pi Z r^2 \left(2\pi Z \zeta = \frac{K'}{2}(2\pi r Z)^2 \right).$$ (5.53)

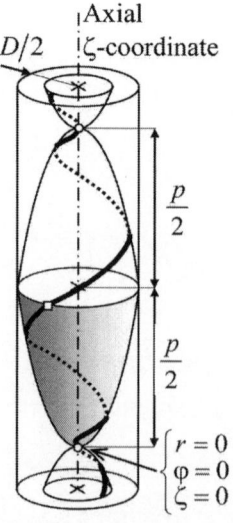

Figure 5.7 Fiber path on paraboloids

(The integrating quantities were renamed by superscript * to avoid any confusion with the upper limits of the integrals.)

Note: If especially $r = D/2$ then $\zeta = K'\pi Zr^2 = K'\pi Z(D/2)^2$ and using Equation (5.50) for K' we obtain $\zeta = p/2$, which was expected.

Note: The solution of Equation (5.15), i.e., Equation (5.20), is still valid.

The parabolic relation between r and ζ is valid for the parts of the fiber path according to Equation (5.53). The three-dimensional fiber path is given by Equations (5.20) and (5.53). It is evident that the first part of the fiber path is lying on the surfaces of the highlighted paraboloid shown in Figure 5.7. The other parts of the fiber path (from outside to inside, etc.) are lying on the symmetrical paraboloids as shown in Figure 5.7. (These are constructed analogically according to the discussion related to Figure 5.3.)

Numerical example

Let us choose that $\pi DZ = \tan 25° = 0.46631$ and $K = 20$ (as well as in Figure 5.3a). It was also found that $p/D = 4.25305$ by means of Equation (5.23) and the same was considered in case of Figure 5.3. The fiber path of this original model stated by Equation (5.21) and the curve shown in Figure 5.3a are displayed as a thin line A in Figure 5.8. The approximated equation with the same value of p/D must use the modified parameter K' determined from Equation (5.50). So, we obtain $K' = \dfrac{2p}{D}\dfrac{1}{\pi DZ} = 2\dfrac{4.25305}{0.46631} = 18.24138$. This value when we used in Equation (5.53) we obtained the first part of the approximated function – thick line B in Figure 5.8. It is evident that the differences between both curves are imperceptible in this case.

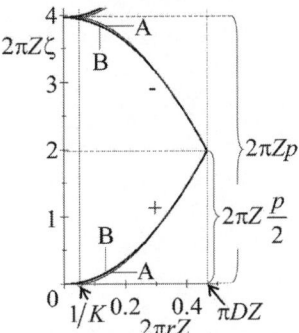

Figure 5.8 Comparison of fiber paths in $\zeta - r$ (i.e. $2\pi Z\zeta - 2\pi Z r$) graphs.
Parameters used: $\pi DZ = \tan\beta_D = \tan 25° = 0.46631$, A: Equation (5.21),
$K = 20$, B: Equation (5.53), $K' = 18.24138$ (Common $p/D = 4.25305$)

Note: The differences observed are even smaller for the higher values of K and K'. The smaller values of K and K' provide a little higher difference, but so small values (so small period of migration) are usually not realistic in real yarns.

Fiber length

By substituting $\tan^2 \alpha$ from Equation (5.51) to (5.2) and using Equation (4.1) for $\tan \beta$, we find the following differential equation:

$$d^2 l = \frac{\tan^2 \alpha + \tan^2 \beta + 1}{\tan^2 \alpha} d^2 r = \left(1 + K'^2 \tan^4 \beta + K'^2 \tan^2 \beta\right) d^2 r,$$

$$dl = \pm\sqrt{1 + K'^2 \tan^4 \beta + K'^2 \tan^2 \beta} \ dr = \pm\sqrt{1 + K'^2 \left(2\pi r Z\right)^4 + K'^2 \left(2\pi r Z\right)^2} \ dr.$$

$$(5.54)$$

The one part of fiber path (e.g., sign +, from inside to outside) has length $l_1/2$ and it lies on the yarn of length $p/2$. (Fiber length l_1 occupies the yarn length of period p.) Thus,

$$l_1 = 2\frac{l_1}{2} = 2 \int_{r=0}^{r=D/2} dl = 2 \int_0^{D/2} \sqrt{1 + K'^2 \left(2\pi r Z\right)^4 + K'^2 \left(2\pi r Z\right)^2} \ dr,$$

Substitution: $2\pi r Z = x, \ dr = dx/(2\pi Z).$

$$l_1 = \frac{1}{\pi Z} \int_0^{\pi D Z} \sqrt{1 + K'^2 x^4 + K'^2 x^2} \ dx. \qquad (5.55)$$

The integral at the right-hand side leads to an elliptical integral which can be solved by using a suitable numerical method.

Yarn retraction – introductory equation

It is valid that $\delta = 1 - p/l_1$ – see Equation (5.30)[13]. We use Equation (5.55) in this expression to obtain the following expression:

---

13   Yarn retraction δ is defined in Section 4.3 by Equation (4.30) and Figure 4.5. In place of symbol (starting length) ζ_0, we use l_1 and in place of symbol (final length) ζ, we use p in Equation (4.30).

$$\delta = 1 - \frac{p}{l_1} = 1 - \frac{\pi Z p}{\int_0^{\pi DZ} \sqrt{1 + K'^2 x^4 + K'^2 x^2}\ dx} = 1 - \frac{p}{D} \frac{\pi DZ}{\int_0^{\pi DZ} \sqrt{1 + K'^2 x^4 + K'^2 x^2}\ dx}.$$

$$(5.56)$$

The yarn retraction in the previous equation is shown as a function of three quantities: K', p/D, and $\kappa = \tan\beta_D = \pi DZ$. The relation among these quantities determined by Equation (5.50) is illustrated in Figure 5.9.

Note: Figure 5.9 is very similar to Figure 5.4; however, the lines are precisely straight now. The thick line corresponds to the value $\pi DZ = \tan 25° = 0.46631$, used in the examples illustrated in Figures 5.3 and 5.8.

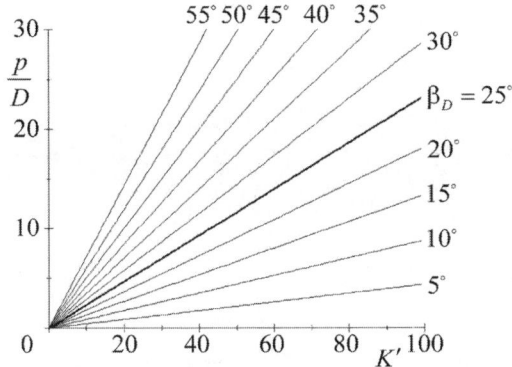

Figure 5.9 Relationship among K', p/D and
β_D ($\tan\beta_D = \pi DZ$) according to Equation (5.50)

It then raises a question that is stated as follows. Which of the relation between K' and p/D is really valid for yarns of different levels of twist intensity? We can again derive only two borderline concepts of yarn retraction: (1) idea of constant value of K' and (2) idea of constant value of p/D.

Yarn retraction for common constant K'

Let us assume that all yarns have the common constant value of parameter K'. It means that the relation between angles α and β is same for all yarns with different twists according to Equation (5.51).

By using the ratio p/D from Equation (5.50) to (5.56), we obtain the following equation:

$$\delta = 1 - \frac{K'\pi DZ}{2} \frac{\pi DZ}{\int_0^{\pi DZ} \sqrt{1 + K'^2 x^4 + K'^2 x^2}\ dx} = 1 - \frac{K'(\pi DZ)^2}{2\int_0^{\pi DZ} \sqrt{1 + K'^2 x^4 + K'^2 x^2}\ dx}$$

Substitution: $y = x/(\pi DZ)$, $dx = \pi DZ\ dy$,

$$\delta = 1 - \frac{K'\pi DZ}{2\int_0^1 \sqrt{1 + K'^2 y^4 (\pi DZ)^4 + K'^2 y^2 (\pi DZ)^2}\ dy}.$$

(5.57)

If $\pi DZ \to 0$, then

$$\lim_{\pi DZ \to 0} \delta = 1 - \frac{K' \cdot 0}{2\int_0^1 dy} = 1.$$

(5.58)

If $K' \to \infty$, then by using Equation (A4.14) from Appendix 4 we find

$$\lim_{K' \to \infty} \delta = 1 - \lim_{K' \to \infty} \frac{K'\pi DZ}{2\int_0^1 \sqrt{1 + K'^2 y^4 (\pi DZ)^4 + K'^2 y^2 (\pi DZ)^2}\ dy}$$

$$= 1 - \lim_{K' \to \infty} \frac{1}{2\int_0^1 \sqrt{\dfrac{1}{K'^2 (\pi DZ)^2} + y^4 (\pi DZ)^2 + y^2}\ dy}$$

$$= 1 - \frac{1}{2\int_0^1 y\sqrt{y^2 (\pi DZ)^2 + 1}\ dy} = 1 - \frac{1}{2\,\pi DZ \dfrac{1}{3}\left[\left(\dfrac{1}{(\pi DZ)^2} + 1\right)^{3/2} - \dfrac{1}{(\pi DZ)^3}\right]},$$

$$\lim_{K' \to \infty} \delta = 1 - \frac{3/2}{\pi DZ \left\{\left[1 + 1/(\pi DZ)^2\right]^{\frac{3}{2}} - 1/(\pi DZ)^3\right\}}.$$

(5.59)

The relation between yarn retraction and twist intensity according to Equation (5.57), including the limits according to Equations (5.58) and (5.59), is illustrated graphically in Figure 5.10. This graph is similar to Figure 5.5. Nevertheless, as opposed to Figure 5.5, all curves possess a finite value of K' and start at a same point $(0,1)$ according to Equation (5.58). The trends of the curves shown in Figure 5.10 are also similar to those displayed in Figure 5.5 and include quite unrealistic decreasing trend at smaller values of twist intensity πDZ. The limit relation $(K' \to \infty)$ leads to the retraction of helical yarn according to Equation (5.34) or (4.56). However, the analogical curve [Equation (5.59) shown by small circular points in Figure 5.10] differs itself from the above-mentioned helical relation mainly at higher values of πDZ.

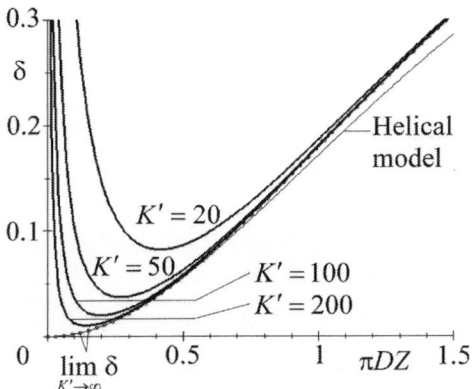

Figure 5.10 Yarn retraction by constant value of K' –
Equation (5.57). Small circular points – Equation (5.59). Thin line – Equation (4.56)

Yarn retraction for common constant p/d

Let us assume that all yarns have the common constant value of parameter p/D. Then by using Equation (5.50), i.e., $K'\pi DZ = 2p/D$, in (5.56), we obtain

$$\delta = 1 - \frac{p}{D} \frac{\pi DZ}{\int_0^{\pi DZ} \sqrt{1 + 4\left(\frac{p}{D}\right)^2 \frac{x^4}{(\pi DZ)^2} + 4\left(\frac{p}{D}\right)^2 \frac{x^2}{(\pi DZ)^2}}\, dx}$$

Substitution: $y = x/(\pi DZ)$, $dx = \pi DZ\, dy$,

$$= 1 - \frac{p}{D} \frac{1}{\int_0^1 \sqrt{1 + 4\left(\frac{p}{D}\right)^2 (\pi DZ)^2 y^4 + 4\left(\frac{p}{D}\right)^2 y^2}\, dy}. \tag{5.60}$$

If $\pi DZ \to 0$, then by using Equation (A4.13) from Appendix 4 we find

$$\lim_{\pi DZ \to 0} \delta = 1 - \frac{p}{D} \lim_{\pi DZ \to 0} \frac{1}{\int_0^1 \sqrt{1 + 4\left(\frac{p}{D}\right)^2 (\pi DZ)^2 y^4 + 4\left(\frac{p}{D}\right)^2 y^2}\, dy}$$

$$= 1 - \frac{p}{D} \frac{1}{\int_0^1 \sqrt{1 + 4\left(\frac{p}{D}\right)^2 y^2}\, dy} = 1 - \frac{p}{D} \frac{1}{2\frac{p}{D}\int_0^1 \sqrt{\frac{1}{4(p/D)^2} + y^2}\, dy}$$

$$= 1 - \frac{1}{2\int_0^1 \sqrt{\frac{1}{4(p/D)^2} + y^2}\, dy}$$

$$= 1 - \frac{1}{2\left\{\frac{1}{2}\frac{1}{2(p/D)}\left[\sqrt{1 + 4(p/D)^2} + \frac{1}{2(p/D)}\ln\left(2(p/D) + \sqrt{1 + 4(p/D)^2}\right)\right]\right\}},$$

$$\lim_{\pi DZ \to 0} \delta = 1 - \frac{2p/D}{\sqrt{1 + 4(p/D)^2} + \frac{1}{2(p/D)}\ln\left(2(p/D) + \sqrt{1 + 4(p/D)^2}\right)}. \tag{5.61}$$

Further, let us find out yarn retraction from Equation (5.60) when $p/D \to \infty$. This is given as follows:

$$\lim_{p/D \to \infty} \delta = 1 - \lim_{p/D \to \infty} \left\{ \frac{p}{D} \frac{1}{\int_0^1 \sqrt{1 + 4\left(\frac{p}{D}\right)^2 (\pi DZ)^2 y^4 + 4\left(\frac{p}{D}\right)^2 y^2}\, dy} \right\}$$

$$= 1 - \lim_{p/D \to \infty} \left\{ \frac{1}{\int_0^1 \sqrt{\frac{1}{(p/D)^2} + 4(\pi DZ)^2 y^4 + 4y^2}\, dy} \right\}$$

$$= 1 - \frac{1}{\int_0^1 \sqrt{4(\pi DZ)^2 y^4 + 4y^2}\, dy} = 1 - \frac{1}{2\int_0^1 y\sqrt{(\pi DZ)^2 y^2 + 1}\, dy}. \tag{5.62}$$

The integral stated in the last expression is identical to that mentioned in Equation (5.59) so that the solution of Equation (5.62) takes the following form:

$$\lim_{p/D \to \infty} \delta = 1 - \frac{3/2}{(\pi DZ)\left\{\left[1 + 1/(\pi DZ)^2\right]^{\frac{3}{2}} - 1/(\pi DZ)^3\right\}}. \tag{5.63}$$

Note: If $p/D \to \infty$, then $K' \to \infty$ according to Equation (5.50). Therefore, the identity of the functions mentioned in Equations (5.59) and (5.63) is evident.

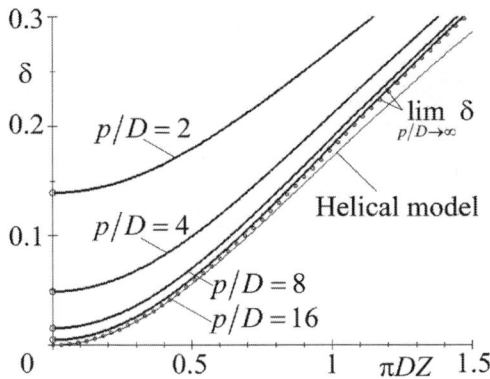

Figure 5.11 Yarn retraction by constant value of p/D –

Equation (5.60). Small circular points – Equation (5.62). Thin line – Equation (4.56)

Figure 5.11 illustrates Equations (5.60), (5.61) and (5.63). This is similar to the graph shown in Figure 5.6. The functions shown in Equation (5.60) represent a set of thick curves – e.g., $p/D = 2, 4, 8, 16$. The values of $\lim\limits_{\pi DZ \to 0} \delta$, according to Equation (5.61), show the rings along δ-axis. These express yarn retractions for untwisted yarn, e.g., like an entangled filament yarn, etc. The function of $\lim\limits_{p/D \to \infty} \delta$ according to Equation (5.62), shown by small circular points in Figure 5.11, differs itself from the helical relation mainly at higher values of πDZ.

Local packing density

The original model of Treloar's migration based, among others, on the assumptions 6 and 7 stated at the starting part of Section 5.3, i.e., based on the constant value of period $p(r) = p$ and constant value of local packing density $\mu_r = \mu$ at all radii r inside the yarn. On the contrary, the approximate model of Treloar's migration, described in this section, accords to the changes in fiber path to the shape described by Equation (5.47) 'ad hoc', while the above-mentioned assumption 6, i.e., $p(r) = p$, remains always valid. Naturally, such approximation must bring a change in the earlier constant value $\mu_r = \mu$.

By using Equations (5.3) and (4.1) and $p(r) = p$ in Equation (5.4), we can write

$$\mu_r = \frac{2NsZ}{p}\sqrt{\frac{\tan^2\alpha + \tan^2\beta + 1}{\tan^2\alpha\tan^2\beta}} = \frac{2NsZ}{p(2\pi rZ)}\frac{dl}{dr}. \tag{5.64}$$

Nevertheless, we can rearrange the last expression using Equations (5.13) and (5.50).

Note: Let us remind that the symbol μ represents the packing density of the whole yarn. Equation (4.7), derived in Chapter 4, i.e., $\mu = 8\int_0^{D/2}\mu_r r\, dr / D^2$, has general validity so that it is also valid now.

So, we obtain the expression

$$\frac{2NsZ}{p} = \frac{2\left[(1-\delta)\mu\dfrac{\pi D^2}{4}\right]Z}{D\dfrac{K'}{2}(\pi DZ)} = \frac{(1-\delta)\mu}{K'}. \tag{5.65}$$

Let us substitute Equation (5.65) to (5.64)

$$\mu_r = \frac{(1-\delta)\mu}{K'(2\pi rZ)}\frac{dl}{dr}, \quad \frac{\mu_r}{\mu} = \frac{1-\delta}{K'(2\pi rZ)}\frac{dl}{dr}. \tag{5.66}$$

Now, we substitute $1-\delta$ and dl/dr from Equations (5.54)[14] and (5.57)

$$\frac{\mu_r}{\mu} = \frac{1}{K'(2\pi rZ)}\frac{K'\pi DZ}{2\int_0^1\sqrt{1+K'^2y^4(\pi DZ)^4 + K'^2y^2(\pi DZ)^2}\,dy}$$

$$= \frac{\pi DZ}{2\pi rZ}\frac{\sqrt{1+K'^2(2\pi rZ)^4 + K'^2(2\pi rZ)^2}}{2\int_0^1\sqrt{1+K'^2y^4(\pi DZ)^4 + K'^2y^2(\pi DZ)^2}\,dy},$$

$$\frac{\mu_r}{\mu} = \frac{\sqrt{1+K'^2y^4(\pi DZ)^4 + K'^2y^2(\pi DZ)^2}}{2y\int_0^1\sqrt{1+K'^2y^4(\pi DZ)^4 + K'^2y^2(\pi DZ)^2}\,dy}, \quad \text{where } y = \frac{2\pi rZ}{\pi DZ} = \frac{r}{D/2}. \tag{5.67}$$

---

14   We interpret the derivative dl/dr as a positive value, now.

The ratio μ_r/μ can be called relative local packing density. This changes in relation to the radial position $y = r/(D/2)$ as shown in case of examples in Figure 5.12. (It was obtained by using a suitable method of numerical integration.)

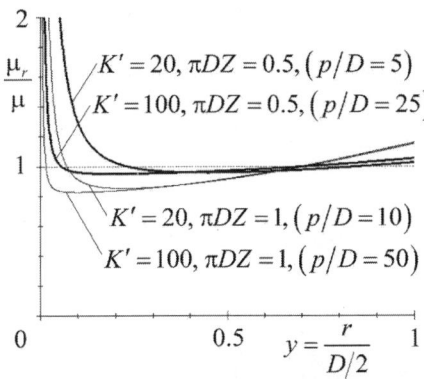

Figure 5.12 Relative local packing density in relation to radial position by $K' = 20, 100$, $\pi DZ = 0.5, 1$ – Equation (5.67),

[p/D according Equation (5.50)]

It is evident that the relative local packing density is limited (non-real) to infinity at yarn axis. Further, in the main part of the yarn body, the relative local packing density shows a slightly increasing trend. As a whole, the calculated curves are not too realistic.

Experimental verification

To verify the theoretical model, the internal structure of a set of carded ring and rotor yarns with a wide range of count and twist produced mainly from 100% viscose staple fibers and a few from cotton fibers was studied by employing the OMEST system developed by Stejskal and Kašpárek [2]. This system used an optical instrument and the tracer fiber technique [3], and gave two perpendicular views of the fiber path in yarn. The fiber paths were then analyzed by a special computer program[15].

Nečkář and Soni [4] evaluated the (mean) values of $|\tan\alpha|$ and $\tan\beta$ at different radii r in case of a lot of yarns. (See also Ref. [5].) The experimental dependence between the quantity $|\tan\alpha|\tan\beta$ and radius $r \in (0, D/2)$ or

---

15   See the next sections for more information.

$\tan \beta = 2\pi r Z \in (0, \pi D Z)$ was determined for different yarns. Such experimental trends have increasing and slightly convex character and all are lying inside the hatched area as shown in Figure 5.13. Also, the model curves are shown in Figure 5.13. It is evident that the quantity $|\tan \alpha| \tan \beta$ is a constant according to Equation (5.51) for Treloar's approximated model. We obtain another function from Equation (5.12) for Treloar's original model. It is valid to write that

$$|\tan \alpha| \tan \beta = \frac{\sqrt{1 + \tan^2 \beta}}{\sqrt{K^2 \tan^2 \beta - 1}} \tan \beta = \frac{\sqrt{1 + (2\pi r Z)^2}}{\sqrt{K^2 (2\pi r Z)^2 - 1}} 2\pi r Z. \quad (5.68)$$

Such functions are limited to infinity for $2\pi r Z = 1/K$ [compare it with Equation (5.17)] and for higher values they follow the thick lines shown in Figure 5.13.

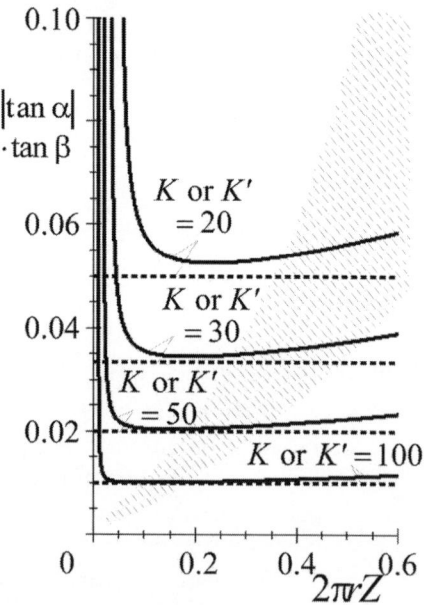

Figure 5.13 Comparison of experimental and model trends, hatched lines – experimental trends, thick lines – initial model, Equation (5.68), dashed lines – approx., Equation (5.51)

It is evident from Figure 5.13 that the theoretical and experimental trends are completely different. It means that the theoretical models are in disproportion with the experimental results.

Note: The original theoretical results derived by Treloar [1] were not compared with experimental results because of unavailability of special instruments and methods and quick computer technique at that time.

5.5   Equidistant radial fiber migration model

Introduction

The basic idea that the regular fibers going from yarn axis to yarn periphery and then returning to yarn axis, which is presented by assumption 6 in the first paragraph of Section 5.3, is not too real. Let us think about layers (of small thickness δr) in a yarn as shown in Figure 5.14. A layer situated at a small radius has a small (dotted) area of cross section and therefore a small total volume. In spite of the fact that the packing density is relatively high, the fiber volume needs to be small. Therefore, a fiber can seldom intersect such layer.

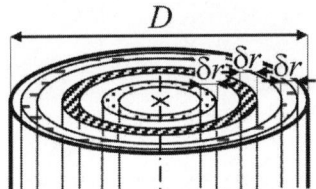

Figure 5.14 Layers inside yarns

A layer situated at a high radius, near to yarn surface, offers a large (dashed) area of cross section and therefore a high total volume. But the packing density is usually (mainly by a real staple yarns) small there, so the fiber volume is also small, and therefore a fiber can seldom intersect such a layer too.

A 'middle' (hatched) layer offers a 'medium' level of cross-sectional area and therefore a medium level of total volume. However, its packing density remains relatively high so that the necessary fiber volume is also relatively high. Therefore, a fiber often intersects such a layer.

It follows from previous ideas that the mean period p of migration changes itself in relation to its radius, $p = p(r)$. Naturally, the fiber path cannot be fully regular according to this concept.

Note: One fiber in $r - \zeta$ graph shown in Figure 5.15 illustrates the local periods $p_1, p_2, \ldots, p_8$, of migration at radius r (vertical dashed line). The arithmetic mean of such local periods determine the mean period $p(r)$ at a given radius.

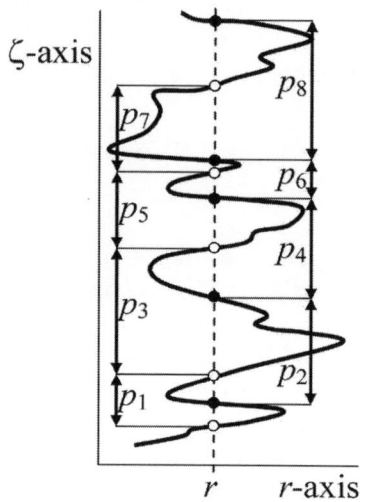

Figure 5.15 Local periods of migration

Assumptions

Let us formulate the assumptions for the concept of ideal equidistant migration. Assumptions from 1 to 5 from Section 5.1 and assumption 7 from Section 5.3 may be valid. However, we introduce the following alternative assumption 6* in place of earlier assumption 6 as follows.

6*. Each fiber path changes its radial position randomly, but in such a manner that the mean number of intersections at radius r is proportional to the fiber volume in the vicinity around this radius.

Basic equation of ideal equidistant migration

Let us think about a long length L of yarn. The cross section of a very thin layer situated at radius r, i.e., the area of a very thin annulus of (medium level) radius r, is $2\pi r \delta r$ (Figure 5.14). The total volume of this layer is $2\pi r \delta r\, L$. The fiber volume inside this layer is

$$\delta V = 2\pi r \delta r\, L \mu_r = (2\pi \delta r\, L) r \mu_r, \tag{5.69}$$

where μ_r is the packing density at the given radius r. [See also Figure 4.3 and Equation (4.3)]

The intersection at radius r by one fiber is (on an average) repeated by half-period $p(r)/2$ (from inside to outside or from outside to inside). Then, the total number of intersections per fiber is

$$\upsilon(r) = \frac{L}{p(r)/2}.\tag{5.70}$$

Note: This equation is identical to Equation (5.5) and also Equation (3.173)[16].

Now, according to assumption 6* and Equations (5.69) and (5.70), we can write

$$\upsilon(r) = C\delta V, \quad \frac{L}{p(r)/2} = C(2\pi\delta r\, L)r\mu_r, \quad p(r) = \frac{2}{C(2\pi\delta r)r\mu_r},$$

$$p(r) = \frac{c}{r\mu_r}, \text{ where } c = \frac{2}{C(2\pi\delta r)}.\tag{5.71}$$

(Parameter C having dimension inverse of cube of length and parameter c having dimension square of length are considered as proportionality constants.)

By using Equations (5.71) and (5.1) in the generally valid fundamental Equation (5.6) we obtain the relation as follows:

$$\tan^2\alpha = \frac{1+\tan^2\beta}{\left[\dfrac{\mu_r}{2NsZ}\dfrac{c}{r\mu_r}\right]^2 \tan^2\beta - 1} = \frac{1+\tan^2\beta}{\left[\dfrac{\pi c}{Ns(2\pi rZ)}\right]^2 \tan^2\beta - 1}$$

$$= \frac{1+\tan^2\beta}{\left[\dfrac{\pi c}{Ns\tan\beta}\right]^2 \tan^2\beta - 1} = \frac{1+\tan^2\beta}{\left[\dfrac{\pi c}{Ns}\right]^2 - 1},$$

$$\tan^2\alpha = \frac{1+\tan^2\beta}{Q^2-1}, \quad \text{(or alternativelly } (Q^2-1)\tan^2\alpha = 1+\tan^2\beta),\tag{5.72}$$

where the common dimensionless parameter is

$$Q = \frac{\pi c}{Ns}.\tag{5.73}$$

Equation (5.72) is the basic equation of equidistant migration. [Compare it with Equations (5.12) and (5.51).]

---

16   Do not forget that the symbol Q in Chapter 3 was renamed as symbol $\upsilon(r)$ in this chapter – see the note following assumption 4 in Section 5.1.

The basic equation of equidistant migration is defined for all radii $r \in (0, D/2)$ and for all parameters $Q > 1$.

Differential equation

Let us use Equations (5.1) and (4.1) in (5.72). Thus, we obtain

$$\tan \alpha = \frac{dr}{d\zeta} = \pm \sqrt{\frac{1 + \tan^2 \beta}{Q^2 - 1}} = \pm \sqrt{\frac{1 + (2\pi r Z)^2}{Q^2 - 1}},$$

$$d\zeta = \pm \sqrt{\frac{Q^2 - 1}{1 + (2\pi r Z)^2}} \, dr. \tag{5.74}$$

Note: The second differential Equation (5.15), i.e., $d\varphi = 2\pi Z \, d\zeta$, is still valid.

Meaning of parameter Q

Because of our assumption 7 (Section 5.3), the packing density $\mu(r) = \mu$ is a constant in ideal equidistant model. Then, Equation (5.13), i.e., $Ns = (1 - \delta)\mu \pi D^2/4$, is fully valid. By applying this in Equation (5.73), we get an alternative expression as follows:

$$Q = \frac{\pi c}{(1 - \delta)\mu \pi D^2/4} = \frac{c}{(1 - \delta)\mu(D/2)^2}. \tag{5.75}$$

Fiber path

Let us solve the differential Equation (5.74) by using Equation (A4.15) from Appendix 4 as follows:

$$\int d\zeta = \pm \int \sqrt{\frac{Q^2 - 1}{1 + (2\pi r Z)^2}} \, dr,$$

$$\frac{\pm \zeta}{\sqrt{Q^2 - 1}} = \int \frac{dr}{\sqrt{1 + (2\pi r Z)^2}} = \frac{1}{2\pi Z} \ln \left| 2\pi r Z + \sqrt{1 + (2\pi r Z)^2} \right| + k,$$

$$\pm 2\pi Z \zeta = \sqrt{Q^2 - 1} \ln \left(\sqrt{1 + (2\pi r Z)^2} + 2\pi r Z \right) + k, \tag{5.76}$$

where k is a constant of integration.

The three schemes shown in Figure 5.16 illustrate the solution according to Equation (5.76). The thin lines are very flat but they are not precisely straight as it is shown from mathematical structure of Equation (5.76). The increasing lines correspond to + sign and the decreasing lines correspond to − sign and the different lines in one graph (creating the above-mentioned scheme) correspond to the different values of the constant k of integration. The thick line in each graph in Figure 5.16 resembles an example of fiber path whose directions from inside to outside and from outside to inside change randomly.

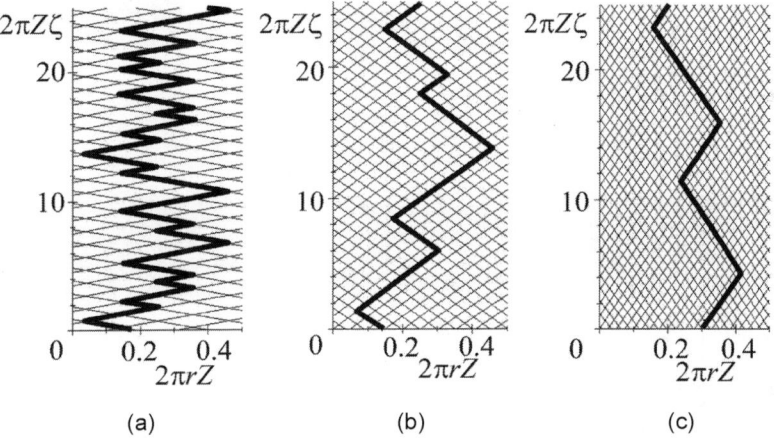

(a)   (b)   (c)

Figure 5.16 Random fiber path. Set of thin lines: Equation (5.76) by
(a) $Q = 5$, (b) $Q = 20$, (c) $Q = 40$.
Thick line: Example of randomly changed fiber path

Idea of equidistance

By applying Equation (5.72) in (5.3), we obtain

$$\frac{d^2r}{d^2l} = \frac{\tan^2\alpha}{\tan^2\alpha + \tan^2\beta + 1} = \frac{\tan^2\alpha}{\tan^2\alpha + \left(Q^2 - 1\right)\tan^2\alpha} = \frac{1}{1 + \left(Q^2 - 1\right)}, \quad \frac{d^2r}{d^2l} = \frac{1}{Q^2}.$$

(5.77)

Thinking about positive elementary increments dr and dl, we can write

$$\frac{dr}{dl} = \left|\frac{\pm 1}{Q}\right| = \frac{1}{Q}, \quad dl = Q\,dr.$$

(5.78)

It is visible that the length dl of each fiber element lying in each differential layer of thickness dr is same, independent of radius r. Otherwise, the length of fiber path equidistantly increases by a constant step of radius r.

Number of intersections in differential layer

One fiber intersects the differential layer at radius r in the length L of yarn $\upsilon(r)$-times. [This quantity is mentioned also by Equation (5.70).] Each fiber element lying inside the differential layer have the length dl and fiber cross-sectional area s so that its volume is $s\,dl$. The number of all such elements from one fiber is $\upsilon(r)s\,dl$ and the total fiber volume in the differential layer from all N fibers [17] is

$$dV = N\upsilon(r)s\,dl .\tag{5.79}$$

The total volume of the above-mentioned differential layer is

$$dV_c = 2\pi r\,dr\,L .\tag{5.80}$$

We obtain the following equation for the packing density in the differential layer:

$$\mu = \frac{dV}{dV_c} = \frac{N\upsilon(r)s\,dl}{2\pi r\,dr\,L} = \frac{N\upsilon(r)s}{2\pi r\,L}\frac{dl}{dr}.\tag{5.81}$$

By applying Equation (5.78) in the last expression, we obtain

$$\mu = \frac{N\upsilon(r)sQ}{2\pi r\,L}, \quad \upsilon(r) = \frac{\mu 2\pi r\,L}{NsQ} .\tag{5.82}$$

Yarn retraction for common parameter Q

We interpret the elementary increments dr and   as positive quantity and then we write Equation (5.74) in the following form:

$$d\zeta = \left| \pm\sqrt{\frac{Q^2-1}{1+(2\pi rZ)^2}} \right| dr = \sqrt{\frac{Q^2-1}{1+(2\pi rZ)^2}}\,dr .\tag{5.83}$$

---

17   Let us remember assumption 4 in Section 5.1 (an ergodic idea).

Each fiber element has an axial height $d\zeta$ in the yarn. The 'sum' (integral) of all elements (integral over all radii) of a given fiber must create the yarn length L; $L = \int_{r=0}^{r=D/2} \upsilon(r)d\zeta$. By using Equations (5.82), (5.83), the relation $NsQ = \pi c$ form Equation (5.73) and Equation (A4.16) from Appendix 4, we obtain the following relation:

$$L = \int_{r=0}^{r=D/2} \upsilon(r)d\zeta = \int_{0}^{D/2} \frac{\mu 2\pi r\, L}{NsQ} \frac{\sqrt{Q^2 - 1}}{\sqrt{1 + (2\pi r Z)^2}}\, dr,$$

$$1 = \frac{\mu 2\pi \sqrt{Q^2 - 1}}{NsQ} \int_{0}^{D/2} \frac{r\, dr}{\sqrt{1 + (2\pi r Z)^2}} = \frac{\mu 2\sqrt{Q^2 - 1}}{c} \int_{0}^{D/2} \frac{r\, dr}{\sqrt{1 + (2\pi r Z)^2}}$$

$$= \frac{\mu 2\sqrt{Q^2 - 1}}{c} \frac{1}{(2\pi Z)^2}\left[\sqrt{1 + (\pi D Z)^2} - 1\right],$$

$$\frac{c}{\mu} = \frac{2\sqrt{Q^2 - 1}}{(2\pi Z)^2}\left[\sqrt{1 + (\pi D Z)^2} - 1\right]. \tag{5.84}$$

Replacing the last expression in Equation (5.75), we finally get the following expression for yarn retraction in case of ideal equidistant model:

$$1 - \delta = \frac{c}{\mu} \frac{4}{Q D^2} = \frac{2\sqrt{Q^2 - 1}}{(2\pi Z)^2}\left(\sqrt{1 + (\pi D Z)^2} - 1\right)\frac{4}{Q D^2},$$

$$\delta = 1 - \frac{\sqrt{Q^2 - 1}}{Q} \frac{2}{(\pi D Z)^2}\left(\sqrt{1 + (\pi D Z)^2} - 1\right). \tag{5.85}$$

If Q is very high, i.e., $Q \to \infty$, then

$$\lim_{Q \to \infty} \delta = 1 - \lim_{Q \to \infty}\left[\sqrt{1 - \frac{1}{Q^2} \frac{2}{(\pi D Z)^2}\left(\sqrt{1 + (\pi D Z)^2} - 1\right)}\right]$$

$$= 1 - \frac{2}{(\pi D Z)^2}\left(\sqrt{1 + (\pi D Z)^2} - 1\right) = \frac{\sqrt{1 + (\pi D Z)^2} - 1}{\sqrt{1 + (\pi D Z)^2} + 1}. \tag{5.86}$$

[See Equation (4.56) for rearrangement.] This result is identical to Braschler's Equation (4.56) for yarn retraction in case of helical model.

If $\pi DZ \to 0$ (untwisted yarn), then according to Equation (5.85) and applying L'Hospital's rule we find

$$\lim_{\pi DZ \to 0} \delta = 1 - \frac{\sqrt{Q^2 - 1}}{Q} \lim_{\pi DZ \to 0} \left[\frac{2}{(\pi DZ)^2} \left(\sqrt{1 + (\pi DZ)^2} - 1 \right) \right]$$

$$= 1 - 2 \frac{\sqrt{Q^2 - 1}}{Q} \lim_{\pi DZ \to 0} \left[\frac{\pi DZ / \sqrt{1 + (\pi DZ)^2}}{2 (\pi DZ)} \right]$$

$$= 1 - \frac{\sqrt{Q^2 - 1}}{Q} \lim_{\pi DZ \to 0} \left[\frac{1}{\sqrt{1 + (\pi DZ)^2}} \right], \quad \lim_{\pi DZ \to 0} \delta = 1 - \frac{\sqrt{Q^2 - 1}}{Q}. \qquad (5.87)$$

Yarn retraction in case of ideal equidistant model according to Equation (5.85) together with special cases of Equations (5.86) and (5.87) are graphically illustrated in Figure 5.17. This graph is partly similar to those shown in Figures 5.6 and 5.11. Some positive values of yarn retraction also exist in case of untwisted yarn, i.e., when $\pi DZ = 0$. It evokes an image of entangled yarn, which is analogous to Figures 5.6 and 5.11.

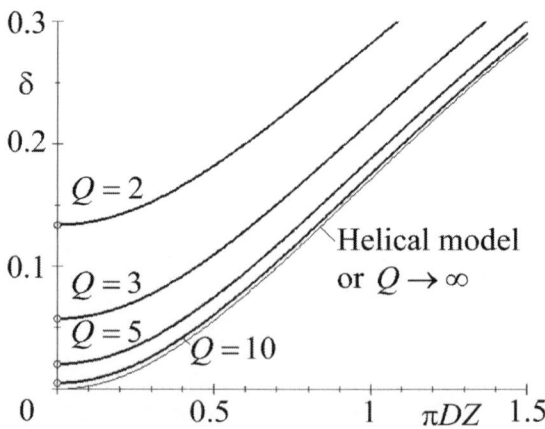

Figure 5.17 Yarn retraction for constant value of Q. Thick lines – Equations (5.85). Thin line – Equations (4.56) and (5.86). Small circular points – Equation (5.87)

Period of migration

The period of migration $p(r)$ changes its value with radius, as we mentioned at the introductory part of this section. Because we assume a constant value of packing density, i.e., $\mu_r = \mu$, Equation (5.71) can be expressed in the form $p(r) = c/(r\mu)$, where the ratio c/μ follows from Equation (5.75). Thus,

$$p(r) = \frac{1}{r}\frac{c}{\mu} = \frac{1}{r}Q(1-\delta)\left(\frac{D}{2}\right)^2,$$

$$\frac{p(r)}{D} = \frac{1}{2}\frac{\pi DZ}{2\pi rZ}Q(1-\delta), \tag{5.88}$$

where yarn retraction δ is given by Equation (5.85). Equation (5.88) takes the following form at yarn surface ($r = D/2$)

$$\frac{p(D/2)}{D} = \frac{1}{2}Q(1-\delta). \tag{5.89}$$

Figure 5.18 illustrates the character of Equations (5.88) and (5.89) for values $Q = 20$ and 40.

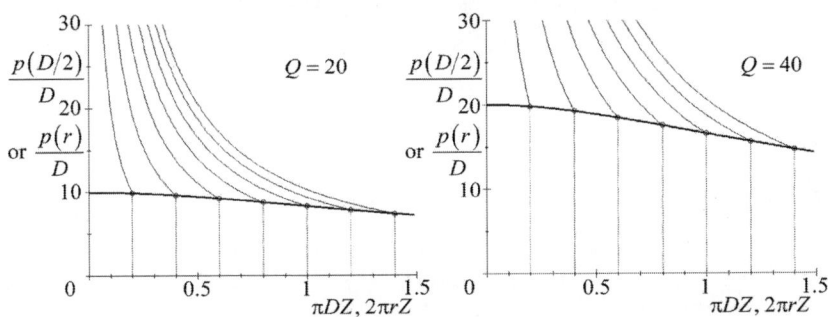

Figure 5.18 Period of migration in ideal equidistant model.

Thick line – $p(D/2)/D$ as the function of πDZ , Equations (5.89) and (5.85).

Thin lines – $p(r)/D$ as the function of $2\pi rZ$, Equations (5.87) and (5.85).

O – the end point $2\pi rZ = \pi DZ$

It is visible that the shortest period $p(D/2)$ [or relative value $p(D/2)/D$] is slightly decreasing with an increase of twist intensity πDZ , according to Equations (5.89) and (5.85). The local period $p(r)$ [or its relative value $p(r)/D$] is infinitely high at the yarn axis and hyperbolically decreases with radius r to the minimum value at yarn surface.

5.6   Approximation of equidistant radial migration model

Assumption

Usually, the relation $\tan^2 \beta \ll 1$ is valid for all radii r in yarns. So, let us assume that $1 + \tan^2 \beta \cong 1$. Then, Equation (5.72) obtains the following form:

$$\tan^2 \alpha = \frac{1}{Q^2 - 1}. \tag{5.90}$$

Differential equation

By applying Equation (5.1) in (5.90), we obtain the differential equation as follows:

$$\frac{\mathrm{d}^2 r}{\mathrm{d}^2 \zeta} = \frac{1}{Q^2 - 1}, \quad \mathrm{d}\zeta = \pm\sqrt{Q^2 - 1}\,\mathrm{d}r. \tag{5.91}$$

Note: The second differential Equation (5.15), i.e., $\mathrm{d}\varphi = 2\pi Z\,\mathrm{d}\zeta$, as well as its solution according to Equation (5.20) is still valid.

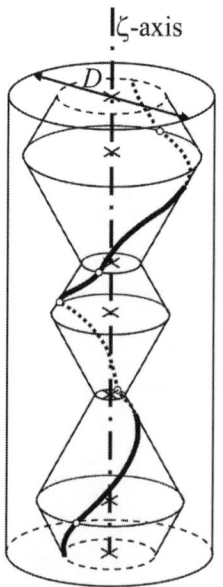

Figure 5.19 Fiber path on a cone surfaces

Fiber path

Let us solve the differential Equation (5.91) as follows:

$$\int d\zeta = \pm \int \sqrt{Q^2 - 1}\, dr, \quad \zeta = \pm\sqrt{Q^2 - 1}\, r + k,$$

$$2\pi Z\zeta = \pm\sqrt{Q^2 - 1}\, 2\pi rZ + k.$$

(5.92)

(k is an integral constant.)

The graphical interpretation of the last equation is very near to that shown in Figure 5.16. Nevertheless, the lines are now quite straight. (They were not so straight in Figure 5.16.) The linear relation according to Equation (5.92) represents conical surfaces on which one fiber is 'wound', as it is shown in Figure 5.19. (Compare it with paraboloids in Figure 5.7.)

Yarn retraction by common parameter Q

Let us use Equations (5.1) and (5.90) in (5.3) as follows:

$$\frac{d^2 l}{d^2 r} = \frac{\tan^2 \alpha + \tan^2 \beta + 1}{\tan^2 \alpha} = 1 + \frac{\tan^2 \beta}{\tan^2 \alpha} + \frac{1}{\tan^2 \alpha}$$

$$= 1 + \left(Q^2 - 1\right)\left(2\pi rZ\right)^2 + \left(Q^2 - 1\right),$$

$$\frac{d^2 l}{d^2 r} = Q^2 + \left(Q^2 - 1\right)\left(2\pi rZ\right)^2.$$

(5.93)

By interpreting the elementary increments dr and dl as positive quantity, we can write Equation (5.93) in the following form:

$$\frac{dl}{dr} = \left|\pm\sqrt{Q^2 + \left(Q^2 - 1\right)\left(2\pi rZ\right)^2}\right| = \sqrt{Q^2 + \left(Q^2 - 1\right)\left(2\pi rZ\right)^2}\,.$$

(5.94)

Equations (5.79) to (5.81) are also valid in this case. By using Equation (5.94) in (5.81), we can express the packing density μ and then the number of intersections $\upsilon(r)$ – in the differential layer at radius r in the yarn length L per fiber – as follows:

$$\mu = \frac{N\upsilon(r)s}{2\pi r\, L}\frac{dl}{dr} = \frac{N\upsilon(r)s}{2\pi r\, L}\sqrt{Q^2 + \left(Q^2 - 1\right)\left(2\pi rZ\right)^2}\,,$$

$$\upsilon(r) = \frac{\mu 2\pi r\, L}{Ns}\frac{1}{\sqrt{Q^2 + \left(Q^2 - 1\right)\left(2\pi rZ\right)^2}}\,.$$

(5.95)

Again, by interpreting the elementary increments dr and $d\zeta$ as positive quantity, we can write Equation (5.91) in the form $d\zeta = \left|\pm\sqrt{Q^2-1}\right|dr = \sqrt{Q^2-1}\,dr$. By using this expression and Equation (5.95), we derive an expression, which is analogical to the first expression of Equation (5.84). By applying Equation (5.13) in the form $Ns/(\pi\mu) = (1-\delta)D^2/4$ and Equation (A4.17) from Appendix 4, we can write

$$L = \int_{r=0}^{r=D/2} \upsilon(r)\,d\zeta = \int_0^{D/2} \frac{\mu 2\pi r\,L}{Ns} \frac{\sqrt{Q^2-1}}{\sqrt{Q^2+\left(Q^2-1\right)\left(2\pi r Z\right)^2}}\,dr,$$

$$1 = \frac{\mu 2\pi}{Ns} \int_0^{D/2} \frac{r\,dr}{\sqrt{\dfrac{Q^2}{Q^2-1}+\left(2\pi r Z\right)^2}} = \frac{\mu 2\pi}{Ns} \frac{1}{\left(2\pi Z\right)^2}\left[\sqrt{\frac{Q^2}{Q^2-1}+\left(\pi DZ\right)^2}-\sqrt{\frac{Q^2}{Q^2-1}}\right]$$

$$= \frac{\mu 2\pi}{Ns} \frac{1}{\left(2\pi Z\right)^2} \sqrt{\frac{Q^2}{Q^2-1}}\left[\sqrt{1+\frac{Q^2-1}{Q^2}\left(\pi DZ\right)^2}-1\right],$$

$$1 = \frac{2}{(1-\delta)\left(\pi DZ\right)^2} \sqrt{\frac{Q^2}{Q^2-1}}\left[\sqrt{1+\frac{Q^2-1}{Q^2}\left(\pi DZ\right)^2}-1\right].$$

$$(5.96)$$

We express yarn retraction δ from the last equation. So, we finally obtain the following relation:

$$\delta = 1 - \frac{2}{\left(\pi DZ\right)^2} \sqrt{\frac{Q^2}{Q^2-1}}\left[\sqrt{1+\frac{Q^2-1}{Q^2}\left(\pi DZ\right)^2}-1\right]. \qquad (5.97)$$

If $Q \to \infty$, then

$$\lim_{Q\to\infty}\delta = 1 - \frac{2}{\left(\pi DZ\right)^2} \lim_{Q\to\infty} \sqrt{\frac{Q^2}{Q^2-1}}\left[\sqrt{1+\frac{Q^2-1}{Q^2}\left(\pi DZ\right)^2}-1\right]$$

$$= 1 - \frac{2}{\left(\pi DZ\right)^2}\left[\sqrt{1+\left(\pi DZ\right)^2}-1\right]. \qquad (5.98)$$

This is equal to Equation (5.86), i.e., identical to Braschler's Equation (4.56) for helical model.

If $\pi DZ \to 0$ (untwisted yarn), then according to Equation (5.97) and applying L'Hospital's rule we find

$$\lim_{\pi DZ \to 0} \delta = 1 - \lim_{\pi DZ \to 0} \left\{ \frac{2}{(\pi DZ)^2} \sqrt{\frac{Q^2}{Q^2 - 1}} \left[\sqrt{1 + \frac{Q^2 - 1}{Q^2}(\pi DZ)^2} - 1 \right] \right\}$$

$$= 1 - 2\sqrt{\frac{Q^2}{Q^2 - 1}} \lim_{\pi DZ \to 0} \frac{\dfrac{Q^2 - 1}{Q^2} 2\pi DZ \bigg/ \left[2\sqrt{1 + \dfrac{Q^2 - 1}{Q^2}(\pi DZ)^2} \right]}{2\pi DZ}$$

$$= 1 - \sqrt{\frac{Q^2}{Q^2 - 1}} \lim_{\pi DZ \to 0} \frac{\dfrac{Q^2 - 1}{Q^2}}{\sqrt{1 + \dfrac{Q^2 - 1}{Q^2}(\pi DZ)^2}}, \qquad \lim_{\pi DZ \to 0} \delta = 1 - \sqrt{\frac{Q^2 - 1}{Q^2}}.$$

$$(5.99)$$

This expression is identical to Equation (5.87).

The behaviours of Equations (5.97), (5.98) and (5.99) are illustrated in Figure 5.20. It seems that this graph is very similar to the one shown in Figure 5.17.

Period of migration

Equation (5.70) is valid always and Equation (5.95) is valid too. From the equivalency of the right-hand sides of these two equations and using Equation (5.13) ($Ns/(\pi\mu) = (1 - \delta)D^2/4$), we obtain the expression for the local period of migration as follows:

$$\frac{L}{p(r)/2} = \frac{\mu 2\pi r L}{Ns} \frac{1}{\sqrt{Q^2 + (Q^2 - 1)(2\pi rZ)^2}},$$

$$p(r) = \frac{Ns}{\mu\pi} \frac{1}{r} \sqrt{Q^2 + (Q^2 - 1)(2\pi rZ)^2}$$

$$= (1 - \delta)\frac{D^2}{4} \frac{1}{r} \sqrt{Q^2 + (Q^2 - 1)(2\pi rZ)^2},$$

$$\frac{p(r)}{D} = \frac{1}{2}(1 - \delta)\frac{\pi DZ}{2\pi rZ} \sqrt{Q^2 + (Q^2 - 1)(2\pi rZ)^2}. \qquad (5.100)$$

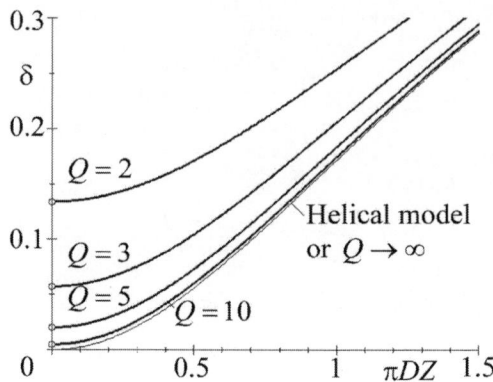

Figure 5.20 Yarn retraction by constant value of Q. Thick lines –
Equations (5.97). Thin line – Equations (4.56) and (5.98).
Small circular points – Equation (5.99)

This equation obtains a special shape at yarn surface, i.e., when $r = D/2$ such that

$$\frac{p(D/2)}{D} = \frac{1}{2}(1-\delta)\sqrt{Q^2 + (Q^2 - 1)(\pi DZ)^2} .$$ (5.101)

[Yarn retraction shown by Equations (5.100) and (5.101) is determined according to Equation (5.97) in the last two equations.]

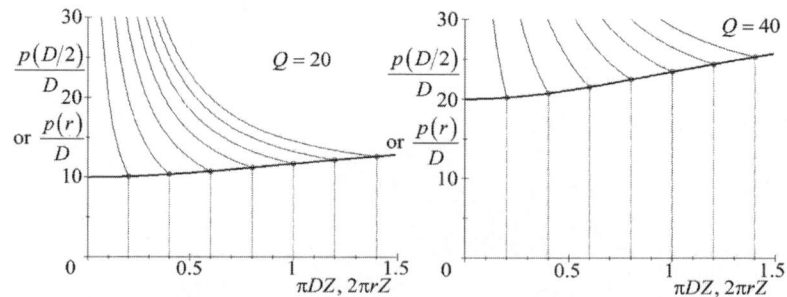

Figure 5.21 Period of migration in case of approximation of equidistant model.
Thick line – $p(D/2)/D$ as the function of πDZ , Equations (5.101) and (5.97).
Thin lines – $p(r)/D$ as the function of $2\pi rZ$, Equations (5.100) and (5.97).
$\bigcirc$ – the end point $2\pi rZ = \pi DZ$

Figure 5.21 illustrates the character of Equations (5.100) and (5.101) at $Q = 20$ and $Q = 40$, this is same as we did in case of Figure 5.18. A comparison of these two figures reveals that the periods of migration change to some degree as a result of approximation. In opposite to the previous case (Figure 5.17), the shortest period $p(D/2)$ [or relative value $p(D/2)/D$] is slightly increasing with twist intensity πDZ. Nevertheless, the local period $p(r)$ [or its relative value $p(r)/D$] is infinitely high along the yarn axis and hyperbolically decreases with the increase in radius r to the minimum value at yarn surface.

Experimental verification

To verify the theoretical model, the internal structure of a set of carded ring and rotor yarns with a wide range of count and twist produced mainly from 100% viscose staple fibers and a few from cotton fibers was studied.

Note: Read the comments stated at the time of discussion of Figure 5.13.

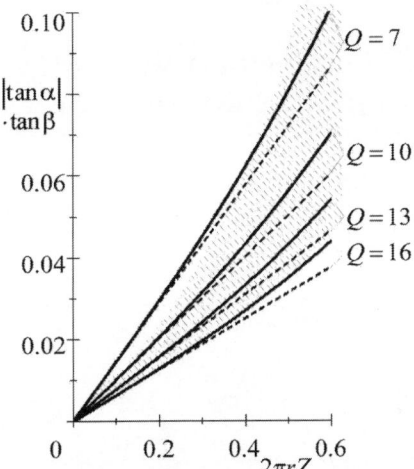

Figure 5.22 Comparison of experimental and model trends.
Hatched – experimental trends, Thick lines – initial model, Equation (5.102),
Dashed lines – approx., Equation (5.103)

The same hatched experimental trends are also shown in Figure 5.22; however, the theoretical results illustrate the directional character of equidistant migration. The following expression is then valid to write from Equations (5.72) and (5.90) by using Equation (4.1):

$$|\tan\alpha|\tan\beta = \sqrt{\frac{1+\tan^2\beta}{(Q^2-1)}}\,\tan\beta = \sqrt{\frac{1+(2\pi rZ)^2}{(Q^2-1)}}\,(2\pi rZ), \qquad (5.102)$$

$$|\tan\alpha|\tan\beta = \frac{\tan\beta}{\sqrt{Q^2-1}} = \frac{2\pi rZ}{\sqrt{Q^2-1}}.$$ (5.103)

We observe that – in comparison with Treloar's model – the trends obtained from the equidistant model are in acceptable accordance with the experimental results.

Local packing density

The approximation of equidistant model is carried out with an assumption of constant local packing density; this is shown in Equation (5.95). On the contrary, we discarded the assumption about the constant value of period of migration – see, e.g., Figure 5.21.

5.7   Characteristics of migration proposed by Hearle and coworkers

Characteristics of Treloar's approximated model

Hearle et al. [6–8] worked on Treloar's approximated model of radial migration, described in Section 5.4, and proposed the following characteristics of migration.

Let us rearrange the differential Equation (5.52) with K' form Equation (5.50) as follows:

$$r\,dr\frac{2}{\left(D/2\right)^2} = \pm\frac{d\zeta}{K'2\pi Z}\frac{2}{\left(D/2\right)^2} = \pm\frac{d\zeta}{\dfrac{2p}{D}\dfrac{1}{\pi DZ}2\pi Z}\frac{2}{\left(D/2\right)^2} = \pm\frac{d\zeta}{p/2}.$$ (5.104)

By integrating the last differential equation, we find the following expression:

$$\frac{r^2}{2}\frac{2}{\left(D/2\right)^2} = \pm\frac{\zeta}{p/2}+c, \quad \frac{r^2}{\left(D/2\right)^2} = \pm\frac{\zeta}{p/2}+c,$$ (5.105)

where c is an integral constant, the positive sign is related to the increasing part (from inside to outside), and the negative sign is related to the decreasing part (from outside to inside); see also Figures 5.7 and 5.8. The ratio $r/(D/2)$ characterizes the 'radial position' of fiber inside the yarn.

Hearle et al. [6, 7] introduced the 'quadratic radial position' as a dimensionless quantity Y as follows:

$$Y = \frac{r^2}{\left(D/2\right)^2}.$$ (5.106)

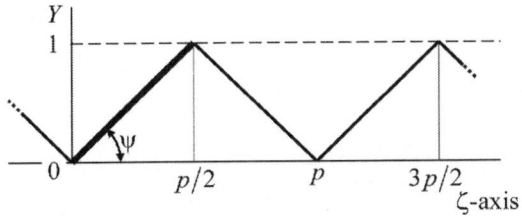

Figure 5.23 $\zeta - Y$ graph according to Equation (5.107) ('complete migration')

By using Equation (5.106), Equation (5.105) can be written as follows:

$$Y = \pm\zeta/(p/2) + c \,. \tag{5.107}$$

The behaviour of Equation (5.107) is illustrated graphically in Figure 5.23. [We used the starting point according to Equation (5.48).]

The following expression is valid for the first (thick) fiber portion shown in Figure 5.23:

$$Y = \zeta/(p/2) \,. \tag{5.108}$$

Note: The following expressions are valid for the other fiber portions: for second portion $Y = -\zeta/(p/2) + 2$, for third portion: $Y = +\zeta/(p/2) - 2$, for fourth portion $Y = -\zeta/(p/2) + 4$, for fifth portion $Y = +\zeta/(p/2) - 4$, etc.

Let us note that all these equations correspond to the general expression stated in Equation (5.107).

The first (thick) line representing the fiber portion is sufficient for determination of three characteristics: fiber mean position, root mean square (RMS) deviation and Intensity of migration.

Fiber mean position

This quantity $\bar{Y}$ expresses the mean value of Y. So, it is valid to write that

$$\bar{Y} = \int\limits_0^{p/2} Y \, d\zeta \bigg/ \int\limits_0^{p/2} d\zeta. \tag{5.109}$$

By using Equation (5.5.108) in (5.109), we find the value of fiber mean position according to Treloar's approximated model of radial migration as follows:

$$\bar{Y} = \dfrac{\displaystyle\int\limits_0^{p/2} Y \, d\zeta}{\displaystyle\int\limits_0^{p/2} d\zeta} = \dfrac{\displaystyle\int\limits_0^{p/2} \dfrac{\zeta}{p/2} \, d\zeta}{\displaystyle\int\limits_0^{p/2} d\zeta} = \dfrac{\dfrac{1}{p/2}\left[\dfrac{\zeta^2}{2}\right]_0^{p/2}}{[\zeta]_0^{p/2}} = \dfrac{\dfrac{1}{p/2}\dfrac{(p/2)^2}{2}}{p/2} = \dfrac{1}{2} \,. \tag{5.110}$$

RMS deviation

The quantity D_{RMS} expresses the root mean square deviation of Y. So, it is valid to write that

$$D_{\text{RMS}} = \sqrt{\int_0^{p/2} \left(Y - \bar{Y}\right)^2 \, d\zeta \Big/ \int_0^{p/2} d\zeta}. \tag{5.111}$$

By using Equations (5.108) and (5.110) in (5.111), we find the following expression for the RMS deviation according to Treloar's approximated model of radial migration:

$$D_{\text{RMS}} = \sqrt{\int_0^{p/2} \left(\frac{\zeta}{p/2} - \frac{1}{2}\right)^2 d\zeta \Big/ \int_0^{p/2} d\zeta} = \sqrt{\int_0^{p/2} \left[\left(\frac{2\zeta}{p}\right)^2 - \frac{2\zeta}{p} + \frac{1}{4}\right] d\zeta \Big/ [\zeta]_0^{p/2}}$$

$$= \sqrt{\int_0^{p/2} \left[\frac{4}{p^2}\zeta^2 - \frac{2}{p}\zeta + \frac{1}{4}\right] d\zeta \Big/ (p/2)} = \sqrt{\int_0^{p/2} \left[\frac{8}{p^3}\zeta^2 - \frac{4}{p^2}\zeta + \frac{2}{4p}\right] d\zeta}$$

$$= \sqrt{\frac{8}{p^3}\left[\frac{\zeta^3}{3}\right]_0^{p/2} - \frac{4}{p^2}\left[\frac{\zeta^2}{2}\right]_0^{p/2} + \frac{2}{4p}[\zeta]_0^{p/2}}$$

$$= \sqrt{\frac{8}{p^3}\frac{(p/2)^3}{3} - \frac{4}{p^2}\frac{(p/2)^2}{2} + \frac{2}{4p}(p/2)}$$

$$= \sqrt{\frac{1}{3} - \frac{1}{2} + \frac{1}{4}} = \frac{1}{2\sqrt{3}} = 0.289. \tag{5.112}$$

Intensity of migration

This quantity J represents the square root of mean square of tangent of angle ψ shown in Figure 5.22. So, it is valid to write that

$$J = \sqrt{\int_0^{p/2} \left(\frac{dY}{d\zeta}\right)^2 d\zeta \Big/ \int_0^{p/2} d\zeta}. \tag{5.113}$$

It is evidently valid from Equation (5.108) that $dY/d\zeta = 2/p$. Thus

$$J = \sqrt{\left(\frac{2}{p}\right)^2 \int_0^{p/2} d\zeta \Big/ \int_0^{p/2} d\zeta} = \frac{2}{p}. \tag{5.114}$$

Incomplete migration

The major part of the experimental results led to significantly other values of fiber mean position and RMS deviation than they were predicted by Equations (5.110) and (5.112). Therefore, Hearle et al. [6, 7] introduced the idea of 'incomplete migration', illustrated in Figure 5.24. They imagined an ideal fiber moving linearly and periodically only between a couple of values $Y_{min} > 0$ and $Y_{max} < 1$. [Equation (5.106) generally defines the quantity Y.]

The individual part of (thick) the function shown in Figure 5.24 corresponds to the following general equation:

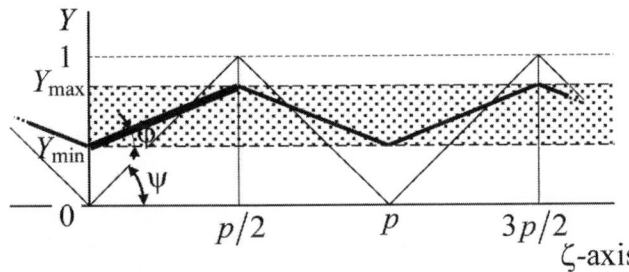

Figure 5.24 $\zeta - Y$ graphs. Thin line – 'complete migration', see Figure 5.23. Thick line – 'incomplete migration', Equation (5.115)

$$Y = \pm\left(Y_{max} - Y_{min}\right)\frac{\zeta}{p/2} + Y_{min} + c. \tag{5.115}$$

(c is a suitable parameter of actual part.) The following expression is valid for the first (thickest) fiber portion:

$$Y = \left(Y_{max} - Y_{min}\right)\frac{\zeta}{p/2} + Y_{min}. \tag{5.116}$$

Note: The following expressions are valid for other fiber portions:

(a) for second portion:
$$Y = -\left(Y_{max} - Y_{min}\right)\zeta/\left(p/2\right) + Y_{min} + 2\left(Y_{max} - Y_{min}\right),$$

(b) for third portion: $Y = \left(Y_{max} - Y_{min}\right)\zeta/\left(p/2\right) + Y_{min} - 2\left(Y_{max} - Y_{min}\right),$

(c) for fourth portion: $Y = -\left(Y_{max} - Y_{min}\right)\zeta/\left(p/2\right) + Y_{min} + 4\left(Y_{max} - Y_{min}\right),$
etc.

Let us note that all these equations correspond to the general Equation (5.115).

The first (thickest) part is enough for determination of the three charac-teristics of incomplete migration proposed by Hearle et al. [6, 7].

The fiber mean position follows the following expression obtained by using Equations (5.109) and (5.116):

$$
\bar{Y} = \frac{\displaystyle\int_0^{p/2}\left[\left(Y_{\max} - Y_{\min}\right)\frac{\zeta}{p/2} + Y_{\min}\right]d\zeta}{\displaystyle\int_0^{p/2} d\zeta}
$$

$$
= \frac{\dfrac{Y_{\max} - Y_{\min}}{p/2}\dfrac{(p/2)^2}{2} + Y_{\min}\, p/2}{p/2} = \frac{Y_{\max} - Y_{\min}}{2} + Y_{\min}. \tag{5.117}
$$

We can obtain the following expression from Equations (5.116) and (5.117):

$$
\left[Y - \bar{Y}\right]^2 = \left[\left(Y_{\max} - Y_{\min}\right)\frac{\zeta}{p/2} + Y_{\min} - \frac{Y_{\max} - Y_{\min}}{2} - Y_{\min}\right]^2
$$

$$
= \left(Y_{\max} - Y_{\min}\right)^2\left[\frac{\zeta}{p/2} - \frac{1}{2}\right]^2 = \left(Y_{\max} - Y_{\min}\right)^2\left[\frac{\zeta^2}{(p/2)^2} - \frac{\zeta}{p/2} + \frac{1}{4}\right]. \tag{5.118}
$$

So, RMS deviation is expressed by using Equations (5.111) and (5.118) as follows:

$$
D_{\mathrm{RMS}} = \sqrt{\frac{\displaystyle\int_0^{p/2}\left(Y_{\max} - Y_{\min}\right)^2\left[\frac{\zeta^2}{(p/2)^2} - \frac{\zeta}{p/2} + \frac{1}{4}\right]d\zeta}{\displaystyle\int_0^{p/2} d\zeta}},
$$

$$
D_{\mathrm{RMS}} = \left(Y_{\max} - Y_{\min}\right)\sqrt{\frac{\dfrac{(p/2)^3}{3(p/2)} - \dfrac{(p/2)^2}{2(p/2)} + \dfrac{(p/2)}{4}}{p/2}} = \sqrt{\frac{1}{3} - \frac{1}{2} + \frac{1}{4}} = \frac{Y_{\max} - Y_{\min}}{2\sqrt{3}}. \tag{5.119}
$$

Finally, the derivative of Y is $dY/d\zeta = \tan\varphi = \left(Y_{\max} - Y_{\min}\right)/(p/2)$, according to Equation (5.116) and Figure 5.24, so that the intensity of migration can be expressed from Equation (5.113) as follows:

$$J = \sqrt{\frac{\int_0^{p/2}\left(\frac{dY}{d\zeta}\right)^2 d\zeta}{\int_0^{p/2} d\zeta}} = \sqrt{\frac{\left[\frac{Y_{max}-Y_{min}}{p/2}\right]^2 \int_0^{p/2} d\zeta}{p/2}} = \frac{Y_{max}-Y_{min}}{p/2}. \qquad (5.120)$$

If we know the characteristics $\overline{Y}, D_{RMS}, J$ experimentally[18], then it is possible to determine the parameters Y_{min}, Y_{max}, p. From Equations (5.117) and (5.119), we can easily find that

$$Y_{min} = \overline{Y} - \sqrt{3}D_{RMS}, \qquad Y_{max} = \overline{Y} + \sqrt{3}D_{RMS}, \qquad (5.121)$$

and from Equations (5.119) and (5.120), we obtain the (equivalent) period of migration as follows:

$$p = 4\sqrt{3}\, D_{RMS}/J. \qquad (5.122)$$

Note: Let us note that Hearle's idea of incomplete migration is very artificial, very far away from a logical image of internal yarn geometry. It is because above all the fibers must pass also through small radii $[Y = r^2/(D/2)^2 < Y_{min}]$ as well as through high radii $[Y = r^2/(D/2)^2 > Y_{max}]$.

Characteristics of approximated equidistant model

We cannot think about the yarn length equal to a half-period, as shown in previous paragraph, because the local periods are different at different radii in the case of approximated equidistant model. So, let us now think about a general yarn length L.

A fiber intersects a general radius r with a frequency $\upsilon(r)$ on yarn length L. This quantity is described in Equation (5.95). Nevertheless, it is also valid from Equation (5.13) that $Ns/(\pi\mu) = (1-\delta)D^2/4$. So we can express Equation (5.95) as follows:

---

18 See References [3, 8, 9] for one possible method of experimental determination of characteristics of migration, this is based on the so-called 'tracer fiber technique'.

$$\upsilon(r) = \frac{8L}{(1-\delta)D^2} \frac{r}{\sqrt{Q^2 + (Q^2-1)(2\pi r Z)^2}}$$

$$= \frac{8L}{(1-\delta)D^2\sqrt{Q^2-1}} \frac{r}{\sqrt{\dfrac{Q^2}{Q^2-1} + (2\pi r Z)^2}}, \tag{5.123}$$

where yarn retraction δ, according to the approximated equidistant model, is described in Equation (5.97).

Mean radial position $\bar{y}$

Similar to Hearle's definition (5.106), let us define the dimensionless radial position y and also mean radial position $\bar{y}$ as follows:

$$y = \frac{r}{D/2}, \quad \bar{y} = \frac{\bar{r}}{D/2}, \tag{5.124}$$

where $\bar{r}$ denotes the mean radius of fiber elements in yarn[19]. It is valid to write that

$$\bar{r} = \frac{\displaystyle\int_{r=0}^{r=D/2} r\,\upsilon(r)\,d\zeta}{\displaystyle\int_{r=0}^{r=D/2} \upsilon(r)\,d\zeta}. \tag{5.125}$$

The denominator of the last equation is evidently equal to yarn length L, as it is written in the first equivalency of Equation (5.96). The numerator can be rearranged to the following form: $\int_0^{D/2} r\,\upsilon(r)\,|d\zeta/dr|\,dr$.

Note: Because we think about positive increments $d\zeta$ and dr, we used the absolute value in the previous expression.

We substitute Equations (5.123), (5.91) and Equation (A4.20) from Appendix 4 to Equation (5.125) and obtain the mean radius as follows:

$$\bar{r} = \frac{1}{L}\int_0^{D/2} r\,\upsilon(r)\left|\frac{d\zeta}{dr}\right|dr = \frac{1}{L}\int_0^{D/2} r\,\upsilon(r)\sqrt{Q^2-1}\,dr, \tag{5.126}$$

---

19 Let us note that the relations described in Equation (5.124) are linear, not quadratic as described in Equation (5.109).

$$\bar{r} = \frac{1}{L}\int_0^{D/2} r\,\frac{8L}{(1-\delta)D^2\sqrt{Q^2-1}}\,\frac{r}{\sqrt{\dfrac{Q^2}{Q^2-1}+(2\pi r Z)^2}}\sqrt{Q^2-1}\,dr,$$

$$\bar{r} = \frac{2}{(1-\delta)(\pi DZ)^2}\int_0^{D/2}\frac{(2\pi r Z)^2\,dr}{\sqrt{\dfrac{Q^2}{Q^2-1}+(2\pi r Z)^2}}$$

$$= \frac{2}{(1-\delta)(\pi DZ)^2}\cdot\frac{1}{2\pi Z}\frac{1}{2}\sqrt{\frac{Q^2}{Q^2-1}}(\pi DZ)\left[\sqrt{1+x^2}-\frac{1}{x}\ln\left(x+\sqrt{1+x^2}\right)\right],$$

$$\bar{y} = \frac{\bar{r}}{D/2} = \frac{1}{(1-\delta)(\pi DZ)^2}\sqrt{\frac{Q^2}{Q^2-1}}\left[\sqrt{1+x^2}-\frac{1}{x}\ln\left(x+\sqrt{1+x^2}\right)\right],$$

$$\text{where } x = \sqrt{\frac{Q^2-1}{Q^2}(\pi DZ)^2}.$$

$$(5.127)$$

Equation (5.97) can be written in the following form:

$$1-\delta = \frac{2}{(\pi DZ)^2}\sqrt{\frac{Q^2}{Q^2-1}}\left[\sqrt{1+x^2}-1\right],\ \text{where } x = \sqrt{\frac{Q^2-1}{Q^2}(\pi DZ)^2}.\ (5.128)$$

We apply Equation (5.128) in (5.127) to obtain the final expression for $\bar{y}$ as follows:

$$\bar{y} = \frac{\bar{r}}{D/2} = \frac{\dfrac{1}{(\pi DZ)^2}\sqrt{\dfrac{Q^2}{Q^2-1}}\left[\sqrt{1+x^2}-\dfrac{1}{x}\ln\left(x+\sqrt{1+x^2}\right)\right]}{\dfrac{2}{(\pi DZ)^2}\sqrt{\dfrac{Q^2}{Q^2-1}}\left[\sqrt{1+x^2}-1\right]},$$

$$\bar{y} = \frac{\bar{r}}{D/2} = \frac{1}{2}\frac{\sqrt{1+x^2}-\dfrac{1}{x}\ln\left(x+\sqrt{1+x^2}\right)}{\sqrt{1+x^2}-1},\ \text{where } x = \sqrt{\frac{Q^2-1}{Q^2}(\pi DZ)^2}.$$

$$(5.129)$$

Figure 5.25 presents a graphical interpretation of the mean radial position $\bar{y}$.

Let us express a common starting point of all curves, i.e., the point at $\pi DZ = 0$ now. The easiest way for solving this case consists of using Equation (5.123) such that it relates to the untwisted yarn ($Z \to 0$). By applying Equation (5.87) to (5.123), we obtain

$$\lim_{Z \to 0} \upsilon(r) = \lim_{Z \to 0} \left[\frac{8L}{(1-\delta)D^2 \sqrt{Q^2-1}} \frac{r}{\sqrt{\dfrac{Q^2}{Q^2-1} + (2\pi r Z)^2}} \right] = \frac{2L}{\left(1 - \lim\limits_{Z \to 0}\delta\right)(D/2)^2} \frac{r}{Q}$$

$$= \frac{2L}{\dfrac{\sqrt{Q^2-1}}{Q}(D/2)^2} \frac{r}{Q} = 2L \frac{r}{\sqrt{Q^2-1}(D/2)^2}.$$

(5.130)

By applying this in Equation (5.126), we can write

$$\lim_{Z \to 0} \bar{r} = \frac{1}{L} \lim_{Z \to 0} \int_0^{D/2} r \lim_{Z \to 0}\left[\upsilon(r)\right] \sqrt{Q^2-1}\, dr$$

$$= \frac{1}{L} \int_0^{D/2} r \cdot 2L \frac{r}{\sqrt{Q^2-1}(D/2)^2} \cdot \sqrt{Q^2-1}\, dr$$

$$= \frac{2}{(D/2)^2} \int_0^{D/2} r^2\, dr = \frac{2}{(D/2)^2}\frac{1}{3}(D/2)^3, \quad \lim_{Z \to 0} \bar{y} = \frac{\lim\limits_{Z \to 0}\bar{r}}{D/2} = \frac{2}{3}. \quad (5.131)$$

We also observe that the relation shows a slightly decreasing trend and is almost independent for each reasonable values of Q in Figure 5.25. (A constant value is obtained only when $Q \to 1$.) The reason for such 'decreasing effect' is evident from Figure 5.26.

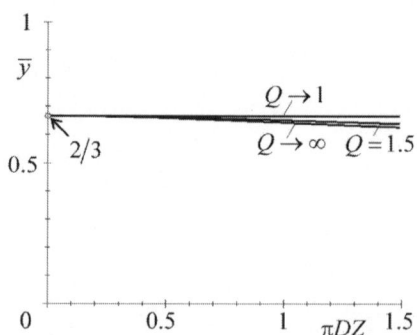

Figure 5.25 Mean radial position according to Equation (5.129)

That is, we assume a constant packing density in each differential layer (see assumption 7 in Section 5.3, also used in the starting part of Section 5.5).

However, we analyzed the radial position of fiber elements only in relation to ζ-coordinate – see Equation (5.125), i.e., without influence of angle β from yarn twisting.

The angle $\tan\alpha = \mathrm{d}r/\mathrm{d}\zeta$ is constant for all fiber elements in this model, but the lengths of fiber elements in the differential layers are not same. These elementary lengths UV are smaller at smaller radii (they have smaller angle β) and higher at higher radii (having higher value of angle β); it is shown in Figure 5.26. So, each fiber element UV at higher radius brings more fiber volume to its differential layer than a fiber element at smaller radius. In consequence of this image, the fiber does not need to 'visit' the peripheral layers in the twisted yarn (high value of πDZ) so often than in the case of untwisted yarn ($\pi DZ \to 0$). Therefore, the mean radial position $\bar{y}$ must decrease with an increase of πDZ.

Mean square value of radial position

In analogy to Equation (5.125) the mean square value of radius r, i.e., $\overline{r^2}$, is given by

$$\overline{r^2} = \int_{r=0}^{r=D/2} r^2 \upsilon(r)\,\mathrm{d}\zeta \Big/ \int_{r=0}^{r=D/2} \upsilon(r)\,\mathrm{d}\zeta. \tag{5.132}$$

The denominator is always equal to yarn length L, according to the first equivalency of Equation (5.96). The numerator can be expressed as $\int_0^{D/2} r^2 \upsilon(r)|\mathrm{d}\zeta/\mathrm{d}r|\,\mathrm{d}r$.

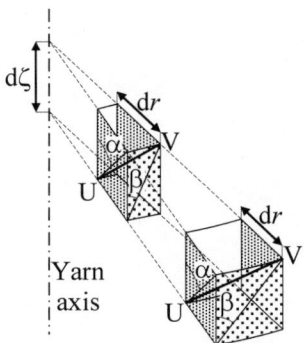

Figure 5.26 Two fiber elements at different radii

Note: Because we think about the positive increments dζ and dr, we used the absolute value in the previous expression.

We apply the aforesaid expressions and Equations (5.123), (5.91) and (A4.21) from Appendix 4 to obtain the following expression:

$$\overline{r^2} = \frac{1}{L} \int_0^{D/2} r^2 \upsilon(r) \left| \frac{d\zeta}{dr} \right| dr$$

$$= \frac{1}{L} \int_0^{D/2} r^2 \frac{8L}{(1-\delta)D^2 \sqrt{Q^2-1}} \frac{r}{\sqrt{\dfrac{Q^2}{Q^2-1} + (2\pi r Z)^2}} \sqrt{Q^2-1}\, dr$$

$$= \frac{D}{(1-\delta)(D/2)^3} \int_0^{D/2} \frac{r^3}{\sqrt{\dfrac{Q^2}{Q^2-1} + (2\pi r Z)^2}}\, dr$$

$$= \frac{D}{(1-\delta)(\pi D Z)^3} \int_0^{D/2} \frac{(2\pi r Z)^3}{\sqrt{\dfrac{Q^2}{Q^2-1} + (2\pi r Z)^2}}\, dr$$

$$= \frac{D}{(1-\delta)(\pi D Z)^3} \frac{1}{2\pi Z} \left(\frac{Q^2}{Q^2-1} \right)^{3/2} \left[\frac{1}{3}(1+x^2)^{3/2} - \sqrt{1+x^2} + \frac{2}{3} \right],$$

$$\overline{y^2} = \overline{\left(\frac{r}{D/2} \right)^2} = \frac{2}{(1-\delta)(\pi D Z)^4} \left(\frac{Q^2}{Q^2-1} \right)^{3/2} \left[\frac{1}{3}(1+x^2)^{3/2} - \sqrt{1+x^2} + \frac{2}{3} \right],$$

where $x = \sqrt{\dfrac{Q^2-1}{Q^2}}(\pi D Z)^2$.

(5.133)

Further, by using Equation (5.128), it is valid to write that

$$\overline{y^2} = \overline{\left(\frac{r}{D/2} \right)^2} = \frac{\dfrac{2}{(\pi D Z)^4} \left(\dfrac{Q^2}{Q^2-1} \right)^{3/2} \left[\dfrac{1}{3}(1+x^2)^{3/2} - \sqrt{1+x^2} + \dfrac{2}{3} \right]}{\dfrac{2}{(\pi D Z)^2} \sqrt{\dfrac{Q^2}{Q^2-1}} \left[\sqrt{1+x^2} - 1 \right]}$$

$$= \frac{\dfrac{Q^2-1}{Q^2}(\pi D Z)^2 \left[\dfrac{1}{3}(1+x^2)^{3/2} - \sqrt{1+x^2} + \dfrac{2}{3} \right]}{\left[\sqrt{1+x^2} - 1 \right]},$$

$$\overline{y^2} = \overline{\left(\frac{r}{D/2}\right)^2} = \frac{\dfrac{1}{x^2}\left[\dfrac{1}{3}\left(1+x^2\right)^{3/2} - \sqrt{1+x^2} + \dfrac{2}{3}\right]}{\left[\sqrt{1+x^2} - 1\right]}, \quad \text{where } x = \sqrt{\frac{Q^2-1}{Q^2}}\left(\pi DZ\right)^2.$$

$$(5.134)$$

Let us express the starting point curves, i.e., the point at $\pi DZ = 0$, now. The easiest way for solving this consists of using Equations (5.130) and (5.91) in (5.132). This is shown as follows:

$$\lim_{Z\to 0}\overline{r^2} = \frac{\displaystyle\lim_{Z\to 0}\int_{r=0}^{r=D/2} r^2 \upsilon(r)\,d\zeta}{\displaystyle\int_{r=0}^{r=D/2} \upsilon(r)\,d\zeta}$$

$$= \frac{1}{L}\int_{r=0}^{r=D/2} r^2 \lim_{Z\to 0}\upsilon(r)\left|\frac{d\zeta}{dr}\right|dr = \frac{1}{L}\int_{r=0}^{r=D/2} r^2 \frac{2L\,r}{\sqrt{Q^2-1}\left(\dfrac{D}{2}\right)^2}\sqrt{Q^2-1}\,dr$$

$$= \frac{2}{\left(D/2\right)^2}\int_{r=0}^{r=D/2} r^3\,dr = \frac{2}{\left(D/2\right)^2}\frac{1}{4}\left(D/2\right)^4 = \frac{1}{2}\left(D/2\right)^2.$$

$$\lim_{Z\to 0}\overline{y^2} = \frac{\displaystyle\lim_{Z\to 0}\overline{r^2}}{\left(D/2\right)^2} = \frac{1}{2}. \qquad (5.135)$$

Note: Because we think about the positive increments $d\zeta$ and dr, we used absolute value in the previous expression.

Standard deviation of radial position

It is generally known that the variance (dispersion) of quantity y is

$$V_y = \overline{y^2} - \left(\overline{y}\right)^2. \qquad (5.136)$$

The standard deviation σ_y of quantity y is then expressed as a square root of variance, i.e.,

$$\sigma_y = \sqrt{V_y} = \sqrt{\overline{y^2} - \left(\overline{y}\right)^2}. \qquad (5.137)$$

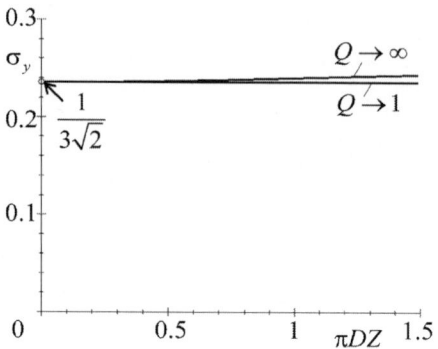

Figure 5.27 Standard deviation σ_y

Note: We must use Equations (5.129) and (5.134) in the last two expressions for numerical solution of standard deviation.

Especially for untwisted yarn, we need to use Equations (5.131) and (5.135). So, the following equation is obtained

$$\lim_{\pi DZ \to 0} \sigma_y = \sqrt{\lim_{\pi DZ \to 0} \overline{y^2} - \left(\lim_{\pi DZ \to 0} \overline{y}\right)^2}$$

$$= \sqrt{\frac{1}{2} - \left(\frac{2}{3}\right)^2} = \sqrt{\frac{1}{18}} = \frac{1}{3\sqrt{2}}. \tag{5.138}$$

The graphical interpretation of standard deviation σ_y, according to Equation (5.137), utilizing Equations (5.129), (5.134) and (5.138), is shown in Figure 5.27. The curves show a negligibly increasing trend for each values of $Q > 1$. [Only for $Q = 1$ the value of the studied function is equal to $1/(3\sqrt{2})$.] The reason for slightly increasing trend is logically the same as the slightly decreasing trend of mean radial position. (See Figure 5.25 and the interpretation according to the scheme shown in Figure 5.26.)

Intensity of migration

In analogy to Hearle's concept – Equation (5.113), let us define the intensity of migration f in case of approximated equidistant model according to the following equation:

$$f = \sqrt{\int_{r=0}^{r=D/2} \left(\frac{dy}{d\zeta}\right)^2 \upsilon(r)\,d\zeta \bigg/ \int_{r=0}^{r=D/2} \upsilon(r)\,d\zeta}. \tag{5.139}$$

Nevertheless, it is valid from Equations (5.124) and (5.91) to write that

$$\frac{dy}{dr} = \frac{1}{(D/2)}, \quad \frac{1}{(D/2)^2} = \left(\frac{dy}{dr}\right)^2 = \left(\frac{dy}{d\zeta}\right)^2 \left(\frac{d\zeta}{dr}\right)^2 = \left(\frac{dy}{d\zeta}\right)^2 (Q^2 - 1),$$

$$\left(\frac{dy}{d\zeta}\right)^2 = \frac{1}{(D/2)^2 (Q^2 - 1)}. \tag{5.140}$$

By using Equation (5.140) in (5.141), we find the intensity of migration f as follows:

$$f = \sqrt{\frac{1}{(D/2)^2 (Q^2 - 1)}} \int_{r=0}^{r=D/2} \upsilon(r) d\zeta \bigg/ \int_{r=0}^{r=D/2} \upsilon(r) d\zeta = \frac{1}{(D/2)\sqrt{Q^2 - 1}}. \tag{5.141}$$

Let us note that the dimension of intensity of migration f is the inverse of length, i.e., same as Hearle's intensity of migration J as expressed in Equation (5.114).

Equation (5.101) describes the (smallest) local period of migration on yarn surface. If yarn twist is limited to zero then the following expression is valid for such a period:

$$\lim_{\pi DZ \to 0} \frac{p(D/2)}{D} = \lim_{\pi DZ \to 0} \left[\frac{1}{2}(1-\delta)\sqrt{Q^2 + (Q^2 - 1)(\pi DZ)^2} \right] = \frac{1}{2}\left(1 - \lim_{\pi DZ \to 0} \delta\right)Q. \tag{5.142}$$

By applying Equation (5.99) to the above-mentioned expression, we find the following:

$$\lim_{\pi DZ \to 0} \frac{p(D/2)}{D} = \sqrt{\frac{Q^2 - 1}{Q^2}} Q = \frac{1}{2}\sqrt{Q^2 - 1}, \quad \lim_{\pi DZ \to 0} p(D/2) = (D/2)\sqrt{Q^2 - 1}. \tag{5.143}$$

However, the last expression is identical to the denominator expressed in Equation (5.141). Thus,

$$f = \frac{1}{(D/2)\sqrt{Q^2 - 1}} = \frac{1}{\lim\limits_{\pi DZ \to 0} p(D/2)}. \tag{5.144}$$

So, the intensity of migration f is a reciprocal value of the limit period of migration on yarn surface in case of untwisted yarn ($\pi DZ \rightarrow 0$); it corresponds to the frequency of migration in this case.

Comments to the characteristics of migration

There exist two triplets of characteristics of migration here: the quantities $\overline{Y}, D_{\mathrm{RMS}}, J$, based on Hearle's concept of characterization of approximated Treloar's model of radial migration and the quantities $\overline{y}, \sigma_y, f$, based on the concept of characterization of approximated equidistant model of radial migration.

The advantage of Hearle's characteristics $\overline{Y}, D, J$ lies in their invariability to twist intensity πDZ . Nevertheless, it is based on structurally very improbable concept. Let us note that the generalization by a way of 'incomplete migration' is very artificial. It would rather better to think about the effects of non-constant packing density in a yarn – see next section.

On the contrary, the characteristics $\overline{y}, \sigma_y, f$ of approximated equidistant model depend on twist intensity, but these dependences are only marginal (see, e.g., Figure 5.25 and Figure 5.27). It can be sufficient to use the values at $\pi DZ = 0$ as characteristics of migration for majority of the existing yarns. Nevertheless, we can expect other values than the model quantities – see Equations (5.131), (5.138), (5.144) – in case of real yarns. It could be due to the effect of non-constant packing density in a yarn, too – see next section.

5.8   Radial migration in case of non-constant packing density

The previous models are based, among others, on assumption 6 (constant period of migration) and/or assumption 7 (constant local packing density) introduced ad hoc in Section 5.3. However, the behaviour of radial packing density – as a result of compression of fibers inside the yarn – is studied experimentally and this phenomenon should be one of the inputs to the migration models. The current section makes an attempt to solve this idea for the approximated models of radial fiber migration.

Common valid relations

The local packing density μ_r at radius r is determined by fiber volume dV and total volume dV_c in the differential layer at a radius r. By using general Equations (5.79) and (5.80), we can write

$$\mu_r = \frac{dV}{dV_c} = \frac{N\upsilon(r)s\,dl}{2\pi r\,dr\,L} = \frac{N\upsilon(r)s}{2\pi r\,L}\frac{dl}{dr}. \tag{5.145}[20]$$

One fiber intersects the differential layer at radius r in the yarn length L with a frequency of $\upsilon(r)$.

The quantity $\upsilon(r)$ – the number of intersections by one fiber to the differential layer at radius r in the yarn length L – can be explicitly expressed as follows:

$$\upsilon(r) = \frac{\mu_r 2\pi r\,L}{Ns}\left|\frac{dr}{dl}\right|. \tag{5.146}[21]$$

One fiber leaves $\upsilon(r)$ number of elements in the differential layer, each with axial distance $d\zeta$. (See, e.g., Figure 5.2.) The sum (integral) of all elements $d\zeta$ (integral over all radii) of the given fiber must evidently create the yarn length L such that $L = \int_{r=0}^{r=D/2}\upsilon(r)d\zeta$. Then,

$$L = \int_{r=0}^{r=D/2}\upsilon(r)d\zeta = \int_{r=0}^{r=D/2}\frac{\mu_r 2\pi r\,L}{Ns}\left|\frac{dr}{dl}\right|d\zeta = \frac{L}{Ns}\int_0^{D/2} 2\pi r\mu_r\left|\frac{dr}{dl}\right|\left|\frac{d\zeta}{dr}\right|dr,$$

$$NsZ = \int_0^{D/2} 2\pi r Z\,\mu_r\left|\frac{dr}{dl}\right|\left|\frac{d\zeta}{dr}\right|dr. \tag{5.147}$$

The next expression follows from Equations (5.1) and (5.3):

$$\left|\frac{dr}{dl}\right|\left|\frac{d\zeta}{dr}\right| = \frac{|\tan\alpha|}{\sqrt{\tan^2\alpha + \tan^2\beta + 1}}\frac{1}{|\tan\alpha|} = \frac{1}{\sqrt{\tan^2\alpha + \tan^2\beta + 1}}. \tag{5.148}$$

Using Equation (5.148) in (5.147), we find

$$NsZ = \int_0^{D/2}\frac{2\pi r Z\mu_r\,dr}{\sqrt{\tan^2\alpha + \tan^2\beta + 1}}. \tag{5.149}$$

---

20 This expression is practically same as Equation (5.81), but, in opposite to the present idea, there is a constant value $\mu_r = \mu$ assumed there.

21 We use absolute value in this equation, because the number of intersections must not be negative.

The fiber length in a length L of yarn is evidently $\int_{r=0}^{r=D/2} \upsilon(r)\,dl$, so that the yarn retraction is $\delta = 1 - L\Big/\int_{r=0}^{r=D/2} \upsilon(r)\,dl$. By substituting Equation (5.146) in the last expression, we obtain

$$\delta = 1 - \frac{L}{\int\limits_{r=0}^{r=D/2} \upsilon(r)\,dl} = 1 - \frac{L}{\int\limits_{r=0}^{r=D/2} \frac{\mu_r\, 2\pi r\, L}{Ns}\left|\frac{dr}{dl}\right| dl} = 1 - \frac{NsZ}{\int\limits_0^{D/2} 2\pi r Z\,\mu_r dr}, \qquad (5.150)$$

Substitution: $y = \dfrac{r}{D/2}, \quad dy = \dfrac{dr}{D/2}$,

$$\delta = 1 - \frac{NsZ}{\int\limits_0^1 2\pi y \frac{D}{2} Z\,\mu_y \frac{D}{2}\,dy} = 1 - \frac{NsZ}{\frac{(\pi DZ)^2}{2\pi Z}\int\limits_0^1 y\mu_y dy}, \quad \text{where } \mu_y = \mu_{r=yD/2} \;\; (5.151)$$

By substituting Equations (5.146) and (5.3) in (5.70), we find the following expression for the period of migration

$$p(r) = \frac{2L}{\upsilon(r)} = \frac{2L}{\frac{\mu_r\, 2\pi r\, L}{Ns}\left|\frac{dr}{dl}\right|} = \frac{2NsZ}{\mu_r\, 2\pi r Z}\frac{\sqrt{\tan^2\alpha + \tan^2\beta + 1}}{|\tan\alpha|}. \qquad (5.152)$$

Variable packing density in case of approximated Treloar's model

The fiber paths follow Equations (4.1) and (5.52) for determination of angles β and α, i.e., Equation (5.53) for their trajectories. If these equations are valid and the nature of radial local packing density is known from the experimental results, then a suitable fiber path must (randomly) change the 'directions', i.e., from inside to outside and from outside to inside in the yarn. This is illustrated in Figure 5.28. (A similar idea was used in case of equidistant migration – see Figure 5.16.)

By substituting Equations (4.1) and (5.52) in (5.149), we find the following expression for this case:

$$NsZ = \int\limits_0^{D/2} \frac{2\pi r Z \mu_r\, dr}{\sqrt{\dfrac{1}{K'^2\,(2\pi rZ)^2} + (2\pi rZ)^2 + 1}}$$

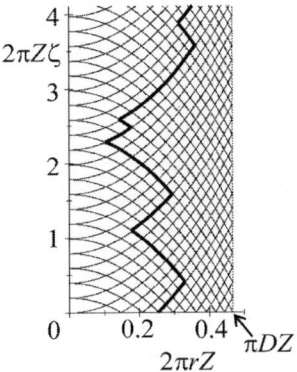

Figure 5.28 Random fiber paths in $\zeta - r$ (i.e. $2\pi Z\zeta - 2\pi rZ$) graph.

Parameters of net: $\pi DZ = \tan 25° = 0.46631$, $K' = 18.24138$.

Curves of net: Equation (5.53). [Equation (5.50): $p/D = 4.25305$]

$$= \int_0^{D/2} \frac{K'(2\pi rZ)^2 \, \mu_r \, dr}{\sqrt{1 + K'^2 (2\pi rZ)^4 + K'^2 (2\pi rZ)^2}}, \qquad (5.153)$$

Substitution: $y = \dfrac{r}{D/2}$, $dy = \dfrac{dr}{D/2}$,

$$NsZ = \int_0^1 \frac{K'\left(2\pi y \dfrac{D}{2} Z\right)^2 \mu_y \dfrac{D}{2} dy}{\sqrt{1 + K'^2 y^4 (\pi DZ)^4 + K'^2 y^2 (\pi DZ)^2}} = \frac{K'(\pi DZ)^3}{2\pi Z} I, \qquad (5.154)$$

where

$$I = \int_0^1 \frac{y^2 \mu_y \, dy}{\sqrt{1 + K'^2 y^4 (\pi DZ)^4 + K'^2 y^2 (\pi DZ)^2}}, \qquad \text{expression } \mu_y = \mu_{r=yD/2}.$$

$$(5.155)$$

The yarn retraction is given by Equation (5.151). When we substitute the term NsZ of this equation by Equation (5.154), we find

$$\delta = 1 - \frac{\dfrac{K'(\pi DZ)^3}{2\pi Z} I}{\dfrac{(\pi DZ)^2}{2\pi Z} \displaystyle\int_0^1 y\mu_y dy} = 1 - \frac{K'(\pi DZ) I}{\displaystyle\int_0^1 y\mu_y dy}. \qquad (5.156)$$

By applying Equations (4.1), (5.52) and later Equation (5.154) in (5.152), we obtain

$$p(r) = \frac{2NsZ}{\mu_r 2\pi rZ} \frac{\sqrt{\dfrac{1}{K'^2(2\pi rZ)^2} + (2\pi rZ)^2 + 1}}{\left| \dfrac{1}{K'(2\pi rZ)} \right|}$$

$$= 2NsZ \frac{\sqrt{1 + K'^2(2\pi rZ)^4 + K'^2(2\pi rZ)^2}}{\mu_r 2\pi rZ},$$

Substitution: $y = \dfrac{r}{D/2}$,

$$\frac{p(y)}{D} = \frac{2NsZ}{D} \frac{\sqrt{1 + K'^2 y^4 (\pi DZ)^4 + K'^2 y^2 (\pi DZ)^2}}{\mu_r y (\pi DZ)}$$

$$= \frac{K'(\pi DZ)^3}{\pi DZ} I \frac{\sqrt{1 + K'^2 y^4 (\pi DZ)^4 + K'^2 y^2 (\pi DZ)^2}}{\mu_y y (\pi DZ)},$$

$$\frac{p(y)}{D} = K'(\pi DZ) I \frac{\sqrt{1 + K'^2 y^4 (\pi DZ)^4 + K'^2 y^2 (\pi DZ)^2}}{\mu_y y},$$

where $p(y) = p(r = yD/2)$, $\mu_y = \mu_{r=yD/2}$. \hfill (5.157)

Variable packing density in case of approximated equidistant model

The fiber paths follow Equations (4.1) and (5.90) for determination of angles β and α. If these equations are valid and the nature of radial local packing density is known from the experimental results then a suitable fiber path changes (randomly) the directions, i.e., from inside to outside and from outside to inside in the yarn. This was illustrated earlier in Figure 5.16.

By substituting Equations (4.1) and (5.90) in (5.149), we find the following expression for this case:

$$NsZ = \int_0^{D/2} \frac{2\pi rZ \mu_r \, dr}{\sqrt{\dfrac{1}{Q^2 - 1} + (2\pi rZ)^2 + 1}} = \int_0^{D/2} \frac{2\pi rZ \mu_r \, dr}{\sqrt{\dfrac{Q^2}{Q^2 - 1} + (2\pi rZ)^2}}, \hfill (5.158)$$

Substitution: $y = \dfrac{r}{D/2}$, $dy = \dfrac{dr}{D/2}$,

$$NsZ = \int_0^1 \frac{2\pi y \dfrac{D}{2} Z \mu_y \dfrac{D}{2} dy}{\sqrt{\dfrac{Q^2}{Q^2 - 1} + y^2 (\pi DZ)^2}} = \frac{(\pi DZ)^2}{2\pi Z} \int_0^1 \frac{y \mu_y \, dy}{\sqrt{\dfrac{Q^2}{Q^2 - 1} + y^2 (\pi DZ)^2}},$$

$$NsZ = \frac{(\pi DZ)^2}{2\pi Z} J, \quad J = \int_0^1 \frac{y \mu_y \, dy}{\sqrt{\dfrac{Q^2}{Q^2 - 1} + y^2 (\pi DZ)^2}}. \tag{5.159}$$

Note: If $\mu_r = \mu$ is constant, then Equation (5.158) is same as Equation (5.96), derived earlier.

The yarn retraction is obtained in the following manner after substituting the term NsZ of Equations (5.151) to (5.159):

$$\delta = 1 - \frac{\dfrac{(\pi DZ)^2}{2\pi Z} J}{\dfrac{(\pi DZ)^2}{2\pi Z} \int_0^1 y \mu_y dy} = 1 - \frac{J}{\int_0^1 y \mu_y dy}, \quad \text{where } \mu_y = \mu_{r = yD/2}. \tag{5.160}$$

By applying Equations (4.1), (5.90) and later Equation (5.159) in (5.152), we obtain the expression for the period of migration as follows:

$$p(r) = \frac{2NsZ}{\mu_r 2\pi r Z} \frac{\sqrt{\dfrac{1}{Q^2 - 1} + (2\pi r Z)^2 + 1}}{\dfrac{1}{\sqrt{Q^2 - 1}}} = \frac{2NsZ\sqrt{Q^2 - 1}}{\mu_r 2\pi r Z} \sqrt{\frac{Q^2}{Q^2 - 1} + (2\pi r Z)^2}$$

$$= \frac{2 \dfrac{(\pi DZ)^2}{2\pi Z} J \sqrt{Q^2 - 1}}{\mu_r 2\pi r Z} \sqrt{\frac{Q^2}{Q^2 - 1} + (2\pi r Z)^2},$$

$$\frac{p(r)}{D} = \frac{\pi DZ \, J\sqrt{Q^2 - 1}}{\mu_r 2\pi r Z} \sqrt{\frac{Q^2}{Q^2 - 1} + (2\pi r Z)^2}, \quad \text{Substitution: } y = \frac{r}{D/2},$$

$$\frac{p(y)}{D} = \sqrt{Q^2 - 1} \, J \frac{\sqrt{\dfrac{Q^2}{Q^2 - 1} + y^2 (\pi DZ)^2}}{\mu_y y},$$

where $p(y) = p(r = yD/2)$, $\mu_y = \mu_{r = yD/2}$. \tag{5.161}

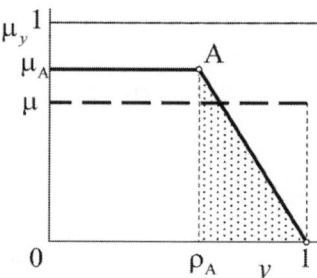

Figure 5.29 *Trapezoidal shape of function* μ_y *against the variable* $y = r/(D/2)$

Note: Let us remind that the numerical solution of integrals I and J, i.e., Equations (5.155) and (5.159), must use a suitable numerical method of integration in each practical case.

Numerical examples

One simplified case of non-constant local packing density – 'trapezoidal' function[22] μ_r – was proposed in Chapter 4 while discussing the helical model; see Figure 4.15 and Equations (4.94) to (4.103).

However, we are working with the quantity $y = r/R = r/(D/2)$ in the current section, so that the local packing density μ_y is the trapezoidal function of y, now – see Figure 5.29. Using symbol $\rho_A = r_A/R$ according to Equation (4.96) and $\xi = \mu/\mu_A$ according to Equation (4.98), we can write in accordance with Equation (4.99)

$$\mu_y = \mu_A \text{ for } y \in \langle 0, \rho_A \rangle,$$

$$\mu_y = -y\frac{\mu_A}{1-\rho_A} + \frac{\mu_A}{1-\rho_A} = \frac{\mu_A}{1-\rho_A}(1-y) \text{ for } y \in (0, \rho_{A,1}), \qquad (5.162)$$

---

22 The stated behaviour of local packing density is naturally a very rough approximation of the general behaviour of radial packing density. Nevertheless, it respects the generally observed phenomenon in case of staple yarns – namely that the packing density in the internal layers of yarn is relatively higher and approximately constant, whereas in the peripheral layers of yarn, the packing density is decreasing to a very small value (near to zero) in a sphere of yarn hairiness. On the contrary, this simplification can express also the case of constant local packing density (twisted filament yarns).

Applying by turns the second Equation in (4.101), and Equations (4.103) and (4.102), we can also write

$$\xi = \frac{\mu}{\mu_A} = \frac{1 + \rho_A + \rho_A^2}{3} \in \langle 1/3, 1 \rangle, \quad \rho_A = -\frac{1}{2} + \sqrt{3}\sqrt{\xi - \frac{1}{4}}. \qquad (5.163)$$

For the yarn packing density μ, it is also valid to write from Equation (4.101) and/or (5.162)

$$\mu = \frac{8}{D^2} \int_0^{D/2} \mu_r r \, dr = 2 \int_0^1 \mu_y y \, dy, \quad \text{substitution:} \, y = \frac{r}{D/2}, \quad dy = \frac{dr}{D/2}. \qquad (5.164)$$

A few examples show the graphical interpretation of approximated Treloar's model of migration in Figures 5.30 and 5.31.

Note: For both figures, we selected packing density μ (used 0.5), parameter K' (used 20 or 100), values μ_A (used from 0.5 to 0.8), we calculated $\xi = \mu/\mu_A$, and verified if the condition $\xi \in \langle 1/3, 1 \rangle$ is valid – Equation (5.163). Then we evaluated ρ_A from Equation (5.163). For Figure 5.30, we calculated the integral in denominator in Equation (5.156) by using of μ_y from Equation (5.162). Subsequently, we gradually selected (increasing) values πDZ, calculated definite integrals I according to Equation (5.155) by using μ_y from Equation (5.162), and evaluated δ-values from Equation (5.156). For Figure 5.31, we selected more the value πDZ (used 0.5 or 1) and calculated the definite integral I according to Equation (5.155) by using μ_y from Equation (5.162). Subsequently, we gradually selected (increasing) values y and, for each value, we calculated $p(y)/D$ from Equation (5.157).

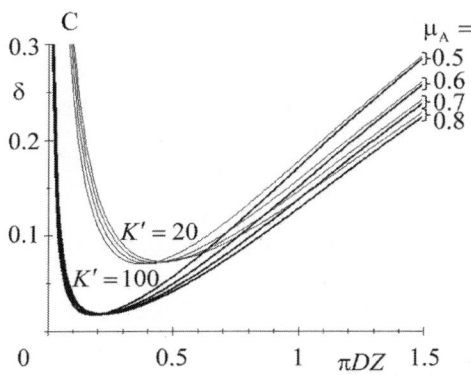

Figure 5.30 Yarn retraction according to Equation (5.156) with Equation (5.155). Common parameter $\mu = 0.5$. Variable μ_y by Equations (5.162) and (5.163)

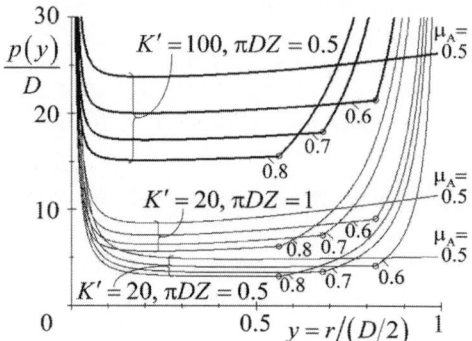

Figure 5.31 Period of migration according to Equation (5.157) with (5.155). Common parameter $\mu = 0.5$. Variable μ_y by Equations (5.162), (5.163). $\bigcirc$: the points of $y = \rho_A$

The behaviour of yarn retraction, shown in Figure 5.30, remains non-real for the smaller values of twist intensity $\tan\beta_D = \kappa = \pi DZ$. A similar result was seen in Figure 5.10. Nevertheless, we observe the decreasing trend of yarn retraction with the increase of μ_A, as shown in Figure 5.30 (more markedly trapezoidal shape of function μ_y in case of Figure 5.29).

The relative value of local period of migration $p(y)/D$ possesses continually decreasing trend with y only at $\mu_A = \mu = 0.5$, i.e., at a constant value of local packing density in the whole yarn body, as shown in Figure 5.31. In all other cases ($\mu_A > \mu$), the quantity $p(y)/D$ follows a U-shaped function, i.e., this is decreasing and then increasing with the increase of y. Such behaviour immediately follows Equation (5.157); because, if $y \to 0$ (at yarn axis) or $\mu_y \to 0$ [when $y = r/(D/2) \to 1$] then evidently $p(y)/D \to \infty$.

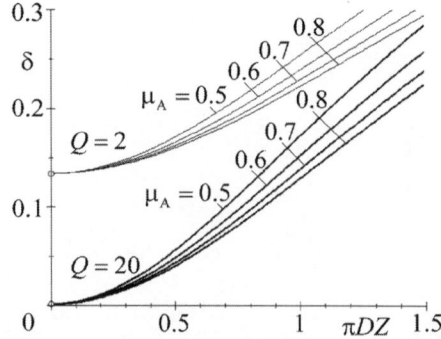

Figure 5.32 Yarn retraction according to Equation (5.160) with Equation (5.159). Common parameter $\mu = 0.5$. Variable μ_y by Equations (5.162), (5.163).
$\bigcirc$: the points by Equation (5.99)

It is also shown that the aforesaid curves are higher (i.e., the frequency of migration is smaller) with the increase of parameter K' (less intensity of migration) and also with the increase of twist intensity $\tan \beta_D = \kappa = \pi DZ$.

Another few examples of the graphical interpretation of approximated equidistant model of migration are shown in Figures 5.32 and 5.33.

Note: Figures 5.32 and 5.33 follow the Figures 5.30 and 5.31 likewise but we select parameter Q (used 2 or 8 or 20) on the place of K'. For Figure 5.32 we apply Equation (5.160) on the place of Equation (5.156) and integral J according to Equation (5.159) on the place of I. For Figure 5.33 we select the value πDZ (used 0.5), apply mentioned integral J on the place of I, and Equation (5.161) on the place of (5.157). [Equations (5.162) and (5.163) are still valid.]

The yarn retraction, shown D in Figure 5.32, is relatively real. The curves at $\mu_A = \mu = 0.5$, i.e., at a constant value of local packing density in the whole yarn body, are identical with the corresponding curves shown in Figure 5.20. Nevertheless, the decreasing trends of yarn retraction are shown with the increase of μ_A in Figure 5.32 (more markedly trapezoidal shape of function μ_y in case of Figure 5.29).

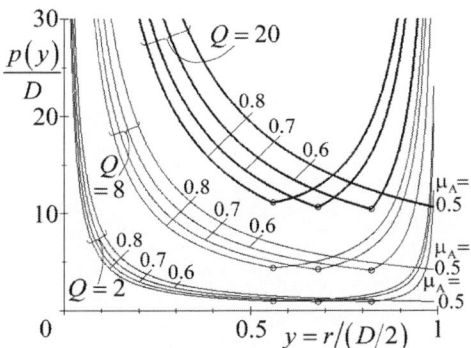

Figure 5.33 Period of migration according to Equation (5.161)
with Equation (5.159). Common parameters: $\mu = 0.5$, $\pi DZ = 0.5$.
Variable μ_y by Equations (5.162), (5.163). O: the points of $y = \rho_A$

The relative value of local period of migration $p(y)/D$ shows continuously decreasing trend with y only at $\mu_A = \mu = 0.5$ as shown in Figure 5.33; these behaviours are the same as those shown in Figure 5.21. In all other cases ($\mu_A > \mu$), the quantity $p(y)/D$ follows the above-mentioned U-shape as in the previous case. Such behaviours follow immediately Equation

(5.161); because, if $y \to 0$ (at yarn axis) or $\mu_y \to 0$ [at $y = r/(D/2) \to 1$], then evidently $p(y)/D \to \infty$.

It is also shown that the aforesaid curves are higher (i.e., the frequencies of migration are smaller) with an increase of parameter Q (less intensity of migration).

5.9   Two examples of experimental results

We studied the internal structure of staple yarns experimentally several years ago by employing the so-called 'tracer fiber technique'[23] according to Morton and Yen [3] using Omest instrument developed by Kasparek [10]. We measured the spatial curves of fibers and then we determined the quantities (angles $\vartheta_r, \vartheta_\varphi, \vartheta_\zeta$ and α, β, γ – see Figure 5.2) of mean directional characteristics of the short fiber portions at differential radii.

The research report [4] contains results of several staple fiber yarns. Some results related to two carded rotor-spun viscose yarns (37 mm staple length) are presented in this section.

Note: The mean characteristics were determined on each small class interval of radius. These mean values are shown in the following graphs without a special marking of 'mean'. The determined values are marked by small rings and/or squares.

Local packing density

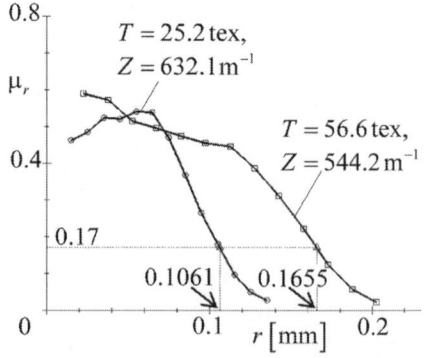

Figure 5.34 Behaviour of local packing densities

---

23   It follows preparation of transparent fibers in yarn and using special immersion liquid so that some black fibers, blended with white fibers, can be geometrically analyzed in three directions.

The behaviour of μ_r is illustrated in Figure 5.34[24]. It is shown that a trapezoidal shape of these functions, according to Figure 5.29, can be taken as a very rough approximation of the measured trends. Based on our experimental experience, we determined yarn radius $D/2$ as the radius at which the local packing density is $\mu_r = 0.17$. So, we found $D/2 = 0.1061\,mm$ for the yarn with 25.2 tex count and $D/2 = 0.1655\,mm$ for the yarn with 56.6 tex count – triangular points shown in Figure 5.34. (The hairiness effect is prevailing outside such yarn radius.)

Note: The yarn radius, found in the previous paragraph, was used for determination of radial position $y = r/(D/2)$ in all other graphs.

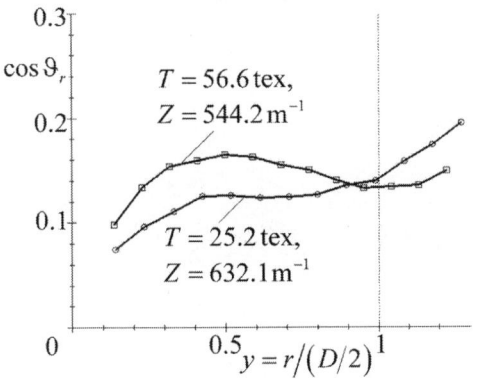

Figure 5.35 Behaviour of mean quantities $\cos \vartheta_r$

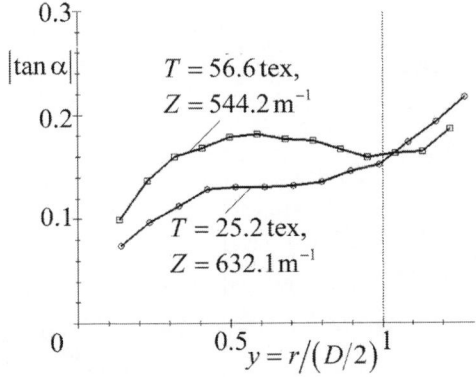

Figure 5.36 Behaviour of mean quantities $|\tan \alpha|$

---

24   The values near to yarn axis were very much variable in a consequence to the method of measurement and evaluation. These values are therefore not presented in our graphs.

Angular characteristics $\cos \vartheta_r$ **and** $|\tan \alpha|$

Figures 5.35 and 5.36 show the behaviour of these quantities. The variable $\cos \vartheta_r$ is not quite a constant, as assumed in the equidistant model – Equations (5.78) and (5.3) – and $|\tan \alpha|$ is also not a constant, as assumed in the approximated equidistant model – Equation (5.90). Nevertheless, the behaviour of the curves – except the region near to yarn axes – is relatively flat. So, the equidistant model and the approximated equidistant model are roughly acceptable from the point of view of (mean) fiber directions at different radii.

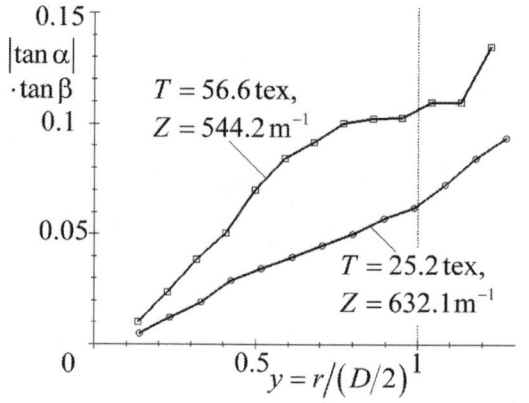

Figure 5.37 Behaviour of mean quantities $|\tan \alpha| \tan \beta$

Note: A characteristic value of $\cos \vartheta_r$ and/or $|\tan \alpha|$ is around 0.14 as shown in Figures 5.35 and 5.36. This value corresponds to the value of parameter $Q \doteq 7$ according to Equations (5.78) with (5.3) and/or Equation (5.90).

Quantity $|\tan \alpha| \tan \beta$

The behaviour of these functions is markedly increasing as shown in Figure 5.37. It fully corresponds with Figure 5.22 on one hand, and it is in contradiction with the theoretical curves of Treloar's models shown in Figure 5.13 on the other hand.

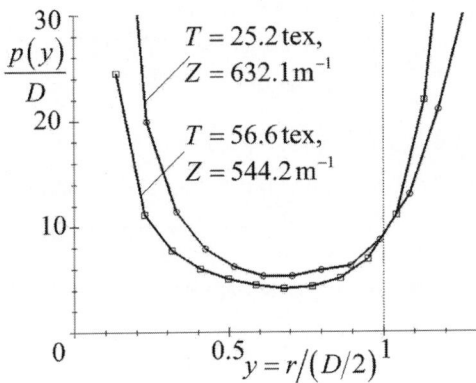

Figure 5.38 Behaviour of relative values of local period $p(y)/D$

Relative values of local period $p(y)/D$

The nature of these functions is illustrated in Figure 5.38. A typical U-shaped curve which was assumed theoretically in Figures 5.31 and 5.33 for cases of local yarn packing density (trapezoidal shape) was obtained.

5.10   Creation of radial migration

There exist a lot of different mechanisms that lead to creation of radial migration of fibers in yarns during the spinning process. These can be, for example:

1. Mechanism of equalizing of fiber lengths
2. Mechanism of bring-in migration in
   (a) regular alternative and
   (b) random alternative
3. Mechanism of pre-twisting in spinning triangle
4. Mechanism of angular-radial instability

Let us characterize these mechanisms briefly.

Equalizing of fiber lengths

Morton [11] formulated probably the oldest idea of creation of radial migration in yarn. He noticed different lengths of fibers in the helical models. The helixes situated at higher radii needed longer length than the helixes lying at small radii, although the starting lengths of all the fibers were (ideally) same. Therefore, the fibers lying at higher radii were tensioned so that they had a tendency to push themselves inside the yarn, i.e., to smaller radii. On the contrary, the fibers lying at smaller radii had relatively surplus of lengths so that they 'could' go to higher radii. The fiber lengths were partly equalized in this way.

The described mechanism is in principle real, but this is probably not too significant; because other fibers in the surrounding do obstruct such type of fiber movement.

Regular bring-in migration

Hearle et al. [6, 7] observed radial migration also in case of twisted filament yarns. They found that the periods of migration were roughly same as the reciprocal values of the protective twists of the starting yarns (i.e., yarns without following technological twisting). Figure 5.39 illustrates the explanation from Hearle et al.

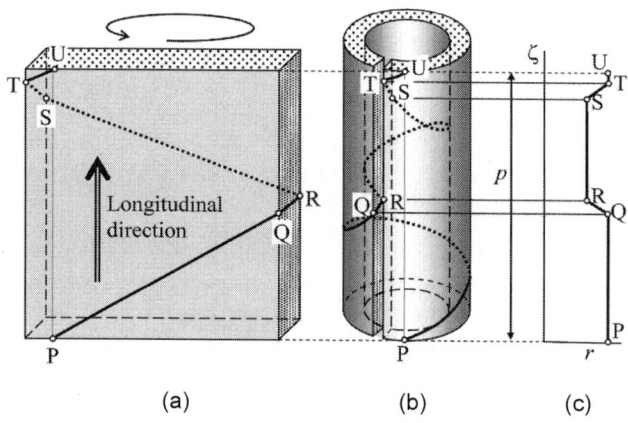

(a)   (b)   (c)

Figure 5.39 Principle of explanation of regular bring-in migration

The starting filament yarn is flattened between rollers to the shape of a ribbon and a fiber with protective twist has the (idealized) shape according the path PQRSTU shown in Figure 5.39a.

The following 'ribbon twisting' causes the so-called wrapping of ribbon to a 'tube' – Figure 5.39b – together with twisting of this tube[25]. So, the fiber periodically changes outer and inner radii of the tube, i.e., migrate, with regular period p, equal to coil height of protective twist – see Figure 5.39b and 5.39c.

Random bring-in migration

The thin fiber bundle is a starting product for spinning of staple yarn. However, the staple fibers creating such bundle are usually not perfectly parallel; they are wavy, randomly crimped, etc. This non-parallelism is translated into

---

25   For simplicity, the technological twisting of creating tube is not illustrated in Figure 5.39b.

yarn structure by the process of yarn twisting similarly as in previous (regular) case. The phenomenon of fiber migration into yarns originates just by means of such mechanism. Neckář [9] proposed that this mechanism is most probably the dominant factor, creating fiber migration in staple yarns.

Pre-twisting in spinning triangle

The yarn starts forming just at the end of the spinning triangle due to twisting in ring frame. In the spinning triangle, we observe a thin partial bundles consisting of few fibers, which are twisting themselves together. This pre-twisting is illustrated schematically in Figure 5.40.

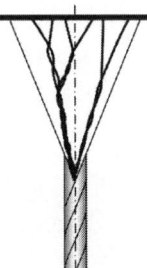

Figure 5.40 Pre-twisting in spinning triangle

Although such thin bundles can enter into the yarn structure as a fully regular helical model, however, in reality, the thin bundles are one of the sources of fiber migration in yarn.

Note: The so-called 'solo-spun' technology supports the pre-twisting effect by means of a specific fine denticulation of lower (hard) roller. So, the fibers are probably more linked or interlaced mutually.

Angular-radial instability

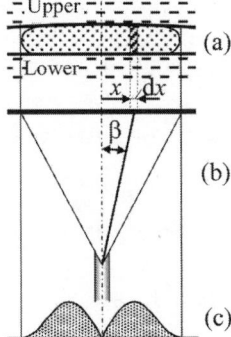

Figure 5.41 Principle of angular-radial instability

The flat fiber bundle, comes out from the clamping line of the front rollers of a ring frame, is shown in Figure 5.41a. Some (differential) number of fibers are lying in the (differential) part dx shown at a distance x in the hatched area. All these fibers make (ideally) an angle β to the common top of the spinning triangle – see Figure 5.41b, and 'desire' to lie in the yarn body just with this angle.

However, the angle β corresponds to the (differential) layer at radius r ($\tan\beta = 2\pi rZ$) where the local packing density is given by μ_r. It means that the corresponding layer consists of another number of fibers than the (differential) number which comes from the hatched part shown in Figure 5.41a. If the number of the incoming fibers is too high then it is not possible to 'jam' all the fibers at a suitable layer so that some of them must cross themselves to another radius where the angle β is 'incorrect'. On the contrary, if the number of fibers coming to a radial layer is too small, then the other fibers with incorrect angle β also come to this layer. The result of this mechanism represents radial angular instability where the fibers with given angle β permanently move from their radial position – so they migrate.

In an abstract case, we can obtain a non-migrating (helical) structure if the profile of the flat fiber bundle, comes out from the clamping line, will be 'double-top profile' according the scheme shown in Figure 5.41c.

Note: The so-called 'siro-spun' technology substitutes the above-mentioned theoretical idea of 'double-top profile' by the couple of small bundles used for creation of one yarn.

5.11   Closing notes to fiber migration in yarns

In the past, Treloar's original and approximate models of radial migration of fibers in yarns [1] created the first non-helical concept of yarn structure – see Sections 5.3 and 5.4. Then, Hearle et al. [6, 7] – going out from Treloar's approximation variant – introduced the characteristics of migration – see Section 5.7. Later on, a large number of attempts were seen to evaluate Hearle's characteristics of radial migration for many different yarns, mostly without analyzing the real theoretical basis.

We discover that the Treloar's original and approximated models do not reflect natural rules of yarn structure – see, e.g., Figure 5.13. Nevertheless, Treloar's unique concept and methodology are generally thought as excellent initial basis for construction of the modified model of radial migration, called equidistant model – see Sections 5.5 and 5.6. It was enough to change only one starting assumption (we changed assumption 6) to a new assumption 6* and loosen the strict 'regularity' of fiber path. It emerged that such model corresponds much better with the structure of real yarns.

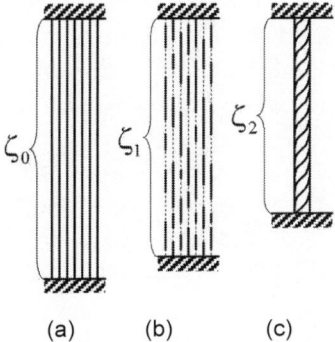

(a)   (b)   (c)

Figure 5.42 Yarn retraction from entangled bundle:
(a) parallel fiber bundle, (b) entangled bundle, (c) twisted yarn

The yarn retraction is also analyzed from migration models in this chapter. Traditionally, the yarn retraction compared the final length of twisted yarn with ideal starting length of bundle of parallel fiber. Nevertheless, this is not fully right, especially for staple yarns, because the starting fiber bundle (e.g., issuing from the front rollers in ring frame) is mostly created from wavy, randomly crimped fibers. We can speak about 'entangled' bundle. Then we usually need to know the yarn retraction as a comparison of the final entangled length and real starting entangled length.

The models of equidistant migration allow us to calculate traditionally defined yarn retraction also for the above-mentioned entangled bundle (yarn). By using Equation (4.30), we can write the following relation from Figure 5.42:

$$\left.\begin{array}{l} \delta_1 = \dfrac{\zeta_0 - \zeta_1}{\zeta_0}, \quad \zeta_1 = \zeta_0\left(1 - \delta_1\right), \\[2mm] \delta_2 = \dfrac{\zeta_0 - \zeta_2}{\zeta_0}, \quad \zeta_2 = \zeta_0\left(1 - \delta_2\right). \end{array}\right\} \tag{5.165}$$

Now, we are able to formulate yarn retraction δ_Y as a comparison of the final entangled length and the real starting entangled length:

$$\delta_Y = \frac{\zeta_1 - \zeta_2}{\zeta_1} = \frac{\zeta_0\left(1 - \delta_1\right) - \zeta_0\left(1 - \delta_2\right)}{\zeta_0\left(1 - \delta_1\right)} = \frac{\delta_2 - \delta_1}{1 - \delta_1}, \tag{5.166}$$

where δ_1, δ_2 can be studied based on the idea of equidistant model.

The variants of equidistant models are better than those of Treloar's models, but unfortunately, the former are also not too perfect. The future models should consider the random character of directions of short fiber segments as well as random character of local packing density. Further, fundamental research needs to be carried out to understand the mechanisms of creation of radial migration. Some more details on this are available in the book [9], but our understanding and theoretical knowledge of these processes are still limited.

Besides radial migration, fundamental research also needs to be carried out with other types of migration, i.e., axial and twisted migration. A few ideas on these topics are presented in the book [9], but these problems are in general still unknown to a considerable extent.

5.12   References

[1]   Treloar, L. R. G., A Migration Filament Theory of Yarn Properties, *Journal of Textile Institute,* 56, T359–T380, 1965.

[2]   Stejskal, A., and Kašpárek, J., CSSR Patent No. 117179 (In Czech).

[3]   Morton, W. E., and Yen, K. C., The Arrangement of Fibers in Fibro Yarns, *Journal of Textile Institute*, 43, T60, 1952.

[4]   Neckář, B., and Soni, M. K., Internal Structure of OE-Yarns from Viscose Fibers, Research Report No. PT2-XI.79, State Textile Research Institute, Liberec, 1979.

[5]   Neckář, B., Soni, M. K., and Das, D., Modelling of radial migration of fibers in yarns, *Textile Research Journal*, 76, 486–491, 2006.

[6]   Hearle, J. W. S., Gupta, B. S., and Merchant, V. B., Migration of Fibers in Yarns, Part I: Characterization and Idealization of Migration Behavior, *Textile Research Journal,* 35, 329–334, 1965.

[7]   Hearle, J. W. S., and Gupta, B. S., Migration of Fibers in Yarns, Part II: A Study of Migration in Staple Fiber Rayon Yarns, *Textile Research Journal*, 35, 788, 1965.

[8]   Hearle, J. W. S., Grosberg, P., and Backer, S., Structural Mechanics of Fibers, Yarns, and Fabrics, Volume 1, John Wiley & Sons, New York, 1969.

[9]   Neckář, B., Příze: Tvorba, struktura, vlastnosti, (Yarns: Creation, structure, properties) SNTL Prague, 1990 (In Czech).

[10]   Kasparek, J., Geometric and Mechanical Properties of Open-end Yarn in "Open-end Spinning, Rohlena, V. (Ed.), Elsevier Scientific Publishing Company, Amsterdam, 214, 1975.

[11]   Morton, W. E., The Arrangement of Fibers in Single Yarns, *Textile Research Journal*, 26, 325–331, 1956.

Mass unevenness of staple fiber yarns

6.1　　General ideas of irregularity

Quite generally, a fibrous assembly can be geometrically divided into many equal parts and the selected properties of these parts can be investigated. If the properties of all parts are same, then we speak about regularity with regard to the given division (dimension) and the chosen properties. If the whole assembly has the same properties in all possible divisions, then it is completely regular in an absolute sense.

Nevertheless, a real yarn is never absolutely regular. The expression regular or 'practically regular' may be adopted with regard to some type of division and some particular properties. (For example, the filament yarns made up of synthetic fibers are practically regular with regard to local yarn fineness.)

If a yarn is not regular we say that it is 'irregular' or 'uneven'. In such case, it is required to understand the mechanism that brings the irregularity or variation in properties among the individual parts of the yarn. At the same time, it is also important to define the relation between the way of dividing the yarn into many parts and the variability of properties of the parts.

The methods of investigation of irregularity depend mainly on:

1. *Type of fibrous assembly*. One-dimensional (linear) textiles, e.g., slivers, yarns, etc., are usually divided into length sections and the irregularity of those sections is evaluated.

2. *Observed properties*. Generally, material, geometrical and other physical properties of fibrous assembly can be investigated. From the practical point of view, the irregularity of mass m along the yarn is most frequently observed. In the case of constant fiber density ρ, the mass irregularity also expresses the irregularity of fiber volume V, because $V = m/\rho$.

3. *Case of variation*. The variability of properties can be deterministic (e.g., regular periodic change of local yarn fineness), fully random (this is dominant in case of mass irregularity of staple fiber yarns) and a combination of the two (most frequent situation in case of staple fiber yarns).

4. *Depth of understanding*. Depending on the purpose, sometimes it is sufficient to describe the irregularity in a simple manner, and for

some other purposes, it is required to deeply investigate the particular causes of irregularity. The descriptive models of irregularity use the general (usually statistical) tools without deeply knowing the unique features of creation and structure of the yarns. The explanatory models, considering the unique features of creation and structure of the yarn, can logically explain the irregularity.

5. *Longitudinal relationships*. The yarn irregularity can be described by suitable mathematical structures with and/or without relation to linear character of the yarn. For example, the distribution functions and their characteristics do not take this linear character into consideration, whereas the autocorrelation function, spectral function, etc., do consider this linear character.

6.2   Martindale's model of sliver irregularity

The simplest example of a linear textile possessing random nature of irregularity is a sliver produced from staple fibers. This is a semi-finished product[1], which is used in spinning mills for production of yarns. Let us now focus our mind to the problem of mass irregularity of slivers.

Ideal sliver

Martindale [1] proposed an explanatory model of mass irregularity of slivers. He assumed a sliver produced from straight and parallel fibers of equal length *l*. The graphical illustration of his model is displayed in Figure 6.1. The right-hand ends of the fibers are marked by the small circles. According to Martindale's idea, the fibers are deposited to form a sliver in the direction of short arrows (a) individually, (b) parallel to the direction of sliver axis, (c) mutually independently and (d) quite randomly.

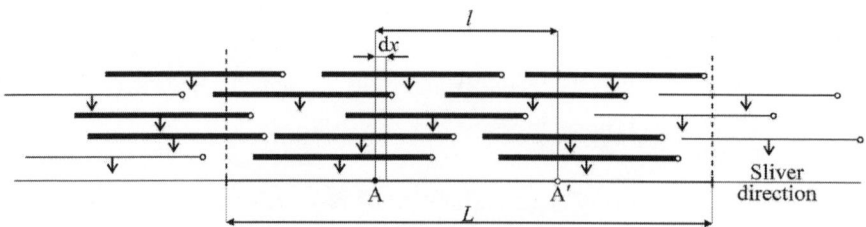

Figure 6.1 Random deposition of individual fibers, parallel to the direction of sliver axis

---

1   Let us note that the yarn can be thought of a thin sliver after twisting.

As shown, the right-hand ends of thick fibers are lying in a given length L, other (thin) fibers have their right-hand ends lying outside the length L. If the fibers have passed the line drawn perpendicular from point A to the axis of the sliver, then their right-hand ends must lie in-between the lines drawn perpendicular from point A and A′, respectively, to the axis of the sliver, i.e., in the section of length l. So, the probability that a fiber, randomly selected from the fibers having their right ends lying in length L, passes the line drawn perpendicular from point A to the axis of the sliver is

$$p = l/L .\tag{6.1}$$

It follows the geometrical definition of probability.

Binomial sliver

Let us assume that there are N fibers whose right-hand ends are lying in length L. It is then valid to write

$$N = N_1 L ,\tag{6.2}$$

where N_1 is the mean number of fibers per unit length of section L.

Because of the assumption of independency, each fiber can pass the line drawn perpendicular from point A to the axis of the sliver with the same probability p. The probability $B(n)$, that just n fibers, $n \le N$, pass the aforesaid line, evidently follows the binomial distribution[2]

$$B(n) = \binom{N}{n} p^n (1-p)^{N-n} .\tag{6.3}$$

Such an imaginatively created sliver is referred to as binomial sliver. The mean value of number n of fibers in the binomial sliver is

$$\bar{n} = Np ,\tag{6.4}$$

their variance and standard deviation are, respectively,

$$\sigma_n^2 = Np(1-p), \quad \sigma_n = \sqrt{Np(1-p)} = \sqrt{\bar{n}(1-p)} ,\tag{6.5}$$

---

2   The readers may like to revise the expressions mentioned in Equations (6.3), (6.4), (6.5) from a handbook of theory of probability.

and the coefficient of variation[3] of number of fibers in the binomial sliver is

$$v(n) = \frac{\sigma_n}{\bar{n}} = \frac{\sqrt{\bar{n}(1-p)}}{\bar{n}} = \sqrt{\frac{1-p}{\bar{n}}} .$$ (6.6)

The derivation of the coefficient of variation of local mass of binomial sliver is rather complicated because (1) the number n of fibers present in the cross sections of binomial sliver is a random variable, and (2) the fiber fineness t varies from fiber to fiber; therefore, it is a random variable too.

We denote the fibers in the sliver cross section serially by the subscript $i = 1, 2, \ldots, n$. The corresponding fiber finenesses are denoted by $t_1, t_2, \ldots, t_n$, where each of them is a random variable with common statistical characteristics, e.g., mean value $\bar{t}$ and coefficient of variation $v(t)$.

The number of fibers n passing through the elementary section of length dx at point A is shown in Figure 6.1. Considering the i-th fiber, having mass $t_i\, dx$ in the elementary section, the overall mass dm of all fibers is

$$dm = 0 \qquad\qquad \text{for } n = 0,$$

$$\sum_{i=1}^{n} dm = \sum_{i=1}^{n}(t_i dx) = dx\sum_{i=1}^{n} t_i \quad \text{for } n = 1, 2, \ldots,$$ (6.7)

and the fineness of this elementary section is $T = dm/dx$. Thus

$$T = dm/dx = 0 \qquad \text{for } n = 0,$$

$$T = dm/dx = \sum_{i=1}^{n} t_i \quad \text{for } n = 1, 2, \ldots$$ (6.8)

Note: This expression is true for a very short section, i.e., an elementary part of sliver. Therefore, this is called as local sliver fineness.

The value of T according to Equation (6.8) satisfies all the requirements of variable type y according to Equation (A5.15) as stated in Appendix A5. Note that Equation (A5.15) utilizes alternative symbols: y instead of T, x_i instead of t_i and m instead of n. Because the discrete variable n in Equation (6.8) follows the binomial distribution with a mean value $\bar{n}$ and maximum value N, it is then valid to write Equation (A5.23) from Appendix 5 for

---

3   We express the coefficient of variation as a dimensionless quantity, not in percentage.

coefficient of variation $v(T)$ of local sliver fineness; the symbol m_{max} is used in Appendix 5 in the place of the present symbol N. By using the present symbolism and applying Equation (6.4), we find

$$v^2(T) = \frac{1}{\overline{n}}\left[v^2(t) + \left(1 - \frac{\overline{n}}{N}\right)\right] = \frac{1}{\overline{n}}\left[v^2(t) + (1-p)\right],$$

$$v(T) = \frac{1}{\sqrt{\overline{n}}}\sqrt{v^2(t) + (1-p)}. \qquad (6.9)$$

The mean number of fibers in the sliver cross section is $\overline{n} = \tau k_n$, where $\tau = \overline{T}/\overline{t}$ and $k_n = 1$, because we think here about a sliver prepared from parallel fibers[4] – see Figure 6.1. So, it is valid to write that

$$\overline{n} = \overline{T}/\overline{t}, \qquad (6.10)$$

and by using Equation (6.10) in (6.9), we can write the following expression

$$v^2(T) = \frac{\overline{t}}{\overline{T}}\left[v^2(t) + (1-p)\right], \quad v(T) = \sqrt{\frac{\overline{t}}{\overline{T}}}\sqrt{v^2(t) + (1-p)}. \quad (6.11)$$

[The probability p can be expressed from Equation (6.1) and/or from Equations (6.4) to (6.10) as $p = l/L = \overline{n}/N = \overline{T}/(N\overline{t})$.]

Note: Let us note that, according to Equation (6.11), the higher is the probability p, the lower is the coefficient of variation.

Poisson sliver

The deposition of fibers explained previously can be extended to a longer section L, now, to the limit value $L \to \infty$. The number N of fibers increases in such a manner that the mean number N_1 of fibers per unit length, introduced by Equation (6.2), remains constant. According to Equation (6.2), it is valid that $\lim_{L\to\infty} N = \lim_{L\to\infty}(N_1 L) = \infty$ and according to Equation (6.1), $\lim_{L\to\infty} p = \lim_{L\to\infty}(l/L) = 0$. The limit of mean value $\overline{n}$ is found after applying Equations (6.1) and (6.2) into (6.4) as follows:

---

4   See the commentaries made after Equation (1.43) in Chapter 1.

$$\bar{n} = \lim_{L \to \infty} (Np) = \lim_{L \to \infty} \left(N_1 L \frac{l}{L} \right) = N_1 l .$$

(6.12)

The variance is found by applying Equations (6.1), (6.2) and (6.12) into (6.5) as follows:

$$\sigma_n^2 = \lim_{L \to \infty} \left[Np(1-p) \right] = \lim_{L \to \infty} \left[N_1 L \frac{l}{L} \left(1 - \frac{l}{L} \right) \right] = \lim_{L \to \infty} \left(N_1 l - N_1 \frac{l^2}{L} \right) = N_1 l = \bar{n} .$$

(6.13)

Evidently, the variance is equal to the mean value in this distribution. It is well known that the binomial distribution is limited to Poisson distribution

$$P(n) = \frac{\bar{n}^n}{n!} e^{-\bar{n}} ,$$

(6.14)

in this case. The sliver model in which the number of fibers in sliver cross section follows the Poisson distribution is usually called Poisson sliver.

By using Equation (6.13), the coefficient of variation of number of fibers present in the cross section of the Poisson sliver is now

$$v(n) = \frac{\sqrt{\sigma_n^2}}{\bar{n}} = \frac{\sqrt{\bar{n}}}{\bar{n}} = \frac{1}{\sqrt{\bar{n}}} .$$

(6.15)

Equation (A5.26) in Appendix 5 is valid for Poisson distribution, utilizing y instead of T, x_i instead of t_i and m instead of n. Therefore, it is valid now

$$v^2(T) = \frac{1}{\bar{n}} \left[v^2(t) + 1 \right], \quad v(T) = \frac{1}{\sqrt{\bar{n}}} \sqrt{v^2(t) + 1} .$$

(6.16)

Further, by substituting Equation (6.10) in (6.16), we can write

$$v^2(T) = \frac{\bar{t}}{\bar{T}} \left[v^2(t) + 1 \right], \quad v(T) = \sqrt{\frac{\bar{t}}{\bar{T}}} \sqrt{v^2(t) + 1} .$$

(6.17)

Note: Let us remark that the last expression results in a little higher value of the coefficient of variation $v(T)$ than that obtained from Equation (6.11). It is because a negative value of p is absent in Equation (6.17). That is, a total N number of fibers is surely present in a given length L in case of binomial sliver, whereas the number N of fibers is randomly different, i.e., it can vary from zero to infinity in case of Poisson sliver.

Equation (6.17) is often modified for practical purpose. The fiber fineness $t = (\pi\rho/4)d^2$, according to Equation (1.6), is a square function of the equivalent fiber diameter which corresponds to Equation (A5.27) from Appendix 5 – we consider the quantity $\pi\rho/4$ as a constant. If the equivalent diameter of the individual fibers is not highly varied, Equation (A5.31) in Appendix 5 can be approximately used. According to the present symbolism, the coefficient of variation can be expressed as

$$v^2(t) = 4v^2(d).$$ (6.18)

Traditionally, the coefficient of variation of fiber diameter is denoted by

$$CV_d \equiv v(d).$$ (6.19)

The coefficient of variation of local yarn fineness $v(T)$ in the Poisson sliver is traditionally represented as

$$CV_{\lim} \equiv v(T).$$ (6.20)

In literature, this term is known as 'limit irregularity'.

By applying Equations (6.18), (6.19) and (6.20) into (6.17), we find the following expressions for the limit irregularity:

$$CV_{\lim} = \sqrt{\frac{\overline{t}}{\overline{T}}}\sqrt{1 + v^2(t)} = \sqrt{\frac{\overline{t}}{\overline{T}}}\sqrt{1 + 4CV_d^2}.$$ (6.21)

Note: Most often, this relation use the quantities expressed in percentage, i.e., $CV_{\lim[\%]} = 100\sqrt{\overline{t}_{[\text{tex}]}/\overline{T}_{[\text{tex}]}}\sqrt{1 + 0,0004\,CV_{d[\%]}^2}$. It is believed that the limit irregularity is the least possible irregularity for a random sliver. Nevertheless, this is valid only for Poisson sliver (not for binomial sliver). Let us remind that Zellweger Uster [2] uses Equation (6.21) for slivers as well as for twisted yarns.

6.3   Influences of fiber direction and fiber length on sliver irregularity

Poissons sliver from oblique fibers

Let us solve an abstract case of sliver, which is randomly created from a set of oblique fibers of length *l* that make an angle ϑ to the longitudinal direction of the sliver – see Figure 6.2.

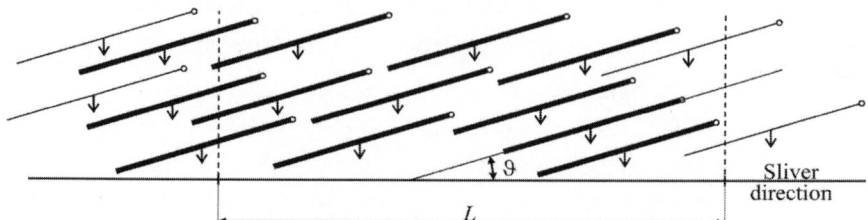

Figure 6.2 Random deposition of individual and mutually
parallel fibers,oblique to the direction of sliver

One (magnified) oblique fiber of length l, which is making an angle ϑ to the direction of the sliver axis, is also shown in Figure 6.3. The area of fiber cross section is s and the (larger) area of fiber section perpendicular to the sliver direction is s^*.

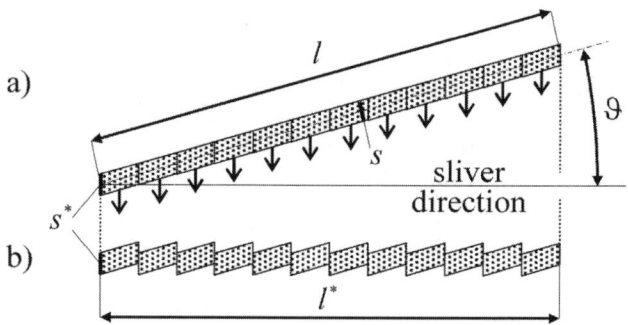

Figure 6.3 Oblique and effective fiber

Let us imaginatively divide this fiber into many short segments (infinite number of infinitesimally small segments) according to Figure 6.3a, i.e., perpendicular to sliver direction. Further, let us move all these individual segments (in the direction of arrows) to the sliver axis. So we obtain a saw-toothed object as shown in Figure 6.3b. If the number of segments is infinitely large, then the 'teeth' are infinitely small and then we obtain a so-called effective parallel fiber.

Such an effective fiber is parallel to the direction of the sliver and has length $l^* = l\cos\vartheta$ according to Figure 6.3, cross-sectional area $s^* = s/\cos\vartheta$ according to Equation (1.40), and fineness $t^* = s^*\rho$ according to Equation (1.5). (The original oblique fiber has fineness $t = s\rho$.) Thus,

$$t^* = s^*\rho = \frac{s\rho}{\cos\vartheta} = \frac{t}{\cos\vartheta}. \tag{6.22}$$

Because the angle ϑ is same for all the fibers, it must be valid for the mean fineness of the effective fibers that

$$\bar{t}^* = \bar{t}/\cos\vartheta. \tag{6.23}$$

The coefficient of variation of the effective fiber fineness that does not change, according to Equation (A5.7) shown in Appendix 5, is as follows:

$$v\left(t^*\right) = v\left(\frac{t}{\cos\vartheta}\right) = v(t). \tag{6.24}$$

(Here, it is true that $1/\cos\vartheta = \text{constant}$ for all fibers.)

The coefficient of variation $v(T)$ of local sliver fineness T is found by applying Equations (6.22) and (6.24) into (6.17) as follows:

$$v^2(T) = \frac{\bar{t}^*}{\bar{T}}\left[1 + v^2\left(t^*\right)\right] = \frac{\bar{t}}{\bar{T}\cos\vartheta}\left[1 + v^2(t)\right], \quad v(T) = \sqrt{\frac{\bar{t}}{\bar{T}\cos\vartheta}}\sqrt{1 + v^2(t)}. \tag{6.25}$$

It is obvious that the sliver produced from obliquely arranged fibers has higher coefficient of variation of local fineness than that formed from parallel fibers. Equation (6.25) will be identical to Equation (6.17) by substituting $\vartheta = 0$. On the contrary, if the angle $\vartheta \rightarrow \pi/2$ then the coefficient of variation of local fineness $v(T) \rightarrow \infty$. It is because the fibers (fiber portions) are perpendicular to the longitudinal direction of sliver in this case so that the finite (or infinity long) length of fibers are lying on an elementary length of sliver.

The number of fibers in the cross section of the sliver with obliquely arranged fibers is found from Equation (6.10) as $\bar{n} = \bar{T}/\bar{t}^*$, and we get the following expression by substituting Equation (6.23) into the last expression:

$$\bar{n} = \frac{\bar{T}}{\bar{t}^*} = \frac{\bar{T}}{\bar{t}}\cos\vartheta. \tag{6.26}$$

Equation (6.25) can also be obtained by applying Equation (6.26) into (6.16).

Note: If we compare Equations (6.26) and (1.44) with (1.36), we can conclude that the coefficient k_n is

$$k_n = \cos\vartheta. \tag{6.27}$$

It corresponds to Equation (1.43) for this special case.

Effect of doubling on sliver irregularity

Figure 6.4 shows a scheme of doubling of $j = 1, 2, \ldots, m$ number of independent partial slivers. Each j-th partial sliver has mean fineness $\overline{T}_j$, mean number of fibers $\overline{n}_j$ in cross section, and coefficient $k_{n,j}$. The fibers in the j-th partial sliver have their mean fineness $\overline{t}_j$, coefficient of variation of fiber fineness $v(t_j)$, constant fiber length l_j and fiber density ρ_j. In section A of the sliver – see Figure 6.4 – the random local fineness is T_j and the random local number of fibers is n_j. By doubling these partial slivers a so-called doubled sliver[5] is produced.

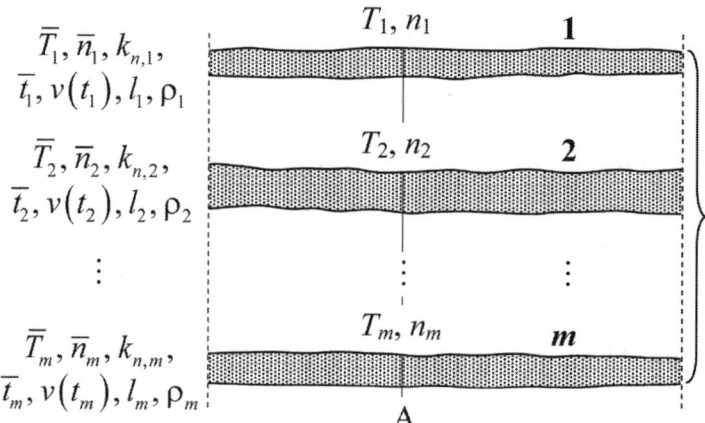

Figure 6.4 Doubling of partial slivers

General characteristics of fiber blend

The doubled sliver can be considered to be a case of fiber blend that is created from fibers of partial slivers. The following relations are then valid to write for fiber blends based on systematic derivations that are shown in Reference [3,4].

The mean value of fineness of the doubled sliver is

$$\overline{T} = \sum_{j=1}^{m} \overline{T}_j .$$
(6.28)

---

5   The quantities corresponding to the doubled sliver are denoted without any subscript.

The mass portion of fibers from the j-th partial sliver is

$$g_j = \overline{T}_j / \overline{T}, \quad \left(\sum_{j=1}^{m} g_j = 1 \right). \tag{6.29}$$

The mean number of fibers present in the cross section of the doubled sliver is

$$\overline{n} = \sum_{j=1}^{m} \overline{n}_j . \tag{6.30}$$

The mean fiber fineness in the doubled sliver is

$$\overline{t} = 1 \bigg/ \sum_{j=1}^{m} \frac{g_j}{\overline{t}_j} . \tag{6.31}$$

The length portion of fibers from the j-th partial sliver, i.e., the cumulative length of all fibers in the j-th partial sliver in relation to the total length of all fibers in the doubled sliver is

$$\lambda_j = g_j \frac{\overline{t}}{\overline{t}_j}, \quad \left(\sum_{j=1}^{m} \lambda_j = \overline{t} \sum_{j=1}^{m} \frac{g_j}{\overline{t}_j} = 1 \right). \tag{6.32}$$

The mean number $\overline{n}_j$ of fibers present in the cross section of the j-th partial sliver and the mean number $\overline{n}$ of fibers present in the cross section of the doubled sliver can be expressed by Equation (1.44) where τ is given by Equation (1.36). Thus

$$\overline{n}_j = \left(\overline{T}_j / \overline{t}_j \right) k_{n,j}, \tag{6.33}$$

$$\overline{n} = \left(\overline{T} / \overline{t} \right) k_n. \tag{6.34}$$

(Generally, the coefficient $k_{n,j}$ of the partial slivers and the coefficient k_n of the doubled sliver are not required to be equal to 1.)

By using the last two equations in Equation (6.30) and rearranging such expression by using Equations (6.29) and (6.33), we obtain the expression for k_n as follows:

$$\frac{\overline{T}}{\overline{t}} k_n = \sum_{j=1}^{m} \left(\frac{\overline{T}_j}{\overline{t}_j} k_{n,j} \right), \quad k_n = \sum_{j=1}^{m} \left[\left(\frac{\overline{T}_j}{\overline{T}} \frac{\overline{t}}{\overline{t}_j} \right) k_{n,j} \right] = \sum_{j=1}^{m} \left[\left(g_j \frac{\overline{t}}{\overline{t}_j} \right) k_{n,j} \right] = \sum_{j=1}^{m} \left(\lambda_j k_{n,j} \right). \tag{6.35}$$

As shown, the coefficient k_n is the arithmetic mean of $k_{n,j}$ values, weighted by the fiber length portions λ_j. Nevertheless, we will use also the harmonic mean $k_{n\,\mathrm{HARM}}$ as

$$k_{n\,\mathrm{HARM}} = \frac{1}{\displaystyle\sum_{j=1}^{m} \frac{\lambda_j}{k_{n,j}}} \quad \left(\frac{1}{k_{n\,\mathrm{HARM}}} = \sum_{j=1}^{m} \frac{\lambda_j}{k_{n,j}} \right). \tag{6.36}$$

Note: The coefficient k_n, according to Equation (6.35), is the actual coefficient k_n of the doubled sliver. It satisfies the relationship $k_n = s/\overline{s}^*$ stated in Equation (1.39). In opposite to this, the harmonic mean $k_{n\,\mathrm{HARM}}$, given by Equation (6.36), is only a formal mathematical variable. If $k_{n,j}$ are different for different partial slivers, then the inequality $k_{n\,\mathrm{HARM}} < k_n$ is valid. (This is derived in Appendix P7 of the book [3].)

Further, let us express two local quantities. The local fineness of the doubled sliver is

$$T = \sum_{j=1}^{m} T_j, \tag{6.37}$$

and the local number of fibers present in the cross section of the doubled sliver is

$$n = \sum_{j=1}^{m} n_j. \tag{6.38}$$

Coefficient of variation of mass irregularity of doubled sliver

From the general definition of coefficient of variation, we can express the variance of the j-th partial sliver as follows:

$$D(T_j) = \overline{T}_j^2\, v^2(T_j). \tag{6.39}$$

According to the basic statistical rules, the variance $D(T)$ of the doubled sliver is the summation of the variances of the partial slivers. So, by using Equation (6.39), it is valid to write that

$$D(T) = \sum_{j=1}^{m} D(T_j) = \sum_{j=1}^{m} \left[\overline{T}_j^2\, v^2(T_j) \right]. \tag{6.40}$$

Then, the square of coefficient of mass variation of the doubled sliver is

$$v^2(T) = \frac{D(T)}{\overline{T}^2} = \frac{1}{\overline{T}^2}\sum_{j=1}^{m}\left[\overline{T}_j^2 v^2(T_j)\right], \quad v(T) = \frac{1}{\overline{T}}\sqrt{\sum_{j=1}^{m}\left[\overline{T}_j^2 v^2(T_j)\right]}, \quad (6.41)$$

and/or by using Equation (6.29)

$$v^2(T) = \sum_{j=1}^{m}\left[\left(\frac{\overline{T}_j}{\overline{T}}\right)^2 v^2(T_j)\right] = \sum_{j=1}^{m}\left[g_j^2 v^2(T_j)\right], \quad v(T) = \sqrt{\sum_{j=1}^{m}\left[g_j^2 v^2(T_j)\right]}.$$

$$(6.42)$$

Note: Zellweger Uster [2] uses the symbol $CV_{\lim}$ in place of $v(T)$ according to Equation (6.20) and similarly the symbol $CV_{\lim j}$ in place of $v(T_j)$. Then, Equation (6.41) can be represented as follows:

$$CV_{\lim[\%]} = \sqrt{\sum_{j=1}^{m}\left[\overline{T}_{j[\text{tex}]}^2 CV_{\lim j[\%]}^2\right]}\Big/\overline{T}_{[\text{tex}]}. \quad (6.43)$$

Note: We may observe that while deriving Equations (6.41) and (6.42), there was no need to consider the individual slivers either as binomial sliver or as Poisson sliver. These expressions are valid for any doubled sliver composed of independent partial slivers.

Doubling of Poisson slivers with obliquely arranged fibers

Let us assume that the doubled sliver is created from a set of partial Poisson slivers. The different partial slivers have different finenesses, differently arranged oblique fibers, different lengths of fibers and different finenesses of fibers. Accordingly, the coefficient of variation of the local fiber fineness of the j-th sliver $v(T_j)$ can be written according to Equation (6.25) as follows:

$$v^2(T_j) = \frac{\overline{t}_j}{\overline{T}_j \cos\vartheta_j}\left[1 + v^2(t_j)\right], \quad v(T_j) = \sqrt{\frac{\overline{t}_j}{\overline{T}_j \cos\vartheta_j}}\sqrt{1 + v^2(t_j)}.$$

$$(6.44)$$

The following expression can be obtained by substituting Equation (6.44) into (6.41) and also applying Equations (6.29) and (6.32):

$$v^2\left(T\right)=\frac{1}{\overline{T}^2}\sum_{j=1}^{m}\left\{\overline{T}_j^2\,v^2\left(T_j\right)\right\}=\frac{1}{\overline{T}^2}\sum_{j=1}^{m}\left\{\overline{T}_j^2\,\frac{\overline{t}_j}{\overline{T}_j\,\cos\vartheta_j}\left[1+v^2\left(t_j\right)\right]\right\}$$

$$=\frac{1}{\overline{T}^2}\sum_{j=1}^{m}\left\{\frac{\overline{T}_j\,\overline{t}_j}{\cos\vartheta_j}\left[1+v^2\left(t_j\right)\right]\right\}=\sum_{j=1}^{m}\left\{\overbrace{\left(\frac{\overline{T}_j}{\overline{T}}\,\frac{\overline{t}}{\overline{t}_j}\right)}^{=\lambda_j}\left(\frac{\overline{t}_j}{\overline{t}}\right)^2\frac{\overline{t}}{\overline{T}}\frac{1}{\cos\vartheta_j}\left[1+v^2\left(t_j\right)\right]\right\},$$

$$v^2\left(T\right)=\frac{\overline{t}}{\overline{T}}\sum_{j=1}^{m}\left\{\frac{\lambda_j}{\cos\vartheta_j}\left(\frac{\overline{t}_j}{\overline{t}}\right)^2\left[1+v^2\left(t_j\right)\right]\right\}.$$

$$(6.45)$$

The last expression is the universal equation that is used for expressing the coefficient of variation of the local fineness of Poisson sliver with obliquely oriented fibers.

Let us note that Equation (6.45) does not contain fiber length l_j. It means that the fiber lengths (and their distribution in the doubled sliver) do not influence on the coefficient of variation of local fineness of (doubled) sliver.

Special cases of Poisson slivers

Let us think about three special cases:

1. The easiest case:

    if $\overline{t}_j=\overline{t}=$ constant, $v\left(t_j\right)=v\left(t\right)=$ constant, $\vartheta_j=\vartheta=0$, then, by

    using Equation (6.32), Equation (6.45) takes the following form:

$$v^2\left(T\right)=\frac{\overline{t}}{\overline{T}}\sum_{j=1}^{m}\left\{\frac{\lambda_j}{\cos\vartheta}\left(\frac{\overline{t}}{\overline{t}}\right)^2\left[1+v^2\left(t\right)\right]\right\}=\frac{\overline{t}}{\overline{T}}\left[1+v^2\left(t\right)\right]\sum_{j=1}^{m}\lambda_j=\frac{\overline{t}}{\overline{T}}\left[1+v^2\left(t\right)\right].$$

$$(6.46)$$

This expression is identical to Equation (6.17).

2. If $\overline{t}_j=\overline{t}=$ constant, $v\left(t_j\right)=v\left(t\right)=$ constant, $\vartheta_j=\vartheta=$ constant, then, by using Equation (6.32), Equation (6.45) obtains the following form:

$$v^2\left(T\right)=\frac{\overline{t}}{\overline{T}}\sum_{j=1}^{m}\left\{\frac{\lambda_j}{\cos\vartheta}\left(\frac{\overline{t}}{\overline{t}}\right)^2\left[1+v^2\left(t\right)\right]\right\}=\frac{\overline{t}}{\overline{T}\cos\vartheta}\left[1+v^2\left(t\right)\right]\sum_{j=1}^{m}\lambda_j$$

$$=\frac{\overline{t}}{\overline{T}\cos\vartheta}\left[1+v^2\left(t\right)\right].$$

$$(6.47)$$

This expression is identical to Equation (6.25). It shows that the coefficient of variation of local fineness of the doubled sliver, formed from a set of partial slivers of different finenesses, is the same as the coefficient of variation of each partial sliver in this case.

According to Equation (6.27), each individual sliver has the same coefficient $k_{n,j} = \cos \vartheta$ and according to Equation (6.35), the doubled sliver has the same coefficient

$$k_n = \sum_{j=1}^{m} (\lambda_j \, k_{n,j}) = \sum_{j=1}^{m} (\lambda_j \cos \vartheta) = \cos \vartheta \sum_{j=1}^{m} \lambda_j = \cos \vartheta.$$ Equation (6.47) can also be expressed as follows:

$$v^2(T) = \frac{\overline{t}}{T \, k_n} \left[1 + v^2(t) \right].$$ (6.48)

3. If only $\overline{t}_j = \overline{t} = \text{constant}$ and $v(t_j) = v(t) = \text{constant}$, then Equation (6.45) obtains the following form by using Equation (6.32):

$$v^2(T) = \frac{\overline{t}}{T} \sum_{j=1}^{m} \left\{ \frac{\lambda_j}{\cos \vartheta_j} \left(\frac{\overline{t}}{\overline{t}} \right)^2 \left[1 + v^2(t) \right] \right\} = \frac{\overline{t}}{T} \left[1 + v^2(t) \right] \sum_{j=1}^{m} \frac{\lambda_j}{\cos \vartheta_j},$$

$$v^2(T) = \frac{\overline{t}}{T} \left[1 + v^2(t) \right] \xi, \quad \text{where} \quad \xi = \sum_{j=1}^{m} \frac{\lambda_j}{\cos \vartheta_j}.$$ (6.49)

Note: The quantity ξ represents an 'increasing factor' of 'traditional' square of unevenness of Poisson sliver.

Equation (6.27) is valid for each partial sliver and then $\lambda_j / \cos \vartheta_j = \lambda_j / k_{n,j}$. Using Equation (6.36), we can write $\sum_{j=1}^{m} (\lambda_j / k_{n,j}) = 1 / k_{n,\text{HARM}}$ and the alternative expression for Equation (6.49) is

$$v^2(T) = \frac{\overline{t}}{T} \left[1 + v^2(t) \right] \frac{1}{k_{n,\text{HARM}}} = \frac{\overline{t}}{T \, k_n} \left[1 + v^2(t) \right] \frac{k_n}{k_{n,\text{HARM}}}.$$ (6.50)

As the arithmetic mean of k_n is higher than the harmonic mean of $k_{n,\text{HARM}}$, it is obvious that $v^2(T)$, according to Equation (6.50), is higher than $v^2(T)$, according to Equation (6.48). In other words, the sliver produced from fibers of different orientations has higher irregularity than the slivers having the same coefficient k_n, but formed from the fibers oriented only at one direction, even in oblique.

Continuous distribution of angle ϑ

Equation (6.49) characterizes the doubling of partial slivers that differ them-selves by different angles ϑ_j and finenesses $\overline{T}_j$. [Because, according to Equations (6.28), (6.29) and (6.32), it is valid that $\lambda_j = g_j = \overline{T}_j / \sum_{j=1}^{m} \overline{T}_j$.]
The mass unevenness of the doubled sliver, according to Equation (6.49), is the product of two components: traditional unevenness of Poisson sliver from parallel fibers according to Equation (6.17) and the expression $\xi = \sum_{j=1}^{m} (\lambda_j / \cos \vartheta_j)$.

Now, let us imagine that we are doubling very (infinitely) high number of partial slivers ($m \to \infty$) when the length portion $\lambda_j = g_j$ of fibers in each partial sliver is very (differentially) small ($\lambda_j \to d\lambda$). In the limit case, we use infinite number of elementary partial slivers where each such sliver is identified by its angle ϑ and the elementary length portion is given by

$$d\lambda = u(\vartheta)d\vartheta, \quad \left(\int_{\vartheta=0}^{\vartheta=\pi/2} d\lambda = \int_{0}^{\pi/2} u(\vartheta)d\vartheta = 1 \right), \tag{6.51}$$

where the function $u(\vartheta)$ denotes the probability density function of non-ori-ented[6] angle $\vartheta \in \langle 0, \pi/2 \rangle$ in the doubled sliver. Then the expression for ξ is changed from Equation (6.49) to

$$\xi = \sum_{j=1}^{m} \frac{\lambda_j}{\cos \vartheta_j} \to \xi = \int_{\vartheta=0}^{\vartheta=\pi/2} \frac{d\lambda}{\cos \vartheta} = \int_{0}^{\pi/2} \frac{u(\vartheta)d\vartheta}{\cos \vartheta}. \tag{6.52}$$

Unevenness of 'helical' yarn

The helical model was earlier described in detail in Chapter 4. A helical as well as an ideal helical model is a fully theoretical concept of an absolutely regular yarn. Nevertheless, the real yarn, which is created through random processes using staple fibers, shows a significant level of mass unevenness and its directional distribution of fibers is more or less similar to the theo-retical distribution that can be derived from the equations of helical model. We, therefore, can apply this ideal distribution for a rough estimation of the influence of fiber slopes on yarn mass unevenness.

---

6   Because there are cosines in Equation (6.49) and it is valid that $\cos \vartheta = \cos(-\vartheta)$, we can work with non-oriented angles.

According to assumption 5 in Section 4.1, it is valid that the height of each coil of each fiber is a constant in the helical model. This height is equal to $1/Z$ and the corresponding fiber length in the coil at yarn radius r is

$$l_1 = \frac{1/Z}{\cos\beta} = \frac{1}{Z\cos\beta}. \qquad (6.53)^7$$

This also follows immediately from Figure 4.2.

The slope of fiber is marked by angle β in this ideal helical model and such angle takes the values only from the interval $\beta \in \langle 0, \beta_D \rangle$, where the maximum angle β_D corresponds to the slope of the surface fiber in the yarn – see, e.g., Figure 1.9 and Equation (1.52). Now, the expressions according to Equations (6.51) and (6.52) take the special forms as follows:

$$d\lambda = u(\beta)d\beta, \quad \left(\int_{\vartheta=0}^{\vartheta=\beta_D} d\lambda = \int_0^{\beta_D} u(\beta)d\beta = 1 \right), \qquad (6.54)$$

$$\xi = \int_{\beta=0}^{\beta=\beta_D} \frac{d\lambda}{\cos\beta} = \int_0^{\beta_D} \frac{u(\beta)d\beta}{\cos\beta}. \qquad (6.55)$$

The number dn of fibers present in the cross section of the differential layer[8] at radius r is expressed by Equation (4.5). By using Equations (4.2) and (4.3), we can also write

$$dn = \frac{2\pi}{s}\cos\beta\,\mu_r r\,dr = \frac{dS}{s}\cos\beta, \qquad (6.56)$$

where dS denotes the elementary increment of yarn substance cross-sectional area S of the differential layer. The length dl_f of all dn fibers in the differential layer is then

$$dl_f = l_1 dn = \frac{1}{Z\cos\beta}\frac{2\pi}{s}\cos\beta\,\mu_r r\,dr = \frac{1}{Z\cos\beta}\frac{dS}{s}\cos\beta = \frac{dS}{Zs}. \qquad (6.57)$$

The total length l_f of all fibers in the yarn portion length $1/Z$ is

---

7   Let us remind that Z denotes yarn twist and β represents the slope of the fiber at radius r. [Equation (4.2) is valid.]

8   The differential layer is shown in Figure 4.3. The sectional area of fibers inside the differential annulus is dS, the total sectional area of all fibers in the yarn cross section – substance cross-sectional area of yarn – is S.

$$l_f = \int\limits_{r=0}^{r=D/2} dl_f = \frac{1}{Zs} \int\limits_{r=0}^{r=D/2} dS = \frac{S}{Zs}.$$ (6.58)

[Equation (4.6) was used here for rearrangement.]

The elementary length portion $d\lambda$ of fibers in the differential layer is then

$$d\lambda = \frac{dl_f}{l_f} = \frac{dS/Zs}{S/Zs} = \frac{dS}{S}.$$ (6.59)

By using Equations (4.3) and (4.6) in (6.59), it is also valid to write that

$$d\lambda = \frac{dS}{S} = \frac{2\pi\mu_r r\,dr}{2\pi \int\limits_0^{D/2} \mu_r r\,dr} = \frac{\mu_r r\,dr}{\int\limits_0^{D/2} \mu_r r\,dr},$$ (6.60)

where μ_r represents the local packing density of the differential layer at radius r.

By using Equation (4.1), i.e., $\tan\beta = 2\pi rZ$, we can write

$$r = \frac{\tan\beta}{2\pi Z}, \quad dr = \frac{d\beta}{2\pi Z \cos^2\beta},$$

$$r\,dr = \frac{\tan\beta\,d\beta}{(2\pi Z)^2 \cos^2\beta} = \frac{\sin\beta\,d\beta}{(2\pi Z)^2 \cos^3\beta} = \frac{D^2 \sin\beta\,d\beta}{4(\pi DZ)^2 \cos^3\beta} = \frac{D^2}{4\tan^2\beta_D} \frac{\sin\beta\,d\beta}{\cos^3\beta}.$$ (6.61)

[Equation (1.52), i.e., $\tan\beta_D = \pi DZ$ was used here for rearrangement.]

We can also express the local packing density of the differential layer at radius r as a function of the corresponding angle β; we use the symbols as follows:

$$\mu_r = \mu_{r=\tan\beta/(2\pi Z)} = \mu_\beta.$$ (6.62)

By substituting Equations (6.61) and (6.62) in the numerator of Equation (6.60) and applying Equation (6.61) as an integral substitution to the denominator, we obtain the following expression:

$$d\lambda = \frac{\mu_\beta \dfrac{D^2}{4\tan^2\beta_D} \dfrac{\sin\beta\,d\beta}{\cos^3\beta}}{\dfrac{D^2}{4\tan^2\beta_D} \int\limits_0^{\beta_D} \mu_\beta \dfrac{\sin\beta\,d\beta}{\cos^3\beta}} = \frac{\mu_\beta \dfrac{\sin\beta}{\cos^3\beta}\,d\beta}{\int\limits_0^{\beta_D} \mu_\beta \dfrac{\sin\beta}{\cos^3\beta}\,d\beta}.$$ (6.63)

The probability density function $u(\beta)$ is originated after substituting Equation (6.63) in (6.54) as follows:

$$d\lambda = u(\beta)d\beta = \frac{\mu_\beta \dfrac{\sin\beta}{\cos^3\beta}d\beta}{\displaystyle\int_0^{\beta_D}\mu_\beta \dfrac{\sin\beta}{\cos^3\beta}d\beta}, \quad u(\beta) = \frac{\mu_\beta \dfrac{\sin\beta}{\cos^3\beta}}{\displaystyle\int_0^{\beta_D}\mu_\beta \dfrac{\sin\beta}{\cos^3\beta}d\beta}. \tag{6.64}$$

Finally, by applying Equation (6.64) to (6.55), we obtain the following expression for ξ:

$$\xi = \int_0^{\beta_D}\mu_\beta \frac{\sin\beta}{\cos^4\beta}d\beta \Big/ \int_0^{\beta_D}\mu_\beta \frac{\sin\beta}{\cos^3\beta}d\beta, \tag{6.65}$$

which can be used in Equation (6.49).

Unevenness of 'ideal helical' yarn

We need to know μ_β as a function of β and/or μ_r as a function of r – see Equation (6.62) – for calculation of the quantity ξ by using Equation (6.65). However, the quantity ξ can be calculated also for the special case – ideal helical model – where $\mu_\beta = \mu_r = $ constant . Then,

$$\xi = \int_0^{\beta_D}\frac{\sin\beta}{\cos^4\beta}d\beta \Big/ \int_0^{\beta_D}\frac{\sin\beta}{\cos^3\beta}d\beta = \left(-\int_1^{\cos\beta_D}\frac{dt}{t^4}\right)\Big/\left(-\int_1^{\cos\beta_D}\frac{dt}{t^3}\right) = \left[\frac{1}{3t^3}\right]_1^{\cos\beta_D}\Big/\left[\frac{1}{2t^2}\right]_1^{\cos\beta_D}$$

Substitution: $\cos\beta = t, \; -\sin\beta\, d\beta = dt,$

$$= \left[\frac{1}{3}\left(\frac{1}{\cos^3\beta_D}-1\right)\right]\Big/\left[\frac{1}{2}\left(\frac{1}{\cos^2\beta_D}-1\right)\right] = \left[\frac{1}{3}\frac{1-\cos^3\beta_D}{\cos^3\beta_D}\right]\Big/\left[\frac{1}{2}\frac{1-\cos^2\beta_D}{\cos^2\beta_D}\right]$$

$$= \frac{2}{3}\frac{1-\cos^3\beta_D}{\cos^3\beta_D}\frac{\cos^2\beta_D}{\sin^2\beta_D} = \frac{2}{3}\frac{1-\cos^3\beta_D}{\cos\beta_D\sin^2\beta_D}. \tag{6.66}$$

$$\sqrt{\xi} = \sqrt{\frac{2}{3}\frac{1-\cos^3\beta_D}{\cos\beta_D\sin^2\beta_D}}. \tag{6.67}$$

The quantity ξ, according to Equation (6.66), and/or the quantity $\sqrt{\xi}$, according to Equation (6.67), correspond to the angular distribution of fibers in the ideal helical model. Nevertheless, by using this quantity in Equation (6.49), we can roughly estimate the significance of increase of the coefficient of variation of local fineness $v(T)$, resulting from the angular distribution of fibers due to twist. According to Equation (6.49), the coefficient of mass variation $v(T)$ is increased by a multiplication of $\sqrt{\xi}$. This factor is illustrated in Figure 6.5.

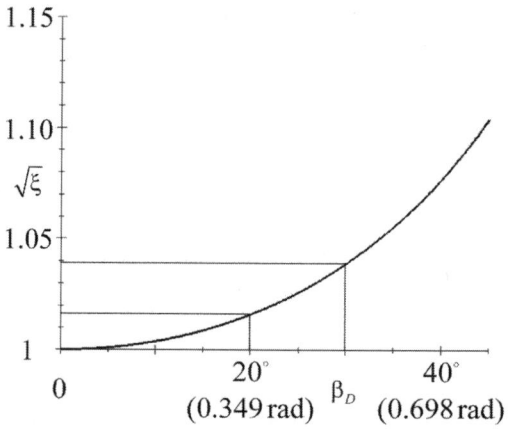

Figure 6.5 Factor $\sqrt{\xi}$ according to Equation (6.67)

Note: If the angle β_D of the surface fiber lies between $20°$ and $30°$, then the value $\sqrt{\xi}$ lies roughly between 1.015 and 1.04. Such an increase is not really too significant. Besides, the real yarn usually has smaller values of the packing densities μ_r in the layers near to the yarn surface, where the fiber slopes are higher. It means that the influence of real value $\sqrt{\xi}$ can even be smaller than that shown in Figure 6.5.

Unevenness of sliver

Let us consider that the fiber portions are inclined at different angles ϑ drawn from the longitudinal direction of the sliver. The following probability density function $u(\vartheta)$ was derived in the book [4] for such structures:

$$u(\vartheta) = \frac{2}{\pi} \frac{C}{C^2 - \left(C^2 - 1\right)\cos^2 \vartheta}, \quad \vartheta \in \langle 0, \pi/2 \rangle. \tag{6.68}$$

The behaviour of the function is illustrated in Figure 6.6.

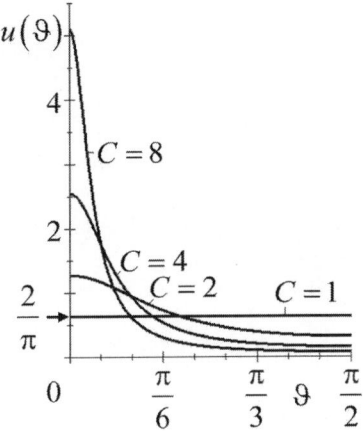

Figure 6.6 Probability density function $u(\vartheta)$, according to Equation (6.68)

Note: The parameter $C \geq 1$ expresses the degree of preference of the fibers to be inclined to the longitudinal direction of the sliver. The higher is the parameter C, the more is the fibers oriented near to the longitudinal direction; $C = 1$ corresponds to isotropic orientation.

Because the probability density function $u(\vartheta)$ has its minimum $u(\vartheta) = u(\pi/2) = 2/(C\pi)$ (i.e., when the fiber portion is lying perpendicular to the longitudinal direction of the sliver) let us rearrange the function $u(\vartheta)$, expressed in Equation (6.68), as follows:

$$
\begin{aligned}
u(\vartheta) &= \frac{2}{\pi}\left\{\frac{1}{C}+\left[\frac{C}{C^2-\left(C^2-1\right)\cos^2\vartheta}-\frac{1}{C}\right]\right\} \\
&= \frac{2}{\pi}\left\{\frac{1}{C}+\frac{C^2-C^2+\left(C^2-1\right)\cos^2\vartheta}{C\left[C^2-\left(C^2-1\right)\cos^2\vartheta\right]}\right\} \\
&= \frac{2}{\pi}\left\{\frac{1}{C}+\frac{\left(C^2-1\right)/C}{C^2/\cos^2\vartheta-\left(C^2-1\right)}\right\}=\frac{2}{C\pi}+\frac{2}{\pi}\frac{\left(C^2-1\right)/C}{C^2\tan^2\vartheta+1}, \quad \vartheta\in\langle 0,\pi/2\rangle.
\end{aligned}
$$

(6.69)

Then, according to Equations (6.52) and (6.69), it is valid to write that

$$\xi = \int_0^{\pi/2} \frac{u(\vartheta)\,d\vartheta}{\cos\vartheta} = \frac{2}{C\pi} \int_0^{\pi/2} \frac{d\vartheta}{\cos\vartheta} + \frac{2(C^2-1)}{C\pi} \int_0^{\pi/2} \frac{d\vartheta}{\cos\vartheta\left(C^2\tan^2\vartheta+1\right)}.$$

(6.70)

Let us solve the first integral in the last expression as follows:

$$\int \frac{d\vartheta}{\cos\vartheta} = \int \frac{dz}{z} = \ln|z| = \ln\left|\frac{1}{\cos\vartheta} + \tan\vartheta\right|,$$

Substitution: $\dfrac{1}{\cos\vartheta} + \tan\vartheta = z$, $\dfrac{1}{\cos\vartheta}\left(\tan\vartheta + \dfrac{1}{\cos\vartheta}\right)d\vartheta = dz$, $\dfrac{d\vartheta}{\cos\vartheta} = \dfrac{dz}{z}$,

$$\int_0^{\pi/2} \frac{d\vartheta}{\cos\vartheta} = \ln\left|\frac{1}{0} + \infty\right| - \ln\left|\frac{1}{1} + \tan 0\right| = \infty.$$

(6.71)

The second integral, expressed by Equation (6.70), takes only the positive values (or values equal to zero) so that this integral cannot take a negative value. Then, by using these results in Equation (6.70) and according to Equation (6.49), we can write the relations as follows:

$$\xi = \infty, \quad v^2(T) = \infty, \quad v(T) = \infty .$$

(6.72)

Why did we obtain so 'strange' result? It is because the probability density function as expressed in Equation (6.68) has the positive value $u(\vartheta) = u(\pi/2) = 2/(C\pi)$ and the local fineness $v(T)$ according to Equation (6.25) is limited to infinity when $\vartheta = \pi/2$. [See the comments given in the context of Equation (6.25).]

Note: Nevertheless, in practice, such strange outcome is not observed. It is because the theoretically calculated result according to Equation (6.71) is related to the coefficient of variation of local mass unevenness, i.e., variation among the infinitesimally small portions of the sliver. The practically measured unevenness always characterizes the coefficient of mass variation of the portions of sliver and/or yarn having lengths of several millimetres; which is a quite other problem. (Let us remind that one millimetre is usually considered to be a very long length in relation to fiber dimension. It usually represents more than fifty times of fiber diameter.) The relation between the chosen length of the measured portion of the sliver and/or yarn and the corresponding coefficient of mass variation is known by the term 'variance-length curve', described in References [5, 6].

6.4   Some alternative ideas of yarn irregularity

Huberty's index of irregularity

Usually, it is recommended to compare the calculated value of yarn irregularity with the measured one of a real yarn.

The calculated result $v(T)$ is usually determined according to Equation (6.17). (This follows the idea of Poisson sliver made up of parallel fibers.) Nevertheless, the symbol CV_{lim}, according to Equation (6.21), is often preferred to be expressed in percentage.

On the contrary, we call $v_{\text{eff}}(T)$ and/or CV_{eff} (dimensionless or in percentage) the really measured value[9] – i.e., 'effective' value – of yarn irregularity.

Huberty [7] introduced the well-known index of irregularity I as the ratio between the measured and calculated values

$$I = \frac{v_{\text{eff}}(T)}{v(T)} = \frac{CV_{\text{eff}}}{CV_{\text{lim}}} = \frac{CV_{\text{eff}[\%]}}{CV_{\text{lim}[\%]}}.$$ (6.73)

The practically observed values of Huberty's index of yarn irregularity are relatively high; they often range from 1.2 to more than 2. Such high values cannot be explained by the slopes of fibers due to twist (compare with Figure 6.5) or by another small inaccuracy of the theoretical model. It seems to be resulting from a quite new significant phenomenon.

Uster irregularity

The generally known 'Uster statistics', produced by Zellweger Uster Company, do not find any logical reason for the above-mentioned disproportion between the measured and calculated yarn irregularity. In opposite to Equation (6.17), they often use the following empirical relation

$$v_{\text{USTER}}(T) = \frac{a}{T^b} \quad \left(\ln v_{\text{USTER}}(T) = \ln a - b \ln \overline{T} \right).$$ (6.74)

Here, a and b are empirical parameters, related to the material and technological characteristics of the yarns. Let us note that this relation is linear in logarithmical scale (see the expression in brackets) as it is often shown in the popular graphs of Uster statistics.

---

9   This quantity is often used by Zellweger Uster.

Bornet's idea of clusters

Bornet [8] made an attempt to investigate the causes of so high values of Huberty's index of irregularity. He claimed that the sliver is not formed by random deposition of individual fibers but from random deposition of fiber clusters. It contrary to Martindale's assumption [Section 6.2, assumptions (a) and (c)], the fiber clusters consist of several fibers depositing together during the formation of sliver and they behave like individual 'thick' fiber. The number of clusters is naturally less than the number of individual fibers present in the sliver. The mean fineness of fiber clusters $\overline{t}_{\text{BORNET}}$, as suggested by Bornet, is given by the following empirical equation:

$$\overline{t}_{\text{BORNET}} = \frac{1}{4}\overline{T}^{1/3}\,\overline{t}^{2/3}\,, \text{ i.e., } \frac{\overline{t}_{\text{BORNET}}}{\overline{t}} = \frac{1}{4}\left(\frac{\overline{T}}{\overline{t}}\right)^{1/3}. \tag{6.75}$$

The mean value of fineness of fiber cluster depends on mean fiber fineness $\overline{t}$ and relative sliver fineness $\tau = \overline{T}/\overline{t}$ – see Equation (1.36). A thick sliver is formed from a 'high' number of clusters. Bornet's empirical Equation (6.75) is plotted in Figure 6.7.

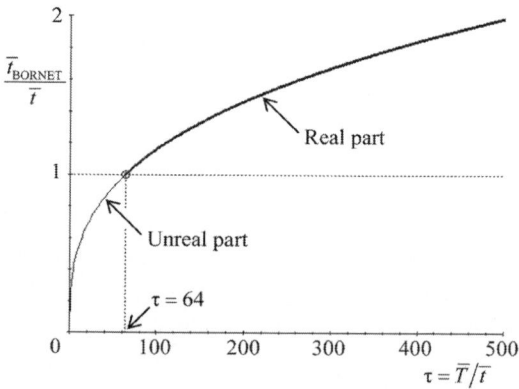

Figure 6.7 Bornet's empirical Equation (6.75)

Note: It is obvious that if the ratio $\tau = \overline{T}/\overline{t} < 64$ then $\overline{t}_{\text{BORNET}} < \overline{t}$, which is evidently illogical. The mean number of fibers in a sliver cross section, formed from parallel fibers ($k_n = 1$), can be estimated from Equation (1.44), i.e., $n = \tau$ now. Therefore, Bornet's Equation (6.75) cannot be valid for slivers or yarns having less than 64 fibers in the cross sections.

It is also valid to write from Equation (6.75) that

$$\frac{\overline{t}_{\text{BORNET}}}{\overline{T}} = \frac{1}{4}\left(\frac{\overline{t}}{\overline{T}}\right)^{2/3}.$$

(6.76)

By applying this ratio in place of $\overline{t}/\overline{T}$ in Equation (6.17), we obtain the following Bornet's expression for sliver and/or yarn irregularity

$$v_{\text{BORNET}}(T) = \sqrt{\frac{\overline{t}_{\text{BORNET}}}{\overline{T}}}\sqrt{v^2(t)+1} = \frac{1}{2}\left(\frac{\overline{t}}{\overline{T}}\right)^{1/3}\sqrt{v^2(t)+1}.$$

(6.77)

Hierarchical structure of random fiber aggregates

The answer to the question why the Huberty's index of yarn irregularity is usually so high based on the Bornet's idea of fiber clusters is right in principle. The movement of fibers resulting in creation of various fiber groups was experimentally observed many times in different spinning processes.

Note: For example, a tooth of an opening roller often throws many groups of fibers (more than one) to the transport channel in rotor spinning process. Also, in ring-spinning machine, it is well known that the movement of fibers in the drawing process takes place in groups (mainly the so-called 'floating' fibers).

Nevertheless, Bornet's excellent idea was mathematically described only by empirically with the help of Equations (6.75) to (6.77).

Nečkář[10] [10] suggested that the main cause of higher values of Huberty's index of yarn irregularity is due to random deposition of different fiber aggregates while forming a sliver. This probabilistic model introduces several structural units as shown in Figure 6.8[11].

The basic, smallest and easiest structural unit is a fiber – level 4 in Figure 6.8.

---

10   The initial idea of this probabilistic model was already published in the book [9].

11   Originally, we thought about only one type of fiber groups (clusters). However, our experimental experience showed the necessity to work with two levels of fiber aggregates – bundles and clusters.

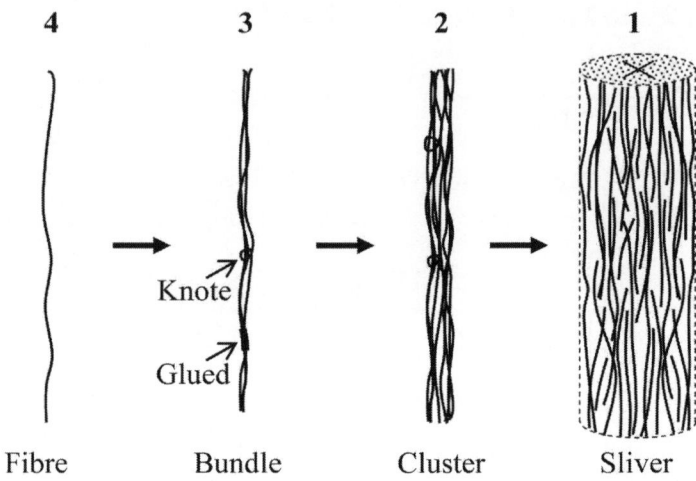

Figure 6.8 Structural units of sliver

A bundle – level 3 – is created from several fibers which are bonded together by glued parts and/or by tight knots. The separation of these fibers into individual one is practically impossible by a given technological process. The fiber bundle behaves always as one thick fiber.

The higher structural unit is a cluster – level 2 – shown in Figure 6.8. The clusters are formed by loose binding of several bundles. The clusters can be separated (under certain circumstances) by application of a given technological process.

Finally, a sliver – level 1 – is formed from fiber clusters as shown in Figure 6.8.

Note: The raw material is opened in several stages to produce fiber flocks in a real technological process. So, clusters, bundles and individual fibers arise in a sequence 'opposite' to the sequence of arrows shown in Figure 6.8. Nevertheless, we can imagine (without loosing any generality) that the creation of bundles from individual fiber, creation of clusters from bundles and creation of slivers from clusters.

Symbols and basic relations

All structural units have different levels of fineness. Their variabilities are characterized by the coefficients of variation of fineness. The symbols from Table 6.1 will be used.

Table 6.1 Symbols of fineness

Level	Structural unit	Individual fineness	Coefficient of variation
1	Sliver	$t_1 \equiv T$	$v_1 \equiv v(T)$
2	Cluster	t_2	v_2
3	Bundle	t_3	v_3
4	Fiber	$t_4 \equiv t$	$v_4 \equiv v(t)$

The higher units are created by aggregation of lower units (fiber → bundle → cluster → sliver) in the present model. The number of lower units in a higher unit will be marked by the symbols given in Table 6.2.

Table 6.2 Number of lower units in higher units

No. of lower units	Present in higher unit		
	1 – Sliver cross section	2 – Cluster	3 – Bundle
2 – Clusters	q_{21}	1	–
3 – Bundles	q_{31}	q_{32}	1
4 – Fibers	$q_{41} \equiv n$	q_{42}	q_{43}

Note: We use the new symbols t_1, t_4, v_1, v_4 in place of formerly used symbols $T, t, v(T), v(t)$ in Table 6.1, and the new symbol q_{41} in place of formerly used symbol n in Table 6.2. This is because of easier development of the algorithms.

The local sliver fineness t_1 is the summation of finenesses of the constituent clusters

$$t_1 = 0 \text{ for } q_{21} = 0, \quad t_1 = \sum_{i=1}^{q_{21}} (t_2)_i \text{ for } q_{21} = 1,2,\dots . \tag{6.78}$$

The individual cluster fineness t_2 is the summation of finenesses of the constituent bundles

$$t_2 = 0 \text{ for } q_{32} = 0, \quad t_2 = \sum_{i=1}^{q_{32}} (t_3)_i \text{ for } q_{32} = 1,2,\dots . \tag{6.79}$$

Finally, the individual bundle fineness t_3 is the summation of finenesses of the constituent fibers

$$t_3 = 0 \text{ for } q_{43} = 0, \quad t_3 = \sum_{i=1}^{q_{43}} (t_4)_i \text{ for } q_{43} = 1, 2, \ldots. \tag{6.80}$$

The mean number of the lower units in the higher unit will be denoted by the symbols given in Table 6.2, but with a bar; e.g., the mean number of fibers in the cluster is $\overline{q}_{42}$, etc. According to the statistical rules, the mean number of fibers present in the cluster can be expressed as follows:

$$\overline{q}_{42} = \overline{q}_{43}\overline{q}_{32} . \tag{6.81}$$

The mean number of fibers present in the cross section of the sliver is

$$\overline{q}_{41} = \overline{q}_{42}\overline{q}_{21} = \overline{q}_{43}\overline{q}_{32}\overline{q}_{21} \tag{6.82}$$

and the mean number of bundles present in the cross section of the sliver is

$$\overline{q}_{31} = \overline{q}_{32}\overline{q}_{21} . \tag{6.83}$$

The maximum number of the lower units present in the higher units will be denoted by the symbols given in Table 6.2, but with the subscript 'max'; e.g., the maximum number of fibers present in the cluster is $q_{42\max}$, etc. The maximum number of fibers present in the cluster can be expressed as follows:

$$q_{42\max} = q_{43\max} q_{32\max} . \tag{6.84}$$

The maximum number of fibers present in the cross section of the sliver is

$$q_{41\max} = q_{42\max} q_{21\max} = q_{43\max} q_{32\max} q_{21\max} \tag{6.85}$$

and the maximum number of bundles present in the cross section of the sliver is

$$q_{31\max} = q_{32\max} q_{21\max} . \tag{6.86}$$

It is obvious from the above equations that the maximum number of fibers present in a bundle does not exceed the number of fibers present in a cluster, and the latter does not exceed the number of fibers present in the cross section of the sliver. Hence, it is valid to write that

$$q_{43max} \leq q_{42max} \leq q_{41max} \,. \qquad (6.87)$$

Similarly, the following expression is also valid to write

$$q_{32max} \leq q_{31max} \,. \qquad (6.88)$$

Sliver formation

While creating the above-mentioned hierarchical model of yarn irregularity, many assumptions of Martindale's ideal sliver, described in Section 6.2, are not followed. We assume that a sliver is created from straight staple fibers of equal length l. Also, we assume that from a higher structural unit is formed from lower structural unit (fiber $\rightarrow$ bundle, bundle $\rightarrow$ cluster, cluster $\rightarrow$ sliver) and the structural units are (a) mutually independent, (b) completely random and (c) parallel to the longitudinal direction of the sliver.

According to the above assumptions, the number of fibers present in a bundle q_{43} is considered as a random variable following a suitable distribution in a defined interval $q_{43} \in \langle 0, q_{43max} \rangle$. Similarly, the number of bundles present in a cluster q_{32} follows a suitable distribution in a defined interval $q_{32} \in \langle 0, q_{32max} \rangle$. Finally, the number of clusters present in a sliver q_{21} follows a suitable distribution in a defined interval $q_{21} \in \langle 0, q_{21max} \rangle$.

6.5   Mass irregularity of structural units following binomial and Poisson distributions

Variant of binomial distributions

Let us imagine that the number of fibers in a bundle q_{43} follows a suitable binomial distribution, the number of bundles in a cluster q_{32} follows a suitable binomial distribution, and the number of clusters in the cross section of a sliver q_{21} follows also a suitable binomial distribution. All three distributions have their individual parameters, i.e., $\bar{q}_{43}, q_{43max}$, further $\bar{q}_{32}, q_{32max}$ and finally $\bar{q}_{21}, q_{21max}$.

Coefficient of variation of local sliver fineness

The finenesses t_1, t_2, t_3, t_4 are random variables too. Their mutual relationships are given by Equations (6.78) to (6.80). If we compare them with Equation (A5.15) stated in Appendix 5, we will clearly recognize that the variables t_1, t_2, t_3 (fineness of sliver, cluster and bundle) are the random variable like variable type y as mentioned in Appendix 5. In such case, the coefficient of variation of the random variable y is given by Equation (A5.23) in Appendix 5. By applying the present symbolism, we find the following triplet of expressions:

$$v_1^2 = \frac{1}{\overline{q}_{21}} \left\{ v_2^2 + \left[1 - \frac{\overline{q}_{21}}{q_{21\,max}} \right] \right\}, \tag{6.89}$$

$$v_2^2 = \frac{1}{\overline{q}_{32}} \left\{ v_3^2 + \left[1 - \frac{\overline{q}_{32}}{q_{32\,max}} \right] \right\}, \tag{6.90}$$

$$v_3^2 = \frac{1}{\overline{q}_{43}} \left\{ v_4^2 + \left[1 - \frac{\overline{q}_{43}}{q_{43\,max}} \right] \right\}. \tag{6.91}$$

By applying Equation (6.91) into (6.90), and the resulting expression into (6.89), we find the coefficient of variation of local sliver fineness. Moreover, we find the following expression using Equations (6.81) and (6.82):

$$v_1^2 = \frac{\dfrac{\dfrac{\dfrac{v_4^2 + \left[1 - \dfrac{\overline{q}_{43}}{q_{43\,max}} \right]}{\overline{q}_{43}} + \left[1 - \dfrac{\overline{q}_{32}}{q_{32\,max}} \right]}{\overline{q}_{32}} + \left[1 - \dfrac{\overline{q}_{21}}{q_{21\,max}} \right]}{\overline{q}_{21}}$$

$$= \frac{v_4^2 + \left[1 - \dfrac{\overline{q}_{43}}{q_{43\,max}} \right] + \left[1 - \dfrac{\overline{q}_{32}}{q_{32\,max}} \right] \overline{q}_{43} + \left[1 - \dfrac{\overline{q}_{21}}{q_{21\,max}} \right] \overline{q}_{43}\overline{q}_{32}}{\overline{q}_{43}\overline{q}_{32}\overline{q}_{21}},$$

$$v_1^2 = \frac{v_4^2 + \left[1 - \dfrac{\overline{q}_{43}}{q_{43\,max}} \right] + \left[1 - \dfrac{\overline{q}_{32}}{q_{32\,max}} \right] \overline{q}_{43} + \left[1 - \dfrac{\overline{q}_{21}}{q_{21\,max}} \right] \overline{q}_{42}}{\overline{q}_{41}}. \tag{6.92}$$

Mean number of fibers present in a cluster

The average size of a fiber cluster depends, to some extent, on the number of fibers present in the cross section of the sliver.

In a drafting system of drawframe and/or ring-spinning machine, the fibers, as it is known, move in groups, i.e., as fiber clusters. If we produce a coarser sliver and/or yarn, a relatively high number of fibers will pass through the drafting system. Such fibers may create clusters and move together in the drafting zone. In opposite to this, if a finer sliver and/or yarn is produced, a smaller number of fibers will pass through the drafting system. Thus, a relatively small number of fibers create clusters which will be found in the drafting zone. The proper setting of the drafting system will minimize the number of fibers in clusters in the drafting zone. The proper process parameters should be maintained to obtain the desired sliver or yarn fineness.

The fiber strand in an open-end spinning machine is formed from fiber clusters deposited in the groove of a rotor. The individual cluster is combed out by the opening roller. The production of coarser yarn necessitates deposition of a coarser strand in the rotor groove. This can be realized by feeding a coarser sliver and increasing the speed of the opening roller. Accordingly, the number of fibers per unit tooth of the opening roller should be relatively high than that in case of production of finer yarns.

Therefore, it is clear that the mean number of fibers in a cluster increases with the mean number of fibers in the cross section of a sliver, but the exact form of this dependence is still not known. It may be possible that the mean cluster size $\overline{q}_{42}$ increases proportionally with the increase in mean number of fibers $\overline{q}_{41}$ in the cross section of the sliver and/or the yarn. This assumption can be symbolically expressed as follows:

$$\overline{q}_{42} = P\overline{q}_{41}.$$
(6.93)

The introduced parameter P can be thought to express the rate of individualization of fibers by the technological processes. For a given material, P can be considered as a rate of 'quality' of the technological process. (The smaller is it the better is the processing of fiber material.)

Coefficient of variation of local sliver fineness

Returning to the traditional symbols, i.e., $\overline{n} \equiv \overline{q}_{41}$, $v(T) \equiv v_1$, and $v(t) \equiv v_4$ (see Tables 6.1 and 6.2), and by applying Equations (6.93) and (6.10) – into (6.92) – we obtain the following expressions:

$$v_1^2 = \frac{v_4^2 + \left(1 - \dfrac{\overline{q}_{43}}{q_{43\,\text{max}}}\right) + \left(1 - \dfrac{\overline{q}_{32}}{q_{32\,\text{max}}}\right)\overline{q}_{43} + \left(1 - \dfrac{\overline{q}_{21}}{q_{21\,\text{max}}}\right)P\overline{q}_{41}}{\overline{q}_{41}}$$

$$= \frac{v_4^2 + 1 + \overline{q}_{43}\left(1 - \dfrac{1}{q_{43\,\text{max}}} - \dfrac{\overline{q}_{32}}{q_{32\,\text{max}}}\right) + P\overline{q}_{41}\left(1 - \dfrac{\overline{q}_{21}}{q_{21\,\text{max}}}\right)}{\overline{q}_{41}},$$

$$v^2(T) = a\frac{v^2(t) + 1 + \overline{q}_{43}\left(1 - \dfrac{1}{q_{43\,\text{max}}} - \dfrac{\overline{q}_{32}}{q_{32\,\text{max}}}\right) + P\overline{n}\left(1 - \dfrac{\overline{q}_{21}}{q_{21\,\text{max}}}\right)}{\overline{n}}.$$
(6.94)

We can also rewrite the above expression in the following form:

$$A = 1 + v^2(t) + \bar{q}_{43}\left(1 - \frac{1}{q_{43\,max}} - \frac{\bar{q}_{32}}{q_{32\,max}}\right), \quad B = P\left(1 - \frac{\bar{q}_{21}}{q_{21\,max}}\right), \tag{6.95}$$

and then

$$v^2(T) = \frac{1}{n}(A + B\bar{n}), \quad v(T) = \frac{1}{\sqrt{n}}\sqrt{A + B\bar{n}} = \sqrt{\frac{\bar{t}}{T}}\sqrt{A + B\frac{\bar{T}}{t}}. \tag{6.96}$$

The parameters A and B are used for characterization of the material and the processing technology, respectively.

Note: Let us note that if especially $A = 1 + v^2(t)$, i.e., $\bar{q}_{43}\left(1 - 1/q_{43\,max} - \bar{q}_{32}/q_{32\,max}\right) = 0$, and $B = 0$ then Equation (6.96) obtains a form which is identical to the traditional Martindale's Equation (6.17). Thus, Equation (6.96) can be thought as a generalization of traditional Martindale's equation.

Discussion of parameters A and B

The ratio $\bar{q}_{32}/q_{32\,max}$ mentioned in the first expression of Equation (6.95) expresses the ratio of the mean to the maximum number of bundles in the cluster. If the number of bundles present in a cluster q_{32} follows the binomial distribution, then this ratio expresses the probability that a randomly chosen bundle 'falls' into a given cluster [see Equation (A5.22) in Appendix 5]. Similarly, the ratio $\bar{q}_{21}/q_{21\,max}$ mentioned in the second expression of Equation (6.95) expresses the probability that a randomly chosen cluster falls into a given sliver cross section.

Under certain conditions, the maximum number of fibers present in a bundle $q_{43\,max}$ will be relatively small (e.g., $q_{43\,max} = 2$). Consequently, the probability $\bar{q}_{32}/q_{32\,max}$ (bundle to cluster) may approach to one. The maximum number of bundles present in a cluster $q_{32\,max}$ will be rather higher than the mean number of bundles $\bar{q}_{32}$ present in the cluster. So, the value expressed within the brackets in the definition of A in Equation (6.95) can also take a negative value (naturally, always more than -1). If the mean value of the number of fibers present in a bundle $\bar{q}_{43}$ is greater than 0 then the expression $\bar{q}_{43}\left(1 - 1/q_{43\,max} - \bar{q}_{32}/q_{32\,max}\right)$ included in Equation (6.95) can be negative under certain conditions. In this case, $A < \left[v^2(t) + 1\right]$.

The probability $\bar{q}_{21}/q_{21\,max}$ that a randomly chosen cluster falls into a given sliver cross section can be close to one. (Such case occurs, e.g., if some fiber clusters are present in a very short length of the sliver.) Then, the value stated within the brackets in the definition of B in Equation (6.95) approaches to zero, $B \to 0$. If $A < \left[1 + v^2(t)\right]$ and $B \to 0$, then the following inequality can be written from Equation (6.96) for this special case

$$v(T) = \sqrt{\frac{\bar{t}}{\bar{T}}}\sqrt{A + B\bar{n}} < \sqrt{\frac{\bar{t}}{\bar{T}}}\sqrt{v^2(t)+1}\,. \tag{6.97}$$

Under certain conditions, the coefficient of variation of local sliver fineness, calculated from Equation (6.96), might be smaller than that obtained from Equation (6.17) in case of Poisson sliver. In terms of Zellweger Uster, this is smaller than the so-called 'limit' irregularity.

Note: The coefficient of variation $v(T)$, according to Equation (6.96), could be smaller than the 'limiting' value of Equation (6.17), because the distribution is considered to be binomial, i.e., the upper value is limited to a certain maximum value. A similar relationship was also found in case of binomial sliver [see the text following Equation (6.17)]. Of course, in real cases, the coefficient of variation $v(T)$ is always higher than that of Equation (6.17), as this was explained before in the context of Huberty's index of irregularity.

Input and output values

We need to specify the following four groups of input values for numerical determination of the model of sliver and/or yarn irregularity:

1. Mean fineness of fiber $\bar{t}$ and mean fineness of yarn $\bar{T}$. The determination of these quantities is usually easy. By applying Equation (6.10), i.e., $\bar{n} = \bar{T}/\bar{t}$, we can determine also the mean number of fibers present in the cross section of the sliver; this is $\bar{q}_{41} \equiv \bar{n}$ according to Table 6.2.

2. Coefficient of variation $v(t)$ of fiber fineness. Usually, the coefficient of variation $v(d)$ of fiber diameter can be measured and then the coefficient of variation $v(t)$ of fiber fineness can be calculated by using Equation (6.18).

3. Parameters A and B. Having a very large dataset of measured values of sliver irregularities (with slivers of different finenesses), the values

A and B can be estimated from Equation (6.96) by using statistical regression technique.

Note: Equation (6.96) can be rearranged as a linear expression $y = Ax + B$, where $y = v^2(T)$ and $x = 1/\overline{n} = \overline{t}/\overline{T}$.

The parameters A and B can also be expressed by applying Equation (6.81) to (6.95) as follows:

$$A - 1 - v^2(t) = \overline{q}_{43}\left(1 - \frac{1}{q_{43\,max}}\right) - \frac{\overline{q}_{43}\overline{q}_{32}}{q_{32\,max}} = \overline{q}_{43}\left(1 - \frac{1}{q_{43\,max}}\right) - \frac{\overline{q}_{42}}{q_{32\,max}},$$

$$\overline{q}_{43} = \frac{A - 1 - v^2(t) + \dfrac{\overline{q}_{42}}{q_{32\,max}}}{1 - \dfrac{1}{q_{43\,max}}}. \tag{6.98}$$

By using the expression for B from Equation (6.95) together with (6.93) and rearranging (6.82), we can obtain the following expression:

$$B = \frac{\overline{q}_{42}}{\overline{q}_{41}}\left(1 - \frac{\overline{q}_{21}}{q_{21\,max}}\right), \quad B\overline{q}_{41} = \overline{q}_{42}\left(1 - \frac{\overline{q}_{21}}{q_{21\,max}}\right) = \overline{q}_{42} - \frac{\overline{q}_{42}\overline{q}_{21}}{q_{21\,max}} = \overline{q}_{42} - \frac{\overline{q}_{41}}{q_{21\,max}},$$

$$\overline{q}_{42} = B\overline{q}_{41} + \frac{\overline{q}_{41}}{q_{21\,max}}. $$

$$\tag{6.99}$$

4. Maximum number of fibers present in a bundle q_{43max}, in a cluster q_{42max} and in the cross section of a sliver q_{41max}. This triplet of input quantities is extremely difficult to determine. We do not know a possible way for assessment of them till now. We can only make some intuitive estimation at most. (Therefore, the exact determination of mass irregularity of binomial sliver/yarn based on our model has been so far inconvenient for application.)

Generally, we need to know eight input values $[\overline{q}_{41}, v(t), A, B, q_{43\,max}, q_{42\,max}, q_{41\,max}]$. Then, we are able to calculate all other parameters step by step:

- determine the equation for the yarn irregularity $v(T)$ according Equation (6.96),
- calculate q_{32max} from Equation (6.84),

- calculate $q_{21\mathrm{max}}$ from Equation (6.85) and
- calculate $q_{31\mathrm{max}}$ from Equation (6.86).

Note: The resulting values must be cardinal numbers. If they are not, then the input values $q_{43\,\mathrm{max}}, q_{42\,\mathrm{max}}, q_{41\,\mathrm{max}}$ cannot be fully correct.

Further let us

- calculate $\overline{q}_{42}$ from Equation (6.99),
- calculate $\overline{q}_{43}$ from Equation (6.98),
- calculate $\overline{q}_{32}$ from Equation (6.81),
- calculate $\overline{q}_{21}$ from Equation (6.82) and
- calculate $\overline{q}_{31}$ from Equation (6.83).

Note: If some of the calculated mean numbers are higher than the corresponding maximum numbers then the triplet of input quantities $q_{43\,\mathrm{max}}, q_{42\,\mathrm{max}}, q_{41\,\mathrm{max}}$ cannot be correct.

Example (hypothetical): We consider a strand (yarn) of fineness $\overline{T} = 29.5\,\mathrm{tex}$ made from fibers of fineness $\overline{t} = 0.17\,\mathrm{tex}$, so that the mean number of fibers in the cross section of the yarn is $\overline{q}_{41} = \overline{n} = \overline{T}/\overline{t} = 173.529$. Further,

$v(t) = 0.18$ (18%), $A = 1.36$, $B = 0.028$, $q_{43\mathrm{max}} = 3$, $q_{42\mathrm{max}} = 30$, $q_{41\mathrm{max}} = 900$.

Then, we obtain $v(T) = 0.189$ (18.9%), $q_{32\mathrm{max}} = 10$, $q_{21\mathrm{max}} = 30$, $q_{31\mathrm{max}} = 300$.

Furthermore, $\overline{q}_{42} = 10.643$, $\overline{q}_{43} = 2.088$, $\overline{q}_{32} = 5.098$, $\overline{q}_{21} = 16.304$, $\overline{q}_{31} = 83.113$.

Mean value of the number of fibers in non-empty bundles and clusters

The binomial model of the distribution of number of fibers in the bundles and clusters considers also the empty bundles and clusters (without fibers). Nevertheless, a fibrous assembly cannot be practically considered without any fiber.

Until now, we have worked with the mean number of fibers present in the bundles $\overline{q}_{43}$ and the mean number of fibers present in the clusters $\overline{q}_{32}$. Let us now introduce the mean number of fibers present in the non-empty bundles and clusters.

In Appendix A6.5, the mean number of fibers present in the non-empty bundles is derived and defined by Equation (A6.14) as follows:

$$\bar{q}_{43}^{*} = \frac{\bar{q}_{43}}{1-\left(1-\bar{q}_{43}/q_{43\,max}\right)^{q_{43\,max}}}.$$ (6.100)

In the same appendix, the mean number of fibers present in the non-empty clusters is given by Equation (A6.14) as follows:

$$\bar{q}_{42}^{*} = \frac{\bar{q}_{42}}{1-\left(1-\dfrac{\bar{q}_{32}}{q_{32\,max}}\dfrac{\bar{q}_{43}}{\bar{q}_{43}^{*}}\right)^{q_{32\,max}}}.$$ (6.101)

Example: We consider the same fiber strand (yarn) as considered in the previous example. We calculated $q_{43max}=3$, $q_{32max}=10$, $\bar{q}_{43}=2.088$, $\bar{q}_{32}=5.098$ and $\bar{q}_{42}=10.643$. By applying Equations (6.100) and (6.101), we obtain $\bar{q}_{43}^{*}=2.148$ and $\bar{q}_{42}^{*}=10.654$. The difference between $\bar{q}_{43}=2.088$ and $\bar{q}_{43}^{*}=2.148$ and the difference between $\bar{q}_{42}=10.643$ and $\bar{q}_{42}^{*}=10.654$ are not significant in this case. Nevertheless, in other cases, they might be significant.

Variant of Poisson distributions

We assumed binomial distributions in the previous sections. Now, let us assume that the aforesaid binomial distribution is approximated to Poisson distribution. Then, the maximum number of the lower units in the higher units ($q_{43\,max}$, $q_{42\,max}$, $q_{41\,max}$, $q_{32\,max}$, $q_{31\,max}$ and $q_{21\,max}$) approaches to infinity[12].

Accordingly, we obtain the parameters A and B from Equation (6.95) as follows:

$$A = \lim_{\substack{q_{43\,max}\to\infty \\ q_{32\,max}\to\infty}} \left\{1+v^{2}\left(t\right)+\bar{q}_{43}\left(1-\frac{1}{q_{43\,max}}-\frac{\bar{q}_{32}}{q_{32\,max}}\right)\right\}=1+v^{2}\left(t\right)+\bar{q}_{43},$$ (6.102)

$$B = \lim_{q_{21\,max}\to\infty} \left\{P\left(1-\frac{\bar{q}_{21}}{q_{21\,max}}\right)\right\}=P.$$ (6.103)

Because the mean number of fibers present in the bundles $\bar{q}_{43}\geq 0$, and also $P\geq 0$, it is evident that the coefficient of variation of local fineness according to Equation (6.96) is higher or at least equal to the value given by Equation (6.17) for the Poisson sliver.

---

12 Compare it with Equations (6.12) to (6.14) and read the text written before the equations.

Input and output values

We need to specify the following input values for numerical determination of the mass irregularity of Poisson sliver/yarn based on our model

1. Mean fineness of fiber $\bar{t}$, mean fineness of yarn $\bar{T}$, hence $\bar{q}_{41} = \bar{n} = \bar{T}/\bar{t}$.

2. Coefficient of variation $v(t)$ of fiber fineness.

3. Parameters A and B.

Note: In opposite to the binomial distributions, there is no need to find out maximum numbers of fibers; as they are limited to infinity in this model.

The mean number of fibers present in the bundle follows Equation (6.102):

$$\bar{q}_{43} = A - 1 - v^2(t),\qquad\qquad(6.104)$$

and the mean number of fibers present in the cluster follows Equations (6.93) and (6.103)

$$\bar{q}_{42} = B\bar{q}_{41}.\qquad\qquad(6.105)$$

Further, as in the previous case, we can:

- calculate $\bar{q}_{32}$ from Equation (6.81),
- calculate $\bar{q}_{21}$ from Equation (6.82) and
- calculate $\bar{q}_{31}$ from Equation (6.83).

Example (hypothetical): As in the previous examples, we consider a strand (yarn) of fineness $\bar{T} = 29.5\,\text{tex}$ from fibers of fineness $\bar{t} = 0.17\,\text{tex}$, so that the mean number of fibers present in the cross section of the yarn is $\bar{q}_{41} = \bar{n} = \bar{T}/\bar{t} = 173.529$. Moreover, $v(t) = 0.18\,(18\%)$, $A = 1.36$, $B = 0.028$, so that $v(T) = 0.189\,(19.9\%)$ according to Equation (6.96). Now, $\bar{q}_{43} = 0.3276$ according to Equation (6.104), $\bar{q}_{42} = 4.858$ according to Equation (6.105), and $\bar{q}_{32} = 14.832$, $\bar{q}_{21} = 35.714$ and $\bar{q}_{31} = 529.70$ according to Equations (6.81) to (6.83).

Mean value of number of fibers in non-empty bundles and clusters

In the Poisson model of sliver formation, the mean number of fibers present in the non-empty bundle is found from the limit of Equation (6.100) as follows:

$$\overline{q}_{43}^{*} = \lim_{q_{43\,max} \to \infty} \left[\frac{\overline{q}_{43}}{1 - \left(1 - \overline{q}_{43} / q_{43\,max}\right)^{q_{43\,max}}} \right] = \frac{\overline{q}_{43}}{1 - e^{-\overline{q}_{43}}}. \qquad (6.106)^{13}$$

The mean number of fibers present in the non-empty clusters is found from the limit of Equation (6.101) by using Equations (6.106) and (6.81) as follows:

$$\overline{q}_{42}^{*} = \lim_{\substack{q_{32\,max} \to \infty \\ \frac{\overline{q}_{43}}{\overline{q}_{43}^{*}} \to 1-\exp(-\overline{q}_{43})}} \left\{ \frac{\overline{q}_{42}}{1 - \left[1 - \dfrac{\overline{q}_{32}}{q_{32\,max}} \dfrac{\overline{q}_{43}}{\overline{q}_{43}^{*}}\right]^{q_{32\,max}}} \right\}$$

$$= \lim_{q_{32\,max} \to \infty} \left\{ \frac{\overline{q}_{42}}{1 - \left[1 - \dfrac{\overline{q}_{32}}{q_{32\,max}}\left(1 - e^{-\overline{q}_{43}}\right)\right]^{q_{32\,max}}} \right\}$$

$$= \frac{\overline{q}_{42}}{1 - e^{-\overline{q}_{32}\left(1-e^{-\overline{q}_{43}}\right)}} = \frac{\overline{q}_{42}}{1 - e^{-\overline{q}_{32}\overline{q}_{43}\frac{\left(1-e^{-\overline{q}_{43}}\right)}{\overline{q}_{43}}}} = \frac{\overline{q}_{42}}{1 - e^{-\overline{q}_{42}\frac{\left(1-e^{-\overline{q}_{43}}\right)}{\overline{q}_{43}}}} = \frac{\overline{q}_{42}}{1 - e^{\frac{-\overline{q}_{42}}{\overline{q}_{43}^{*}}}}. \qquad (6.107)$$

The relation between the mean number of fibers in the complete bundles $\overline{q}_{43}$ and in the non-empty bundles $\overline{q}_{43}^{*}$ according to Equation (6.106) is shown in Figure 6.9a. Similarly, the relation between the mean number of fibers in the complete clusters $\overline{q}_{42}$ and in the non-empty clusters $\overline{q}_{42}^{*}$ according to Equation (6.107) is shown in Figure 6.9b.

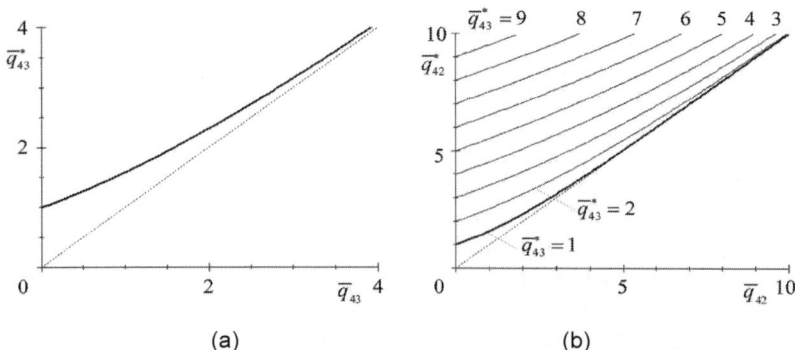

(a)   (b)

Figure 6.9 Mean number of fibers present in (a) non-empty bundles and (b) non-empty clusters according to Poisson's variants of distributions – Equations (6.106) and (6.107)

---

13   Let us remind that it is generally valid to write $\lim_{x \to \infty}\left(1 - a/x\right)^{x} = e^{-a}$

Example: We consider the same fiber strand (yarn) as in the previous example. Besides other values, we calculated $\overline{q}_{43} = 0.3276$ and $\overline{q}_{32} = 14.832$. Now, by applying Equations (6.106) and (6.107), we obtain the mean number of fibers present in the non-empty bundles $\overline{q}_{43}^* = 1.173$ and the mean number of fibers present in the non-empty clusters $\overline{q}_{42}^* = 4.937$.

The summary of all results obtained from (hypothetical) examples in this section is reported in Table 6.3. We can see that the differences between binomial and Poisson models can be very significant.

Interpretation of results

To understand the occurrence of irregularity in a better way, we need to know about the 'size' of bundles and clusters and their locations in the sliver. The bundle size determines the extent of 'inseparability' of the textile fibers, regardless of the fineness of the sliver. In opposite to this, the size of the clusters is significantly influenced by the quality of the technology employed to make the sliver.

Let us assume that we know the coefficient of variation $v(t)$ of actual fiber fineness and the parameters A and B evaluated from long years of observation of slivers (yarns) in the laboratory. [See the note in the context of Equation (6.98) for estimation of A and B.]

The interpretation is easier in the case of Poisson model. The mean number of fibers $\overline{q}_{43}$ in the bundle (non-separable unit) is followed from Equation (6.102). If this value is very high, then it is required to consider other materials for processing, which contains fibers that can be opened easily.

Note: If we obtain $\overline{q}_{43} < 0$, then the Poisson model is evidently unacceptable. In such cases, the mass irregularity probably follows the binomial model.

The mean number of fibers $\overline{q}_{42}$ in the cluster (separable unit) follows Equation (6.93), where $P = B$ according to Equation (6.103). If this value is very high, then, according to Equation (6.105), $\overline{q}_{42}/\overline{q}_{41}$ is also very high. This means that the number of clusters is also very high in relation to the total number of fibers in the cross section of the sliver and/or yarn. In such case, it is required to check the setting of the drafting system, card clothing, etc. In short, the high value of parameter A indicates mainly unsuitability of the raw material, whereas the high value of parameter B is associated with the poor quality of processing technology.

Table 6.3 Summary of the values calculated in the examples used (Input values are written in boldface)

Mean sliver fineness T	**29.5 tex**
Mean fiber fineness $\bar{t}$	**0.17 tex**
CV of fiber fineness $v(t)$	**0.18 (18 %)**
Parameter A	**1.36**
Parameter B	**0.028**
Irregularity $v(T)$	0.189 (18.9 %)

Model		Binomial		Poisson	
Values of	Maximum	Mean value		Mean value	
		All	Non-empty	All	Non-empty
No. of fibers in bundle q_{43}	**3**	2.088	2.148	0.3276	1.173
No. of fibers in cluster q_{42}	**30**	10.643	10.654	4.858	4.937
No. of fibers in sliver q_{41}	**900**	173.529		173.529	
No. bundles in cluster q_{32}	10	5.098		14.832	
No. of bundles in sliver q_{31}	300	83.113	Not calculated	529.70*	Not calculated
No. of clusters in sliver q_{21}	30	16.304		35.714	

* At first sight, the relation $\bar{q}_{31} \gg \bar{q}_{41}$ looks improbable. However, we must consider that many bundles can be empty. (See that $\bar{q}_{43}$ is only 0.3276 in this example.)

The variant of the binomial model does not permit the values to be exceeded to the definite maximum values $q_{43\,max}$, $q_{42\,max}$, $q_{41\,max}$. Nevertheless, these input values are not only very difficult to obtain but also difficult to guess. Therefore, the practical meaning of the parameters A and B is not very clear. However, it seems that a high value of parameter A indicates unsuitability of the raw material and a high value of parameter B indicates poor quality of the technological process also in case of binomial model.

Comparison with Uster Statistics

The previous theoretical results were compared in the book [3] with the yarn irregularity[14] according to Uster Statistics [11]. Here, we would like to present such comparisons.

Note: On the one hand, the Uster Statistics [11] report relatively old results of yarn irregularity (1997) according to Zellweger Uster Company. On the other hand, such results are probably not too much influenced by the contemporary regulations of the drafting process, so that they better represent the idea of random creation of yarns.

Table 6.4 Irregularity parameters of cotton carded ring-spun yarns [3]

Uster Statistics			Corresponding parameters of Equation (6.96)		Poisson model, mean number of fibers in bundles	
Curve designation (%)	Parameters of Equation (6.74)					
	a $[\text{tex}^b]$	b	A	B	$\bar{q}_{43}$	$\bar{q}_{43}^*$
95	0.35340	0.20419	2.30832	0.01615	1.268	1.765
75	0.34617	0.22119	2.12841	0.01287	1.088	1.641
50	0.34479	0.24224	1.99331	0.01008	0.953	1.551
25	0.32511	0.24750	1.74470	0.00844	0.705	1.393
5	0.31036	0.25255	1.56561	0.00726	0.526	1.286

In Uster Statistics, the values $v(T)$ are marked as $CV\%$, $\bar{T}$ is expressed in 'tex' and the results are plotted in logarithmic graph $\ln \bar{T} - \ln v(T)$. The lines of 95%, 75%, 50%, 25% and 5% indicate the yarn production worldwide. The relations are often approximated by means of Equation (6.74) which follows a linear equation in logarithmic coordinates.

Table 6.4 and Figure 6.10 illustrate the yarn irregularity based on the example of cotton carded ring-spun yarns. (The linear scales are used here in contrary to the original logarithmic coordinates.)

The thin lines [Equation (6.74) with a, b from Table 6.4] represent the original Uster Statistics. (These curves lie 'in the middle' of the thick red coloured lines in the original Uster Statistics.) The thick lines in Figure 6.10 represent our model, i.e., Equation (6.96) with A and B taken from Table 6.4. [We considered $\bar{t} = 0.18\,\text{tex}$ and $v(t) = 0.2\,(20\%)$ for calculations.]

---

14   Yarn is considered as a fibrous assembly of sliver type, where it is assumed that there is no significant effect of twist on the irregularity.

As shown, both types of curves show similar behaviour. Let us also consider that the curves from Uster Statstics are originated empirically from a 'cloud of many points' representing the laboratory results of a lot of spinning mills worldwide.

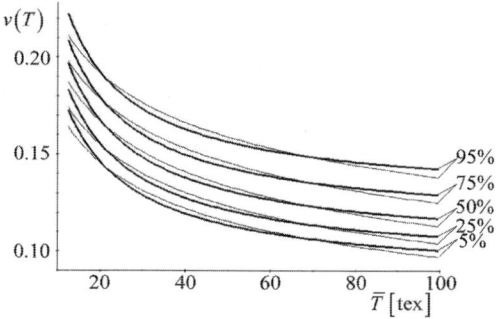

Figure 6.10 Cotton carded ring-spun yarns according to Reference [3]. —: Equation (6.74), : Equation (6.96)

Because in all cases $A > 1 + v^2(t)$, the values $\overline{q}_{43}$ and $\overline{q}_{43}^*$ can be determined from Equations (6.106) and (6.107) for the Poisson model of formation of slivers – see Table 6.4.

Note: This is not true always. For example, the values determined for woolen combed ring-spun yarns do not allow us to determine the parameters for the Poisson model because $A < 1 + v^2(t)$ in all cases – see Reference [3] for more details.

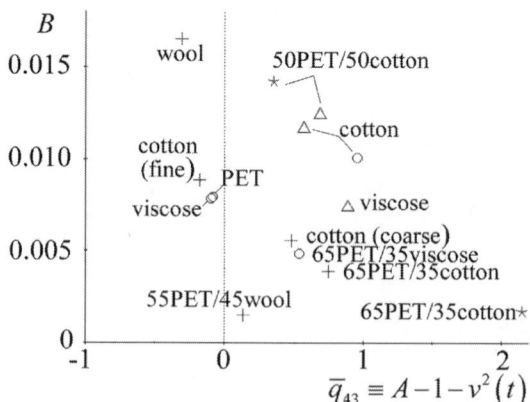

Figure 6.11 Parameters of 50% Uster Statistics lines [11] – different material and technologies: ○: carded, +: combed (worsted), △: open end, *: air jet

Figure 6.11 characterizes the relation mentioned earlier. There are many values of B and $\overline{q}_{43} \equiv A - 1 - v^2(t)$ determined for many different yarns from the lines represented by 50 % in Uster Statistics. It is shown that the mono-component yarns, having highly oriented fibers (finer combed and/or worsted yarns from cotton or wool or yarns from 100% of synthetic fibers) follow $A < 1 + v^2(t)$ so that the Poisson model is not applicable for such yarns.

It is also interesting to note that the most of the blended yarns have relatively small values of B, but relatively high values of $\overline{q}_{43}$ as shown in Figure 6.11. It points out to the existence of significant (probably mono-component) bundles in the yarns, whereas their mutual blending in the clusters and in the cross section of the yarns is relatively good.

6.6   References

[1]   Martindale, J. G., A new method of measuring the irregularity of yarns with some observations on the origin of irregularities in worsted slivers and yarns, *Journal of Textile Institute*, 36, T39–T47, 1945.

[2]   Anwendungshandbuch für Gleichmässigkeitsprüfer (Manual of irregularity tester) No. 240344-14100 b. Zellweger, Uster 1986. (In German).

[3]   Neckář, B. and Ibrahim, S., Structural Theory of Fibrous Assemblies and Yarns, Technical University of Liberec, Liberec, 2003.

[4]   Neckář, B. and Das, D., Theory of Structure and Mechanics of Fibrous Assemblies, Woodhead Publishing India Pvt. Ltd., New Delhi, 2012.

[5]   Townsend, M. W. H. and Cox, D. R., The analysis of yarn irregularity, *Journal of Textile Institute*, 42, P107–P113, 1951.

[6]   Breny, H., The calculations of the variance-length curves from the length distribution of fibers, *Journal of Textile Institute*, 44, P1–P9, 1953.

[7]   Huberty, A., Première étude des Paramètres Caractérisant la Régularité des Fils, IWTO Techn. Comm. Proc. 1, 55, 1947 (In French).

[8]   Bornet, G. M., The rating of yarns for short-term unevenness, *Textile Research Journal*, 34, 381–400, 1964.

[9]   Neckář, B., Příze Příze: Tvorba, struktura, vlastnosti (Yarns: Creation, Structure and Properties), SNTL, Prague, 1990 (In Czech).

[10] Neckář, B., Neuere Erkenntisse zur Garnungleichmässigkeit. (Contribution to the yarn unevenness.) *Melliand Textileberichte,* 70, 480–486, 1989 (In German).

[11] Uster Statistics. Zellweger, Uster 1997.

Hairiness of staple fiber yarns

7.1   Introductory images and experience

Preliminary ideas of yarn hairiness

The fiber portions, like fiber ends and fiber loops, protrude from the relatively constricted basic body of staple fiber yarns. This phenomenon is known by the term 'yarn hairiness'. The hairs in the outer layers of staple fiber yarns show certain behaviours during usage. For example, the textile products prepared from staple fiber yarns offer 'mossy' handle (very different from products made from filament yarns), specific cover effect and distinct fluid (heat, moisture, etc.) transport properties. On the contrary, the hairs are responsible for various complications arising in many technological processes, especially weaving and also imparting poor look of the fabrics. To remove the hairs from yarns or fabrics, a technological operation called singeing is often carried out. Nevertheless, the hairiness remains as one of the significant structural characteristics of staple fiber yarns.

Figure 7.1 illustrates a part of the cross section of a staple fiber yarn. It typically shows different spheres of the yarn, perceived more or less intuitively. The inner sphere is lying very close to the yarn axis. It has relatively high level of fiber packing density. In this sphere, the fibers are mutually compressed by means of twist. Further, the mechanical behaviour of the fibrous assemblies is mostly dictated by this sphere.

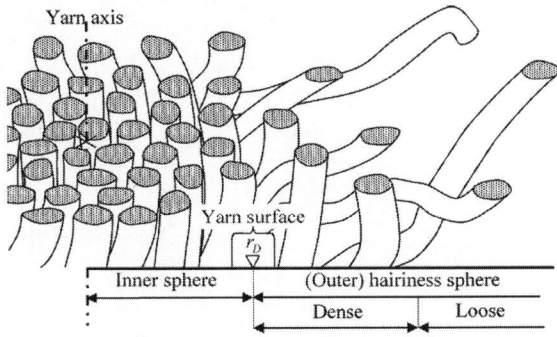

Figure 7.1 Spheres of yarn

The (outer) hairiness sphere adjoins to the previous inner region. In this sphere, the fibers and/or fiber ends and loops are relatively 'free' so that the fiber material is not significantly influenced by the mechanical actions imparted to the yarn. The structure and behaviour of this sphere are mostly determined by the probabilistic laws.

The sphere of hairiness may be consisting of two kinds of hairs – dense hairs and loose hairs. The dense hairs can be perceived as free fibers that do not have any significant mutual interaction; however, the density of fiber-to-fiber contacts cannot be negligible for such hairs. The dense hairs create a certain 'soft and flexible' cover onto the inner sphere. Usually, the thickness of this layer is only a few tenth of a millimetre. On the contrary, the loose hairs can be thought of long individually separated 'flying' fibers. Such hairs are mostly removed by means of singeing process.

The transition of inner sphere to hairiness sphere cannot be perceived as a 'jumping' phenomenon. There exists a small compound area between them which we feel as 'yarn surface'. Usually, we locate the yarn radius r_D [1] (one-half of yarn diameter D) in this area so that

$$r_D = D/2 . \tag{7.1}$$

Observation of hairiness

The methods adopted for studying yarn hairiness were developed mostly in the second half of the 20th century. There exist various experimental methods and instruments, whereas the theoretical approaches, based on probabilistic concepts, were hardly developed. Krupincova [1] reviewed different methods developed for studying yarn hairiness experimentally.

The easiest experimental method suggests us to determine the weights of fibers before and after singeing process and estimate yarn hairiness from the difference between them [2, 3]. Evidently, this method is very subjective and quite vague.

There is a group of experimental methods where yarn hairiness is analyzed from the microscopic images of the longitudinal views of the yarns. In some of the methods, we use standard yarn boards and grade spun yarns for appearance [4], while in other methods, we count the number of hairs (at different distances from the yarn body) and/or the dimension of the hairs [2–4]. The newer instruments analyze similar quantities electronically [4–10]. This

---

1   The determination of yarn diameter D and/or yarn radius r_D is the question of a chosen convention in each case. See later on for more details on this.

category of instruments analyzes hairs at a distance from the yarn body, e.g., 1 mm, indicating that the sphere of loose hairs is not usually studied well by them.

Another principle of measurement is used by Uster Tester [11]. In this instrument, a suitable polarized light 'switches on' the fibers in the sphere of hairs. (In this case, the inner sphere remains dark in the picture of the yarns.) Then, the light intensity of the luminous fibers shows a cumulative intensity of the hairs. This method also measures the sphere of dense hairs.

The present principles of image analysis technique allow us to deeply evaluate the microscopic images of the yarns. They also allow us to develop and validate the probabilistic models of yarn hairiness.

7.2   Probabilistic model of yarn hairiness – general equations

Initial assumptions

The theoretical model reported in this chapter is based on the ideas described in References [12–14].

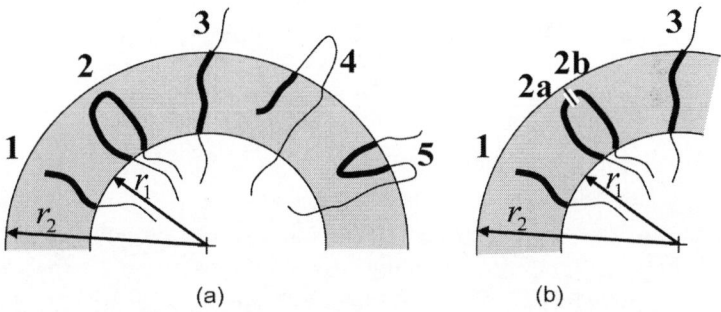

(a)   (b)

Figure 7.2 Shapes of fiber portions between two radii

Let us have a couple of two radii r_1 and r_2 in the hairiness sphere of yarn ($r_2 > r_1 \geq r_D$), and different fiber portions between them – see Figure 7.2a. In this region, the fibers may have different shapes, such as

- free ends – type 1,
- loops – type 2,
- protruding segments – type 3,
- reversal ends – type 4 and
- reversal loops – type 5.

Nevertheless, we can assume that reversal ends 4 and reversal loops 5 occur very seldom so that they can be neglected. Moreover, each loop (type 2) can be (imaginatively) divided into two parts 2a and 2b, whose shapes correspond to a couple of free ends. So, we can only think of fiber ends and protruding fibers in the sphere of hairiness; these are shown in Figure 7.2b.

Hairiness sphere

The inner (compact) sphere of yarn of a given length has yarn diameter D and yarn radius $r_D = D/2$, according to Equation (7.1). This sphere is represented by a dark grey coloured cylinder in Figure 7.3.

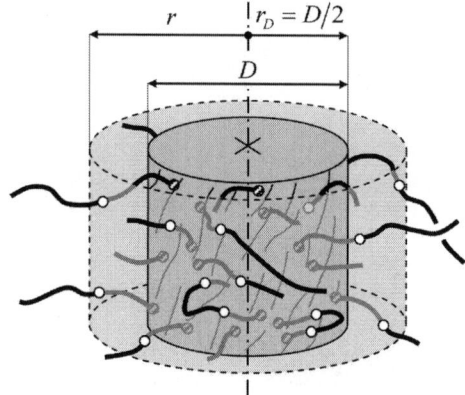

Figure 7.3 Protruding fibers in hairiness sphere

All hairs start from the surface of the inner sphere (small hatched rings). Some of them are ending at the light grey coloured part of the hairiness sphere, i.e., between radii r_D and r. The other hairs pass the depicted general radius $r \geq r_D$ (shown by small white coloured rings) and continue to move to higher radii. We call r_{max} as the most distant radius where the longest hair ends, so that $r_D \leq r \leq r_{max}$.

Number of protruding fibers

As all hairs are starting at radius r_D, the number m_D of protruding fibers at a given length of yarn – small hatched rings – is the highest at this radius. The number m of protruding fibers at a general radius $r \geq r_D$ is usually smaller than m_D, because some hairs are ended in-between r and r_D; $m \leq m_D$. In general, the number m of protruding fibers is a decreasing function of yarn radius.

Let us introduce also the (local) fiber packing density μ_D at a specific radius r_D and the (local) fiber packing density μ_r at a general radius r.

Note: If $r > r_{max}$, then evidently $m = 0$ and $\mu_r = 0$, because it is the empty space lying at the above-mentioned radii.

The number of protruding fibers at radius r is m, the number of protruding fibers at a higher radius $r + dr$ is $m + dm$. However, m is a decreasing function of r so that the elemental quantity dm must be negative. It means that the (positive) number of fiber ends in the differential layer[2] between radii r and $r + dr$ is given by the elementary quantity $(-dm)$.

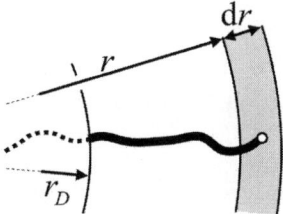

Figure 7.4 End of hair in a differential layer

Let us assume that a fiber (hair) surely passes through the radius r as shown in Figure 7.4. Then, what is the conditional probability that this fiber is ending into a (grey) differential layer, immediately adjoining the above-mentioned radius r? This is evidently an elementary value and in general varies with radius r. Such an elementary conditional probability is introduced hereunder

$$dp = g(r)dr ,\tag{7.2}$$

where $g(r)$ is a suitable positive function of radius r. We call this as a 'probabilistic function'.

Now, the number of fiber (hair) ends in the differential layer $(-dm)$ is possible to express as a product of all fibers m protruding from radius r and the conditional probability dp that they are ending immediately in the differential layer. Then,

---

2   See Figure 4.3 for the differential layer.

$$\underset{\substack{\text{total number}}}{-dm} = \overset{\substack{\text{total number}}}{m} \cdot \overset{\substack{\text{probability}}}{dp} = m\, g(r)\, dr .\tag{7.3}$$

After rearranging the last expression, we derive the following expression for the number of protruding fibers m:

$$\frac{dm}{m} = -g(r)\, dr, \quad \int_{m_D}^{m} \frac{dy}{y} = -\int_{r_D}^{r} g(t)\, dt, \quad \ln\frac{m}{m_D} = -\int_{r_D}^{r} g(t)\, dt,$$

$$m = m_D \exp\left[-\int_{r_D}^{r} g(t)\, dt \right].\tag{7.4}$$

(The integration variables are renamed from r to t and from m to y, because r and m are already used as the upper limits of the definite integrals.)

Note: The probabilistic function g is always a positive function [see its meaning according to Equation (7.2)], so that $-\int_{r_D}^{r} g(t)\, dt$ is a negative function of r, and $\exp\left[-\int_{r_D}^{r} g(t)\, dt \right]$ is a decreasing function of r. Therefore, m is a decreasing function of r in each case.

Distribution of fiber (hair) ends

The total number of all fiber ends in the hairiness sphere is equal to the number m_D of fibers, protruding from the starting yarn radius r_D. All the fiber ends are lying randomly in the interval (r_D, r_{max}) which is consequently the interval of the random variable r of the fiber (hair) ends. The number of fiber ends lying in the general radial interval (r_D, r) is then $m_D - m$ and the relative frequency of radii of fiber ends in this interval is $\Psi(r) = (m_D - m)/m_D$. By applying Equation (7.4), we can write

$$\left. \begin{aligned} \Psi(r) &= \frac{m_D - m}{m_D} = 1 - \exp\left[-\int_{r_D}^{r} g(t)\, dt \right], \\ 1 - \Psi(r) &= \exp\left[-\int_{r_D}^{r} g(t)\, dt \right]. \end{aligned} \right\}\tag{7.5}$$

It is evident that $\Psi(r)$ must be a distribution function of radii of fiber ends[3].

The corresponding probability density function is

$$\psi(r) = \frac{d\Psi(r)}{dr} = -\exp\left[-\int_{r_D}^{r} g(t)\,dt\right]\cdot\left[-g(r)\right] = g(r)\exp\left[-\int_{r_D}^{r} g(t)\,dt\right],$$

(7.6)

and by applying Equation (7.5) in the last expression, we can also write

$$\psi(r) = g(r)\left[1 - \Psi(r)\right].$$

(7.7)

It is valid from Equation (7.5) that $\Psi(r_D) = 0$. So, the probability density function $\psi(r_D)$, according to Equation (7.7), is equal to the probabilistic function $g(r_D)$ at this starting radius r_D.

Especially, the value of the distribution function must be

$$\Psi(r_{max}) = 1$$

(7.8)

at the maximum radius r_{max} of the above-stated interval. By using Equation (7.8) in (7.5), we obtain the following expression:

$$1 - \Psi(r_{max}) = 0 = \exp\left[-\int_{r_D}^{r_{max}} g(t)\,dt\right], \quad \int_{r_D}^{r_{max}} g(t)\,dt = \infty.$$

(7.9)

There are generally two possibilities to solve the last equation.

1. The (non-negative) probabilistic function $g(r)$ must be limited to infinity at some points in the interval $\langle r_D, r_{max}\rangle$. [Such case is illogical in relation to the definition of $g(r)$, according to Equation (7.2).]

2. The maximum r_{max} of the random variable r of fiber (hair) ends must be limited to infinity. (We will use this assumption later on.)

---

3   The interval for the random variable r of fiber (hair) ends is (r_D, r_{max}). $\Psi(r)$ is the relative frequency and/or the probability that a randomly chosen fiber (hair) end has its radius lying in the interval (r_D, r), where $r \in (r_D, r_{max})$.

Let us note that the validity of some of these possibilities is necessary, but not sufficient.

Fiber elements

One general fiber element in the sphere of hairiness, lying at a differential layer between radii r and $r + dr$, is represented by the thick line in Figure 7.5. The cylindrical coordinates of this element are $r \geq r_D, \varphi, \zeta$, the local Cartesian axes are x_r, x_φ, x_ζ, elementary increments are $dr, d\varphi, d\zeta$, and the inclination angles are $\vartheta_r, \vartheta_\varphi, \vartheta_\zeta$. The length of the element is $dl = dr / \cos \vartheta_r$.

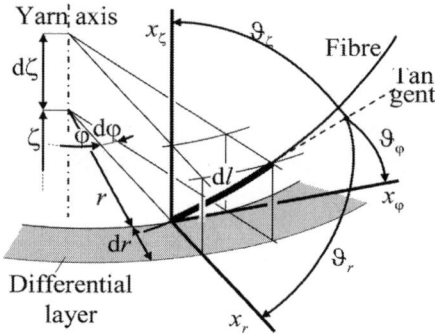

Figure 7.5 Fiber element in hairiness sphere

On a given (very long) yarn length, a lot of different fiber elements, marked by subscripts $i = 1, 2, \ldots, m$, lie in the above-mentioned differential layer (at radius r) with their individual inclination angles $\vartheta_{r,i}, \vartheta_{\varphi,i}, \vartheta_{\zeta,i}$ and individual lengths dl_i. Thus,

$$dl_i = dr / \cos \vartheta_{r,i} . \tag{7.10}$$

The mean length of these fiber elements is

$$\overline{dl} = \frac{1}{m} \sum_{i=1}^{m} dl_i = \frac{1}{m} \sum_{i=1}^{m} \frac{dr}{\cos \vartheta_{r,i}} , \tag{7.11}$$

or

$$\overline{dl} = dr \, \lambda(r), \quad \text{where} \quad \lambda(r) = \frac{1}{m} \sum_{i=1}^{m} \frac{1}{\cos \vartheta_{r,i}} . \tag{7.12}$$

The radial function $\lambda(r)$ is the mean of reciprocal values of $\cos \vartheta_{r,i}$.

In analogy to Equation (7.12), we also introduce the radial function $\sigma(r)$ as the mean of reciprocal values of $\cos \vartheta_{\zeta,i}$

$$\sigma(r) = \frac{1}{m} \sum_{i=1}^{m} \frac{1}{\cos \vartheta_{\zeta,i}} . \tag{7.13}$$

Note: The functions $\lambda(r)$ and $\sigma(r)$ characterize the specific directional distribution of fiber elements at a given radius. They are the characteristics of orientation of fiber elements in the given differential layer.

Packing density

By using Equations (7.10), (7.12) and (7.4), the total length of all fiber elements in the differential layer is

$$dL = \sum_{i=1}^{m} dl_i = \sum_{i=1}^{m} \frac{dr}{\cos \vartheta_{r,i}} = m \, \overline{dl} = m \lambda(r) dr$$

$$= m_D \exp\left[-\int_{r_D}^{r} g(t) dt \right] \lambda(r) dr. \tag{7.14}$$

In the yarn of (a very long) length ζ, the total volume of differential layer between radii r and $r + dr$ is given by the expression $dV_c = 2\pi r \, dr \, \zeta$ [4]. The fiber volume inside this differential layer is $dV = s \, dL$, where s represents the area of fiber cross section. In the sphere of hairiness, the (local) packing density of the fibers in the differential layer is possible to express by using Equation (7.14) as follows:

$$\mu_r = \frac{dV}{dV_c} = \frac{s \, dL}{2\pi r \, dr \, \zeta} = \frac{s \, m_D \exp\left[-\int_{r_D}^{r} g(t) dt \right] \lambda(r) dr}{2\pi r \, dr \, \zeta}$$

$$= \frac{s \, m_D}{2\pi \zeta} \frac{\lambda(r)}{r} \exp\left[-\int_{r_D}^{r} g(t) dt \right]. \tag{7.15}$$

---

4   In principle, Figure 4.3 is applicable also to the sphere of hairiness. The differential annulus is the sectional area of the differential layer in yarn cross-section; its area is $2\pi r \, dr$ – read the text written with reference to Figure 4.3.

Especially, the packing density $\mu_r = \mu_D$ at the starting radius $r = r_D$. Thus,

$$\mu_D = \frac{s\,m_D}{2\pi\zeta}\frac{\lambda(r_D)}{r_D}\exp\left[-\int_{r_D}^{r_D}g(t)\,dt\right] = \frac{s\,m_D}{2\pi\zeta}\frac{\lambda(r_D)}{r_D}, \tag{7.16}$$

$$\frac{\mu_D r_D}{\lambda(r_D)} = \frac{s\,m_D}{2\pi\zeta}. \tag{7.17}$$

By using the last expression in Equation (7.15), we find the packing density of fibers in the differential layer in the following form:

$$\mu_r = \frac{\mu_D r_D}{\lambda(r_D)}\frac{\lambda(r)}{r}\exp\left[-\int_{r_D}^{r}g(t)\,dt\right] = \mu_D\frac{r_D}{r}\frac{\lambda(r)}{\lambda(r_D)}\exp\left[-\int_{r_D}^{r}g(t)\,dt\right]. \tag{7.18}$$

Number of fibers in the hairiness sphere

There are $i = 1, 2, \ldots, m$ different fiber elements with different angles $\vartheta_{\zeta,i}$ in the differential layer under consideration. An oblique i-th fiber element has the sectional area

$$s_i^* = s/\cos\vartheta_{\zeta,i} \tag{7.19}$$

in the yarn cross section. [See also the derivation in Chapter 1, Figure 1.8 and Equation (1.40).] By using Equations (7.19) and (7.13), the mean value of fiber sectional areas s_i^* of the fiber elements is

$$\overline{s^*} = \frac{1}{m}\sum_{i=1}^{m}s_i^* = s\frac{1}{m}\sum_{i=1}^{m}\frac{1}{\cos\vartheta_{\zeta,i}} = s\sigma(r). \tag{7.20}$$

The total sectional area of fibers inside the differential layer is

$$dS = 2\pi r\,dr\,\mu_r. \tag{7.21}$$

[See also Equation (4.3).] Then, the number of fibers in the differential layer is

$$dn_H = \frac{dS}{\overline{s^*}} = \frac{2\pi r\,dr\,\mu_r}{s\sigma(r)}. \tag{7.22}$$

Note: A very similar idea was used for derivation of Equation (4.5) in the helical model.

The number of fibers in the whole hairiness region of yarn cross section can be obtained by integrating the previous expression. Thus,

$$n_{\mathrm{H}} = \int_{r=r_D}^{r=r_{max}} \mathrm{d}n_{\mathrm{H}} = \int_{r_D}^{r_{max}} \frac{2\pi r\, \mathrm{d}r\, \mu_r}{s\,\sigma(r)} = \frac{2\pi}{s} \int_{r_D}^{r_{max}} \frac{r\,\mu_r}{\sigma(r)}\, \mathrm{d}r \,. \tag{7.23}$$

7.3   Exponential model of yarn hairiness

Assumptions

Besides a few parameters, we need to know three radial functions, i.e., $g(r)$, $\lambda(r)$ and $\sigma(r)$, for calculation of all previous equations. Unfortunately, we do not have a special theory about the character of these functions. So it is necessary to consider some assumptions on ad hoc basis.

Let us introduce the following two assumptions for the hairiness sphere in our exponential model.

1. The probability that the fiber passing through radius r has its end lying in the differential layer $(r, r + \mathrm{d}r)$ – see Figure 7.4 – does not depend on r. It means that each protruding fiber (hair) has the same chance to end at each radius. Then, according to Equation (7.2), the probabilistic function can be expressed as follows:

   $$g(r) = g \ldots \text{constant} \,. \tag{7.24}$$

   Because Equation (7.9) was derived, the following expression must be valid now:

   $$\int_{r_D}^{r_{max}} g\, \mathrm{d}t = g \int_{r_D}^{r_{max}} \mathrm{d}t = g\left(r_{max} - r_D\right) = \infty, \quad r_{max} = \infty \,. \tag{7.25}$$

   It means that (theoretically) the yarn hairiness spreads out from the starting radius r_D to infinity in this model.

2. The distribution of orientation of fiber elements is independent of radius r. In others words, the distribution of directions of fiber elements is the same at all radii. Then, according to Equations (7.12) and (7.13), we can write the following expressions:

   $$\lambda(r) = \lambda \ldots \text{constant}, \quad \sigma(r) = \sigma \ldots \text{constant} \,. \tag{7.26}$$

Number of protruding fibers

By using Equation (7.24) in (7.4), we obtain the following expression:

$$m = m_D \exp\left[-\int_{r_D}^{r} g \, dt \right] = m_D e^{-g(r-r_D)} . \tag{7.27}$$

As shown, the number of protruding fibers is exponentially decreasing with the increase of radius r. Therefore, we call this model as exponential model.

Half-decrease interval

At another (higher) radius r^*, the number of protruding fibers is $m^* = m_D e^{-g(r^*-r_D)}$. At a special radius $r^* = r + h$, the number of protruding fibers decreases to one half, $m^* = m/2$. Then,

$$\frac{m}{2} = m_D e^{-g(r+h-r_D)} = \left[m_D e^{-g(r-r_D)} \right] e^{-gh} = m e^{-gh},$$

$$\frac{1}{2} = e^{-gh}, \quad -\ln 2 = -gh, \quad h = \frac{\ln 2}{g} \quad (g = \ln 2/h). \tag{7.28}$$

The introduced quantity h is a parameter which is independent of yarn radius. We call this as half-decrease interval[5] of number of protruding fibers in the hairiness sphere. By applying Equation (7.28) in (7.27), we obtain the following modified expression for the number of protruding fibers:

$$m = m_D e^{-\ln 2 \frac{r-r_D}{h}} = m_D 2^{-\frac{r-r_D}{h}} = m_D 2^{\frac{r_D}{h}} 2^{-\frac{r}{h}} . \tag{7.29}$$

Packing density

By using Equations (7.24) and (7.26) in (7.18), we express the (local) packing density as a function of radius r in the following form:

---

5   It is similar to the well-known half-life decay of radioactive elements.

$$\mu_r = \mu_D \frac{r_D}{r} \frac{\lambda}{\lambda} \exp\left[-\int_{r_D}^{r} g\,dt\right] = \mu_D \frac{r_D}{r} e^{-g(r-r_D)}$$

$$= \mu_D \frac{r_D}{r} e^{-\ln 2 \frac{r-r_D}{h}} = \mu_D \frac{r_D}{r} 2^{-\frac{r-r_D}{h}} = \mu_D r_D 2^{\frac{r_D}{h}} \frac{2^{-\frac{r}{h}}}{r}. \tag{7.30}$$

Let us mention that the expressions $m_D 2^{r_D/h}$ and $\mu_D r_D 2^{r_D/h}$ are considered to be the yarn parameters in Equations (7.29) and (7.30).

Number of fibers in the hairiness sphere

Equations (7.26) and (7.30) are valid in our exponential model. So, according to Equation (7.22), we obtain the following expression for the number of fibers in the differential layer:

$$dn_H = \frac{2\pi r\,dr\,\mu_D r_D 2^{\frac{r_D}{h}} \frac{2^{-\frac{r}{h}}}{r}}{s\,\sigma} = \frac{2\pi\mu_D r_D 2^{\frac{r_D}{h}}}{s\,\sigma} 2^{-\frac{r}{h}}\,dr. \tag{7.31}$$

By using Equations (7.25) and (7.31), the number of fibers in the whole hairiness region is expressed by

$$n_H = \int_{r=r_D}^{r=r_{max}=\infty} dn_H = \frac{2\pi\mu_D r_D 2^{\frac{r_D}{h}}}{s\,\sigma} \int_{r_D}^{\infty} 2^{-\frac{r}{h}}\,dr = \frac{2\pi\mu_D r_D 2^{\frac{r_D}{h}}}{s\,\sigma} \int_{r_D}^{\infty} 2^{-\frac{r}{h}}\,dr$$

$$= \frac{2\pi\mu_D r_D 2^{\frac{r_D}{h}}}{s\,\sigma} \int_{r_D}^{\infty} e^{-\frac{r}{h}\ln 2}\,dr = \frac{2\pi\mu_D r_D 2^{\frac{r_D}{h}}}{s\,\sigma}\left[-\frac{h}{\ln 2} e^{-\frac{r}{h}\ln 2}\right]_{r_D}^{\infty},$$

$$n_H = \frac{2\pi\mu_D r_D 2^{\frac{r_D}{h}}}{s\,\sigma} \frac{h}{\ln 2} e^{-\frac{r_D}{h}\ln 2} = \frac{2\pi\mu_D r_D 2^{\frac{r_D}{h}}}{s\,\sigma} \frac{h}{\ln 2} 2^{-\frac{r_D}{h}} = \frac{2\pi\mu_D r_D h}{s\,\sigma\ln 2}. \tag{7.32}$$

Parallel light beams

In image analysis technique, we usually observe the yarn images under parallel light beams, perpendicular to yarn axis. Such images allow us to analyze the character of yarn hairiness.

The light beams, passing beside a yarn at a distance $x \geq r_D$ (in the hairiness sphere), are illustrated in Figure 7.6. Some light beams (a set of arrows) can pass at a distance x without any problem. Some other light beams are 'hindered' by the hairs (see small white squares in Figure 7.6). This phenomenon is evidently random.

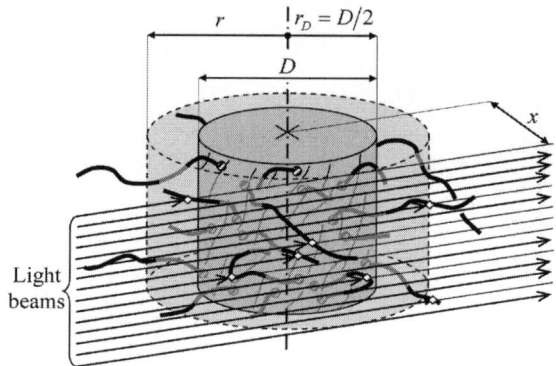

Figure 7.6 Light beams in the sphere of hairiness

Light beam in a yarn cross section

The fiber sections are randomly distributed in the hairiness region as shown in Figure 7.7. The elementary number of fiber sections (centres of fiber sections), lying in the differential annulus[6] at radii from r to $r + dr$, is dn_H. The total number of all fiber sections in the whole hairiness region is n_H. So, the probability that a randomly chosen fiber section from the hairiness region has its centre just lying inside the above-mentioned differential annulus is dn_H / n_H.

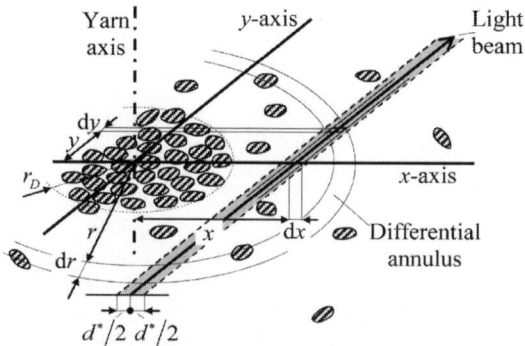

Figure 7.7 Light beam in a yarn cross section

---

6   The differential annulus is the section of the differential layer in a yarn cross section.

Further, let us imagine that an infinitesimally small rectangle $dx\,dy$ on the Cartesian coordinates x, y is lying 'inside' the differential annulus – see Figure 7.7. The conditional probability that the randomly chosen fiber centre is laying inside the infinitesimal rectangle $dx\,dy$, when it surely lies in the differential annulus, is then $dx\,dy/(2\pi r\,dr)$.

Note: Because the area of the differential annulus is $2\pi r\,dr$ – see Figure 4.3 and the text written before Equation (4.3) in Chapter 4.

The complete probability that the centre of a randomly chosen fiber from the hairiness region is lying in the rectangle $dx\,dy$ is then

$$dq_{xy} = \frac{dn_{\mathrm{H}}}{n_{\mathrm{H}}}\frac{dx\,dy}{2\pi r\,dr}. \tag{7.33}$$

By using Equation (7.31) and the generally known relation $r = \sqrt{x^2 + y^2}$, we rearrange Equation (7.33) as follows:

$$dq_{xy} = \frac{2\pi\mu_D r_D\,2^{\frac{r_D}{h}}}{s\,\sigma}2^{-\frac{r}{h}}\,dr\,\frac{1}{n_{\mathrm{H}}}\frac{dx\,dy}{2\pi r\,dr} = \frac{\mu_D r_D\,2^{\frac{r_D}{h}}}{s\,\sigma r}2^{-\frac{r}{h}}\frac{dx\,dy}{n_{\mathrm{H}}}$$

$$= \frac{\mu_D r_D\,2^{\frac{r_D}{h}}}{s\,\sigma\sqrt{x^2 + y^2}}2^{-\frac{\sqrt{x^2+y^2}}{h}}\frac{dx\,dy}{n_{\mathrm{H}}}. \tag{7.34}$$

The probability dq_{xy} is now a function of Cartesian coordinates x, y.

Equivalent diameter

To tackle with the problem of fiber 'hindering' the light beam during the measurement of yarn hairiness, we replace different sectional areas of fibers in the hairiness region (shown in Figure 7.7) by the circles of mean sectional area.

Let the diameter of the substituted circle, the so-called equivalent diameter, be d^*. Then, the area of such circle is $\pi d^{*2}/4$. This is equal to the mean sectional area of fibers $\overline{s^*}$. This mean value can be determined from Equation (7.20), but the general function $\sigma(r)$ is a constant σ according to the assumption stated in Equation (7.26). So, we can write $\overline{s^*} = s\sigma$ and then

$$\overline{s^*} = s\sigma = \frac{\pi d^{*2}}{4}, \quad d^* = \sqrt{\frac{4s}{\pi}} = d\sqrt{\sigma}. \tag{7.35}$$

[Equation (1.6) was used for fiber diameter d.]

Note: σ represents the arithmetic mean of the reciprocal values of $\cos \vartheta_\zeta$ according to Equation (7.13), so that $\sigma \geq 1$. It means that the equivalent diameter d^* of fiber section is usually a little higher than the fiber diameter d. (Because the area of oblique fiber section is higher than the area of fiber cross section.)

The grey area surrounding the light beam, shown in Figure 7.7, is displayed in Figure 7.8. The above-mentioned grey strip has thickness d^* ($d^*/2$ on both sides from the light beam). The circles from A to D in Figure 7.8 represent circular fiber sections with equivalent diameter d^*.

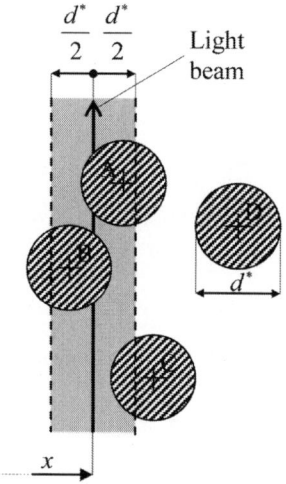

Figure 7.8 'Fiber' circles in relation to the light beam

If the centre of a fiber section lies inside the grey strip – e.g., section A and/or B – then such fiber is evidently hindering the light beam. Otherwise, if the centre of a fiber section lies outside the grey strip – e.g., section C and/ or D – then such fiber does not hinder the light beam.

Probability of hindering

What is the probability that a fiber section, randomly chosen from the hairiness region, hinders the light beam at a distance x (Figure 7.7)? This is equal to the probability that the centre of such (simplified) fiber section lies in a grey strip. It happens when the x-coordinate of such a fiber centre lies in the interval $(x - d^*/2, x + d^*/2)$ and the y-coordinate lies in the interval $(-\infty, \infty)$. By using Equation (7.34), the probability of such an event can be written as follows:

$$Q(x) = \iint\limits_{\substack{x \in (x-d^*/2,\, x+d^*/2) \\ y \in (-\infty,\infty)}} dq_{xy} = \frac{\mu_D r_D\, 2^{\frac{r_D}{h}}}{s\sigma n_{\mathrm{H}}} \iint\limits_{\substack{t \in (x-d^*/2,\, x+d^*/2) \\ y \in (-\infty,\infty)}} \frac{2^{-\frac{\sqrt{t^2+y^2}}{h}}}{\sqrt{t^2+y^2}}\, dt\, dy$$

$$= \frac{\mu_D r_D\, 2^{\frac{r_D}{h}}}{s\sigma n_{\mathrm{H}}} \int\limits_{x-d^*/2}^{x+d^*/2} \left[\int\limits_{-\infty}^{\infty} \frac{2^{-\frac{\sqrt{t^2+y^2}}{h}}}{\sqrt{t^2+y^2}}\, dy \right] dt = \frac{2\mu_D r_D\, 2^{\frac{r_D}{h}}}{s\sigma n_{\mathrm{H}}} \int\limits_{x-d^*/2}^{x+d^*/2} \left[\int\limits_{0}^{\infty} \frac{2^{-\frac{\sqrt{t^2+y^2}}{h}}}{\sqrt{t^2+y^2}}\, dy \right] dt.$$

$$(7.36)$$

Note: The integration variable was renamed from x to t because x is now used as the limits of integration. The integral function expressed in the square brackets is an even function so that the integral from $-\infty$ to ∞ is numerically equal to twice the integral from 0 to ∞.

The last equation can be expressed also in the following manner:

$$Q(x) = \frac{\mu_D r_D\, 2^{\frac{r_D}{h}+1}}{s\sigma n_{\mathrm{H}}} \int\limits_{x-d^*/2}^{x+d^*/2} f(t)\, dt, \quad \text{where} \quad f(t) = \int\limits_{0}^{\infty} \frac{2^{-\frac{\sqrt{t^2+y^2}}{h}}}{\sqrt{t^2+y^2}}\, dy. \quad (7.37)$$

The introduced function $f(t)$ can be rearranged as follows:

$$f(t) = \int\limits_{0}^{\infty} \frac{2^{-\frac{\sqrt{t^2+y^2}}{h}}}{\sqrt{t^2+y^2}}\, dy = \int\limits_{0}^{\pi/2} \frac{2^{-\frac{\sqrt{t^2+t^2\tan^2 u}}{h}}}{\sqrt{t^2+t^2\tan^2 u}} \frac{t\, du}{\cos^2 u} = \int\limits_{0}^{\pi/2} \frac{2^{-\frac{t\sqrt{1+\tan^2 u}}{h}}}{\sqrt{1+\tan^2 u}} \frac{du}{\cos^2 u}$$

Substitution: $y = t \tan u$, $dy = \dfrac{t\, du}{\cos^2 u}$

$$= \int\limits_{0}^{\pi/2} 2^{-\frac{t}{h\cos u}} \cos u \frac{du}{\cos^2 u} = \int\limits_{0}^{\pi/2} 2^{-\frac{t}{h\cos u}} \frac{du}{\cos u}. \quad (7.38)$$

Probability of passing the light beam

The quantity $Q(x)$ is the probability that one randomly chosen fiber section hinders the light beam at a distance x. The probability that a randomly chosen fiber section does not hinder the light beam is $1 - Q(x)$, and the probability that none of the fiber sections (from the hairiness region) hinders the light beam at the distance x is evidently

$$P(x) = \{1 - Q(x)\}^{n_H} = \left\{ 1 - \frac{1}{n_H} \left[\frac{\mu_D r_D \, 2^{\frac{r_D}{h}+1}}{s\sigma} \int_{x-d^*/2}^{x+d^*/2} f(t)\,dt \right] \right\}^{n_H} . \quad (7.39)[7]$$

[Equation (7.37) was used in the previous expression.] Let us give a name to the function $P(x)$ as 'show-through' function.

It is possible to assume that the number of fiber sections in the whole hairiness region n_H is relatively high in case of most of the real yarns. Then, we can write the following expressions from Equation (7.39) by using (7.38):

$$P(x) \approx \lim_{n_H \to \infty} \left\{ 1 - \frac{1}{n_H} \left[\frac{\mu_D r_D \, 2^{\frac{r_D}{h}+1}}{s\sigma} \int_{x-d^*/2}^{x+d^*/2} f(t)\,dt \right] \right\}^{n_H}$$

$$= \exp\left[-\frac{\mu_D r_D \, 2^{\frac{r_D}{h}+1}}{s\sigma} \int_{x-d^*/2}^{x+d^*/2} f(t)\,dt \right], \quad (7.40)[8]$$

$$-\ln P(x) = \frac{\mu_D r_D \, 2^{\frac{r_D}{h}+1}}{s\sigma} \int_{x-d^*/2}^{x+d^*/2} f(t)\,dt = \frac{\mu_D r_D \, 2^{\frac{r_D}{h}+1}}{s\sigma} \int_{x-d^*/2}^{x+d^*/2} \left(\int_0^{\pi/2} 2^{-\frac{t}{h\cos u}} \frac{du}{\cos u} \right) dt .$$

$$(7.41)$$

The sequence of integration in the last double integral can be interchanged. Thus, we obtain

$$-\ln P(x) = \frac{\mu_D r_D \, 2^{\frac{r_D}{h}+1}}{s\sigma} \int_0^{\pi/2} \left[\int_{x-d^*/2}^{x+d^*/2} 2^{-\frac{t}{h\cos u}} \, dt \right] \frac{du}{\cos u}$$

$$= \frac{\mu_D r_D \, 2^{\frac{r_D}{h}+1}}{s\sigma} \int_0^{\pi/2} \left[\int_{x-d^*/2}^{x+d^*/2} e^{-\frac{t\ln 2}{h\cos u}} \, dt \right] \frac{du}{\cos u}$$

---

7   There are in totally n_H number of randomly distributed fiber sections in the hairiness region of yarn cross sec-tion.

8   Let us remind that it is valid to write $\lim_{x \to \infty}(1 - a/x)^x = e^{-a}$ – see a suitable handbook of mathematics.

$$= \frac{\mu_D r_D \, 2^{\frac{r_D}{h}+1}}{s\sigma} \int_0^{\pi/2} \left[\left(-\frac{h\cos u}{\ln 2} e^{-\frac{t\ln 2}{h\cos u}} \right)_{x-d^*/2}^{x+d^*/2} \right] \frac{du}{\cos u}$$

$$= \frac{h\mu_D r_D \, 2^{\frac{r_D}{h}+1}}{s\sigma \ln 2} \int_0^{\pi/2} \left[\cos u \left(-e^{-\frac{x+d^*/2}{h\cos u}\ln 2} + e^{-\frac{x-d^*/2}{h\cos u}\ln 2} \right) \right] \frac{du}{\cos u}$$

$$= \frac{h\mu_D r_D \, 2^{\frac{r_D}{h}+1}}{s\sigma \ln 2} \int_0^{\pi/2} \left(-e^{-\frac{x+d^*/2}{h\cos u}\ln 2} + e^{-\frac{x-d^*/2}{h\cos u}\ln 2} \right) du$$

$$= \frac{h\mu_D r_D \, 2^{\frac{r_D}{h}+1}}{s\sigma \ln 2} \left(\int_0^{\pi/2} 2^{-\frac{x-d^*/2}{h\cos u}} du - \int_0^{\pi/2} 2^{-\frac{x+d^*/2}{h\cos u}} du \right). \tag{7.42}$$

By applying Equation (7.35) in (7.42), we can write the following expression for the show-through function

$$-\ln P(x) = \frac{h\mu_D r_D \, 2^{\frac{r_D}{h}+1}}{\left(\pi d^{*2}/4\right)\ln 2} \left(\int_0^{\pi/2} 2^{-\frac{x-d^*/2}{h\cos u}} du - \int_0^{\pi/2} 2^{-\frac{x+d^*/2}{h\cos u}} du \right),$$

$$-\ln P(x) = \frac{8h\mu_D r_D}{\pi d^{*2}\ln 2} 2^{\frac{r_D}{h}} \left(\int_0^{\pi/2} 2^{-\frac{x-d^*/2}{h\cos u}} du - \int_0^{\pi/2} 2^{-\frac{x+d^*/2}{h\cos u}} du \right). \tag{7.43}$$

We can introduce the parameters q, H and a 'helping' function $z(x)$ as follows:

$$q = \frac{8h\mu_D r_D}{\pi d^{*2}\ln 2} 2^{\frac{r_D}{h}}, \tag{7.44}$$

$$H = \frac{8h}{\pi d^{*2}\ln 2}, \quad \text{i.e.,} \quad \frac{q}{H} = \mu_D r_D 2^{\frac{r_D}{h}}, \tag{7.45}$$

$$z(x) = \int_0^{\pi/2} 2^{-\frac{x-d^*/2}{h\cos u}} du - \int_0^{\pi/2} 2^{-\frac{x+d^*/2}{h\cos u}} du. \tag{7.46}$$

The following expression for the show-through function arises after substitution of Equations (7.44) and (7.46) into (7.43):

$$-\ln P(x) = q\,z(x). \tag{7.47}$$

Also, the packing density can be rearranged by applying Equation (7.45) in (7.30) in the following manner:

$$\mu_r = \mu_D r_D\, 2^{\frac{r_D}{h}}\, \frac{2^{-\frac{r}{h}}}{r} = \frac{q}{H}\, \frac{2^{-\frac{r}{h}}}{r}. \tag{7.48}$$

Note: If especially $r = r_D$, then we obtain the following expression according to the first expression stated in Equation (7.48):

$$\mu_{r=r_D} = \mu_D r_D\, 2^{\frac{r_D}{h}}\, \frac{2^{-\frac{r_D}{h}}}{r_D} = \mu_D. \tag{7.49}$$

Number of fibers in the whole hairiness region

By rearranging Equation (7.32) for the number of fiber section in the whole hairiness region by means of Equations (7.35) and (7.44), we can also write

$$n_H = \frac{2\pi\mu_D r_D\, h}{\left(\pi d^{*2}/4\right)\ln 2} = q\, \frac{\pi}{2^{\frac{r_D}{h}}}. \tag{7.50}$$

Blackening function

The probability that at least one fiber section in the yarn cross section hinders the light beam at a distance x is then written in the following manner:

$$Z(x) = 1 - P(x), \quad \text{i.e.,} \quad -\ln\left[1 - Z(x)\right] = -\ln P(x). \tag{7.51}$$

The quantity $Z(x)$ also informs us 'how much' amount of light beams is 'stopped' by the hairs situated at a distance x from the yarn axis (see Figure 7.6).

Note: By applying digital image analysis technique, the quantity $Z(x)$, representing the portion of black among all the pixels lying on the image of the yarn at a distance x ($x \geq r_D$) from the yarn axis, can be evaluated. Let us therefore call the function $Z(x)$ as the blackening function.

By using Equation (7.43) and/or (7.47) in (7.51), we can write the blackening function as follows:

$$Z(x) = 1 - \exp\left[-\frac{8h\mu_D r_D}{\pi d^{*2}\ln 2}\, 2^{\frac{r_D}{h}} \left(\int_0^{\pi/2} 2^{-\frac{x - d^*/2}{h\cos u}}\, du - \int_0^{\pi/2} 2^{-\frac{x + d^*/2}{h\cos u}}\, du \right) \right]$$

$$= 1 - \exp\left[-q\, z(x) \right]. \tag{7.52}$$

7.4   Double-exponential model of yarn hairiness

Past experience

The experimental analysis of different types of yarns showed that the derived single-exponential model does not precisely correspond to the real hairiness spheres (see more in the next section, e.g., Figure 7.12). This experience leads us to an idea of two types of hairs present in the hairiness sphere, whose distributions are mutually independent and randomly blended in the hairiness sphere. One type of hairs is composed mostly of shorter fiber portions and they are concentrated mainly around the yarn surface. The second type of hairs is composed mostly from relatively long flying fibers. The introduced idea would be put more precisely in this section under the name 'double-exponential model'[9].

Note: The idea of two types of hairs also expresses our long-time intuitive experience from the subjective observations of yarn images using microscopy. We usually observed a 'dense' layer of short fiber ends and different fiber loops (something like 'moose') immediately on yarn surface. At longer distances, we observed only relatively free flying hairs creating an outer 'loose' layer of hairiness sphere. (See also Figure 7.1.)

Assumptions

We use the following two assumptions for the double-exponential model:

1. There are two mutually independent types of hairs present in the hairiness sphere of yarn. The quantities related to the first and second types of hairs are denoted with the subscripts 1 and 2, respectively. The common parameters and variables are shown without this subscript.

2. Each type of hairs follows its own exponential model (generally with different parameters, of course) as formulated in the previous section.

Common parameters and variables

We use the following quantities:

---

9   To differentiate between the previous exponential model from Section 7.3 and this exponential model, the former is sometimes called single-exponential model.

- Common parameters: Equivalent diameter of fiber section d^* and yarn radius r_D.

- Common variables: General radius $r \geq r_D$ and distance from yarn axis $x \geq r_D$.

Survey of partial quantities

We use the following set of partial quantities (quantities related to hairiness type 1 and hairiness type 2):

- Half-decrease intervals h_1, h_2 – parameters.

- Initial values of packing density $\mu_{D,1}, \mu_{D,2}$ – parameters.

- Parameters q_1, q_2 – see Equation (7.44):

$$q_1 = \frac{8h_1\mu_{D,1}r_D}{\pi d^{*2}\ln 2}2^{\frac{r_D}{h_1}}, \quad q_2 = \frac{8h_2\mu_{D,2}r_D}{\pi d^{*2}\ln 2}2^{\frac{r_D}{h_2}}. \tag{7.53}$$

- Parameters H_1, H_2 – see Equation (7.45):

$$H_1 = \frac{8h_1}{\pi d^{*2}\ln 2}, \text{ i.e., } \frac{q_1}{H_1} = \mu_{D,1}r_D2^{\frac{r_D}{h_1}}, \quad H_2 = \frac{8h_2}{\pi d^{*2}\ln 2}, \text{ i.e., } \frac{q_2}{H_2} = \mu_{D,2}r_D2^{\frac{r_D}{h_2}}. \tag{7.54}$$

- Packing densities $\mu_{r,1}, \mu_{r,2}$ at general radius – see Equation (7.48):

$$\mu_{r,1} = \frac{q_1}{H_1}\frac{2^{-\frac{r}{h_1}}}{r}, \quad \mu_{r,2} = \frac{q_2}{H_2}\frac{2^{-\frac{r}{h_2}}}{r}; \tag{7.55}$$

especially for $r = r_D$, it is valid to write

$$\mu_{r=r_D,1} = \mu_{D,1} = \frac{q_1}{H_1}\frac{2^{-\frac{r_D}{h_1}}}{r_D}, \quad \mu_{r=r_D,2} = \mu_{D,2} = \frac{q_2}{H_2}\frac{2^{-\frac{r_D}{h_2}}}{r_D}. \tag{7.56}$$

- Total number of fiber sections in the hairiness region of yarn cross section $n_{H,1}, n_{H,2}$ – see Equation (7.50):

$$n_{H,1} = q_1\frac{\pi}{2^{\frac{r_D}{h_1}}}, \quad n_{H,2} = q_2\frac{\pi}{2^{\frac{r_D}{h_2}}}. \tag{7.57}$$

- Probabilities that no one fiber section hinders the light beam, i.e., show-through functions $P_1(x), P_2(x)$ – see Equation (7.47):

$$-\ln P_1(x) = q_1 z_1(x), \quad -\ln P_2(x) = q_2 z_2(x), \tag{7.58}$$

where the following expressions can be written in agreement with Equation (7.46),

$$z_1(x) = \int_0^{\pi/2} 2^{-\frac{x-d^*/2}{h_1 \cos u}} du - \int_0^{\pi/2} 2^{-\frac{x+d^*/2}{h_1 \cos u}} du, \quad z_2(x) = \int_0^{\pi/2} 2^{-\frac{x-d^*/2}{h_2 \cos u}} du - \int_0^{\pi/2} 2^{-\frac{x+d^*/2}{h_2 \cos u}} du.$$

$$\tag{7.59}$$

- The blackening functions $Z_1(x), Z_2(x)$ – see Equation (7.52):

$$Z_1(x) = 1 - \exp\left[-q_1 z_1(x)\right], \quad Z_2(x) = 1 - \exp\left[-q_2 z_2(x)\right].^{[10]} \tag{7.60}$$

Total quantities

Besides the previous partial quantities, we also evaluate the total quantities determined from both types of hairs together.

The total value of packing density at a general radius r is evidently the summation of such values from both types of hairs. By using Equation (7.55), the following equation is valid to write

$$\mu_r = \mu_{r,1} + \mu_{r,2} = \frac{1}{r}\left(\frac{q_1}{H_1} 2^{-\frac{r}{h_1}} + \frac{q_2}{H_2} 2^{-\frac{r}{h_2}}\right). \tag{7.61}$$

Especially for $r = r_D$, it is valid to write that

$$\mu_D = \mu_{D,1} + \mu_{D,2} = \frac{1}{r_D}\left(\frac{q_1}{H_1} 2^{-\frac{r_D}{h_1}} + \frac{q_2}{H_2} 2^{-\frac{r_D}{h_2}}\right). \tag{7.62}$$

[Compare it with Equation (7.56).]

The total number of fiber sections in the hairiness region of yarn cross section is evidently the summation of number of fiber sections from both types of hairs as given by Equations (7.57). Thus,

$$n_H = n_{H,1} + n_{H,2} = \pi\left(\frac{q_1}{2^{\frac{r_D}{h_1}}} + \frac{q_2}{2^{\frac{r_D}{h_2}}}\right). \tag{7.63}$$

---

10   They are hypothetical functions that are arising out from an abstract picture of only one type of hairiness (type 1 or type 2) in the hairiness sphere, while the second type is eliminated.

The probability that none of the fiber sections (from hairiness region) hinders the light beam at a distance x from the yarn axis means that none of the fiber sections of type 1 and simultaneously none of the fiber sections of type 2 hinder the light beam. Both events must occur at the same time so that we must multiply both probabilities as stated in Equation (7.58). Thus,

$$P(x) = P_1(x) P_2(x), \quad -\ln P(x) = -\ln P_1(x) - \ln P_2(x),$$
$$-\ln P(x) = q_1 z_1(x) + q_2 z_2(x). \tag{7.64}$$

Equation (7.60) determines the partial blackening function. The total blackening function results from the definition according to Equations (7.51) and (7.64)

$$Z(x) = 1 - P(x) = 1 - \exp\left[-q_1 z_1(x) - q_2 z_2(x)\right]. \tag{7.65}$$

7.5   Experimental blackening function – evaluation of hairiness

Experimental blackening function

The blackening function can be analyzed experimentally by using, e.g., a suitable image analysis technique. Then we obtain yarn pictures as shown in Figure 7.9.

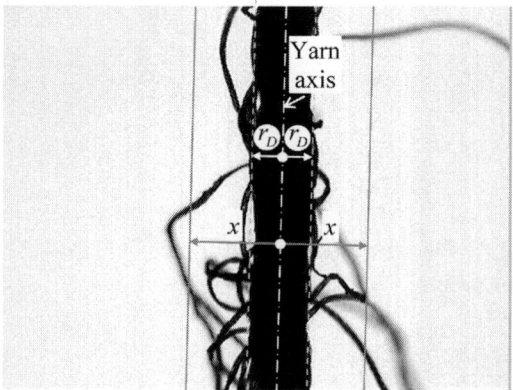

Figure 7.9 Yarn picture in the parallel light beams

Note: Practically, the magnification of a yarn by means of an optical (microscopic) system and then using an image analysis technique bring many difficult problems, which must be solved. Of them, the suitable transformation of the original picture (Figure 7.9) to the black-and-white size together with increasing sharpness[11] brings enormous difficulty.

At first, the yarn axis must be determined as a 'middle line' in the compact black strip as shown in the yarn image. Then, the ratio of 'black' and 'white' pixels is evaluated at each distance x, starting from the yarn axis. So we obtain (evaluating a lot of yarn pictures together) the experimental blackening function $Z_E(x)$ – e.g., the saw-tooted curve as shown in Figure 7.10. (It is $Z_E(x) = 1$ near to yarn axis always.)

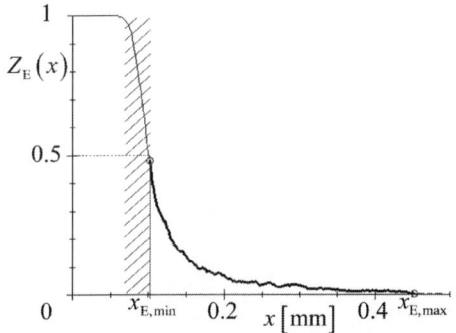

Figure 7.10 Behaviour of experimental blackening function $Z_E(x)$
(Example-yarn – cotton carded ring yarn of 20 tex count)

Note: In the following, we will illustrate how to evaluate yarn hairiness. For this, we use a cotton carded ring-spun yarn of 20 tex count. We call this yarn as 'example-yarn'.

At this point, an obvious question arises, i.e., where is the required yarn radius r_D lying in the yarn? As of now, we do not know about it. (Maybe, the yarn radius is lying somewhere in the visible hatched interval as shown in Figure 7.10. We shall discuss this problem later on.) However, we are certain that the value of the blackening function $Z_E(x = r_D)$ must be higher than the value 0.5[12]. So, we can think that if $x \geq x_{E,\min}$, where $Z_E(x_{E,\min}) \leq 0.5$, then we are certainly talking about the hairiness sphere.

---

11   The light used is not perfectly parallel so that the hairs on the yarn surface are not often enough sharp.

12   The value 0.5 was chosen empirically. The fibers are half-transparent at this distance and it is not usually imaginable that the fibers are significantly compressed in this space.

Note: The experimental blackening function is practically determined by a set of points $\left[x, Z_E(x)\right]$. We select the x-coordinate of the first point having $Z_E(x) \leq 0.5$ at the distance $x_{E,min}$.

On the contrary, the experimental blackening function is practically equal to zero where x is enough high. [For example, it is practically appearing when $Z_E(x) \leq 0.005$.] The corresponding borderline radius of the example-yarn is denoted by $x_{E,max}$ in Figure 7.10.

So, we utilize the experimentally found blackening function $Z_E(x)$ only on the interval $\left(x_{E,min}, x_{E,max}\right)$ – the thick part in Figure 7.10. This thick part of the experimental blackening function can be expressed by means of the derived theoretical models, i.e., single-exponential model and/or double-exponential model.

Further, from the definition according to Equation (7.51), it is valid to write that $-\ln P(x) = -\ln\left[1 - Z(x)\right]$. By using experimental function $Z_E(x)$, we can also write that

$$-\ln P_E(x) = -\ln\left[1 - Z_E(x)\right], \tag{7.66}$$

where $P_E(x)$ is the corresponding experimental show-through function (i.e., the function determined by means of experimental blackening function). This is required for the following evaluation.

Equivalent diameter of fiber

Equation (7.35) defines the equivalent diameter $d^* = d\sqrt{\sigma}$, where σ represents the arithmetic mean of the reciprocal values of $\cos \vartheta_\zeta$ – see Figure 7.5 and Equation (7.13). Unfortunately, we do not have any information about the directional arrangement of fibers (hairs) in the hairiness sphere. A rough estimation, e.g., as shown in Appendix 7, can then be used.

By applying Equation (A7.4) from Appendix 7, we use the value

$$\sqrt{\sigma} = 1.216, \quad d^* = d \cdot 1.216, \tag{7.67}$$

for the following calculation.

By assuming that the equivalent diameter of cotton fibers is $d = 0.012\,\text{mm}$, we obtain $d^* = 0.012 \cdot 1.216 = 0.0146 \doteq 0.015\,\text{mm}$. We will use this value for evaluation of our example-yarn.

Notes to integration

The integrals used in Equation (7.59) can be rearranged as follows:

$$\int_0^{\pi/2} 2^{-\frac{x\pm d^*/2}{h\cos u}}\,du = \int_0^{\pi/2} e^{-\frac{(x\pm d^*/2)\ln 2}{h\cos u}}\,du = e^{-\frac{(x\pm d^*/2)\ln 2}{h\cos u}} \int_0^{\pi/2} e^{-\frac{(x\pm d^*/2)\ln 2}{h}\left(\frac{1}{\cos u}-1\right)}\,du$$

$$= e^{-k}\int_0^{\pi/2} e^{\left(-\frac{k}{\cos u}+k\right)}\,du, \quad \text{where} \quad k=\frac{\left(x\pm d^*/2\right)\ln 2}{h}. \qquad (7.68)$$

Note: Here, k is a positive value, because $x > d^*/2$ in the hairiness sphere. Let us also mention that the quantities h_1 and/or h_2 mentioned in the double-exponential model are used in place of h in the previous expression.

The curves shown in Figure 7.11 display the behaviour of the integral function in the large interval of parameter k. The values of the integrals (i.e., areas under the curves) take the values from 0 (when $k\to\infty$) to $\pi/2$ (when $k\to 0$).

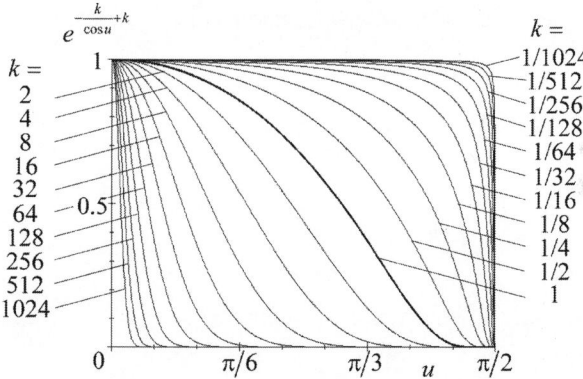

Figure 7.11 Behaviour of the integral in Equation (7.68)

The type of integral, according to Equation (7.68), always results in a finite value, but it must be solved by using a suitable numerical method of integration.

Evaluation of single-exponential model

Here, the initial parameters are obtained by comparing the experimental blackening function $Z_E(x)$ (its thick part from $x_{E,\min}$ to $x_{E,\max}$) and the single-exponential model function $Z(x)$ – according to Equation (7.52) in conjunction with Equation (7.46). When we know the parameter d^*, we need to estimate two suitable parameters – q and h. [Parameter h uses the function $z(x)$ according to Equation (7.46).] For this, a numerical method of statistical regression can be used. The result obtained for the example yarn is shown in Figure 7.12. Needless to say, this is a typical result.

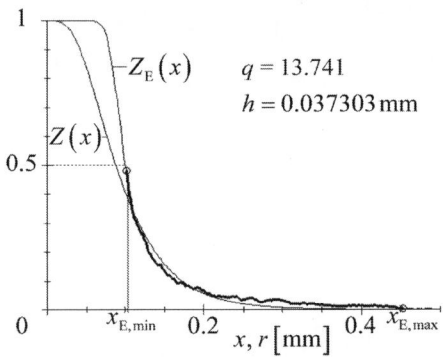

Figure 7.12 Experimental blackening function $Z_E(x)$ and calculated blackening function $Z(x)$ by single-exponential model (example-yarn, $d^* = 0.0150\,\text{mm}$)

Note: It was used that $d^* = 0.015\,\text{mm}$. The model parameters ($h = 0.037303$ mm and $q = 13.741$) were determined by numerical method of statistical regression carried out on the experimental data of the example-yarn (same as Figure 7.10).

It can be seen that the experimentally determined curve $Z_E(x)$ and the curve $Z(x)$ of single-exponential model are not in good agreement. We, therefore, do not prefer the single exponential model.

Evaluation of double-exponential model

Here, the initial parameters follow from a suitable comparison between the experimental blackening function $Z_E(x)$ (from $x_{E,\min}$ to $x_{E,\max}$) and the double-exponential model function $Z(x)$ according to Equation (7.65) with Equation (7.59). When we know the parameter d^*, we need to estimate four suitable parameters – q_1, q_2 and h_1, h_2 . [The parameters h_1, h_2 use the function $z_1(x), z_2(x)$, according to Equation (7.59).] A numerical method of statistical regression is used.

In case of the example-yarn (Figure 7.10), we determined that $h_1 = 0.014404\,\text{mm}$, $q_1 = 139,76$, $h_2 = 0.096642\,\text{mm}$, $q_2 = 2.598289$. This (typical) result is illustrated graphically in Figure 7.13.

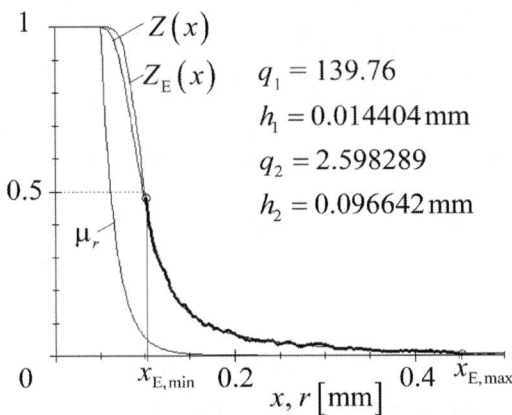

Figure 7.13 Experimental blackening function $Z_E(x)$, calculated blackening function $Z(x)$, and packing density μ_r in case of double-exponential model (example-yarn, $d^* = 0.015\,\text{mm}$)

Note: The model parameters h_1, h_2 and q_1, q_2 were determined by using the numerical method of statistical regression carried out on the experimental data (thick line). Let us note that a numerical method is necessary in relation to the half decrease intervals h_1, h_2, while the 'best' values q_1, q_2 can be calculated analytically for each couple h_1, h_2 (method of least squares).

It can be seen that the experimentally determined curve $Z_E(x)$ and the curve $Z(x)$ of double-exponential model show very good correspondence in the hairiness sphere.

We prefer the double-exponential model, because such a very good agreement is typical for numerous staple yarns studied by us.

Quantities of double-exponential model

Having known the value d^* and the determined values h_1, h_2 and q_1, q_2, we can further calculate the followings step by step:

- Parameters H_1, H_2 from Equation (7.54).

- Partial packing density functions $\mu_{r,1}, \mu_{r,2}$ from Equation (7.55) and function μ_r from Equation (7.61). (The behaviour of the function μ_r for the example yarn is shown in Figure 7.13.)
- Functions $z_1(x), z_2(x)$ from Equation (7.59).
- Partial blackening functions $Z_1(x), Z_2(x)$ from Equation (7.60) and blackening function $Z(x)$ from Equation (7.65). (The behaviour of the function $Z(x)$ for the example-yarn is shown in Figure 7.13.)

For the determination of other characteristics, it is required to specify the yarn radius.

Yarn cover radius

The distance $x_{E,min}$ at which the experimental blackening function is equal to 0.5 was presented in the introductory part of this section. (More precisely, the distance $x_{E,min}$ represents the first point of the experimental blackening function $Z_E(x)$, where $Z_E(x = x_{E,min}) \le 0.5$.)

After evaluation of the parameters as mentioned earlier, we can find the (model) value x_{cover} at which the (model) blackening function $Z(x_{cover})$ is just equal to the value $1/2$. By using Equations (7.65) and (7.59), we can write

$$\frac{1}{2} = Z(x_{cover}) = 1 - \exp\left[-q_1 z_1(x_{cover}) - q_2 z_2(x_{cover})\right],$$

$$\frac{1}{2} = \exp\left[-q_1 z_1(x_{cover}) - q_2 z_2(x_{cover})\right], -\ln 2 = -q_1 z_1(x_{cover}) - q_2 z_2(x_{cover}),$$

$$\ln 2 = q_1 \left(\int_0^{\pi/2} 2^{-\frac{x_{cover}-d^*/2}{h_1 \cos u}} du - \int_0^{\pi/2} 2^{-\frac{x_{cover}+d^*/2}{h_1 \cos u}} du \right)$$

$$+ q_2 \left(\int_0^{\pi/2} 2^{-\frac{x_{cover}-d^*/2}{h_2 \cos u}} du - \int_0^{\pi/2} 2^{-\frac{x_{cover}+d^*/2}{h_2 \cos u}} du \right). \tag{7.69}$$

The numerically calculated root x_{cover} of the last equation determines the distance at which $Z(x_{cover}) = 1/2$.

The distance x_{cover} can be interpreted as a characteristic of yarn radius – called cover radius $r_{D,\text{cover}}$ – of yarn. (The double-value represents the cover diameter D_{cover} of yarn.) This is shown as follows:

$$r_{D,\text{cover}} = x_{\text{cover}}, \quad D_{\text{cover}} = 2r_{D,\text{cover}} = 2x_{\text{cover}}. \tag{7.70}$$

Note: This cover radius or cover diameter has practical consequences. At radii $r < r_{D,\text{cover}}$, we see more hairs than 'light windows'; we 'feel' that this is primarily the yarn body at such radii. On the other hand, at radii $r > r_{D,\text{cover}}$, we see more light windows than hairs; we do not feel it as a (compact) yarn but more as some individual hairs only. Therefore, the cover radius and cover diameter are important for solving the so-called cover factors in woven or knitted fabrics, etc.

Note: The other quantities mentioned with subscript 'cover' will use the radius $r_D = r_{D,\text{cover}}$ for their calculations.

Yarn density radius

At the beginning of this section, we discussed an idea of inner sphere, hairiness sphere and a small area of yarn surface – see Figure 7.1. We imagined that the probabilistic regulations are dominant in the hairiness sphere, mechanical regulations determine the inner sphere and some mixed influences are typical for the above-mentioned area of yarn surface.

It is extremely difficult to determine a borderline for the inner sphere. Based on a very subjective estimation (our visual experience with the microscopic images of different cross sections of yarns), we can characterize this borderline in accordance with a density radius $r_{D,\text{dens}}$ at which the packing density $\mu_r = \mu_{D,\text{dens}}$ is equal to the value 0.1.

Accordingly, we can write Equation (7.62) in the following special form:

$$0.1 = \mu_{D,\text{dens}} = \frac{1}{r_{D,\text{dens}}} \left(\frac{q_1}{H_1} 2^{-\frac{r_{D,\text{dens}}}{h_1}} + \frac{q_2}{H_2} 2^{-\frac{r_{D,\text{dens}}}{h_2}} \right). \tag{7.71}$$

The density radius $r_{D,\text{dens}}$ is the root of the last equation that can be solved by using a suitable numerical method.

Let us mention that the yarn density diameter D_{dens} is twice the yarn density radius. Further, the yarn density radius can be used as a characteristic distance x_{dens} for calculation of the blackening functions. So, we write

$$x_{\text{dens}} = r_{D,\text{dens}}, \qquad D_{\text{dens}} = 2r_{D,\text{dens}}.$$ (7.72)

Note: This density radius or density diameter estimates a cylinder having significantly compressed fibers. The internal structure and the internal mechanics of such cylinder have a certain deterministic character to a large degree.

Note: The other quantities mentioned with subscript 'dens' will use the radius $r_D = r_{D,\text{dens}}$ for their calculations.

Quantities making use of yarn radius

By applying yarn radius $r_D = r_{D,\text{cover}}$ and $r_D = r_{D,\text{dens}}$ [or x_{cover} and/or x_{dens} – Equations (7.70), (7.72)], we can further calculate the followings step by step:

- Partial values of packing density $\mu_{D1,\text{cover}}$, $\mu_{D2,\text{cover}}$ and/or $\mu_{D1,\text{dens}}$, $\mu_{D2,\text{dens}}$ from Equation (7.56) and the total value $\mu_{D,\text{cover}}$ from Equation (7.62). [The value $\mu_{D,\text{dens}} = 0.1$ was used as the input value – see Equation (7.71).]

- Partial values $n_{H1,\text{cover}}$, $n_{H2,\text{cover}}$ and/or $n_{H1,\text{dens}}$, $n_{H2,\text{dens}}$ from Equation (7.57) and the total values $n_{H,\text{cover}}$ and $n_{H,\text{dens}}$ from Equation (7.63).

- Partial blackening values $Z_1(x = x_{\text{cover}})$, $Z_2(x = x_{\text{cover}})$ and/or $Z_1(x = x_{\text{dens}})$, $Z_2(x = x_{\text{dens}})$ from Equation (7.60) and the total value $Z(x = x_{\text{dens}})$ from Equation (7.65). [The value $Z(x = x_{\text{cover}}) = 0.5$ was used as the input value – see the first equivalency in Equation (7.69).]

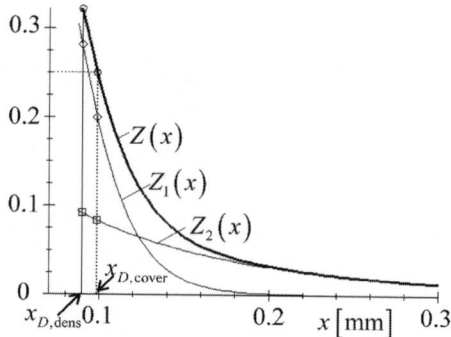

Figure 7.14 Blackening and partial blackening functions (example-yarn, $d^* = 0.015\,\text{mm}$)

Note: The blackening function $Z(x)$, the partial blackening functions $Z_1(x), Z_2(x)$, the values of the distances $x_{\text{cover}}, x_{\text{dens}}$ and the special blackening values $Z(x_{\text{cover}})$, $Z_1(x_{\text{cover}})$, $Z_2(x_{\text{cover}})$ and $Z(x_{\text{dens}})$, $Z_1(x_{\text{dens}})$, $Z_2(x_{\text{dens}})$ are illustrated graphically in Figure 7.14 for the example yarn (cotton carded ring-spun yarn of 20 tex count, Figure 7.10).

Integral characteristics of hairiness

Sometimes, an overall scalar characteristic of yarn hairiness is required. Therefore, it is possible to define the following integral characteristic:

$$I_{\text{cover}} = \int_{r_{D,\text{cover}}=x_{\text{cover}}}^{\infty} Z(x)\,dx, \quad I_{\text{dens}} = \int_{r_{D,\text{dens}}=x_{\text{dens}}}^{\infty} Z(x)\,dx. \qquad (7.73)$$

Characteristics of the example – yarn

The summary of the determined characteristics is shown in Table 7.1.

Table 7.1 Quantities evaluated on the basis of experimental blackening function (Figure 7.10) for example-yarn (cotton carded ring-spun yarn of 20 tex count)[*]

Values from regression	$h_1 = 0.014404\,\text{mm}$, $\quad q_1 = 139.76$			$h_2 = 0.096642\,\text{mm}$, $\quad q_2 = 2.598289$		
	(calculated at $d^* = \mathbf{0.015}$ mm)					

Radii (distances)	Cover radius (distance) $r_{D,\text{cover}} = x_{\text{cover}} = 0.09847\,\text{mm}$			Density radius (distance) $r_{D,\text{dens}} = x_{\text{dens}} = 0.08935\,\text{mm}$		
Hairs	$Z(x_{\text{cover}})$	$\mu_{D,\text{cover}}$	$n_{H,\text{cover}}$	$Z(x_{\text{dens}})$	$\mu_{D,\text{dens}}$	$n_{H,\text{dens}}$
Subscript 1	0.39957	0.05282	3.8426	0.56331	0.09029	5.9605
Subscript 2	0.16733	0.00825	4.0282	0.18412	0.00971	4.3006
All	**0.5**	0.06107	7.8708	0.64371	**0.1**	10.2611
Integrals:	$\int_{x_{\text{cover}}}^{\infty} Z(x)\,dx = 0.02236$ mm			$\int_{x_{\text{dens}}}^{\infty} Z(x)\,dx = 0.02756$ mm		

* The bold printed values indicate the input parameters.

Note: Let us note that (1) $h_1 \ll h_2$ ($h_1/h_2 \doteq 0.15$) and (2) $\mu_{D1} \gg \mu_{D2}$ ($\mu_{D1,\text{cover}}/\mu_{D2,\text{cover}} \doteq 6.4$, and $\mu_{D1,\text{dens}}/\mu_{D2,\text{dens}} \doteq 9.3$) in the case of example yarn. Such a difference between the first and the second types of hairs is the characteristic of almost all staple yarns. After all, it is also evident from the behaviours of the partial blackening functions $Z_1(x)$ and $Z_2(x)$ as shown in Figure 7.14.

Relation of radii – a hypothetical case

Let us – hypothetically – imagine that the thick yarns have (1) still the same half-decrease intervals h_1, h_2 and (2) still the same values $\mu_{D1,\text{dens}}, \mu_{D2,\text{dens}}$ (where $\mu_{D1,\text{dens}} + \mu_{D2,\text{dens}} = 0.1$) at their yarn density radii $r_{D,\text{dens}}$. It means that each type of hairs starts at the yarn density radius always with the same density and decreases with the same 'speed'. In such a hypothetical case, we can write the following expression in accordance with Equation (7.53):

$$ q_1 = \left[\frac{8h_1\mu_{D,1}}{\pi d^{*2}\ln 2} \right] r_{D,\text{dens}}\, 2^{\frac{r_{D,\text{dens}}}{h_1}}, \quad q_2 = \left[\frac{8h_2\mu_{D,2}}{\pi d^{*2}\ln 2} \right] r_{D,\text{dens}}\, 2^{\frac{r_{D,\text{dens}}}{h_2}}, \quad (7.74) $$

where the expressions shown within square brackets are still constants. Having two yarns – 'old' (e.g., the example yarn) and 'new' (with other yarn density radius) – we can recalculate the quantities for a new yarn. It is then valid to write that

$$ q_{1\,\text{new}} = q_{1\,\text{old}}\, \frac{r_{D,\text{dens new}}}{r_{D,\text{dens old}}}\, 2^{\frac{r_{D,\text{dens new}} - r_{D,\text{dens old}}}{h_1}}, \quad q_{2\,\text{new}} = q_{2\,\text{old}}\, \frac{r_{D,\text{dens new}}}{r_{D,\text{dens old}}}\, 2^{\frac{r_{D,\text{dens new}} - r_{D,\text{dens old}}}{h_2}}. $$

$$ (7.75) $$

By using the old values from the example-yarn (Table 7.1), we can always evaluate $q_{1\,\text{new}}, q_{2\,\text{new}}$ for several new yarns having several other new values of $r_{D,\text{dens new}}$. Further, we can calculate all other characteristics of hairiness, including the corresponding radii $r_{D,\text{cover new}}$. Figure 7.15 illustrates the relation between $r_{D,\text{dens}}$ and $r_{D,\text{cover}}$ for several new hypothetical yarns at these conditions. It can be seen that the ratio of radii is changed with the yarn fineness.

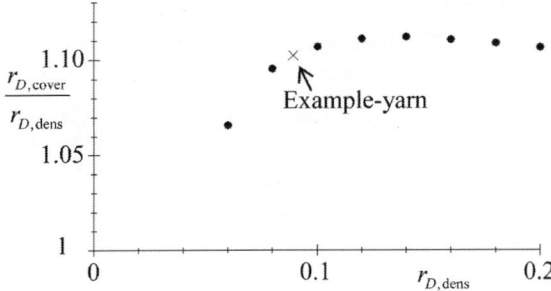

Figure 7.15 Relation between $r_{D,\text{dens}}$ and $r_{D,\text{cover}}$ for constant values of h_1, h_2 and $\mu_{D1,\text{dens}}, \mu_{D2,\text{dens}}$ used from Table 7.1

7.6   Experience and observed trends

Experimental results

Three sets of combed and carded ring-spun yarns were prepared from the same cotton material such that each set consists of five levels of yarn fineness, and there were three levels of twist coefficients in each fineness, i.e., in total 15 yarns in each set. The hairiness of these yarns was also tested by Uster Tester. Further, viscose ring yarns were prepared from one type of fiber material in seven levels of fineness. Table 7.2 characterizes the fineness and twist of the experimental yarns [15].

Table 7.2 Overview of yarns

Type of yarns	Yarn finenesses T [tex]	Koechlin's twist coefficients α [m^{-1} ktex$^{1/2}$]*	H_{USTER}
Cotton combed ring-spun yarns**	14.76, 19.68, 24.60, 29.53, 36.91	110...△, 120...○, 130...□	Yes
Cotton carded ring-spun yarns**	14.76, 19.68, 24.60, 29.53, 36.91	110...△, 120...○, 130...□	Yes
Viscose ring-spun yarns	7.4, 10, 14.5, 20, 29.5, 38, 50	Ordinary...○	No

*The marked graphical symbols are used in the following graphs.
**Krupincova [1] prepared these yarns and carried out the initial evaluation (parameters h_1, h_2 and q_1, q_2) of these yarns.

Half-decrease intervals

Figures 7.16–7.18 show the half-decrease intervals of the aforesaid yarns.

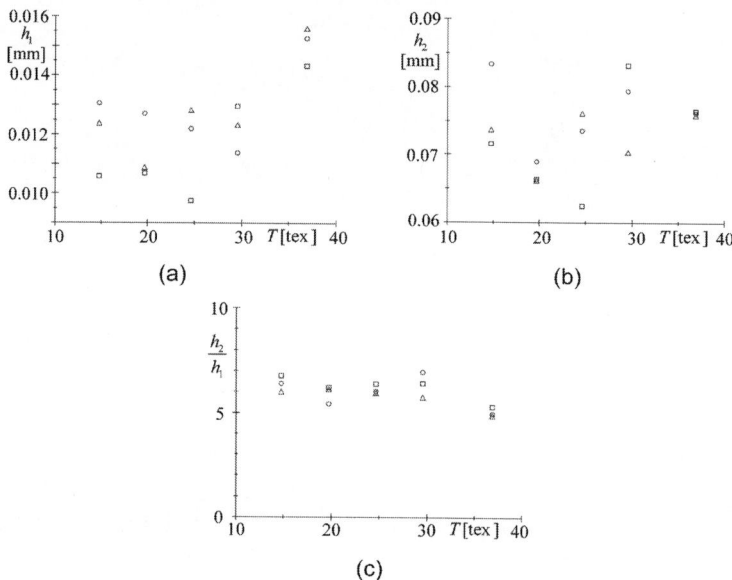

Figure 7.16 Half-decrease intervals h_1, h_2 of cotton combed ring-spun yarns

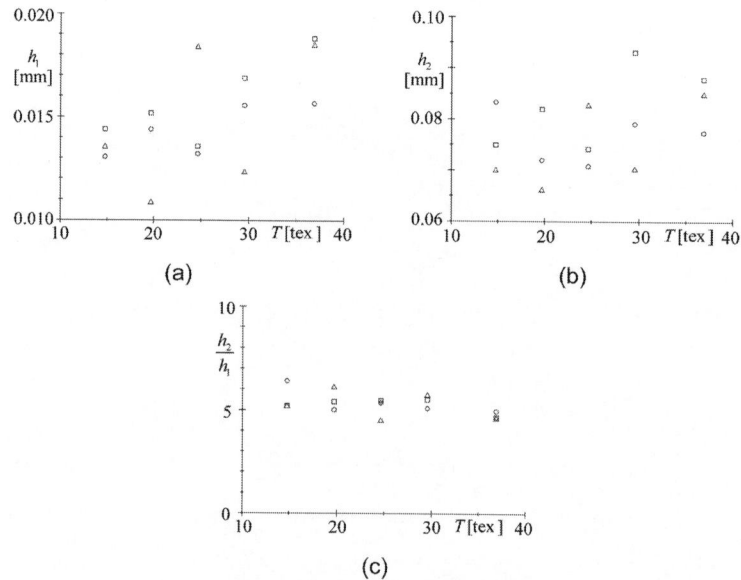

Figure 7.17 Half-decrease intervals h_1, h_2 of cotton carded ring-spun yarns

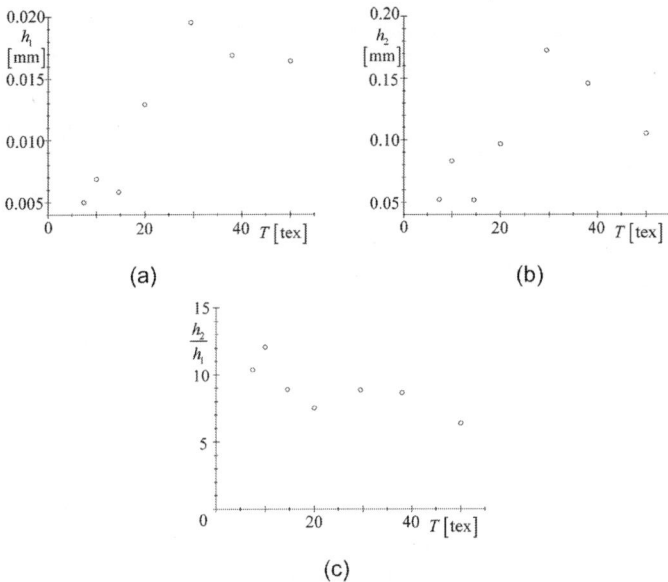

Figure 7.18 Half-decrease intervals h_1 h_2 of viscose ring-spun yarns

The following observations can be made from the aforesaid graphs.

(a) The values h_1 and h_2 are different for different yarns. Nevertheless, they – mostly h_1 – display a slightly increasing tendency for the coarser yarns. Why is it so? It is shown in Figure 7.19 that a fiber loop has the same length l in the thin and thick yarns. Let us assume that the distances of starting points A and B along the yarn circumferences are same in both cases ($\alpha_a r_{Da} = \alpha_b r_{Db}$). Then – as it is shown – the distance of the loop from the yarn surface is higher with the coarser yarn than with the finer yarn. The influence of twist factor to the half-decrease intervals (cotton yarns) is probably not too significant as shown in the graphs.

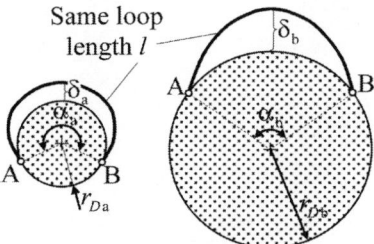

Figure 7.19 Fiber loop on the small and big yarn cylinder

Note: Another experiment showed a certain loss of h_1 with the increase of yarn twist. It can be ascribed to the fact that the higher twist resulted in drawing of the hairs towards the yarn body.

(b) The differences between h_1 and h_2 of the half-decrease intervals are very high. The ratio h_2/h_1 is a little higher than 5 in case of both groups of cotton yarns and this is around 8 in case of viscose yarns. It shows that these two types of hairs are very different from each other. (Compare with the 'speeds of decreasing' of the partial blackening functions as shown in Figure 7.14 – the earlier example yarn.)

The first type of hairs decreases quickly to (practically) zero with the increase in yarn radius. These hairs create something like a moos on the yarn surface which is often good for us. It brings a pleasant handle, fullness of the fabric, etc.

On the contrary, the second type of hairs decreases very slowly with the increase in yarn radius. These hairs are creating mostly from 'long flying fibers' which often bring difficulty in the subsequent technological processes namely weaving as well as make the fabric looking poor. This is precisely why the singeing operation is often carried out in textile technology.

Yarn density diameter and yarn cover diameter

Figures 7.20–7.22 show the results of yarn diameters; they are twice the corresponding yarn radii – see Equations (7.70) and (7.72).

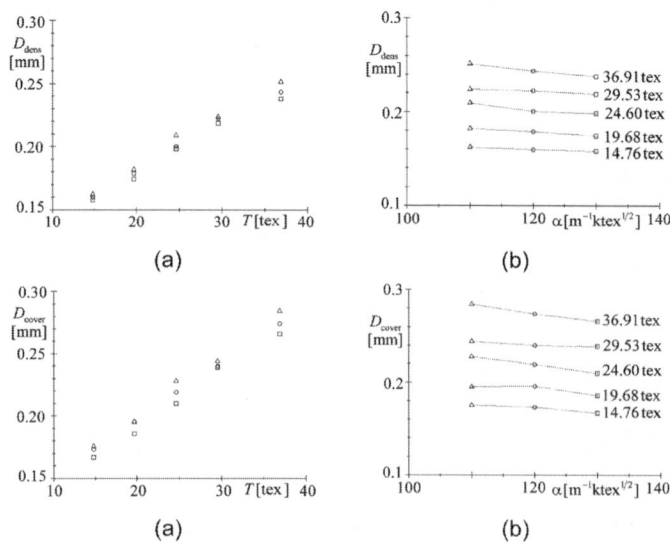

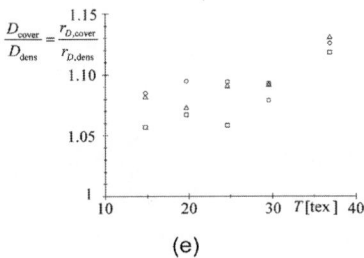

(e)

Figure 7.20 Yarn density and cover diameters
of cotton combed ring-spun yarns

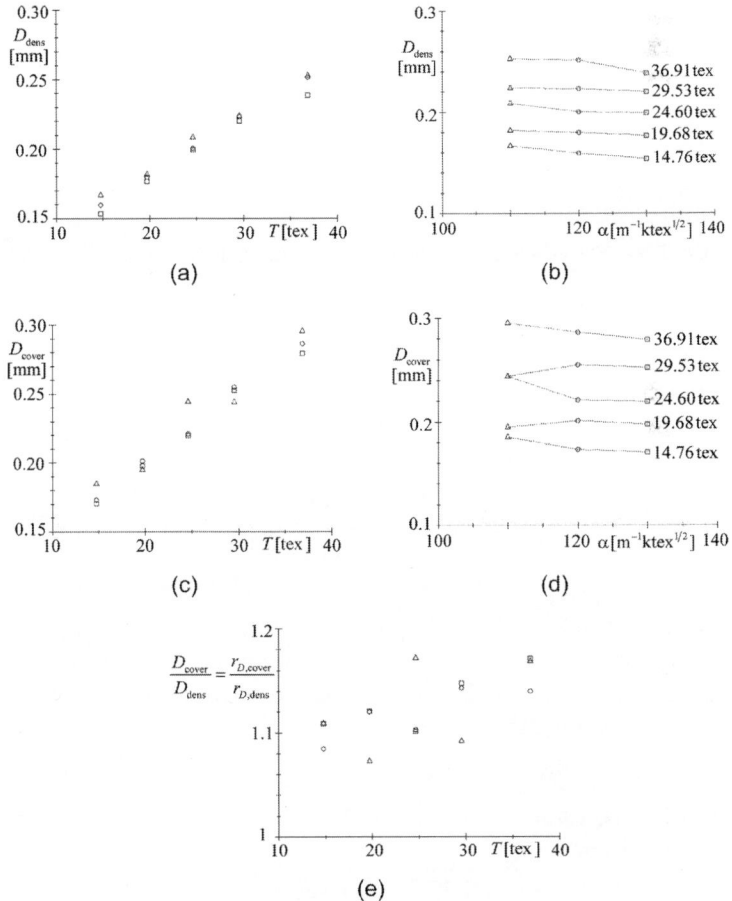

Figure 7.21 Yarn density and cover diameters of
cotton carded ring-spun yarns

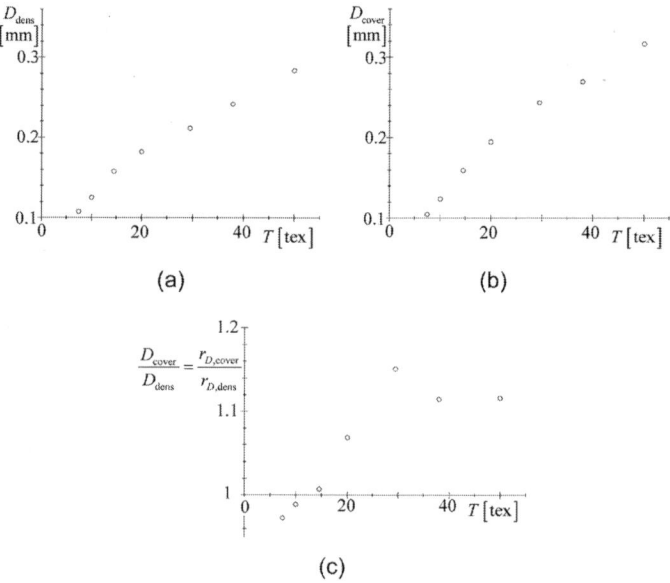

Figure 7.22 Yarn density and cover diameters of viscose ring-spun yarns

The following observations can be made from the aforesaid graphs.

(a) It is shown in Figure 7.20 that both diameters D_{dens} and D_{cover} increase with the increase in yarn fineness T. It also proves that the diameters are slightly decreasing by increasing yarn twist (twist coefficient). This is shown in Figure 7.21. This effect is most prominent at D_{dens}, i.e., at the level of packing density $\mu_{D,\text{dens}} = 0.1$ – see Equation (7.71). The diameter D_{cover} [at the level of blackening function $Z(x_{\text{cover}}) = Z(D_{\text{cover}}/2) = 0.5$ – see Equations (7.69) and (7.70)] decreases clearly with the increase of twist coefficients in case of cotton combed yarns, while a similar trend is a little 'defocused' in case of cotton carded yarns, probably as a consequence of a lower degree of fiber arrangement in these yarns.

(b) The values of the ratio $D_{\text{cover}}/D_{\text{dens}}$ are around 1.07 in case of cotton combed yarns and around 1.11 in case of cotton carded yarns. This ratio is slightly increasing with the increase in yarn fineness in both types of cotton yarns – see Figures 7.20e and 7.21e. It corresponds to the first part of the trend line shown in Figure 7.15, derived for 'hypothetical yarns'. The graph shown in Figure 7.22c is somewhat different from the previous case. There is a clearly increasing trend of the ratio $D_{\text{cover}}/D_{\text{dens}}$ with the increase in yarn fineness here, but we obtained the values smaller than 1 for finer yarns; it means that $D_{cover} < D_{dens}$ there. However, a similar trend is also shown in the graph of Figure 7.15 (hypothetical yarns) at its extrapolation to the left, i.e., to a very small yarn radii. We can correctly imagine that the extremely fine yarns can be so much transparent on their edges that the cover diameters at the blackening value $Z\left(D_{cover}/2\right) = 0.5$ need a higher value of packing density than $\mu_{D,dens} = 0.1$.

Packing densities

Figures 7.23–7.25 show the calculated values of packing densities.

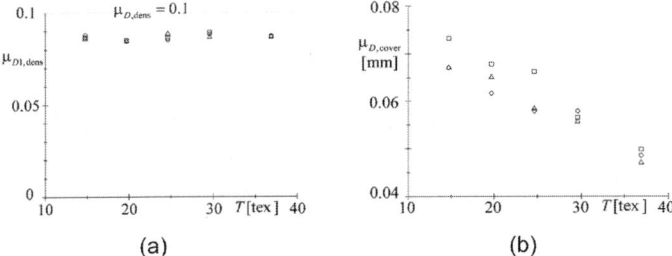

(a)   (b)

Figure 7.23 Packing densities in cotton combed ring-spun yarns

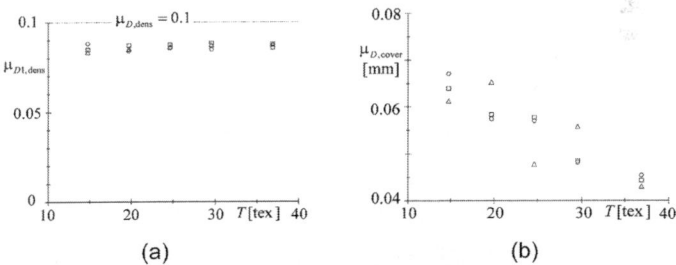

(a)   (b)

Figure 7.24 Packing densities in cotton carded ring-spun yarns

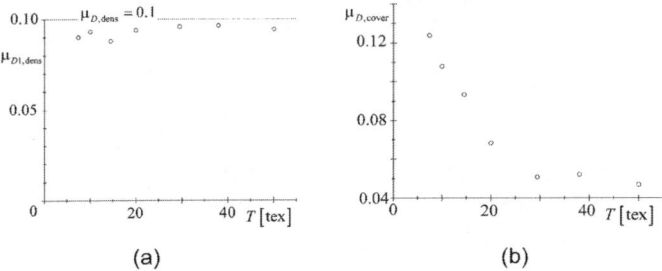

(a)   (b)

Figure 7.25 Packing densities in viscose ring-spun yarns

The following observations are possible to make from the above-mentioned graphs.

(a) The packing density $\mu_{D,\text{dens}} = 0.1$ was selected as one of the input parameters and, based on this, the radius $r_{D,\text{dens}} = D_{\text{dens}}/2$ was determined – see Equations (7.71) and (7.72). It is also valid to write according to Equation (7.62) that $0.1 = \mu_{D,\text{dens}} = \mu_{D1,\text{dens}} + \mu_{D2,\text{dens}}$. The values $\mu_{D1,\text{dens}}$ show graphs (a) in Figures 7.23–7.25; their subtraction from 0.1 evidently represent the values of $\mu_{D2,\text{dens}}$.

At first, we can see that the above-mentioned values $\mu_{D1,\text{dens}}$ and $\mu_{D2,\text{dens}}$ are practically constants. This is approximately 0.86 for the cotton combed as well as for the cotton carded yarns and 0.92 for the viscose yarns. It means that about 90% of the hairs starting at diameter D_{dens} belong to the first type of hairiness, i.e., 'moos hairs' and only about 10% belong to the second type of hairiness, i.e., long flying fibers (even less in case of viscose yarns). Together with the knowledge about the half-decrease intervals we can say that

- The most of the hairs start at diameter D_{dens} as a first type of hairs, nevertheless, the density of these hairs decreases very quickly with the increase in radius. We call this hairiness (moos) as the dense hairiness.

- A small part of the fibers start at diameter D_{dens} as a second type of hairs. The density of these hairs decreases slowly with the increase in radius. We call this hairiness (long flying fibers) as the loose hairiness.

(b) The (total) packing density $\mu_{D,\text{cover}}$ at diameter D_{cover} (blackening value 0.5) is shown in the graph b of Figures 7.23–7.25. These packing densities show markedly decreasing trend with the increase in yarn fineness. This is probably because

- In case of coarse yarns (thick yarn cylinder), a light beam is passing through the hairiness sphere beside the yarn body (see Figure 7.6) on a 'very long length' so that the probability of hindering the light (during measurement of hairiness) by a fiber is relatively high. To obtain the value of blackening function just equal to 0.5, the light beam must pass beside the yarn body in a region of relatively small value of packing density; $\mu_{D,\text{cover}}$ should be small.

- On the contrary, in case of fine yarns (thin yarn cylinder), a light beam is passing through the hairiness sphere beside the yarn body at 'only short length' so that the probability of hindering the light beam by a fiber is relatively small. To obtain the value of blackening function just equal to 0.5, the light beam must pass beside the yarn body in a region of relatively high value of packing density; $\mu_{D,\text{cover}}$ can be high.

(c) A little strange result is obtained regarding the nature of decreasing tendency of $\mu_{D,\text{cover}}$ in case of viscose yarn – Figure 7.25b. Maybe, in case of coarser yarns, the effect of 'thick yarn cylinder' is not so important. (Yarn of 30 tex fineness can be rather anomalous – see also Figures 7.18 and 7.22.)

Integrals I_{dens} and Uster hairiness

Yarn hairiness of both types of cotton yarns was measured also by using Uster Tester instrument. The comparison between Uster hairiness H_{Uster} and integral[13] $I_{\text{dens}} = \int_{r_{D,\text{dens}}=x_{\text{dens}}}^{\infty} Z(x)\,\mathrm{d}x$ according to Equation (7.33) is shown in Figures 7.26 and 7.27.

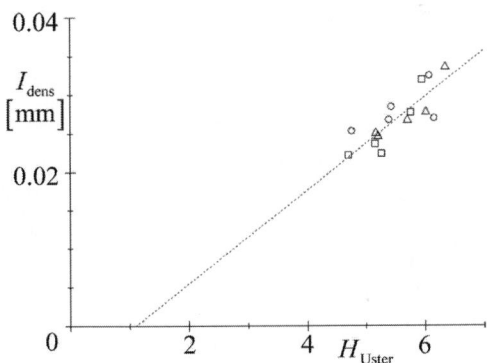

Figure 7.26 Comparison of H_{Uster} with integral value I_{dens} in case of cotton combed ring yarns

---

13   Uster Tester measures cumulative yarn harness, i.e., including the moos of the yarn body – see Section 7.1. According to Equation (7.33), it logically corresponds to the integral I_{dens}.

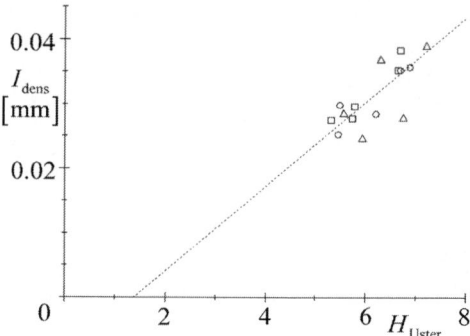

Figure 7.27 Comparison of H_{Uster} with integral value I_{dens} in case of cotton carded ring yarns

The following observations can be made from the above-mentioned graphs.

(a) The positive high correlation between H_{Uster} and I_{dens} is evident. A slightly better result is shown in case of combed yarns, because these yarns do offer a little better directional arrangement of the fibers.

(b) The dashed straight lines start from points of $H_{\text{Uster}} > 0.1$ in both graphs. Why do we obtain such result? An attempt is made here to find out a tentative explanation for this behaviour. Let us think about a limit case, e.g., about a twisted filament yarn. Evidently, this does not have any hairiness, i.e., $I_{\text{dens}} = 0$. Nevertheless, such a yarn shows a value $H_{\text{Uster}} \in (1.2, 3)$ according to our experience. It is probably because the edges of such a yarn are always slightly shining under the polarized light of Uster Tester instrument (including mono-filament yarn). It thus results that Uster Tester does not measure the real hairiness of such yarns.

7.7   References

[1] Krupincova, G., Chlupatost přízí (Hairiness of yarns), PhD Thesis, Technical University of Liberec, Czech Liberec, 2012 (In Czech).

[2] Barella, A., Yarn hairiness, *Textile Progress*, 43 (1), The Textile Institute, Manchester, UK, 1983.

[3] Barella, A., The hairiness of yarn: The review of the literature and a survey of the present position, *Journal of Textile Institute*, 57, T461–T489, 1966.

[4]   Barella, A., La velosidad de los hilos, historia de un parámetro y de la influencia de las nuevas tecnologia sobre la evolución de su mediciós, *Tarabajo de divulagión. Boletin intexter*, 8, 118, 2000.

[5]   Barella, A., Yarn hairiness: A survey of recent literature and a description of a new instrument for measuring yarn hairiness, *Journal of Textile Institute*, 61, 438–447, 1970.

[6]   Suh, M. W., Jasper, W., Gunay, M., 3-D electronic imaging of fabric qualities by on line yarn data, National Textile Center, Project S01-NS12, Annual Report, 2004 (http://www.ntcresearch.org/).

[7]   Barella, A., Recent developments in yarn – hairiness studies, *Journal of Textile Institute*, 64, 558–564, 1973.

[8]   Barella, A., Torn, J., Vigo, J. P., Application of a new hairiness meter to the study of sources of yarn hairiness. *Textile Research Journal*, 41, 126–133, 1972.

[9]   Steadman, R. G., Cotton testing, *Textile Progress*, 27 (1), The Textile Institute, Manchester, UK, 1997.

[10] Saville, B. P., Physical Textile Testing. Woodhead Publishing Ltd, UK, 1999.

[11] Zellweger Uster – Handbook of Uster Tester 4, No. 5, 2002 (http://www.uster.com).

[12] Neckář, B., Příze: Tvorba, struktura, vlastnosti (Yarns: Creation, Structure, Properties). SNTL Publisher, Prague, 1990 (In Czech).

[13] Neckář, B., Yarn hairiness, Part 1: Theoretical model of yarn hairiness, 7th International Conference STRUTEX, Liberec, Czech Republic, 2000.

[14] Neckář, B., Das, D., and Krupincová, G., Hairiness of staple fiber yarns, Part I: Mathematical modeling, *Journal of Textile Institute*, 3, 332–337, 2015.

[15] Krupincová, G., Neckář, B., and Das, D., Hairiness of staple fiber yarns, Part II: Model validation, *Journal of Textile Institute*, 3, 338–345, 2015.

Internal mechanics of twisted yarns

8.1 Differential equation of radial equilibrium

A group of assumptions on continuity is often considered for solving the problems of internal yarn mechanics. For example, if an object consists of enormous number of particles then an individual particle does not play a too important role for the whole body and such object can be interpreted by a continuum and can be studied by using the tools of mathematical analysis. In this chapter, we will study internal mechanics of yarns using such idea of continuum, though we are aware that the yarns do not fully correspond to this idea[1].

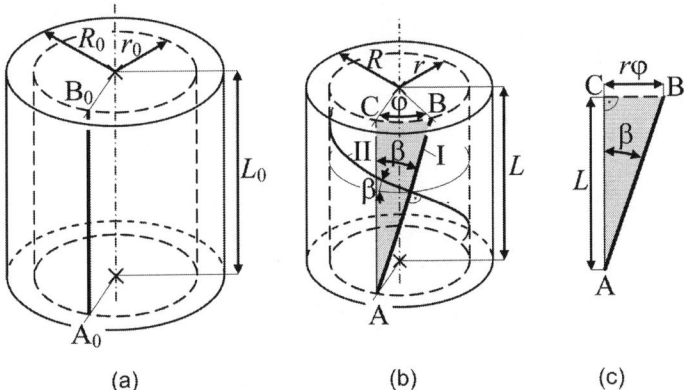

Figure 8.1 Twisting of continuous cylindrical body: (a) non-twisted case; (b) twisted case: I – fiber helix, II – perpendicular helix; (c) geometry of grey triangle

---

1 A yarn is considered to lie 'in-between' a continuous model and a discontinuous model. On one hand, it has too many different particles (i.e., fibers or relatively independently functioning fiber portions) for solving according to a discontinuous model. On the other hand, it has too little particles for 'carefree' solving according to a continuous model. Unfortunately, a special mathematical 'language' does not exist for a type of object like our yarn.

Ideal fiber helix

The easiest case is quasi-static twisting of a cylindrical continuum yarn body. It introduces the concept of the so-called ideal fiber[2] inside a cylinder. At the initial non-twisted position, the ideal fiber creates a straight line A_0B_0 at radius r_0. Such a fiber is parallel to the axis of the yarn cylinder having radius R_0 and length L_0 – see Figure 8.1a.

This ideal fiber creates the fiber helix AB after twisting – see Figure 8.1b, curve I. Simultaneously, the radius r_0 decreases to a smaller value r (compression of material) and the length of cylinder L_0 gets shortened to a value L (yarn retraction – see Chapter 4, Section 4.3). The twist angle φ expresses the angular rotation of cross sections of yarn cylinder and the helix angle β[3] characterizes the slope of the fiber (see also Chapter 4, Figure 4.2).

One coil of fiber represents the twist angle $\varphi = 2\pi$. The number of coils in a unit length of yarn is called twist Z (see Section 1.3) so that the twist angle φ at length L is

$$\varphi = 2\pi ZL. \tag{8.1}$$

The length of the arch CB is $r\varphi$, as shown in Figure 8.1b. We obtain a triangle ABC after unrolling the grey area ABC to the plane. This is illustrated in Figure 8.1c. By using Equation (8.1), we can write

$$\tan\beta = \frac{r\varphi}{L} = \frac{r\,2\pi ZL}{L} = 2\pi rZ. \tag{8.2}$$

Note: The last equation is identical to Equation (4.1) in Section 4.1; see also Figure 4.2. Nevertheless, Equation (8.2) is now valid for an ideal fiber, i.e., independent of real fibers.

---

2   The image of an ideal fiber does not need to be identical with that of a real fiber in a yarn. This term is used in continuum mechanics.

3   In other chapters of this book, we sometimes use angle β to denote 'twist angle' with a sense of spinning process, i.e., 'yarn twist'. However, in this chapter, we follow the terminology of general mechanics where the term twist angle is denoted by angle φ and the angle β is called 'helix angle'.

It is generally known that the first and second curvatures[4] are constants for all elements of a helix. The first and second curvatures k_1 and k_2 of the ideal fiber helix are, respectively[5],

$$k_1 = \frac{\sin^2 \beta}{r}, \tag{8.3}$$

$$k_2 = \frac{\sin \beta \cos \beta}{r}. \tag{8.4}$$

Perpendicular helix

It is possible to create also another helix at radius r whose helix angle is $(\beta - \pi/2)$ – see helix II in Figure 8.1b. Such a 'perpendicular' helix is cutting the earlier fiber helix at an angle $\pi/2$ [6] so that it forms an angle β to the tangent of cross-sectional ring of the cylinder[7]. The first and second curvatures c_1 and c_2 of perpendicular helix are, respectively,

$$c_1 = \frac{\sin^2 (\beta - \pi/2)}{r} = \frac{\cos^2 \beta}{r}, \tag{8.5}$$

$$c_2 = \frac{\sin (\beta - \pi/2) \cos (\beta - \pi/2)}{r} = \frac{-\cos \beta \sin \beta}{r}. \tag{8.6}$$

Elements of yarn body

If twisting is a quasi-static process, then the conditions of static equilibrium must be satisfied at each place and with each moment of twisting.

---

4   The first curvature (flexion) is the reciprocal of radius of curvature, i.e., the radius of the osculating circle. The second curvature (torsion) expresses the torsion of curve (e.g., torsion of ideal fiber, now).

5   Expressions are presented or derived in different handbooks or text books of mathematics.

6   More precisely, the tangents to both helixes create an angle of 90° at the point of intersection.

7   More precisely, the tangents to the perpendicular helix and the cross-sectional ring of cylinder (both curves at same radius r) create an angle β at the point of intersection.

Let us think about a general element of a cylinder on axial coordinate ζ with elementary height $\mathrm{d}\zeta$ lying at radius r with elementary thickness $\mathrm{d}r$ – prism ABCDEFGH according to Figure 8.2[8]. Let us assume

$$r' = r - \mathrm{d}r/2, \quad r'' = r + \mathrm{d}r/2. \tag{8.7}$$

(The same superscripts can also be used with angle β that corresponds to the above-mentioned radii.)

The edges

1. AE, BF, DH, CG and also the mid-lines KN, LM and PQ are elements of fiber helixes,
2. AB, EF, DC, HG and also the mid-lines KL, UV and NM are elements of perpendicular helixes and
3. DA, CB, HE, GF as well as the mid-line RS are segments of radial straight lines.

It is evident that all angles among (non-parallel) edges are right-angles, i.e., the general element is an elementary prism.

The lengths of the edges and the mid-lines can be found from Figure 8.2. It is valid to write that

$$\left.\begin{aligned}
&DH = CG = \mathrm{d}\zeta/\cos\beta', \quad KN = PQ = LM = \mathrm{d}\zeta/\cos\beta, \quad AE = BF = \mathrm{d}\zeta/\cos\beta'', \\
&DC = HG = r'\,\mathrm{d}\varphi\cos\beta', \quad KL = UV = NM = r\,\mathrm{d}\varphi\cos\beta, \quad AB = EF = r''\,\mathrm{d}\varphi\cos\beta'', \\
&DA = CB = RS = HE = GF = \mathrm{d}r.
\end{aligned}\right\} \tag{8.8}$$

The following surfaces can then be expressed by using the previous relations:

$$\left.\begin{aligned}
DCGH &= DC \cdot CG = r'\,\mathrm{d}\varphi\,\mathrm{d}\zeta, \\
ABFE &= AB \cdot AE = r''\,\mathrm{d}\varphi\,\mathrm{d}\zeta, \\
AEHD &= BFGC = AD \cdot KN = \mathrm{d}r\,\mathrm{d}\zeta/\cos\beta, \\
ABCD &= EFGH = DA \cdot KL = r\cos\beta\,\mathrm{d}r\,\mathrm{d}\varphi.
\end{aligned}\right\} \tag{8.9}$$

Note: The mid-line KN was used for calculation of surfaces AEHD and BFGC. The mid-line KL was used for calculation of surfaces ABCD and EFGH.

---

8   In Figure 8.2, the object ABCDEFGH is shown relatively big. This is because a lot of geometrical parameters were necessary to illustrate in this scheme. Nevertheless, we must keep in mind that we always think about an infinitely small element whose dimensions are also described in the same figure.

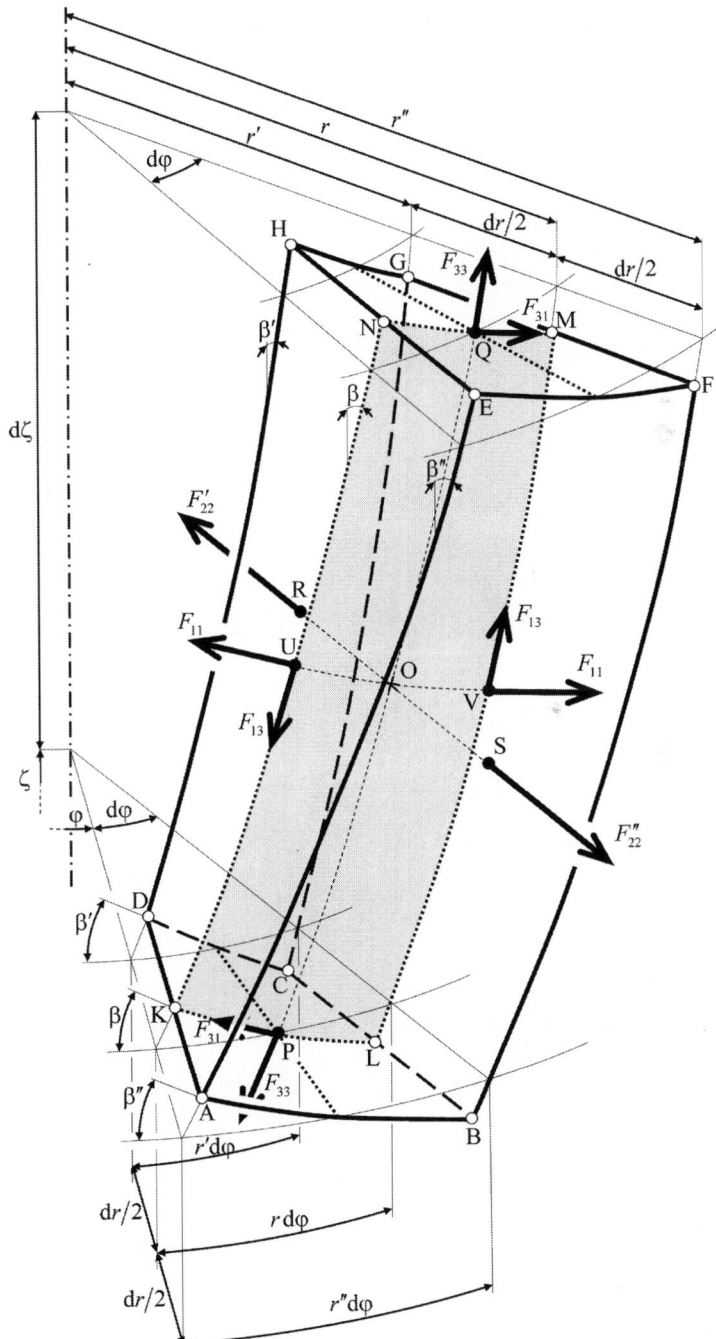

Figure 8.2 Rectangular yarn element with a general moment of twisting

In accordance with Figure 8.3, the character of deformation decides the character of force. Initially, let us think about an element $A_0B_0C_0D_0E_0F_0G_0H_0$[9] of initial, i.e., non-twisted cylinder at radius r_0, axial length $d\zeta_0$, radial thickness dr_0, angular width $d\varphi$. Its all (non-parallel) edges intersect themselves perpendicularly.

This element changes itself to a new position ABCDEFGH due to turning of cross sections of cylinder (i.e., yarn) in consequence of twisting; however, both cross sections remain parallel to each other. The radius of this element decreases to a value r, length decreases to $d\zeta$, radial thickness decreases to dr; only the angular width $d\varphi$ must stay same[10]. The edges AD, BC, EH, FG still remain perpendicular to the 'walls' ABFE and DCGH. It means that no shear forces can be acting on the stated areas ABFE and DCGH. In opposite to them, the angle EFB or HGC (i.e., also angle FEA or GHD) does not stay perpendicular after twisting. Generally, it indicates an existence of shear forces on all other surfaces of the twisted element.

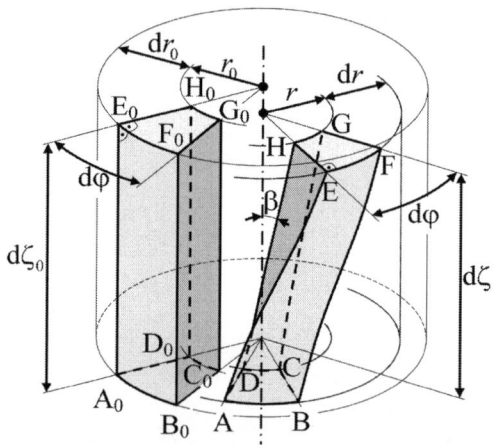

Figure 8.3 Deformation of starting element due to twist

We will study the state of stress acting on the earlier rectangular element shown in Figure 8.2. Let us interpret the forces and stresses using the directions of following local Cartesian (rectangular) coordinates:

---

9 The elements shown in Figure 8.3 are specified differently than the element introduced in Figure 8.2.

10 Yarn cylinder under consideration must remain as a continuous body also after twisting.

- direction '1' – direction of line OV, i.e., normal line to ideal helix,
- direction '2' – direction of radial line OS, i.e., binormal line to ideal helix and
- direction '3' – direction of line OQ, i.e., tangential line to ideal helix.

We think about points of force actions (•) acting on the middle of elemental surface areas.

Let us consider the symbols of normal forces as F_{11}, F'_{22}, F''_{22}, F_{33} and the symbols of shear forces as F_{13}, F_{31}. The normal stresses σ_{11}, σ'_{22} and σ''_{22}, σ_{33} and the shear stress σ_{13} [11] belong to the forces stated earlier.

The state of stress may be changed only in the radial direction 2 because of axial symmetry of twisting process. The stress σ'_{22} acts on the area DCGH (radius r'), the stress σ_{22} acts on the mid-area KLMN (radius r) and the stress σ''_{22} acts on the area ABFE (radius r''). We introduce two partial incremental changes $d\sigma'_{22}$ and $d\sigma''_{22}$ that correspond to the radial increase form r' to r (increment $dr/2$) and from r to r'' (also increment $dr/2$)

$$d\sigma'_{22} = \sigma_{22} - \sigma'_{22}, \quad d\sigma''_{22} = \sigma''_{22} - \sigma_{22}. \tag{8.10}$$

(Formally, we introduce positive increment of stress by increasing the radius.) We also introduce the whole increment of stress by the total increment of radius from r' to r'' (increment dr).

$$d\sigma_{22} = \sigma''_{22} - \sigma'_{22} = d\sigma''_{22} + d\sigma'_{22}. \tag{8.11}$$

Each force is a product of stress and the corresponding area so that

$$\left.\begin{aligned}
F_{11} &= \sigma_{11}\text{AEHD} = \sigma_{11}\,dr\,d\zeta/\cos\beta, \\
F'_{22} &= \sigma'_{22}\text{DCGH} = \sigma'_{22}r'\,d\varphi\,d\zeta, \\
F''_{22} &= \sigma''_{22}\text{ABFE} = \sigma''_{22}r''\,d\varphi\,d\zeta, \\
F_{33} &= \sigma_{33}\text{ABCD} = \sigma_{33}r\cos\beta\,dr\,d\varphi, \\
F_{13} &= \sigma_{13}\text{AEHD} = \sigma_{13}\,dr\,d\zeta/\cos\beta, \\
F_{31} &= \sigma_{13}\text{ABCD} = \sigma_{13}r\cos\beta\,dr\,d\varphi.
\end{aligned}\right\} \tag{8.12}$$

---

11   It is generally known that $\sigma_{13} = \sigma_{31}$ because each stress tensor must be symmetrical.

Radial equilibrium

It is necessary to determine the projections of all the forces onto the radial direction 2 for formulation of radial equilibrium. The partial resultant of a couple of forces F_{22}'' and F_{22}' (to the centrifugal direction) can simply be expressed by their difference $d(F_{22})_2$ [12]. Then, we can write the following expression by using Equations (8.10) to (8.12) and (8.7):

$$d(F_{22})_2 = F_{22}'' - F_{22}' = \sigma_{22}'' r'' \, d\varphi \, d\zeta - \sigma_{22}' r' \, d\varphi \, d\zeta$$

$$= \left[(d\sigma_{22}'' + \sigma_{22}) \left(r + \frac{dr}{2} \right) - (\sigma_{22} - d\sigma_{22}') \left(r - \frac{dr}{2} \right) \right] d\varphi \, d\zeta$$

$$= \left[\begin{array}{c} d\sigma_{22}'' r + d\sigma_{22}'' \dfrac{dr}{2} + \sigma_{22} r + \sigma_{22} \dfrac{dr}{2} - \sigma_{22} r \\[2mm] + \sigma_{22} \dfrac{dr}{2} + d\sigma_{22}' r - d\sigma_{22}' \dfrac{dr}{2} \end{array} \right] d\varphi \, d\zeta$$

$$= \left[d\sigma_{22}'' r + \overbrace{d\sigma_{22}'' \frac{dr}{2}}^{\substack{\text{differential} \\ \text{of higher order}}} + \sigma_{22} \frac{dr}{2} + \sigma_{22} \frac{dr}{2} + d\sigma_{22}' r - \overbrace{d\sigma_{22}' \frac{dr}{2}}^{\substack{\text{differential} \\ \text{of higher order}}} \right] d\varphi \, d\zeta$$

$$= \left[(d\sigma_{22}'' + d\sigma_{22}') r + \sigma_{22} dr \right] d\varphi \, d\zeta = \left[d\sigma_{22} r + \sigma_{22} dr \right] d\varphi \, d\zeta,$$

$$d(F_{22})_2 = \sigma_{22} dr \, d\varphi \, d\zeta + r \frac{d\sigma_{22}}{dr} dr \, d\varphi \, d\zeta.$$

$$(8.13)^{[13]}$$

All other forces work on the cylindrical surface with radius r as shown in Figure 8.2 The 'grey' coloured elementary mid-area KLMN, shown in Figure 8.2, is separately illustrated also in Figure 8.4a.

The following three ideas are used for determination of contributions from centripetal forces as shown in Figure 8.4a:

A. Because the grey area is differentially small we can interpret the arches of helixes (arch PQ of ideal helix I and arch UV of perpendicular helix II) as the arches of osculating circles. It is well known that the radii of such osculating circles are reciprocal of the corresponding first curvatures. In a nutshell, the radius of the elementary arch

---

12   The subscript stated outside the brackets indicates the direction onto which the forces are projected.

13   It is known that the differentials of higher order can be neglected.

PQ, according to Equation (8.3), is $1/k_1$ and the radius of the elementary arch UV, according to Equation (8.5), is $1/c_1$.

B. Quite generally, if an elementary arch (dashed), having radius a and the elementary angular distance $d\alpha$, is stressed by a couple of tangential forces F then the resulting centripetal force is $F\,d\alpha$ – see Figure 8.4b. It is because the component of each force F (to the 'vertical' direction as shown in Figure 8.4b is $F\sin(d\alpha/2) = F\,d\alpha/2$ [14].

C. Without loosing a sense of universality, we can 'vertically' [15] shift' each couple of equally high parallel shear forces to the common line of direction of the same force because the grey area is differentially small.

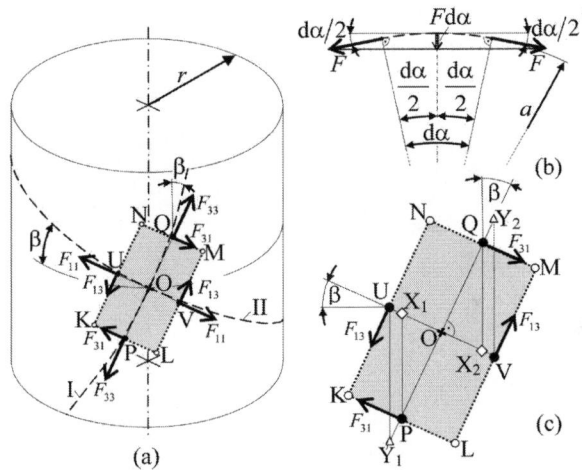

Figure 8.4 Forces acting on cylindrical surface with radius r. (a) Elementary grey surface KLMN with functioning forces: I – helix of ideal fiber, II – perpendicular helix; (b) general creation of centripetal force; (c) shear forces on grey surface KLMN

Let us at first determine the contribution of the forces F_{11} as shown in Figure 8.4a. The length of the elementary arch UV is $(1/c_1)d\alpha$, where $1/c_1$ is the radius of the osculating circle of perpendicular helix and $d\alpha$ is the elementary angular distance of the arch UV (all in the plane of the osculating circle). By using Equations (8.5) and (8.8), we can write

---

14   It is well known that $\sin x \rightarrow x$ when $x \rightarrow 0$. (The angle is expressed in radian, of course.)

15   It means the direction of yarn axis.

$$\text{UV} = (1/c_1)\,d\alpha,$$

$$r\,d\varphi\cos\beta = \frac{r}{\cos^2\beta}\,d\alpha, \qquad (8.14)$$

$$d\alpha = \cos^3\beta\,d\varphi.$$

The couple of forces F_{11}, as shown in Figure 8.4a, creates a centripetal force $d(F_{11})_2$. According to the previous idea B, it is valid to write the following expression by using Equations (8.12) and (8.14):

$$d(F_{11})_2 = F_{11}d\alpha = \sigma_{11}\,dr\,d\zeta/\cos\beta\cdot\cos^3\beta\,d\varphi = \sigma_{11}\cos^2\beta\,dr\,d\varphi\,d\zeta. \qquad (8.15)$$

Similarly, we can determine the contribution of the forces F_{33} as shown in Figure 8.4a. The length of the elementary arch PQ is $(1/k_1)\,d\gamma$, where $1/k_1$ is the radius of the osculating circle of ideal helix and $d\gamma$ is the elementary angular distance of the arch PQ (all in the same plane of the osculating circle). By using Equations (8.3) and (8.8), we can write

$$\text{UV} = (1/k_1)\,d\gamma, \qquad \frac{d\zeta}{\cos\beta} = \frac{r}{\sin^2\beta}\,d\gamma, \quad d\gamma = \frac{1}{r}\frac{\sin^2\beta}{\cos\beta}\,d\zeta. \qquad (8.16)$$

The couple of the forces F_{33}, as shown in Figure 8.4a, introduces a centripetal force $d(F_{33})_2$. According to the previous idea B, it is valid to write the following expression by using Equations (8.12) and (8.16):

$$d(F_{33})_2 = F_{33}d\gamma = \sigma_{33}r\cos\beta\,dr\,d\varphi\cdot\frac{1}{r}\frac{\sin^2\beta}{\cos\beta}\,d\zeta = \sigma_{33}\sin^2\beta\,dr\,d\varphi\,d\zeta. \qquad (8.17)$$

A little more complicated problem is to determine the contribution of the shear forces F_{31}. We will vertically shift the couple of these forces to the line UV – see the previous idea C. As a result, new (rhombic) points X_1 and X_2 are generated as shown in Figure 8.4c. According to the triangles OX_2Q and/or OX_1P, the length $OX_1 = OX_2 = (PQ/2)\tan\beta$, so that the following expression is valid to write by using Equation (8.8):

$$X_1X_2 = OX_1 + OX_2 = PQ\tan\beta = (d\zeta/\cos\beta)\tan\beta. \qquad (8.18)$$

(The arch X_1X_2 as well as the arch UV lies on the perpendicular helix.) The length of the elementary arch X_1X_2 is $(1/c_1)\,d\xi$, where $1/c_1$ is the radius of the first curvature of perpendicular helix and $d\xi$ is the elementary angular distance of the arch X_1X_2 (all in the same plane of the osculating circle). By using Equations (8.5) and (8.18), we can write

$$X_1X_2 = (1/c_1)d\xi, \quad \frac{d\zeta}{\cos\beta}\tan\beta = \frac{r}{\cos^2\beta}d\xi, \quad d\xi = \frac{1}{r}\tan\beta\cos\beta\,d\zeta. \quad (8.19)$$

The couple of the shear forces F_{31}, as shown in Figure 8.4a, brings a centripetal force $d(F_{31})_2$. According to the previous idea B, it is valid to write the following expression by using Equations (8.12) and (8.19):

$$d(F_{31})_2 = F_{31}d\xi = \sigma_{13}r\cos\beta\,dr\,d\varphi\cdot\frac{1}{r}\tan\beta\cos\beta\,d\zeta = \sigma_{13}\sin\beta\cos\beta\,dr\,d\varphi\,d\zeta.$$

$$(8.20)$$

The contribution of the shear forces F_{13} can be determined in a similar way. We will vertically shift the couple of these forces to the line PQ – see previous idea C. As a result of this action, the new (triangular) points Y_1 and Y_2 are generated as shown in Figure 8.4c. It is then valid to write from triangle OUY_1 and/or OVY_2 that $OY_1 = OY_2 = (UV/2)/\tan\beta$, so that it is further valid to write the following expression by using Equation (8.8):

$$Y_1Y_2 = OY_1 + OY_2 = \frac{UV}{\tan\beta} = \frac{r\,d\varphi\cos\beta}{\tan\beta}. \quad (8.21)$$

(The arch Y_1Y_2 and the arch PQ lie on the ideal helix.) The length of the elementary arch Y_1Y_2 is $(1/k_1)d\psi$, where $1/k_1$ is the radius of the first curvature of ideal helix and $d\psi$ is the elementary angular distance of the arch Y_1Y_2 (all lie in the plane of the osculating circle). By using Equations (8.3) and (8.21), we can write

$$Y_1Y_2 = (1/k_1)d\psi, \quad \frac{r\,d\varphi\cos\beta}{\tan\beta} = \frac{r}{\sin^2\beta}d\psi, \quad d\psi = \sin\beta\cos^2\beta\,d\varphi. \quad (8.22)$$

The couple of the shear forces F_{13}, as shown in Figure 8.4a, introduces the centripetal force $d(F_{13})_2$. According to the previous idea (B), it is valid to write the following expression by using Equations (8.12) and (8.22):

$$d(F_{13})_2 = F_{13}d\psi = \sigma_{13}\frac{dr\,d\zeta}{\cos\beta}\cdot\sin\beta\cos^2\beta\,d\varphi = \sigma_{13}\sin\beta\cos\beta\,dr\,d\varphi\,d\zeta. \quad (8.23)$$

The radial equivalency of forces follows the partial contributions that can be formulated according to Equations (8.13), (8.15), (8.17), (8.20) and (8.23). The contribution of $d(F_{22})_2$, according to Equation (8.13), represents the increment of the radial force with radius; therefore, this will bear a plus sign in the equation of equivalency. All other contributions – the stated decrements – i.e., the centripetal forces oriented in the opposite direction to the increase of radius, must bear the negative sign in the equation of equivalency. It is then valid that

$$d(F_{22})_2 - d(F_{33})_2 - d(F_{31})_2 - d(F_{11})_2 - d(F_{13})_2 = 0, \qquad (8.24)$$

$$\sigma_{22} dr \, d\varphi \, d\zeta + r \frac{d\sigma_{22}}{dr} dr \, d\varphi \, d\zeta - \sigma_{33} \sin^2 \beta \, dr \, d\varphi \, d\zeta - \sigma_{13} \sin \beta \cos \beta \, dr \, d\varphi \, d\zeta$$
$$- \sigma_{11} \cos^2 \beta \, dr \, d\varphi \, d\zeta - \sigma_{13} \sin \beta \cos \beta \, dr \, d\varphi \, d\zeta = 0,$$

$$\sigma_{22} + r \frac{d\sigma_{22}}{dr} - \sigma_{33} \sin^2 \beta - \sigma_{11} \cos^2 \beta - \sigma_{13} 2 \sin \beta \cos \beta = 0. \qquad (8.25)$$

The last expression is the differential equation of radial equilibrium in a twisted yarn cylinder. We can express this differential equation also by using Equations (8.3), (8.4) and (8.5), i.e., through curvatures of ideal and perpendicular helixes. Thus,

$$\sigma_{22} + r \frac{d\sigma_{22}}{dr} - \sigma_{33} rk_1 - \sigma_{11} rc_1 - \sigma_{13} 2 rk_2 = 0. \qquad (8.26)$$

In literature, a few attempts were made earlier to determine the differential equations for radial equilibrium. Budnikov [1] was probably the first to work on this research problem. This was also solved by Hearle [2] with a correction of the work done by Treloar and Hearle [3]. A more general solution was given by Cheng et al. [4]. Further, White et al. [5] and Neckář [6] presented quite general solution according to Equation (8.25) at the same time. (See it also in Reference [7]).

8.2   Deformations and packing density

Packing density

The unstressed element, shown in Figure 8.3, lies at radius r_0 and it has edges $A_0D_0 = dr_0$, $D_0C_0 = r_0 d\varphi$ and $D_0H_0 = d\zeta_0$. The volume of this element is $dV_0 = (A_0D_0)(D_0C_0)(D_0H_0)$, $dV_0 = (A_0D_0)(D_0C_0)(D_0H_0) = r_0 \, dr_0 \, d\varphi \, d\zeta_0$. The density of the whole element is ρ_0^*. The constant density of fiber mass is still designated as ρ. The packing density of the above-mentioned element, according to Equation (1.27) of Chapter 1, is $\mu_0 = \rho_0^*/\rho$, and the mass of it is $dm = dV_0 \rho_0^* = dV_0 \mu_0 \rho$.

After twisting, the element changes itself to a new position. Its edges are $AD = dr$, $DC = r\,d\varphi$ and $DH = d\zeta/\cos\beta$. According to the geometry shown in Figure 8.3, its volume is $dV = (AD)(DC)(DH \cdot \cos\beta) = r\,dr\,d\varphi\,d\zeta$. The density of the stressed element is marked by ρ^* so that its packing density is $\mu = \rho^*/\rho$ and the mass of the stressed element is $dm = dV\rho^* = dV\mu\rho$. By using the previous relations, the following expression is then valid to write

$$dm = dV_0\,\mu_0\,\rho = dV\,\mu\rho, \quad \frac{\mu}{\mu_0} = \frac{dV_0}{dV} = \frac{r_0\,dr_0\,d\varphi\,d\zeta_0}{r\,dr\,d\varphi\,d\zeta} = \frac{r_0\,dr_0\,d\zeta_0}{r\,dr\,d\zeta}. \quad (8.27)$$

Ideal phases of deformation

The visible objects were drawn relatively large as shown in Figure 8.3 to display a good quality image. Nevertheless, we think about elementary objects. Therefore, we can interpret them also as elementary prisms. Figure 8.5 shows several ideal phases of deformation of elementary prisms from the initial position $A_0B_0C_0D_0E_0F_0G_0H_0$ to the final position ABCDEFGH. (The directions of local coordinates are shown in the schemes above the elementary prisms in Figure 8.5.)

The initial elementary prism $A_0B_0C_0D_0E_0F_0G_0H_0$ – scheme (a), changes itself to prism $A_1B_1C_1D_1E_1F_1G_1H_1$ – scheme (b), by only normal deformation determined by strains $\varepsilon_1, \varepsilon_2, \varepsilon_3$ in the directions of local coordinates. Thus,

$$1 + \varepsilon_1 = A_1B_1/A_0B_0 = A_1B_1/r_0\,d\varphi, \quad (8.28)$$

$$1 + \varepsilon_2 = A_1D_1/A_0D_0 = A_1D_1/dr_0, \quad (8.29)$$

$$1 + \varepsilon_3 = A_1E_1/A_0E_0 = A_1E_1/d\zeta_0. \quad (8.30)$$

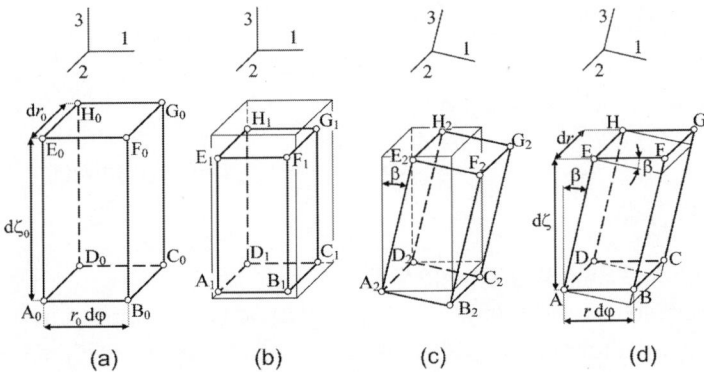

Figure 8.5 Ideal phases of deformation of element: (a) initial position, (b) normal deformation, (c) rotation of element, (d) shear deformation

Then, the elementary prism turns itself around axis A_1D_1 at angle β to the oblique direction $A_2B_2C_2D_2E_2F_2G_2H_2$ – scheme (c). However, no additional deformation originates due to this turning. Finally, it follows shear deformation with an angle β according to scheme (d) and so the resulting element ABCDEFGH is originated. The edge A_2B_2 changes its length to AB and similarly also edges D_2C_2, E_2F_2, H_2G_2 change their lengths to DC, EF, HG, respectively. The lengths of all other edges remain unchanged.

By using Figure 8.5(d), we can write the following expressions:

$$A_1B_1 = A_2B_2 = AB \cdot \cos\beta = r\,d\varphi\cos\beta, \tag{8.31}$$

$$A_1D_1 = A_2D_2 = AD = dr, \tag{8.32}$$

$$A_1E_1 = A_2E_2 = AE = d\zeta/\cos\beta. \tag{8.33}$$

By applying the previous triplet of expressions in Equations (8.28) to (8.30), we find

$$1+\varepsilon_1 = \frac{r\,d\varphi\cos\beta}{r_0\,d\varphi} = \frac{r}{r_0}\cos\beta, \tag{8.34}$$

$$1+\varepsilon_2 = \frac{dr}{dr_0} = \frac{1}{dr_0/dr}, \tag{8.35}$$

$$1+\varepsilon_3 = \frac{d\zeta/\cos\beta}{d\zeta_0} = \frac{d\zeta}{d\zeta_0}\frac{1}{\cos\beta}. \tag{8.36}$$

The last three equations allow us to express the packing density, according to Equation (8.27), in the following form

$$\frac{\mu}{\mu_0} = \frac{r_0\,dr_0\,d\zeta_0}{r\,dr\,d\zeta} = \left(\frac{r_0}{r\cos\beta}\right)\left(\frac{dr_0}{dr}\right)\left(\frac{d\zeta_0}{d\zeta}\cos\beta\right), \quad \mu = \frac{\mu_0}{(1+\varepsilon_1)(1+\varepsilon_2)(1+\varepsilon_3)}. \tag{8.37}$$

Note: The packing density is not now immediately influenced by the helix angle β.

Rearrangement of differential equation

The instantaneous values of twist Z – see Equation (8.2) – and yarn retraction $d\zeta_0/d\zeta$ are common parameters for all elements in a yarn with a given moment of twisting. Further, a function $r_0 = r_0(r)$, showing the change from initial radius r_0 to an instantaneous radius r, must also exist for each moment of twisting.

The deformation of the element is fully described by four quantities $\varepsilon_1, \varepsilon_2, \varepsilon_3, \beta$ which are only functions of variable r at a given moment of twisting. Therefore, we can express derivatives of $\varepsilon_1, \varepsilon_2, \varepsilon_3, \beta$ with respect to r.

Note: The following couple of expressions obtained from Equation (8.2) will be used for derivations:

$$\cos\beta = \frac{1}{\sqrt{1+\tan^2\beta}} = \frac{1}{\sqrt{1+(2\pi r Z)^2}}, \tag{8.38}$$

$$\frac{d(\cos\beta)}{dr} = -\frac{1}{2}\left[1+(2\pi r Z)^2\right]^{-3/2} 2(2\pi r Z)2\pi Z = -\left[1+(2\pi r Z)^2\right]^{-3/2}\frac{(2\pi r Z)^2}{r}$$

$$= -\frac{1}{r}\cos^3\beta\tan^2\beta = -\frac{1}{r}\cos\beta\sin^2\beta. \tag{8.39}$$

By using Equation (8.34), we determine the following derivative by the help of previous two equations and Equation (8.35):

$$\frac{d\varepsilon_1}{dr} = \frac{\partial\varepsilon_1}{\partial r} + \frac{\partial\varepsilon_1}{\partial(\cos\beta)}\frac{d(\cos\beta)}{dr} + \frac{\partial\varepsilon_1}{\partial r_0}\frac{dr_0}{dr}$$

$$= \frac{\cos\beta}{r_0} + \frac{r}{r_0}\left(-\frac{1}{r}\cos\beta\sin^2\beta\right) + r\cos\beta\frac{-1}{r_0^2}\frac{dr_0}{dr}$$

$$= \frac{\cos\beta}{r_0}\left[1-\sin^2\beta-\frac{r}{r_0}\frac{dr_0}{dr}\right] = \frac{\cos\beta}{r_0}\left[\cos^2\beta - \frac{\dfrac{r}{r_0}\cos\beta}{\dfrac{dr}{dr_0}\cos\beta}\right], \tag{8.40}$$

$$\frac{d\varepsilon_1}{dr} = \frac{1+\varepsilon_1}{r}\left[\cos^2\beta - \frac{1+\varepsilon_1}{(1+\varepsilon_2)\cos\beta}\right].$$

From Equation (8.35), we can derive

$$\frac{d\varepsilon_2}{dr} = \frac{d}{dr}\left(\frac{1}{dr_0/dr}\right) = \frac{-d^2 r_0/dr^2}{(dr_0/dr)^2} = -(1+\varepsilon_2)^2\frac{d^2 r_0}{dr^2}. \tag{8.41}$$

Further, applying Equations (8.36) and (8.39), we can write

$$\frac{d\varepsilon_3}{dr} = \frac{d\zeta}{d\zeta_0}\frac{(-1)d(\cos\beta)/dr}{\cos^2\beta} = \frac{d\zeta}{d\zeta_0}\frac{\frac{1}{r}\cos\beta\sin^2\beta}{\cos^2\beta} = \frac{d\zeta}{d\zeta_0}\frac{\tan\beta\sin\beta}{r}. \quad (8.42)$$

Finally, we find the following expression from Equation (8.2):

$$\beta = \arctan(2\pi rZ),$$

$$\frac{d\beta}{dr} = \frac{1}{1+(2\pi rZ)^2}2\pi Z = \frac{1}{1+\tan^2\beta}\frac{\tan\beta}{r} = \cos^2\beta\frac{\tan\beta}{r} = \frac{\cos\beta\sin\beta}{r}. \quad (8.43)$$

We assume that a common deformation law stands well when all the elements in the yarn cylinder are twisted; we think about a homogenous material. This law expresses the mechanical characteristic of the fibrous assembly from which the yarn is created. So, each of the following stresses $\sigma_{11}, \sigma_{22}, \sigma_{33}, \sigma_{13}$ is generally a function of four deformations $\varepsilon_1, \varepsilon_2, \varepsilon_3, \beta$ [16] in this case. For example, $\sigma_{22} = \sigma_{22}(\varepsilon_1, \varepsilon_2, \varepsilon_3, \beta)$, thus also

$$\frac{d\sigma_{22}}{dr} = \frac{\partial\sigma_{22}}{\partial\varepsilon_1}\frac{d\varepsilon_1}{dr} + \frac{\partial\sigma_{22}}{\partial\varepsilon_2}\frac{d\varepsilon_2}{dr} + \frac{\partial\sigma_{22}}{\partial\varepsilon_3}\frac{d\varepsilon_3}{dr} + \frac{\partial\sigma_{22}}{\partial\beta}\frac{d\beta}{dr}. \quad (8.44)$$

The following expression is obtained after substitution of the last expression in the differential equation according to Equation (8.25) and also using Equation (8.41) for rearrangement:

$$\sigma_{22} + r\left[\frac{\partial\sigma_{22}}{\partial\varepsilon_1}\frac{d\varepsilon_1}{dr} + \frac{\partial\sigma_{22}}{\partial\varepsilon_2}\frac{d\varepsilon_2}{dr} + \frac{\partial\sigma_{22}}{\partial\varepsilon_3}\frac{d\varepsilon_3}{dr} + \frac{\partial\sigma_{22}}{\partial\beta}\frac{d\beta}{dr}\right]$$

$$- \sigma_{33}\sin^2\beta - \sigma_{11}\cos^2\beta - \sigma_{13}2\sin\beta\cos\beta = 0,$$

$$\frac{\sigma_{22}}{r} + \frac{\partial\sigma_{22}}{\partial\varepsilon_1}\frac{d\varepsilon_1}{dr} + \frac{\partial\sigma_{22}}{\partial\varepsilon_2}\left[-(1+\varepsilon_2)^2\frac{d^2r_0}{dr^2}\right] + \frac{\partial\sigma_{22}}{\partial\varepsilon_3}\frac{d\varepsilon_3}{dr} + \frac{\partial\sigma_{22}}{\partial\beta}\frac{d\beta}{dr}$$

$$- \frac{\sigma_{33}}{r}\sin^2\beta - \frac{\sigma_{11}}{r}\cos^2\beta - 2\frac{\sigma_{13}}{r}\sin\beta\cos\beta = 0,$$

$$\frac{\partial\sigma_{22}}{\partial\varepsilon_2}\left[-(1+\varepsilon_2)^2\frac{d^2r_0}{dr^2}\right] = -\frac{\sigma_{22}}{r} - \frac{\partial\sigma_{22}}{\partial\varepsilon_1}\frac{d\varepsilon_1}{dr} - \frac{\partial\sigma_{22}}{\partial\varepsilon_3}\frac{d\varepsilon_3}{dr} - \frac{\partial\sigma_{22}}{\partial\beta}\frac{d\beta}{dr}$$

$$+ \frac{\sigma_{33}}{r}\sin^2\beta + \frac{\sigma_{11}}{r}\cos^2\beta + 2\frac{\sigma_{13}}{r}\sin\beta\cos\beta,$$

---

[16] More generally, such equations express the relation between stress tensor and strain tensor of fibrous material.

$$\frac{d^2 r_0}{dr^2} = \frac{-1}{\left(1+\varepsilon_2\right)^2 \partial\sigma_{22}/\partial\varepsilon_2} \left[-\frac{\partial\sigma_{22}}{\partial\varepsilon_1}\frac{d\varepsilon_1}{dr} - \frac{\partial\sigma_{22}}{\partial\varepsilon_3}\frac{d\varepsilon_3}{dr} - \frac{\partial\sigma_{22}}{\partial\beta}\frac{d\beta}{dr} \right.$$

$$\left. -\frac{\sigma_{22}}{r} + \frac{\sigma_{33}}{r}\sin^2\beta + \frac{\sigma_{11}}{r}\cos^2\beta + 2\frac{\sigma_{13}}{r}\sin\beta\cos\beta \right].$$

$$(8.45)$$

Let us think that except the common parameters (e.g., Z, $d\zeta/d\zeta_0$):

- according to Equation (8.2), helix angle β is the function of only radius r,
- according to Equations (8.34) to (8.36), deformations $\varepsilon_1, \varepsilon_2, \varepsilon_3$ are functions of only $r, r_0, dr_0/dr$ and
- each stress $\sigma_{11}, \sigma_{22}, \sigma_{33}, \sigma_{13}$ is a given function of $\varepsilon_1, \varepsilon_2, \varepsilon_3, \beta$ and so the stresses are functions of only $r, r_0, dr_0/dr$.

Therefore, Equation (8.45) is the differential equation of second order with explicitly expressed second derivatives, as follows:

$$d^2 r_0/dr^2 = f\left(r, r_0, dr_0/dr\right). \qquad (8.46)$$

The solution of it determines the function $r_0 = r_0\left(r\right)$.

8.3   Solution of differential equation of radial equilibrium

Initial conditions

The nature of stresses and strains must be smooth[17] at each inner point of continuum. A point on the axis of yarn cylinder is also an inner point so that in the surrounding of it the nature of the strain ε_2 must be smooth; however, the yarn cylinder must also be axially symmetrical. Both assumptions are satisfied by the following relation:

$$\lim_{r\to 0}\left(d\varepsilon_2/dr\right) = 0 . \qquad (8.47)$$

Because $\left(1+\varepsilon_2\right) \neq 0$, according to Equation (8.41), it must be valid that

$$\lim_{r\to 0}\left(d^2 r_0/dr^2\right) = 0. \qquad (8.48)$$

---

17   Such a character is continuous in function and also in first derivative.

It means that the relation

$$\lim_{r \to 0} \left(dr_0 / dr \right) = k \ldots \text{parameter} \tag{8.49}$$

is valid in the (differentially small) surrounding of point $r = 0$. (It is known that $r_0 = 0$ when $r = 0$.) Now, the following expression results from Equations (8.2) and (8.34) to (8.36) by the use of Equation (8.49):

$$\lim_{r \to 0} \beta = 0, \tag{8.50}$$

$$\lim_{r \to 0} \left(1 + \varepsilon_1 \right) = \lim_{r \to 0} \left(1 + \varepsilon_2 \right) = \frac{1}{k}, \tag{8.51}[18]$$

$$\lim_{r \to 0} \left(1 + \varepsilon_3 \right) = d\zeta / d\zeta_0 . \tag{8.52}$$

By applying Equations (8.51) and (8.52) in (8.37), the packing density around the axial point of yarn cylinder takes the following form:

$$\mu_{r=0} = \lim_{r \to 0} \mu = \frac{\mu_0 k^2}{d\zeta / d\zeta_0}, \quad \text{i.e. also} \quad k = \sqrt{\frac{\mu_{r=0} \, d\zeta / d\zeta_0}{\mu_0}} . \tag{8.53}$$

It is possible to consider the following couple of conditions as the initial conditions:

1. The function $r_0 = r_0(r)$ is passing through the initial point $r = 0$ at $r_0 = 0$.

2. Value of the parameter $\lim_{r \to 0} \left(dr_0 / dr \right) = k$.

Other relations are only a result of continuum behaviour.

Because stresses must be smooth at each inner point of continuum and yarn cylinder is axially symmetrical, the following expression is valid in analogy to Equation (8.47):

$$\lim_{r \to 0} \left(d\sigma_{22} / dr \right) = 0 . \tag{8.54}$$

Further, we can write

$$\lim_{r \to 0} \left(\sigma_{13} \right) = 0 , \tag{8.55}$$

---

18   Usually, $\lim_{r \to 0} \left(1 + \varepsilon_1 \right) = \lim_{r \to 0} \left(1 + \varepsilon_2 \right) < 1$ (compression around yarn axis) so that $k > 1$.

because no shear deformation is present surrounding the yarn axis. (According to Equation (8.50), $\beta = 0$ there.)

Inputs

The function $r_0 = r_0(r)$, satisfying the initial conditions, is the desirable solutions of differential equation according to Equation (8.45). The necessary inputs for solving this are

1. *Deformation law.* This expresses several stresses $\sigma_{11}, \sigma_{22}, \sigma_{33}, \sigma_{13}$ as suitable functions of deformations $\varepsilon_1, \varepsilon_2, \varepsilon_3, \beta$. The adequate deformation law must describe the so-called 'transversely isotropic continuum' where the 'transverse' plane means the plane perpendicular to the tangential direction 3, i.e., perpendicular to (tangent of) the helix of ideal fiber – e.g., the plane USVR in the element shown in Figure 8.2.[19] (There are also different parameters like different moduli, etc., that can be included in the expressions creating the deformation law.)

2. *Parameter* $d\zeta/d\zeta_0$. This value is determined by the change of axial length of the yarn cylinder (a) in consequence of yarn retraction due to twisting and (b) in consequence of yarn tensioning by twisting.

3. *Twist Z.* This traditional parameter describes the level of twisting of yarn.

4. *Parameter k.* This value represents the derivative dr_0/dr at the point on yarn axis. [See Equations (8.49) and also (8.53).]

Solution

The solution of differential equation of radial equilibrium can be obtained as follows:

A. Individual stresses and their partial derivatives with respect to deformations must be expressed as functions of $\varepsilon_1, \varepsilon_2, \varepsilon_3, \beta$ from the given deformation law.

B. Deformations $\varepsilon_1, \varepsilon_2, \varepsilon_3, \beta$ and their derivatives with respect to r must be expressed as functions of $r, r_0, dr_0/dr$ by means of Equations (8.2), (8.34), (8.35), (8.36) and Equations (8.40), (8.41), (8.42), (8.43).

---

19 The necessity of transverse isotropy also comes when we imagine the element from Figure 8.2 in the limit position on the axis of the yarn cylinder.

C. The differential equation (Equation 8.46) is originated by mutually substituting the previously obtained expressions and then substituting the resulting expression to Equation (8.45).

Usually, we cannot solve such differential equation analytically. Therefore, it is necessary to use a suitable numerical method for solving such type of differential equation[20].

Nevertheless, one more problem must be solved – that is related to the unknown value of k present in the second initial condition. This parameter could be calculated by the second expression of Equation (8.53) (values $d\zeta/d\zeta_0$ and μ_0 are known), but we do not know the value $\mu_{r=0} \in (\mu_0, 1)$. Nevertheless, we can use the following boundary condition in lieu of the present initial condition $\lim\limits_{r \to 0} (dr_0/dr) = k$.

$$\sigma_{22} = 0 \quad \text{when} \quad r = R . \tag{8.56}$$

(R_0 is the radius of non-twisted cylinder and R is the radius of twisted cylinder – see Figure 8.1.) This condition says that no radial stress is allowed to be present on the surface of twisted yarn cylinder. If we suggest a too high value of $\mu_{r=0}$ then k is also too high and σ_{22} is less than 0 at radius $r = R$. In contrary to this, if we suggest a too small value of $\mu_{r=0}$, then k is also too small and σ_{22} is greater than 0 at radius $r = R$. This relation allows us to use the so-called 'shooting method' for numerical solving of our differential equation. By a convergent iterative algorithm (e.g., splitting of intervals), we can suggest different new values of $\mu_{r=0} \in (\mu_0, 1)$ to determine the value of σ_{22} at radius $r = R$ in each step. In this way, we can come near to the desirable relation $\sigma_{22} = 0$.

Possible results

The function $r_0 = r_0(r)$ (and naturally its derivative) is the primary solution of differential equation of radial equilibrium and many other results can be determined from this result. Above all, they are

1. *Diameter D of yarn cylinder.* It is especially valid that $R_0 = r_0(R)$ on the surface of the yarn cylinder, whereas the initial radius R_0 is a

---

20   For example, the Runge–Kutte method was used for first several points and then the predictor–corrector method was applied for other points in [6] and [7]. See Reference [8] for the mentioned methods.

known value. Yarn radius R can be determined from the above-mentioned relation and then the yarn diameter can be determined from $D = 2R$.

2. *Radial nature of deformations* $\varepsilon_1, \varepsilon_2, \varepsilon_3$ – according to Equations (8.34), (8.35), (8.36) by using Equation (8.2).

3. *Radial nature of packing density* μ – according to Equation (8.37).

4. *Radial nature of stresses* $\sigma_{11}, \sigma_{22}, \sigma_{33}, \sigma_{13}$ – according to the law of deformation.

5. *Axial force on yarn cylinder.* Figure 8.6 shows a part of the element from Figure 8.2, which is cut by the plane AYXD, i.e., perpendicular to the axis of yarn cylinder. (Area ABCD can be identical to Figure 8.2.) Also, all the stresses acting on it are illustrated here.

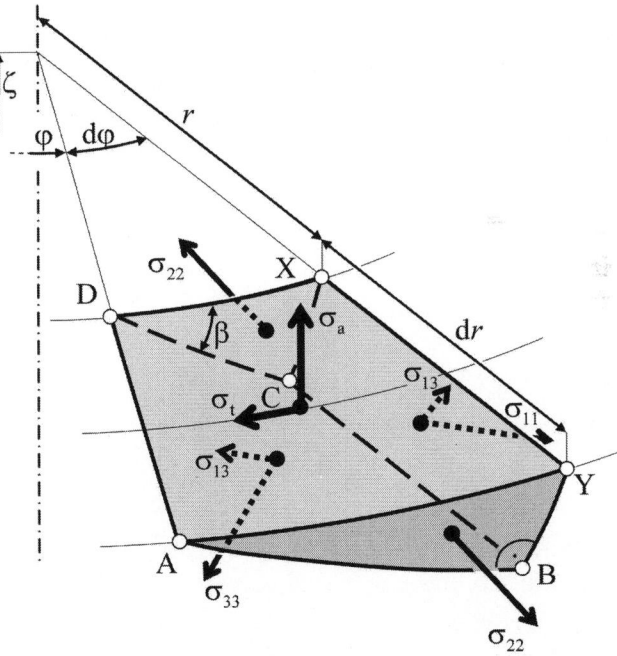

Figure 8.6 Part of element from Figure 8.2 cut by a plane perpendicular to yarn axis

Note: Segments BY and CX are parts of fiber helixes, whereas segments AB and DC are parts of perpendicular helixes.

The following areas are shown in Figure 8.6:

$$AYXD = r\,d\varphi\,dr , \tag{8.57a}$$

$$ABCD = AYXD\cos\beta$$
$$= r\,d\varphi\,dr\cos\beta,$$

$$(8.58a)^{21}$$

$$BYXC = AYXD\sin\beta$$
$$= r\,d\varphi\,dr\sin\beta,$$

$$(8.59a)$$

Besides the earlier stated stresses $\sigma_{11}, \sigma_{22}, \sigma_{33}, \sigma_{13}$, the normal stress σ_a in axial direction and the shear stress σ_t in tangential direction also act on the area AYXD. The normal force dF_Y acts on the area AYXD in axial direction. Then, the following equivalency is valid in consequence of the equilibrium of forces acting on the elementary body shown in Figure 8.6:

$$dF_Y = \sigma_a AYXD = ABCD\left(\sigma_{33}\cos\beta - \sigma_{13}\sin\beta\right) + BYXC\left(\sigma_{11}\sin\beta - \sigma_{13}\cos\beta\right)$$
$$= r\,d\varphi\,dr\cos\beta\left(\sigma_{33}\cos\beta - \sigma_{13}\sin\beta\right) + r\,d\varphi\,dr\sin\beta\left(\sigma_{11}\sin\beta - \sigma_{13}\cos\beta\right)$$
$$= \left(\sigma_{11}\sin^2\beta + \sigma_{33}\cos^2\beta - 2\sigma_{13}\sin\beta\cos\beta\right)r\,dr\,d\varphi,$$
$$dF_Y = \left(\sigma_{11}\sin^2\beta + \sigma_{33}\cos^2\beta - \sigma_{13}\sin 2\beta\right)r\,dr\,d\varphi. \qquad (8.57)$$

Then, the whole axial force acting on a yarn cylinder is

$$F_Y = \iint_{\substack{r\in(0,R)\\ \varphi\in(0,2\pi)}} dF_Y = 2\pi\int_0^R\left(\sigma_{11}\sin^2\beta + \sigma_{33}\cos^2\beta - \sigma_{13}\sin 2\beta\right)r\,dr. \quad (8.58)$$

6. *Yarn retraction.* Force F_Y is changed with different input values of parameter $d\zeta/d\zeta_0$, then we obtain a higher value of F_Y for a higher value of $d\zeta/d\zeta_0$ and vice-versa. It is possible to find also such value $\left(d\zeta/d\zeta_0\right)_{F_Y=0}$ at which force $F_Y = 0$ (by using a convergent iterative algorithm). Such a situation corresponds to 'free' yarn, i.e., non-tensioned yarn. So, the determined value $\left(d\zeta/d\zeta_0\right)_{F_Y=0}$ can be used for finding yarn retraction δ according to Equation (4.30) stated in Chapter 4.

---

21 This equation is identical to the corresponding expression in Equation (8.9), as expected.

Note: We theoretically assume a yarn which is quite uniform along its length so that we can use the value $(d\zeta/d\zeta_0)_{F_Y=0}$ in place of ζ/ζ_0 [22] in equations mentioned in Section 4.3 of Chapter 4. Especially, yarn retraction is expressed as

$$\delta = 1 - (d\zeta/d\zeta_0)_{F_Y=0} \tag{8.59}$$

according to Equation (4.30).

Also, the neutral position x_n in yarn can be expressed from Equation (4.37) and then the neutral radius r_n can be found from Equation (4.36).

7. *Force–strain and stress–strain curve of yarn.* Let us imagine a differentially small length $d\zeta_0$ as a part of non-twisted yarn cylinder. The length of this part after twisting (free, non-tensioned) is $d\zeta_0(1-\delta)$ – see the definition of yarn retraction δ given in Chapter 4. The length of the part of yarn is $d\zeta$ after applying tension to the twisted yarn. Then, yarn strain ε_Y can be expressed as follows:

$$\varepsilon_Y = \frac{d\zeta - d\zeta_0(1-\delta)}{d\zeta_0(1-\delta)} = \frac{d\zeta/d\zeta_0}{1-\delta} - 1. \tag{8.60}$$

If we calculate yarn force F_Y according to Equation (8.58) and yarn strain ε_Y according to Equation (8.60) repeatedly for lots of input values $d\zeta/d\zeta_0 \geq (d\zeta/d\zeta_0)_{F_Y=0}$, then we obtain lots of points of force–strain relation of the yarn. We can also find specific stress–strain relation by dividing each yarn force F_Y by yarn fineness T as it will be given in Chapter 9, Equation (9.1).

8. *Torsional moment.* The elementary contribution dM_Y to torsional moment of yarn applied on the elementary area AYXD, as shown in Figure 8.6, is

---

22 Let us remind that ζ_0 is the length of untwisted yarn and ζ is the length of twisted but non-tensioned ('free') yarn, as stated in Section 4.3.

$$dM_Y = \left[\overbrace{AYXD\sigma_t}^{\text{shear force}} \right] r$$

$$= \left[ABCD \left(\sigma_{33} \sin\beta + \sigma_{13} \cos\beta \right) + BYXC \left(-\sigma_{11} \cos\beta - \sigma_{13} \sin\beta \right) \right] r$$

$$= \left[r\, d\varphi\, dr \cos\beta \left(\sigma_{33} \sin\beta + \sigma_{13} \cos\beta \right) + r\, d\varphi\, dr \sin\beta \left(-\sigma_{11} \cos\beta - \sigma_{13} \sin\beta \right) \right] r$$

$$= \left[\left(\sigma_{33} - \sigma_{11} \right) \sin\beta \cos\beta + \sigma_{13} \left(\cos^2\beta - \sin^2\beta \right) \right] r^2 d\varphi\, dr,$$

$$dM_Y = \left[\frac{\left(\sigma_{33} - \sigma_{11} \right)}{2} \sin 2\beta + \sigma_{13} \cos 2\beta \right] r^2 d\varphi\, dr. \tag{8.61}$$

[Equations (3.58a) and (3.59a) were used for rearrangement.] The total torsional moment is then

$$M_Y = \iint\limits_{\substack{r\in(0,R) \\ \varphi\in(0,2\pi)}} dM_Y = 2\pi \int_0^R \left[\frac{\left(\sigma_{33} - \sigma_{11} \right)}{2} \sin 2\beta + \sigma_{13} \cos 2\beta \right] r^2\, dr. \tag{8.62}$$

Let us state that all the items of information (point 1 to point 7 in this section) can be evaluated for each moment of twisting process by means of increasing input values Z as well as for each moment of tensioning yarn cylinder by means of increasing input values $d\zeta/d\zeta_0$. So, creation and tensioning of each yarn can be viewed as specific processes theoretically.

8.4   Some possibilities of solution of deformation law

The determination of stresses $\sigma_{11}, \sigma_{22}, \sigma_{33}, \sigma_{13}$ as suitable functions of deformations $\varepsilon_1, \varepsilon_2, \varepsilon_3, \beta$ requires basic inputs for solving the differential equation of radial equilibrium in a yarn cylinder, described in the previous section. Unfortunately, there is not enough knowledge available on this problem as of now. Nevertheless, we would like to present some relations which can provide a help to obtain a solution to above-mentioned problem in future.

Note: We will work with the two definitions of stresses in this section. One is the engineering (nominal) stress defined as a force applied per unit area of non-deformed body. (This force will be initially applied on the non-deformed area.) Such stress is marked with superscript * here; e.g., σ_{11}^*. The second definition is stress – the true (Cauchy's) stress – defined as a force acting on the area of the deformed body. (This force acted on the final area, i.e., the deformed area.) Such stress is marked without superscript here; e.g., σ_{11}.

Initial relations

The initial unitary cube of fibrous material, shown in Figure 8.7, is displayed in the Cartesian system of rectangular coordinates x_1, x_2 and x_3. The volume of fibers in this cube is V and so the initial packing density is $\mu_0 = V/(1 \cdot 1 \cdot 1)$. Thus,

$$\mu_0 = V .\tag{8.63}$$

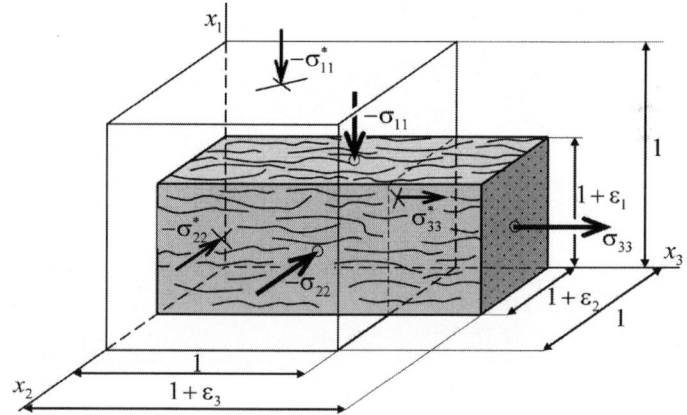

Figure 8.7 Deformations of fibrous material by normal stresses

We assume that

- the fibers are largely oriented to the direction of x_3 -axis and
- the fibrous material is isotropic in the plane perpendicular to the direction of x_3 -axis, i.e., in the plane of axes x_1, x_2 (transverse isotropy).

Now, let us think that only the normal forces $\sigma_{11}^*, \sigma_{22}^*, \sigma_{33}^*$ are acting on the surfaces of the initial unitary cube[23]. Of them, forces $\sigma_{11}^*, \sigma_{22}^*$ take minus sign, because they are shown in the opposite direction, i.e., compressional direction shown in Figure 8.7 (the typical situation in yarns). Each square area is equal to 1 so that forces $\sigma_{11}^*, \sigma_{22}^*, \sigma_{33}^*$ are also the engineering (nominal) normal stresses. (The shear stresses are not considered here.)

---

23 In Figure 8.7, the points, where the stated forces are applied, are marked by $\times$ symbol.

The initial cube deforms itself to the grey prism in consequence of the initial strains $\varepsilon_1, \varepsilon_2, \varepsilon_3$ so that the dimensions of the deformed body are $(1+\varepsilon_1), (1+\varepsilon_2), (1+\varepsilon_3)$. We assume that the fiber volume V remains always same in the process of deformation[24]. Then, the total volume of this prism is $V_c = (1+\varepsilon_1)(1+\varepsilon_2)(1+\varepsilon_3)$ and – using Equation (8.63) – the packing density after deformation is

$$\mu = \frac{V}{V_c} = \frac{\mu_0}{(1+\varepsilon_1)(1+\varepsilon_2)(1+\varepsilon_3)}. \tag{8.64}$$

Further, let us introduce the following triplet of quantities

$$\left.\begin{array}{ll} \mu_1 = \dfrac{\mu_0}{1+\varepsilon_1}, & 1+\varepsilon_1 = \dfrac{\mu_0}{\mu_1}, \\[3mm] \mu_2 = \dfrac{\mu_0}{1+\varepsilon_2}, & 1+\varepsilon_2 = \dfrac{\mu_0}{\mu_2}, \\[3mm] \mu_3 = \dfrac{\mu_0}{1+\varepsilon_3}, & 1+\varepsilon_3 = \dfrac{\mu_0}{\mu_3}. \end{array}\right\} \tag{8.65}$$

Note: The quantities μ_1, μ_2, μ_3 are not generally packing densities; they are only an alternative way to describe strains.

By using Equations (8.64) and (8.65), we obtain

$$\mu_1\mu_2\mu_3 = \frac{\mu_0}{1+\varepsilon_1}\frac{\mu_0}{1+\varepsilon_2}\frac{\mu_0}{1+\varepsilon_3} = \mu\mu_0^2. \tag{8.66}$$

The true (Cauchy's) stresses $\sigma_{11}, \sigma_{22}, \sigma_{33}$ act on the rectangular surfaces of the deformed prism[25]. By using Equation (8.65) and (8.66), it is valid to write that

$$\left.\begin{array}{l} \sigma_{11} = \dfrac{\sigma_{11}^*}{(1+\varepsilon_2)(1+\varepsilon_3)} = \sigma_{11}^*\dfrac{\mu_2\mu_3}{\mu_0^2} = \sigma_{11}^*\dfrac{\mu}{\mu_1}, \\[4mm] \sigma_{22} = \dfrac{\sigma_{22}^*}{(1+\varepsilon_1)(1+\varepsilon_3)} = \sigma_{22}^*\dfrac{\mu_1\mu_3}{\mu_0^2} = \sigma_{22}^*\dfrac{\mu}{\mu_2}, \\[4mm] \sigma_{33} = \dfrac{\sigma_{33}^*}{(1+\varepsilon_1)(1+\varepsilon_2)} = \sigma_{33}^*\dfrac{\mu_1\mu_2}{\mu_0^2} = \sigma_{33}^*\dfrac{\mu}{\mu_3}. \end{array}\right\} \tag{8.67}$$

---

24   It means that the compression of fiber assembly results in the decrease of air volume only.

25   In Figure 8.7, the points, where the above-mentioned forces are applied, are marked by symbol O.

Let us further formulate the following derivatives of expressions stated in Equation (8.65):

$$\left.\begin{aligned}
\frac{d\mu_1}{d\varepsilon_1} &= \frac{-\mu_0}{\left(1+\varepsilon_1\right)^2} = \frac{-\mu_1^2}{\mu_0}, \\[2mm]
\frac{d\mu_2}{d\varepsilon_2} &= \frac{-\mu_0}{\left(1+\varepsilon_2\right)^2} = \frac{-\mu_2^2}{\mu_0}, \\[2mm]
\frac{d\mu_3}{d\varepsilon_3} &= \frac{-\mu_0}{\left(1+\varepsilon_3\right)^2} = \frac{-\mu_3^2}{\mu_0}.
\end{aligned}\right\}
\qquad (8.68)$$

Finally, let us determine the partial derivatives of packing density by using Equations (8.64) and (8.65) as follows:

$$\left.\begin{aligned}
\frac{\partial\mu}{\partial\varepsilon_1} &= \frac{-\mu_0}{\left(1+\varepsilon_1\right)^2\left(1+\varepsilon_2\right)\left(1+\varepsilon_3\right)} = \frac{-\mu\mu_1}{\mu_0}, \\[2mm]
\frac{\partial\mu}{\partial\varepsilon_2} &= \frac{-\mu_0}{\left(1+\varepsilon_1\right)\left(1+\varepsilon_2\right)^2\left(1+\varepsilon_3\right)} = \frac{-\mu\mu_2}{\mu_0}, \\[2mm]
\frac{\partial\mu}{\partial\varepsilon_3} &= \frac{-\mu_0}{\left(1+\varepsilon_1\right)\left(1+\varepsilon_2\right)\left(1+\varepsilon_3\right)^2} = \frac{-\mu\mu_3}{\mu_0}.
\end{aligned}\right\}
\qquad (8.69)$$

Hypothesis of deformation energy and work done

It is necessary to introduce a suitable hypothesis of deformation energy to solve the current problem of deformation. The following additional type of relation seems to be probably the easiest here:

$$E = a_1 F_1\left(\mu_1\right) + a_2 F_2\left(\mu_2\right) + a_3 F_3\left(\mu_3\right) + b F_b\left(\mu\right), \qquad (8.70)$$

where F_1, F_2, F_3, F_b are suitable functions and a_1, a_2, a_3, b are suitable parameters. Nevertheless, as we assumed transverse isotropy in the plane perpendicular to x_3-axis, the common function $F_1 = F_2 = F$ and the common parameter $a_1 = a_2 = a$ must be valid. Thus,

$$E = a F\left(\mu_1\right) + a F\left(\mu_2\right) + a_3 F_3\left(\mu_3\right) + b F_b\left(\mu\right). \qquad (8.71)$$

[Energy can also be perceived as a function of $\varepsilon_1, \varepsilon_2, \varepsilon_3$; therefore, Equations (8.64) and (8.65) are also valid.]

The elementary increment of deformation energy is

$$dE = \frac{\partial E}{\partial \varepsilon_1} d\varepsilon_1 + \frac{\partial E}{\partial \varepsilon_2} d\varepsilon_2 + \frac{\partial E}{\partial \varepsilon_3} d\varepsilon_3 \qquad (8.72)$$

and elementary increment of work done is

$$dA = \sigma_{11}^* d\varepsilon_1 + \sigma_{22}^* d\varepsilon_2 + \sigma_{33}^* d\varepsilon_3. \qquad (8.73)$$

According to the well-known idea of conservative system, it could be assumed that the elementary increment of deformation energy is equal to the elementary increment of work done, i.e., $dE = dA$. By partial generalization of this idea, we can use

$$dE = C\,dA, \quad C \leq 1 \ldots \text{constant},$$

$$\frac{\partial E}{\partial \varepsilon_1} d\varepsilon_1 + \frac{\partial E}{\partial \varepsilon_2} d\varepsilon_2 + \frac{\partial E}{\partial \varepsilon_3} d\varepsilon_3 = C\left(\sigma_{11}^* d\varepsilon_1 + \sigma_{22}^* d\varepsilon_2 + \sigma_{33}^* d\varepsilon_3\right). \qquad (8.74)^{26}$$

[Equations (8.72) and (8.73) were used.] However, the last expression must be valid for every triplet of values μ_1, μ_2, μ_3 so that the following relations of engineering (nominal) stresses must be true

$$\left.\begin{array}{ll} \dfrac{\partial E}{\partial \varepsilon_1} d\varepsilon_1 = C\,\sigma_{11}^* d\varepsilon_1, & \sigma_{11}^* = \dfrac{1}{C}\dfrac{\partial E}{\partial \varepsilon_1}, \\[2ex] \dfrac{\partial E}{\partial \varepsilon_2} d\varepsilon_2 = C\,\sigma_{22}^* d\varepsilon_2, & \sigma_{22}^* = \dfrac{1}{C}\dfrac{\partial E}{\partial \varepsilon_2}, \\[2ex] \dfrac{\partial E}{\partial \varepsilon_3} d\varepsilon_3 = C\,\sigma_{33}^* d\varepsilon_3, & \sigma_{33}^* = \dfrac{1}{C}\dfrac{\partial E}{\partial \varepsilon_3}. \end{array}\right\} \qquad (8.75)$$

---

26 The deformation energy E depends only on the final values of deformations and each increment of the work done is fully 'stored' as the increment of the deformation energy into the deformed body. This is a known theoretical idea that is often quoted under the conservative system. Nevertheless, a part of work done is usually dissipated in the form of thermal energy due to fiber-to-fiber friction, etc., so that $C < 1$. Only for conservative system, it is valid that $C = 1$. Our assumption, given by Equation (8.74), therefore bears a little 'looser' sense than that meant by the strict conservative system.

Let us denote the derivatives of the functions F, F_3, F_b by symbols f, f_3, f_b. Then, by using Equations (8.68) and (8.69), we obtain the general engineering (nominal) stresses from Equations (8.71) and (8.75) as follows:

$$
\begin{aligned}
\sigma_{11}^* &= \frac{1}{C}\left[a\frac{dF(\mu_1)}{d\mu_1}\frac{d\mu_1}{d\varepsilon_1} + b\frac{\partial F_b(\mu)}{\partial\mu}\frac{\partial\mu}{\partial\varepsilon_1} \right] = \frac{-1}{C}\left[af(\mu_1)\frac{\mu_1^2}{\mu_0} + bf_b(\mu)\frac{\mu\mu_1}{\mu_0} \right], \\
\sigma_{22}^* &= \frac{1}{C}\left[a\frac{dF(\mu_2)}{d\mu_2}\frac{d\mu_2}{d\varepsilon_2} + b\frac{\partial F_b(\mu)}{\partial\mu}\frac{\partial\mu}{\partial\varepsilon_2} \right] = \frac{-1}{C}\left[af(\mu_2)\frac{\mu_2^2}{\mu_0} + bf_b(\mu)\frac{\mu\mu_2}{\mu_0} \right], \\
\sigma_{33}^* &= \frac{1}{C}\left[a_3\frac{dF_3(\mu_3)}{d\mu_3}\frac{d\mu_3}{d\varepsilon_3} + b\frac{\partial F_b(\mu)}{\partial\mu}\frac{\partial\mu}{\partial\varepsilon_3} \right] = \frac{-1}{C}\left[a_3f_3(\mu_3)\frac{\mu_3^2}{\mu_0} + bf_b(\mu)\frac{\mu\mu_3}{\mu_0} \right].
\end{aligned}
$$

$$(8.76)$$

The true (Cauchy's) stresses follows Equation (8.67). By applying Equation (8.76) and also Equation (8.66), we find

$$
\begin{aligned}
\sigma_{11} &= \frac{1}{C}\left[af(\mu_1)\frac{-\mu_1^2}{\mu_0}\frac{\mu_2\mu_3}{\mu_0^2} + bf_b(\mu)\frac{-\mu\mu_1}{\mu_0}\frac{\mu_2\mu_3}{\mu_0^2} \right] \\
&= \frac{-1}{C}\left[af(\mu_1)\frac{\mu_1\mu}{\mu_0} + bf_b(\mu)\frac{\mu^2}{\mu_0} \right], \\
\sigma_{22} &= \frac{1}{C}\left[af(\mu_2)\frac{-\mu_2^2}{\mu_0}\frac{\mu_1\mu_3}{\mu_0^2} + bf_b(\mu)\frac{-\mu\mu_2}{\mu_0}\frac{\mu_1\mu_3}{\mu_0^2} \right] \\
&= \frac{-1}{C}\left[af(\mu_2)\frac{\mu_2\mu}{\mu_0} + bf_b(\mu)\frac{\mu^2}{\mu_0} \right], \\
\sigma_{33} &= \frac{1}{C}\left[a_3f_3(\mu_3)\frac{-\mu_3^2}{\mu_0}\frac{\mu_1\mu_2}{\mu_0^2} + bf_b(\mu)\frac{-\mu\mu_3}{\mu_0}\frac{\mu_1\mu_2}{\mu_0^2} \right] \\
&= \frac{-1}{C}\left[a_3f_3(\mu_3)\frac{\mu_3\mu}{\mu_0} + bf_b(\mu)\frac{\mu^2}{\mu_0} \right].
\end{aligned}
$$

$$(8.77)$$

There are different functions and parameters stated in Equation (8.76) and Equation (8.77), which must be strengthened by means of theoretical and experimental ways. A study of some special straining can bring out effective results on this topic. We will now introduce three such cases: (1) non-deformed fibrous material, (2) transverse deformation and (3) uniaxial tensioning in fiber direction.

1. *Non-deformed fibrous material.* If the unitary cube of fibrous material is not deformed, then $\varepsilon_1 = \varepsilon_2 = \varepsilon_3 = 0$. So, according to Equations (8.64) and (8.65), $\mu_1 = \mu_2 = \mu_3 = \mu = \mu_0$, and evidently $\sigma_{11}^* = \sigma_{22}^* = \sigma_{33}^* = 0$ as well as $\sigma_{11} = \sigma_{22} = \sigma_{33} = 0$. Then, according to Equations (8.76) and (8.77), the following expressions are valid in this special case:

$$\left.\begin{aligned}
\frac{C\sigma_{11}^*}{-\mu_0} = \frac{C\sigma_{11}}{-\mu_0} = \frac{C\sigma_{22}^*}{-\mu_0} = \frac{C\sigma_{22}}{-\mu_0} = a\,f(\mu_0) + b\,f_b(\mu_0) = 0, \\
\frac{C\sigma_{33}^*}{-\mu_0} = \frac{C\sigma_{33}}{-\mu_0} = a_3\,f_3(\mu_0) + b\,f_b(\mu_0) = 0.
\end{aligned}\right\} \tag{8.78}$$

The above expression is right when

$$f(\mu_0) = f_3(\mu_0) = f_b(\mu_0) = 0. \tag{8.79}$$

2. *Transverse deformation.* Let us imagine that $\varepsilon_2 = \varepsilon_3 = 0$ and only $\varepsilon_1 < 0$ as shown in Figure 8.7. This is the case of uniaxial compressive deformation in transverse isotropic direction. Then, according to Equations (8.64) and (8.65),

$$\mu_2 = \mu_3 = \mu_0, \quad \mu = \mu_1. \tag{8.80}$$

In this special case, the true (Cauchy's) stresses, presented generally in Equation (8.77), takes the following forms:

$$\left.\begin{aligned}
\sigma_{11} &= \frac{-1}{C}\left[a\,f(\mu)\frac{\mu^2}{\mu_0} + b\,f_b(\mu)\frac{\mu^2}{\mu_0}\right] = \frac{-\mu^2}{C\mu_0}\left[a\,f(\mu) + b\,f_b(\mu)\right], \\
\sigma_{22} &= \frac{-1}{C}\left[a\,f(\mu_0)\frac{\mu_0\mu}{\mu_0} + b\,f_b(\mu)\frac{\mu^2}{\mu_0}\right] = \frac{-\mu^2}{C\mu_0}b\,f_b(\mu), \\
\sigma_{33} &= \frac{-1}{C}\left[a_3 f_3(\mu_0)\frac{\mu_0\mu}{\mu_0} + b\,f_b(\mu)\frac{\mu^2}{\mu_0}\right] = \frac{-\mu^2}{C\mu_0}b\,f_b(\mu).
\end{aligned}\right\} \tag{8.81}$$

[Equation (8.79) was used for rearrangement.]

Hearle and El-Behery [9], for the first time, studied the ratio $\omega = \sigma_{22}/\sigma_{11}$, i.e., the transverse stress σ_{22} divided by the stress σ_{11}. (Both stresses were measured experimentally.) Roughly speaking, this ratio is a constant, except for very small pressure (very small packing density). Similar results were obtained later on by Gurová

[10]. In this work, a set of different materials was experimentally taken and the ratio ω was evaluated as an experimental function of packing density μ. An example is shown in Figure 8.8. It can be observed that the ratio ω does not change after packing density reaches around $\mu \cong 0.05$. (According to Figure 8.8, the ratio ω becomes a constant at about 0.57.) So, we can write

$$\omega = \sigma_{22}/\sigma_{11} \ldots \text{constant} . \tag{8.82}$$

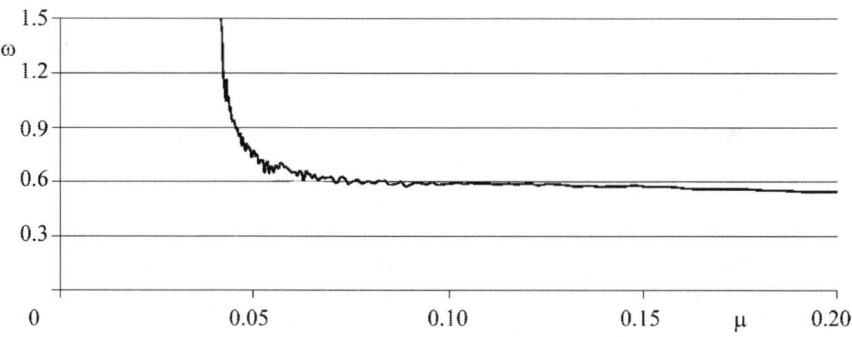

Figure 8.8 Experimental behaviour of ω as a function of packing density μ according to Gurová [10]. Material: Polyester fibers of 80 mm cut length and 6.7 dtex fineness

By using Equation (8.81) in (8.82), the following expression can be obtained:

$$\omega = \frac{\sigma_{22}}{\sigma_{11}} = \frac{b\,f_b(\mu)}{a\,f(\mu) + b\,f_b(\mu)} = \frac{1}{\dfrac{a\,f(\mu)}{b\,f_b(\mu)} + 1} = \text{constant} . \tag{8.83}$$

Thus,

$$\frac{f_b(\mu)}{f(\mu)} = k \ldots \text{constant} , \tag{8.84}$$

and according to Equations (8.83) and (8.84),

$$\omega = \frac{1}{\dfrac{a\,f(\mu)}{b\,k\,f(\mu)} + 1} = \frac{1}{\dfrac{a}{bk} + 1}, \quad bk = a\frac{\omega}{1-\omega}, \quad b = \frac{a}{k}\frac{\omega}{1-\omega} . \tag{8.85}$$

Further, the following expression is valid to write by using Equations (8.84) and (8.85):

$$b f_b (\mu) = \left[\frac{a}{k} \frac{\omega}{1-\omega} \right] \left[k \, f(\mu) \right] = a \frac{\omega}{1-\omega} f(\mu). \tag{8.86}$$

By using the last expression in Equation (8.81), we find

$$\left. \begin{array}{l} \sigma_{11} = \dfrac{-\mu^2}{C\mu_0} \left[a f(\mu) + a \dfrac{\omega}{1-\omega} f(\mu) \right] = \dfrac{-\mu^2 a}{C\mu_0} f(\mu) \dfrac{1}{1-\omega}, \\[3mm] \sigma_{22} = \sigma_{33} = \dfrac{-\mu^2 a}{C\mu_0} f(\mu) \dfrac{\omega}{1-\omega}. \end{array} \right\} \tag{8.87}$$

The uniaxial compressive deformation was earlier studied experimentally as well as theoretically. Probably, the first theoretical model was presented by van Wyk [11]. He derived the following well-known relation between packing density μ and compressive pressure p, which is basically the negative value of the stress σ_{11}.

$$-\sigma_{11} = p = k_p \mu^3. \tag{8.88}$$

(The complete derivation of this expression is given in Reference [12].) Nevertheless, this expression does not correspond to the experimental results at higher values of packing density (approximately over 0.25). Therefore, the following modification was made, which was derived in References [7] and [12] in detail:

$$-\sigma_{11} = p = k_p \left[\varphi(\mu) - \varphi(\mu_0) \right], \tag{8.89}$$

where the function φ of a general argument x is expressed as follows:

$$\varphi(x) = \frac{x^3}{\left[1 - (x/\mu_m)^{2+g} \right]^3}, \quad x \in (0, \mu_m). \tag{8.90}$$

(The parameter μ_m denotes the theoretical maximum packing density; usually $\mu_m \to 1$. Based on experimental experience, the parameter g is usually very near to 1 [7, 12].)

Note: The value of $\varphi(\mu_0)$ is usually negligible, because μ_0 is very small. Nevertheless, this precisely guarantees a zero pressure at $\mu = \mu_0$, i.e., by non-compressed material.

Baljasov [13] studied uniaxial compressive deformation experimentally. He compressed different fiber materials in a rigid (non-deformed) metal box by applying a wide range of pressure and measuring the volume occupied by the fiber materials. Some of his results are illustrated on semi-logarithmic graphs in Figure 8.9. In this figure, two theoretical curves are also shown – van Wyk's relation according to Equation (8.88) – curve 1, and modified variant according to Equations (8.89) and (8.90) – curve 2. It is shown that curve 2 describe the experimental results of Baljasov very well.

Note: The differences between the random organization of fibers and roughly parallel organization of fibers are not too significant, as shown in the graphs of Figure 8.9. It evokes an idea that fiber orientation is not too important in this case. Such idea corresponds to the fact that σ_{22} and σ_{33} are described by the same expression for uniaxial compressive deformation in transverse isotropic direction – see Equation (8.87).

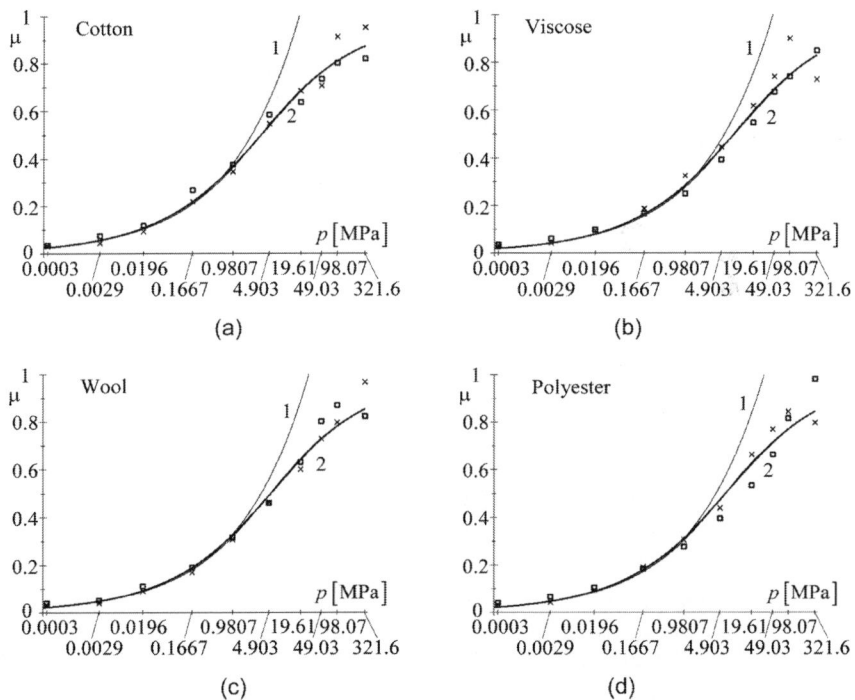

Figure 8.9 Dependence of compressive pressure on packing density
– comparison between theoretical and experimental results.
Experimental points according to Baljasov [13]:
✕: random organization of fibers; ☐: roughly parallel organization of fibers.
Curves calculated using k_p = 18MPa (cotton), k_p = 45MPa (viscose),
k_p = 28MPa (wool), k_p =33MPa (polyester).
1: Original van Wyk's model – Equation (8.88); 2: modified van Wyk's model –
Equations (8.89) and (8.90), by μ_m = 1, g = 1, μ_0 = 0.02

The equivalency of stress σ_{11} obtained from Equations (8.87) and (8.89) by using (8.90) allows us to express the function f in the following manner:

$$\sigma_{11} = -p = -k_p \left[\varphi(\mu) - \varphi(\mu_0) \right] = \frac{-\mu^2 a}{C\mu_0} f(\mu) \frac{1}{1-\omega},$$

$$f(\mu) = \frac{k_p C(1-\omega)}{a} \left[\varphi(\mu) - \varphi(\mu_0) \right] \frac{\mu_0}{\mu^2}, \tag{8.91}$$

or quite generally

$$f(x) = \frac{k_p C(1-\omega)}{a} \left[\varphi(x) - \varphi(\mu_0) \right] \frac{\mu_0}{x^2}. \tag{8.92}$$

(x is a general symbol of argument of function f.)

Let us accept (a) a constant value of ratio ω and (b) earlier character of function f. Then, knowing Equation (8.86) and the previous function according to Equation (8.92), we are able to rearrange the engineering (nominal) stresses from Equation (8.76) as follows:

$$\begin{aligned}
\sigma_{11}^* &= \frac{-1}{C} \left\{ a f(\mu_1) \frac{\mu_1^2}{\mu_0} + a \frac{\omega}{1-\omega} f(\mu) \frac{\mu\mu_1}{\mu_0} \right\} \\
&= \frac{-1}{C} \left\{ a \left[\frac{k_p C(1-\omega)}{a} \left[\varphi(\mu_1) - \varphi(\mu_0) \right] \frac{\mu_0}{\mu_1^2} \right] \frac{\mu_1^2}{\mu_0} \right. \\
&\quad \left. + a \frac{\omega}{1-\omega} \left[\frac{k_p C(1-\omega)}{a} \left[\varphi(\mu) - \varphi(\mu_0) \right] \frac{\mu_0}{\mu^2} \right] \frac{\mu\mu_1}{\mu_0} \right\}, \\
&= -k_p \mu_1 \left\{ (1-\omega) \frac{\varphi(\mu_1) - \varphi(\mu_0)}{\mu_1} + \omega \frac{\varphi(\mu) - \varphi(\mu_0)}{\mu} \right\}.
\end{aligned} \tag{8.93}$$

$$\begin{aligned}
\sigma_{22}^* &= \frac{-1}{C} \left\{ a f(\mu_1) \frac{\mu_2^2}{\mu_0} + a \frac{\omega}{1-\omega} f(\mu) \frac{\mu\mu_2}{\mu_0} \right\} \\
&= \frac{-1}{C} \left\{ a \left[\frac{k_p C(1-\omega)}{a} \left[\varphi(\mu_2) - \varphi(\mu_0) \right] \frac{\mu_0}{\mu_2^2} \right] \frac{\mu_2^2}{\mu_0} \right. \\
&\quad \left. + a \frac{\omega}{1-\omega} \left[\frac{k_p C(1-\omega)}{a} \left[\varphi(\mu) - \varphi(\mu_0) \right] \frac{\mu_0}{\mu^2} \right] \frac{\mu\mu_2}{\mu_0} \right\}, \\
&= -k_p \mu_2 \left\{ (1-\omega) \frac{\varphi(\mu_2) - \varphi(\mu_0)}{\mu_2} + \omega \frac{\varphi(\mu) - \varphi(\mu_0)}{\mu} \right\}.
\end{aligned} \tag{8.94}$$

$$\sigma_{33}^* = \frac{-1}{C}\left\{a_3 f_3(\mu_3)\frac{\mu_3^2}{\mu_0} + a\frac{\omega}{1-\omega}f(\mu)\frac{\mu\mu_3}{\mu_0}\right\}$$

$$= \frac{-1}{C}\left\{a_3 f_3(\mu_3)\frac{\mu_3^2}{\mu_0} + a\frac{\omega}{1-\omega}\left[\frac{k_p C(1-\omega)}{a}\left[\varphi(\mu)-\varphi(\mu_0)\right]\frac{\mu_0}{\mu^2}\right]\frac{\mu\mu_3}{\mu_0}\right\},$$

$$= -k_p\mu_3\left\{\frac{a_3}{Ck_p}f_3(\mu_3)\frac{\mu_3}{\mu_0} + \omega\frac{\varphi(\mu)-\varphi(\mu_0)}{\mu}\right\}. \tag{8.95}$$

Similarly, it is possible to express also true (Cauchy's) stresses from Equation (8.67) and the triplet of last equations as follows:

$$\sigma_{11} = -k_p\mu\left\{(1-\omega)\frac{\varphi(\mu_1)-\varphi(\mu_0)}{\mu_1} + \omega\frac{\varphi(\mu)-\varphi(\mu_0)}{\mu}\right\}, \tag{8.96}$$

$$\sigma_{22} = -k_p\mu\left\{(1-\omega)\frac{\varphi(\mu_2)-\varphi(\mu_0)}{\mu_2} + \omega\frac{\varphi(\mu)-\varphi(\mu_0)}{\mu}\right\}, \tag{8.97}$$

$$\sigma_{33} = -k_p\mu\left\{\frac{a_3}{Ck_p}f_3(\mu_3)\frac{\mu_3}{\mu_0} + \omega\frac{\varphi(\mu)-\varphi(\mu_0)}{\mu}\right\}. \tag{8.98}$$

3. *Application of uniaxial tension along fiber direction.* Let us accept the previous assumption, i.e., Equations (8.93) to (8.98) and imagine a special straining of the initial unitary cube, as shown in Figure 8.7, by applying the engineering (nominal) stress σ_{33}^* only; while keeping the other stresses $\sigma_{11}^*, \sigma_{22}^*$ at zero. A positive strain ε_3 evokes negative strains ε_1 and ε_2 in consequence of the effect of cross contraction. Both quantities ε_1 and ε_2 have the same value ε_t under the assumed transverse isotropy.

Note: Let us use subscript 't' (like transverse) also by other quantities having the same value in consequence of the transverse isotropy.

Thus, in this case:

$$\sigma_{11}^* = \sigma_{22}^* = 0, \quad \sigma_{11} = \sigma_{22} = 0, \tag{8.99}$$

$$\varepsilon_1 = \varepsilon_2 = \varepsilon_t, \tag{8.100}$$

$$\mu_1 = \mu_2 = \mu_t = \frac{\mu_0}{1+\varepsilon_t}, \tag{8.101}$$

$$\mu = \frac{\mu_t^2\mu_3}{\mu_0^2}, \quad \left(\mu_t = \mu_0\sqrt{\mu/\mu_3}\right). \tag{8.102}$$

[Equations (8.65), (8.66) and (8.67) were used for formulation of the previous relations.]

By applying Equations (8.99), (8.101) and (8.102) in Equations (8.93) and (8.94), we find the following relation in this case:

$$
\left.
\begin{aligned}
\sigma_{11}^* = \sigma_{22}^* &= -k_p \mu_t \left\{ (1-\omega) \frac{\varphi(\mu_t) - \varphi(\mu_0)}{\mu_t} + \omega \frac{\varphi(\mu) - \varphi(\mu_0)}{\mu} \right\} = 0, \\
(1-\omega) \frac{\varphi(\mu_t) - \varphi(\mu_0)}{\mu_t} &+ \omega \frac{\varphi(\mu_t^2 \mu_3 / \mu_0^2) - \varphi(\mu_0)}{\mu_t^2 \mu_3 / \mu_0^2} = 0, \\
\text{symbolically: } \mu_t &= \Phi(\mu_3).
\end{aligned}
\right\}
\tag{8.103}
$$

The previous result, together with Equation (8.90), represents the function Φ assigning μ_t to the quantity μ_3.

Note: If $\mu_t = \mu_0$, then Equation (8.103) takes the following form:

$$
\begin{aligned}
(1-\omega) \frac{\varphi(\mu_0) - \varphi(\mu_0)}{\mu_0} &+ \omega \frac{\varphi(\mu_0^2 \mu_3 / \mu_0^2) - \varphi(\mu_0)}{\mu_0^2 \mu_3 / \mu_0^2} \\
&= \omega \frac{\varphi(\mu_3) - \varphi(\mu_0)}{\mu_3} = 0, \quad \mu_3 = \mu_0.
\end{aligned}
\tag{8.104}
$$

Summarily, the function Φ expressed in Equation (8.103) passes through the point $\mu_3 = \mu_t = \mu_0$.

Also, the contraction ratio – i.e., $\eta = -\varepsilon_t / \varepsilon_{33}$ according to our symbols – follows the description of uniaxial tension. From Equations (8.65), (8.100), (8.101) and (8.103), it is valid to write that

$$
\varepsilon_t = \mu_0 / \mu_t - 1, \quad \varepsilon_3 = \mu_0 / \mu_3 - 1,
$$

$$
\eta = -\frac{\varepsilon_t}{\varepsilon_{33}} = -\frac{\mu_0 / \mu_t - 1}{\mu_0 / \mu_3 - 1} = -\frac{\mu_0 / \Phi(\mu_3) - 1}{\mu_0 / \mu_3 - 1}.
\tag{8.105}
$$

The contraction ratio takes a special value when $\omega = 1$. Then, Equation (8.103) takes the following form:

$$
\frac{\varphi(\mu_t^2 \mu_3 / \mu_0^2) - \varphi(\mu_0)}{\mu_t^2 \mu_3 / \mu_0^2} = 0, \quad \frac{\mu_t^2 \mu_3}{\mu_0^2} = \mu_0, \quad \text{i.e., } \mu = \mu_0.
\tag{8.106}
$$

[Equation (8.102) was used.] It shows that when $\omega = 1$ the initial packing density μ is always a constant and equal to the initial value μ_0. This means that the volume of a fibrous assembly is not changed by applying uniaxial tension. Rearranging Equation (8.106) by using Equations (8.65) and (8.101), we find the contraction ratio in this special case as follows:

$$\frac{\mu_t^2 \mu_3}{\mu_0^2} = \mu_0, \quad \frac{\mu_0^2}{\left(1+\varepsilon_t\right)^2} \frac{\mu_0}{1+\varepsilon_3} = \mu_0^3,$$

$$\left(1+\varepsilon_t\right)^2 \left(1+\varepsilon_3\right) = 1, \quad \varepsilon_t = 1/\sqrt{1+\varepsilon_3} - 1, \tag{8.107}$$

$$\eta = -\frac{\varepsilon_t}{\varepsilon_3} = -\frac{1/\sqrt{1+\varepsilon_3} - 1}{\varepsilon_3} \quad \text{when } \omega = 1.$$

We found the relation between contraction ratio η and strain ε_3 by applying Equation (8.90) in (8.103), solving this equation, and then applying result in Equation (8.105). [If $\omega = 1$, then Equation (8.107) is valid.] Figure 8.10 illustrates the trends of this relation.

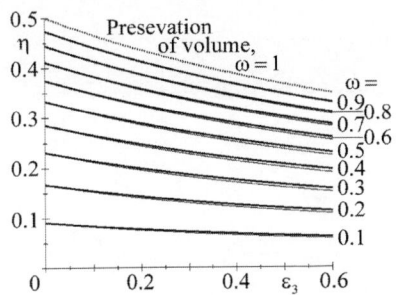

Figure 8.10 Contraction ratio by uniaxial tensioning in fiber direction. Common parameters: $\mu_m = 1$, $g = 1$; curves: thick – $\mu_0 = 0.02$, thin – $\mu_0 = 0.3$, dotted – with $\omega = 1$, Equation (8.107)

All curves follow a decreasing trend, as expected. In one hand, the parameter ω influences on the value of contraction ratio significantly, but, on the other hand, the initial value μ_0 of packing density is not too important; all differences between thick curve (usual value $\mu_0 = 0.02$) and thin curve (extremely high, perhaps unrealistic value $\mu_0 = 0.3$) are very small.

The engineering (nominal) stress σ_{33}^* is obtained from Equation (8.95)[27] by using Equations (8.102) and (8.103) as follows:

---

27 Let us remind that σ_{11}^* and σ_{22}^* are still equal to zero according to Equation (8.89).

$$\sigma_{33}^* = -k_p \mu_3 \left\{ \frac{a_3}{Ck_p} f_3(\mu_3) \frac{\mu_3}{\mu_0} + \omega \frac{\varphi\left[\mu_t^2 \mu_3/\mu_0^2\right] - \varphi(\mu_0)}{\mu_t^2 \mu_3/\mu_0^2} \right\}$$

$$= -k_p \mu_3 \left\{ \frac{a_3}{Ck_p} f_3(\mu_3) \frac{\mu_3}{\mu_0} + \omega \frac{\varphi\left[\Phi^2(\mu_3)\mu_3/\mu_0^2\right] - \varphi(\mu_0)}{\Phi^2(\mu_3)\mu_3/\mu_0^2} \right\}. \qquad (8.108)$$

This equation describes the relation $\sigma_{33}^* - \mu_3$, i.e., function $\sigma_{33}^*(\mu_3)$, by the application of uniaxial tension along fiber direction.

Sometimes, it is possible to obtain the function $\sigma_{33}^* = \sigma_{33\,exp}^*(\mu_3)$ experimentally[28]. Then, the following term can be expressed from Equation (8.108):

$$\frac{a_3}{Ck_p} f_3(\mu_3) \frac{\mu_3}{\mu_0} = \frac{\sigma_{33\,exp}^*(\mu_3)}{-k_p \mu_3} - \omega \frac{\varphi\left[\Phi^2(\mu_3)\mu_3/\mu_0^2\right] - \varphi(\mu_0)}{\Phi^2(\mu_3)\mu_3/\mu_0^2}. \qquad (8.109)$$

(Expressions are functions of μ_3 only.) However, this term is present in Equation (8.98) too. Thus, the final triplet of equations, determined the engineering (nominal) stresses, is obtained from Equations (8.93), (8.94) and (8.95) with (8.109) as follows:

$$\sigma_{11}^* = -k_p \mu_1 \left\{ (1-\omega) \frac{\varphi(\mu_1) - \varphi(\mu_0)}{\mu_1} + \omega \frac{\varphi(\mu) - \varphi(\mu_0)}{\mu} \right\},$$

$$\sigma_{22}^* = -k_p \mu_2 \left\{ (1-\omega) \frac{\varphi(\mu_2) - \varphi(\mu_0)}{\mu_2} + \omega \frac{\varphi(\mu) - \varphi(\mu_0)}{\mu} \right\},$$

$$\sigma_{33}^* = -k_p \mu_3 \left\{ \frac{\sigma_{33\,exp}^*(\mu_3)}{-k_p \mu_3} - \omega \frac{\varphi\left[\Phi^2(\mu_3)\mu_3/\mu_0^2\right] - \varphi(\mu_0)}{\Phi^2(\mu_3)\mu_3/\mu_0^2} + \omega \frac{\varphi(\mu) - \varphi(\mu_0)}{\mu} \right\}.$$

$$(8.110)$$

Similarly, the final triplet of equations, determined the true (Cauchy's) stresses, is obtained from Equations (8.96), (8.97) and (8.98) with (8.109) as follows:

---

28 Primarily, we obtain the stress–strain relation $\sigma_{33}^* - \varepsilon_3$ experimentally; then each ε_3 can be recalculated to μ_3, according to Equation (8.65).

$$\sigma_{11} = -k_p \mu \left\{ (1-\omega) \frac{\varphi(\mu_1) - \varphi(\mu_0)}{\mu_1} + \omega \frac{\varphi(\mu) - \varphi(\mu_0)}{\mu} \right\},$$

$$\sigma_{22} = -k_p \mu \left\{ (1-\omega) \frac{\varphi(\mu_2) - \varphi(\mu_0)}{\mu_2} + \omega \frac{\varphi(\mu) - \varphi(\mu_0)}{\mu} \right\},$$

$$\sigma_{33} = -k_p \mu \left\{ \frac{\sigma_{33\,\text{exp}}^*(\mu_3)}{-k_p \mu_3} - \omega \frac{\varphi\left[\Phi^2(\mu_3)\mu_3/\mu_0^2\right] - \varphi(\mu_0)}{\Phi^2(\mu_3)\mu_3/\mu_0^2} + \omega \frac{\varphi(\mu) - \varphi(\mu_0)}{\mu} \right\}.$$

$$(8.111)$$

[Function φ is determined from Equation (8.90), function Φ is determined from Equation (8.103) and function $\sigma_{33\,\text{exp}}^*$ is determined experimentally by applying uniaxial tension along the fiber direction of a fiber bundle.]

Concluding notes

The resulting expressions, according to Equation (8.110) and/or (8.111), represent one possible model of relations among the normal stresses and the corresponding strains. Nevertheless, this is 'not completed', because it did not solve the shear stresses and the corresponding deformations. However, Hearle et al. [14] reasoned – probably well – that the shear stresses σ_{13} are minimal in relation to normal stresses so that they can be neglected.

The model of deformation law, presented in this section, can be but need not to be valid for an actual fiber material in a yarn. But, it creates problems even if we accept this theoretical model. Especially, the experimental determination of ω-value and the function $\sigma_{33\,\text{exp}}^*$ bring problems for the actually used fiber material above all. It is necessary to have a special equipment in a laboratory[29] for measurement of ω-value, i.e., evaluation of stresses σ_{11} and σ_{22} by uniaxial compressive deformation in transverse isotropic direction – see Equation (8.82).

To determine the function $\sigma_{33\,\text{exp}}^*$ by the application of uniaxial tension along the fiber direction is still more complicated. It is relatively easy to find out the stress–strain relation of (straight) fibers and/or fiber bundles in a

---

29 Usually, a fiber material is compressed by a suitable piston in a nondeformable metal box. The pressure evoked by the piston is measured together with pressure worked on the side-wall of the box by a special sensor.

laboratory, nevertheless, a lot of other factors significantly affect the mechanical behaviour of a fiber material. They are, e.g., crimp of fibers, variability of fiber breaking strain and tenacity (see Chapter 9), fiber-to-fiber slippage and fiber-to-fiber friction (see Chapter 11), diverse behaviour resulting from the applications of positive and negative (compressive) tensions along fiber direction, etc. Solving such complicated problem could probably require some suitable combination of experimental experiences together with images of some theoretical models.

A long ago, Neckář [6, 7] used a semi-empirical model of deformation law by solving the differential equation of radial equilibrium of forces in a yarn body as stated in Sections 8.1 to 8.3 and obtained an acceptable results; see one example of behaviour of radial packing density displayed in Figure 8.11.

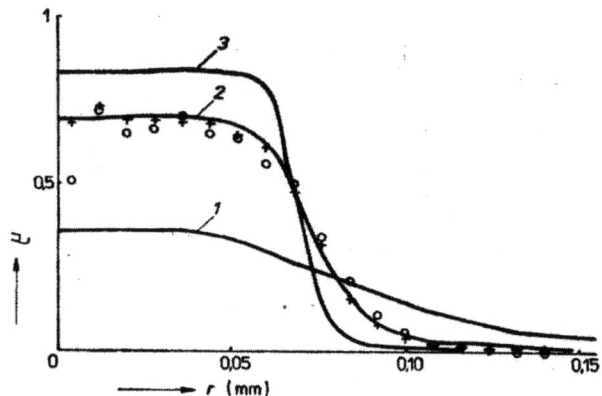

Figure 8.11 Radial packing density in yarn – one example from [7].
Experimental ring yarn: 19.6 tex, twist 669 m^{-1}, from viscose fibers 39.2 mm,
0.162 tex; O, +: results measured (two different experimental methods);
curve 2: calculated results; other curves: curve 1 – calculation by twist 400 m^{-1},
curve 3 – calculation by twist 1000 m^{-1}

Note: The packing densities at different radii were experimentally analyzed by means of the evaluation of spatial fiber curves in the yarn (Morton's tracer fiber techniques – see points O) and also by the evaluation of yarn cross sections (see points +), shown in Figure 8.10. Curve 2 represents the calculated result. Furthermore, two other curves – curve 1 and curve 3 were also obtained for illustrating the influence of yarn twist. (See Reference [7] for full description of the model used.)

Figure 8.10 shows that the way, introduced in previous section, can bring enough real results, corresponding to experimental experiences. Nevertheless, this way is relatively complicated and primarily requires many inputs (deformation law above all) which are not enough known by us as of now. Therefore, some fully empirical models were derived [e.g., Equation (1.69) with the example according to Equation (1.70)] or some semi-empirical models were obtained for easier use. One concept of semi-empirical model is shown in the following section.

8.5 A semi-empirical equation for yarn diameter

Let us think about yarn structure corresponding to the helical model – see Chapter 4, Section 4.1. (We use the same symbols as stated in Chapter 1, i.e., yarn fineness T, yarn twist Z, yarn diameter D, general radius r, angle of fiber inclination β, etc.) The first curvature of the fiber helix at radius r is $k_1 = \sin^2 \beta / r$ – see Equation (8.3). The radius of curvature r_c is reciprocal of the first curvature, so that

$$r_c = 1/k_1 = r / \sin^2 \beta .$$

(8.112)

(It is generally known that the radius of curvature represents the radius of an osculating circle which can 'replace' the elementary arch of the helical curve.)

Centripetal force per unit volume of fiber

Let us consider a part of yarn (diameter D, length l) that contains a helical fiber at a general radius r – see the thick line shown in Figure 8.12a. Further, let us think about an elementary fiber portion UV in the stated fiber. Its curvature apparently describes radius r_c. The angular distance between points U and V is $d\varphi$. The stated fiber element is shown in the plane of the osculating circle (Figure 8.12b).

An axial force F imparts a tension in the element and it creates a centripetal force dP in consequence of the curvature of the element. The following expression is then valid from the geometry, shown in Figure 8.12b:

$$dP = 2\left[F \sin \frac{d\varphi}{2} \right] = 2F \frac{d\varphi}{2} = F d\varphi .$$

(8.113)[30]

---

30   It is generally known that if $\varphi \to 0$, then $\sin \varphi \to \varphi$.

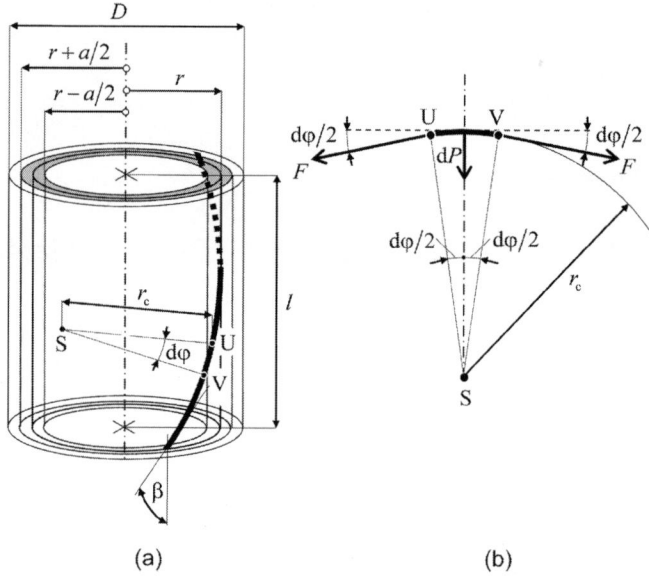

(a)   (b)

Figure 8.12 Centripetal force of helical fiber

The length of element UV is $r_c d\varphi$ and the fiber cross-sectional area is s. Then the volume of the element is

$$dV = sr_c d\varphi = \frac{sr\,d\varphi}{\sin^2\beta}.$$ (8.114)

[Equation (8.112) was used here for rearrangement.]
The centripetal force per unit volume of fiber is then

$$P_1 = \frac{dP}{dV} = \frac{F d\varphi}{sr\,d\varphi/\sin^2\beta} = \frac{F\sin^2\beta}{sr}.$$ (8.115)

Note: Equation (4.1), i.e., $\tan\beta = 2\pi r Z$, determines the angle β generated due to yarn twist Z.

Compressing zone

The derived model is based on the following two ideas:

1. Around the yarn surface, i.e., at radii near to $D/2$, the local packing density, number of fiber-to-fiber contacts, frictional forces and consequently also axial forces F and centripetal forces per unit volume of fiber P_1 are probably very small.

2. Around the yarn axis, i.e., at radii near to zero, the angle β is very small so that the centripetal force per unit volume of fiber P_1 is also very small – see Equation (8.115).

Based on the above-stated ideas, we can assume that the significant centripetal forces are present on the fibers that are lying somewhere 'in-between' the above-mentioned regions; the influences of other fibers can be neglected. Let this 'compressing zone' be lying in-between $r - a/2$ to $r + a/2$ as shown in Figure 8.12a; the mean radius of this zone is called r and its thickness is a.

Pressure developed in the compressing zone

The cross-sectional area of the compressing zone is $\pi(r + a/2)^2 - \pi(r - a/2)^2 = 2\pi r a$ (grey area shown in Figure 8.12a), the total volume of it is $V_{c,a} = 2\pi r a l$ and the fiber volume in the compressing zone is

$$V_a = 2\pi r a l \mu, \tag{8.116}$$

where μ is the packing density of fibers in the compressing zone. By using Equations (8.115) and (8.116), the total centripetal force in the compressing zone is

$$P = P_1 V_a = \frac{F \sin^2 \beta}{sr} 2\pi r a l \mu = F \sin^2 \beta \, 2\pi a l \mu / s. \tag{8.117}$$

For simplification, let us further assume that the centripetal force P acts on the cylinder at a (mean) radius r of the compressing zone. (We centralize all partial forces at radius r.) The surface area of the cylinder with this radius is $A = 2\pi r l$ – see Figure 8.12a. So, the pressure developed in the compressing zone is

$$p = \frac{P}{A} = \frac{F \sin^2 \beta \, 2\pi a l \mu / s}{2\pi r l} = \frac{F \sin^2 \beta}{rs} a\mu. \tag{8.118}$$

There is a fiber portion from a yarn, with angle β and axial force F, as shown in Figure 8.13. This force can be interpreted as a vector sum of component F_a, parallel to the direction of yarn axis and perpendicular component F_t. It is valid to write that

$$F_a = F \cos\beta. \tag{8.119}$$

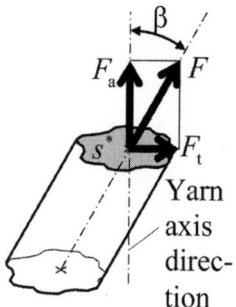

Figure 8.13 Resolution of axial force on fiber

The grey sectional area of fiber is $s^* = s/\cos\beta$ [see also Chapter 1, Equation (1.40) and Figure 1.8]. The normal stress σ acting on the sectional area of the fiber is then

$$\sigma = \frac{F_a}{s^*} = \frac{F\cos\beta}{s/\cos\beta} = \frac{F}{s}\cos^2\beta, \quad F = \frac{\sigma s}{\cos^2\beta}. \tag{8.120}$$

By applying the second expression from Equation (8.120) in (8.118) and then rearranging it by using Equations (4.1), (1.51), (1.47), (1.53), (1.38), we get

$$p = \frac{\sigma s}{\cos^2\beta} \frac{\sin^2\beta}{rs} a\mu = \frac{\sigma a\mu}{r}\tan^2\beta = \frac{\sigma a\mu}{r}(2\pi r Z)^2 = \frac{2\sigma a\mu}{D\dfrac{2r}{D}}(\pi D Z)^2 \left(\frac{2r}{D}\right)^2$$

$$= \frac{2\sigma a\mu}{D}\kappa^2\left(\frac{2r}{D}\right) = \frac{2\sigma a\mu}{D_S/\sqrt{\mu}}\kappa^2\left(\frac{2r}{D}\right) = 2\sigma a\left(\frac{2r}{D}\right)\sqrt{\mu}\frac{1}{D_S}\left(\frac{\mu\kappa^2}{4\pi}\right)4\pi$$

$$= 8\pi\sigma a\left(\frac{2r}{D}\right)\sqrt{\mu}\frac{\alpha_S^2}{D_S} = 8\pi\sigma\left(\frac{a}{d}\right)\left(\frac{2r}{D}\right)\sqrt{\mu}\frac{\alpha_S^2}{D_S/d}$$

$$= 8\pi\sigma\left(\frac{a}{d}\right)\left(\frac{2r}{D}\right)\sqrt{\mu}\frac{\alpha_S^2}{\sqrt{\tau}}, \tag{8.121}$$

or

$$p = C\sqrt{\mu}\frac{\alpha_S^2}{\sqrt{\tau}}, \quad \text{where } C = 8\pi\sigma\left(\frac{a}{d}\right)\left(\frac{2r}{D}\right). \tag{8.122}$$

Note: Besides the earlier-stated quantities, the following symbols occur again in last two expressions: κ for twist intensity, D_S for substance diameter of yarn, α_s for substance type of Koechlin's twist coefficient, τ for relative fineness of yarn and d for equivalent fiber diameter. (See Chapter 1 for a detailed explanation.)

Quantity C

The value of C depends on three quantities: σ, $2r/D$ and a/d. The axial stress σ, acting on the yarn substance cross-sectional area, is determined by the centrifugal forces and applying tension along the fibers during twisting. This stress could be perhaps a constant. The relative position of the compressing zone $2r/D$ may be the same with regard to yarn surface cylinder. So this quantity could also be perhaps a constant. To a certain extent, it is most difficult to comment on the relative thickness of the compressing zone a/d because we have not yet got enough clear physical imagination about this. But, based on our practical experience, this value could be considered to be a constant, too[31]. Summarily, we can assume that the quantity C is a suitable constant and the first expression in Equation (8.122) represents a characteristic of the pressure inside a yarn, for a given yarn structure.

Homogenous compression

We can roughly think about a yarn as a transverse isotropic fiber bundle which is compressed by a same pressure from all sides – see Figure 8.14. This is so-called homogenous compression of fiber bundle, which was already solved earlier – see Equations (5.142) and (5.55) in Reference [12]. It was derived that

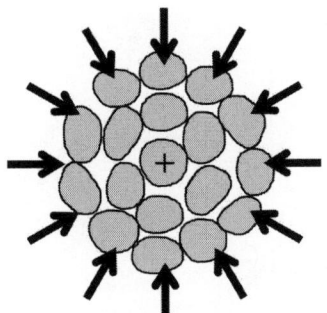

Figure 8.14 Homogenous compression of fiber bundle

---

31   This is an open problem for future research.

$$p = k_p b \frac{\mu^3}{\left[1-(\mu/\mu_m)^{2+g}\right]^3} \, . \qquad (8.123)^{32}$$

where μ_m is the maximum (limit) packing density and k_p, b and g are parameters.

Packing density

We have two expressions for the same pressure – the pressure according to Equation (8.122) derived from yarn structure and the pressure according to Equation (8.123) derived from compression of the fiber assembly. The equivalency of the right-hand sides of both equations results in the following expression:

$$k_p b \frac{\mu^3}{\left[1-(\mu/\mu_m)^{2+g}\right]^3} = C\sqrt{\mu}\,\frac{\alpha_S^2}{\sqrt{\tau}} \, . \qquad (8.124)$$

The right-hand side of the last equation can be further rearranged by using Equations (1.53), (1.36), (1.33), (1.5) and (1.7) from Chapter 1 as follows:

$$C\sqrt{\mu}\,\frac{\alpha_S^2}{\sqrt{\tau}} = C\sqrt{\mu}\,\frac{\left(Z\sqrt{S}\right)^2}{\sqrt{T/t}} = C\sqrt{\mu}\,\frac{\sqrt{t}}{\sqrt{T}}\left(Z\sqrt{T/\rho}\right)^2 = C\sqrt{\mu}\,\frac{\sqrt{s\rho}}{\sqrt{T}}\left(Z\sqrt{T/\rho}\right)^2$$

$$= C\sqrt{\mu}\,\frac{\sqrt{\dfrac{\pi d^2}{4}\rho}}{\sqrt{T}}\,\frac{Z^2 T}{\rho} = C\sqrt{\mu}\,\frac{d\sqrt{\pi}}{2\sqrt{\rho}}Z^2\sqrt{T} = \sqrt{\mu}\left(C\frac{d\sqrt{\pi}}{2\sqrt{\rho}}\right)\left(ZT^{1/4}\right)^2 .$$

$$(8.125)$$

(The last expression includes, besides the earlier-stated quantities, fiber fineness t, yarn fineness T and fiber density ρ; see Chapter 1 for a detailed explanation.)

Hence Equation (8.124) can be expressed in the following form:

---

32 Note that the symbol β substitutes the symbol b and the symbol $-\sigma$ replaces the symbol p in the original work [12].

$$k_p b \frac{\mu^3}{\left[1-\left(\mu/\mu_m\right)^{2+g}\right]^3} = \sqrt{\mu}\left(C\frac{d\sqrt{\pi}}{2\sqrt{\rho}}\right)\left(ZT^{1/4}\right)^2,$$

$$\frac{\mu^{2.5}}{\left[1-\left(\mu/\mu_m\right)^{2+g}\right]^3} = \left(C\frac{d\sqrt{\pi}}{2k_p b\sqrt{\rho}}\right)\left(ZT^{1/4}\right)^2. \qquad (8.126)$$

The quantity

$$Q = Cd\sqrt{\pi}\Big/\left(2k_p b\sqrt{\rho}\right) \qquad (8.127)$$

is a common parameter depending on the fiber materials used and the spinning technology employed to prepare the yarns. So, the final equation, which determines the yarn packing density, is

$$\frac{\mu^{2.5}}{\left[1-\left(\mu/\mu_m\right)^{2+g}\right]^3} = Q\left(ZT^{1/4}\right)^2. \qquad (8.128)$$

The packing density μ is a constant when the expression $ZT^{1/4}$ is a constant. On the contrary, Koechlin's theory assumes that the packing density is a constant when $ZT^{1/2}$ is a constant.

Note: In the Koechlin's theory, the packing density μ is a constant – Equation (1.61) – when the Koechlin's twist coefficient α is a constant – Equation (1.63). However, this twist coefficient is defined, according to Equation (1.54), as $\alpha = Z\sqrt{T}$.

Parameters

It is necessary to determine the parameters μ_m, g and Q for practical application of Equation (8.128).

The parameter μ_m, i.e., the maximum (limit) packing density, is generally near to one. But, this value is a little smaller in case of yarn. Let us (theoretically) imagine a 'limit' staple yarn, having its packing density near to one in its central region. Nevertheless (1) this limit yarn also has a smaller concentration of fibers in the layers near to yarn surface, because the fibers cannot be enough compressed there and (2) small gaps are always present among the fibers in yarns. Therefore, the average value of yarn packing density shall be a little smaller than one. Based on our experience, a suitable value of the maximum (limit) packing density lies around $\mu_m = 0.9$ for staple yarns.

(It also corresponds to the limit structure, i.e., $\mu = \mu_{lim} = \pi / (2\sqrt{3}) = 0.907$. See the first Chapter of Reference [12] for more general consideration.)

Parameter g is usually equal to one, according to the experimental results. (See the comments given on page number 150 in Reference [12].)

By using usual physical dimensions, Equation (8.128) takes the following form:

$$\frac{\mu^{2.5}}{\left[1-\left(\mu/\mu_{m}\right)^{2+g}\right]^{3}} = Q_{[m^{2}\,tex^{-1/2}]}\left(Z_{[m^{-1}]}T_{[tex]}^{1/4}\right)^{2}. \tag{8.129}$$

Parameter Q changes its values in relation to fiber material and type of spinning technology. Table 8.1 reports on the values of this parameter for different fiber materials and different spinning technologies.

Table 8.1 Values of parameter Q

Typical values Q [m² tex$^{-1/2}$], (μ_m = 0.9, g = 1)				
Fibrous material		Spinning technology		
Type	Density ρ [kg m^{-3}]	Combed	Carded	OE, type BD
Cotton	1520	$1.46 \cdot 10^{-7}$	$9.61 \cdot 10^{-8}$	$6.18 \cdot 10^{-8}$
Viscose, cotton type	1500	$4.12 \cdot 10^{-7}$		$1.76 \cdot 10^{-7}$
Polyester, cotton type	1360	$2.98 \cdot 10^{-7}$		$1.29 \cdot 10^{-7}$
Wool	1310	$2.16 \cdot 10^{-7}$	$1.20 \cdot 10^{-7}$	$6.49 \cdot 10^{-8}$

Note: The values of parameter Q, presented in Table 8.1, can be taken as examples only. However, the actual values must be selected according to the actual experience with specific fiber materials used and spinning technologies employed to prepare the yarns.

Example 1

For a cotton carded yarn with $T = 29.5\,tex$, $Z = 737\,m^{-1}$, $\rho = 1520\,kg\,m^{-3}$, the right-hand side of Equation (8.129), by using $Q = 9.61 \cdot 10^{-8}\,m^{2}\,tex^{-1/2}$ from Table 8.1, is 0.2835. It corresponds to $\mu = 0.489$ on the left-hand side (calculated by numerical method). Then, according to Equation (1.47), yarn diameter is $D = 0.225\,mm$.

Approximation

A practical problem in applying Equation (8.129) is lying in searching for the root of this equation. However, it is possible to use the following approximation for solving the packing density. Let us think that all packing densities of different yarns are lying in a region around some value μ^{*} [33]. (For example, we can estimate that the packing densities of cotton carded yarns are not too far from the value $\mu^{*} = 0.46$ according to our limited experience.) Then, the following approximated equation, surroundings the value $\mu = \mu^{*}$, can be found [12]:

$$A = 3\frac{1+(1+g)\left(\mu^{*}/\mu_{\mathrm{m}}\right)^{2+g}}{1-\left(\mu^{*}/\mu_{\mathrm{m}}\right)^{2+g}}, \quad B = \frac{1}{\left[1-\left(\mu^{*}/\mu_{\mathrm{m}}\right)^{2+g}\right]^{3}\left(\mu^{*}\right)^{A-3}} \quad (8.130)$$

$$\frac{\mu^{3}}{\left[1-\left(\mu/\mu_{\mathrm{m}}\right)^{2+g}\right]^{3}} \cong B\mu^{A}. \quad (8.131)[34]$$

By using this approximation in Equation (8.128), we obtain

$$\frac{B\mu^{A}}{\sqrt{\mu}} = Q\left(ZT^{1/4}\right)^{2},$$

$$\mu = \left(\frac{Q}{B}\right)^{\frac{1}{A-0.5}}\left(ZT^{1/4}\right)^{\frac{2}{A-0.5}} = \left(\frac{Q}{B}\right)^{\frac{1}{A-0.5}} Z^{\frac{2}{A-0.5}} T^{\frac{1}{2A-1}}. \quad (8.132)$$

By the use of physical dimensions, Equation (8.132) can be expressed as follows:

$$\mu = \left(\frac{Q_{[\mathrm{m}^{2}\mathrm{tex}^{-1/2}]}}{B}\right)^{\frac{1}{A-0.5}}\left(Z_{[\mathrm{m}^{-1}]}T_{[\mathrm{tex}]}^{1/4}\right)^{\frac{2}{A-0.5}} = \left(\frac{Q_{[\mathrm{m}^{2}\mathrm{tex}^{-1/2}]}}{B}\right)^{\frac{1}{A-0.5}} Z_{[\mathrm{m}^{-1}]}^{\frac{2}{A-0.5}} T_{[\mathrm{tex}]}^{\frac{1}{2A-1}}. \quad (8.133)$$

(A and B are dimensionless parameters.)

---

33   We can also – only once – evaluate the packing density $\mu = \mu^{*}$ for a selected yarn, which is lying 'in the middle' of the produced and/or used set of yarns, from numerical solving of Equation (8.129).

34   This approximation is presented in the fifth chapter of the book [12] in the context of compression of fibrous assemblies. However, there are symbols b and c used in place of symbols A and B, respectively.

Example 2

Let us use in case of a cotton carded yarn, according to Table 8.1, $g = 1$, $\mu_m = 0.9$, $Q = 9.61 \cdot 10^{-8}\,\mathrm{m^2\,tex^{-1/2}}$, $\rho = 1520\,\mathrm{kg\,m^{-3}}$ and let us choose $\mu^* = 0.46$. Then, according to Equation (8.130), we get $A = 4.3869$, $B = 4.5126$, and the approximated expression for packing density is expressed as

$$\mu = 0.010622\left(Z_{[m^{-1}]}\,T^{1/4}_{[tex]}\right)^{0.51456} = 0.010622\,Z^{0.51456}_{[m^{-1}]}\,T^{0.12864}_{[tex]} \qquad (8.134)$$

according to Equation (8.132). Equation (8.134) represents the thin line in Figure 8.15; the thick line reflects the original Equation (8.129). A good accordance between both curves is visible in a region around the chosen value of μ^* (roughly from $ZT^{1/4} = 1000$ to $2500\,\mathrm{m^{-1}\,tex^{1/4}}$).

The experimental values of packing densities in two sets of cotton carded ring yarns were determined [by means of measurement of yarn diameters – see Equation (1.47)] in a large range of fineness (from 16.5 to 100 tex) and twist coefficient (from 83 to 140 $\mathrm{m^{-1}\,ktex^{-1/2}}$). The points shown in Figure 8.15 display a good accordance between both curves in the region under consideration.

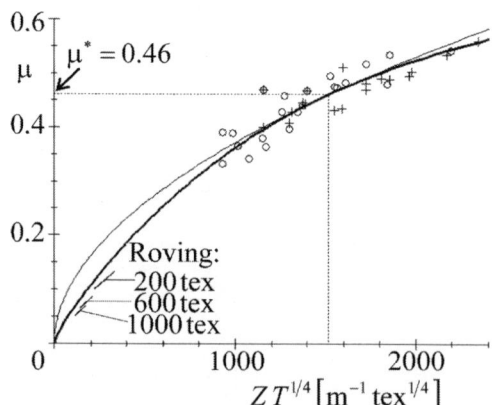

Figure 8.15 Packing density μ as the function of variable $ZT^{1/4}$ by carded cotton ring yarns. Thick line: Equation (8.129); thin line: Equation (8.134); points O,+: two sets of experimental results; roving: Salaba's empirical expressions

Salaba [15] studied the packing densities of rovings by means of measurement of their diameters. The short curves displayed in Figure 8.15 illustrate the results. While the original Equation (8.129) describes the experimental results more or less acceptable, the approximated curve according to Equation (8.132) (being agreed with yarns) is quite away from the experimental results. It illustrates a limited scope of validity of each approximated equation, however, a relatively large validity of the original expression.

Concluding remarks

There are two 'borderline' conceptions how to determine the packing density of yarn: The 'simplest' way according to Koechlin's idea, described in Section 1.4 of this book and, on the other hand, the 'most difficult' conception, based on the solution of differential equation of radial equilibrium in yarn, described in Sections 8.1 to 8.4 in this chapter[35]. Both the concepts bring a lot of troubles. Koechlin's yarn diameter $D = K\sqrt{T}$ – Equation (1.67) – is generally known and widely used, but it is not sufficiently precise. On the other hand, the practical utilization of the differential equation – Equation (8.45) – brings different difficulties with determination of necessary inputs, mostly the deformation law of fiber material – see Section 8.4. The way, introduced in this section, can be viewed as an in-between model, which is relatively simple and simultaneously precise for common practical usage.

8.6   Suitable yarn twist

Equation (8.128) allows us to solve the problem of suitable yarn twist by modification of the first and second special assumptions of Koechlin, described in Section 1.4.

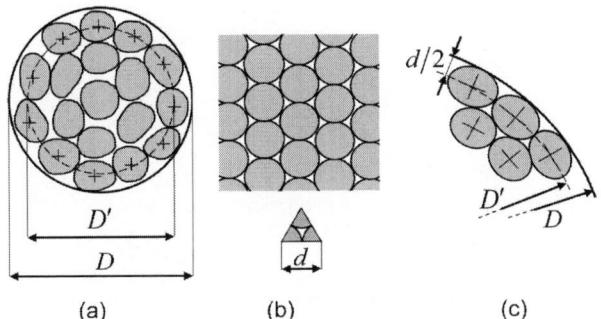

(a)   (b)   (c)

Figure 8.16 Derivation of Schwarz constant: (a) yarn diameter and modified yarn diameter, (b) limit structure of parallel fiber bundle, (c) peripheral layers of yarn by limit structure

---

35   On the top of it, a lot of quite empirical equations can be found in literature, e.g., Zurek's book [16] contains a set of formerly published expressions of such type.

Schwarz constant

We shall modify the idea of geometrical similarity (second special assumption of Koechlin) by the idea of modified yarn diameter. There are two diameters shown in the scheme of yarn cross section, displayed in Figure 8.16a. The cylindrical surface with yarn diameter D encircles the whole (idealized) yarn while the cylindrical surface with modified diameter D' is passing among (i.e., near to) the centres of the fibers lying on the surface layer of the yarn. It is evident that we need to use fiber angle at the modified diameter D' when we are thinking more precisely about the slope of fibers in the peripheral layer. The ratio

$$C_D = \frac{D'}{D} \tag{8.135}$$

is known as Schwarz constant [17].

At first, let us think about the Schwarz constant in the (theoretical) case of the so-called limit structure of a yarn. This idea worked with the closest bundle of cylindrical fibers with diameter d, whose cross section is shown in Figure 8.16b. The equilateral triangle, creating the structural unit of this configuration, has total area $d \cdot d \cos 30°/2 = d^2 \sqrt{3}/4$. The area of fibers occupied inside this triangle is $\pi d^2/8$. The limit packing density of such structure is then

$$\mu_{\text{lim}} = \frac{\pi d^2/8}{d^2/4} = \frac{\pi}{2\sqrt{3}} = 0.907. \tag{8.136}$$

(See Chapter 1 of the book [12] for a more general consideration.)

In the limit structure, the peripheral layers of fibers in the yarn can be imaginable from the scheme shown in Figure 8.16c. Here, $D - D' = d$. By using this relation in Equation (8.135) and subsequently rearranging the expression by using Equations (1.47), (1.35), (8.136), (1.38) and (1.36), we obtain the following expression for the Schwarz constant:

$$C_D = \frac{D-d}{D} = 1 - \frac{d}{\sqrt{4S/(\pi\mu_{\text{lim}})}} = 1 - \frac{d}{\sqrt{4(\pi D_S^2/4)/(\pi\mu_{\text{lim}})}} = 1 - \sqrt{\mu_{\text{lim}}}\frac{d}{D_S}$$

$$= 1 - \sqrt{\frac{\pi}{2\sqrt{3}}}\frac{1}{\sqrt{\tau}} = 1 - 0.952\sqrt{\frac{t}{T}}. \tag{8.137}$$

(In this expression, there are symbols of substance cross-sectional area S, substance diameter D_S and relative fineness τ; see Chapter 1 for more details.)

Because the constant in the last expression is very near to one, we can use the following simplified approximated expression:

$$C_D = 1 - \sqrt{\frac{t}{T}}.$$ (8.138)

The real yarns have packing densities $\mu < \mu_{lim}$, i.e., a higher value for D as well as for D'. Nevertheless, we assume that the mentioned 'magnification' of both quantities is, roughly speaking, proportional. Then, the Schwarz constant C_D is also a constant, independent of the packing density of the yarn.

Modified angle β'

The slope of helical fiber axes, lying at radius $D'/2$, follows Equation (4.1); see also Figure 4.2. It is valid to write that

$$\tan\beta' = 2\pi(D'/2)Z = \pi DZ \cdot D'/D = \kappa C_D.$$ (8.139)

[Equation (1.52) determines the twist intensity $\kappa = \pi DZ$.]

Note: Equation (8.139) is generally known; see, e.g., Equation (2.46) in the book [14].

Packing density of yarn

We shall modify also the first special assumption of Koechlin about the packing density, which is a function of twist intensity only (see Section 1.4). We will here use Equation (8.128) in place of earlier Equation (1.55).

Equation (8.128) can be rearranged by using Equations (1.52), (1.47), (8.139), (8.138) as follows:

$$\frac{\mu^{2.5}}{\left[1 - (\mu/\mu_m)^{2+g}\right]^3} = Q\left(\frac{\pi DZ}{\pi D}T^{1/4}\right)^2$$

$$= Q\left(\frac{\kappa}{\pi\sqrt{4T/(\pi\mu\rho)}}T^{1/4}\right)^2 = Q\frac{(\tan\beta'/C_D)^2\,\mu\rho}{4\pi\sqrt{T}},$$

$$\frac{\mu^{1.5}}{\left[1 - (\mu/\mu_m)^{2+g}\right]^3} = Q\frac{\tan^2\beta'\,\rho}{4\pi\sqrt{T}\,C_D^2} = \left(Q\frac{\tan^2\beta'\,\rho}{4\pi}\right)\frac{1}{\sqrt{T}}\frac{1}{\left(1 - \sqrt{t/T}\right)^2},$$ (8.140)

or

$$\frac{\mu^{1.5}}{\left[1-\left(\mu/\mu_{\mathrm{m}}\right)^{2+g}\right]^{3}} = \frac{R}{\sqrt{T}\left(1-\sqrt{t/T}\right)^{2}}, \tag{8.141}$$

where

$$R = Q\frac{\tan^{2}\beta'\rho}{4\pi}. \tag{8.142}$$

Let us think about yarns prepared from same (staple) fibrous material and for same (and/or analogical) end-use – assumptions 1 and 3 of Koechlin's concept in Section 1.4.

Further, let us imagine that such yarns are produced by employing the same type of technology but with different fineness. The question is: how would you twist such yarns? It is probably so that the modified angle β' remains always the same (an improved idea of geometrical similarity of yarns). Then, the parameter R is constant for all such yarns according to Equation (8.142).

Alternatively, let us imagine that the yarns are always having the same fineness but produced by different types of technologies (combed ring, carded ring, rotor, etc.). Such yarns must be twisted probably so intensively as to achieve the same level of packing density (i.e., same density of fiber-to-fiber contacts, same compression of fibers, same effects of fiber-to-fiber friction and fiber-to-fiber slippage, etc.). Then, the left-hand side of Equation (8.141) becomes a constant, the denominator on the right-hand side is also a constant (because of the assumption of constant yarn fineness) so that the quantity R must be a constant, too.

Thus, the value of R shall be a constant in both, mutually quite different cases. It leads to generalization of the hypothesis that the value of R should be a constant for all yarns prepared from the same material and designed for same (and/or analogical) purpose of end-use. Accepting this hypothesis, Equation (8.141) is a new relation, independent to Equation (8.128).

We can use Equation (8.141) with the following physical dimensions:

$$\frac{\mu^{1.5}}{\left[1-\left(\mu/\mu_{\mathrm{m}}\right)^{2+g}\right]^{3}} = \frac{R_{[\mathrm{tex}^{1/2}]}}{\sqrt{T_{[\mathrm{tex}]}}\left(1-\sqrt{t_{[\mathrm{tex}]}/T_{[\mathrm{tex}]}}\right)^{2}}, \quad R_{[\mathrm{tex}^{1/2}]} = \text{constant}. \tag{8.143}$$

Table 8.2 illustrates the most common values of R, especially for warp yarns used in weaving.

Table 8.2 Typical values of R

Material	$\rho\left[\text{kgm}^{-3}\right]$	$R\ [\text{tex}^{1/2}]$
Cotton – long staple	1520	2.145
Cotton – medium staple	1520	2.500
Viscose, cotton type	1500	4.589
Polyester, cotton type	1360	3.563
Wool	1310	2.341

Note: Let us remind that R does not depend on spinning technology, but its value depends on the purpose of end-use. (For example, yarns for knitting need smaller values of R, etc.) The parameter R reminds us something as a generalized expression of 'twist coefficient'.

Common method of application

Usually, we know the type of fiber material including fiber fineness $t_{[\text{tex}]}$, type of spinning technology, and yarn fineness $T_{[\text{tex}]}$. Assuming suitable values μ_m and g (commonly $\mu_m = 0.9$ and $g = 1$), it is possible to determine also parameters $Q_{[\text{m}^2\text{tex}^{-1/2}]}$ and $R_{[\text{tex}^{1/2}]}$, e.g., from Tables 8.1 and 8.2. Now, we find packing density μ of the yarn as a root of Equation (8.143) by using a suitable numerical method. Further, we substitute this value μ to the left-hand side of Equation (8.129) and then we express very easily the desired suitable twist-value Z. [The corresponding value of Koechlin's twist coefficient $\alpha_{[\text{m}^{-1}\text{ktex}^{1/2}]}$ can be then found out from Equation (1.66).]

Two equations for packing density – rearrangement

Equations (8.128) and (8.141) create a couple of (mutually independent) equations for determination of the packing density of yarn. We can also divide Equation (8.128) by Equation (8.141) and then we get the expression as follows:

$$\mu = Q\left(ZT^{1/4}\right)^2 \frac{\sqrt{T}\left(1-\sqrt{t/T}\right)^2}{R} = \frac{Q}{R}Z^2 T\left(1-\sqrt{t/T}\right)^2. \tag{8.144}$$

By using the physical dimensions that are commonly used, we can write

$$\mu = \frac{Q_{[m^2tex^{-1/2}]}}{R_{[tex^{1/2}]}} Z^2_{[m^{-1}]} T_{[tex]} \left(1 - \sqrt{t_{[tex]}/T_{[tex]}}\right)^2.$$

(8.145)

Note: Let us remind that Equation (8.144), including Equation (8.145), was derived from original (i.e., non-approximated) Equations (8.128) and (8.141). By means of a common method of application as stated in the previous paragraph, the last equation can be utilized for determination of suitable twist in place of Equation (8.129). [Calculation of Equation (8.145) is formally a little easier than by using Equation (8.129).]

Approximation of Equation (8.144)

Let us think that

1. The last factor in Equation (8.144) is possible to approximate according to the following expression:

$$1 - \sqrt{t/T} \doteq U T^V.$$

(8.146)

By differentiating the previous equation with respect to T, we obtain

$$0.5\sqrt{t}\, T^{-1.5} \doteq UV\, T^{V-1}.$$

(8.147)

2. The 'typical' or 'average' yarn, which is lying 'in the middle' of the produced and/or used set of yarns, has – besides its packing density μ^* – also the known yarn fineness T^*. (We will later denote also other quantities of this 'middle' yarn with superscript *.) We denote the values of the approximated expression and its first derivative identical to the original expressions by $T = T^*$. Thus

$$1 - \sqrt{t/T^*} = U \cdot \left(T^*\right)^V, \quad 0.5\sqrt{t}\, \left(T^*\right)^{-1.5} = UV \cdot \left(T^*\right)^{V-1}.$$

(8.148)

We obtain the following expressions after rearrangement of the last expressions

$$0.5\sqrt{t}\, \left(T^*\right)^{-1.5} = UV \cdot \left(T^*\right)^{V-1} = \frac{V}{T^*}\left[U \cdot \left(T^*\right)^V\right] = \frac{V}{T^*}\left[1 - \sqrt{t/T^*}\right],$$

$$V = \frac{0.5\sqrt{t}\, \left(T^*\right)^{-0.5}}{1 - \sqrt{t/T^*}} = \frac{0.5\sqrt{t/T^*}}{1 - \sqrt{t/T^*}}, \quad U = \frac{1 - \sqrt{t/T^*}}{\left(T^*\right)^V}.$$

(8.149)

We can also write the last two expressions in terms of physical dimensions that are commonly used in the following manner:

$$V = \frac{0.5\sqrt{t_{[\mathrm{tex}]}/T^*_{[\mathrm{tex}]}}}{1-\sqrt{t_{[\mathrm{tex}]}/T^*_{[\mathrm{tex}]}}}, \quad U_{[\mathrm{tex}^{-V}]} = \frac{1-\sqrt{t_{[\mathrm{tex}]}/T^*_{[\mathrm{tex}]}}}{\left(T^*_{[\mathrm{tex}]}\right)^V}. \tag{8.150}$$

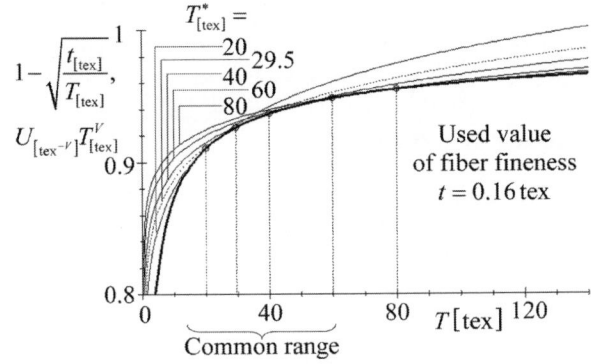

Figure 8.17 Comparison between original and approximated factors
in case of cotton carded yarns.
Thick line: original factor; thin lines: approximations with different values of T^*;
O: contact points of original and approximated factor; dotted line: – approximation
with $T^* = 29.5$ tex

Figure 8.17 illustrates an example (cotton carded yarns) for comparison between the original factor $1-\sqrt{t_{[\mathrm{tex}]}/T_{[\mathrm{tex}]}}$ and its approximation $U_{[\mathrm{tex}^{-V}]}T^V_{[\mathrm{tex}]}$ – see Equations (8.146) and (8.149). It is shown that the differences among these two expressions are quite insignificant in the common range chosen for yarn fineness.

By applying the approximated Equation (8.146) to Equation (8.144), we find

$$\mu = \frac{Q}{R}Z^2T\left(UT^V\right)^2 = \frac{Q}{R}U^2Z^2T^{2V+1}, \tag{8.151}$$

where Equation (8.149) determines parameters U and V.

Twist exponent and method of twisting

Two independent approximated equations were derived for the same packing density of yarn – Equations (8.132) and (8.151). The next equation follows from the equivalency of the right-hand sides of these equations:

$$\frac{Q}{R}U^2Z^2T^{2V+1} = \left(\frac{Q}{B}\right)^{\frac{1}{A-0.5}}Z^{\frac{2}{A-0.5}}T^{\frac{1}{2A-1}}, \quad Z^{2-\frac{2}{A-0.5}}T^{2V+1-\frac{1}{2}\frac{1}{A-0.5}} = \frac{R}{Q^{1-\frac{1}{A-0.5}}U^2B^{\frac{1}{A-0.5}}},$$

$$Z^{2\frac{A-1.5}{A-0.5}} T^{2V+\frac{A-1}{A-0.5}} = \frac{R}{Q^{\frac{A-1.5}{A-0.5}} U^2 B^{\frac{1}{A-0.5}}}, \quad ZT^{2\frac{A-1.5}{A-0.5}} = \frac{R^{\frac{2V+\frac{A-1}{A-0.5}}{2\frac{A-1.5}{A-0.5}}}}{Q^{\frac{\frac{A-1.5}{A-0.5}}{2\frac{A-1.5}{A-0.5}}} U^{\frac{2}{2\frac{A-1.5}{A-0.5}}} B^{\frac{\frac{1}{A-0.5}}{2\frac{A-1.5}{A-0.5}}}},$$

$$ZT^{\frac{V(A-0.5)}{A-1.5}+\frac{A-1}{2(A-1.5)}} = \frac{R^{\frac{A-0.5}{2(A-1.5)}}}{Q^{\frac{1}{2}} U^{\frac{A-0.5}{A-1.5}} B^{\frac{1}{2(A-1.5)}}}. \tag{8.152}$$

Let us introduce the twist exponent as follows:

$$q = \frac{V(A-0.5)}{A-1.5} + \frac{A-1}{2(A-1.5)}, \tag{8.153}$$

and the general twist coefficient as follows:

$$\alpha_q = \frac{R^{\frac{A-0.5}{2(A-1.5)}}}{Q^{\frac{1}{2}} U^{\frac{A-0.5}{A-1.5}} B^{\frac{1}{2(A-1.5)}}}. \tag{8.154}$$

Then, we can write

$$\alpha_q = ZT^q. \tag{8.155}$$

Let us note that q and α_q are not functions of yarn fineness; they are only parameters of a given type of yarns; the last equation is then identical to the definition of general twist coefficient[36], stated in Chapter 1, Table 1.8.

We can also write the last two expressions in terms of physical dimensions that are commonly used as follows:

$$\alpha_{q[m^{-1}tex^q]} = Z_{[m^{-1}]} T^q_{[tex]}, \quad \alpha_{q[m^{-1}tex^q]} = \frac{R^{\frac{A-0.5}{2(A-1.5)}}_{[tex^{1/2}]}}{Q^{1/2}_{[m^2 tex^{-1/2}]} U^{\frac{A-0.5}{A-1.5}}_{[tex^{-V}]} B^{\frac{1}{2(A-1.5)}}}. \tag{8.156}$$

[The quantities, A, B, V and q are dimensionless according to Equations (8.130), (8.149) and (8.153).]

---

36 Symbol α is used in Chapter 1, in contrary to symbol α_q which is used now.

A common method of application

There are two possible methods for finding the suitable yarn twist – the original method of the derived model (needed numerical solving for the root of equation) and the approximated alternative for it. Both methods require the followings to determine:

- fiber fineness t in tex,
- fiber density ρ in $kg\,m^{-3}$,
- maximum packing density μ_m (recommended value $\mu_m = 0.9$),
- parameter g (recommended value $g = 1$),
- parameter Q in $m^2\,tex^{-1/2}$, selected according to the fiber material and the spinning technology (like Table 8.1) and
- parameter R in $tex^{1/2}$, selected according to the fiber material and the intended end-use[37] (like Table 8.2).

A. The original method consists of the following steps:

(a) Determination of actual yarn fineness T in tex.

(b) Calculation of right-hand side value of Equation (8.143) and then finding the corresponding value of packing density μ (by using a numerical method, e.g., 'half-interval method', or a table of values prepared beforehand).

(c) Calculation of left-hand side value of Equation (8.129) or (8.145) by using the previous value of packing density, and then finding the corresponding value of yarn twist Z in m^{-1} from the stated equation.

(d) If necessary, it is possible to determine the actual values of Koechlin's twist coefficient (denoted by $\alpha_{Koechlin}$ now) in $m^{-1}ktex^{1/2}$ and Phrix's twist coefficient (denoted by a_{Phrix} now) in $m^{-1}ktex^{2/3}$ by using Equations (1.66) and (1.68) from Chapter 1.

Note: By repeating the previous procedure for different yarn finenesses, we get the original relation between yarn fineness and suitable yarn twist.

B. The approximated method consists of the following steps:

---

37   We mentioned earlier that the parameter R is something like a 'generalized twist coefficient'.

(a) Selection of yarn fineness $T = T^*$ in tex for a typical or average yarn, which is lying in the middle of the produced and/or used set of yarns. (The subjective step accords to our previous experience.).

(b) Calculation of packing density value $\mu = \mu^*$, by using point (b) according to previous original method A.

(c) Calculation of dimensionless parameters A and B according to Equation (8.130).

(d) Calculation of dimensionless parameter V and parameter U in tex^{-V} according to Equation (8.150).

(e) Calculation of dimensionless general exponent q of twist according to Equation (8.153).

(f) Calculation of α_q in $\text{m}^{-1}\text{tex}^q$ according to the second expression of Equation (8.156).

(g) Selection of actual yarn fineness T in tex.

(h) Determination of twist Z in m^{-1} for actual yarn, according to the first expression of Equation (8.156).

(i) If necessary, it is possible to determine the actual values of Koechlin's twist coefficient α_{Koechlin} in $\text{m}^{-1}\text{ktex}^{1/2}$ and Phrix's twist coefficient a_{Phrix} in $\text{m}^{-1}\text{ktex}^{2/3}$ by using Equations (1.66) and (1.68)

Note: By repeating the previous procedure from (g) to (i) for different yarn finenesses, we obtain the approximated relation between yarn fineness and suitable yarn twist.

Example

Let us illustrate the previous methods of application on the following example of cotton carded ring-spun yarns. Two yarns were selected – one yarn for weaving with higher R-value (higher twisting) and another yarn for knitting with smaller R-value (smaller twisting). Table 8.3 reports on the input parameters and evaluated quantities for approximated relationships.

Table 8.3 Calculated data by approximation

Common input parameters:			
$t = 0.16\,\text{tex}$, $\rho = 1520\,\text{kg}\,\text{m}^{-3\,a}$, $\mu_m = 0.9^{\,a}$, $g = 1^{a}$, $Q = 9.61 \cdot 10^{-8}\,\text{m}^2\,\text{tex}^{-1/2\,a}$			

Quantity	Dimension	Middle yarn, $T^* = 29.5\,\text{tex}$	
		Warp yarn	**Knitted yarn**
R (input)	$\text{tex}^{1/2}$	2.500[b]	2.000
μ^*	–	0.47760[c]	0.44206[c]
A	–	4.5812	4.2099
B	–	5.2283	3.9198
V	–	0.039750	0.039750
U	tex^{-V}	0.80975	0.80975
q	–	0.63379	0.64667
α_q	$\text{m}^{-1}\text{tex}^q$	5984.2	5379.0
Values of middle yarn:			
Z^*	m^{-1}	700.57	602.85
$\alpha^*_{\text{Koechlin}}$	$\text{m}^{-1}\text{ktex}^{1/2}$	120.33	103.54
α^*_{Phrix}	$\text{m}^{-1}\text{ktex}^{2/3}$	66.89	57.56

[a]See Table 8.1.
[b]See Table 8.2.
[c]The root found by numerical method.

The following expression is valid to write from Equations (1.66) and (8.156)

$$Z_{[\text{m}^{-1}]} = \frac{\alpha_{q[\text{m}^{-1}\text{tex}^q]}}{T^q_{[\text{tex}]}} = \frac{31.623\,\alpha_{\text{Koechlin}[\text{m}^{-1}\text{ktex}^{1/2}]}}{\sqrt{T_{[\text{tex}]}}}, \quad \alpha_{\text{Koechlin}[\text{m}^{-1}\text{ktex}^{1/2}]} = \frac{\alpha_{q[\text{m}^{-1}\text{tex}^q]}}{31.623\,T^{q-0.5}_{[\text{tex}]}},$$

(8.157)

we can write the following expressions using the data given in Table 8.3

$$\alpha_{\text{Koechlin}[m^{-1}\text{ktex}^{1/2}]} = \frac{5984.2}{31.623\,T_{[\text{tex}]}^{0.13379}} = \frac{189.24}{T_{[\text{tex}]}^{0.13379}} \cdots \text{warp yarn,}$$

$$\alpha_{\text{Koechlin}[m^{-1}\text{ktex}^{1/2}]} = \frac{5379.0}{31.623\,T_{[\text{tex}]}^{0.14667}} = \frac{170.10}{T_{[\text{tex}]}^{0.14667}} \cdots \text{knitted yarn.}$$

(8.158)

Similarly, it is valid to write from Equations (1.68) and (8.156)

$$Z_{[m^{-1}]} = \frac{\alpha_{q[m^{-1}\text{tex}^q]}}{T_{[\text{tex}]}^q} = \frac{100\,a_{\text{Phrix}[m^{-1}\text{ktex}^{2/3}]}}{T_{[\text{tex}]}^{2/3}}, \quad a_{\text{Phrix}[m^{-1}\text{ktex}^{2/3}]} = \frac{\alpha_{q[m^{-1}\text{tex}^q]}}{100\,T_{[\text{tex}]}^{q-2/3}}. \quad (8.159)$$

Now, we can write the following expressions using the data given in Table 8.3

$$a_{\text{Phrix}[m^{-1}\text{ktex}^{2/3}]} = \frac{5984.2}{100\,T_{[\text{tex}]}^{0.63379-2/3}} = 59.842\,T_{[\text{tex}]}^{0.032876} \cdots \text{warp yarn,}$$

$$a_{\text{Phrix}[m^{-1}\text{ktex}^{2/3}]} = \frac{5379.0}{100\,T_{[\text{tex}]}^{0.64667-2/3}} = 53.790\,T_{[\text{tex}]}^{0.019996} \cdots \text{knitted yarn.}$$

(8.160)

Finally, let us remind that the general twist coefficient $\alpha_{q[m^{-1}\text{tex}^q]}$ is a constant for a given type of yarns. Thus, according to Table 8.3,

$$\alpha_{q[m^{-1}\text{tex}^{0.63379}]} = 5984.2\ldots \text{warp yarn,}$$

$$\alpha_{q[m^{-1}\text{tex}^{0.64667}]} = 5379.0\ldots \text{knitted yarn.}$$

(8.161)

By using the approximated Equations (8.158), (8.160) and (8.161), yarn twist was calculated in accordance with the original method of solving. (See algorithm A stated before this example.) Then, the behaviours of $\alpha_{\text{Koechlin}[m^{-1}\text{ktex}^{1/2}]}$, $a_{\text{Phrix}[m^{-1}\text{ktex}^{2/3}]}$ and $\alpha_{q[m^{-1}\text{tex}^q]}$ were analogically determined by using Equations (1.66), (1.68) and the first expression of Equation (8.156).

The derived relations illustrate triplet of pictures shown in Figure 8.18. The thick curves show the behaviour obtained by following the original method of solving, the thin curves show analogical behaviours, as found by approximated method of solving. It then stands to reason that

- The original and approximated behaviours are in very good accordance, especially in the common range of yarn finenesses. (Keep it in mind that the approximated ones were constructed 'around' the typical yarn fineness of 29.5 tex lying inside the common range; therefore, this accordance must be the best in this range.)

- The suitable twist coefficient α_{Koechlin} decreases with increasing yarn fineness T, as it is empirically very well known from the industrial practices of spinning yarns. Thus, these empirical relations result from the regulations of internal mechanics of yarns.
- The behaviour of suitable twist coefficient α_{Phrix} as shown in Figure 8.18b is very flat and comes near to the idea of a constant value. (It corresponds to the industrial spinning practices of yarns in Czech Republic and in some other countries.)
- The behaviour of suitable generalized twist coefficient α_q as shown in Figure 8.18c is constant, which results immediately from Equation (8.161). Table 8.3 determines the twist exponents ($q = 0.63379$ and 0.64667), in this example. It is in good accordance with the empirical experience, reported in Table 1.9 of Chapter 1.

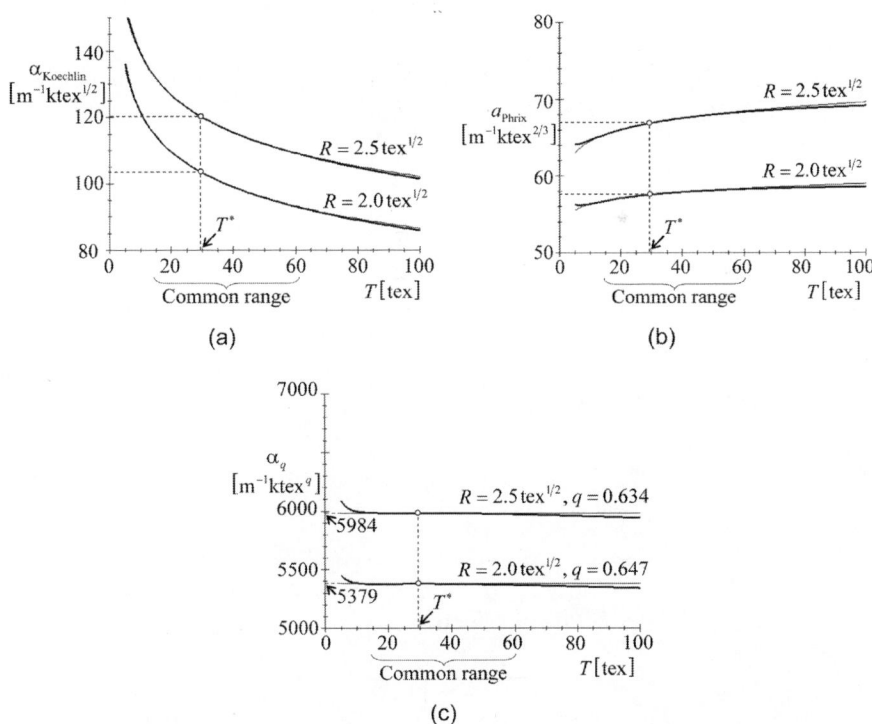

Figure 8.18 Suitable twist coefficients of cotton carded ring yarns – examples: Parameters used: Table 8.3. $R = 2.5\,\text{tex}^{1/2}$ – twisting of warp yarn for weaving; $R = 2.0\,\text{tex}^{1/2}$ – twisting of yarns for knitting; thick lines – original solving (by numerical method); thin lines – approximated equations; O – contact points of original and approximated curves

8.7   References

[1]   Budnikov, V.I., O strukture prjazi i rovnicy i o naprjazeniach vozni-kajuscich pri krucenii voloknistovo produkta (About yarn and roving structure and about stresses generated by twisting of fiber product). Naucno-issledovatelskie trudy, Tom X (Scientific and Research Works, Tome X), Moskovskij tekstilnyj institut (Moscow Textile Institute), Gizlegprom Publisher, 29–66, 1948 (in Russian).

[2]   Hearle, J.W.S., The mechanics of twisted yarns: The influence of transverse forces on textile behaviour, *Journal of Textile Institute*, 49, T383–T408, 1958.

[3]   Treloar, L.R.G. and Hearle, J.W.S., The mechanics of twisted yarns: A correction, *Journal of Textile Institute*, 53, T446–T448, 1962.

[4]   Cheng, C.C., White, J.L., and Duckett, K.E., A continuum mechanics approach to twisted yarns, *Textile Research Journal*, 44, T798–T803, 1974.

[5]   White, J.L., Cheng, C.C., Spruiell, J.E., Some aspects of mechanics of continuous filament twisted yarns and the deformation of fibers, Applied Polymer Symposium, No. 27, 275–294, 1975.

[6]   Neckář, B., Fyzikální model vnitřní struktury příze (Physical model of internal yarn structure), PhD thesis, Technical University of Liberec, 1975 (In Czech).

[7]   Neckář, B., Příze: tvorba, struktura, vlastnosti (Yarns: Creation, structure, properties). SNTL Publisher, Prague, 1990 (In Czech).

[8]   Korn, G.A. and Korn, T.M., Mathematical Handbook for Scientists and Engineers, McGraw Hill Book Company, New York, 1968.

[9]   Hearle, J.W.S. and El-Behery, H.M.A., The transmission of transverse stresses in fiber assemblies, *Journal of Textile Institute*, 51, T164–T171, 1960.

[10]   Gurová, M., Jednoosé stlačování vlákenného materiálu (Uniaxial compression of fibrous assemblies), MSc Thesis, Technical University of Liberec, 2008 (In Czech).

[11]   van Wyk, C.M., Note on the compressibility of wool, *Journal of Textile Institute*, 37, T285–T291, 1949.

[12]   Neckář, B. and Das, D., Theory of Structure and Mechanics of Fibrous Assemblies, Woodhead Publishing India Pvt. Ltd., New Delhi, 2012.

[13]   Baljasov, P.D., Szatie textilnych volokon v mase I technologija testilnovo proizvodstva (Compression of textile fiber mass and technology

of textile manufacture), Legkaya promyshlenost, Moscow, 1976 (in Russian).

[14] Hearle, J.W.S, Grosberg, P., Backer, S. Structural Mechanics of Fibers, Yarns, and Fabrics, Vol. 1, Wiley Interscience, New York, London, Sydney, Toronto, 1969.

[15] Salaba, J., Geometrie a vlastnosti staplových přízí (Geometry and Properties of Staple Yarns), PhD Thesis, Technical University of Liberec, 1975 (in Czech).

[16] Zurek, W., Struktura przedzy (Yarn Structure), WNT Publishing, Warszawa, 1971 (in Polish).

[17] Schwarz, E.R., An introduction to the micro-analyses of yarn twist, *Journal of Textile Institute*, 24, T105–T118, 1933.

Tensile behaviour of yarns

9.1 Stress–strain relation

Introduction

The tensile behaviour of staple fiber yarns is considered to be one of the most difficult topics of theoretical research in the area of structural mechanics of yarns. Several attempts were made to develop satisfactory theoretical basis onto which our practical understanding can be built upon. Gegauff [1] studied the stress–strain relation in twisted yarn by taking into account of tensile characteristics of fibers. An analytical expression was derived establishing the relationship among fiber strain, yarn strain and twist angle. Further, the tensile force utilization coefficient in twisted yarns was predicted as a function of yarn twist angle. This model, however, did not take into account of change in yarn diameter during extension of yarn. Considering this Platt [2, 3] and Hearle [4] attempted to develop more accurate models of stress–strain relation in twisted yarns. Another pioneering work in the area of structural mechanics of yarns included Peirce's contribution to the weak link theory for twisted yarns [5]. (This problem will be solved in Chapter 10 of this book.) All theoretical work besides many other empirical researches led to establish a fact that the tensile behaviour of twisted yarns are by and large determined by the constituent fiber properties, internal structure of yarns and external factors such as testing parameters, etc.

Terms and definitions

The tensile behaviour of yarn includes the relationship between the tensile (axial) force acting on a yarn and its deformation and/or destruction. If a tensile force F_Y is acting on a yarn with fineness T then the (engineering[1]) tensile stress of the yarn is expressed by $\sigma'_Y = F_Y/S$, where $S = T/\rho$ denotes

---

1   Engineering stress is defined by the force acting per unit initial cross-sectional area of a specimen. You may like to compare it with Equation (1.21). For more details, see Appendix 3 of our earlier book [6].

the substance cross-sectional area of yarn – see Figure 1.6 and Equations (1.32), (1.33) and ρ is fiber density.

The (engineering) specific tensile stress of the yarn is expressed by

$$\sigma_Y = \frac{F_Y}{T} = \frac{\sigma_Y'}{\rho}. \tag{9.1}$$

(In textile practice, the preferred unit of specific stress in yarn is $N\,tex^{-1}$.)

Similarly, if a tensile force F_f is acting on a fiber with fineness t, then the (engineering) tensile stress of this fiber is $\sigma_f' = F_f/s$ ($s = t/\rho$ denotes the cross-sectional area of the fiber) and specific tensile stress of the fiber is

$$\sigma_f = \frac{F_f}{t} = \frac{F_f}{s\rho} = \frac{\sigma_f'}{\rho}. \tag{9.2}$$

[The last equation is identical to Equation (1.21).]

Note: In this chapter, we prefer to work with specific tensile stresses of fibers and yarns.

The tensile force F_Y or the tensile stress σ_Y' or the specific tensile stress σ_Y causes a tensile deformation in yarn. As a result, the initial length $l_{Y,0}$ of the yarn is increased to a longer length l_Y. The difference $\Delta l_Y = l_Y - l_{Y,0}$ is called tensile elongation of yarn and the ratio

$$\varepsilon_Y = \frac{\Delta l_Y}{l_{Y,0}} = \frac{l_Y - l_{Y,0}}{l_{Y,0}} \tag{9.3}$$

is called tensile strain of yarn.

Note: The strain is usually expressed in percentage, however, in this book, we prefer to use dimensionless strain according to Equation (9.3).

Similar quantities are also used for tensile strain of fibers. If the initial fiber length $l_{f,0}$ is increased to a longer length l_f, then the difference $\Delta l_f = l_f - l_{f,0}$ is known as tensile elongation of fiber and the ratio

$$\varepsilon_f = \frac{\Delta l_f}{l_{f,0}} = \frac{l_f - l_{f,0}}{l_{f,0}} \tag{9.4}$$

is called tensile strain of fiber.

In case of yarns as well as in case of fibers, the specific stresses are functions of strains. Such functions are called specific stress–strain functions. The behaviours of such functions are shown by the so-called specific stress–strain curves. A scheme of such curves is illustrated in Figure 9.1 [7].

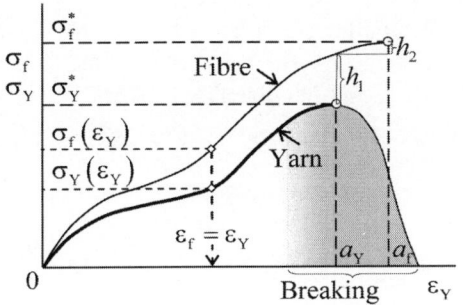

Figure 9.1 Scheme of specific stress–strain curves of (mean) fiber and yarn. ○: points of tenacity, i.e., specific breaking stress (specific strength) and breaking strain; ◇: a general point at same strain; grey area: significant breakage of fibers in yarn

Let us imagine that the specific stress–strain functions start from origin (0,0) as an increasing function (surely in the first part). The maximum points of specific stresses on the specific stress–strain curves determine the specific breaking stress σ_f^* of fiber (specific strength of fiber or fiber tenacity) and the specific breaking stress σ_Y^* of yarn (specific strength of yarn or yarn tenacity). The corresponding strains a_f and a_Y are known as breaking strains of fiber and yarn, respectively – see Figure 9.1.

Note: In this chapter, we prefer to use subscript 'f' for fiber quantities and subscript 'Y' for yarn quantities. However, in the subsequent sections, we use these subscripts only if the difference between fiber quantities and yarn quantities is not clear from the context.

Coefficient of utilization of specific fiber strength

Let us imagine a yarn prepared from one type of fibrous material. The ratio of yarn tenacity to (mean) fiber tenacity is called 'coefficient of utilization of fiber tenacity'[2]. This is shown as follows:

---

2   This type of coefficient of utilization was introduced by Johannsen [9]. He used the German term 'Ausnützung der Substanz', where 'substance' indicates 'fibers'. Following Johannsen's terminology, we use the term 'coefficient of utilization of fiber tenacity'.

$$\varphi_\sigma^* = \frac{\sigma_Y^*}{\sigma_f^*} . \qquad (9.5)$$

This coefficient is smaller than one in all practical cases.

Coefficient of utilization of fiber breaking strain

Similarly, we define 'coefficient of utilization of fiber breaking strain' in the following manner:

$$\varphi_a^* = \frac{a_Y}{a_f} . \qquad (9.6)$$

This coefficient is generally smaller than one, but it can be sometimes greater than one too.

Figure 9.1 illustrates the specific stress–strain curve of a twisted staple fiber yarn and the similar curve of the corresponding (average) fiber. The specific stress–strain curve of yarn usually follows more or less similar shape as the specific stress–strain curve of fiber (in case of twisted staple fiber yarn); however, the yarn curve lies under the fiber curve. It is in consequence of specific fiber arrangement and fiber-to-fiber interaction in yarn. The corresponding difference in specific stress is h_1 at yarn breaking strain $\varepsilon_f = a_Y$. However, the breaking strain of yarn a_Y is mostly smaller than the breaking strain of fiber a_f. So, the specific stress of fiber usually increases more – by a distance h_2 from a_Y to a_f. This is mostly governed by the breakages of individual fiber and the internal structure of yarn. Thus, the coefficient of utilization of fiber tenacity is the result of two different influences: (1) changes in yarn structure by its tensile deformation (without significant fiber breakage) and (2) process of fiber breakage in yarn.

Note: There are two regions shown under the specific stress–strain curve of yarn in Figure 9.1. The original structural character of yarn remains relatively stable and practically all fibers are mechanically stressed in the first (white-colour) region. However, the yarn structure is significantly changed and more fibers break in the second (grey-colour) region. Of course, the borderline between these two regions is not too clear; the first one is continuously changed to the second one. (Figure 9.1 illustrates it.) Nevertheless, the idea of 'separation' of both regions makes it easier to develop theoretical models.

Note: The non-twisted flat filament yarns (parallel fiber bundles), discussed later on, can show a different behaviour than that displayed in Figure 9.1. Nevertheless, the coefficient of utilization of filament tenacity is usually smaller than one here too.

Coefficient of utilization of specific fiber stress

Let us now study the region of specific stress–strain curve of yarn before breakages of fiber and yarn. At a same strain $\varepsilon_f = \varepsilon_Y$, we find a specific fiber stress $\sigma_f(\varepsilon_Y)$ and a specific yarn stress $\sigma_Y(\varepsilon_Y)$ – see Figure 9.1. We call the ratio

$$\varphi(\varepsilon_Y) = \frac{\sigma_Y(\varepsilon_Y)}{\sigma_f(\varepsilon_Y)} \tag{9.7}$$

as 'coefficient of utilization of specific fiber stress'. Generally, this expression is not a constant but a function of yarn strain.

Types of models

The theoretical models of tensile behaviour of yarns are mostly solved for non-twisted flat filament yarns (parallel fiber bundles), twisted filament yarns and twisted staple fiber yarns. In some of these models, the specific tensile stress is studied in relation to the tensile strain; while in others, the process of yarn breakage is solved along with prediction of yarn tenacity and yarn breaking strain.

9.2   Notes to parallel fiber bundles – fiber blends

The mathematical modeling of tensile behaviour of a non-twisted yarn, interpreted as a parallel fiber bundle prepared from many fiber components, is systematically derived with all details in Chapter 6 of our earlier book [6]. Here, we want only to remind you about the principles used and results obtained there.

General image

We think about a fiber bundle with (initial) fineness T, which is consisting of a set of straight, parallel and mutually independent fibers, where all the fibers are clamped in-between two jaws A and B of a tensile tester, i.e., set with a gauge length h – Figure 9.2a. A tensile force F_Y is applied to the bundle, which corresponds to the specific stress $\sigma_Y = F_Y/T$. The jaw B moves to a new position B′, as a result, each fiber is elongated to a new length $h(1+\varepsilon)$, where ε means yarn strain as well as fiber strain ($\varepsilon_f = \varepsilon_Y = \varepsilon$) – Figure 9.2b.

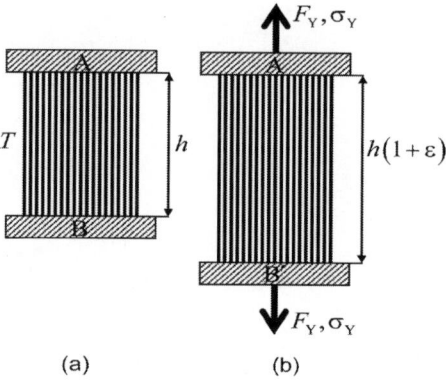

(a)   (b)

Figure 9.2 Bundle of parallel fibers between jaws

Trivial case

If all fibers are absolutely identical (theoretical case), then the specific stress–strain curves of fiber and bundle are same, $\sigma_Y(\varepsilon) = \sigma_f(\varepsilon)$, bundle tenacity is equal to fiber tenacity, $\sigma_Y^* = \sigma_f^*$, and the bundle breaking strain is equal to the fiber breaking strain, $a_Y = a_f$. Thus, all coefficients of utilization [φ_σ^*, φ_a^*, and $\varphi(\varepsilon_Y)$] are equal to one – see Equations (9.5) to (9.7).

Bi-component fiber bundle

Let us imagine a blended fiber bundle prepared from two types of fibers. All the fibers of the same type have absolutely same properties, but the properties of the fiber from the first and the second types are (can be) quit different. Hamburger [7], probably for the first time, solved the problem of tenacity of such bi-component fiber bundle. In accordance with the book [6], the following symbols are used (see also Figure 9.3):

- Type of fiber – subscript[3] 1 for fibers having smaller breaking strain, i.e., $a_1 \le a_2$, and subscript 2 for fibers having higher breaking strain, i.e., $a_2 \ge a_1$.

---

3   Here, we use subscripts 1 and/or 2 in the place of earlier used subscript f.

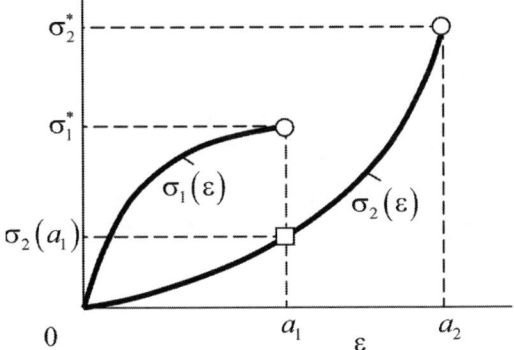

Figure 9.3 Scheme of specific stress–strain curves of first and second type of fibers

- g_1, g_2 are the mass fractions, $g_1 + g_2 = 1$. ($g_1 \cdot 100$ and $g_2 \cdot 100$ yield the 'percentages of components', as used in industry.)
- $\sigma_1^* \ \sigma_2^*$ are the fiber-specific breaking stresses.
- $\sigma_2(a_1)$ is the specific stress of fiber type 2 when its strain is a_1.

Then, as stated in the book [6], the yarn tenacity is

$$\sigma_Y^* = \max\left[g_1\sigma_1^* + g_2\sigma_2(a_1), g_2\sigma_2^* \right] \tag{9.8}$$

and the yarn breaking strain is

$$\begin{aligned}
a_Y &= a_1 & \text{if } \sigma_Y^* = \sigma_Y(a_1) > \sigma_Y(a_2), \\
a_Y &= a_2 & \text{if } \sigma_Y^* = \sigma_Y(a_2) > \sigma_Y(a_1), \\
a_Y &\text{ is undefined} & \text{if } \sigma_Y^* = \sigma_Y(a_1) = \sigma_Y(a_2).
\end{aligned} \tag{9.9}$$

The graphical interpretation of Equations (9.8) and (9.9) is shown by the broken lines in Figure 9.4.

Let us note that the minimum bundle tenacity can be even smaller than the smaller of the two fiber tenacities σ_1^* and σ_2^*. This is shown by black point C (mass fractions $g_{1,C}, g_{2,C}$) in scheme (a) or (b). This mass fraction should be avoided in practice. The corresponding mass fraction at point C is

$$g_{2,C} = \frac{\sigma_1^*}{\sigma_1^* + \sigma_2^* - \sigma_2(a_1)}, \quad g_{1,C} = 1 - g_{2,C} = \frac{\sigma_2^* - \sigma_2(a_1)}{\sigma_1^* + \sigma_2^* - \sigma_2(a_1)}, \tag{9.10}$$

and the corresponding bundle tenacity at point C is

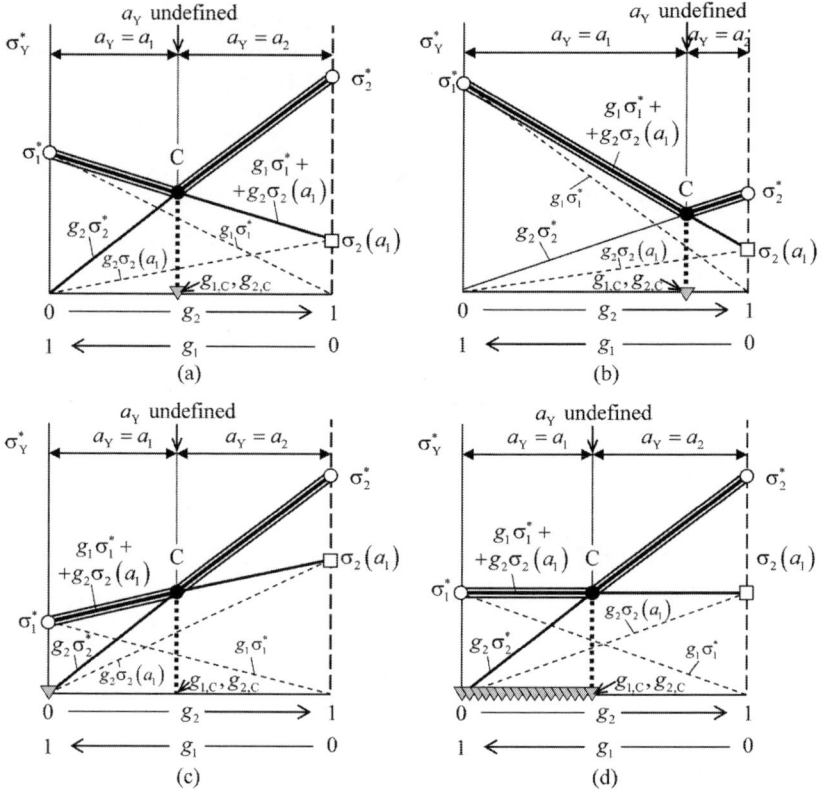

Figure 9.4 Behaviour of tenacity and breaking strain of fiber bundle in relation to mass fraction of fibers in the bundle. (a) $\sigma_1^* < \sigma_2^*$ and $\sigma_1^* > \sigma_2(a_1)$, (b) $\sigma_1^* > \sigma_2^*$ and $\sigma_1^* > \sigma_2(a_1)$, (c) $\sigma_1^* < \sigma_2(a_1)$, (d) $\sigma_1^* = \sigma_2(a_1)$.
━━━ : bundle tenacity, ▽ : points of minimum bundle tenacity

$$\sigma_{Y,C}^* = g_{2,C}\sigma_2^* = \frac{\sigma_1^*\sigma_2^*}{\sigma_1^* + \sigma_2^* - \sigma_2(a_1)}.\qquad(9.11)$$

Note: The equations, presented here, were derived for parallel fiber bundle prepared from two components. Nevertheless, these results are often used also for rough estimation of tenacity of twisted staple yarn prepared from two components. In such case, the corresponding yarn parameters are required to be used in place of fiber parameters – see the book [6].

Multi-component fiber bundle

Hamburger's idea can be generalized to the fiber bundles prepared from more than one component. Let us imagine that a fiber bundle consists of $K \geq 2$

components whose stress–strain curves are illustrated in Figure 9.5. (We assume that all fibers of the same type have absolutely same properties.)

We arranged components on the basis of fiber breaking strain in ascending order from subscript[4] 1 – smallest breaking strain (a_1) – over general subscripts i, j (a_i, a_j), to the final subscript K (a_K) – highest breaking strain. If the bundle strain lies in the interval $\varepsilon \in (a_{i-1}, a_i)$, then the fibers from subscript 1 to $i-1$ are broken and the force in the bundle is taken by fibers with subscript $j \geq i$ – see Figure 9.5.

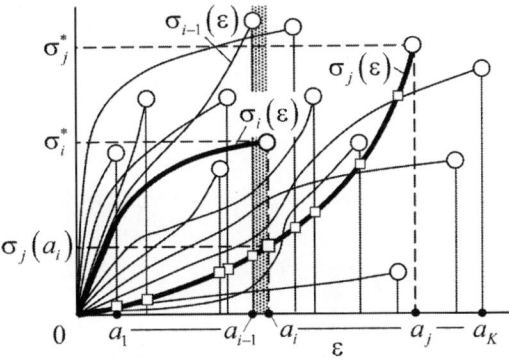

Figure 9.5 Scheme of stress–strain curves of fibers from individual components

Let us assume that we know

- Tenacities σ_i^* (○) of all fibers.

- Specific stress (□) of each j-th fiber; the stress–strain curve $\sigma_j(\varepsilon)$ is related to the breaking stress of all 'previous' fiber types, $i < j$ – see Figure 9.5.

- All mass fractions g_i ($\sum_{i=1}^{K} g_i = 1$).

Then, as stated in the book [6], the tenacity of the parallel fiber bundle consisting of $K \geq 2$ fiber components is

$$\sigma_Y^* = \max \left\{ \sum_{j=i}^{K} g_j \sigma_j(a_i) \right\}_{i=1}^{K}. \tag{9.12}$$

---

4   Here, we use subscripts 1, 2,... for fiber quantities in place of earlier used subscript f and/or 1 and 2.

The minimum bundle tenacity can be smaller than the smallest tenacity of the fibers also in this case. The corresponding 'worst' mass fraction can be calculated according to our special, rather complicated algorithm derived and published in the book [6].

9.3   Notes to parallel fiber bundles – variable fiber properties

The mathematical modeling of tensile behaviour of non-twisted parallel fiber bundle consisting of fibers with variable properties is fully derived in Chapter 6 of our book [6]. Here, we wish to remind you about the principles used and results obtained there.

Tension of one fiber

We often use mono-component fiber bundle prepared from one type of fibers with variable properties. The following properties are important in this regard

- actual fiber crimp and its variability,
- fiber tenacity and its variability and
- fiber breaking strain and its variability.

Figure 9.6 illustrates (from left to right) the straining of a (general) fiber. The distance d of separation between the jaws is equal to the gauge length h at the initial position. In general, the starting fiber length is greater than the gauge length, i.e., $l_0 \geq h$. The following parameter

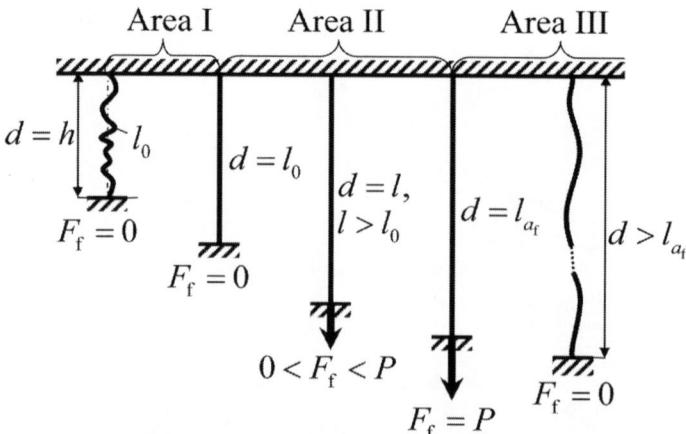

Figure 9.6 Straining of fiber between jaws

$$\lambda = \frac{l_0 - h}{h} = \frac{l_0}{h} - 1 \tag{9.13}$$

characterizes the waviness (so-called crimp) of fiber.

The distance d of separation between the jaws and the gauge length h determine the jaw displacement in the following manner:

$$\varepsilon_Y = \frac{d - h}{h} = \frac{d}{h} - 1. \tag{9.14}$$

It characterizes the movement of jaws and – because a fiber is a part of fiber bundle – it is also the bundle strain. The tensile force in fiber $F_f = 0$ at all distances between the jaws $d \in (0, l_0)$, i.e., for all displacements of the jaws $\varepsilon_Y \in \left(0, \frac{l_0 - h}{h}\right)$. (The fiber is only straightened in Area I.)

Because of further straining, the actual fiber length l is increasing more and more than the initial fiber length l_0 so that the fiber strain

$$\varepsilon_f = \frac{l - l_0}{l_0} = \frac{l}{l_0} - 1 \tag{9.15}$$

increases from zero to its maximum value $\varepsilon_f = a_f$ – fiber breaking strain. (Fiber breaking length is l_{a_f}.) Then, Equation (9.15) takes the following form:

$$a_f = \frac{l_{a_f} - l_0}{l_0} = \frac{l_{a_f}}{l_0} - 1. \tag{9.16}$$

The tensile force in fiber F_f increases from zero to the breaking force $F_f = P$ in Area II. Finally, the fiber breaks due to more straining so that the fiber tensile force takes a value equal to zero[5]. Summarily, a fiber takes a positive value of tensile force only in Area II, shown in Figure 9.6. The expression $d = l_0$ determines the lower limit and the expression $d = l_{a_f}$ determines the upper limit of Area II. After obtaining an expression for d from Equation (9.14) and for l_0 from Equation (9.13), the lower limit of jaw displacement is obtained as follows:

$$h(1 + \varepsilon) = h(1 + \lambda), \quad \varepsilon_Y = \lambda. \tag{9.17}$$

---

5   This assumption is used in Section 6.4 of our book [6].

By substituting the expression $d = l_{a_f}$ into Equation (9.14) and then using Equations (9.13) and (9.16), we obtain the following condition for the upper limit of jaw displacement:

$$\varepsilon_Y = \frac{l_{a_f}}{h} - 1 = \frac{l_{a_f}}{l_0}\frac{l_0}{h} - 1 = (1 + a_f)(1 + \lambda) - 1 = a_f\lambda + a_f + \lambda. \qquad (9.18)$$

If the fiber length is same 'inside' Area II, then the distance between the jaws is equal to fiber length, i.e., $l = d$. By substituting this expression into Equation (9.14) and then by using Equations (9.13) and (9.15) we obtain the expression for fiber strain in area II as follows

$$\varepsilon_Y = \frac{l}{h} - 1 = \frac{l}{l_0}\frac{l_0}{h} - 1 = (1 + \varepsilon_f)(1 + \lambda) - 1 = \varepsilon_f\lambda + \varepsilon_f + \lambda, \qquad \varepsilon_f = \frac{\varepsilon_Y - \lambda}{1 + \lambda}.$$
$$(9.19)$$

The tensile forces F_f acting on the single fibers can be expressed as an increasing function of fiber strains ε_f, i.e., $F_f = F_f(\varepsilon_f)$. The argument ε_f can be expressed according to Equation (9.19). Thus,

$$F_f = F_f(\varepsilon_f) = F_f\left(\frac{\varepsilon_Y - \lambda}{1 + \lambda}\right). \qquad (9.20)$$

So, the tensile force F_f in a fiber (if it lies in Area II) is an individual function of F_f at a common jaw displacement ε and individual fiber crimp λ.

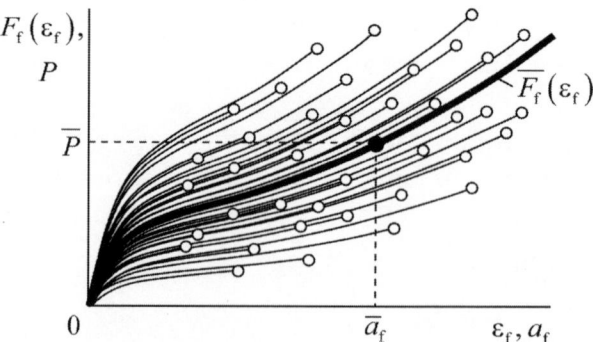

Figure 9.7 Force–strain curves of fibers: —: curves of single fibers; O: breaking points of single fibers; ▬: average function; ●: middle point of break (mean fiber breaking force $\overline{P}$, mean fiber breaking strain $\overline{a}_f$)

The scheme shown in Figure 9.7 illustrates force–strain curves of fibers. Usually, the shapes of force–strain curves of fibers are very similar to each other. We then think of an average function $\overline{F_f}(\varepsilon_f)$ such that each single force–strain curve (before fiber break) is a multiple of this one. We locate the average function $\overline{F_f}(\varepsilon_f)$ in such a manner that it is passing through the 'middle point' of fiber breakage, i.e., through mean fiber breaking force $\overline{P}$ and mean fiber breaking strain $\overline{a_f}$ – see the thick line in Figure 9.7.

The introduced 'theorem of similarity' says that the force–strain function of each fiber (before breakage) is proportional to the average force–strain function. Then, $F_f = F_f(\varepsilon_f) = K\,\overline{F_f}(\varepsilon_f)$, where K is a suitable constant of proportionality for a given single fiber. If $F_f = P$ (fiber breaking force) then $\varepsilon_f = a_f$ (fiber breaking strain), and the expressions $P = K\,\overline{F_f}(a_f)$, $K = P/\overline{F_f}(a_f)$ are valid. By applying the last expression in the first equation, we obtain the relation as follows:

$$F_f = \frac{P}{\overline{F_f}(a_f)}\,\overline{F_f}(\varepsilon_f) = \frac{P}{\overline{F_f}(a_f)}\,\overline{F_f}\!\left(\frac{\varepsilon_Y - \lambda}{1+\lambda}\right). \tag{9.21}$$

[Equation (9.19) was used here.] In this case, the breaking force P and breaking strain a_f are only two individual scalar characteristics of single fiber and $\overline{F_f}(\varepsilon_f)$ is a common function for all fibers[6].

Variation in fibers

According to the 'theorem of similarity' – Equation (9.21), the individual fiber is described by a triplet of scalar values: λ, P, a_f [and one common function $\overline{F_f}(\varepsilon_f)$, of course]. Let us think that the distribution of fiber crimp λ in a large fiber bundle is independent and described by the probability density function $v(\lambda)$. The distribution of pairs of P (fiber breaking force) and a_f (fiber breaking strain) in this large fiber bundle is described by the conjugate probability density function $u(P, a_f)$.

---

6   If we know the average function $\overline{F_f}(\varepsilon_f)$, then only two scalar quantities P and a_f determine the complete force–strain curve of any fiber.

Mean force per fiber in bundle

The mean force per fiber in the bundle is generally described by the following expression:

$$\bar{F} = \iiint\limits_{\substack{P\in(0,\infty)\\ a_f\in(0,\infty)\\ \lambda\in(0,\infty)}} F_f\, u\left(P,a_f\right) v(\lambda)\,\mathrm{d}a_f\,\mathrm{d}P\,\mathrm{d}\lambda . \tag{9.22}$$

(The total force in a bundle is the product of $\bar{F}$ and the number of fibers present in the cross section of the bundle.)

Nevertheless, the force F_f acting on a general fiber in a fiber bundle is greater than zero only when it is lying in Area II, shown in Figure 9.6. In all other cases, the fiber force is equal to zero. Therefore, Equation (9.22) was rearranged in the book [6] by accepting the borderlines of Area II according to Equations (9.17) and (9.18). So we obtained

$$\bar{F} = \int\limits_{0}^{\varepsilon_Y}\left\{\int\limits_{\frac{\varepsilon_Y-\lambda}{1+\lambda}}^{\infty}\left[\int\limits_{0}^{\infty}P u\left(a_f,P\right)\mathrm{d}P\right]\frac{1}{\overline{F_f}\left(a_f\right)}\,\mathrm{d}a_f\right\}\overline{F_f}\left(\frac{\varepsilon_Y-\lambda}{1+\lambda}\right)v(\lambda)\,\mathrm{d}\lambda . \tag{9.23}$$

This is a general[7] expression for calculation of force in a fiber bundle.

Special case 1 – fibers with variable crimp only

By using Equation (9.23), the following expressions were derived in the book [6] for this case

$$\left.\begin{array}{l}
\text{if } \varepsilon_Y \le a_f, \text{ then } \bar{F} = \displaystyle\int\limits_{0}^{\varepsilon_Y} \overline{F_f}\left(\frac{\varepsilon_Y-\lambda}{1+\lambda}\right)v(\lambda)\,\mathrm{d}\lambda, \\[3mm]
\text{if } \varepsilon_Y > a_f, \text{ then } \bar{F} = \displaystyle\int\limits_{\frac{\varepsilon_Y-a_f}{1+a_f}}^{\varepsilon_Y} \overline{F_f}\left(\frac{\varepsilon_Y-\lambda}{1+\lambda}\right)v(\lambda)\,\mathrm{d}\lambda.
\end{array}\right\} \tag{9.24}$$

The following (theoretical) example was considered under the following conditions:

---

7   Only the ideas of large fiber bundle and the 'theorem of similarity', according to Equation (9.21), are used.

1. $v(\lambda)$ refers to the lognormal probability density function of fiber crimp; mean value is denoted by $\overline{\lambda}$ and coefficient of variation is indicated by CV_{λ}.

2. Common function $\overline{F_f}(\varepsilon)$ is linear.

Figure 9.8 illustrates the resulting curves obtained from Equation (9.24).

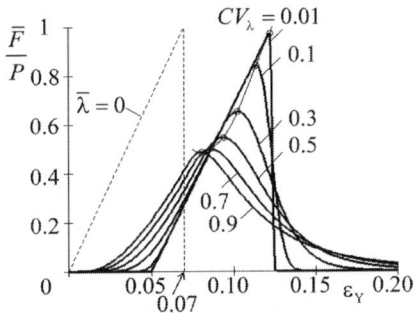

Figure 9.8 Relation between ε and $\overline{F}/P$

for different values of CV_{λ} (—); parameters used:

$a_f = 0.07\ (7\%),\ \overline{\lambda} = 0.05$; O: breaking points; ---: quite straight fibers

Note: The mean force per fiber in the bundle $\overline{F}$ is divided by (constant) fiber breaking force P and this ratio is plotted in Figure 9.8. The ratio $\overline{F}/P$ is de facto the coefficient of utilization of fiber tenacity in this case.

It is shown that

* The maximum mean force per fiber in the bundle ($\overline{F}$ at the breaking points O) is smaller than the fiber breaking force P in all cases. The higher is the coefficient of variation CV_{λ}, the smaller is the magnitude of this ratio.

* The breaking strains of the bundles (ε at breaking points O) are higher than the breaking strain a_f of fiber in each case. Nevertheless, the higher is the coefficient of variation CV_{λ} the smaller is the breaking strain of the bundle.

Special case 2 – Completely straight fibers

In this special case, the variability of fiber breaking force P and the variability of fiber breaking strain a_f are relevant only. We also used the assumption of 'symmetrical strength' stated in References [6, 9], i.e., we supposed that the breaking points of fibers (points ○ in Figure 9.7) are lying 'symmetrical-ly'[8] around the average function $\overline{F_f}(\varepsilon_f)$.

By using Equation (9.23), the following expression was derived in Reference [6] for this special case:

$$\overline{F} = \overline{F_f}(\varepsilon_Y)\left[1 - G(\varepsilon_Y)\right], \text{ where } G(\varepsilon_Y) = \int_0^{\varepsilon_Y} g(a_f) \, da_f . \qquad (9.25)$$

[The marginal probability density function of fiber breaking strain $g(a_f) = \int_0^\infty u(a_f, P) \, dP .]$

By using Equation (9.25), the following (theoretical) example was solved under the following conditions:

1. $g(a_f)$ refers to the lognormal probability density function of fiber breaking strain; the mean value is denoted by $\overline{a_f}$ and the coefficient of variation is indicated by CV_{a_f}.

2.   Common function $\overline{F_f}(\varepsilon_f)$ is linear.

Figure 9.9 illustrates the resulting curves.

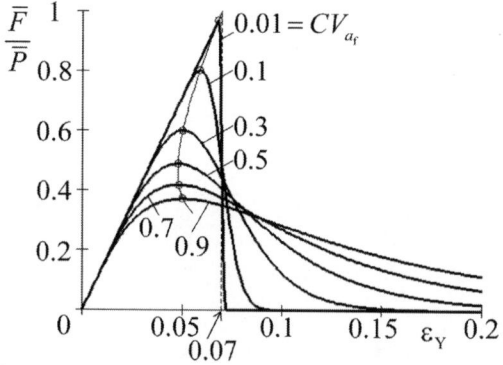

Figure 9.9 Relation between ε_Y and $\overline{F}/\overline{P}$ for different values of CV_{a_f}

(—) by $\overline{a}_f = 0.07$ (7%) ; ○: breaking points;

---: bundle consisting of identical fibers ($CV_{a_f} = 0$)

---

8   More precisely, the conditional mean value of fiber breaking forces P at $\varepsilon_Y = \varepsilon_f = a_f$ is lying just on the average curve $\overline{F_f}$. See Reference [6] for a detailed discussion.

Note: The mean force per fiber in the bundle $\overline{F}$ is divided by the mean fiber breaking force $\overline{P}$ and this ratio is plotted in Figure 9.9. The ratio $\overline{F}/\overline{P}$ is de facto the coefficient of utilization of fiber tenacity in this case.

It is shown that

- The maximum mean forces per fiber in the bundle ($\overline{F}$ at breaking points $\bigcirc$) are smaller than the mean fiber breaking force $\overline{P}$ in all cases. The higher is the coefficient of variation CV_{a_f}, the smaller is the magnitude of this ratio.

- The breaking strains of the bundles (ε_Y at breaking points $\bigcirc$) are smaller than the mean fiber breaking strain $\overline{a}_f$ for all practical values of CV_{a_f}. (The smallest value lies near to the value $CV_{a_f} = 0.5$.)

Note: Also, in many other cases, the decrease of bundle tenacity and bundle breaking strain depend (theoretically) only on the coefficient of variation of fiber breaking strain CV_{a_f}. Therefore, the materials with high variability of fiber breaking strain can bring forth the risk of too small bundle tenacity.

Special case 3 – Most general case

By considering the variability of all three variables (λ, P, a_f), using the aforementioned assumption of 'symmetrical strength', and accepting linear common function $\overline{F}_f(\varepsilon_f)$, Equation (9.23) was rearranged in Reference [6] as follows:

$$\frac{\overline{F}}{\overline{P}} = \frac{1}{\overline{a}_f} \int_0^{\varepsilon_Y} \left[1 - G\left(\frac{\varepsilon_Y - \lambda}{1+\lambda}\right)\right] \frac{\varepsilon_Y - \lambda}{1+\lambda} v(\lambda) \mathrm{d}\lambda,$$

$$\text{where } G\left(\frac{\varepsilon_Y - \lambda}{1+\lambda}\right) = \int_0^{\frac{\varepsilon_Y - \lambda}{1+\lambda}} g(a_f) \mathrm{d}a_f. \tag{9.26}$$

We calculated three different curves, shown in Figure 9.10 according to Equation (9.26), assuming further that

1. $v(\lambda)$ refers to the lognormal probability density function of fiber crimp, where the mean value is denoted by $\overline{\lambda}$ and the coefficient of variation is indicated by CV_λ.

2. $g(a_f)$ is the lognormal probability density function of fiber breaking strain, where the mean value is indicated by $\overline{a}_f$, coefficient of variation is denoted by CV_{a_f}.

3. The common function $\overline{F_f}(\varepsilon_f)$ is linear.

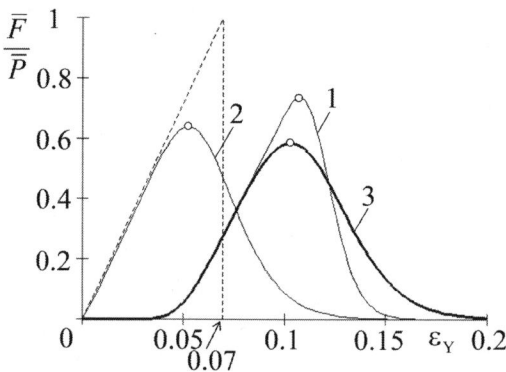

Figure 9.10 Relation between $\overline{F}/\overline{P}$ and ε

by $\overline{a}_f = 0.07$, $CV_\lambda = 0.2$. Curve 1: $CV_{a_f} = 0.01\,(\rightarrow 0)$, $\overline{\lambda} = 0.05$; curve 2: $CV_{a_f} = 0.25$, $\overline{\lambda} = 0.001\,(\rightarrow 0)$; curve 3: $CV_{a_f} = 0.25$, $\overline{\lambda} = 0.05$; O: breaking points; ---: bundle consisting of identical straight fibers

Curve 1 corresponds to Figure 9.8 ($CV_{a_f} \approx 0$), curve 2 corresponds to Figure 9.9 ($\overline{\lambda} \approx 0$) and curve 3 is the function according to Equation (9.26).

Concluding remarks to parallel fiber bundles

The last two sections report briefly on the logical bases of mathematical models and the corresponding results. Nevertheless, these themes were solved with all details in our earlier book [6]. Especially, for specialists, we recommend to study our earlier book [6].

Nevertheless, the regulations applying for different types of parallel fiber bundles are assumed to be more or less applicable to the non-parallel fiber structures, particularly in different types of twisted yarns. However, the new phenomenon of spatial fiber structure must be considered at the same time. In this regard, the tensile behaviour of twisted yarn needs to be modelled first.

9.4   Tensile behaviour of twisted yarns – general solution

Introduction

In Section 9.1, we commented on the two regions of specific stress–strain curve of yarn (Figure 9.1). In the first (white-colour) region, the original structure of the yarn remained relatively stable and practically all fibers were mechanically stressed, i.e., practically no fiber was broken.

In the following sections, the regulations that are valid in the 'white-colour region' will be discussed. As such the borderline of this region is not very clear; hence the validity of the derived results must be restricted intuitively based on different experience. The second (grey-colour) region is relatively mostly significant in case of staple yarns.

The problem of tensile behaviour of twisted yarns is usually solved by using two assumptions for simplification:

1. All fibers have same properties.
2. Yarn structure corresponds to the helical model described in Chapter 4.

(Geagauff [1] published the first version of such model.)

Fiber element

Figure 9.11 displays a general (thick) element of a helical fiber of length dl and angle β, which is lying at its initial radius r. (This is the position before tensile testing.) This is determined by an elementary (dark grey) cylindrical surface with dimensions $r\,d\varphi$ and $d\zeta$.

After straining of yarn, this element shifts to a new position at a smaller radius r' with a new angle β' and a new set of dimensions $r'\,d\varphi$ and $d\zeta'$ of a new elementary (light grey) cylindrical surface. (The elementary cylindrical surfaces[9] are shown once more – separately – on the right-hand side of Figure 9.11.)

---

9   The 'horizontal dimensions' of the elementary cylindrical surfaces, shown in Figure 9.11, are circular bows with a common elementary angle $d\varphi$ (measured in radians). Therefore, the corresponding lengths are $r\,d\varphi$ and $r'\,d\varphi$.

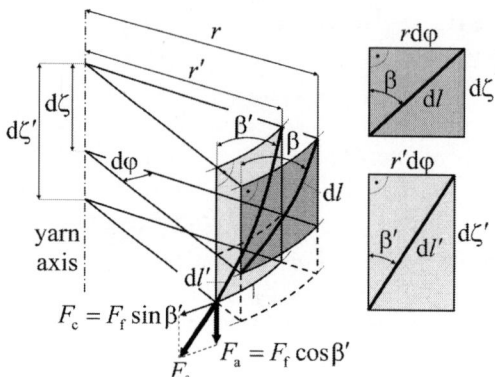

Figure 9.11 Coordinates of fiber element in a yarn before and after deformation

The following relations are then valid to write

$$\tan\beta = r\,d\phi/d\zeta,$$ (9.27)

$$\tan\beta' = r'\,d\phi/d\zeta'.$$ (9.28)

Relative characteristics of shifting

Let us introduce the following relative (dimensionless) quantities for characterization of the above-mentioned shifting of fiber element.

We assume that the axial strain ε_Y of each fiber element is same as the axial strain of the yarn. This is expressed as follows:

$$\varepsilon_Y = \frac{d\zeta' - d\zeta}{d\zeta} = \frac{d\zeta'}{d\zeta} - 1, \quad \text{i.e.,} \quad d\zeta' = (1 + \varepsilon_Y)d\zeta.$$ (9.29)

The radial strain ε_r of fiber element takes the following form:

$$\varepsilon_r = \frac{r' - r}{r} = \frac{r'}{r} - 1, \quad \text{i.e.,} \quad r' = (1 + \varepsilon_r)r.$$ (9.30)

(The radius r' in almost all cases is less than the radius r, hence $\varepsilon_r < 0$.)

The contraction ratio η_r, often known as Poisson's ratio, is defined by

$$\eta_r = -\varepsilon_r/\varepsilon_Y.$$ (9.31)

Here, $\eta_r \geq 0$ because $\varepsilon_r \leq 0$. We assume that the values of radial strains ε_r as well as contraction ratios η_r are same for fiber elements lying at the same radius but can be generally different for fibers lying at different radii; they are functions of radius r.

Note: It is well known from continuum mechanics that $\eta_r \in \langle 0, 0.5 \rangle$ for small deformations. (Contraction ratio 0.5 means the constant volume of strained body.) In another cases, contraction ratios can obtain other values (e.g., higher than 0.5).

At last, the fiber strain ε_f is defined as follows:

$$\varepsilon_f = \frac{dl' - dl}{dl} = \frac{dl'}{dl} - 1, \quad \text{i.e.,} \quad dl' = (1 + \varepsilon_f) dl . \tag{9.32}$$

Geometrical expression of fiber strain

Based on the Pythagorean theorem, the following relations can be written with a view to the elementary cylindrical surfaces shown in Figure 9.11:

$$d^2l = d^2\zeta + (r\, d\varphi)^2 , \tag{9.33}$$

$$d^2l' = d^2\zeta' + (r'\, d\varphi)^2 . \tag{9.34}$$

Let us rearrange Equation (9.34) by using Equations (9.29) to (9.32) as follows:

$$\left[(1 + \varepsilon_f) dl \right]^2 = \left[(1 + \varepsilon_Y) d\zeta \right]^2 + \left[(1 + \varepsilon_r) r' d\varphi \right]^2 ,$$

$$(1 + \varepsilon_f)^2 d^2l = (1 + \varepsilon_Y)^2 d^2\zeta + (1 + \varepsilon_r)^2 (r\, d\varphi)^2$$

$$= (1 + \varepsilon_Y)^2 d^2\zeta + (1 - \eta_r \varepsilon_Y)^2 (r\, d\varphi)^2 ,$$

$$(1 + \varepsilon_f)^2 = \frac{(1 + \varepsilon_Y)^2 d^2\zeta + (1 - \eta_r \varepsilon_Y)^2 (r\, d\varphi)^2}{d^2l} . \tag{9.35}$$

By applying Equation (9.27) in (9.35), we obtain

$$(1 + \varepsilon_f)^2 = \frac{(1 + \varepsilon_Y)^2 d^2\zeta + (1 - \eta_r \varepsilon_Y)^2 (r d\varphi)^2}{d^2\zeta + (r\, d\varphi)^2}$$

$$= \frac{(1 + \varepsilon_Y)^2 + (1 - \eta_r \varepsilon_Y)^2 (r d\varphi/d\zeta)^2}{1 + (r d\varphi/d\zeta)^2}$$

$$
= \frac{\left(1+\varepsilon_Y\right)^2 +\left(1-\eta_r\varepsilon_Y\right)^2 \tan^2\beta}{1+\tan^2\beta}
$$

$$
= \frac{1+2\varepsilon_Y +\varepsilon_Y^2 + \tan^2\beta - 2\eta_r\varepsilon_Y \tan^2\beta + \eta_r^2\varepsilon_Y^2 \tan^2\beta}{1+\tan^2\beta}
$$

$$
=1+\varepsilon_Y \frac{2 - 2\eta_r \tan^2\beta}{1/\cos^2\beta} + \varepsilon_Y^2 \frac{1+\eta_r^2 \tan^2\beta}{1/\cos^2\beta},
$$

$$
\left(1+\varepsilon_f\right)^2 =1+2\varepsilon_Y\left(\cos^2\beta - \eta_r \sin^2\beta\right) + \varepsilon_Y^2\left(\cos^2\beta + \eta_r^2 \sin^2\beta\right), \qquad (9.36)
$$

$$
\varepsilon_f = \sqrt{1+2\varepsilon_Y\left(\cos^2\beta - \eta_r \sin^2\beta\right) + \varepsilon_Y^2\left(\cos^2\beta + \eta_r^2 \sin^2\beta\right)} - 1. \qquad (9.37)
$$

Note: The highest fiber strain is obtained when $\beta = 0$ (central fiber); then $\varepsilon_f = \varepsilon_Y$ – see Equation (9.37). Moreover, $\eta_r <1$ and if $\beta>0$, then $\cos^2\beta - \eta_r \sin^2\beta <1$, $\cos^2\beta + \eta_r^2 \sin^2\beta <1$, and so $\varepsilon_f < \varepsilon_Y$ according to Equation (9.37). The oblique fiber elements have their strain values smaller than the axial strain values of yarn. The higher is the angle β, the smaller is the fiber strain.

Relation between angles

We can use Equations (9.29) to (9.32) and Equation (9.27) in (9.28) to obtain the following relationship between the angles β' and β:

$$
\tan\beta' = \frac{\left[(1+\varepsilon_r)r\right]d\varphi}{\left[(1+\varepsilon_Y)d\zeta\right]} = \frac{1+\varepsilon_r}{1+\varepsilon_Y}\tan\beta = \frac{1-\eta_r\varepsilon_Y}{1+\varepsilon_Y}\tan\beta. \qquad (9.38)
$$

Nevertheless, the relationship expressed in Equation (4.1), i.e., $\tan\beta = 2\pi rZ$, is valid for (initial) angle β in the case of helical model, where Z signifies yarn twist. So we can write

$$
\tan\beta = 2\pi rZ, \quad \tan\beta' = \frac{1-\eta_r\varepsilon_Y}{1+\varepsilon_Y} 2\pi rZ . \qquad (9.39)
$$

$$
\left.
\begin{array}{l}
\cos\beta = \dfrac{1}{\sqrt{1+\tan^2\beta}} = \dfrac{1}{\sqrt{1+(2\pi rZ)^2}}, \quad \left(\sin\beta = \dfrac{2\pi rZ}{\sqrt{1+(2\pi rZ)^2}}\right), \\[4mm]
\cos\beta' = \dfrac{1}{\sqrt{1+\tan^2\beta'}} = \dfrac{1}{\sqrt{1+\dfrac{(1-\eta_r\varepsilon_Y)^2}{(1+\varepsilon_Y)^2}(2\pi rZ)^2}}.
\end{array}
\right\} \qquad (9.40)
$$

[The first equivalency in Equation (9.40) is identical with Equation (4.2).] Thus,

$$\frac{\cos\beta'}{\cos\beta} = \frac{\sqrt{1+\tan^2\beta}}{\sqrt{1+\dfrac{\left(1-\eta_r\varepsilon_Y\right)^2}{\left(1+\varepsilon_Y\right)^2}\tan^2\beta}} = \frac{\sqrt{1+\left(2\pi rZ\right)^2}}{\sqrt{1+\dfrac{\left(1-\eta_r\varepsilon_Y\right)^2}{\left(1+\varepsilon_Y\right)^2}\left(2\pi rZ\right)^2}} . \quad (9.41)$$

Fiber force and stress

The tensile force F_f in fiber is an increasing function of fiber strain ε_f. This is mathematically expressed as follows:

$$F_f = F_f\left(\varepsilon_f\right) . \quad (9.42)$$

Nevertheless, we characterize the stress–strain relation of fiber more frequently. We denote the specific engineering tensile stress[10] in fiber as follows:

$$\sigma_f = \sigma_f\left(\varepsilon_f\right) . \quad (9.43)$$

The following expression is valid to write from Equation (1.21):

$$\sigma_f\left(\varepsilon_f\right) = \frac{F_f\left(\varepsilon_f\right)}{t} = \frac{F_f\left(\varepsilon_f\right)}{s\rho} , \quad (9.44)$$

where t is fiber fineness, s is cross-sectional area of fiber, ρ is fiber density, and ε_f is fiber strain that can be determined from Equation (9.37).

Component of force

The force F_f can be resolved into two perpendicular components, as shown in Figure 9.11. The component F_a (in the 'axial direction' of yarn) is parallel to yarn axis and the component F_c (in 'cross direction' of yarn) is perpendicular to yarn axis. So,

$$F_a = F_f\left(\varepsilon_f\right)\cos\beta' . \quad (9.45)$$

---

10   Let us remind that the tensile stress is the axial force in fiber per unit area of fiber cross section and the specific stress is the above-mentioned stress divided by fiber density – see the text written before Equation (1.21). The specific stress can also be expressed as a ratio of tensile force to fiber fineness (e.g., N/tex), as shown in Equation (1.21).

The force F_a contributes to the total axial force in yarn. (The cross-directional force $F_c = F_f(\varepsilon_f)\sin\beta'$ – together with all other fibers – creates a torsional moment in yarn.)

Differential layer

We introduced the image of a so-called differential layer of (initial) yarn (Figure 4.3) in Section 4.1 and found that the number dn of fibers present in the cross section of the differential layer, according to Equation (4.5), is $dn = 2\pi r\, dr\, \mu_r/(s/\cos\beta)$. After straining of yarn, the differential layer changes its position from initial radius r to a new (smaller) radius r'. Such a differential layer participates in determining the total axial force in yarn with a contribution amounting to

$$dF_Y = F_a\, dn.\tag{9.46}$$

The total axial force F_Y in yarn is then expressed as an integral function over all differential layers, i.e., form radius $r = 0$ to radius $r = D/2$, where D stands for yarn diameter.

Note: All fibers lie inside a yarn cylinder with diameter D in the case of helical model.

Force in yarn

By using step by step Equations (9.46), (9.45), (4.5), (9.44) and applying Equations (9.41) and (9.40), we can express the total axial force F_Y in yarn as follows:

$$F_Y = F_Y(\varepsilon_Y) = \int_{r=0}^{r=D/2} dF_Y = \int_{r=0}^{r=D/2} F_a\, dn = \int_0^{D/2} F_f(\varepsilon_f)\cos\beta'\, \frac{2\pi r\, dr\, \mu_r}{s/\cos\beta}$$

$$= \int_0^{D/2} s\rho\, \sigma_f(\varepsilon_f)\cos\beta'\, \frac{2\pi r\, dr\, \mu_r}{s/\cos\beta} = 2\pi\rho \int_0^{D/2} \sigma_f(\varepsilon_f)\cos\beta'\cos\beta\,\mu_r r\, dr,$$

$$F_Y(\varepsilon_Y) = 2\pi\rho \int_0^{D/2} \sigma_f(\varepsilon_f)\left(\frac{\cos\beta'}{\cos\beta}\right)\cos^2\beta\,\mu_r r\, dr,\tag{9.47}$$

where ε_f can be determined from Equation (9.37). [The angle β and the ratio $\cos\beta'/\cos\beta$ are given by Equations (9.39), (9.40) and (9.41).]

Note: The force F_Y is a function of ε_Y, because this variable is present in Equation (9.37) that determines ε_f as a function of ε_Y.

Specific stress–strain relation in yarn

In the case of helical model, the substance cross-sectional area of yarn[11], according to Equation (4.6), is $S = 2\pi\int_0^{D/2} \mu_r r\, dr$. So, the (engineering) stress of yarn is $\sigma'_Y = F_Y/S$ and the (engineering) specific stress is then $\sigma_Y = \sigma'_Y/\rho = F_Y/(S\rho)$[12]. By applying Equations (9.47) and (4.6), we can write the expression as follows:

$$\sigma_Y\left(\varepsilon_Y\right) = \frac{F_Y\left(\varepsilon_Y\right)}{S\rho} = \frac{2\pi\rho\int_0^{D/2} \sigma_f\left(\varepsilon_f\right)\left(\dfrac{\cos\beta'}{\cos\beta}\right)\cos^2\beta\,\mu_r r\, dr}{2\pi\rho\int_0^{D/2} \mu_r r\, dr},$$

$$\sigma_Y\left(\varepsilon_Y\right) = \frac{\int_0^{D/2} \sigma_f\left(\varepsilon_f\right)\left(\dfrac{\cos\beta'}{\cos\beta}\right)\cos^2\beta\,\mu_r r\, dr}{\int_0^{D/2} \mu_r r\, dr}.$$

(9.48)

Note: It is always valid that σ_Y is a function of ε_Y, because this variable is present in Equation (9.37) that determines ε_f as a function of ε_Y.

Non-twisted bundle

The angle $\beta = 0$ for all radii in a non-twisted bundle of fibers. Then, according to Equation (9.40), $\cos\beta = 1$ and $\sin\beta = 0$; according to Equation (9.41), $\cos\beta'/\cos\beta = 1$; and according to Equation (9.37), $\varepsilon_f = \varepsilon_Y$. Then, Equation (9.48) can be expressed in the following form:

---

11 Let us remind that the substance cross-sectional area of yarn is the sum of sectional areas of all fibers present in the cross section of the yarn – see Figure 1.6 and Equation (1.32).

12 Let us remind that the value $S\rho$ is equal to yarn fineness T as shown in Equation (1.33).

$$
\sigma_Y\left(\varepsilon_Y\right) = \frac{\displaystyle\int_0^{D/2} \sigma_f\left(\varepsilon_f\right)\mu_r r\, dr}{\displaystyle\int_0^{D/2} \mu_r r\, dr} = \frac{\sigma_f\left(\varepsilon_f\right)\displaystyle\int_0^{D/2} \mu_r r\, dr}{\displaystyle\int_0^{D/2} \mu_r r\, dr} = \sigma_f\left(\varepsilon_f\right) = \sigma_f\left(\varepsilon_Y\right). \quad (9.49)
$$

Note: We place the expression $\sigma_f\left(\varepsilon_f\right)$ before the integral, because we assumed that all fibers have same properties – it also means that the fibers are present in different radii. (The same result was presented in the paragraph entitled 'Trivial case' in Section 9.2.)

Coefficient of utilization of specific fiber stress

The coefficient of utilization of specific fiber stress in yarn is defined by Equation (9.7). By applying Equation (9.48) in (9.7), we find

$$
\varphi\left(\varepsilon_Y\right) = \frac{\sigma_Y\left(\varepsilon_Y\right)}{\sigma_f\left(\varepsilon_Y\right)} = \frac{\displaystyle\int_0^{D/2} \sigma_f\left(\varepsilon_f\right)\left(\frac{\cos\beta'}{\cos\beta}\right)\cos^2\beta\,\mu_r r\, dr}{\sigma_f\left(\varepsilon_Y\right)\displaystyle\int_0^{D/2} \mu_r r\, dr}. \quad (9.50)
$$

Note: The coefficient of utilization of specific fiber stress is equal to 1 in case of a non-twisted fiber bundle, because Equation (9.49) is valid in this special case.

Calculations and their practical problems

The total axial force $F_Y\left(\varepsilon_Y\right)$ in yarn, the specific (engineering) stress $\sigma_Y\left(\varepsilon_Y\right)$ in yarn and the coefficient of utilization of specific fiber stress $\varphi\left(\varepsilon_Y\right)$ in yarn can be practically calculated as the functions of axial strain of yarn ε_Y, if we know:

(a) Specific engineering stress–strain function of fiber $\sigma_f\left(\varepsilon_f\right)$.

(b) Radial function of packing density μ_r and radial function of contraction ratio η_r.

(c) Parameters – yarn twist Z, fiber density ρ and yarn diameter D.

At each selected values of axial strain of yarn ε_Y and radius r, we can apply the following algorithm for calculation of the integral function stated in Equations (9.47), (9.48) and/or (9.50):

1. Input – a value of ε_Y.

2. Input – a value of r ($r \in \langle 0, D/2 \rangle$).

3. Calculation of $\cos\beta$ and $\sin\beta$ from Equation (9.40).

4. Calculation of the ratio $\cos\beta'/\cos\beta$ from Equation (9.41).

5. Calculation of value ε_f from Equation (9.37).

6. Calculation of the integral function in Equations (9.47), (9.48), and/
    or (9.50) for the given input values.

By repeating this algorithm from point 2 to point 6 (for a same value of
ε_Y) for a set of $r \in \langle 0, D/2 \rangle$, we obtain an integral function as a function of
radius. Then, after (numerical) integration, we obtain the resulting integral
value which corresponds to the input value of ε_Y. Further, by repeating the
whole previous procedure for a set of values of ε_Y (point 1), we obtain the
above-mentioned integral as a function of yarn strain ε_Y. In this way, the tri-
plet of functions $F_Y(\varepsilon_Y)$, $\sigma_Y(\varepsilon_Y)$, $\varphi(\varepsilon_Y)$ can be obtained. [However, we
must calculate the integral $\int_0^{D/2} \mu_r r \, dr$ expressed in the denominator of Equations (9.48) and (9.50).]

The described procedure of calculation is possible in principle, but practically extraordinarily difficult. It is because we usually do not know the necessary input values of radial functions μ_r and η_r. (The experimental
determination of the third function, i.e., specific stress–strain function of fiber
$\sigma_f(\varepsilon_f)$, is well known and it can be obtained in a well-equipped fiber – and/
or textile laboratory.).

The radial function η_r could be possible to derive based on, e.g., experimental knowledge of (initial) radial function of packing density μ_r and
(final) radial function of packing density μ_r' in a yarn under tension – see
Appendix 8.

9.5   Tensile behaviour of twisted yarns – simplified solution

The complete solution of Equations (9.47), (9.48) and/or (9.50) is not usually
possible, because we usually do not have the necessary input information –
mainly the radial function of packing density μ_r and the radial function of
contraction ratio η_r. Nevertheless, we can make this problem easier by

means of a certain simplified assumptions. (Compare it also with the part 'Special cases' in Appendix 8.)

Small strains

At first, let us assume that the yarn strain ε_Y is small. Then, the corresponding strain ε_f of fiber at a (initial) radius r is even smaller than ε_Y – see the note under Equation (9.37). It is possible to express ε_f from Equation (9.37) as follows:

$$1 + \varepsilon_f = \sqrt{1 + 2\varepsilon_Y \left(\cos^2 \beta - \eta_r \sin^2 \beta \right) + \varepsilon_Y^2 \left(\cos^2 \beta + \eta_r^2 \sin^2 \beta \right)},$$

$$\left(1 + \varepsilon_f\right)^2 = 1 + 2\varepsilon_f + \overset{\to 0}{\varepsilon_f^2} = 1 + 2\varepsilon_Y \left(\cos^2 \beta - \eta_r \sin^2 \beta \right) + \overset{\to 0}{\varepsilon_Y^2} \overbrace{\left(\cos^2 \beta + \eta_r^2 \sin^2 \beta \right)}^{\text{smaller than 1}},$$

$$\varepsilon_f \approx \varepsilon_Y \left(\cos^2 \beta - \eta_r \sin^2 \beta \right). \tag{9.51}[13]$$

Note: Gegauff [1] considered $\eta_r = 0$ and obtained the well-known expression $\varepsilon_f = \varepsilon_Y \cos^2 \beta$.

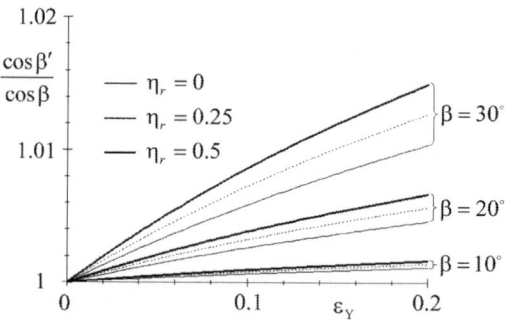

Figure 9.12 Ratio $\cos\beta'/\cos\beta$ as a function of ε_Y

Equation (9.41) determines the numerical value of the ratio $\cos\beta'/\cos\beta$ for different values of ε_Y. Figure (9.12) displays this relation in a typical twisted staple fiber yarn.

Note: The majority of fibers in a yarn has usually angle β smaller than $30°$, the contraction ratio of yarns is usually not too far away from 0.5, and the axial strain ε_Y of yarn is used to be smaller than 0.2 (20%). Then, the

---

13   A higher power of a small value is a much smaller quantity.

quantity $\cos^2\beta'/\cos^2\beta$ differs from the value 1 by no more than about 3% in such situation.

We observe that

$$\cos\beta'/\cos\beta \approx 1 \qquad (9.52)$$

can be evidently a good approximation of this ratio.

Independence of radius

In spite of the fact that the packing density μ_r and the contraction ratio η_r of differential layer generally depend on the radius, we assume (for simplification) that this quantities are independent of the radius r. Thus,

$$\mu_r = \mu \ldots \text{constant}, \qquad (9.53)^{14}$$

$$\eta_r = \eta \ldots \text{constant}. \qquad (9.54)$$

Simplified equations

By using Equations (9.52) and (9.53), we will formulate the following equations for the total axial force in yarn, specific engineering stress in yarn and coefficient of utilization of specific fiber stress in yarn.

The total axial force $F_Y(\varepsilon_Y)$ in yarn follows Equation (9.47). It is valid to write that

$$F_Y(\varepsilon_Y) = 2\pi\rho\mu \int_0^{D/2} \sigma_f(\varepsilon_f)\cos^2\beta\, r\, dr,$$

Substitution: $r = \dfrac{2\pi rZ}{2\pi Z} = \dfrac{\tan\beta}{2\pi Z} = \dfrac{D\tan\beta}{2\pi DZ} = \dfrac{D\tan\beta}{2\tan\beta_D}, \quad dr = \dfrac{D}{2\tan\beta_D}\dfrac{d\beta}{\cos^2\beta},$

$$r\, dr = \dfrac{D\tan\beta}{2\tan\beta_D} \cdot \dfrac{D}{2\tan\beta_D}\dfrac{d\beta}{\cos^2\beta} = \left(\dfrac{D}{2\tan\beta_D}\right)^2 \dfrac{\sin\beta}{\cos^3\beta}d\beta,$$

$$F_Y(\varepsilon_Y) = 2\pi\rho\mu \int_0^{\beta_D} \sigma_f(\varepsilon_f)\cos^2\beta\left(\dfrac{D}{2\tan\beta_D}\right)^2 \dfrac{\sin\beta}{\cos^3\beta}d\beta$$

$$= 2\pi\rho\mu\left(\dfrac{D}{2\tan\beta_D}\right)^2 \int_0^{\beta_D} \sigma_f(\varepsilon_f)\tan\beta\, d\beta. \qquad (9.55)$$

---

14 By using Equation (9.53), the helical model turns to the so-called ideal helical model, as determined by assumption 6 stated in Section 4.1 in Chapter 4.

[Equation (1.51), i.e., $\tan\beta_D = \pi DZ$, and Equation (4.1), i.e., $\tan\beta = 2\pi rZ$, were used for rearrangement.]

The specific engineering stress $\sigma_Y(\varepsilon_Y)$ in yarn follows Equation (9.48). It is valid to write that

$$
\sigma_Y(\varepsilon_Y) = \frac{\mu \int_0^{D/2} \sigma_f(\varepsilon_f)\cos^2\beta\, r\, dr}{\mu \int_0^{D/2} r\, dr}
$$

$$
= \frac{\left(\dfrac{D}{2\tan\beta_D}\right)^2 \int_0^{\beta_D} \sigma_f(\varepsilon_f)\tan\beta\, d\beta}{\left(\dfrac{D}{2\tan\beta_D}\right)^2 \int_0^{\beta_D} \dfrac{\sin\beta}{\cos^3\beta}\, d\beta} = \frac{\int_0^{\beta_D} \sigma_f(\varepsilon_f)\tan\beta\, d\beta}{\int_0^{\beta_D} \dfrac{\sin\beta}{\cos^3\beta}\, d\beta}. \quad (9.56)
$$

[Here, Equations from (9.51) to (9.54) and the substitution as shown in Equation (9.55) were used.] Let us solve the denominator of Equation (9.56) as follows:

$$
\int_0^{\beta_D} \frac{\sin\beta}{\cos^3\beta}\, d\beta = -\int_1^{\cos\beta_D} \frac{dt}{t^3} = \left[\frac{1}{2t^2}\right]_1^{\cos\beta_D} = \frac{1}{2}\left(\frac{1}{\cos^2\beta_D} - 1\right) = \frac{\tan^2\beta_D}{2}. \quad (9.57)
$$

Substitution: $\cos\beta = t,\ -\sin\beta\, d\beta = dt$.

By applying Equation (9.57) in (9.56), we finally obtain

$$
\sigma_Y(\varepsilon_Y) = \frac{\int_0^{\beta_D} \sigma_f(\varepsilon_f)\tan\beta\, d\beta}{\int_0^{\beta_D} \dfrac{\sin\beta}{\cos^3\beta}\, d\beta} = \frac{2}{\tan^2\beta_D} \int_0^{\beta_D} \sigma_f(\varepsilon_f)\tan\beta\, d\beta. \quad (9.58)
$$

The coefficient of utilization of specific fiber stress $\varphi(\varepsilon_Y)$ in yarn follows Equation (9.50). By using the previous equation, it is valid to write that

$$\varphi(\varepsilon_Y) = \frac{\sigma_Y(\varepsilon_Y)}{\sigma_f(\varepsilon_Y)} = \frac{\dfrac{2}{\tan^2 \beta_D} \displaystyle\int_0^{\beta_D} \sigma_f(\varepsilon_f) \tan \beta \, d\beta}{\sigma_f(\varepsilon_Y)}$$

$$= \frac{2}{\sigma_f(\varepsilon_Y)\tan^2 \beta_D} \int_0^{\beta_D} \sigma_f(\varepsilon_f) \tan \beta \, d\beta. \tag{9.59}$$

Note: The fiber strain, determining the fiber specific stress $\sigma_f(\varepsilon_f)$, should be calculated from Equation (9.51), i.e., $\varepsilon_f \approx \varepsilon_Y(\cos^2 \beta - \eta \sin^2 \beta)$ – in case of small deformation.

Simpliest equation like Gegauff's equation

Besides all the previous simplifications, Gegauff [1] further assumed that the specific tensile stress generated in a fiber is a linear function of fiber strain. This is stated as follows:

$$\sigma_f(\varepsilon_f) = K\varepsilon_f, \quad K = E/\rho \ldots \text{constant} . \tag{9.60}$$

Note: The constant K is the ratio of fiber (Young) modulus E to fiber density ρ.

By using Equations (9.51) and (9.60), the expression $\int_0^{\beta_D} \sigma_f(\varepsilon_f) \tan \beta \, d\beta$, present in Equations (9.55), (9.58) and (9.59), takes the following form:

$$\int_0^{\beta_D} \sigma_f(\varepsilon_f) \tan \beta \, d\beta = K \int_0^{\beta_D} \varepsilon_f \tan \beta \, d\beta = K\varepsilon_Y \int_0^{\beta_D} (\cos^2 \beta - \eta \sin^2 \beta) \tan \beta \, d\beta. \tag{9.61}$$

The last definite integral can be solved by using Equation (A9.5) in Appendix 9. So, it is valid

$$\int_0^{\beta_D} \sigma_f(\varepsilon_f) \tan \beta \, d\beta = K\varepsilon_Y \int_0^{\beta_D} (\cos^2 \beta - \eta \sin^2 \beta) \tan \beta \, d\beta$$

$$= K\varepsilon_Y \frac{\tan^2 \beta_D}{2} \left[(1+\eta)\cos^2 \beta_D + \eta \frac{\ln \cos^2 \beta_D}{\tan^2 \beta_D} \right]. \tag{9.62}$$

By using Equation (9.62) in (9.55), we obtain the total axial force $F_Y(\varepsilon_Y)$ in yarn in the following manner:

$$F_Y(\varepsilon_Y) = 2\pi\rho\mu\left(\frac{D}{2\tan\beta_D}\right)^2 \int_0^{\beta_D} \sigma_f(\varepsilon_f)\tan\beta\, d\beta$$

$$= 2\pi\rho\mu\left(\frac{D}{2\tan\beta_D}\right)^2 K\varepsilon_Y \frac{\tan^2\beta_D}{2}\left[(1+\eta)\cos^2\beta_D + \eta\frac{\ln\cos^2\beta_D}{\tan^2\beta_D}\right],$$

$$F_Y(\varepsilon_Y) = \pi\rho\mu\left(\frac{D}{2}\right)^2 K\varepsilon_Y\left[(1+\eta)\cos^2\beta_D + \eta\frac{\ln\cos^2\beta_D}{\tan^2\beta_D}\right]. \tag{9.63}$$

By using Equation (9.62) in (9.58), we similarly obtain the specific engineering stress $\sigma_Y(\varepsilon_Y)$ in yarn as follows:

$$\sigma_Y(\varepsilon_Y) = \frac{2}{\tan^2\beta_D}\int_0^{\beta_D}\sigma_f(\varepsilon_f)\tan\beta\, d\beta$$

$$= \frac{2}{\tan^2\beta_D} K\varepsilon_Y \frac{\tan^2\beta_D}{2}\left[(1+\eta)\cos^2\beta_D + \eta\frac{\ln\cos^2\beta_D}{\tan^2\beta_D}\right],$$

$$\sigma_Y(\varepsilon_Y) = K\varepsilon_Y\left[(1+\eta)\cos^2\beta_D + \eta\frac{\ln\cos^2\beta_D}{\tan^2\beta_D}\right]. \tag{9.64}$$

Finally, by using Equation (9.62) in (9.59) and $\sigma_f(\varepsilon_Y) = K\varepsilon_Y$ from Equation (9.60), we obtain the coefficient of utilization of specific fiber stress $\varphi(\varepsilon_Y)$ in yarn as follows:

$$\varphi(\varepsilon_Y) = \frac{2}{\sigma_f(\varepsilon_Y)\tan^2\beta_D}\int_0^{\beta_D}\sigma_f(\varepsilon_f)\tan\beta\, d\beta$$

$$= \frac{2}{K\varepsilon_Y\tan^2\beta_D} K\varepsilon_Y \frac{\tan^2\beta_D}{2}\left[(1+\eta)\cos^2\beta_D + \eta\frac{\ln\cos^2\beta_D}{\tan^2\beta_D}\right],$$

$$\varphi(\varepsilon_Y) = (1+\eta)\cos^2\beta_D + \eta\frac{\ln\cos^2\beta_D}{\tan^2\beta_D}. \tag{9.65}$$

Note: Equation (1.51), i.e., $\tan\beta_D = \pi DZ$ is always valid in all of these equations.

It follows from Equation (9.65) that the coefficient of utilization of specific fiber stress $\varphi(\varepsilon_Y)$ in yarn is independent of yarn strain ε_Y in this special (Gegauf's) case; this coefficient is a function of only angle β_D and contraction ratio η.

The graphical interpretation of Equation (9.65) is illustrated in Figure 9.13. It stands to reason that the coefficient of utilization of specific fiber stress in yarn is decreasing with increasing angle β_D (the angle of helical fiber present on yarn surface). In other words, a more twisted yarn is mechanically 'softer' that a less twisted yarn. A question can be raised now: how one can logically imagine such result?

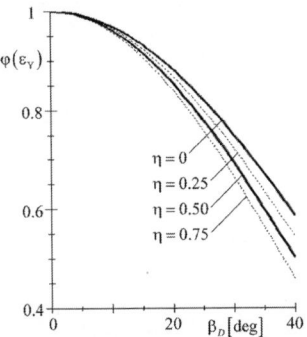

Figure 9.13 Gegauff's coefficient of utilization of specific fiber stress $\varphi(\varepsilon_Y)$ in yarn according to Equation (9.65)

Consider that there are two fibers lying vertically from yarn axis as shown in Figure 9.14. Let us imagine that the same vertical component F_a (parallel to yarn axis) acts on both of the fibers. The relatively smaller resultant force F_f acts on the direction of fiber axis, as shown in Figure 9.14a, because the angle β is small in this case. On the contrary, the analogical resultant force F_f, shown in Figure 9.14b, is evidently higher due to the reason that the angle β is higher for this fiber. Consequently, the component of vertical force generated during the process of applying tension to yarn causes higher resultant forces by the oblique fibers. [See also the note expressed in parentheses after Equation (9.45).]

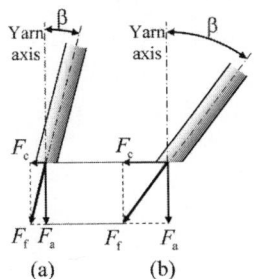

Figure 9.14 Resolution of the forces acting on the fibers

9.6　　　Tensile behaviour of generalized helical model – simplified solution

Idea of generalization

Up to now, all the previous equations were based on the image of helical and/ or ideal helical model. It means according to Equation (4.1) that all fibers have the same angle β at a given radius. Now, the 'classical' helical model will be partly generalized.

Let us imagine that all fibers at a radius r (in the differential layer at radius r) are permanently lying on the corresponding cylindrical surface[15] (alike helical model), but they do not have the same angle β; the fibers form different angles ϑ to the direction of yarn axis.

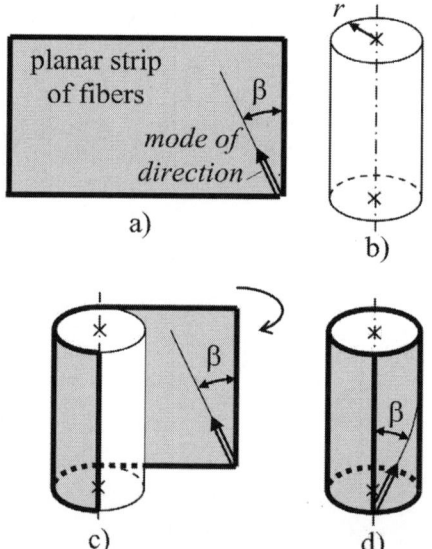

Figure 9.15 Image of originating of (differential) fiber layer in yarn

The origin of (differentially) thin fiber layer can be imagined according to the scheme shown in Figure 9.15. Let us have (a) (grey) planar strip a of fibers at the beginning. The fibers have different directions ϑ there, but the modal[16] direction follows the angle $\vartheta = \beta$. Further, let us have a cylinder

---

15　Quite generally, fiber paths are changed at different radii, e.g., in case of models of fiber migration discussed in Chapter 5. Nevertheless, we assume that this phenomenon is not significant here.

16　This is true in a statistical sense.

(b) around yarn axis at a general radius r. Now, let us roll the initial planar strip of the cylindrical surface as shown in Figure 15(c). As a result, we obtain a (differential) layer of yarn (d) where fibers have different angles ϑ and the modal angle $\vartheta = \beta$ follows the direction to the yarn axis.

Distribution of angle ϑ

The distribution of fiber directions for planar orientation of fibers was derived in Chapter 3 of our book [6]. It was found that the unimodal probability density function of non-oriented[17] angle $\vartheta \in (0, \pi/2)$ follows:

$$u(\vartheta) = \frac{1}{\pi} \frac{C}{C^2 - (C^2 - 1)\cos^2(\vartheta + \beta)} + \frac{1}{\pi} \frac{C}{C^2 - (C^2 - 1)\cos^2(\vartheta - \beta)}, \qquad (9.66)^{18}$$

$C \geq 1$...constant, a 'measure' of preference of modal direction.

The graphical interpretation of this probability density function is illustrated in Figure 9.16. As it is shown, the most frequent (modal) direction follows the angle β; the higher is the parameter C, the higher is the concentration of fiber directions near to the modal angle β. Let us now discuss the limit cases.

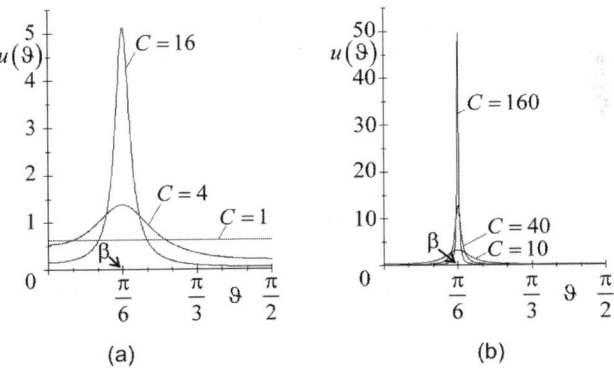

(a)   (b)

Figure 9.16 Graphical interpretation of non-oriented angle ϑ; example by
$\beta = \pi/6$ (30 deg) : (a) smaller values of C, (b) higher values of C

---

17   It means that we describe directions of fibers only by positive angles, independently if they are inclined 'right' or 'left' from the 'vertical' direction of yarn axis.

18   This is practically same as Equation (3.49) stated in the book [6]. The only difference is that in the place of earlier preferential angle α now the modal angle β is used.

1. If $C = 1$, then the probability density function $\lim_{C \to 1} u(\vartheta) = 2/\pi = \text{constant}$. This borderline case is the case of isotropic orientation – see the corresponding curve in Figure 9.16a.

2. If $C \to \infty$, i.e., C is unlimitedly higher, then the fiber directions are more concentrated in a smaller vicinity around the modal direction given by angle β – see the curves in Figure 9.16b. Finally, all the fibers follow the same angle $\vartheta = \beta$ as a limit case of such process. The probability density function $u(\vartheta)$ is then limited to the so-called Dirac delta-function $\delta(\vartheta - \beta)$. It is valid that $\delta(\vartheta - \beta) = \infty$ if $\vartheta = \beta$, $\delta(\vartheta - \beta) = 0$ for $\vartheta \neq \beta$, and $\int_{-\infty}^{\infty} \delta(\vartheta - \beta) d\vartheta = 1$. Finally, for each general function $f(\vartheta)$, the following expression is valid

$$\int_{-\infty}^{\infty} f(\vartheta) \delta(\vartheta - \beta) d\vartheta = f(\beta). \tag{9.67}$$

In our case, the probability density function $u(\vartheta)$, according to Equation (9.66), represents also the probability density function of fiber inclination in differential layers of yarn.

The distribution of fiber directions in yarn cross section differs from the distribution of fiber directions in the whole planar strip of fibers. The expression for the probability density function of such angles ϑ is derived in our book [6] as follows:

$$u^*(\vartheta) = \frac{\cos \vartheta \, u(\vartheta)}{k_n}, \tag{9.68}$$

where

$$k_n = \int_0^{\pi/2} \cos \vartheta \, u(\vartheta) \, d\vartheta. \tag{9.69}[19]$$

Now, $u^*(\vartheta)$ represents also the probability density function of fiber inclination in the cross section of differential layers of yarn.

---

19 Equations (9.68) and (9.69) are practically same as Equations (3.20) and (3.18) in the book [6].

Number of fibers in sections of differential layer

A section of very thin planar strip, whose thickness is δh and length is 1, is shown in Figure 9.17[20]. The number of sectioned fibers (hatched fiber 'islands') per unit sectional length was derived in the book [6] as follows:

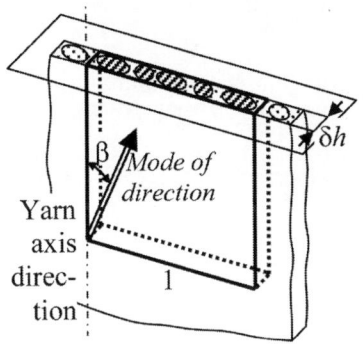

Figure 9.17 Section of a planar fiber assembly

$$\upsilon = \delta h \frac{\mu \rho}{t} k_n ,$$
(9.70)[21]

where t is fiber fineness, ρ is fiber density, and μ is packing density of fibers in the above-mentioned planar strip. [Fiber cross-sectional area, according to Equation (1.5), is $s = t/\rho$.] Further, if the thickness of the above-mentioned planar strip is elementary, $\delta h \to dh$, then the number of fibers per unit sectional length is also elementary, $\upsilon \to d\upsilon$, and we can write from Equation (9.70) that

$$d\upsilon = \frac{dh\, \mu\, k_n}{s} .$$
(9.71)

After rolling of the initial planar strip of cylindrical surface – Figure 9.15 – the elementary thickness dh is changed to the elementary increment dr of radius. Moreover, the length of the differential layer at radius r is $2\pi r$, so that the number dn of sectioned fibers in the afore-mentioned differential layer of yarn cross section is

---

20   This figure is very similar to Figure 3.18 in the book [6].

21   This expression is identical to Equation (3.57) in the book [6].

$$dn = d\upsilon\, 2\pi r = dr\, \mu \frac{k_n}{s}\, 2\pi r = \frac{2\pi r \mu\, dr}{s}\, k_n.$$ (9.72)

Note: The mean sectional area of fiber in the differential layer $\overline{s}^* = s/k_n$ is determined by Equation (1.39). So, it is also valid that $dn = 2\pi r \mu\, dr/\overline{s}^*$. This expression is analogous to Equations (4.4) and (4.5) derived in case of helical model; only the constant sectional area $s/\cos\beta$ was used in the place of current mean sectional area of fiber $\overline{s}^* = s/k_n$.

Simplified assumptions
We will use the simplified assumptions, known from the previous section, i.e.,

1. Quantities $\mu_r = \mu$ and $\eta_r = \eta$ are constants, independent of radius [see also Equations (9.53) and (9.54)].
2. Small deformations. It allows us to approximately use the initial geometry of yarn also in the place of geometry of the yarn under tension. Especially, $\vartheta' \doteq \vartheta$, $\beta' \doteq \beta$, $\beta'_D \doteq \beta_D$, $r' \doteq r$, $D' \doteq D$, etc.
3. The yarn strain ε_Y and the fiber strain ε_f are also small and correspond to

$$\varepsilon_f \doteq \varepsilon_Y \left(\cos^2 \vartheta - \eta \sin^2 \vartheta \right).$$ (9.73)

[This is very similar to Equation (9.51).]
Note: There exits an upper limit of angle $\vartheta = \vartheta_u$ at which the fiber strain is equal to zero, $\varepsilon_f = 0$. It is then valid to write that

$$0 = \varepsilon_Y \left(\cos^2 \vartheta_u - \eta \sin^2 \vartheta_u \right),$$
$$0 = \cos^2 \vartheta_u - \eta \sin^2 \vartheta_u = \left(1 - \sin^2 \vartheta_u \right) - \eta \sin^2 \vartheta_u$$
$$= 1 - \sin^2 \vartheta_u \left(1 + \eta \right), \quad \sin^2 \vartheta_u = \frac{1}{\left(1 + \eta \right)},$$
$$\vartheta_u = \arcsin \frac{1}{\sqrt{1 + \eta}}.$$ (9.74)

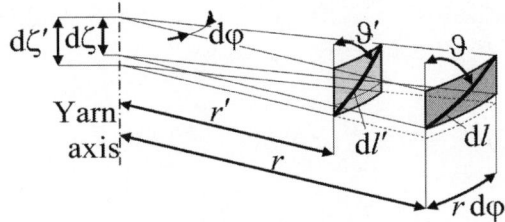

Figure 9.18 Shortening of fiber element by yarn straining, if $\vartheta > \vartheta_u$

If $\vartheta \in (\vartheta_u, \pi/2)$ [22], then, according to Equation (9.73), the fiber strain is negative, i.e., the fiber should be axially 'compressed'. Such a situation is illustrated in Figure 9.18, where $\vartheta > \vartheta_u$ so that $dl' < dl$ and $\varepsilon_f = (dl' - dl)/dl < 0$ according to Equation (9.32). (Compare Figure 9.18 with Figure 9.11.) Nevertheless, we can assume that

- the number of such fiber portions is very small so that their influence to the mechanical properties of yarn is not too significant,
- an axial compression of fiber (fiber portion) is difficult to imagine; it maybe more probable to imagine fiber slippage, fiber crimp and/or another similar process in this situation.

Based on the previous assumptions, if $\vartheta \in (\vartheta_u, \pi/2)$ then the fiber's contribution to the axial yarn force (specific stress) is zero.

Force in yarn

According to Equation (9.44), the specific stress of fiber $\sigma(\varepsilon_f)$ determines the force in fiber, i.e., $F_f(\varepsilon_f) = \sigma_f(\varepsilon_f) s \rho$. A fiber is inclined at an angle ϑ to the yarn axis so that the component of this force in the yarn direction is $F_a = F_f(\varepsilon_f) \cos \vartheta$ – compare it with Equation (9.45) (and/or $F_a = 0$ if $\vartheta \in (\vartheta_u, \pi/2)$ according to Equation (9.74) – compare it with the previous note). Nevertheless, the fibers are inclined at many different angles ϑ at a given radius. In the cross section of the differential layer, the distribution of

---

22   The fibers with angles ϑ near to $\pi/2$ are found seldom in yarn structure; however, they were identified by experimental analysis of internal structure of yarns by Neckář and Soni [12].

their angles ϑ is described according to Equation (9.68). So, the mean of force per fiber in the direction of yarn is

$$
\overline{F}_a = \int_0^{\vartheta_u} F_a u^*(\vartheta)\,d\vartheta = \int_0^{\vartheta_u} F_f(\varepsilon_f)\cos\vartheta\, u^*(\vartheta)\,d\vartheta
$$

$$
= \int_0^{\vartheta_u} F_f(\varepsilon_f)\cos\vartheta\,\frac{\cos\vartheta\, u(\vartheta)}{k_n}\,d\vartheta = \frac{s\rho}{k_n}\int_0^{\vartheta_u}\sigma_f(\varepsilon_f)\cos^2\vartheta\, u(\vartheta)\,d\vartheta. \qquad (9.75)
$$

As this force is carried by the whole differential layer, it is valid to write that

$$
dF_Y = dn\,\overline{F}_a = \frac{2\pi r\mu\,dr}{s}\,k_n\,\frac{s\rho}{k_n}\int_0^{\vartheta_u}\sigma_f(\varepsilon_f)\cos^2\vartheta\, u(\vartheta)\,d\vartheta
$$

$$
= 2\pi r\mu\rho\,dr\int_0^{\vartheta_u}\sigma_f(\varepsilon_f)\cos^2\vartheta\, u(\vartheta)\,d\vartheta. \qquad (9.76)
$$

[Equations (9.72) and (9.75) were used for rearrangement.]
The total axial force in yarn is then

$$
F_Y(\varepsilon_Y) = \int_{r=0}^{r=D/2} dF_Y = 2\pi\mu\rho\int_{r=0}^{r=D/2}\left[\int_0^{\vartheta_u}\sigma_f(\varepsilon_f)\cos^2\vartheta\, u(\vartheta)\,d\vartheta\right] r\,dr . \qquad (9.77)
$$

The following statements can be made with a view to the last equation.

- The helical angle β determines Equation (4.1), i.e., $\tan\beta = 2\pi r Z$, where Z is yarn twist. Similarly, the twist intensity is $\tan\beta_D = \pi D Z$, according to Equation (1.52).
- The fiber strain $\varepsilon_f \doteq \varepsilon_Y(\cos^2\vartheta - \eta\sin^2\vartheta)$, according to Equation (9.73), is a function of ϑ. Because of the presence of ε_f, the axial force in yarn depends on yarn strain ε_Y.
- The upper limit of angle ϑ, i.e., angle ϑ_u follows Equation (9.74) as a function of cross-contraction ratio η.
- The probability density function $u(\vartheta)$ is determined by Equation (9.66) as a function of ϑ and modal value β.

Equation (9.77) can be rearranged as follows:

$$F_Y(\varepsilon_Y) = \int_{r=0}^{r=D/2} dF_Y = 2\pi\mu\rho \int_0^{D/2} \left[\int_0^{\vartheta_u} \sigma_f(\varepsilon_f)\cos^2\vartheta\, u(\vartheta)d\vartheta \right] r\, dr$$

Substitution: $r = \dfrac{2\pi r Z}{2\pi Z} = \dfrac{\tan\beta}{2\pi Z} = \dfrac{D\tan\beta}{2\pi DZ} = \dfrac{D\tan\beta}{2\tan\beta_D}$, $\quad dr = \dfrac{D}{2\tan\beta_D}\dfrac{d\beta}{\cos^2\beta}$,

$$r\, dr = \frac{D\tan\beta}{2\tan\beta_D} \cdot \frac{D}{2\tan\beta_D}\frac{d\beta}{\cos^2\beta} = \left(\frac{D}{2\tan\beta_D}\right)^2 \frac{\sin\beta}{\cos^3\beta}d\beta,$$

$$F_Y(\varepsilon_Y) = 2\pi\mu\rho \left(\frac{D}{2\tan\beta_D}\right)^2 \int_0^{\beta_D} \left[\int_0^{\vartheta_u} \sigma_f(\varepsilon_f)\cos^2\vartheta\, u(\vartheta)d\vartheta \right] \frac{\sin\beta}{\cos^3\beta}d\beta. \qquad (9.78)$$

Note: Let us imagine that all angles ϑ are limited to angle β in a special case. Then, the probability density function $u(\vartheta)$ is limited to the Dirac delta-function $\delta(\vartheta-\beta)$ and Equation (9.67) is valid. Let us especially select the function $f(\vartheta) = \sigma_f(\varepsilon_f)\cos^2\vartheta$ [according to the expression in Equation (9.78)], where $\varepsilon_f \doteq \varepsilon_Y(\cos^2\vartheta - \eta\sin^2\vartheta)$.

Then, $\int_{-\infty}^{\infty} \sigma_f(\varepsilon_f)\cos^2\vartheta\, \delta(\vartheta-\beta)d\vartheta = \sigma_f(\varepsilon_f)\cos^2\beta$,

where $\varepsilon_f \doteq \varepsilon_Y(\cos^2\beta - \eta\sin^2\beta)$. Thus,

$$F_Y(\varepsilon_Y) = 2\pi\mu\rho \left(\frac{D}{2\tan\beta_D}\right)^2 \int_0^{\beta_D} \sigma_f(\varepsilon_f)\cos^2\beta\frac{\sin\beta}{\cos^3\beta}d\beta$$

$$= 2\pi\mu\rho \left(\frac{D}{2\tan\beta_D}\right)^2 \int_0^{\beta_D} \sigma_f(\varepsilon_f)\cos^2\beta\frac{\sin\beta}{\cos^3\beta}d\beta$$

$$= 2\pi\mu\rho \left(\frac{D}{2\tan\beta_D}\right)^2 \int_0^{\beta_D} \sigma_f(\varepsilon_f)\tan\beta\, d\beta,$$

where $\varepsilon_f = \varepsilon_Y(\cos^2\beta - \eta\sin^2\beta)$.

This expression is identical to the earlier derived Equation (9.55).

Specific stress–strain relation in yarn

Equation (1.45) determines the packing density of yarn in the form $\mu = 4T/(\pi D^2\rho)$, where T is yarn fineness. By using this expression in Equation (9.78), we obtain

$$F_Y(\varepsilon_Y) = 2\pi \cdot \frac{4T}{\pi D^2 \rho} \cdot \rho \frac{D^2}{4\tan^2\beta_D} \int_0^{\beta_D} \left[\int_0^{\vartheta_u} \sigma_f(\varepsilon_f) \cos^2\vartheta\, u(\vartheta)\mathrm{d}\vartheta \right] \frac{\sin\beta}{\cos^3\beta}\mathrm{d}\beta$$

$$= T \frac{2}{\tan^2\beta_D} \int_0^{\beta_D} \left[\int_0^{\vartheta_u} \sigma_f(\varepsilon_f) \cos^2\vartheta\, u(\vartheta)\mathrm{d}\vartheta \right] \frac{\sin\beta}{\cos^3\beta}\mathrm{d}\beta. \tag{9.79}$$

However, Equation (1.21) or (9.1) defines the specific stress also as a ratio of force and fineness. Then, by using the previous expression, we obtain the following expression:

$$\sigma_Y(\varepsilon_Y) = \frac{F_Y(\varepsilon_Y)}{T} = \frac{2}{\tan^2\beta_D} \int_0^{\beta_D} \left[\int_0^{\vartheta_u} \sigma_f(\varepsilon_f) \cos^2\vartheta\, u(\vartheta)\mathrm{d}\vartheta \right] \frac{\sin\beta}{\cos^3\beta}\mathrm{d}\beta. \tag{9.80}$$

Coefficient of utilization of specific fiber stress

The coefficient of utilization of specific fiber stress in yarn is defined according to Equation (9.7). By applying Equation (9.80) in (9.7), we find

$$\varphi(\varepsilon_Y) = \frac{\sigma_Y(\varepsilon_Y)}{\sigma_f(\varepsilon_Y)} = \frac{2}{\sigma_f(\varepsilon_Y)\tan^2\beta_D} \int_0^{\beta_D} \left[\int_0^{\vartheta_u} \sigma_f(\varepsilon_f) \cos^2\vartheta\, u(\vartheta)\mathrm{d}\vartheta \right] \frac{\sin\beta}{\cos^3\beta}\mathrm{d}\beta, \tag{9.81}$$

where Equations (9.66) and (9.73) are used.

Calculations

The total axial force $F_Y(\varepsilon_Y)$ in yarn, the specific (engineering) stress in yarn $\sigma_Y(\varepsilon_Y)$ and the coefficient of utilization of specific fiber stress $\varphi(\varepsilon_Y)$ in yarn can be practically calculated as functions of axial strain of yarn ε_Y, when we know:

(a) Specific (engineering) stress–strain function of fiber $\sigma_f(\varepsilon_f)$.

(b) Parameters – yarn twist Z, fiber density ρ, yarn diameter D, yarn packing density μ, contraction ratio η and the 'measure' of preference of fiber orientation in modal direction C.

The calculation according to Equations (9.78), (9.80) and (9.81) also needs to use Equation (9.73) for ε_f, Equation (9.74) for ϑ_u and Equation (1.52) for β_D. The double integral in the previous equations can be solved by means of a suitable numerical method.

Note: The   numerical   integration   of   the   expression $\int_0^{9_u} \sigma_f(\epsilon_f)\cos^2 9\, u(9)\,d9$, being in Equation (9.81), can be a little complicated. One possible method of solving this is reported in Appendix 10.

Theoretical example

Let us (theoretically) assume that the specific stress–strain function $\sigma_f(\epsilon_f)$ of fiber is linear. Then, the coefficient of utilization of specific fiber stress $\varphi(\epsilon_Y)$ in yarn is independent of yarn strain ϵ_Y. Nevertheless, the degree of utilization relates to the variability of fiber orientation, which is described according to parameter C – see Figure 9.19.

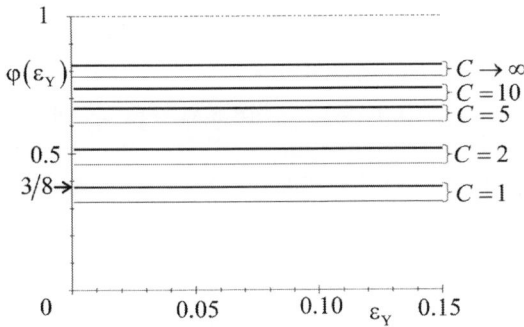

Figure 9.19 Theoretical example of Equation (9.81) considering linear function of $\sigma_f(\epsilon_f)$. $\beta_D = 0.43633\,(25\,\text{deg})$; thick lines: $\eta = 0$; thin lines: $\eta = 0.5$

If $C \rightarrow \infty$, then Equation (9.81) is limited to Equation (9.65), i.e., the equation like Gegauff's equation [see also the text written in relation to Figure 9.16 and Equation (9.67)]. On the contrary, the initial yarn possesses completely isotropic orientation of fiber portions with $C = 1$. If additionally the yarn cross contraction is not influenced by yarn straining, i.e., $\eta = 0$, then the coefficient of utilization of specific fiber stress is equal to $3/8$. [Let us remind that this value was derived for isotropic orientation in Equation (7.125) in Chapter 7 of the book [6].]

Figure 9.20 illustrates the dependence of the coefficient of utilization of specific fiber stress $\varphi(\epsilon_Y)$ in yarn on the peripheral angle β_D of fibers in the yarn.

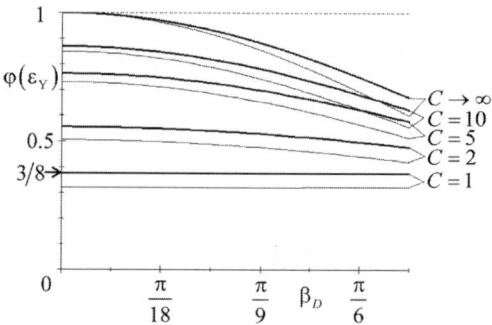

Figure 9.20 Theoretical example of Equation (9.81) by linear function of $\sigma_f(\varepsilon_f)$; thick lines: $\eta = 0$; thin lines: $\eta = 0.5$ (Curves are independent of ε_Y)

Note: The curves are independent of yarn strain ε_Y in this case; see also the comments made in relation to the previous Figure 9.19.

If $C \to \infty$, then Equation (9.81) is limited to Equation (9.65) – Gegauff's equation. The curves shown in Figure 9.20 are same as the corresponding curves displayed in Figure 9.13. If $C = 1$, then the initial yarn is something like 'isotropically oriented cylindrical bundle of fibers' where the inclination of fibers due to twist does not naturally play a role (because they are isotropic). Therefore, when $\eta = 0$, the coefficient of utilization of specific fiber stress is a constant with a value $3/8$, as mentioned in the previous figure.

Note: The curves shown in Figure 9.20 can have a real sense only at enough high values of angle β_D in case of staple fiber yarns. There is because, at a low twist, i.e., at small peripheral angle β_D, the fibers do not stick together markedly, rather they slip over each other with a small level of friction so that the actual value of the coefficient of utilization of specific fiber stress is very small. However, we have not thought about this phenomenon in our models.

9.7   Experimental results

Except the most general variant of the coefficient of utilization of specific fiber stress according to Equation (9.50), the other more or less simplified coefficients $\varphi(\varepsilon_Y)$, according to Equations (9.59), (9.65) and (9.81), were compared with experimental results. This section illustrates some examples of such comparison [15].

Coefficient of utilization of specific fiber stress according to Gegauff's equation

The formula according to Equation (9.65) is the simplest one among all the derived coefficients of utilization of specific fiber stress. The model assumed mainly ideal helical yarn, small deformation, constant contraction coefficient, and a linear stress–strain relation of fiber. Especially in this model, the coefficient of utilization of specific fiber stress in yarn is same for all yarn strains, i.e., $\varphi(\varepsilon_Y)$ is independent to ε_Y. But, as known, this coefficient decreases with an increase of angle β_D, i.e., with an increase of twist intensity, because twist intensity is expressed by $\tan\beta_D = \pi D Z$, according to Equation (1.51)) – see Figure 9.13.

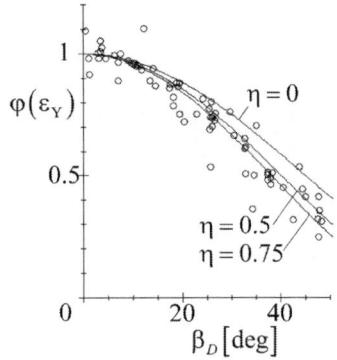

Figure 9.21 Comparison of experimental values according to Hearle et al. [10] with Equation (9.65)

Hearle et al. [10] determined experimentally the relation between fiber strength and yarn tenacity by using a lot of different types of twisted flat filament yarns[23]. Their results are shown in Figure 9.21, along with the model curves according to Equation (9.65).

The flat filament yarns usually display very small 'grey' region, shown in Figure 9.1. Therefore, the presented experimental results of coefficient of utilization of fiber tenacity φ_σ^* in yarn – Equation (9.5) – can be (roughly) compared to the coefficient of utilization of specific fiber stress $\varphi(\varepsilon_Y)$ in yarn. The curves shown in Figure 9.21 follow Equation (9.65) with different values

---

23   The following flat filament yarns were used: viscose 83/75 (yarn fineness in dtex/number of fibers), 111/24, 333/100; Tenasco 444/180, 1833/750; Acetate 111/28,111/48, 333/78; Nylon 111/34, 933/136; Terylen 111/48, 278/48.

of contraction ratio. It is shown that the curve with $\eta = 0$ (without yarn contraction) does not correspond to the experimental results satisfactorily. In contrary, the best accordance is obtained with a value of contraction ratio η lying in-between 0.5^{24} and 0.75.

Note: The value of contraction ratio greater than 0.5 ($\eta > 0.5$) accords to the tensile behaviour of filament yarns quite well. Their packing densities probably increase markedly during tensile testing.

At all events, the quantity $\varphi(\varepsilon_Y)$, according to Equation (9.65), can by and large substitute the coefficient of utilization of fiber tenacity φ_σ^*, defined by Equation (9.5).

Coefficient of utilization of specific fiber stress in viscose staple yarns

The curves of the coefficient of utilization of specific fiber stress in relation to yarn strain generally represent a result of relatively complicated processes. We will illustrate it here with the help of two examples.

A relatively 'good' and 'regular' trends of coefficient of utilization of specific fiber stress was observed in case of viscose staple yarns. The standard viscose fibers of 38 mm cut length and 1.3 dtex fineness were used to prepare viscose yarns. Table 9.1 reports on the characteristics of the yarns produced. Figure 9.22 displays the measured average stress–strain curves.

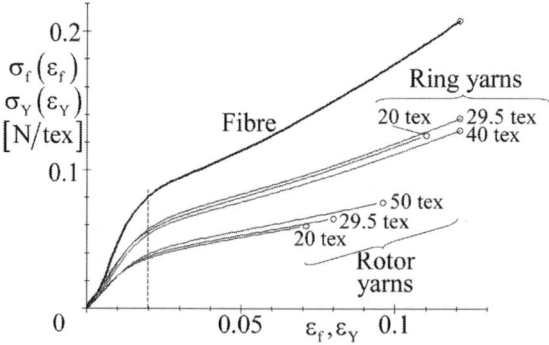

Figure 9.22 Average experimental specific stress–strain curves: curve $\sigma_f(\varepsilon_f)$ of viscose fiber, 1.3 dtex, 38 mm, and curves $\sigma_Y(\varepsilon_Y)$ of yarns. O: end-points

---

24   Cases of constant volume and small deformations in continuum mechanics
      – see also Equation (A8.13) and/or Equation (A8.24) in Appendix 8.

Table 9.1 Parameters of 100% viscose staple yarns

Type of yarn	T [tex]	Z [m^{-1}]	α [m^{-1}ktex$^{1/2}$]	μ^*	D [mm]**
Ring	20	840	119	0.593	0.169
	29.5	650	112	0.570	0.210
	45	515	109	0.550	0.263
Rotor	20	900	127	0.539	0.177
	29.5	680	117	0.507	0.222
	50	550	123	0.492	0.294

* Calculated values according to Equation (8.129), $Q = 4.12 \cdot 10^{-7}$ m^{2}tex$^{-1/2}$ for ring yarns and $Q = 1.76 \cdot 10^{-7}$ m^{2}tex$^{-1/2}$ for rotor yarns - see Chapter 8.

** Calculated values according to Equation (1.47) by using $\rho = 1500$ kg m^{-3}.

The topmost position in Figure 9.22 is occupied by the average specific stress–strain curve of fiber. Further, it can be observed that all the points on the specific stress–strain curve of yarn lie below those on the specific stress–strain curve of fiber. Further, the specific stress–strain curve of rotor yarns lies below that of ring yarns.

Note: The earlier theoretical relationships are not fully valid in the area of very small strains (probably as a consequence of relative imperfections during measurements.) Therefore, we recommend to analyze the mutual relations, starting from the strain value of 0.02 (2%).

Note: The curves shown in Figure 9.22 are the average curves obtained from a series of individual measurements. Nearly 10% of individual curves, having smallest and highest breaking strains, were excluded as outliers while obtaining the average curves. Each average curve was then determined to the value of smallest breaking strain from the series of (remaining) individual curves; see end-points ◯ in Figure 9.22.

There are six graphs shown in Figure 9.23. Three of them (a, b, c) are related to ring yarns, and the remaining three (d, e, f) are related to rotor yarns. The experimental curve, marked by φ_E, expresses the ratios of yarn specific stress to fiber specific stress at same strain values. This is shown hereunder.

Both functions – in the numerator as well as in the denominator – are determined experimentally.

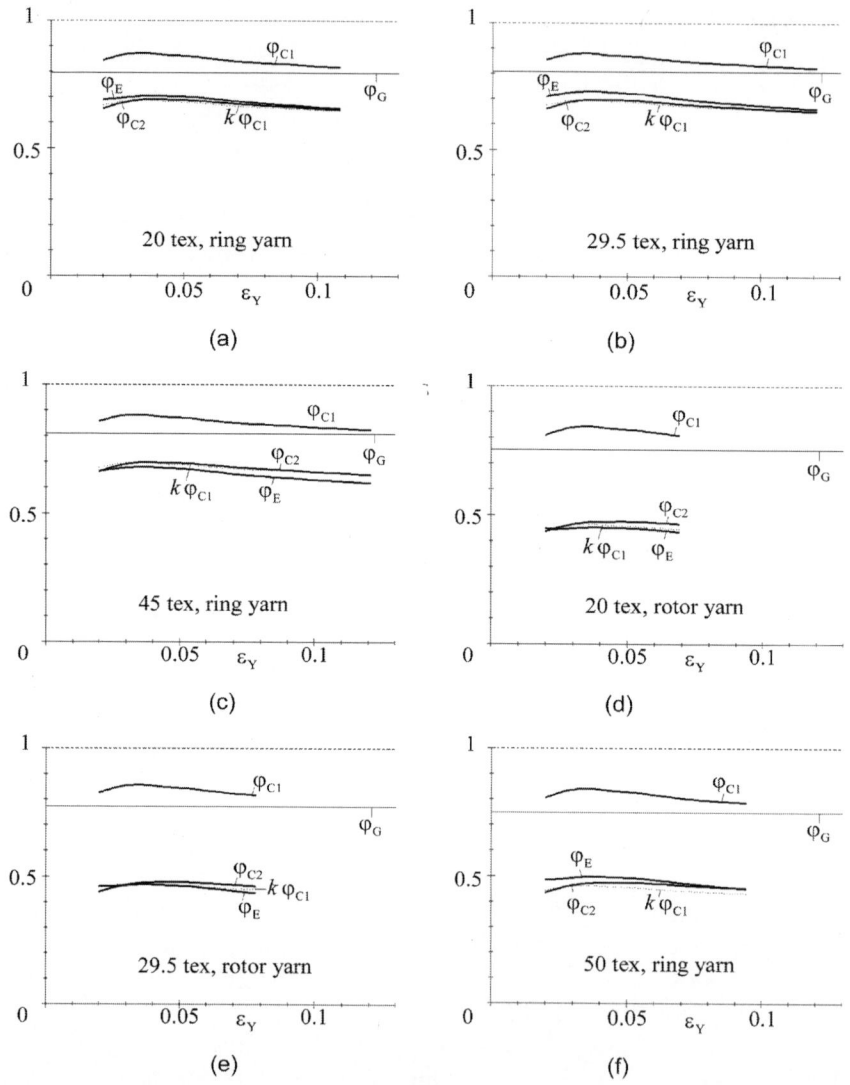

Figure 9.23 Coefficient of utilizations of specific fiber stress in viscose staple yarns. φ_E : fully experimental curve; φ_G : calculated according to Equation (9.65); $\varphi_{C1}, k\varphi_{C1}$: calculated according to Equation (9.59); φ_{C2} : calculated according to Equation (9.81); parameters used: $\eta = 0.5$; ring yarns a, b, c: $C = 4.9$, $k = 0.79$; rotor yarns d, e, f: $C = 1.6$, $k = 0.55$

$$\varphi_E = \frac{\sigma_{Y\,experim}\left(\varepsilon_Y\right)}{\sigma_{f\,experim}\left(\varepsilon_Y\right)}. \tag{9.82}$$

Note: The similar (flat) experimental curves of coefficient of utilization of specific fiber stress in viscose staple yarns were published also in the old research reports [12, 13]. Nevertheless, the level of curves of φ_E was a little higher there – around 0.75 for ring yarns, 0.55 for rotor yarns, and 0.4 for woolen yarns prepared from same viscose (cotton-type) fibers.

The calculated Gegauff's model, i.e., $\varphi_G \equiv \varphi(\varepsilon_Y)$, according to Equation (9.65), is represented by thin straight lines in the graphs shown in Figure 9.23. The curves $\varphi_{C1} \equiv \varphi(\varepsilon_Y)$ correspond to Equation (9.59) ('single-integral equation'). Also k-multiples of φ_{C1}, i.e., the dotted lines $k\varphi_{C1}$, work naturally with the same equation. Finally, the curves $\varphi_{C2} \equiv \varphi(\varepsilon_Y)$ correspond to Equation (9.81) ('double-integral equation'). The parameters from Table 9.1 and $\eta = 0.5$ were used for calculations. In case of all ring yarns, the values of C and k were determined as follows: $C = 4.9$ and $k = 0.79$. And, in case of all rotor yarns, they were determined as follows: $C = 1.6$ and $k = 0.55$.

It can be observed from Figure 9.23 that

- Gegauff's straight lines φ_G are relatively far away from the experimental curves φ_E in terms of their positions as well as their shapes.
- The curves φ_{C1} have similar shapes as the experimental ones φ_E, but the former are lying 'too high' (near Gegauff's lines φ_G) in relation to experimental curves φ_E.
- The curves with suitable k-multiple of φ_{C1} are very close to the experimental curves φ_E. It is interesting to note that a common value (empirically determined parameter $k = 0.79$) satisfies to all ring yarns and another common value (empirically determined parameter $k = 0.55$) satisfies to all rotor yarns.

An obvious question therefore arises: What is the reason for the last observed phenomenon in the case of viscose yarns?

The interpretation offers an idea of fiber orientation distribution[25], expressed by the curves φ_{C2}; these curves lie close to the curves $k\varphi_{C1}$ in all graphs. It is interesting to note the suitable values of the characteristic C of

---

25 First of all, a 'suitable' parameter C takes into account of the influence of fiber orientation in yarn. But, naturally, it can also cover a set of other different or unknown or probably less important influences.

fiber directional distribution (see Figure 9.16). A relatively high common value of C ($C = 4.9$) was necessary to choose for the ring yarns where – in consequence of intensive drafting process – fiber directions were concentrating mostly around the 'helical' (modal) angle β. In case of rotor yarns, we must choose a much smaller common value $C = 1.6$ which typically corresponds to the level of orientation of short fiber segments in a carded web – see the book [6]. In other words, the free flying or moving fibers in the air channel of spinning unit of rotor technology create a structure like carded 'micro-web' on the internal surface of the rotor.

Coefficient of utilization of specific fiber stress in polyester staple yarns

We obtained difficultly in explaining the results of analysis of polyester fibers and yarns. The standard polyester fibers of 38 mm cut length and 1.5 dtex fineness were used to prepare polyester yarns. Table 9.2 reports on the characteristics of the yarns produced and Figure 9.24 displays the measured average stress–strain curves of polyester fiber and yarns.

Table 9.2 Parameters of 100% polyester staple yarns

Type of yarn	T [tex]	Z [m^{-1}]	α [m^{-1}ktex$^{1/2}$]	μ^*	D [mm]**
	20	836	118	0.569	0.181
Ring	29.5	661	114	0.548	0.225
	45	543	115	0.532	0.281
	33	720	131	0.494	0.250
Rotor	42	615	126	0.475	0.288
	50	593	133	0.477	0.313

* Calculated values according to Equation (8.129), $Q = 2.98 \cdot 10^{-7}$ m^{2}tex$^{-1/2}$ for ring yarns and $Q = 1.29 \cdot 10^{-7}$ m^{2}tex$^{-1/2}$ for rotor yarns.

** Calculated values according to Equation (1.47) using $\rho = 1360\,\mathrm{kg\,m}^{-3}$.

The topmost position in Figure 9.24 is occupied by the average specific stress–strain curve of fiber. Further, all the points on the specific stress–strain curve of yarn lie below those on the specific stress–strain curve of fiber. Furthermore, the specific stress–strain curve of rotor yarns lie slightly below the specific stress–strain curve of ring yarns. The former note related to Figure 9.22 is fully valid also to Figure 9.24.

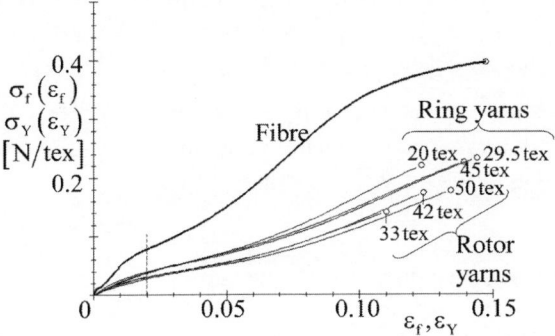

Figure 9.24 Average experimental specific stress–strain curves: curve $\sigma_f(\varepsilon_f)$ of polyester fiber, 1.5 dtex, 38 mm, and curves $\sigma_Y(\varepsilon_Y)$ of yarns. O: end-points

Figure 9.25 presents six graphs related to individual yarns. The symbols used here are the same as those described in Figure 9.23 (see the paragraphs for viscose yarns).

These graphs follow more or less similar logic as the previous curves of viscose yarns. Of course, it was necessary to use another parameters that were specific for these polyester yarns – i.e., values from Table 9.2 and $\eta = 0.5$, $C = 2.0$ and $k = 0.60$ for ring yarns, and $C = 1.17$ and $k = 0.51$ for rotor yarns. Similarly, as in the previous case of viscose yarns, it is also now valid that

- Gegauff's curves φ_G are relatively far away from the experimental curves φ_E.

- The curves φ_{C1} follow roughly similar shapes as the experimental curves φ_E, but the former are lying 'too high' (near to Gegauff's curves) in relation to the experimental curves.

- The curves with suitable k-multiple of φ_{C1} are very close to the experimental curves φ_E, especially at smaller values of yarn strain ε_Y.

- The curves φ_{C2} are lying in close contact with the curves $k\varphi_{C1}$.

Nevertheless, there are two new significant phenomena that we can observe from Figure 9.25.

1. A "suitable" value of parameter C, i.e., $C = 2.0$ for ring yarns, corresponds roughly to a carded web and $C = 1.17$ represents practically isotropic orientation of fiber portions (see the book [6]). Such a high variability of fiber directions is not imaginable, based on our

experience with microscopic evaluations of yarns. It means that there exists another significant influence – besides fiber orientation – that causes the 'fall' of experimental curves φ_E.

2. Besides it, we observe that the experimental curves φ_E increase more rapidly at higher values of breaking strain (from about 0.07 or 0.08) in relation to the curves $k\varphi_{C1}$ and/or φ_{C2}.

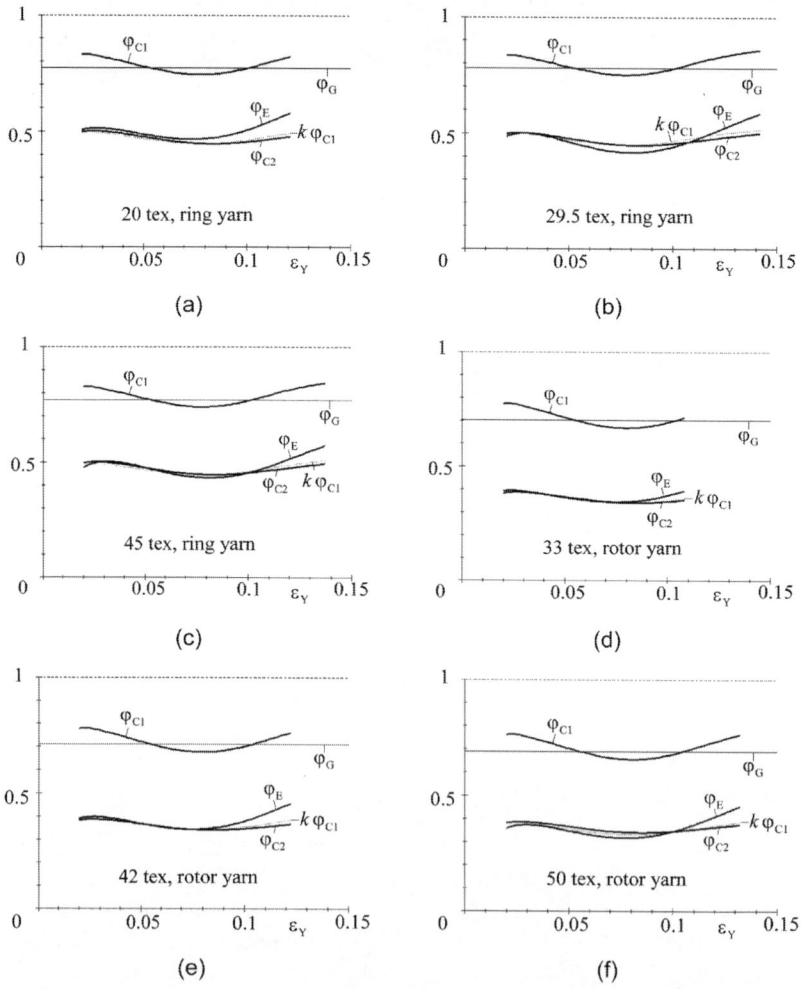

Figure 9.25 Coefficient of utilization of specific fiber stress in polyester staple yarn: φ_E: fully experimental curve; φ_G: calculated according to Equation (9.65); $\varphi_{C1}, k\varphi_{C1}$: calculated according to Equation (9.59); φ_{C2}: calculated according to Equation (9.81).

Parameters used: $\eta = 0.5$; ring yarns a, b, c: $C = 2.0$, $k = 0.60$; rotor yarns d, e, f: $C = 1.17$, $k = 0.51$

Unfortunately, we currently do not have mathematical models for describing the above-mentioned two phenomena. Nevertheless, we imagine that the crimp distribution[26] of short fiber segments (between fiber-to-fiber contacts) and fiber-to-fiber slippage, occurring mostly at higher values of yarn strains, can probably be the main reasons for the observed trends.

9.8   References

[1]   Gegauff, G., Strength and elasticity of cotton threads, *Bulletin de la Society Industrielle de Mulhouse*, 77, 153–176, 1907 (In French).

[2]   Platt, M. M., Mechanics of Elastic Performance of Textile Materials: Part III: Some Aspects of Stress Analysis of Textile Structures – Continous-Filament Yarns, *Textile Research Journal*, 20, 1–15, 1950.

[3]   Platt, M. M., Mechanics of Elastic Performance of Textile Materials: Part VI: Influence of yarn twist on modulus of elasticity, *Textile Research Journal*, 20, 665–667, 1950.

[4]   Hearle, J. W. S., The mechanics of twisted yarns: The influence of transverse forces on tensile behavior, *Journal of Textile Institute*, 49, T389–T408, 1958.

[5]   Peirce F. T., Tensile Tests for Cotton Yarns v. 'The Weakest Link' Theorems on The Strength of Long and of Composite Specimens, *Journal of Textile Institute* 17, T355–T368, 1926.

[6]   Neckář, B. and Das, D, Theory of Structure and Mechanics of Fibrous Assemblies, Woodhead Publishing India Pvt. Ltd., New Delhi, 2012.

[7]   Hamburger, W. J., The industrial application of the stress–strain relation, *Journal of the Textile Institute*, 40, P700–P718, 1949.

[8]   Neckář, B., Morfologie a strukturní mechanika obecných vlákenných útvarů (Morphology and Structural Mechanics of General Fibrous Assemblies), Technical University of Liberec, Liberec, 1998 (In Czech).

[9]   Johannsen, O., Handbuch der Baumwollspinnerei, Rohweberei und Fabrikanlagen, I Band (Handbook of Cotton Spinning, Weaving and Factory Equipment) Verlag von B. F. Voigt, Leipzig, 1930 (In German).

---

26   In Section 9.3, the influence of fiber crimp variability is discussed in the easiest case of parallel fiber bundles – see Equation (9.24) and Figure 9.8.

[10] Hearle, J. W. S., Grosberg, P., and Backer, S., Structural Mechanics of Fibers, Yarns, and Fabrics, Wiley-Interscience, New York, 1969.

[11] Neckář, B., Fyzikální model vnitřní struktury příze. Část III – Měření a vyhodnocování struktury příze (Physical model of internal yarn structure. Part III – Measuring and evaluation of yarn structures), Research Report K-07-X/77, State Textile Research Institute, Liberec 1977 (In Czech).

[12] Neckář, B., and Soni, M. K., Internal Structure of OE-Yarns from Viscose Fibers, Research Report No. PT2-XI.79, State Textile Research Institute, Liberec, 1979.

[13] Neckář, B., Příčné rozměry a tahová křivka přízí (Cross-dimensions and stress–strain curve of yarns), Research Report SV-10-XII/81, State Textile Research Institute, Liberec, 1981 (In Czech).

[14] Neckář, B. and Das, D., Tensile behavior of staple fiber yarns, Part I: Theoretical models, *Journal of Textile Institute*, 108, 922–930, 2016.

[15] Zubair, M., Neckář, B., and Das, D., Tensile behavior of staple fiber yarns, Part II: Model Validation, *Journal of Textile Institute*, 108, 931–934, 2016.

Yarn strength in relation to gauge length

10.1 Introduction

Strength is the oldest, experimentally easiest and therefore most-frequent characteristic of mechanical behaviour of twisted staple fiber yarns. The basic quantity describing this phenomenon is the maximum force, the so-called breaking force, which a yarn can take during its tensile deformation. Naturally, it is observed that a 'thick' yarn has usually higher numerical value of breaking force then a 'thin' yarn. Therefore, in history, the quantity 'breaking length'[1] was defined by the length R required to break a yarn under its own weight[2]. In the present time, we use the so-called yarn tenacity σ, which is denoted by the maximum specific stress of yarn. It is valid to write that $\sigma_{\left[\mathrm{N\,tex^{-1}}\right]} = 0.00981\,R_{[\mathrm{km}]}$.

The tenacity of staple fiber yarn is a result of many influences, for example, longitudinal variability of yarn structure and its probabilistic characteristics, internal structure of yarns described by twist, fiber length, fiber migration, fiber-to-fiber contacts, fiber-to-fiber friction, fiber-to-fiber slippage, etc. Unfortunately, as of today, we do not fully understand all these influences. Let us mention only two examples describing our lack of knowledge. The longitudinal mass variability is commonly monitored along the yarn by commercial instruments like Uster tester. Nevertheless, this has a very different character than the strength variability of yarn. Further, the behaviour of fibers inside the structure of the yarn is not fully known. The fiber-to-fiber contact, fiber-to-fiber friction and fiber-to-fiber slippage are complicated, which are not enough precisely known till date.

---

1 See also comments stated after Equation (1.21) in Chapter 1.

2 Müller [1] stated that the term 'breaking length' suggested, for the first time, by Releaux for calculation of ropes in the first edition of his book Constructeur (Braunschweig, 1861). Later on, Rankin generalized this term for fiber products in Mechanics Magazine in 1866.

This chapter focuses on yarn strength as a random process and its relation to gauge length.

10.2   Yarn strength in relation to gauge length – Peirce's theory [2]

Initial experience and ideas

Usually, yarn strength measurement is carried out at 500 mm gauge length. However, in practice, yarns experience stresses at different lengths. For example, yarns of very small length experience tremendous stress during weaving process. On the contrary, yarns of much longer than 500 mm length are stressed during the post-spinning operations. It is therefore often stated that the standard measurement of yarn strength only at 500 mm gauge length is not sufficient.

The relation between strength and length in yarn has been a topic of interest for many years. There exist many interesting imaginations, concepts and relations regarding yarn strength dependence on gauge length[3]. It is generally observed that the mean strength as well as mean breaking strain of staple fiber yarns decrease with the increase of gauge length. This raises an obvious question: why is it so?

Probably, the first theoretical knowledge regarding strength–length relation in yarn was given by Peirce [2]. His fundamental idea is originated from the weakest link of a chain. Let us imagine that the length of a straightened chain is increased by the application of a tensile force P as shown in Figure 10.1. One link (double line) is the weakest link there. Then this link determines the strength of the whole chain and the strengths of all other (stronger) links do not play any role in determining the strength of the whole chain.

Further let us imagine a long chain which is created from many links such that only a few of them are weak. The strength of such a long chain would correspond to the strength of the weakest link. Alternatively, let us divide this long chain into several short parts and measure the strength of the short parts of the chain. Only one part (containing the weakest link) will have the same strength as the previous long chain; all other short parts will have higher strength. Therefore, the average strength of short parts will be higher than the strength of the long chain.

Peirce [2] made use of this logical principle – 'principle of weakest link' – for deriving a relationship between strength and length in yarn.

---

3   See a critical review published by Das [3] in this regard.

Figure 10.1 Idea of weakest link

Probability of yarn break

Let us imagine a long length l of yarn which is gripped in-between a couple of jaws (Figure 10.2). The yarn will be elongated by means of a tensile force

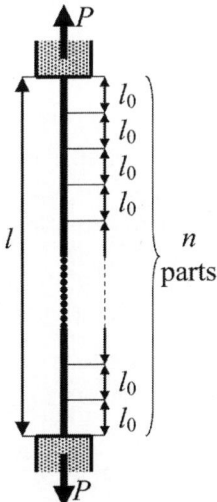

Figure 10.2 'Long' and 'short' parts of a yarn

P. Now, let us (imaginatively) divide the long length l into n (short) parts, all of length l_0; it is then valid that

$$n = l / l_0.$$ (10.1)

Let us introduce the probability $F(P,l)$ that the yarn of a given (long) length l will be broken by a force P. (P is a random variable, whereas l is a parameter.) The introduced function $F(P,l)$ is a non-decreasing function, because if P is higher, then the probability $F(P,l)$ is also higher. Further, $F(P,l) \in \langle 0,1 \rangle$. The yarn strength lies in an interval $\langle P_{min}, P_{max} \rangle$ so that it is valid to write that $F(P \leq P_{min}, l) = 0$ and $F(P \geq P_{max}, l) = 1$. Also, $F(P,l)$ is a distribution function of yarn strengths at length l.

Note: $F(P,l)$ expresses the part of yarn portions of length l, having their strength smaller than the given value P. So, it is evident that the function $F(P,l)$ has the sense of distribution function of P at this gauge length l.

The probability $1 - F(P,l)$ expresses that the yarn of the given length l will not be broken by the force P.

Similarly, let us introduce the probability $F(P,l_0)$ that the yarn of a given (short) gauge length l_0 will be broken by a force P. (P is a random variable, whereas l_0 is a parameter.) The probability that the yarn of the given gauge length l_0 will not be broken by the force P is then $1 - F(P,l_0)$.

In pursuant to Peirce's concept, let us assume that

1. The principle of weakest link is valid.
2. The probability that one (short) part of yarn, length l_0, will be broken is independent to the probabilities of breakage of all other (short) parts. This is shown in Figure 10.2. In other words, the probabilities of breakage of all (short) parts are mutually independent.

Then, each (short) part of yarn of length l_0 has the same probability $1 - F(P,l_0)$ of its non-breakage (survival). If in Figure 10.2 the whole (long) length l is not to be broken, then each (short) part of length l_0 creating length l must not be broken. This idea can be formulated as follows:

$$
\underbrace{1 - F(P,l)}_{\substack{\text{Probability of non-breakage}\\\text{of (long) length } l}} = \overbrace{\underbrace{\left[1 - F(P,l_0)\right]}_{\substack{\text{Probability of non-breakage}\\\text{of 1-st (short) part}}} \cdot \underbrace{\left[1 - F(P,l_0)\right]}_{\substack{\text{Probability of non-breakage}\\\text{of 2-nd (short) part}}} \cdot \cdots \cdot \underbrace{\left[1 - F(P,l_0)\right]}_{\substack{\text{Probability of non-breakage}\\\text{of } n\text{-th (short) part}}}}^{n \text{ identical multiplicants}}
$$

$$
= \left[1 - F(P,l_0)\right]^n = \left[1 - F(P,l_0)\right]^{l/l_0}, \left[1 - F(P,l)\right]^{1/l} = \left[1 - F(P,l_0)\right]^{1/l_0}.
$$

$$(10.2)$$

Note: The couple of lengths l, l_0 was introduced so that their ratio n – Equation (10.1) – is a natural number. Now, we shall generalize this idea to all couples of l, l_0 whose ratio is a positive real number.

Distribution function

Equation (10.2) can also be rearranged as follows:

$$F(P,l) = 1 - \left[1 - F(P,l_0)\right]^{l/l_0} . \tag{10.3}$$

The last equation expresses the distribution function of yarn strength at a (long) gauge length l. If we know the distribution function $F(P,l_0)$ of yarn strength at a (short) gauge length l_0, then we can calculate the distribution function $F(P,l)$ at each other longer gauge length l.

Probability density function

The probability density function $f(P,l)$ of yarn strength at gauge length l can be obtained as the derivative of distribution function with respect to P. By using Equation (10.3), we obtain

$$f(P,l) = \frac{\partial F(P,l)}{\partial P} = \frac{l}{l_0}\left[1 - F(P,l_0)\right]^{\frac{l}{l_0}-1} \overbrace{\partial F(P,l_0)/\partial P}^{=f(P,l_0)},$$

$$f(P,l) = \frac{l}{l_0}\left[1 - F(P,l_0)\right]^{\frac{l}{l_0}-1} f(P,l_0). \tag{10.4}$$

(The expression $f(P,l_0)$ is evidently the probability density function of yarn strength at gauge length l_0.)

Risk function

Equation (10.2) shows that the expressions are not a function of length l and/or l_0. (We thought these two values are fully arbitrary, without any special condition.) These expressions must be a function of P only. (The function $F(P,l)$ must have a form such that we calculate the same value of $\left[1 - F(P,l)\right]^{1/l}$ for each positive value of l, naturally including $l = l_0$.)

Therefore, it is usually introduced a so-called risk function $R(P)$ according to the following expression:

$$\left[1 - F(P,l)\right]^{1/l} = e^{-R(P)}, \quad l > 0 \dots \text{arbitrary.} \tag{10.5}$$

$F(P,l)$ is a non-decreasing (distribution) function ranging from $F(P \le P_{\min}, l) = 0$ to $F(P \ge P_{\max}, l) = 1$. Then, the expression $\left[1 - F(P,l)\right]$ is a non-increasing function ranging from $\left[1 - F(P \le P_{\min}, l)\right] = 1$ to $\left[1 - F(P \ge P_{\max}, l)\right] = 0$. (Naturally, the same is valid also for the expression $\left[1 - F(P,l_0)\right]^{1/l_0} = e^{-R(P)}$.) From these relations, the following properties of risk function are valid:

$$\left. \begin{array}{l} R(P) \in \langle 0, \infty \rangle \dots \text{non-decreasing function,} \\ R(P \le P_{\min}) = R(P_{\min}) = 0, \\ R(P \ge P_{\max}) = R(P_{\max}) = \infty. \end{array} \right\} \tag{10.6}$$

Equation (10.5) is more often used in the following form:

$$F(P,l) = 1 - e^{-l R(P)}, \tag{10.7}$$

which is an alternative expression of the distribution function of yarn strength. So, the distribution function $F(P,l)$ cannot be chosen fully arbitrarily. It must satisfy Equation (10.7), where the risk function corresponds to the relations stated in Equation (10.6).

The alternative expression of the probability density function $f(P,l)$ of yarn strength at gauge length l can be obtained as the derivative of distribution function with respect to P. By using Equation (10.7), we obtain

$$f(P,l) = \frac{\partial F(P,l)}{\partial P} = \frac{dR(P)}{dP} l\, e^{-l R(P)}. \tag{10.8}$$

Normal distribution of yarn strength at (short) gauge length

In addition to the two assumptions described before Equation (10.2), Peirce [2] assumed that

3. The yarn strength P at a (short) gauge length l_0 follows the normal (Gaussian) distribution with mean value $\bar{P}_{l_0}$ and standard deviation σ_{l_0} .

The probability density function is then

$$f(P,l_0) = \frac{1}{\sqrt{2\pi}\,\sigma_{l_0}} \exp\left[-\frac{\left(P-\bar{P}_{l_0}\right)^2}{2\sigma_{l_0}^2}\right], \tag{10.9}$$

and the corresponding distribution function is

$$F(P,l_0) = \int_{-\infty}^{P} f(Q,l_0)\,dQ = \frac{1}{\sqrt{2\pi}\,\sigma_{l_0}} \int_{-\infty}^{P} \exp\left[-\frac{\left(Q-\bar{P}_{l_0}\right)^2}{2\sigma_{l_0}^2}\right] dQ. \tag{10.10}$$

(The integration variable P was renamed to Q in the previous definite integral.) $\bar{P}_{l_0}$ and σ_{l_0} are two parameters.

By applying the previous two expressions in Equations (10.3) and (10.4), we can write the distribution function of yarn strength at gauge length l as follows:

$$F(P,l) = 1 - \left[1 - \frac{1}{\sqrt{2\pi}\,\sigma_{l_0}} \int_{-\infty}^{P} \exp\left[-\frac{\left(Q-\bar{P}_{l_0}\right)^2}{2\sigma_{l_0}^2}\right] dQ\right]^{l/l_0}, \tag{10.11}$$

and the probability density function of yarn strength at gauge length l in the following form:

$$f(P,l) = \frac{l}{l_0}\left[1 - \frac{1}{\sqrt{2\pi}\,\sigma_{l_0}} \int_{-\infty}^{P} \exp\left[-\frac{\left(Q-\bar{P}_{l_0}\right)^2}{2\sigma_{l_0}^2}\right] dQ\right]^{\frac{l}{l_0}-1}$$

$$\cdot \frac{1}{\sqrt{2\pi}\,\sigma_{l_0}} \exp\left[-\frac{\left(P-\bar{P}_{l_0}\right)^2}{2\sigma_{l_0}^2}\right]. \tag{10.12}[4]$$

---

[4]   It is evident that the general probability density function $f(P, l)$, according to Equation (10.12), is not normal (Gaussian) distribution, except for the special case when $l = l_0$.

Linearly transformed yarn strength

It is advantageous to introduce a linearly transformed random quantity u as follows:

$$u = \frac{P - \bar{P}_{l_0}}{\sigma_{l_0}}, \quad P = \sigma_{l_0} u + \bar{P}_{l_0}, \tag{10.13}$$

where P is the random yarn strength, measured on an arbitrary chosen gauge length l, generally different from l_0, and $\bar{P}_{l_0}$ and σ_{l_0}, determined on the (short) length l_0, are simply two parameters. So, the random variable u is evidently a linearly transformed random quantity. (If yarn strength P relates to gauge length $l = l_0$, then u can be termed as Gaussian standardized random quantity.)

Note: By the help of the introduced linearly transformed random quantity u, we can investigate the behaviour of yarn strength distribution in accordance with Equations (10.11) and (10.12) without actual knowledge of the numerical values of parameters $\bar{P}_{l_0}$ and σ_{l_0}.

Distribution of linearly transformed yarn strength

In Appendix 11, the distribution of linearly transformed yarn strength is derived. According to Equation (A11.8) of this appendix, the probability density function related to gauge length l is

$$g(u,l) = \frac{l}{l_0} \left[1 - \Phi(u) \right]^{\frac{l}{l_0} - 1} \varphi(u), \tag{10.14}$$

and according to Equation (A11.9), the corresponding distribution function is

$$G(u,l) = 1 - \left[1 - \Phi(u) \right]^{l/l_0}. \tag{10.15}$$

In the previous two equations, the known functions $\varphi(u)$ and $\Phi(u)$ are probability density function and distribution function of standardized normal (Gaussian) distribution, i.e.,

$$\varphi(u) = \frac{1}{\sqrt{2\pi}} e^{-\frac{u^2}{2}}, \quad \Phi(u) = \frac{1}{\sqrt{2\pi}} \int_{-\infty}^{u} e^{-\frac{v^2}{2}} \, dv = \frac{1}{\sqrt{\pi}} \int_{-\infty}^{u/\sqrt{2}} e^{-t^2} \, dt. \tag{10.16}$$

(The last definite integral is the so-called Laplace–Gaussian integral; which can only be solved numerically.)

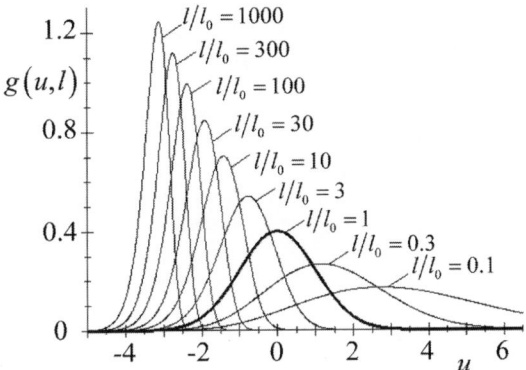

Figure 10.3 Probability density function of linearly transformed yarn strength, (Standardized function by $l/l_0 = 1$)

Figure 10.3 illustrates the behaviour of the probability density functions of the linearly transformed yarn strength as defined according to Equation (10.14) with expressions stated in Equation (10.16). It can be seen that with the increase of gauge length (increase of l/l_0).

- yarn strength decreases (including its mean value),
- strength variability decreases (the function becomes 'narrower'),
- asymmetry of this function slightly increases.

Statistical characteristics of linearly transformed yarn strength

A set of statistical characteristics of transformed yarn strength u is derived in Appendix 11. By using the probability density function, as stated in Equation (10.14), the general non-central m-th statistical moment can be expressed in accordance with Equation (A11.10) as follows:

$$\overline{u^m} = \int_{-\infty}^{\infty} u^m g(u,l)\,du, \quad m = 1,2,3,\ldots.^5 \tag{10.17}$$

According to Equation (A11.11), the mean value of u is

$$\overline{u} = \int_{-\infty}^{\infty} u\, g(u,l)\,du, \tag{10.18}$$

---

5   The symbol $\overline{u^m}$ represents the mean value of the random quantity u^m, while $\overline{u}^m$ denotes m-th power of mean value $\overline{u}$ of the random quantity u. Similar symbols are used in the following equations.

and according to Equation (A11.14), the standard deviation is

$$\sigma_u = \sqrt{\overline{u^2} - \overline{u}^2} .$$

(10.19)

According to Equation (A11.17), the coefficient of variation is then

$$v_u = \frac{\sigma_u}{\overline{u}} = \frac{\sqrt{\overline{u^2} - \overline{u}^2}}{\overline{u}} .$$

(10.20)[6]

Moreover, according to Equation (A11.18), the skewness of linearly transformed yarn strength u is

$$a = \frac{\overline{(u - \overline{u})^3}}{\sigma_u^3} = \frac{\overline{u^3} - 3\overline{u^2}\overline{u} + 2\overline{u}^3}{\left[\overline{u^2} - \overline{u}^2\right]^{3/2}} ,$$

(10.21)

and according to Equation (A11.19), its kurtosis is

$$e = \frac{\overline{(u - \overline{u})^4}}{\sigma_u^4} - 3 = \frac{\overline{u^4} - 4\overline{u^3}\overline{u} + 6\overline{u^2}\overline{u}^2 - 3\overline{u}^4}{\left[\overline{u^2} - \overline{u}^2\right]^2} - 3 .$$

(10.22)

The above-mentioned statistical characteristics (calculated by numerical integration) are reported in Table 10.1.

Table 10.1 Statistical characteristics of linearly transformed yarn strength u according to Peirce's model

Quantity: Calculation: l/l_0	$\overline{u}$ Equation (10.18)	σ_u Equation (10.19)	v_u Equation (10.20)	a Equation (10.21)	e Equation (10.22)
0.1	2.338	1.894	0.810	0.037	−2.690
0.3	1.334	1.480	1.110	0.179	−2.138
0.5	0.704	1.247	1.772	0.135	−1.289
0.7	0.342	1.117	3.268	0.071	−0.504
1.0	**0.000**	**1.000**	**±∞**	**0.000**	**0.000**
2.0	−0.564	0.826	−1.463	−0.137	−1.577

---

6   The dimensionless value is expressed as a simple ratio, not in percentage.

Quantity: Calculation: l/l_0	$\bar{u}$ Equation (10.18)	σ_u Equation (10.19)	v_u Equation (10.20)	a Equation (10.21)	e Equation (10.22)
3.0	−0.846	0.748	−0.884	−0.213	−2.401
5.0	−1.163	0.669	−0.575	−0.303	−2.802
7.0	−1.352	0.626	−0.463	−0.357	−2.898
10.0	−1.539	0.587	−0.381	−0.410	−2.946
20.0	−1.867	0.525	−0.281	−0.501	−2.981
30.0	−2.043	0.496	−0.243	−0.546	−2.989
50.0	−2.249	0.464	−0.207	−0.597	−2.994
70.0	−2.377	0.447	−0.188	−0.627	−2.996
100.0	−2.508	0.429	−0.171	−0.655	−2.997
200.0	−2.746	0.401	−0.146	−0.703	−2.998
300.0	−2.878	0.387	−0.134	−0.726	−2.999
500.0	−3.037	0.370	−0.122	−0.751	−2.999
700.0	−3.138	0.361	−0.115	−0.762	−2.999
1000.0	−3.241	0.352	−0.109	−0.766	−2.999

Statistical characteristics of yarn strength

A set of statistical characteristics of the original yarn strength P is derived in Appendix 11. One can see that the following expressions contain symbols related to the linearly transformed yarn strength u, i.e., $\bar{u}$ according to Equation (10.18), σ_u according to Equation (10.19), and v_u according to Equation (10.20). We then introduce the coefficient of variation – according to Equation (A11.26) – as follows:

$$v_{l_0} = \frac{\sigma_{l_0}}{\bar{P}_{l_0}}, \tag{10.23}$$

which represents the coefficient of variation of yarn strength obtained at (short) gauge length l_0. Finally, the non-central moments $\overline{u^2}$, $\overline{u^3}$ and $\overline{u^4}$ can be defined according to the general expression stated in Equation (10.17).

According to Equation (A11.21), the mean value of P is

$$\bar{P}_l = \sigma_{l_0}\bar{u} + \bar{P}_{l_0}, \tag{10.24}$$

and according to Equation (A11.23), the standard deviation is

$$\sigma_l = \sigma_{l_0} \sigma_u .$$ (10.25)

According to Equation (A11.27), the coefficient of variation is

$$v_l = \frac{v_u}{1 + \dfrac{1}{v_{l_0} \bar{u}}} .$$ (10.26)

Moreover, according to Equations (A11.28) and (A11.29), the skewness a and kurtosis e of probability density function of yarn strength P are

$$a = \frac{\overline{u^3} - 3\overline{u^2}\bar{u} + 2\bar{u}^3}{\left(\overline{u^2} - \bar{u}^2\right)^{3/2}}$$ (10.27)

and

$$e = \frac{\overline{u^4} - 4\overline{u^3}\bar{u} + 6\overline{u^2}\bar{u}^2 - 3\bar{u}^4}{\left[\overline{u^2} - \bar{u}^2\right]^2} - 3 .$$ (10.28)

Let us note that the last two equations are identical with Equations (10.21) and (10.22). It means that the original yarn strength P and the linearly transformed yarn strength u have same skewness and kurtosis.

Peirce's approximation
The numerical calculation of Equation (10.17) [using Equations (10.14) and (10.16)] is usually too impractical. Therefore, Peirce [2] suggested the following approximated expressions:

$$\sigma_u = \frac{\sigma_l}{\sigma_{l_0}} \doteq \left(l/l_0\right)^{-1/5} ,$$ (10.29)

$$\bar{u} = \frac{\bar{P}_l - \bar{P}_{l_0}}{\sigma_{l_0}} \doteq 4.2\left(\sigma_u - 1\right) = 4.2\left[\left(l/l_0\right)^{-1/5} - 1\right] .$$ (10.30)

A comparison between Peirce's original and approximated expressions is displayed in Figure 10.4. It can be seen that the correspondence is evidently very good.

Nevertheless, Peirce's approximation can be rearranged a little. Let us introduce two parameters A and B in relation to (short) gauge length l_0 as stated follows:

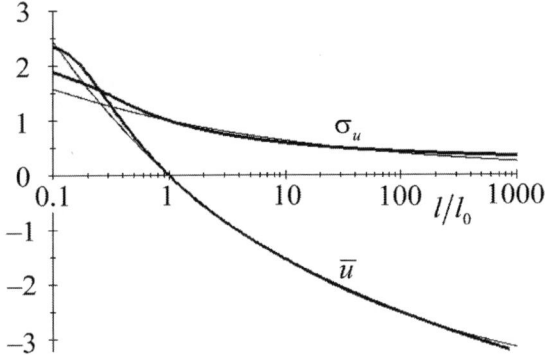

Figure 10.4 Comparison between original (thick) courses – Equations (10.17) to (10.19) – and approximation (thin) functions – Equations (10.29) and (10.30)

$$A = \overline{P}_{l_0} - 4.2\sigma_{l_0}, \qquad B = \sigma_{l_0} l_0^{1/5}. \tag{10.31}$$

By substituting parameter B in Equation (10.29), the following expression is obtained:

$$\sigma_l \doteq \sigma_{l_0} \left(l/l_0 \right)^{-1/5} = B l^{-1/5}. \tag{10.32}$$

By rearranging Equation (10.30) with the help of Equations (10.31) and (10.32), we obtain

$$\overline{P}_l \doteq \overline{P}_{l_0} + 4.2\sigma_{l_0} \left[\left(l/l_0 \right)^{-1/5} - 1 \right] = \overline{P}_{l_0} - 4.2\sigma_{l_0} + 4.2\sigma_{l_0} \left(l/l_0 \right)^{-1/5} = A + 4.2 B l^{-1/5}. \tag{10.33}$$

Note: The parameters A and B, determined according to Equation (10.31), can also be rearranged by using Equation (10.23) as follows:

$$A = \overline{P}_{l_0} - 4.2\sigma_{l_0} = \overline{P}_{l_0} - 4.2v_{l_0}\overline{P}_{l_0} = \overline{P}_{l_0} \left(1 - 4.2v_{l_0} \right), \qquad B = \sigma_{l_0} l_0^{1/5} = v_{l_0}\overline{P}_{l_0} l_0^{1/5}. \tag{10.34}$$

Finally, the coefficient of variation of yarn strength can be approximately expressed from Equations (10.32) and (10.33) as follows:

$$v_l = \frac{\sigma_l}{\overline{P_l}} = \frac{Bl^{-1/5}}{A + 4.2Bl^{-1/5}} = \frac{1}{\dfrac{A}{B}l^{1/5} + 4.2} . \qquad (10.35)$$

We can practically apply the approximated equations in the following way.

1. Determine a short gauge length l_0 [7] and experimentally find out the mean value of yarn strength $\overline{P}_{l_0}$ and its standard deviation σ_{l_0} (or coefficient of variation v_{l_0}) at this gauge length.

2. Calculate parameters A and B according to Equation (10.31) or (10.34).

3. Then, the mean value of yarn strength $\overline{P}_l$ and the standard deviation of yarn strength σ_l are possible to calculate according to Equations (10.33) and (10.32), respectively, for other (usually longer) gauge length l.

One practical example

Figure 10.5 displays a comparison between Peirce's theory and experimental results [4]. A 100% cotton combed ring-spun yarn of 7.4 tex count and 1080 turns per meter (twist) was tested for its strength at different gauge lengths from $l_0 = 50\,\text{mm}$ to $l = 700\,\text{mm}$. The experimental results (circles) were compared with Equation (10.32) and (10.33) where $A = 0.7834\,N$ and $B = 0.0530\,N\,\text{mm}^{1/5}$. The resulting (calculated) curves are shown in Figure 10.5.

It can be observed that Peirce's model reflects the general trends quite well, but it cannot precisely explain the yarn strength at different gauge lengths. (Usually, the experimental mean value of yarn strength $\overline{P}_l$ and the measured value of standard deviation of yarn strength σ_l are slightly higher at higher gauge lengths.)

---

7   This gauge length must be longer than the longest staple fiber in the yarn, otherwise a 'mix' of fiber strength and yarn strength would have been measured. (Idea of weakest link is assumed.)

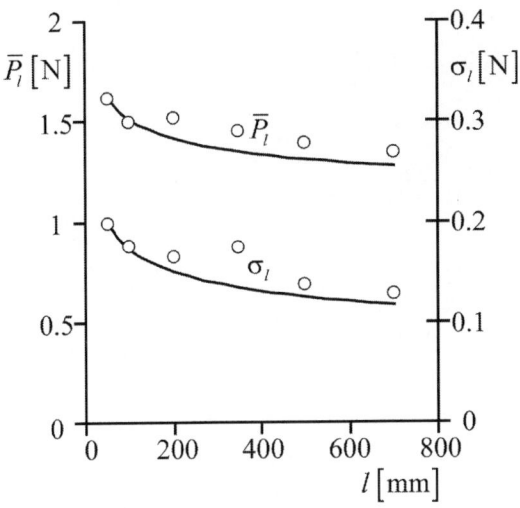

Figure 10.5 One example of comparison between Peirce's model and experimental result: smooth curves – Equations (10.32) and (10.33), circles – experimental values

10.3   Yarn strength in relation to gauge length – the alternative Weibull distribution

Basic idea

Let us imagine that the two assumptions, introduced before Equation (10.2), are still valid and therefore Equations (10.2) to (10.8) are valid too. Nevertheless, let us alternatively determine the third assumption, which was originally introduced before Equation (10.9). It will be valid to restate the (empirically proposed) third assumption as follows:

3. The risk-function $R(P)$ [see Equation (10.5)] of yarn strength P at each general gauge length l is (empirically) described by the following expression:

$$R(P) = \left(\frac{P - P_{min}}{Q} \right)^c, \quad \begin{matrix} P \in (0, \infty) \ldots \text{random quantity,} \\ P_{min} \geq 0, Q \geq 0, c \neq 0 \ldots \text{constants for a yarn.} \end{matrix}$$

$$(10.36)$$

The derivative of previous risk function is stated as follows:

$$\frac{dR(P)}{dP} = \frac{c}{Q} \left(\frac{P - P_{min}}{Q} \right)^{c-1}.$$

$$(10.37)$$

Yarn strength distribution

According to Equations (10.7) and (10.36), the distribution function $F(P,l)$ can be written as follows:

$$F(P,l) = 1 - e^{-lR(P)} = 1 - \exp\left[-l\left(\frac{P - P_{\min}}{Q}\right)^c\right].$$

(10.38)

The corresponding probability density function $f(P,l)$ follows Equations (10.8), (10.36) and (10.37) as follows:

$$f(P,l) = \frac{dR(P)}{dP} l e^{-lR(P)} = \frac{c}{Q}\left(\frac{P - P_{\min}}{Q}\right)^{c-1} l \exp\left[-l\left(\frac{P - P_{\min}}{Q}\right)^c\right].$$

(10.39)

Physical dimensions

Each distribution function, including Equation (10.38), must be dimensionless. (It has the sense of probability.) Because the random variable P and the parameter $P_{\min}$ have the dimension of force (e.g., [N]), the gauge length has the dimension of length (e.g., [m]) and the exponent c must be dimensionless, the constant Q must have the physical dimension of force times length to the power reciprocal value of c (e.g., $[\,N\,m^{1/c}\,]$). In this case, the distribution function, according to Equation (10.38), is dimensionless, as expected. Each probability density function has the physical dimension which corresponds to the reciprocal value of random quantity[8]. In our case, the probability density function has the reciprocal dimension of force (e.g., $[\,N^{-1}\,]$). Then, the quantity $Q f(P,l)$ has dimension $[\,m^{1/c}\,]$ and the quantity $(P - P_{\min})/Q$ has dimension $[\,m^{-1/c}\,]$ in the following graphs.

Graphical interpretation of probability density functions

The graphs shown in Figure 10.6 illustrate the trends of probability density function according to Equation (10.39).

---

8   Generally, a probability density function $f(x)$ has the physical dimension of reciprocal value of x. It is because $f(x)\,dx$ is dimensionless elemental probability and dx is an element of x.

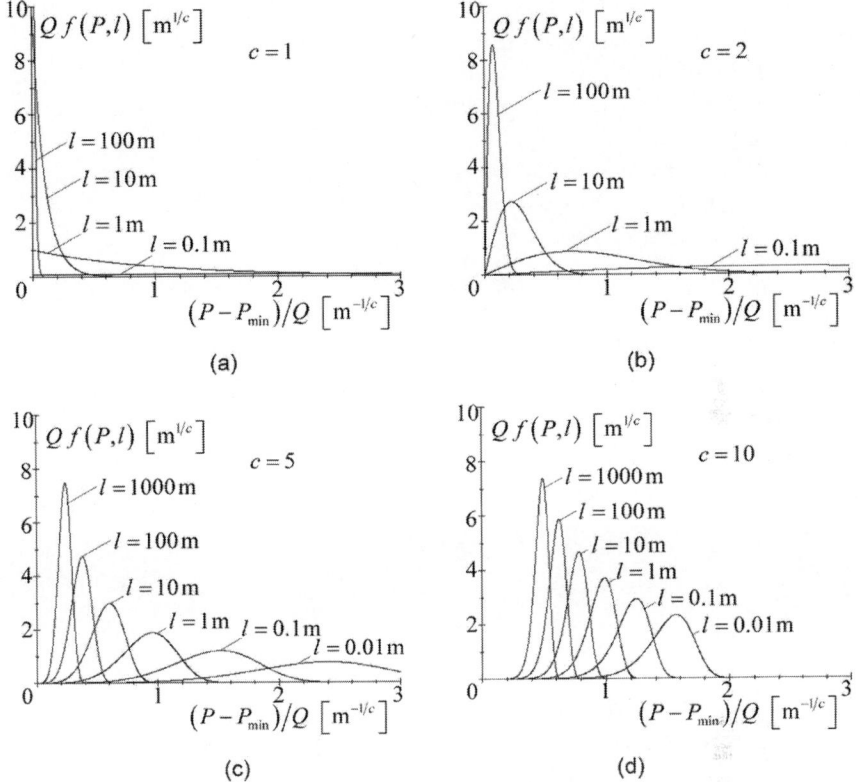

Figure 10.6 Influence of exponent c and gauge lengths l on the probability density function $f(P,l)$ according to Equation (10.39)

It is shown that

1. the probability density functions 'move to your left-hand side' with the increase of gauge length l, irrespective of all values of c.

2. the shapes of probability density functions are significantly changed with the change of exponent c. Especially, when $c = 1$ – Figure 10.6a – the distribution is exponential and the probability density functions are always decreasing with the increase of $(P - P_{min})/Q$.

Parameter of gauge length

Let us introduce a parameter q of gauge length as follows:

$$q = \frac{Q}{l^{1/c}}.$$

(10.40)

Naturally, this parameter depends on the selected gauge length l according to the last equation, when Q and c are constants for a given yarn. (The physical dimension of q is the force, e.g., [N]; compare it with other dimensions of different quantities as mentioned earlier in the paragraph 'Physical dimensions'.)

By applying Equations (10.40) to (10.38) and (10.39), we obtain the distribution function as follows:

$$F(P,l) = 1 - \exp\left[-\left(\frac{P - P_{min}}{q}\right)^c\right]$$

(10.41)

and the probability density function as follows:

$$f(P,l) = \frac{c}{q}\left(\frac{P - P_{min}}{q}\right)^{c-1} \exp\left[-\left(\frac{P - P_{min}}{q}\right)^c\right].$$

(10.42)

The last two expressions determine the distribution of the random quantity P by means of three parameters, namely P_{min}, q and c. This is generally known as Weibull distribution.

Transformed variable

Let us introduce the transformed random quantity as follows:

$$\left.\begin{aligned}
u &= \left(\frac{P - P_{min}}{q}\right)^c, \quad u \in \langle 0, \infty), \\
P &= qu^{1/c} + P_{min}, \quad dP = \frac{q}{c}u^{\frac{1}{c}-1}\,du.
\end{aligned}\right\}$$

(10.43)

Let us use a special symbol $\psi(u)$ for the probability density function of the transformed variable u. Because of the relation $\psi(u)\,du = f(P,l)\,dP$ must be valid, we can derive the probability density function for the transformed variable u by using Equations (10.42) and (10.43) as follows:

$$\psi(u)\,du = \frac{c}{q}\left(\frac{P - P_{min}}{q}\right)^{c-1} \exp\left[-\left(\frac{P - P_{min}}{q}\right)^c\right]\frac{q}{c}u^{\left(\frac{1}{c}-1\right)}\,du = \frac{c}{q}u^{\frac{c-1}{c}}e^{-u}\frac{q}{c}u^{\frac{1}{c}-1}\,du,$$

$$\psi(u) = e^{-u}.$$

(10.44)

The transformed variable u follows exponential distribution with mean value as well as variance equal to one. (Especially, if $c = 1$, $P_{min} = 0$ and $q = 1$, then $u = P$.)

Statistical characteristics

To determine the statistical characteristics, we need to work with the so-called gamma-function as follows:

$$\Gamma(x) = \int_0^\infty u^{x-1} e^{-u}\, du, \quad u \in \langle 0, \infty \rangle. \tag{10.45}$$

Note: The gamma-function is one of the higher transcendental functions, which can only be solved numerically. However, the following expressions are often found to be helpful. $\Gamma(n+1) = n!$, $\Gamma(x+1) = x\Gamma(x)$. (n is a positive integer number. See more in a standard handbook of mathematics.)

In this case, Appendix 12 shows the mathematical derivation of the usual statistical characteristics of yarn strength.

The mean value follows Equation (A12.12) and is stated as

$$\begin{aligned} \bar{P}_l &= \frac{q}{c}\Gamma\left(\frac{1}{c}\right) + P_{min}, \quad \left(\frac{\bar{P}_l - P_{min}}{q} = \frac{1}{c}\Gamma\left(\frac{1}{c}\right) \right), \\ \bar{P}_l &= l^{-\frac{1}{c}}\frac{Q}{c}\Gamma\left(\frac{1}{c}\right) + P_{min}. \end{aligned} \tag{10.46}$$

The standard deviation follows Equation (A12.13) and is mentioned as

$$\begin{aligned} \sigma_l &= q\sqrt{\frac{2}{c}\Gamma\left(\frac{2}{c}\right) - \frac{1}{c^2}\Gamma^2\left(\frac{1}{c}\right)}, \quad \left(\frac{\sigma_l}{q} = \sqrt{\frac{2}{c}\Gamma\left(\frac{2}{c}\right) - \frac{1}{c^2}\Gamma^2\left(\frac{1}{c}\right)} \right), \\ \sigma_l &= l^{-\frac{1}{c}}Q\sqrt{\frac{2}{c}\Gamma\left(\frac{2}{c}\right) - \frac{1}{c^2}\Gamma^2\left(\frac{1}{c}\right)}. \end{aligned} \tag{10.47}$$

The coefficient of variation follows Equation (A12.14) and is reported as follows:

$$v_l = \frac{\sqrt{\dfrac{2}{c}\Gamma\left(\dfrac{2}{c}\right) - \dfrac{1}{c^2}\Gamma^2\left(\dfrac{1}{c}\right)}}{\dfrac{1}{c}\Gamma\left(\dfrac{1}{c}\right) + \dfrac{P_{min}}{q}} = \frac{\sqrt{\dfrac{2}{c}\Gamma\left(\dfrac{2}{c}\right) - \dfrac{1}{c^2}\Gamma^2\left(\dfrac{1}{c}\right)}}{\dfrac{1}{c}\Gamma\left(\dfrac{1}{c}\right) + \dfrac{P_{min}}{Q}l^{1/c}}. \tag{10.48}$$

The skewness follows Equation (A12.15) and is displayed as

$$a = \frac{\frac{3}{c}\Gamma\left(\frac{3}{c}\right) - \frac{6}{c^2}\Gamma\left(\frac{2}{c}\right)\Gamma\left(\frac{1}{c}\right) + \frac{2}{c^3}\Gamma^3\left(\frac{1}{c}\right)}{\left[\frac{2}{c}\Gamma\left(\frac{2}{c}\right) - \frac{1}{c^2}\Gamma^2\left(\frac{1}{c}\right)\right]^{3/2}}. \tag{10.49}$$

Finally, the kurtosis follows Equation (A12.16) and is as follows:

$$e = \frac{\frac{4}{c}\Gamma\left(\frac{4}{c}\right) - \frac{12}{c^2}\Gamma\left(\frac{3}{c}\right)\Gamma\left(\frac{1}{c}\right) + \frac{12}{c^3}\Gamma\left(\frac{2}{c}\right)\Gamma^2\left(\frac{1}{c}\right) - \frac{3}{c^4}\Gamma^4\left(\frac{1}{c}\right)}{\left[\frac{2}{c}\Gamma\left(\frac{2}{c}\right) - \frac{1}{c^2}\Gamma^2\left(\frac{1}{c}\right)\right]^2} - 3. \tag{10.50}$$

Note: Let us note that the skewness a, kurtosis e, and the quantities $\left(\bar{P}_l - P_{\min}\right)/q$ and σ_l/q depend on one parameter only, which is the exponent c. [The parameter q depends on the gauge length l according to Equation (10.40).]

Comparison of models
It is interesting to compare the present results with equations of Peirce's model (Gaussian distribution) from Section 10.2. Peirce proposed the approximated relations as follows: $\bar{P}_l = A + 4.2\sigma_l$ and $\sigma_l = Bl^{-1/5}$ [see Equation (10.32) and (10.33)], where A and B are suitable constants.

The first of Peirce's expressions can be written in the following form $\left(\bar{P}_l - A\right)/4.2 = \sigma_l$. Let us assume that this relation is valid also when we substitute the present constant $P_{\min}$ in place of symbol A as

$$P_{\min} \equiv A. \tag{10.51}$$

Thus,

$$\frac{\bar{P}_l - \overbrace{A}^{\equiv P_{\min}}}{4.2} = \sigma_l, \quad \frac{1}{4.2}\frac{\bar{P}_l - \overbrace{A}^{\equiv P_{\min}}}{q} = \frac{\sigma_l}{q}. \tag{10.52}$$

By substituting Equation (10.46) and (10.47) in Equation (10.52), we obtain the following expression:

$$\frac{1}{4.2}\frac{1}{c}\Gamma\left(\frac{1}{c}\right) = \sqrt{\frac{2}{c}\Gamma\left(\frac{2}{c}\right) - \frac{1}{c^2}\Gamma^2\left(\frac{1}{c}\right)}. \tag{10.53}$$

The last formula expresses an equation having root c; whose value is $c = 4.8$. In other words, Equation (10.52), known from Peirce's model, is valid also here for $c = 4.8$.

With such a calculated value of c and known constant Q, Equation (10.47) determines the standard deviation σ_l. We can introduce the common constant as follows:

$$B = Q\sqrt{\frac{2}{c}\Gamma\left(\frac{2}{c}\right) - \frac{1}{c^2}\Gamma^2\left(\frac{1}{c}\right)} = Q\sqrt{\frac{2}{4.8}\Gamma\left(\frac{2}{4.8}\right) - \frac{1}{4.8^2}\Gamma^2\left(\frac{1}{4.8}\right)}, \tag{10.54}$$

so that Equation (10.47) takes the following form:

$$\sigma_l = Bl^{-\frac{1}{c}} = Bl^{-\frac{1}{4.8}}. \tag{10.55}$$

Finally, from the first expression in Equations (10.52) and (10.55), we find the equation for the mean value of yarn strength as follows:

$$\overline{P_l} = \overset{\equiv P_{min}}{A} + 4.2\sigma_l. \tag{10.56}$$

By comparing the present Equations (10.55) and (10.56) with the former Equations (10.32) and (10.33), we observe only one difference, that is, the exponent $-1/4.8$ is in place of earlier proposed exponent $-1/5$, but this difference is practically not too significant. So, we can say that the last two expressions are practically equivalent with Peirce's approximation, described in the previous section. (We find more significant differences later on with the coefficients of skewness and kurtosis.)

Table 10.2 shows the numerical values of the presented model based on Weibull distribution of yarn strength.

Table 10.2 Statistical characteristics of yarn strength according to Weibull distribution

Quantity:	$\left(\overline{P}_l - P_{\min}\right)/q$	σ_l/q	a	e
Calculation:	Equation (10.46)	Equation (10.47)	Equation (10.49)	Equation (10.50)
c				
0.5	2	4.472	6.619	84.72
1	1	1	2	6
2.0	0.886	0.463	0.631	0.245
3.0	0.893	0.324	0.173	−0.29
4.0	0.906	0.254	−0.08	−0.24
4.8	0.916	0.218	−0.23	−0.11
5.0	0.918	0.210	−0.26	−0.07
6.0	0.928	0.180	−0.38	0.085
7.0	0.935	0.157	−0.47	0.206
8.0	0.942	0.140	−0.54	0.334
10.0	0.951	0.114	−0.64	0.579
50.0	0.989	0.025	−1.03	1.975

One practical example

Militký and Kovačič [5] studied the strength of a basalt multi-filament yarn and verified Weibull distribution of yarn strength with the following parameters: $l = 0.5\,\text{m}$ (gauge length used), $P_{\min} = 0.5327\,\text{GPa}$, $q = 0.431\,\text{GPa}$ and $c = 6.547$.

By using Equation (10.40), we calculate Q as follows:
$$Q = 0.431_{[\text{GPa}]}\, 0.5^{1/6.547}_{[\text{m}]} = 0.3877\ \text{GPa}\,\text{m}^{1/6.547}.$$

By using the last expression stated in Equation (10.46), we obtain
$$\overline{P}_{l[\text{GPa}]} = l^{-1/6.547}_{[\text{m}]}\ \frac{0.3877_{\left[\text{GPa}\,\text{m}^{1/6.547}\right]}}{6.547}\ \Gamma\!\left(\frac{1}{6.547}\right) + 0.5327_{[\text{GPa}]} = l^{-1/6.547}_{[\text{m}]}\, 0.3613_{\left[\text{GPa}\,\text{m}^{1/6.547}\right]}$$
$$+\, 0.5327_{[\text{GPa}]}.$$

Further, using the last expression in Equation (10.47), we calculate standard deviation as follows:

$$\sigma_{l[\text{GPa}]} = l_{[\text{m}]}^{-1/6.547} 0.3877_{\left[\text{GPa\,m}^{1/6.547}\right]} \sqrt{\frac{2}{6.547}\Gamma\left(\frac{2}{6.547}\right) - \frac{1}{6.547^2}\Gamma^2\left(\frac{1}{6.547}\right)}$$

$$= l_{[\text{m}]}^{-1/6.547} 0.06475_{\left[\text{GPa\,m}^{1/6.547}\right]}.$$

Then, the coefficient of variation is expressed as follows:

$$v_l = \frac{\sigma_{P[\text{GPa}]}}{\bar{P}_{[\text{GPa}]}} = \frac{l_{[\text{m}]}^{-1/6.547} 0.06475_{\left[\text{GPa\,m}^{1/6.547}\right]}}{l_{[\text{m}]}^{-1/6.547} 0.3613_{\left[\text{GPa\,m}^{1/6.547}\right]} + 0.5327_{[\text{GPa}]}}$$

$$= \frac{0.06475_{\left[\text{GPa\,m}^{1/6.547}\right]}}{0.3613_{\left[\text{GPa\,m}^{1/6.547}\right]} + 0.5327_{[\text{GPa}]} l_{[\text{m}]}^{1/6.547}}.$$

[Alternatively, one can use Equation (10.48) in this case.]
Finally, the skewness and kurtosis are calculated according to Equations (10.49) and (10.50), respectively. For a given value $c = 6.547$, we obtain $a = -0.43$ and $e = 0.151$. (All necessary gamma-functions are calculated numerically.) In this way, one can predict the statistical characteristics of yarn strength for different gauge lengths.

10.4   Yarn strengths as SEM stochastic process

Introductory image

Let us imagine a (long) part of a yarn, schematically represented in Figure 10.7 by the thick line. This long part is divided into many short sections of constant length l_0, marked by cardinal numbers $i = 1, 2, \ldots, k$. The upper end of each section belongs to it – see small bows in Figure 10.7. (The lower end becomes the upper end of the following section.)

We measure the position of a section from the 'uppermost starting line' (from the first section) by means of the distance of its upper end. So, the distance of the first section is $x = 0 \cdot l_0 = 0$, the distance of the second section is $x = 1 \cdot l_0$, the distance of the third section is $x = 2 \cdot l_0$, etc. Generally,

$$x = (i-1)l_0, \quad l_0 > 0, \quad i = 1, 2, \ldots. \tag{10.57}$$

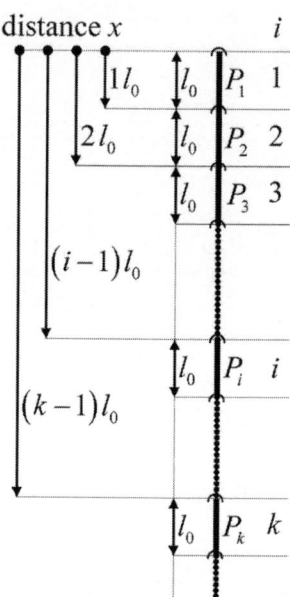

Figure 10.7 Schematic representation of a long length
of yarn divided into many small sections

The (measured) set of yarn strengths P_i, $i = 1, 2, \ldots$ belongs to the tested yarn sections. This methodology of determination of yarn strength is described in a more detail by Das and Neckář [6].

Symbols

We use the symbols like P_i, P_{i+k}, etc., i.e., with a subscript of $1, 2, \ldots$ entirely for the strength of short yarn section of length l_0. The mean value, standard deviation, and coefficient of variation of P_i are $\overline{P}_{l_0}$, σ_{l_0} and v_{l_0}, respectively.

One realization

The set of strengths P_i, $i = 1, 2, \ldots$, shown in Figure 10.7, is the so-called one realization of a stochastic (random) process. (Figure 10.8 shows a schematic graph of one random realization.)

Now, we can repeat the previous process on another (long) part of the same yarn. Then, we probably obtain another (measured) set of strengths P_i, $i = 1, 2, \ldots$, because they are random quantities.

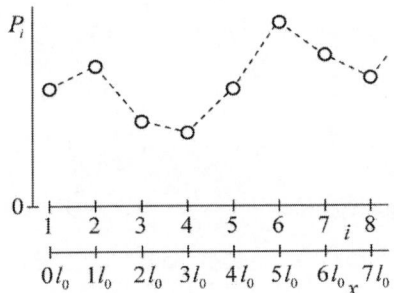

Figure 10.8 Graphical interpretation of one realization of yarn strength on (short) length l_0

By repeating a lot of such realizations, we finally obtain a lot of strength values P_i for each i (and/or x) which together create the so-called a stochastic (random) process with discrete argument i (and/or x).

Probability density functions

Yarn strengths at gauge length l_0 are described by probability density functions. Each set of random quantities P_i, obtained from a lot of realizations shown by the same subscript i, has its own probability density function $f(P_i)$.

The distribution of a couple of values P_i, P_{i+a}, $i = 1, 2, \ldots, a = 0, 1, 2, \ldots$, describes a conjugate probability density function $f(P_i, P_{i+a})$. Finally, the distribution of sequence $P_i, P_{i+a_1}, P_{i+a_2}, \ldots, P_{i+a_n}$ $(i = 1, 2, \ldots,$ where cardinal numbers $a_1 < a_2 < \cdots < a_n$) describes a conjugate probability density function $f(P_i, P_{i+a_1}, P_{i+a_2}, \ldots, P_{i+a_n})$.

Stationary stochastic process

A stationary stochastic process[9] has a behaviour such that each conjugate probability density function $f(P_i, P_{i+a_1}, P_{i+a_2}, \ldots, P_{i+a_n})$ is independent of the initial subscript i. Then, it is especially valid that $f(P_i)$, where $P_i \in (P_{\min}, P_{\max})$, is independent of subscript i. (These probability density functions are independent of the variable which we select at first.) Thus[10]

---

9   See a special handbook on stochastic or random processes for more about stationary process.

10   We use the traditional operators, i.e., E for mean value and D for variance.

1. Mean value $E(P_i) = \int_{P_{min}}^{P_{max}} P_i f(P_i) dP_i = \bar{P}_{l_0}$, same (constant) for each i.

2. Variance $D(P_i) = E\left[\left(P_i - \bar{P}_{l_0}\right)^2\right] = \int_{P_{min}}^{P_{max}} \left(P_i - \bar{P}_{l_0}\right)^2 f(P_i) dP_i = \sigma_{l_0}^2$,

    same (constant) for each i.

3. Standard deviation $\sqrt{\sigma_{l_0}^2} = \sigma_{l_0}$, same (constant) for each i.

    For a given value of k, the conjugate probability density function $f(P_i, P_{i+k})$ is independent of i, so that the following expressions are valid to write

4. Covariance $\mathrm{cov}(P_i, P_{i+k}) = E\left[\left(P_i - \bar{P}_{l_0}\right)\left(P_{i+k} - \bar{P}_{l_0}\right)\right]$

    $$= \int_{P_i = P_{min}}^{P_{max}} \int_{P_{i+k} = P_{min}}^{P_{max}} \left(P_i - \bar{P}_{l_0}\right)\left(P_{i+k} - \bar{P}_{l_0}\right) dP_i\, dP_{i+k},$$

    same (constant) for each i for a given value of k. [When $k = 0$, evidently $\mathrm{cov}(P_i, P_i) = \sigma_{l_0}^2$.]

5. Correlation   coefficient $\rho(P_i, P_{i+k}) = \mathrm{cov}(P_i, P_{i+k})/\sigma_{l_0}^2$,   same (constant) for each i for a given value of k. [When $k = 0$, evidently $\rho(P_i, P_i) = 1$.]

Ergodic stochastic process

Ergodicity[11] allows us to calculate the statistical parameters such as mean and variance (and also standard deviation, covariance and correlation coefficient) from a single (enough 'long') realization of a stochastic process. In case of ergodic random process, the probability and statistical characteristics obtained from a single realization do not change with the increase in number of measurements in the realization. For a very long realization, the statistics of the ergodic stochastic process are mentioned as follows:

1. Mean value $\bar{P}_{l_0} = \lim_{k \to \infty} \frac{1}{k} \sum_{i=1}^{k} P_i$.

---

11   See a special handbook on random process for more about ergodic stochastic process.

2. Variance $\sigma_{l_0}^2 = \lim\limits_{k \to \infty} \dfrac{1}{k} \sum\limits_{i=1}^{k} \left(P_i - \bar{P}_{l_0} \right)^2$.

3. Standard deviation $\sigma_{l_0} = \sqrt{\sigma_{l_0}^2}$.

4. Covariance $\mathrm{cov}(P_i, P_{i+k}) = \lim\limits_{k \to \infty} \sum\limits_{i=1}^{k} \left[\left(P_i - \bar{P}_{l_0} \right) \left(P_{i+k} - \bar{P}_{l_0} \right) \right]$.

5. Correlation coefficient $\rho(P_i, P_{i+k}) = \mathrm{cov}(P_i, P_{i+k}) \big/ \sigma_{l_0}^2$.

Marcovian stochastic process

The event 'strength is P_i' or 'strength has a value P_i' means that the strength of i-th (short) section of yarn lies within the elementary interval of $(P_i, P_i + \mathrm{d}P_i)$. Analogously, the event 'strengths are $P_i, P_{i+1}, \ldots, P_{i+k}$' or 'set of strengths is $P_i, P_{i+1}, \ldots, P_{i+k}$' means that the strength of $i, i+1, \ldots, i+k$ sections of yarn lies within the elementary interval of $(P_i, P_i + \mathrm{d}P_i), (P_{i+1}, P_{i+1} + \mathrm{d}P_{i+1}), \ldots, (P_{i+k}, P_{i+k} + \mathrm{d}P_{i+k})$.

Note: The probabilities of just such events are $f(P_i)\mathrm{d}P_i$, $f(P_i, P_{i+1}, P_{i+2}, \ldots, P_{i+k})\mathrm{d}P_i\, \mathrm{d}P_{i+1}\, \mathrm{d}P_{i+2} \ldots \mathrm{d}P_{i+k}$, etc., where f is the corresponding (simple, conjugate, conditional) probability density function. Evidently, they are infinitesimally small. We will not call them 'elementary' probability in this chapter.

If we still do not know any value of yarn strength (at gauge length l_0) then the probability of the value P_1 of the strength of our first section ($i = 1$) will be $f(P_1)\mathrm{d}P_1$, where $f(P_i)$ generally denotes the probability density function of the random variable P_i.

Further, let us assume that we already know the first i-th strengths $P_1, P_2, \ldots, P_i$ and let us solve the problem of strength P_{i+1}.

Note: If yarn strengths are mutually independent (see previous sections in this chapter), then the probability of a given value P_{i+1} is simply $f(P_{i+1})\mathrm{d}P_{i+1}$.

Generally, the probability of yarn strength P_{i+1} depends on all previous values, i.e., $P_1, P_2, \ldots, P_i$. Nevertheless, a simpler assumption which is often satisfied is that only the previous value P_i influences (immediately) the next value P_{i+1}. This is known as Marcovian character of the Marcovian stochastic process. (In case of Marcovian stochastic process, it is assumed that the strength values $P_1, P_2, \ldots, P_{i-1}$ – the 'past' – are not immediately necessary for determination of the probability of the strength P_{i+1} – the 'future' – when only the strength value P_i – the 'present' – is known.)

SEM stochastic process

The following discussion gives us an idea about the probabilistic character-istics and the statistical characteristics of the stationary, ergodic and Marco-vian stochastic process, i.e., in short, SEM stochastic process. The classical concepts of probability density function and conditional probability density function can be implemented to evaluate the distribution of strength values under the SEM stochastic process. The statistical parameters of this stochas-tic process can also be evaluated.

Distribution of a pair of 'neighbouring' strengths P_i, P_{i+1}

The probability density function $f(P_{i+k})$ gives the distribution of strength P_{i+k} without taking the condition of the strength value of other sections into consideration. But, the strength P_{i+k} of all realizations can be in advance pre-dicted from the given strength value P_i of i-th section ('previous') by the conditional probability density function $\varphi(P_{i+k}|P_i)$. The symbol φ is used as an operator of the conditional probability density function. The first quantity within the parentheses corresponds to one random variable on which the distribution is concerned and the second quantity is a (known) parameter. (Therefore, the sequence of symbols must not be changed!)

The independent probability of strength value P_i of i-th section is $f(P_i)\mathrm{d}P_i$. The conditional probability of strength value P_{i+1} of $(i+1)$-th section depends on the condition of strength P_i of the previ-ous section (i.e., i-th section) and is calculated from the conditional probability density function $\varphi(P_{i+1}|P_i)$. Finally, the conjugate probability that i-th section has strength value P_i and simultaneously $(i+1)$-th section has strength value P_{i+1} is $f(P_i, P_{i+1})\mathrm{d}P_i\,\mathrm{d}P_{i+1}$. According to the theory of probability, this expression can be expressed in the following manner:

$$\left.\begin{array}{l} f(P_i, P_{i+1})\mathrm{d}P_i\,\mathrm{d}P_{i+1} = f(P_i)\mathrm{d}P_i \cdot \varphi(P_{i+1}|P_i)\mathrm{d}P_{i+1}, \\[2mm] \varphi(P_{i+1}|P_i) = \dfrac{f(P_i, P_{i+1})}{f(P_i)}. \end{array}\right\} \tag{10.58}$$

Under the assumption of a stationary stochastic process, both functions $f(P_i)$ and $f(P_i, P_{i+1})$ are same for all values of i, hence the conditional probability density function $\varphi(P_{i+1}|P_i)$ is also same for all values of i.

Distribution of strength values $P_i, \ldots, P_{i+k}$

Now the above discussion can be extended to find out the distribution of strength values $P_i, \ldots, P_{i+k}$. The conjugate probability of a given set of such strengths, i.e., from i-th section to $(i+k)$-th section, is expressed by

$$f(P_i, P_{i+1}, P_{i+2}, \ldots, P_{i+k}) dP_i \, dP_{i+1} \, dP_{i+2} \ldots dP_{i+k}.$$

Another expression follows the Marcovian character of a stochastic process. Let us at first think about $k = 2$. Then, the probability of a set of strength values P_i, P_{i+1}, P_{i+2} is given by the expression $f(P_i, P_{i+1}, P_{i+2}) dP_i \, dP_{i+1} \, dP_{i+2}$. However, we can express the same probability also as a product of two probabilities:

1. Probability that sections i-th and $(i+1)$-th have the strength values P_i, P_{i+1}. According to Equation (10.58), this is expressed as
    $$f(P_i, P_{i+1}) dP_i \, dP_{i+1} = f(P_i) dP_i \cdot \varphi(P_{i+1}|P_i) dP_{i+1}.$$

2. Probability that section $(i+2)$-th has the strength value P_{i+2} under the condition that the strength value of $(i+1)$-th section is P_{i+1}. This probability is $\varphi(P_{i+2}|P_{i+1}) dP_{i+2}$.

So, the resulting probability is

$$f(P_i, P_{i+1}, P_{i+2}) dP_i \, dP_{i+1} \, dP_{i+2} = \left[f(P_i) dP_i \cdot \varphi(P_{i+1}|P_i) dP_{i+1} \right] \left[\varphi(P_{i+2}|P_{i+1}) dP_{i+2} \right]$$

$$= f(P_i) \varphi(P_{i+1}|P_i) \varphi(P_{i+2}|P_{i+1}) dP_i \, dP_{i+1} \, dP_{i+2}. \tag{10.59}$$

Analogously, it is possible to obtain the probability of strengths of i-th section to $(i+k)$-th section gradually for $k = 3$, then $k = 4$ and so on till the last repeat. Hence, the following expression can be obtained:

$$f(P_i, P_{i+1}, P_{i+2}, \ldots, P_{i+k}) dP_i \, dP_{i+1} \, dP_{i+2} \ldots dP_{i+k}$$

$$= f(P_i) \varphi(P_{i+1}|P_i) \varphi(P_{i+2}|P_{i+1}) \ldots \varphi(P_{i+k}|P_{i+k-1}) dP_i \, dP_{i+1} \, dP_{i+2} \ldots dP_{i+k},$$

$$f(P_i, P_{i+1}, P_{i+2}, \ldots, P_{i+k}) = f(P_i) \prod_{j=1}^{k} \varphi(P_{i+j}|P_{i+j-1}), \quad k = 1, 2, \ldots. \tag{10.60}$$

Under the assumption of stationary stochastic process, all the functions stated at the right hand side of Equation (10.60) are same for all values of i. Accordingly, the probability density function $f\left(P_i, P_{i+1}, P_{i+2}, \ldots, P_{i+k}\right)$ is also same for all i in case of this stationary stochastic process.

Distribution of a pair of strength values P_i, P_{i+k}

The probability of i-th section having strength value P_i and at the same time $(i+k)$-th section having strength value P_{i+k} is given by the expression $f\left(P_i, P_{i+k}\right) dP_i\, dP_{i+k}$. This probability can also be expressed by two simultaneous probabilities:

1. probability $f\left(P_i\right) dP_i$ of first section having strength value P_i, and

2. probability of $(i+k)$-th section having strength value P_{i+k} under the given condition that first section has strength value P_i. This is expressed by the term $\varphi\left(P_{i+k} \middle| P_i\right) dP_{i+k}$, where $\varphi\left(P_{i+k} \middle| P_i\right)$ is the conditional probability density function.

According to the multiplication rule of probability, the following expression is valid:

$$f\left(P_i, P_{i+k}\right) dP_i\, dP_{i+k} = f\left(P_i\right) dP_i \cdot \varphi\left(P_{i+k} \middle| P_i\right) dP_{i+k},$$

$$f\left(P_i, P_{i+k}\right) = f\left(P_i\right) \varphi\left(P_{i+k} \middle| P_i\right), \quad k = 1, 2, \ldots \tag{10.61}$$

The above distribution can be obtained in a different way on the basis of theory of probability as shown below:

$$f\left(P_i, P_{i+k}\right)$$

$$= \int_{P_{i+1}=P\min}^{P\max} \int_{P_{i+2}=P\min}^{P\max} \cdots \int_{P_{i+k-1}=P\min}^{P\max} f\left(P_i, P_{i+1}, P_{i+2}, \ldots, P_{i+k}\right) dP_{i+1}\, dP_{i+2} \ldots dP_{i+k-1}$$

$$= f\left(P_i\right) \int_{P_{i+1}=P\min}^{P\max} \int_{P_{i+2}=P\min}^{P\max} \cdots \int_{P_{i+k-1}=P\min}^{P\max} \left[\prod_{j=1}^{k} \varphi\left(P_{i+j} \middle| P_{i+j-1}\right)\right] dP_{i+1}\, dP_{i+2} \ldots dP_{i+k-1},$$

$$k = 2, 3, \ldots \tag{10.62}$$

The following expression is obtained by comparing Equations (10.61) and (10.62):

$$\phi\left(P_{i+k}\mid P_i\right)=\int\limits_{P_{i+1}=P\min}^{P_{\max}}\int\limits_{P_{i+2}=P\min}^{P_{\max}}\cdots\int\limits_{P_{i+k-1}=P\min}^{P_{\max}}\left[\prod_{j=1}^{k}\phi\left(P_{i+j}\mid P_{i+j-1}\right)\right]\mathrm{d}P_{i+1}\,\mathrm{d}P_{i+2}\cdots\mathrm{d}P_{i+k-1},$$

$$k=2,3,\ldots.\tag{10.63}$$

Note: Especially, if $k=1$, then Equations (10.62) and (10.63) are reduced to Equation (10.58).

The above integral function is same for all values of i in case of SEM random process. Accordingly, the conditional probability density function $\phi\left(P_{i+k}\mid P_i\right)$ is also same for all values of i in case of this stochastic process.

Distribution functions of yarn strength as SEM stochastic process

The corresponding distribution function related to the initial probability density function $f\left(P_i\right)$ is

$$F\left(P_i\right)=\int\limits_{P_{\min}}^{P_i}f\left(P_i^*\right)\mathrm{d}P_i^*.\tag{10.64}[12]$$

The conjugate distribution function related to the conjugate probability density function $f\left(P_i,P_{i+1},P_{i+2},\ldots,P_{i+k}\right)$ is defined by following expression:

$$F\left(P_i,P_{i+1},P_{i+2},\ldots,P_{i+k}\right)$$

$$=\int\limits_{P_i^*=P_{\min}}^{P_i}\int\limits_{P_{i+1}^*=P_{\min}}^{P_{i+1}}\cdots\int\limits_{P_{i+k}^*=P_{\min}}^{P_{i+k}}f\left(P_i^*,P_{i+1}^*,\ldots,P_{i+k}^*\right)\mathrm{d}P_i^*\mathrm{d}P_{i+1}^*\cdots\mathrm{d}P_{i+k}^*$$

$$=\int\limits_{P_i^*=P_{\min}}^{P_i}\int\limits_{P_{i+1}^*=P_{\min}}^{P_{i+1}}\cdots\int\limits_{P_{i+k}^*=P_{\min}}^{P_{i+k}}f\left(P_i^*\right)\left[\prod_{j=1}^{k}\phi\left(P_{i+j}^*\mid P_{i+j-1}^*\right)\right]\mathrm{d}P_i^*\mathrm{d}P_{i+1}^*\cdots\mathrm{d}P_{i+k}^*.\tag{10.65}$$

Here, Equation (10.60) is used for rearrangement.

The conditional distribution function related to the conditional probability density function $\phi\left(P_{i+k}\mid P_i\right)$ is defined by the following expression:

$$\Phi\left(P_{i+k}\mid P_i\right)=\int\limits_{P_{i+k}^*=P_{\min}}^{P_{i+k}}\phi\left(P_{i+k}^*\mid P_i\right)\mathrm{d}P_{i+k}^*$$

---

12   We renamed the integrating variable as P_i^* (i.e., with star), because the same quantity cannot be written for integrating variable and also for the upper limit of the integral. This is maintained in the following equations.

$$= \int\limits_{P_{i+k}^*=P\min}^{P_{i+k}} \left\{ \int\limits_{P_{i+1}^*=P\min}^{P_{max}} \int\limits_{P_{i+2}^*=P\min}^{P_{max}} \cdots \int\limits_{P_{i+k-1}^*=P\min}^{P_{max}} \left[\varphi\left(P_{i+1}^*\middle|P_i\right)\prod_{j=2}^{k}\varphi\left(P_{i+j}^*\middle|P_{i+j-1}^*\right) \right] \right.$$

$$\left. \cdot dP_{i+1}^* \, dP_{i+2}^* \ldots dP_{i+k-1}^* \right\} dP_{i+k}^*,$$

$$\Phi\left(P_{i+k}\middle|P_i\right) = \int\limits_{P_{i+1}^*=P\min}^{P_{max}} \int\limits_{P_{i+2}^*=P\min}^{P_{max}} \cdots \int\limits_{P_{i+k-1}^*=P\min}^{P_{max}} \int\limits_{P_{i+k}^*=P\min}^{P_{i+k}} \left[\varphi\left(P_{i+1}^*\middle|P_i\right)\prod_{j=2}^{k}\varphi\left(P_{i+j}^*\middle|P_{i+j-1}^*\right) \right]$$

$$\cdot dP_{i+1}^* \, dP_{i+2}^* \ldots dP_{i+k-1}^* dP_{i+k}^* \quad \text{for } k=2,3,\ldots,$$

$$\Phi\left(P_{i+1}\middle|P_i\right) = \int\limits_{P_{i+1}^*=P\min}^{P_{max}} \varphi\left(P_{i+1}^*\middle|P_i\right)dP_{i+1}^* \quad \text{for } k=1,$$

$$\Phi\left(P_i\middle|P_i\right)=1 \quad \text{for } k=0 \quad \text{(a certain event).}$$

$$(10.66)$$

Here, Equation (10.63) is used for rearrangement.

Statistical characteristics of SEM stochastic process

In case of stationary stochastic process, it is possible to express the probability density function $f\left(P_i,P_{i+k}\right)$ and the conditional probability density function $\varphi\left(P_{i+k}\middle|P_i\right)$ by Equations (10.62) and (10.63) for section $i=1,2,\ldots$ and for value $k=1,2,\ldots$, respectively. If these two functions (same for all i) are known, then the statistical characteristics of the stationary stochastic process can be obtained according to the following expressions[13].

Mean value:

$$\bar{P}_{l_0} = E\left(P_i\right) = \int\limits_{P_{min}}^{P_{max}} P_i f\left(P_i\right)dP_i. \tag{10.67}$$

Variance:

$$\sigma_{l_0}^2 = D\left(P_i\right) = E\left[\left(P_i - \bar{P}_{l_0}\right)^2\right] = \int\limits_{P_{min}}^{P_{max}} \left(P_i - \bar{P}_{l_0}\right)^2 f\left(P_i\right)dP_i,$$

$$\sigma_{l_0}^2 = E\left[\left(P_i^2 - 2P_i\bar{P}_{l_0} + \bar{P}_{l_0}^2\right)\right] = E\left(P_i^2\right) - 2\bar{P}_{l_0} E\left(P_i\right) + \bar{P}_{l_0}^2$$

---

13   We use the traditional operators, i.e., E for mean value and D for variance.

$$= E\left(P_i^2\right) - 2\bar{P}_{l_0}^2 + \bar{P}_{l_0}^2 = E\left(P_i^2\right) - \bar{P}_{l_0}^2 = \int_{P_{min}}^{P_{max}} P_i^2\, f\left(P_i\right) \mathrm{d}P_i - \bar{P}_{l_0}^2. \tag{10.68}$$

Then standard deviation takes the following form:

$$\sigma_{l_0} = \sqrt{\sigma_{l_0}^2}\,. \tag{10.69}$$

The covariance is expressed as follows:

$$\mathrm{cov}\left(P_i, P_{i+k}\right) = E\left[\left(P_i - \bar{P}_{l_0}\right)\left(P_{i+k} - \bar{P}_{l_0}\right)\right]$$

$$= \int_{P_i = P_{min}}^{P_{max}} \int_{P_{i+k} = P_{min}}^{P_{max}} \left(P_i - \bar{P}_{l_0}\right)\left(P_{i+k} - \bar{P}_{l_0}\right) f\left(P_i, P_{i+k}\right) \mathrm{d}P_i\, \mathrm{d}P_{i+k},$$

$$\mathrm{cov}\left(P_i, P_{i+k}\right) = E\left(P_i P_{i+k} - P_i \bar{P}_{l_0} - \bar{P}_{l_0} P_{i+k} + \bar{P}_{l_0}^2\right)$$

$$= E\left(P_i P_{i+k}\right) - \bar{P}_{l_0} E\left(P_i\right) - \bar{P}_{l_0} E\left(P_{i+k}\right) + \bar{P}_{l_0}^2 = E\left(P_i P_{i+k}\right) - \bar{P}_{l_0}^2 - \bar{P}_{l_0}^2 + \bar{P}_{l_0}^2$$

$$= E\left(P_i P_{i+k}\right) - \bar{P}_{l_0}^2 = \int_{P_i = P_{min}}^{P_{max}} \int_{P_{i+k} = P_{min}}^{P_{max}} P_i P_{i+k}\, f\left(P_i, P_{i+k}\right) \mathrm{d}P_i\, \mathrm{d}P_{i+k} - \bar{P}_{l_0}^2. \tag{10.70}$$

Note: The term 'covariance' is generalized also for $k = 0$ and according to Equation (10.68), it is valid to write that

$$\mathrm{cov}\left(P_i, P_i\right) = E\left[\left(P_i - \bar{P}_{l_0}\right)\left(P_i - \bar{P}_{l_0}\right)\right] = E\left[\left(P_i - \bar{P}_{l_0}\right)^2\right] = \sigma_{l_0}^2. \tag{10.71}$$

The correlation coefficient is expressed as

$$\rho\left(P_i, P_{i+k}\right) = \frac{\mathrm{cov}\left(P_i, P_{i+k}\right)}{\sigma_{l_0}^2}, \quad \left(\text{especially } \rho\left(P_i, P_i\right) = \frac{\mathrm{cov}\left(P_i, P_i\right)}{\sigma_{l_0}^2} = 1\right). \tag{10.72}$$

Linearly transformed SEM stochastic process

Here, a simple case of linear transformation of stochastic process is illustrated. A linear function of the stochastic process P_i is given by

$$Q_i = aP_i + b, \quad i = 1, 2, \ldots, \quad a \neq 0, b\ldots\text{constants}. \tag{10.73}$$

By using Equations (10.67) to (10.72), the statistics of such linearly transformed stochastic process can be found out as follows.

Mean value:

$$\overline{Q} = E(Q_i) = E(aP_i + b) = a E(P_i) + b = a\overline{P}_{l_0} + b.$$ (10.74)

Variance:

$$\sigma_Q^2 = D(Q_i) = E\left\{\left[Q_i - \overline{Q}\right]^2\right\}$$

$$= E\left\{\left[(aP_i + b) - (a\overline{P}_{l_0} + b)\right]^2\right\} = E\left\{\left[a(P_i - \overline{P}_{l_0})\right]^2\right\}$$

$$= a^2 E\left\{\left[P_i - \overline{P}_{l_0}\right]^2\right\} = a^2 \sigma_{l_0}^2.$$ (10.75)

Standard deviation:

$$\sigma_Q = \sqrt{\sigma_Q^2} = |a|\sigma_{l_0}.$$ (10.76)

Covariance:

$$\text{cov}(Q_i, Q_{i+k}) = E\left[(Q_i - \overline{Q})(Q_{i+k} - \overline{Q})\right]$$

$$= E\left[\left\{(aP_i + b) - (a\overline{P}_{l_0} + b)\right\}\left\{(aP_{i+k} + b) - (a\overline{P}_{l_0} + b)\right\}\right]$$

$$= E\left[a^2 (P_i - \overline{P}_{l_0})(P_{i+k} - \overline{P}_{l_0})\right] = a^2 E\left[(P_i - \overline{P}_{l_0})(P_{i+k} - \overline{P}_{l_0})\right]$$

$$= a^2 \text{cov}(P_i, P_{i+k}).$$ (10.77)

Correlation coefficient:

$$\rho(Q_i, Q_{i+k}) = \frac{\text{cov}(Q_i, Q_{i+k})}{\sigma_Q^2} = \frac{a^2 \text{cov}(P_i, P_{i+k})}{a^2 \sigma_{l_0}^2} = \rho(P_i, P_{i+k}).$$ (10.78)

The above statistical parameters of the linearly transformed stochastic process are same for all values of i. This linearly transformed stochastic process Q_i is also a type of SEM stochastic process.

Centralized SEM stochastic process

Most often, the linearly transformed stochastic process is of two types: centralized stochastic process and standardized (Gaussian/normal) stochastic process. The centralized stochastic process is defined by the following expression:

$$P_i^\circ = P_i - \overline{P}_{l_0}, \quad i = 1, 2, \ldots .$$ (10.79)

Note: The quantities of centralized stochastic process are marked by a small ring as superscript.

Note: The centralized stochastic process can also be obtained by substituting $a = 1$ and $b = -\overline{P}_{l_0}$ in the definition of the linear transformation of stochastic process Q_i, according to Equation (10.73). So, $Q_i = P_i^\circ$ in this special case.

The statistics of the centralized stochastic process can be obtained by using $a = 1$ and $b = -\overline{P}_{l_0}$ in Equations (10.73) to (10.78) as follows:

$$\overline{P^\circ} = \overline{P}_{l_0} - \overline{P}_{l_0} = 0.$$ (10.80)

$$\sigma_{P^\circ}^2 = \sigma_{l_0}^2.$$ (10.81)

$$\sigma_{P^\circ} = \sigma_{l_0}.$$ (10.82)

$$\mathrm{cov}\left(P_i^\circ, P_{i+k}^\circ\right) = \mathrm{cov}\left(P_i, P_{i+k}\right).$$ (10.83)

$$\rho\left(P_i^\circ, P_{i+k}^\circ\right) = \rho\left(P_i, P_{i+k}\right).$$ (10.84)

Standardized SEM stochastic process

The standardized stochastic process is defined by

$$U_i = \frac{P_i - \overline{P}_{l_0}}{\sigma_{l_0}} = \frac{1}{\sigma_{l_0}} P_i - \frac{\overline{P}_{l_0}}{\sigma_{l_0}}, \quad \left(P_i = \sigma_{l_0} U_i + \overline{P}_{l_0}\right).$$ (10.85)

This expression can also be obtained by substituting $a = 1/\sigma_{l_0}$ and $b = -\overline{P}_{l_0}/\sigma_{l_0}$ in the definition of linear transformation of stochastic process $Q_i = aP_i + b$. The statistics of the standardized stochastic process can be obtained in a similar way as discussed in case of linear transformation of stochastic process, i.e., by using Equations (10.73) to (10.78). They are shown as follows:

$$\overline{U} = \left(1/\sigma_{l_0}\right)\overline{P_{l_0}} - \left(\overline{P_{l_0}}/\sigma_{l_0}\right) = 0, \tag{10.86}$$

$$\sigma_U^2 = \left(1/\sigma_{l_0}\right)^2 \sigma_{l_0}^2 = 1, \tag{10.87}$$

$$\sigma_U = \sqrt{\sigma_U^2} = 1, \tag{10.88}$$

$$\mathrm{cov}\left(U_i, U_{i+k}\right) = \left(1/\sigma_{l_0}\right)^2 \mathrm{cov}\left(P_i, P_{i+k}\right) = \rho\left(P_i, P_{i+k}\right), \tag{10.89}$$

$$\rho\left(U_i, U_{i+k}\right) = \rho\left(P_i, P_{i+k}\right). \tag{10.90}$$

As it is already known that P_i characterizes a SEM stochastic process, hence the above equations infer that the centralized stochastic process and the standardized stochastic process are also a SEM stochastic process.

Sum of two SEM stochastic processes

Let us now consider an SEM stochastic process P_i as a summation of two partial and independent (mutually exclusive) SEM stochastic processes $^{(1)}P_i$ and $^{(2)}P_i$. This statement is algebraically written in the following manner:

$$P_i = {}^{(1)}P_i + {}^{(2)}P_i, \tag{10.91}$$

where $^{(1)}P_i$ and $^{(2)}P_i$ represent the first and second stochastic processes of the above type, respectively.

The statistical parameters of this stochastic process are given below. The mean of the sum of two independent stochastic processes is

$$\overline{P_{l_0}} = E\left(P_i\right) = E\left({}^{(1)}P_i + {}^{(2)}P_i\right) = E\left({}^{(1)}P_i\right) + E\left({}^{(2)}P_i\right) = {}^{(1)}\overline{P_{l_0}} + {}^{(2)}\overline{P_{l_0}}. \tag{10.92}$$

As the sum of two independent stochastic processes is also an independent stochastic process, then the following expression, corresponding to the mean of product of two stochastic processes, is valid for all values of $i = 1, 2, \ldots$ and $k = 0, 1, 2, \ldots$.

$$E\left({}^{(1)}P_i\,{}^{(2)}P_{i+k}\right) = E\left({}^{(1)}P_i\right)E\left({}^{(2)}P_{i+k}\right) = {}^{(1)}\overline{P_{l_0}}\,{}^{(2)}\overline{P_{l_0}}. \tag{10.93}$$

The variance of the sum of two independent stochastic processes is

$$\sigma_{l_0}^2 = E\left[P_i^2\right] - \bar{P}_{l_0}^2 = E\left[\left(^{(1)}P_i + {}^{(2)}P_i\right)^2\right] - \bar{P}_{l_0}^2$$

$$= E\left(^{(1)}P_i^2 + 2\,{}^{(1)}P_i\,{}^{(2)}P_i + {}^{(2)}P_i^2\right) - \bar{P}_{l_0}^2$$

$$= E\left(^{(1)}P_i^2\right) + 2E\left(^{(1)}P_i\right)E\left(^{(2)}P_i\right) + E\left(^{(2)}P_i^2\right) - \left(^{(1)}\bar{P}_{l_0} + {}^{(2)}\bar{P}_{l_0}\right)^2$$

$$= E\left(^{(1)}P_i^2\right) + 2\,{}^{(1)}\bar{P}_{l_0}\,{}^{(2)}\bar{P}_{l_0} + E\left(^{(2)}P_i^2\right) - {}^{(1)}\bar{P}_{l_0}^2 - 2\,{}^{(1)}\bar{P}_{l_0}\,{}^{(2)}\bar{P}_{l_0} - {}^{(2)}\bar{P}_{l_0}^2$$

$$= \left[E\left(^{(1)}P_i^2\right) - {}^{(1)}\bar{P}_{l_0}^2\right] + \left[E\left(^{(2)}P_i^2\right) - {}^{(2)}\bar{P}_{l_0}^2\right] = {}^{(1)}\sigma_{l_0}^2 + {}^{(2)}\sigma_{l_0}^2. \qquad (10.94)$$

The standard deviation of the sum of two independent stochastic processes is

$$\sigma_{l_0} = \sqrt{\sigma_{l_0}^2} = \sqrt{{}^{(1)}\sigma_{l_0}^2 + {}^{(2)}\sigma_{l_0}^2}. \qquad (10.95)$$

By using Equations (10.70), (10.92) and (10.93), the covariance of the sum of two independent stochastic processes is

$$\text{cov}\left(P_i, P_{i+k}\right) = E\left[\left(P_i - \bar{P}_{l_0}\right)\left(P_{i+k} - \bar{P}_{l_0}\right)\right] = E\left(P_i P_{i+k}\right) - \bar{P}_{l_0}^2,$$

$$\text{cov}\left(P_i, P_{i+k}\right) = E\left[\left(^{(1)}P_i + {}^{(2)}P_i\right)\left(^{(1)}P_{i+k} + {}^{(2)}P_{i+k}\right)\right] - \bar{P}_0^2$$

$$= E\left[^{(1)}P_i\,{}^{(1)}P_{i+k} + {}^{(1)}P_i\,{}^{(2)}P_{i+k} + {}^{(2)}P_i\,{}^{(1)}P_{i+k} + {}^{(2)}P_i\,{}^{(2)}P_{i+k}\right] - \bar{P}_{l_0}^2$$

$$= E\left(^{(1)}P_i\,{}^{(1)}P_{i+k}\right) + E\left(^{(1)}P_i\,{}^{(2)}P_{i+k}\right) + E\left(^{(2)}P_i\,{}^{(1)}P_{i+k}\right) + E\left(^{(2)}P_i\,{}^{(2)}P_{i+k}\right) - \left(^{(1)}\bar{P}_{l_0} + {}^{(2)}\bar{P}_{l_0}\right)^2$$

$$= E\left(^{(1)}P_i\,{}^{(1)}P_{i+k}\right) + {}^{(1)}\bar{P}_{l_0}\,{}^{(2)}\bar{P}_{l_0} + {}^{(2)}\bar{P}_{l_0}\,{}^{(1)}\bar{P}_{l_0} + E\left(^{(2)}P_i\,{}^{(2)}P_{i+k}\right) - {}^{(1)}\bar{P}_{l_0}^2 - 2\,{}^{(1)}\bar{P}_{l_0}\,{}^{(2)}\bar{P}_{l_0} - {}^{(2)}\bar{P}_{l_0}^2$$

$$= \left[E\left(^{(1)}P_i\,{}^{(1)}P_{i+k}\right) - {}^{(1)}\bar{P}_{l_0}^2\right] + \left[E\left(^{(2)}P_i\,{}^{(2)}P_{i+k}\right) - {}^{(2)}\bar{P}_{l_0}^2\right]$$

$$= \text{cov}\left(^{(1)}P_i,\,{}^{(1)}P_{i+k}\right) + \text{cov}\left(^{(2)}P_i,\,{}^{(2)}P_{i+k}\right). \qquad (10.96)$$

The correlation coefficient of the sum of two independent stochastic processes is

$$\rho\left(P_i, P_{i+k}\right) = \frac{\mathrm{cov}\left(P_i, P_{i+k}\right)}{\sigma_{l_0}^2} = \frac{\mathrm{cov}\left({}^{(1)}P_i, {}^{(1)}P_{i+k}\right) + \mathrm{cov}\left({}^{(2)}P_i, {}^{(2)}P_{i+k}\right)}{\sigma_{l_0}^2}$$

$$= \frac{{}^{(1)}\sigma_{l_0}^2}{\sigma_{l_0}^2} \frac{\mathrm{cov}\left({}^{(1)}P_i, {}^{(1)}P_{i+k}\right)}{{}^{(1)}\sigma_{l_0}^2} + \frac{{}^{(2)}\sigma_{l_0}^2}{\sigma_{l_0}^2} \frac{\mathrm{cov}\left({}^{(2)}P_i, {}^{(2)}P_{i+k}\right)}{{}^{(2)}\sigma_{l_0}^2}$$

$$= \frac{{}^{(1)}\sigma_{l_0}^2}{\sigma_{l_0}^2}\rho\left({}^{(1)}P_i, {}^{(1)}P_{i+k}\right) + \frac{{}^{(2)}\sigma_{l_0}^2}{\sigma_{l_0}^2}\rho\left({}^{(2)}P_i, {}^{(2)}P_{i+k}\right). \tag{10.97}$$

The previous expressions are independent of i, because these two independent stochastic processes are considered as individual SEM stochastic process, hence all the statistical parameters of this stochastic process (sum of two stochastic processes) are also independent of i. Consequently, the sum of two stochastic processes ${}^{(1)}P_i$ and ${}^{(2)}P_i$ is also an SEM stochastic process.

Sum of several SEM stochastic processes
We could add another independent SEM stochastic process to the above-mentioned sum of two stochastic processes and we would find out analogical equations, and so on. In general, the statistical parameters of a stochastic process, which is the sum of M independent SEM stochastic processes, are illustrated in the following equations. The definition of the stochastic process P_i as a sum of M stochastic processes is as follows:

$$P_i = \sum_{m=1}^{M} {}^{(m)}P_i . \tag{10.98}$$

The statistics of this stochastic process are shown in the following equations.

The mean of the sum of M independent stochastic processes is

$$\overline{P_{l_0}} = \sum_{m=1}^{M} {}^{(m)}\overline{P_{l_0}} . \tag{10.99}$$

The variance of the sum of M independent stochastic processes is

$$\sigma_{l_0}^2 = \sum_{m=1}^{M} {}^{(m)}\sigma_{l_0}^2 . \tag{10.100}$$

The standard deviation of the sum of M independent stochastic processes is

$$\sigma_{l_0} = \sqrt{\sum_{m=1}^{M} {}^{(m)}\sigma_{l_0}} \ . \tag{10.101}$$

The covariance of the sum of M independent stochastic processes is

$$\text{cov}\left(P_i, P_{i+k}\right) = \sum_{m=1}^{M} \text{cov}\left({}^{(m)}P_i, {}^{(m)}P_{i+k}\right). \tag{10.102}$$

The correlation coefficient of the sum of M independent stochastic processes is

$$\rho\left(P_i, P_{i+k}\right) = \frac{1}{\sigma_{l_0}^2} \sum_{m=1}^{M} \left[{}^{(m)}\sigma_{l_0}^2 \ \rho\left({}^{(m)}P_i, {}^{(m)}P_{i+k}\right)\right]. \tag{10.103}$$

[Compare Equations (10.98) to (10.103) with previous Equations (10.91), (10.92) and (10.94) to (10.97).] In this case, it is also valid that the sum of M independent is a SEM stochastic process.

Simulation of SEM stochastic process

An algorithm is required to generate the values of the stochastic process P_i, given its probability density function $f(P_i)$, and the conditional probability density function $\varphi\left(P_{i+1}|P_i\right)$. (Each of these functions is same for all i.) The realization of the stochastic process is possible to simulate on computer by the following sequence.

On the basis of functions f and φ, we can generate random variables of these distributions by using computer; they are computer generators (programs) of such random variables. (Generator f is independent to any parameter, but generator φ needs to know the 'previous' value P_i.)

The computer simulation can be realized by the following way:

1. The generator f generates a value that is considered as value P_1.

2. The generator φ generates a value of P_2 when the value of parameter P_1, which was calculated previously, is given to the generator φ. (Thus, the value P_2 is calculated from the conditional probability density function $\varphi\left(P_2|P_1\right)$.)

3. The value of the parameter P_2, which was calculated previously, is given to the generator φ and the value P_3 is generated.

$$\vdots$$

i.   The value of the parameter P_{i-1}, which was calculated previously, is given to the generator φ and the value P_i is generated.

$$\vdots$$

etc.

Thus, all of the simulated realizations of the stochastic process have the same probability and statistical characteristics as do the experimental realizations. (Of course, a huge number of simulated values can be produced within a short time, which is not possible to obtain experimentally. However, it is required to know the probability density function and the conditional probability density function of this particular stochastic process in advance.) In this way, the simulated values corresponding to the stochastic process, which is the sum of several independent stochastic processes, can be generated. The statistical characteristics of the sum of individual independent stochastic processes can be obtained from Equations (10.99) to (10.103).

Correlation and standardized correlation functions

The covariance function $\operatorname{cov}(P_i, P_{i+k})$ and the correlation coefficient $\rho(P_i, P_{i+k})$ are same for all values of i. However, these functions possess different values for different $k = 0, 1, 2, \ldots$. These functions can be expressed in terms of argument k as follows.

The covariance function expressed in terms of argument k is given by

$$\operatorname{cov}(P_i, P_{i+k}) = \operatorname{cov}(k). \tag{10.104}$$

(Sometimes, the above function is called correlation function – non-standardized – of stochastic process.)

The standardized correlation function in terms of argument k is given by

$$\rho(P_i, P_{i+k}) = \rho(k). \tag{10.105}$$

It may be observed that the argument k is actually the difference between the serial numbers used to designate particular yarn sections.

Distance between two yarn sections – a parameter of random process

It is evident from Figure 10.7 and Equation (10.57) that the distance x of i-th section from the first section is given by the value $l_0(i-1)$, similarly the distance of $(i+k)$-th section from the first section is given by the value $l_0(i+k-1)$, and hence the distance between these two sections is given by

$$x = l_0(i+k-1) - l_0(i-1) = l_0 k . \tag{10.106}$$

The covariance (non-standardized correlation) and the (standardized) correlation functions can also be expressed in terms of this distance x by $\text{cov}(x)$ and $\rho(x)$, respectively. It is clear from Equation (10.57) that k takes only discrete values ($k = 0,1,2,...$), hence x also takes only discrete values $x = 0, l_0, 2l_0,...$, but, in real practice, x can take any other values defined by the interval $x \in \langle 0, \infty \rangle$.

Note: Let us remind that the standardized correlation function is same for strengths P_i, standardized quantities U_i as well as centralized quantities P_i° – see Equations (10.84) and (10.90). Simultaneously, this function is marked with argument k according to Equation (105) and/or x – see texts written after Equation (10.106). So altogether we can write

$$\rho(P_i, P_{i+k}) = \rho(U_i, U_{i+k}) = \rho(P_i^{\circ}, P_{i+k}^{\circ}) = \rho(k) = \rho(x) . \tag{10.107}$$

Strength distribution at a general (long) gauge length

Let us think that the (long) gauge length l contains $k + 1$ (short) sections of length l_0 designated by cardinal numbers $i, i+1,...,i+k$. So the gauge length is

$$l = l_0(k+1), \quad \text{i.e.,} \quad k = \frac{l}{l_0} - 1 . \tag{10.108}$$

(The relation between x, which is the distance between the first and the last sections, and the whole gauge length l is $l = x + l_0$ according to Equations (10.108) and (10.106); see also Figure 10.7.) The values $P_i, P_{i+1}, P_{i+2},..., P_{i+k}$ describe strengths of all (short) yarn sections (length l_0) creating together the whole gauge length l.

Let us now choose a value P. The probability that each strength $P_i, P_{i+1}, P_{i+2},..., P_{i+k}$ obtained at each short length l_0, creating the long length l, is smaller than the chosen value P is described by a special conjugate distribution function, and according to Equation (10.65), all upper limits of the integrals are equal to a common value P. This special case is

$$F\left(P_i = P, P_{i+1} = P, P_{i+2} = P,\ldots, P_{i+k} = P\right)$$

$$= \int_{P_i=P_{min}}^{P} \int_{P_{i+1}=P_{min}}^{P} \cdots \int_{P_{i+k}=P_{min}}^{P} f(P_i)\left[\prod_{j=1}^{k} \varphi\left(P_{i+j}\big|P_{i+j-1}\right)\right] dP_i\, dP_{i+1} \ldots dP_{i+k}. \quad (10.109)^{14}$$

Note: The function according to Equation (10.109) is no more 'conjugate'. This is the function of one random quantity P, and the value k is a parameter related to 'long' gauge length l according to Equation (10.108).

On the contrary, the probability that the strength of all short segments (creating the long gauge length l) will be higher than P, can be expressed by $1 - G(P,k)$. This is shown below

$$1 - G(P,k) = \int_{P_i=P}^{P_{max}} \int_{P_{i+1}=P}^{P_{max}} \cdots \int_{P_{i+k}=P}^{P_{max}} f(P_i)\left[\prod_{j=1}^{k} \varphi\left(P_{i+j}\big|P_{i+j-1}\right)\right] dP_i\, dP_{i+1} \ldots dP_{i+k}.$$

$$(10.110)$$

According to the principle of the weakest link, Equation (10.110) expresses the probability that the (long) yarn section of length l is broken by the force P.

The supplementary probability that the (long) length l is not broken by the force P is $1 - \left[1 - G(P,k)\right] = G(P,k)$. Thus,

$$G(P,k) = 1 - \int_{P_i=P}^{P_{max}} \int_{P_{i+1}=P}^{P_{max}} \cdots \int_{P_{i+k}=P}^{P_{max}} f(P_i)\left[\prod_{j=1}^{k} \varphi\left(P_{i+j}\big|P_{i+j-1}\right)\right] dP_i\, dP_{i+1} \ldots dP_{i+k}.$$

$$(10.111)$$

So, the function $G(P,k)$ represents the distribution function of strength P of (long) yarn section of length l.

---

14 The symbols of integration quantities are marked by $P_i^*, P_{i+1}^*,\ldots, P_{i+k}^*$ in Equation (10.65) to avoid any confusion with the upper limits of the integrals. Now, we introduce one common value P for all upper limits so that the 'mixing' does not happen. We, therefore, mark the integration quantities in Equation (10.109) by $P_i, P_{i+1},\ldots, P_{i+k}$, i.e., without star as a superscript.

Further, we derive the corresponding probability density function of the stochastic quantity P as a derivative of Equation (10.111) in the following manner:

$$g(P,k) = \frac{dG(P,k)}{dP}$$

$$= -\frac{d}{dP}\left\{ \int\limits_{P_i=P}^{P_{max}} \int\limits_{P_{i+1}=P}^{P_{max}} \cdots \int\limits_{P_{i+k}=P}^{P_{max}} f(P_i)\left[\prod_{j=1}^{k} \varphi\left(P_{i+j}\,\middle|\,P_{i+j-1}\right)\right] dP_i\, dP_{i+1}\ldots dP_{i+k} \right\}.$$

(10.112)

Note: If $k=0$, then we find the following expression by using Equation (10.66) in (10.112):

$$g(P,0) = -\frac{d}{dP}\left\{ \int\limits_{P}^{P_{max}} f(P_i)dP_i \right\} = f(P),$$

(10.113)

which is logically expected.

10.5   Yarn strengths as SEMG stochastic process

Introductory idea

In Section 10.4, several relations are derived by means of two general symbols:

- probability density function $f(P_i)$ and
- conditional probability density function $\varphi\left(P_{i+1}\,\middle|\,P_i\right)$.

In this section, we assume that the above-mentioned functions have a specific behaviour, corresponding to normal, i.e., Gaussian distribution. Evidently, such stochastic process is stationary, ergodic, Marcovian and Gaussian stochastic process, in short, SEMG stochastic process. The probability and statistical characteristics of this process are stated in the following sections.

Probabilistic characteristics of SEMG stochastic process

The probability density function of yarn strength P_i (at length l_0) is given by the known Gaussian expression as follows:

$$f(P_i) = \frac{1}{\sqrt{2\pi}\sigma_{l_0}} \exp\left\{ -\frac{\left(P_i - \overline{P}_{l_0}\right)^2}{2\sigma_{l_0}^2} \right\},$$

P_i…random variable, $P_i \in (-\infty, \infty)$,

$\overline{P}_{l_0}$…mean value – parameter,

σ_{l_0}…standard deviation – parameter.

(10.114)

The following expression determines the Gaussian conditional probability density function of random quantity P_{i+1} under the condition that the previous 'neighbouring' quantity P_i is a given value (parameter).

$$\varphi\left(P_{i+1}|P_i\right) = \frac{1}{\sqrt{2\pi}\,\sigma_{l_0}\sqrt{1-r^2}}\exp\left\{-\frac{\left(P_{i+1}-\left[\overline{P}_{l_0}+r\left(P_i-\overline{P}_{l_0}\right)\right]\right)^2}{2\sigma_{l_0}^2\left(1-r^2\right)}\right\},\qquad (10.115)$$

$P_{i+1}\ldots$random variable, $P_{i+1}\in\left(-\infty,\infty\right)$,

$P_i,\overline{P}_{l_0},\sigma_{l_0}\ldots$previous random value, mean value and standard deviation $-$ parameters,

$r=\rho\left(P_i,P_{i+1}\right)\ldots$correlation coefficient between P_i and P_{i+1} $-$ parameter.

(The last two equations are well known from the theory of probability.)

   Note: If $r>0$, then we speak about the correlated SEMG stochastic process; in other cases such process is non-correlated (see, e.g., Sections 10.2 and 10.3 in this chapter).

   The corresponding conjugate probability density function is given by Equation (A13.3) in Appendix A13 as follows:

$$f\left(P_i,P_{i+1}\right) = \frac{1}{2\pi\sigma_{l_0}^2\sqrt{1-r^2}}\left\{-\frac{\left(P_i-\overline{P}_{l_0}\right)^2-2r\left(P_i-\overline{P}_{l_0}\right)\left(P_{i+1}-\overline{P}_{l_0}\right)+\left(P_{i+1}-\overline{P}_{l_0}\right)^2}{2\sigma_{l_0}^2\left(1-r^2\right)}\right\}.$$

$$(10.116)$$

Standardized SEMG stochastic process

Equation (10.85) and/or Equation (A13.4) in Appendix A13 describe standardized normal (Gaussian) random quantity U_i. According to Equation (A13.5), (A13.6) and (A13.7), the probability density function of U_i is

$$f\left(U_i\right) = \frac{1}{\sqrt{2\pi}}\exp\left\{-\frac{U_i^2}{2}\right\},\qquad (10.117)$$

the conditional probability density function is

$$\varphi\left(U_{i+1}|U_i\right) = \frac{1}{\sqrt{2\pi}\sqrt{1-r^2}}\exp\left\{-\frac{\left(U_{i+1}-rU_i\right)^2}{2\left(1-r^2\right)}\right\},\qquad (10.118)$$

and the conjugate probability density function is

$$f\left(U_i,U_{i+1}\right) = \frac{1}{2\pi\sqrt{1-r^2}}\exp\left\{-\frac{U_i^2 - 2rU_iU_{i+1} + U_{i+1}^2}{2\left(1-r^2\right)}\right\}. \qquad (10.119)$$

Note: Equations (10.118) and (10.119) express the relations between 'neighbouring' quantities U_i, U_{i+1} of the standardized process. The parameter r expresses correlation coefficient between the mentioned 'neighbouring' quantities.

The relations between 'far' quantities U_i, U_{i+k}, $k = 1, 2, \ldots$ are derived in Appendix 13. According to Equation (A13.27), the conditional probability density function is

$$\varphi\left(U_{i+k}\middle|U_i\right) = \frac{1}{\sqrt{2\pi}\sqrt{1-r^{2k}}}\exp\left[-\frac{\left(U_{i+k}-r^kU_i\right)^2}{2\left(1-r^{2k}\right)}\right], \qquad (10.120)$$

and according to Equation (A13.28), the conjugate probability density function is

$$f\left(U_i,U_{i+k}\right) = \frac{1}{\sqrt{2\pi}}\exp\left[-\frac{U_i^2}{2}\right]\frac{1}{\sqrt{2\pi}\sqrt{1-r^{2k}}}\exp\left[-\frac{\left(U_{i+k}-r^kU_i\right)^2}{2\left(1-r^{2k}\right)}\right]. \qquad (10.121)$$

Note: For $k = 1$, Equations (10.120) and (10.121) become identical to Equations (10.118) and (10.119), respectively.

Statistical characteristics of standardized SEMG stochastic process

It is generally known that the mean value of standardized normal (Gaussian) process U_i is $\bar{U} = 0$ and the standard deviation of this process $\sigma_U = 1$ – see Equations (10.86) to (10.88). Moreover, the covariance (non-standardized correlation function), according to Equation (A13.31) in Appendix A13, is

$$\mathrm{cov}\left(U_i,U_{i+k}\right) = r^k. \qquad (10.122)$$

The definition of correlation coefficient, according to Equation (10.72), is valid quite generally; so in the present case $\rho\left(U_i,U_{i+k}\right) = \mathrm{cov}\left(U_i,U_{i+k}\right)/\sigma_U^2$ and because $\sigma_U = 1$, it is valid that

$$\rho\left(U_i,U_{i+k}\right) = \mathrm{cov}\left(U_i,U_{i+k}\right) = r^k. \qquad (10.123)$$

Note: The covariance as well as correlation coefficient is a function of k, i.e., the difference between cardinal numbers of a pair of values. Therefore, we speak about the covariance function as well as correlation function.

Note: Let us note that the covariance function as well as correlation function is an exponentially decreasing function with argument k.

Statistical characteristics of general SEMG stochastic process

A general SEMG stochastic process P_i is characterized by its mean value $\bar{P}_{l_0}$, variance $\sigma_{l_0}^2$ and/or standard deviation σ_{l_0}, and correlation coefficient r.

According to Equation (A13.32) in Appendix A13, the covariance function is

$$\mathrm{cov}\left(P_i, P_{i+k}\right) = \sigma_{l_0}^2\, r^k\,,\tag{10.124}$$

and according to Equation (A13.33) in Appendix A13, the correlation function is

$$\rho\left(P_i, P_{i+k}\right) = r^k\,.\tag{10.125}$$

Finally, let us remember that the earlier two expressions are valid for all values of i, so that the covariance function and (standardized) correlation function are functions of k only. So, we can mark these functions in accordance with Equations (10.104) and (10.105) as follows:

$$\mathrm{cov}\left(k\right) = \mathrm{cov}\left(P_i, P_{i+k}\right) = \sigma_{l_0}^2\, r^k\,.\tag{10.126}$$

$$\rho\left(k\right) = \rho\left(P_i, P_{i+k}\right) = r^k\,.\tag{10.127}$$

Simulation of SEMG stochastic process

The algorithm for simulation of a stochastic process P_i, described in the previous section, is easier now, because it is enough to know only one generator – the generator of standardized normal (Gaussian) distribution.

Let us introduce a 'helping' random quantity V_{i+1} as follows:

$$V_{i+1} = \frac{U_{i+1} - rU_i}{\sqrt{1-r^2}},\qquad \sqrt{1-r^2}\ \text{and}\ rU_i\ldots\text{parameters},\tag{10.128}$$

$$U_{i+1} = V_{i+1}\sqrt{1-r^2} + rU_i,\qquad \mathrm{d}U_{i+1}/\mathrm{d}V_{i+1} = \sqrt{1-r^2}\,.\tag{10.129}$$

Then, by using Equations (10.118) and (10.129), we can write

$$f\left(V_{i+1}\right) = \varphi\left(U_{i+1}|U_i\right)\frac{\mathrm{d}U_{i+1}}{\mathrm{d}V_{i+1}} = \frac{1}{\sqrt{2\pi}\sqrt{1-r^2}}\exp\left\{-\frac{\left(U_{i+1}-rU_i\right)^2}{2\left(1-r^2\right)}\right\}\sqrt{1-r^2}$$

$$= \frac{1}{\sqrt{2\pi}}\exp\left\{-\frac{V_{i+1}^2}{2}\right\}. \tag{10.130}^{15}$$

However, this is the probability density function of standardized normal (Gaussian) distribution. So, we can determine the random quantity U_{i+1} from Equation (10.129) in which we substitute the random quantity V_{i+1} generated from the standardized normal (Gaussian) distribution.

Let us assume that we know the values of parameters $\bar{P}_{l_0}, \sigma_{l_0}, r$ of a normal distribution of strengths P_i. Then, the computer generation of individual values of P_i can be realized according to the following scheme:

1. By using Equation (10.85) for $i = 1$, we find

$$P_1 = \sigma_{l_0} U_1 + \bar{P}_{l_0}, \tag{i}$$

    whereas we determine the value U_1 from the generator of standardized normal (Gaussian) distribution according to Equation (10.117).

2. By using Equation (10.85) for $i = 2$, then by applying Equation (10.129) for $i = 1$, and once more by applying Equation (10.85) for $i = 1$, we find

$$P_2 = \sigma_{l_0} U_2 + \bar{P}_{l_0} = \sigma_{l_0}\left(V_2\sqrt{1-r^2} + rU_1\right) + \bar{P}_{l_0}$$

$$= \sigma_{l_0}\left(V_2\sqrt{1-r^2} + r\frac{P_1-\bar{P}_{l_0}}{\sigma_{l_0}}\right) + \bar{P}_{l_0} = \sigma_{l_0}V_2\sqrt{1-r^2} + r\left(P_1 - \bar{P}_{l_0}\right) + \bar{P}_{l_0}, \tag{ii}$$

    whereas we know the value P_1 from previous point and the random value V_2 results from the generator of standardized normal (Gaussian) distribution according to Equation (10.130).

3. Similarly, we obtain

---

15   Generally, if a function $y = g(x)$ is valid between two random variables x and y then $f(y) = f(x)\,\mathrm{d}x/\mathrm{d}y$ is valid between their probability density functions $f(y), f(x)$. (See a handbook of probability.)

$$P_3 = \sigma_{l_0} V_3 \sqrt{1-r^2} + r\left(P_2 - \bar{P}_{l_0}\right) + \bar{P}_{l_0}, \tag{iii}$$

whereas we know the value P_2 from previous point and the random value V_3 results from the generator of standardized normal (Gaussian) distribution according to Equation (10.130).

$\vdots$

4. By using Equation (10.85) step by step for $i = k$ and $i = k-1$ and by applying Equation (10.129) for $i = k-1$, we find

$$P_k = \sigma_{l_0} V_k \sqrt{1-r^2} + r\left(P_{k-1} - \bar{P}_{l_0}\right) + \bar{P}_{l_0}, \tag{iv}$$

whereas we know the value P_{k-1} from previous point and the random value V_k results from the generator of standardized normal (Gaussian) distribution according to Equation (10.130).

$\vdots$

etc.

Sum of more SEMG stochastic processes

Let us think about M mutually independent SEMG stochastic processes $^{(m)}P_i$, $m = 1, 2, \ldots, M$, determined by parameters $^{(m)}\bar{P}_{l_0}$, $^{(m)}\sigma_{l_0}$, $^{(m)}r$.

We use Equations (10.99) to (10.101) for calculation of mean value, variance and standard deviation of sum of SEMG stochastic processes. Equations (10.102) and (10.126) allow us to express the covariance function of sum of SEMG stochastic processes as follows:

$$\mathrm{cov}(k) = \mathrm{cov}\left(P_i, P_{i+k}\right) = \sum_{m=1}^{M} \mathrm{cov}\left(^{(m)}P_i, {}^{(m)}P_{i+k}\right) = \sum_{m=1}^{M}\left(^{(m)}\sigma_{l_0}^2 \, {}^{(m)}r^k\right). \tag{10.131}$$

Finally, by using Equations (10.103) and (10.127), we can write the following correlation function of sum of SEMG stochastic processes:

$$\rho(k) = \rho\left(P_i, P_{i+k}\right) = \frac{1}{\sigma_{l_0}^2} \sum_{m=1}^{M}\left[^{(m)}\sigma_{l_0}^2 \, {}^{(m)}r^k\right]. \tag{10.132}$$

[The resulting variance $\sigma_{l_0}^2$ is determined by using Equation (10.100) as the sum of variances $^{(m)}\sigma_{l_0}^2$.]

Strength at a general gauge length as SEMG stochastic process

Equation (10.112) describes yarn strength P at a general (long) gauge length l. This, for normal (Gaussian) distribution, is derived in Appendix 13, Equation (A13.35). Thus,

$$g(P,k) = -\frac{1}{\sqrt{2\pi}\sigma_{l_0}} \left(\frac{1}{\sqrt{2\pi}\,\sigma_{l_0}\sqrt{1-r^2}} \right)^k \frac{d}{dP} \left\{ \int\limits_{P_i=P}^{\infty} \int\limits_{P_{i+1}=P}^{\infty} \cdots \int\limits_{P_{i+k}=P}^{\infty} \exp\left(-\frac{\left(P_i - \bar{P}_{l_0} \right)^2}{2\sigma_{l_0}^2} \right) \right.$$

$$\left. \cdot \prod_{j=1}^{k} \left[\exp\left(-\frac{\left(P_{i+j} - \left[\bar{P}_{l_0} + r\left(P_{i+j-1} - \bar{P}_{l_0} \right) \right] \right)^2}{2\sigma_{l_0}^2 \left(1 - r^2 \right)} \right) \right] dP_i \, dP_{i+1} \ldots dP_{i+k} \right\}, \qquad k = 1, 2, \ldots$$

(10.133)

Note: Let us remind that the (long) gauge length l contains $k+1$ (short) sections of length l_0 with cardinal numbers $i, i+1, \ldots, i+k$.

The probability density function $g(p,k)$ of the following linearly transformed random quantity:

$$p = \frac{P - \bar{P}_{l_0}}{\sigma_{l_0}}, \qquad \left(P = \sigma l_0 p + \bar{P}_{l_0} \right),$$

(10.134)

takes a simpler shape.

Note: Let us remind that the random quantity p is a linearly transformed random quantity P, but it is not a standardized quantity. It is because the mean value $\bar{P}_{l_0}$ and the standard deviation σ_{l_0} are related to strengths at (short) length l_0, while the random value of the strength P relates to a (long) gauge length l. (p can be a standardized quantity only in a special case where $l = l_0$.)

In Appendix 13, the probability density function $g(p,k)$ of linearly transformed random quantity p is derived. According to Equation (A13.42), it is valid to write that

$$g(p,k) = -\frac{1}{\sqrt{2\pi}} \left(\frac{1}{\sqrt{2\pi}\sqrt{1-r^2}} \right)^k \frac{d}{dp} \left\{ \int\limits_{U_i=p}^{\infty} \int\limits_{U_{i+1}=p}^{\infty} \cdots \int\limits_{U_{i+k}=p}^{\infty} \exp\left(-\frac{U_i^2}{2} \right) \right.$$

$$\left. \cdot \prod_{j=1}^{k} \left[\exp\left(-\frac{\left(U_{i+j} - rU_{i+j-1} \right)^2}{2\left(1 - r^2 \right)} \right) \right] dU_i \, dU_{i+1} \ldots dU_{i+k} \right\}.$$

(10.135)

The main advantage of this probability density function $g(p,k)$ of the linearly transformed random quantity p is that it does not depend on the mean value $\bar{P}_{l_0}$ as well as the standard deviation σ_{l_0}. It depends – besides the parameter k – on the correlation coefficient r only.

Numerical calculations (simulations)

The determination of probability density functions according to Equation (10.133) and/or (10.135) is computationally difficult and the computations of statistical characteristics are more difficult. On the contrary, all these information are relatively easy to obtain by using simulation technique.

Let us assume that the experimentally determined strengths P_i at a short gauge length l_0 follow SEM stochastic process or sum of such several processes. Moreover, let us assume that a suitable probability density function f and a conditional probability density function φ are determined for each stochastic process by means of evaluation of experimental data. Then, we can generate strengths P_i using computer in accordance with Section 10.4, under the paragraph 'Simulation of SEM stochastic process'. Especially in the case of normal (Gaussian) process, the strengths P_i can be generated according to paragraph 'Simulation of SEMG stochastic process'.

Each $k+1$ successively generated values $\{P_i, P_{i+1}, \ldots, P_{i+k}\} \equiv \{P_{i+j}\}_{j=0}^{j=k}$ express strengths of successive (short) lengths l_0 creating together the (long) length l. The yarn strength P at (long) gauge length l is the minimum of all values $\{P_{i+j}\}_{j=0}^{j=k}$.

$$P = \min\{P_{i+j}\}_{j=0}^{j=k}.$$

(10.136)

(The principle of the weakest link theory is assumed here.)

A computer can generate a large set of strengths P in a short time. From such a set of P the probability density function $g(P,k)$ as well as all statistical characteristics can be determined. Further, this evaluation can be repeated for different values of $k = 0, 1, 2, \ldots$, to determine $g(P,k)$ for different values of k, i.e., for $l = l_0, 2l_0, \ldots$, and to determine the corresponding statistical characteristics.

Example 1

Neckář [7] studied yarn strength as a stochastic process with a cotton carded ring-spun yarn of $29.5\,\text{tex}$ count and $710\,\text{m}^{-1}$ twist. The basic (short) gauge length l_0 was chosen as 50 mm. (Evidently, all cotton fibers in this yarn were shorter than this gauge length so that the principle of the weakest link could be assumed.) He always marked sixty segments each of 50 mm length, one after another (No. 1, 2, ..., 60), on a randomly selected 3 m long length of this yarn. The strengths P_i of odd-numbered segments (No. 1, 3, 5, ..., 59) were tested, while the even-numbered segments (No. 2, 4, 6, ..., 60) were used for gripping in-between the jaws of the tensile tester, 25 mm on both sides of the tested gauge length – see Figure 10.9. The test was repeated thirty times on a 3-m long length of the yarn so that we obtained 900 strength values altogether. (However, we excluded 129 incorrect values – breakages in jaws, etc. – in this example.) Table 10.3 illustrates the basic statistical characteristics of strength of this yarn.

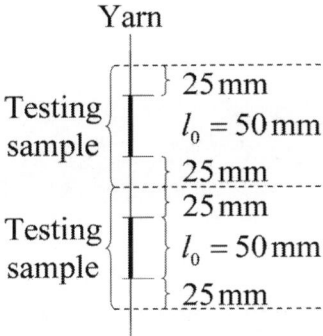

Figure 10.9 Yarn segments for testing. Thick parts – tested segments, thin parts – lengths for gripping

Table 10.3 Statistical characteristics of yarn strength [7]

Quantity	Symbol	Value
Gauge length	$l_0 \left[\text{mm}\right]$	50
No. of valid observations	-	871
Mean	$\bar{P}_{l_0}\left[\text{N}\right]$	4.5179
Standard deviation	$\sigma_{l_0}\left[\text{N}\right]$	0.5670
Coefficient of variation	$v_{l_0}\left[-\right]$	0.1255 (12.55%)
Skewness	$a\left[-\right]$	0.0162
Kurtosis	$e\left[-\right]$	0.3050

The experimentally observed strength values P_i were standardized by using the following expression $U_i = \left(P_i - \overline{P_{l_0}} \right) \Big/ \sigma_{l_0}$ according to Equation (10.85). The experimentally observed probability density function $f\left(U_i\right)$ is expressed by the histogram shown in Figure 10.10; the smooth curve corresponds to the standardized normal (Gaussian) distribution according to Equation (10.117). The character of distribution as shown in Figure 10.10 and the relatively small values of skewness and kurtosis mentioned in Table 10.3 show that this distribution can be approximated by normal (Gaussian) distribution.

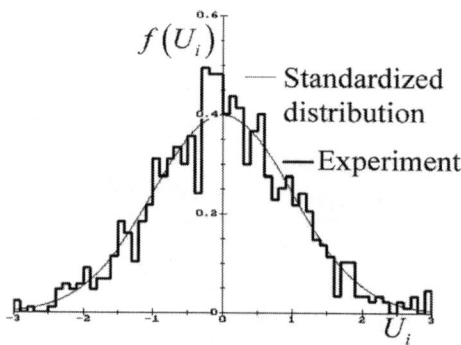

Figure 10.10 Probability density function of standardized quantity U_i for the experimental yarn [7]

Then, the experimental (standardized) correlation function was determined by using a known method[16]. We obtained values $\rho\left(U_i, U_{i+k}\right)$ for $k = 2, 4, 6, \ldots$. However, this correlation function can also be expressed by means distance as follows:

$$x_{[mm]} = l_{0[mm]}k, \quad x_{[mm]} = 50\,\text{mm} \cdot k \tag{10.137}$$

---

16   The 30 sets of values $U_1, U_3, U_5, \ldots, U_{59}$ were available for evaluation. (Even the even-numbered sections were used for gripping by the jaws of tensile tester.) The correlation coefficient $\rho\left(U_i, U_{i+2}\right)$ was determined from all the existing couples U_1, U_3, U_3, U_5, $U_5, U_7, \ldots$. Similarly, the correlation coefficient $\rho\left(U_i, U_{i+4}\right)$ was determined from all of the existing couples U_1, U_5, U_3, U_7, $U_5, U_9, \ldots$, etc. So we obtained a set of coefficients $\rho\left(U_i, U_{i+k}\right), k = 2, 4, 6, \ldots$ which characterized the experimental trend of (standardized) correlation function.

according to Equation (10.106), i.e., $\rho(U_i, U_{i+k}) = \rho(x)$; see also Equation (10.107). Figure 10.11 displays the experimentally obtained correlation coefficients.

Note: The value $\rho(U_i, U_i)$, i.e, when $x = 0$ and/or $k = 0$, is equal to 1 from definition – see Equation (10.72).

Evidently, the experimental results of correlation coefficients cannot be explained by a simple exponential function according to Equation (10.127). On the contrary, the summation of two independent SEMG stochastic processes can be a very good explanation for the experimental results shown in Figure 10.11.

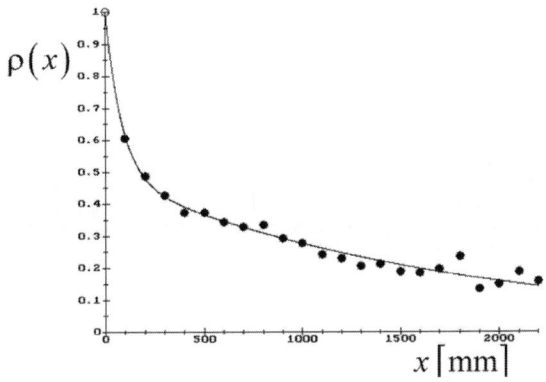

Figure 10.11 (Standardized) correlation function of the experimental yarn [7]: Points – experimentally determined values, smooth curve –regression function, Equation (10.140)

By using Equation (10.132) for $M = 2$, the following expression can be written:

$$\rho(k) = \rho(P_i, P_{i+k})$$
$$= \frac{{}^{(1)}\sigma_{l_0}^2}{\sigma_{l_0}^2}\,{}^{(1)}r^k + \frac{{}^{(2)}\sigma_{l_0}^2}{\sigma_{l_0}^2}\,{}^{(2)}r^k, \tag{10.138}$$

or, by applying Equation (10.137), we find

$$\rho(x) = \rho(P_i, P_{i+k}) = \frac{{}^{(1)}\sigma_{l_0}^2}{\sigma_{l_0}^2}\,{}^{(1)}r^{x_{[mm]}/50} + \frac{{}^{(2)}\sigma_{l_0}^2}{\sigma_{l_0}^2}\,{}^{(2)}r^{x_{[mm]}/50}. \tag{10.139}$$

By applying statistical regression method on the experimental results shown in Figure 10.11, we found

$$\rho(x) = 0.51718\,e^{-0.011913x} + 0.48282\,e^{-0.000055713x}, \tag{10.140}$$

or, by using Equation (10.137), we obtain

$$\rho(k) = 0.51718\,e^{-0.59565k} + 0.48282\,e^{-0.0278565k}. \tag{10.141}$$

By comparing Equation (10.138) with (10.141) and by using σ_{l_0} from Table 10.3, we obtain

$$\left.\begin{array}{l} ^{(1)}\sigma_{l_0}^2 \big/ \sigma_{l_0}^2 = 0.51718, \quad ^{(1)}\sigma_{l_0} = \sqrt{0.51718 \cdot 0.5670} = 0.40776\,\text{N}, \\[4pt] ^{(2)}\sigma_{l_0}^2 \big/ \sigma_{l_0}^2 = 0.48282, \quad ^{(2)}\sigma_{l_0} = \sqrt{0.48282 \cdot 0.5670} = 0.39398\,\text{N}, \\[4pt] ^{(1)}r^k = e^{-0.59565k}, \quad ^{(1)}r = e^{-0.59565} = 0.55120, \\[4pt] ^{(2)}r^k = e^{-0.0278565k}, \quad ^{(2)}r = e^{-0.0278565} = 0.97253. \end{array}\right\} \tag{10.142}$$

Note: Let us note that the differences between the values of correlation coefficients $^{(1)}r$ and $^{(2)}r$ are very significant in this example. This leads us to an idea that two independent and quite different influences are existing in the process of yarn creation. (We obtained analogical results in case of a lot of other yarns [8, 9].) Though these influences are not yet precisely known; however, hypothetically, the mass unevenness of yarn brings one of the influences and the structural unevenness[17] brings the other one.

Further, we used simulation technique for determination of probability density function of yarn strength at different gauge lengths. Unfortunately, we did not know mean values $^{(1)}\overline{P}_{l_0}$ and $^{(2)}\overline{P}_{l_0}$ of the two partial stochastic processes. Therefore, we worked with centralized random quantities

$$^{(1)}P_i^\circ = {}^{(1)}P_i - {}^{(1)}\overline{P}_{l_0}, \quad ^{(2)}P_i^\circ = {}^{(2)}P_i - {}^{(2)}\overline{P}_{l_0}, \tag{10.143}$$

whose mean values are equal to zero according to Equation (10.80).

We used expressions (i) to (iv) [paragraph 'Simulation of SEMG stochastic process' after Equation (10.130)] for the centralized random quantities. This was done after the following modifications were carried out.

---

17 Mass unevenness expresses the variation of fineness of yarn along its length. This is a known phenomenon, described in Chapter 6. The structural unevenness expresses different arrangement of fibers at different places of the yarn. It is also a generally known phenomenon, but there exists a little exact knowledge about it.

Table 10.4 Calculation of sum of two random processes in Example 1

Partial process (1) in [N]	Partial process (2) in [N]	Sum of processes in [N], Equation (10.91)
$^{(1)}P_1^\circ = {}^{(1)}\sigma_{l_0} U + {}^{(1)}\overline{P}^\circ$ $= 0.40776U$	$^{(2)}P_1^\circ = {}^{(2)}\sigma_{l_0} U + {}^{(2)}\overline{P}^\circ$ $= 0.39398U$	$P_1^\circ = {}^{(1)}P_1^\circ + {}^{(2)}P_1^\circ$
$^{(1)}P_2^\circ = {}^{(1)}\sigma_{l_0} U \sqrt{1 - {}^{(1)}r^2}$ $+ {}^{(1)}r\left({}^{(1)}P_1^\circ - {}^{(1)}\overline{P}^\circ\right) + {}^{(1)}\overline{P}^\circ$ $= 0.40776U\sqrt{1 - 0.55120^2}$ $+ 0.55120 \, {}^{(1)}P_1^\circ$ $= 0.34022U + 0.55120 \, {}^{(1)}P_1^\circ$	$^{(2)}P_2^\circ = {}^{(2)}\sigma_{l_0} U \sqrt{1 - {}^{(2)}r^2}$ $+ {}^{(2)}r\left({}^{(2)}P_1^\circ - {}^{(2)}\overline{P}^\circ\right) + {}^{(2)}\overline{P}^\circ$ $= 0.39398U\sqrt{1 - 0.97253^2}$ $+ 0.97253 \, {}^{(2)}P_1^\circ$ $= 0.09171U + 0.97253 \, {}^{(2)}P_1^\circ$	$P_2^\circ = {}^{(1)}P_2^\circ + {}^{(2)}P_2^\circ$
$^{(1)}P_3^\circ = 0.34022U + 0.55120 \, {}^{(1)}P_2^\circ$ $\cdots$	$^{(2)}P_3^\circ = 0.09171U + 0.97253 \, {}^{(2)}P_2^\circ$ $\cdots$	$P_3^\circ = {}^{(1)}P_3^\circ + {}^{(2)}P_3^\circ$ $\cdots$
$^{(1)}P_k^\circ = 0.34022U + 0.55120 \, {}^{(1)}P_{k-1}^\circ$ $\cdots$	$^{(2)}P_k^\circ = 0.09171U + 0.97253 \, {}^{(2)}P_{k-1}^\circ$ $\cdots$	$P_k^\circ = {}^{(1)}P_k^\circ + {}^{(2)}P_k^\circ$ $\cdots$

- in place of $P_1, P_2, P_3, \ldots, P_k,$, we used ${}^{(1)}P_1^\circ, {}^{(1)}P_2^\circ, {}^{(1)}P_3^\circ, \ldots, {}^{(1)}P_k^\circ$, and ${}^{(2)}P_1^\circ, {}^{(2)}P_2^\circ, {}^{(2)}P_3^\circ, \ldots, {}^{(2)}P_k^\circ,$,

- in place of $\overline{P}_{l_0}$, we used ${}^{(1)}\overline{P}^\circ = 0$ and ${}^{(2)}\overline{P}^\circ = 0$,

- in place of σ_{l_0}, we used ${}^{(1)}\sigma_{P^\circ} = {}^{(1)}\sigma_{l_0}$ and ${}^{(2)}\sigma_{P^\circ} = {}^{(2)}\sigma_{l_0}$, according to Equation (10.142),

- in place of r, we used ${}^{(1)}r$ and ${}^{(2)}r$, according to Equation (10.142),

- in place of U_1, U_2 and $V_1, V_2, V_3, \ldots, V_k,$, we used mutually independent values U (and similarly values V) of standardized normal (Gaussian) distribution – Equation (10.117).

By using such modifications, we obtained the following equations stated in Table 10.4 in case of this example:

Finally, by using Equations (10.79), (10.91) and $\overline{P}_{l_0}$ from Table 10.3, we obtained the resulting values of random strengths at short (50 mm) gauge length as follows:

$$P_i = \left(\overbrace{{}^{(1)}P_i^\circ + {}^{(1)}\overline{P}_{l_0}}^{= {}^{(1)}P_i} \right) + \left(\overbrace{{}^{(2)}P_i^\circ + {}^{(2)}\overline{P}_{l_0}}^{= {}^{(2)}P_i} \right) = {}^{(1)}P_i^\circ + {}^{(2)}P_i^\circ + \left({}^{(1)}\overline{P}_{l_0} + {}^{(2)}\overline{P}_{l_0} \right)$$

$$= {}^{(1)}P_i^\circ + {}^{(2)}P_i^\circ + \overline{P}_{l_0},$$

$$P_{i[N]} = {}^{(1)}P_{i[N]}^\circ + {}^{(2)}P_{i[N]}^\circ + 4.5179\,\text{N}, \quad i = 1, 2, \ldots, k. \tag{10.144}$$

We simulated 10,000 5-m long parts of a yarn, each containing 100 segments each of 50 mm length, according to the procedure mentioned above. We, therefore, generated $10{,}000 \cdot 100 = 10^6$ strength values related to (short) gauge length $l_0 = 50\,\text{mm}$. These values were evaluated by analogical way as followed in case of real experimental data.

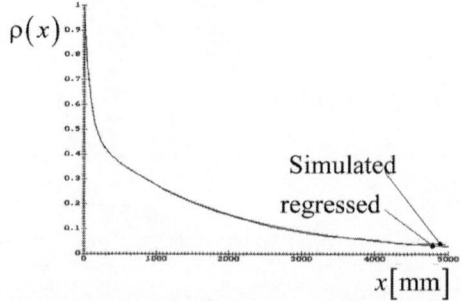

$x[\text{mm}]$

Figure 10.12 Comparison of (standardized) correlation functions, the experimental yarn [7]: Regressed – regression function according to Equation (10.140), Simulated – from computer generated data by using values of Equation (10.142)

Note: Figure 10.12 illustrates the (standardized) correlation function determined according to Equation (10.140), i.e., regression function determined from experimental data, shown in Figure 10.11, and (standardized) correlation function evaluated from the set of simulated values. [Parameters from Equation (10.142) were used during the process of simulation.] The perfect accordance shows that the computer generated data can be really considered as an artificially created 'set of pseudo-experimental data'.

We obtained altogether 10^6 strength values of P related to (short) gauge length $l_0 = 50\,\text{mm}$. (These values have $k = 0, x = 0$.) We generated 10^6 corresponding linearly transformed values p according to Equation (10.134)[18] and illustrated the distribution of thus generated values of p by the histogram shown in Figure 10.13a. This histogram, representing the probability density function $g(p, k = 0)$, creates the standardized normal (Gaussian) probability density function, evidently.

Further, we determined yarn strengths P at different (longer) gauge lengths – according to Equation (10.136) – from the same set of primarily generated values. Then, we calculated the corresponding linearly transformed values p according to Equation (10.134) and created the histograms of probability density functions. Two examples of them (for $l = 500\,\text{mm}$ and $l = 5000\,\text{mm}$) are shown in Figure 10.13b and 10.13c, respectively.

Note: The number of generated yarn strengths decreases with increasing gauge lengths l. We generated 10^6 strength values for $l = l_0 = 50\,\text{mm}$; however, we had to associate several more short segments for creation of each longer gauge lengths every times. (For example, the sequence of 100 short lengths each of $50\,\text{mm}$ creates only one long gauge length $5000\,\text{mm}$.) It explains why the histogram shown in Figure 10.13c is not so 'smooth' as they appeared in the previous two graphs – Figure 10.13a and Figure 10.13b.

The histograms shown in Figure 10.13 show that with increase of gauge length,

- the mean value of yarn strength decreases,
- the standard deviation of yarn strength decreases and
- the asymmetry slightly increases.

---

18   In this case, the gauge length is $l = l_0 = 50\,\text{mm}$, hence the linearly transformed random quantities p according to Equation (10.134) are also the standardized quantities at the same time.

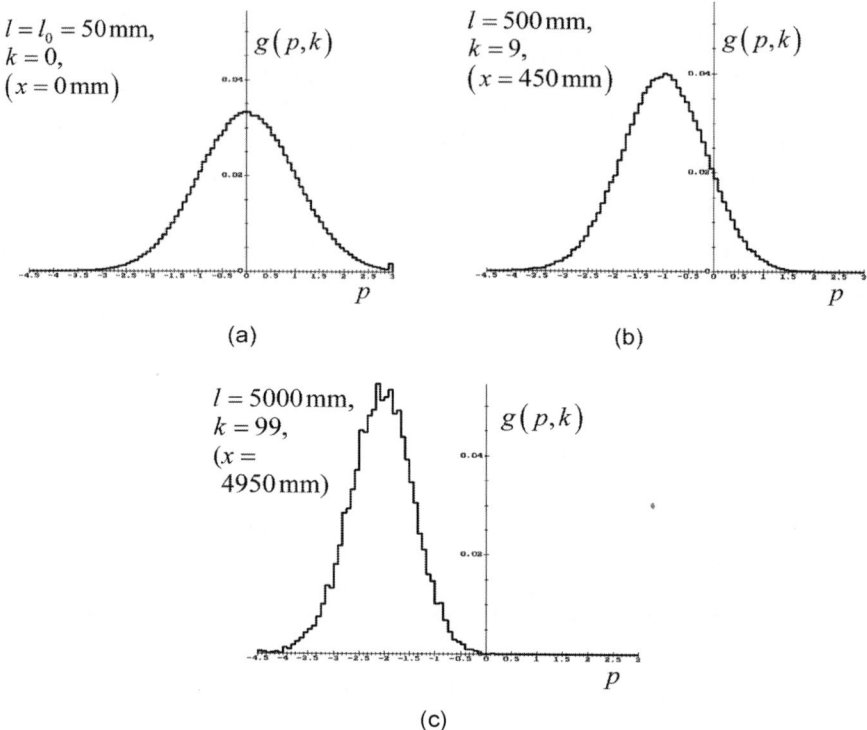

Figure 10.13 Probability density functions of linearly transformed quantity p; the experimental yarn [7]

The mean values and standard deviations of linearly transformed quantities p, evaluated from generated distributions, are shown in Table 10.5 – see the column labelled 'simulation'.

Note: It is interesting to note that the Peirce's type of empirical Equations (10.29) and (10.30) can also be used in this case, but naturally with other values of parameters. We obtained the equations as follows:

$$\left.\begin{aligned}\sigma_p &= \left(l/l_0\right)^{-1/12.5}, \\ \bar{p} &= 6\left[\left(l/l_0\right)^{-1/12.5} - 1\right].\end{aligned}\right\} \tag{10.145}$$

Table 10.5 shows the values according to these two expressions in the column labelled 'approximation'. [Let us note that Equations (10.137), (10.140) to (10.142), (10.144), (10.145) and Tables 10.4 and 10.5) are valid only for the yarn discussed in Example 1.]

Table 10.5 Mean values and standard deviation of linearly transformed quantities p in Example 1

Gauge length l (mm)	Mean value $\bar{p}$		Standard deviation σ_p	
	$\bar{p} = \dfrac{\bar{P}_l - \bar{P}_{l_0}}{\sigma_{l_0}} = \dfrac{\bar{P}_{l[\mathrm{N}]} - 4.5179}{0.5670}$		$\sigma_p = \dfrac{\sigma_l}{\sigma_{l_0}} = \dfrac{\sigma_{l[\mathrm{N}]}}{0.5670}$	
	Simulation	Approximation, Equation (10.145)	Simulation	Approximation, Equation (10.145)
50	0	0	1.00	1
100	−0.27	−0.32	0.96	0.95
150	−0.45	−0.50	0.93	0.92
200	−0.58	−0.63	0.91	0.90
250	−0.68	−0.72	0.89	0.89
300	−0.76	−0.80	0.88	0.87
400	−0.90	−0.92	0.86	0.85
500	−1.00	−1.01	0.84	0.83
600	−1.08	−1.08	0.82	0.82
800	−1.21	−1.19	0.80	0.80
1000	−1.31	−1.28	0.78	0.79
2000	−1.63	−1.53	0.72	0.74
3000	−1.82	−1.68	0.68	0.72
4000	−1.96	−1.77	0.66	0.70
5000	−2.07	−1.85	0.63	0.69

Example 2

A detailed attempt was made by Das [4] to study yarn strength as a stochastic process. One characteristic example – a ring-spun combed cotton yarn of 7.4 tex count and 1080 twist per meter[19] – is discussed here.

---

19   The same yarn was used as an example while discussing Peirce's model (see Figure 10.5).

The same procedure as discussed in Example 1 was used to obtain the experimental strength values.

Note: To realize the yarn strength measurements, a special attachment was devised for semi-automatic feeding of equal length of yarn specimens in-between the jaws of the tensile tester one after another. The (shortest) gauge length was selected as $l_0 = 50\,\text{mm}$ and the testing speed was selected in such a manner that almost all the yarn specimens were broken within 20 ± 3 s.

In addition, the same yarn was also experimentally tested for strength at other gauge lengths viz. $l = 100, 200, 350, 500$ and $700\,\text{mm}$. The basic statistical parameters of strengths measured at different gauge lengths are presented in Table 10.6.

Table 10.6 Experimental results of strengths of 7.4 tex combed ring yarn [4]

$l\,(\text{mm})$	Mean strength $\dfrac{\bar{P}_{l[\text{cN}]}}{7.4\,\text{tex}}$ (cN/tex)	Standard deviation $\dfrac{\sigma_{l[\text{cN}]}}{7.4\,\text{tex}}$ (cN/tex)	Coefficient of variation $\dfrac{\bar{P}_{l[\text{cN}]}}{\sigma_{l[\text{cN}]}} \cdot 100 = v_l \cdot 100\,(\%)$
50*	21.8135	2.6730	12.2548
	$(\bar{P}_{l_0} = 21.8135 \cdot 7.4/100$	$(\sigma_{l_0} = 2.6730 \cdot 7.4/100$	
	$= 1.61420\,\text{N})$	$= 0.1978\,\text{N})$	
100	20.2230	2.3730	11.7553
200	20.4351	2.2284	10.9012
350	19.6149	2.3446	11.9563
500	18.7365	1.8554	9.9056
700	17.1782	1.6420	9.5587

* Here, $l = l_0 = 50\,\text{mm}$. It means that the individually measured values $P = P_i$, mean value $\bar{P}_l = \bar{P}_{l_0}$, standard deviation $\sigma_l = \sigma_{l_0}$ and linearly transformed values $p = \left(P - \bar{P}_{l_0}\right)\!/\sigma_{l_0}$ according to Equation (10.134), are standardized values $U_i = \left(P_i - \bar{P}_{l_0}\right)\!/\sigma_{l_0}$ according to Equation (10.85).

All experimentally measured strengths P at all gauge lengths were linearly transformed to p according to Equation (10.134). The histograms of experimental results – experimental functions $f_{\text{E}}(p)$ – are illustrated in Figure 10.14 together with the (smooth and thin) standardized Gaussian distribution. The relative shifting of the histogram to the left-hand side direction with the increase in gauge length is well noticeable.

Note: The histogram for the shortest gauge length $l = l_0 = 50\,\text{mm}$ is visually in good coherence with the Gaussian standardized distribution. Also, the quantile–quantile plot (Q–Q plot) of strengths evaluated at 50 mm gauge length demonstrates this (see Figure 10.15). Hence, the assumption of Gaussian (normal) distribution is justifiable.

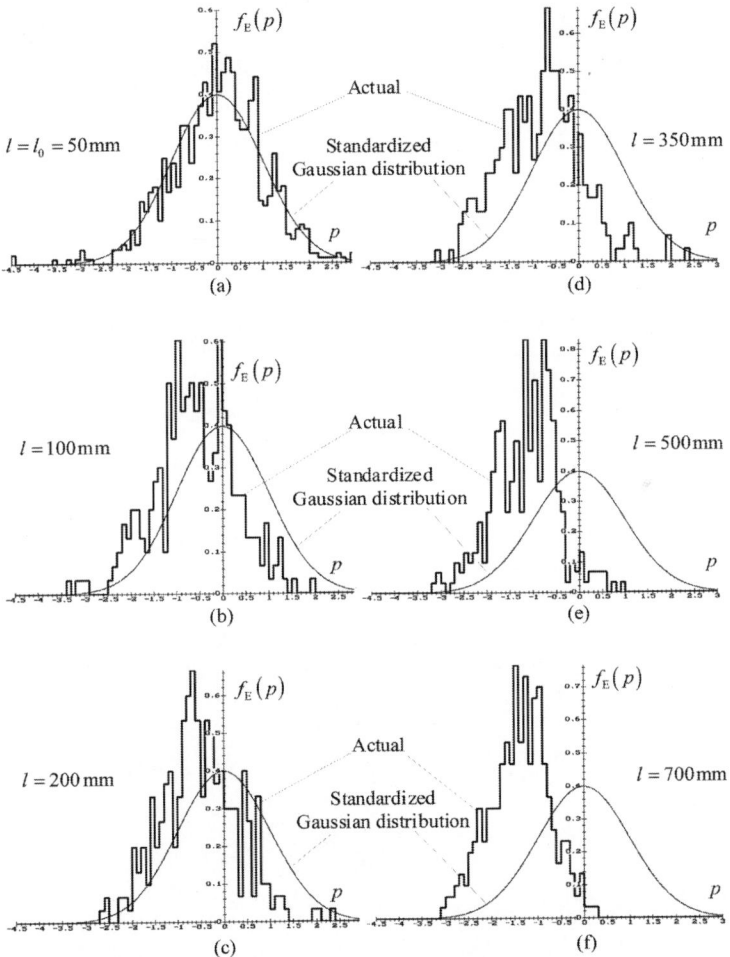

Figure 10.14 Histograms of linearly transformed yarn strengths experimentally measured at different gauge length; the experimental yarn [4]

The experimental values of (standardized) correlation function were determined in the same manner as it was described in Example 1. Subsequently, the sum of two exponential functions, corresponding to Equations (10.138) and (10.139), was determined by using statistical regression technique. We obtained

$$\rho(x) = 0.6604\,e^{-0.014049x} + 0.3396\,e^{-0.000376x},\tag{10.146}$$

or, by using Equation (10.137), we get

$$\rho(k) = 0.6604\,e^{-0.7025k} + 0.3396\,e^{-0.0188k}.\tag{10.147}$$

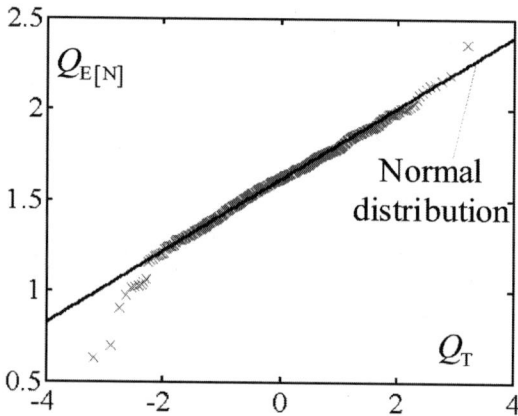

Figure 10.15 Q–Q plot of yarn strength at 50 mm gauge length; the experimental yarn [4]: Quantile values: Q_E …empirical, Q_T …theoretical (Gaussian)

The graphical interpretation of the (standardized) correlation function according to Equation (10.146) and the behaviour of each of the two (steeper fall-off and gradual fall-off) components are shown in Figure 10.16.

By comparing Equation (10.138) with (10.147) and by using σ_{l_0} from Table 10.6, we obtain

$$
\left.
\begin{aligned}
&^{(1)}\sigma_{l_0}^2 / \sigma_{l_0}^2 = 0.6604, \quad {}^{(1)}\sigma_{l_0} = \sqrt{0.6604} \cdot 0.1978 = 0.1607\,\text{N},\\
&^{(2)}\sigma_{l_0}^2 / \sigma_{l_0}^2 = 0.3396, \quad {}^{(2)}\sigma_{l_0} = \sqrt{0.3396} \cdot 0.1978 = 0.1153\,\text{N},\\
&^{(1)}r^k = e^{-0.7025k}, \quad {}^{(1)}r = e^{-0.7025} = 0.4953,\\
&^{(2)}r^k = e^{-0.0188k}, \quad {}^{(2)}r = e^{-0.0188} = 0.9814.
\end{aligned}
\right\}
\tag{10.148}
$$

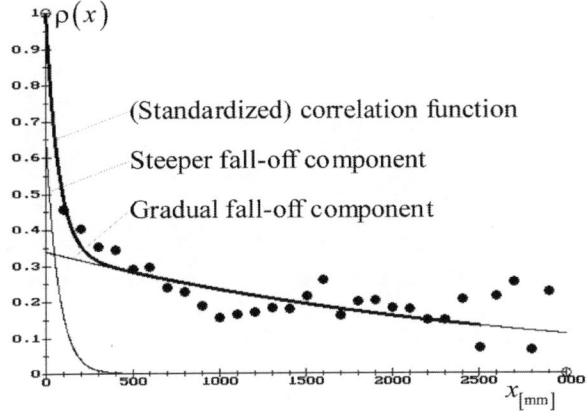

Figure 10.16 (Standardized) correlation function, the experimental yarn [4]:
Points…experimentally determined values

Note: Also, in Example 1, the differences between the values of correlation coefficients $^{(1)}r$ and $^{(2)}r$ are very significant. This leads to an idea of two independent and quite different influences, mentioned as a note after Equation (10.142). The levels of correlation coefficients, namely one value near to 0.5 and second value around 0.9, were observed in Examples 1 and 2 but for different yarns. It may be possible that this phenomenon represents a regulation whose logical interpretation is not yet fully known.

Further, we used the simulation techniques for determination of probability density function of yarn strength at different gauge lengths, in the same way as described in details in Example 1, before and after Equation (10.143).

Finally, by using Equations (10.79), (10.91) and $\overline{P}_{l_0}$ from Table 10.6 (first row), we obtained the resulting values of random strengths at short (50 mm) gauge lengths as follows:

$$P_i = \left(\overbrace{^{(1)}P_i^{\circ} + {}^{(1)}\overline{P}_{l_0}}^{= {}^{(1)}P_i} \right) + \left(\overbrace{^{(2)}P_i^{\circ} + {}^{(2)}\overline{P}_{l_0}}^{= {}^{(2)}P_i} \right) = {}^{(1)}P_i^{\circ} + {}^{(2)}P_i^{\circ} + \left({}^{(1)}\overline{P}_{l_0} + {}^{(2)}\overline{P}_{l_0} \right)$$

$$= {}^{(1)}P_i^{\circ} + {}^{(2)}P_i^{\circ} + \overline{P}_{l_0},$$

$$P_{i[\mathrm{N}]} = {}^{(1)}P_{i[\mathrm{N}]}^{\circ} + {}^{(2)}P_{i[\mathrm{N}]}^{\circ} + 1.61420\,\mathrm{N}, \quad i = 1, 2, \ldots, k. \tag{10.149}$$

Table 10.7 Calculation of sum of two random processes in Example 2

Partial process (1), expressed in [N]	Partial process (2), expressed in [N]	Sum of processes, expressed in [N], Equation (10.91)
$^{(1)}P_1^\circ = {}^{(1)}\sigma_{l_0} U + {}^{(1)}\bar{P}^\circ$ $= 0.1607U$	$^{(2)}P_1^\circ = {}^{(2)}\sigma_{l_0} U + {}^{(2)}\bar{P}^\circ$ $= 0.1153U$	$P_1^\circ = {}^{(1)}P_1^\circ + {}^{(2)}P_1^\circ$
$^{(1)}P_2^\circ = {}^{(1)}\sigma_{l_0} U \sqrt{1 - {}^{(1)}r^2}$ $\quad + {}^{(1)}r\left({}^{(1)}P_1^\circ - {}^{(1)}\bar{P}^\circ\right) + {}^{(1)}\bar{P}^\circ$ $= 0.1607U\sqrt{1 - 0.4953^2}$ $\quad + 0.4953\,{}^{(1)}P_1^\circ$ $= 0.1396U + 0.4953\,{}^{(1)}P_1^\circ$	$^{(2)}P_2^\circ = {}^{(2)}\sigma_{l_0} U \sqrt{1 - {}^{(2)}r^2}$ $\quad + {}^{(2)}r\left({}^{(2)}P_1^\circ - {}^{(2)}\bar{P}^\circ\right) + {}^{(2)}\bar{P}^\circ$ $= 0.1153U\sqrt{1 - 0.9814^2}$ $\quad + 0.9814\,{}^{(2)}P_1^\circ$ $= 0.0221U + 0.9814\,{}^{(2)}P_1^\circ$	$P_2^\circ = {}^{(1)}P_2^\circ + {}^{(2)}P_2^\circ$
$^{(1)}P_3^\circ = 0.1396U + 0.4953\,{}^{(1)}P_2^\circ$	$^{(2)}P_3^\circ = 0.0221U + 0.9814\,{}^{(2)}P_2^\circ$	$P_3^\circ = {}^{(1)}P_3^\circ + {}^{(2)}P_3^\circ$
…	…	…
$^{(1)}P_k^\circ = 0.1396U + 0.4953\,{}^{(1)}P_{k-1}^\circ$	$^{(2)}P_k^\circ = 0.0221U + 0.9814\,{}^{(2)}P_{k-1}^\circ$	$P_k^\circ = {}^{(1)}P_k^\circ + {}^{(2)}P_k^\circ$
…	…	…

Subsequently, we simulated 10^6 strength values at short gauge length (50 mm) in the same style as mentioned in Example 1. (This values have evidently $k = 0, x = 0$.) Then, we determined also yarn strengths P at different (longer) gauge lengths – according to Equation (10.136) – from the same set of primarily generated values, and afterwards we calculated linearly transformed values p according to Equation (10.134) for all yarn strengths at all gauge lengths. (The values $\overline{P}_{l_0}$ and σ_{l_0} are shown in the first row of Table 10.6.) Figure 10.17 shows three histograms of frequency distribution (for $l = l_0 = 50$ mm, $l = 500$ mm, $l = 5000$ mm). Let us note that these histograms are very similar to those shown in Example 1 (Figure 10.13). Also the histograms shown in Figure 10.17 exhibit that with the increase of gauge length, the mean value of yarn strength decreases, the standard deviation of yarn strength decreases, and the asymmetry slightly increases.

Table 10.8 reports on the mean values and standard deviations of linearly transformed quantities p, evaluated from generated distributions (column labelled 'simulation').

The Peirce's type empirical Equations (10.29) and (10.30) can also be obtained in this case, but with other values of parameters. We determined the following relations:

$$\sigma_p = \left(l/l_0 \right)^{-1/9.3}, \quad \overline{p} = 5.33 \left[\left(l/l_0 \right)^{-1/9.3} - 1 \right]. \tag{10.150}$$

Table 10.8 shows the values according to the last expressions (column labelled 'approximation'). An acceptable accordance between simulation and approximation values is evident.

Note: Let us note that Equations (10.137), (10.146) to (10.150), Table 10.6 and Table 10.7 are valid only for the yarn in Example 2.

It is possible to compare three results of mean and standard deviation of strength of this yarn (7.4 tex count and 1080 turns per meter twist):

- experimentally measured (actual) values
- values calculated according to Peirce's model of independent strengths (Figure 10.5) and
- values calculated according to stochastic dependence of yarn strengths (solved now).

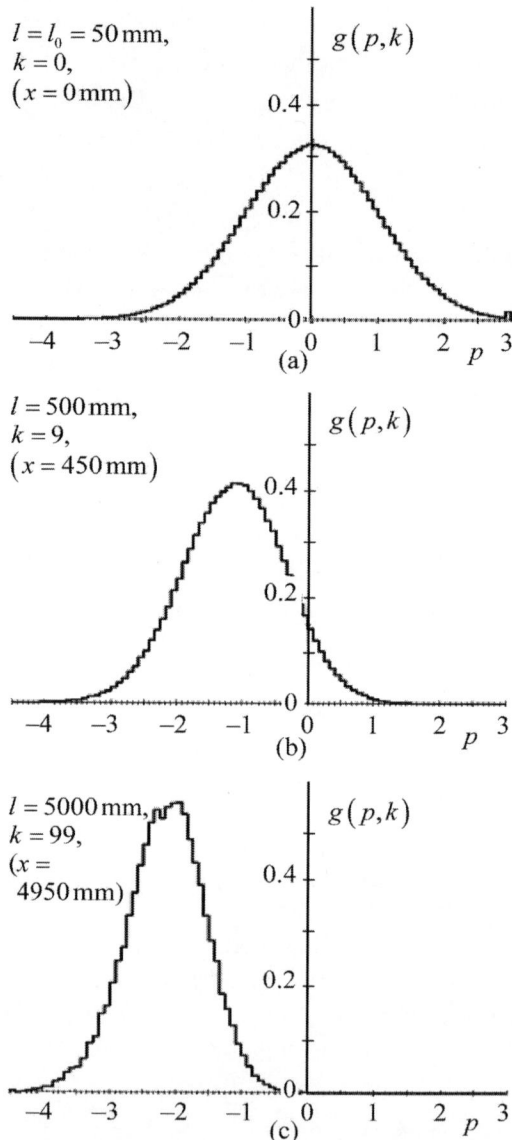

Figure 10.17 Probability density functions of linearly transformed quantity p in the case of experimental yarn [4]

This comparison is illustrated in Figure 10.18. It shows that the last model (stochastic dependence of yarn strengths) represents the experimental results better than Peirce's model of independent strengths.

Table 10.8 Mean values and standard deviation of linearly transformed quantities p in case of Example 2

Gauge length l (mm)	Mean value $\bar{p}$ $$\bar{p} = \frac{\bar{P}_l - \bar{P}_{l_0}}{\sigma_{l_0}} = \frac{\bar{P}_{l[\text{N}]} - 1.61420}{0.1978}$$		Standard deviation σ_p $$\sigma_p = \frac{\sigma_l}{\sigma_{l_0}} = \frac{\sigma_{l[\text{N}]}}{0.1978}$$	
	Simulation	Approximation, Equation (10.150)	Simulation	Approximation, Equation (10.150)
50	0	0	1	1
100	−0.33	−0.38	0.94	0.93
150	−0.53	−0.59	0.91	0.89
200	−0.68	−0.74	0.88	0.86
⋮	⋮	⋮	⋮	⋮
500	−1.13	−1.17	0.79	0.78
⋮	⋮	⋮	⋮	⋮
1000	−1.45	−1.47	0.73	0.72
⋮	⋮	⋮	⋮	⋮
5000	−2.15	−2.08	0.60	0.61

Notes to Examples 1 and 2

The results mentioned in two examples are similar to a certain degree. We can say that evidently:

- The principle of the weakest link is probably an acceptable idea for explanation of decreasing trends of mean value and standard deviation in relation to increasing gauge lengths.

- The sum of two independent SEMG stochastic processes appears to be a satisfactory model to explain the variation of yarn strength along the length of the yarn.

- The significant difference between correlation coefficients $^{(1)}r$ and $^{(2)}r$ (near to 0.5 and 0.9) evokes an idea of two different technological influences (probably mass unevenness and structural unevenness) which are responsible for this observation.

- The main results of stochastic modelling of yarn strength in relation to gauge length can be empirically substituted by Peirce's type of empirical Equations (10.29) and (10.30) but with different parameters. Above all, the exponent $-1/5$ should be usually substituted by a higher value in the denominator (see $-1/12.5$ in Example 1 and $-1/9.3$ in Example 2).

Note: Zurek [10], based on experimental results, recommended an exponent $-1/7$ empirically. Nečkář [11] showed that the logical principle of this empirical recommendation lies in the idea of correlated stochastic processes.

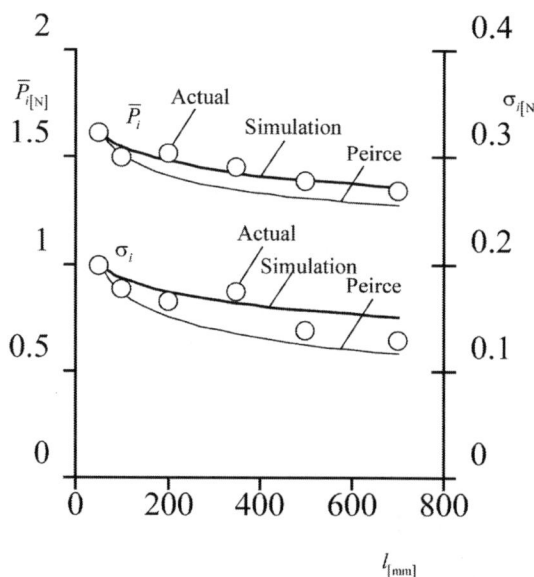

Figure 10.18 Comparison among experimental results, Peirce's model and stochastic model [4]

10.6   References

[1]   Müller, E., Handbuch der Spinnerei (Handbook of Spinning Mill), Berlin, 1892.

[2]   Peirce F. T., Tensile Tests for Cotton Yarns v. – 'The Weakest Link' Theorems on The Strength of Long and of Composite Specimens, *Journal of Textile Institute*, 17, T355–T368, 1926.

[3]   Das, D., Yarn strength as a function of gauge length – a critical review, Vlákna a Textil (Fiber and Textile), 12, 7–12, 2005.

[4]   Das, D., Yarn strength as a stochastic process, PhD Thesis, Technical University of Liberec, Czech Republic, 2005.

[5]   Militký, J., and Kovačič, V., Ultimate mechanical properties of basalt filaments, *Textile Research Journal*, 66, 225–229, 1996.

[6]   Das, D. and Neckář, B., A Methodology for better characterization of yarn strength, Melliand International, 11, 104–106, 2005.

[7]   Neckář, B., Morfologie a strukturní mechanika obecných vlákenných útvarů (Morphology and Structural Mechanics of General Fiber Assemblies) Technical University of Liberec, Liberec 1998 (In Czech).

[8]   Das, D. and Neckář, B., Yarn strength behavior at different gauge length, *Indian Journal of Fiber and Textile Research*, 30, 414–420, 2005.

[9]   Das, D., Mathematical modeling and experimental investigation of yarn strength as a stochastic process, Book of Abstracts of Fiber Society Annual Meeting and Technical Conference, Ithaca, USA, 2004.

[10]   Zurek,W., Struktura liniowych wzrobów włókienniczych (Structure of Linear Textiles), Wydawnictwa Naukovo-Techniczne, Warszava, 1989 (In Polish).

[11]   Neckář, B., Zurek's 1/7 – Yarn Strength as a Stochastic Process, Proceedings of IMWO – Metrology in Textile Engineering (Connected with celebration of 50th Anniversary of Prof. W. Zurek's work), Technical University of Lodz, Lodz, 8–18, 1996.

Constitutive theory of fiber-to-fiber slippage

11.1   Introductory remarks

Yarn tensioning and yarn strength are very complicated phenomena where many partial factors, often not enough known, are playing significant roles in deciding such complex events. A complete model of yarn tensioning and yarn strength is therefore not yet fully known. As of now, one can only think of some influences on yarn tensioning and yarn strength and eventually suggest some empirical or semi-empirical models.

There are three spheres of influence that govern yarn tensioning and yarn strength:

1. Properties of fibers. They include geometrical properties of fibers such as fiber length, fiber fineness, etc., and mechanical properties of fibers such as fiber stress–strain curve, fiber strength, fiber breaking strain, fiber-to-fiber friction, etc.

2. Extrinsic conditions of tensioning and breaking of yarn. They encompass gauge length, rate of straining, etc.

3. Structure of yarn. It comprises of

    (a) macro characteristics of yarn structure, e.g., yarn fineness (yarn count), yarn twist, fiber blend ratio (in case of blended yarns), etc., and

    (b) micro characteristics of yarn structure, e.g., fiber directional arrangement (orientation), fiber crimp, fiber-to-fiber contacts, etc. Of them, a very specific but quantitatively 'mysterious' phenomenon is fiber-to-fiber slippage.

Note: Besides the above-mentioned spheres of influence, we must also take into account of the variability of each partial characteristic that ultimately govern yarn tensioning and yarn strength.

Fiber-to-fiber slippage is a very complex phenomenon, where many partial influences are acting together. They are, for example, fiber geometry in yarn, fiber stress–strain relation, fiber-to-fiber friction, fiber-to-fiber contacts, and above all very significant random variability of microstructure of yarn. This complicated event is not yet fully known theoretically and solved mathematically.

The difficulty that arises in formulating a theory of fiber-to-fiber slippage broadly falls into two categories.

1. Usually, we do not have information on the real inputs that are necessary for solving the problem of fiber-to-fiber slippage. The inputs include the real (partly random) fiber geometry in yarn, the valid law of fiber-to-fiber friction, the pressure imparted on a slipped fiber by the surrounding ones, etc.

2. Nevertheless, even if we (hypothetically) have all input information, we do not know the underlying principles of fiber-to-fiber slippages enough well; for example, we do not know whether the slippage occurs at the ends of the fibers or at the 'middle part' of the fibers.

This chapter is directed to theoretically formulate the regulations of fiber-to-fiber slippages (point number 2). It aims at developing a constitutive (fundamental) set of relations for future theoretical models. (The methods of application of the constitutive theory are illustrated with the help of easy examples.)

11.2   Static equilibrium of fiber elements

Elementary fiber portion

Let us imagine a general fiber in a tensioned or non-tensioned yarn, as shown in Figure 11.1a. A coordinate λ along the fiber path describes individual position on this fiber at its initial (non-tensioned) state. Let us now take an elementary fiber portion of length $d\lambda$, which is lying on coordinate λ. Such an element lies at a radius r and is inclined at an angle ϑ to the direction of yarn axis. Figure 11.1b and 11.1c display a magnified view of this element. (It is hereby assumed that the fiber cross-sectional area s and fiber perimeter p are same throughout its length.) As shown in the figure, the following forces are acting on this element in its length-wise direction[1]:

1. Internal forces – The tensile (normal) forces P and its elementary increment dP are acting on the fiber cross sections with area s.

---

1   Each general force can be interpreted as a couple of length-wise and transverse components that are acting on the fiber. However, we are at this moment studying the length-wise component only.

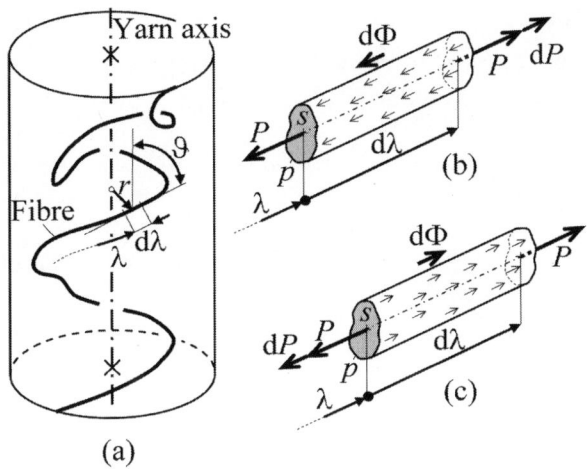

Figure 11.1 General fiber in a yarn and its element

2. Surface forces – The frictional (shear) forces are acting on the fiber surfaces at different places, shown by the small thin arrows in Figure 11.1b and 11.1c. Their resultant force is the elementary frictional force $d\Phi$ which is also acting on the fiber element. This must be always acting in the opposite to the direction of increment of tensile force dP (principle of action and reaction forces) – see Figure 11.1b and 11.1c.

Let us assume that the magnitude of the (elementary) forces is considered to be positive when they are oriented in the direction of the horizontal coordinate λ. In contrary to this, the elementary forces oriented in a direction opposite to the direction of the horizontal coordinate λ are taken as negative. Then, the condition of static equilibrium of forces, as shown in Figure 11.1b and 11.1c, is

$$dP + d\Phi = 0, \quad dP/d\lambda + d\Phi/d\lambda = 0. \tag{11.1}$$

Note: It is evident from Figure 11.1b that $dP > 0$ ($dP/d\lambda > 0$) and $d\Phi < 0$ ($d\Phi/d\lambda < 0$). Further, it is obvious from Figure 11.1c that $dP < 0$ ($dP/d\lambda < 0$) and $d\Phi > 0$ ($d\Phi/d\lambda > 0$).

In a stable situation, the static equilibrium of forces must be valid for all fiber elements in a tensioned yarn as well in a non-tensioned yarn. If this equilibrium is not maintained (e.g., by additional yarn extension), then such elements will promptly move to a new and balanced position for a brief period of time (fiber-to-fiber slippage).

We can express the tensile force $P = \sigma(\lambda)s$ by means of tensile stress $\sigma(\lambda)$[2] and fiber cross-sectional area s. By using Equations (1.5) and (1.7) from Chapter 1, we can write the following expressions:

$$P = \sigma(\lambda)s, \quad dP = s\,d\sigma(\lambda), \quad \frac{dP}{d\lambda} = s\frac{d\sigma(\lambda)}{d\lambda} = \frac{t}{\rho}\frac{d\sigma(\lambda)}{d\lambda} = \frac{\pi d^2}{4}\frac{d\sigma(\lambda)}{d\lambda}, \quad (11.2)$$

where t denotes fiber fineness, ρ indicates fiber density, and d refers to equivalent fiber diameter, according to Equation (1.11) in Chapter 1.

The frictional (shear) force $d\Phi$ can be expressed by using frictional (shear) stress $\varphi(\lambda)$, which represents the actual shear force due to friction related to per unit area of fiber surface along the coordinate λ. Thus, $d\Phi = \varphi(\lambda)\,p\,d\lambda$, where $p\,d\lambda$ is the surface area of (non-tensioned) fiber element – see Figure 11.1b or 11.1c. By applying Equations (1.6) and (1.11) from Chapter 1, we can write

$$d\Phi = \varphi(\lambda)\,p\,d\lambda, \quad \frac{d\Phi}{d\lambda} = \varphi(\lambda)\,p = 2\varphi(\lambda)\sqrt{\frac{t}{\rho}}\sqrt{\pi}(1+q) = \varphi(\lambda)\pi d(1+q),$$

$$(11.3)$$

where p refers to fiber perimeter and q stands for fiber shape factor – see Equation (1.11) in Chapter 1.

By using Equations (11.2) and (11.3) in (11.1), we obtain the following relations for the static equilibrium of forces in the horizontal direction of the fiber element:

$$\frac{\pi d^2}{4}\frac{d\sigma(\lambda)}{d\lambda} + \varphi(\lambda)\pi d(1+q) = 0, \quad \frac{d\sigma(\lambda)}{d\lambda} = -\frac{\varphi(\lambda)}{d}4(1+q), \quad (11.4)$$

and

$$\frac{t}{\rho}\frac{d\sigma(\lambda)}{d\lambda} + 2\varphi(\lambda)\sqrt{\frac{t}{\rho}}\sqrt{\pi}(1+q) = 0, \quad \frac{d\sigma(\lambda)}{d\lambda} = -\frac{\varphi(\lambda)}{\sqrt{t}}2\sqrt{\pi\rho}(1+q). \quad (11.5)$$

---

2   The quantity 'λ' shows that the above-mentioned tensile stress (and later on other quantities) relates to the fiber element lying on coordinate λ. Generally, this tensile stress changes along the fiber.

Note: According to the note stated after Equation (11.1), it is true that $d\sigma(\lambda)/d\lambda > 0$, where $\varphi(\lambda) < 0$ and $d\sigma(\lambda)/d\lambda < 0$, where $\varphi(\lambda) > 0$ – see Equations (11.2) and (11.3).

We can formally rewrite Equations (11.4) and (11.5) – the static equilibrium of forces – also in the following form:

$$\frac{d\sigma(\lambda)}{d\lambda} = -k\varphi(\lambda), \tag{11.6}$$

where k is a suitable positive constant (fiber parameter), which is defined by

$$k = \frac{4}{d}(1+q) = \frac{2}{\sqrt{t}}\sqrt{\pi\rho}(1+q). \tag{11.7}$$

According to the last two expressions, the static equilibrium of the horizontal forces acting on the fiber element depends on $\varphi(\lambda)$ and d or t (besides q and ρ). It has two consequences:

1. According to Equation (11.6), the absolute value of the derivative $|d\sigma(\lambda)/d\lambda|$ increases with an increase of the absolute value of frictional stress $|\varphi(\lambda)|$. In other words, a higher friction implies an ability of the fiber to transfer a higher tensile stress. Of course, this trivial consequence is generally known.

2. According to Equation (11.6), the absolute value of the derivative $|d\sigma(\lambda)/d\lambda|$ also increases with an increase of the value of k. However, k increases with the decrease of equivalent fiber diameter d and/ or fiber fineness t (e.g., in tex) in accordance with Equation (11.7). Thus, a finer fiber exhibits an ability to transfer higher tensile stress. (We often prefer finer fibers when we wish to magnify yarn tenacity, namely in the case of rotor yarns.)

If it is mechanically possible, then the frictional shear stress $\varphi(\lambda)$ will 'automatically' set itself in such a value that would correspond to the static equilibrium of forces, discussed earlier. (Usually, we understand the tensile force P and the tensile stress $\sigma(\lambda)$ as the action force and the action stress, respectively; and the frictional force Φ and the frictional stress $\varphi(\lambda)$ as the reaction force and the reaction stress, respectively.)

In all cases, the actual frictional shear stress $\varphi(\lambda)$ must lie in the interval $\langle -b_\lambda, b_\lambda \rangle$, where a positive parameter b_λ represents a borderline situation[3] at which the maximum frictional effect is evoked. (This value is usually determined by the actual compression of the fiber surface by the other surrounding fibers[4].)

By using the borderline values $-b_\lambda$ and b_λ, in place of $\varphi(\lambda)$ in Equation (11.6), we obtain a couple of 'limit' values for the derivatives $d\sigma_1(\lambda)/d\lambda$ and $d\sigma_2(\lambda)/d\lambda$ as follows:

$$\frac{d\sigma_1(\lambda)}{d\lambda} = -kb_\lambda, \quad \frac{d\sigma_2(\lambda)}{d\lambda} = kb_\lambda, \quad \left(\frac{d\sigma_1(\lambda)}{d\lambda} = -\frac{d\sigma_2(\lambda)}{d\lambda} \right). \quad (11.8)$$

These values create two acceptable borderline situations for the derivative $d\sigma(\lambda)/d\lambda$ in case of static equilibrium of forces on the fiber element.

The actual position of the value $d\sigma(\lambda)/d\lambda$ in the above-mentioned interval $\langle d\sigma_1(\lambda)/d\lambda, d\sigma_2(\lambda)/d\lambda \rangle$ is described by a dimensionless parameter $v_\lambda \in \langle -1, 1 \rangle$ whose logical sense is somewhat like relative frictional loading of fiber element. We introduce

$$\frac{d\sigma(\lambda)}{d\lambda} = \frac{1-v_\lambda}{2} \frac{d\sigma_1(\lambda)}{d\lambda} + \frac{1+v_\lambda}{2} \frac{d\sigma_2(\lambda)}{d\lambda}$$

$$= -v_\lambda \frac{d\sigma_1(\lambda)}{d\lambda} = v_\lambda \frac{d\sigma_2(\lambda)}{d\lambda}, \quad v_\lambda \in \langle -1, 1 \rangle, \quad (11.9)$$

$$\frac{d\sigma(\lambda)}{d\lambda} = \frac{1-v_\lambda}{2}(-kb_\lambda) + \frac{1+v_\lambda}{2}kb_\lambda = kv_\lambda b_\lambda. \quad (11.10)$$

(Here, the expressions stated in Equation (11.8) were used.) By using the last two equations, it can be stated that

---

3   Let us strictly differentiate between the frictional shear stress $\varphi(\lambda)$ and its possible borderline values b_λ and $-b_\lambda$.

4   The well-known Coulomb's model of friction is often used as a most simple variant for determination of b_λ. Unfortunately, this is not very much valid in case of fiber-to-fiber friction.

1. If $v_\lambda = -1$, then $d\sigma(\lambda)/d\lambda = d\sigma_1(\lambda)/d\lambda = -kb_\lambda < 0$. It represents the maximum allowable frictional activity ($\varphi(\lambda) = -b_\lambda$) oriented in a direction opposite to the direction of the horizontal coordinate λ – Figure 11.1c.

2. If $v_\lambda = 1$, then $d\sigma(\lambda)/d\lambda = d\sigma_2(\lambda)/d\lambda = kb_\lambda > 0$. It represents the maximum allowable frictional activity ($\varphi(\lambda) = b_\lambda$) oriented in the direction of the horizontal coordinate λ – Figure 11.1b.

3. If $v_\lambda = 0$, then $d\sigma(\lambda)/d\lambda = \left[d\sigma_1(\lambda)/d\lambda + d\sigma_2(\lambda)/d\lambda\right]/2 = 0$ in accordance with Equation (11.10) and $\varphi(\lambda) = 0$ in accordance with Equation (11.6). It represents zero frictional activity – no frictional force is acting on the fiber element.

Sometimes, the actual value of the derivative $d\sigma(\lambda)/d\lambda$ of tensile stress can lie outside the interval $\langle d\sigma_1(\lambda)/d\lambda, d\sigma_2(\lambda)/d\lambda \rangle$ for a very short (infinitely small) period of time, e.g., during yarn tensioning. (Formally, the relative frictional loading of fiber element $v_\lambda \notin \langle -1,1 \rangle$ characterizes such an instantaneous situation.) Then, the static equilibrium cannot be reached to the original position of fiber element. (Friction cannot balance so high elementary increment or decrement of tensile stress acting on the fiber element.) Therefore, the fiber element must promptly slip to a new statically balanced position. (In general, the fiber element also changes some of its parameters due to this movement.)

Functions of coordinate λ

Generally, the values of positive borderline parameter b_λ and the relative frictional loading v_λ [as well as the frictional shear stress $\varphi(\lambda)$] change along the fiber coordinate λ. We understand b_λ and v_λ [and also $\varphi(\lambda)$] as suitable functions of λ. We assume that b_λ is a continuous function of fiber coordinate λ.

Note: It is important to mention that the coordinate λ identifies the positions of non-tensioned places along the fiber. (Naturally, the final horizontal coordinate will be different, as a consequence of movement, tension, and slippage of fiber and/or fiber portion in yarn.)

Note: Every time, the positive function b_λ is generated mainly through (variable) compression of fibers by other neighbouring fibers. The function

v_λ characterizes the variable nature of relative frictional loading along the fiber. Probably, this function has a markedly random character in consequence of highly variable micro-structure of yarn.

Consequently, the derivatives $d\sigma_1(\lambda)/d\lambda$, $d\sigma_2(\lambda)/d\lambda$ and $d\sigma(\lambda)/d\lambda$ are functions of fiber coordinate λ. By using the expressions stated in Equations (11.8) and (11.10), we can write

$$\frac{d\sigma_1(\lambda)}{d\lambda} = -kb_\lambda, \quad \sigma_1(\lambda) = -k\int b_\lambda \, d\lambda + C_1, \quad b_\lambda \geq 0, \tag{11.11}$$

$$\frac{d\sigma_2(\lambda)}{d\lambda} = kb_\lambda, \quad \sigma_2(\lambda) = k\int b_\lambda \, d\lambda + C_2, \quad b_\lambda \geq 0, \tag{11.12}$$

$$\frac{d\sigma(\lambda)}{d\lambda} = kv_\lambda b_\lambda, \quad \sigma(\lambda) = k\int v_\lambda b_\lambda \, d\lambda + C, \quad b_\lambda \geq 0. \tag{11.13}$$

Note: It is better to work with the constant and non-constant parts of the indefinite integrals separately in this chapter. Therefore, in this chapter, we will interpret the symbols of the indefinite integrals by a simpler ('standardized') form, i.e., without any (non-zero) constant. Using such convention, Equations (11.11), (11.12) and (11.13) can be correctly expressed[5].

The previous borderline functions $\sigma_1(\lambda)$ and $\sigma_2(\lambda)$ represents two single-parametric systems of curves. (The different constants of integration C_1 and C_2 are known to be their parameters.)

The function $\sigma(\lambda)$ of tensile stress according to Equation (11.13) contains – besides the actual common parameter k according to Equation (11.7) – functions v_λ, b_λ, and one suitable parameter C (in the case of continuous function $\sigma(\lambda)$). One set of functions v_λ, b_λ and parameters k and C determine the function $\sigma(\lambda)$.

---

5   Let us introduce one trivial example for more clarity. If, for example, $b_\lambda = b = $ constant then, we find $k\int b \, d\lambda = kb\lambda$ and then, according to Equation (11.12), we obtain $\sigma_2(\lambda) = kb\lambda + C_2$.

Note: The function $\sigma(\lambda)$ can also be a discontinuous function whose graph might have many breaks, especially, at static imbalance which can be appeared immediately after adding more tension to the yarn (see later on).

Let us remind that b_λ is a positive function of λ so that $\int b_\lambda \, d\lambda$ must be an increasing function of λ always. Therefore, each $\sigma_1(\lambda)$ is a monotonously decreasing function and each $\sigma_2(\lambda)$ is a monotonously increasing function of λ, according to Equations (11.11) and (11.12), respectively. In Figure 11.2, the net-like image of thin dashed lines represents a scheme of such borderline systems of curves.

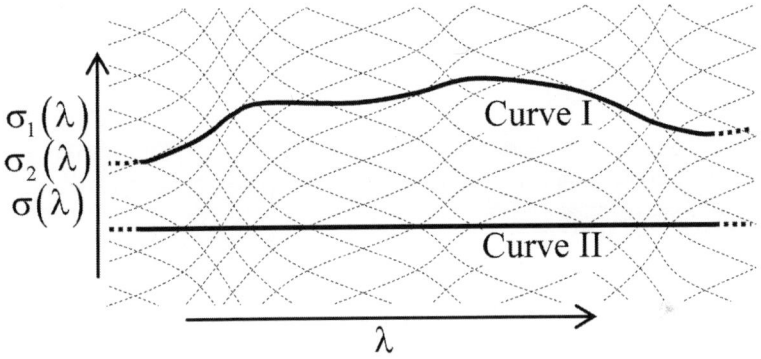

Figure 11.2 Scheme of behaviour of tensile stress and its borderline curves: thin dashed lines – single-parametric systems of borderline curves; $\sigma_1(\lambda)$ – decreasing curves, Equation (11.11); $\sigma_2(\lambda)$ – increasing curves, Equation (11.12); thick continuous lines – possible examples of tensile stresses $\sigma(\lambda)$, Equation (11.13); curve I – possible non-constant behaviour; curve II – straight line behaviour

Note: On the contrary, the function $\sigma(\lambda)$, according to Equation (11.13), can have increasing as well as decreasing parts, because $v_\lambda \in \langle -1,1 \rangle$.

In case of static equilibrium of fiber, the function $\sigma(\lambda)$ of tensile stress must lie between the tangential directions of $\sigma_1(\lambda)$ and $\sigma_2(\lambda)$ on each horizontal coordinate λ. In other words, each function $\sigma(\lambda)$ of tensile stress must intersect to the system of decreasing borderline curves $\sigma_1(\lambda)$ 'from below to above' and to the system of increasing borderline curves $\sigma_2(\lambda)$ 'from above to below' with the increase of λ. Figure 11.2 illustrates the possible behaviour of $\sigma(\lambda)$ – curve I – related to a part of ('infinitely' long) fiber. It is important to believe that the tensile stress $\sigma(\lambda)$ (usually) fluctuates along the fiber in consequence of the variability of frictional conditions generated along the fiber in the yarn. Nevertheless, the constant nature of $\sigma(\lambda)$ – curve II – corresponds to the previous requirements. [Especially, there is $v_\lambda = 0$ in Equation (11.13) so that $\sigma(\lambda) = C$, i.e., a constant.]

Fiber-to-fiber slippages

The nature of tensile stress $\sigma(\lambda)$ changes by applying a tension to the yarn. This tension causes fiber-to-fiber slippage in the yarn. This – together with the gradual breakage of mostly exposed parts of a fiber – leads to the breakage of the whole yarn at the final step of tensile testing of yarn.

There are two basic types of slippages taking place on a fiber:

1. Slippages of fiber ends occur at the end parts of staple (i.e., 'short') fibers in staple yarns and similar products (especially, roving, sliver, etc.). This phenomenon is required to be solved as a statically determinate problem.

2. Slippages of some fiber portions take place on the 'middle' ('internal') parts of fibers in staple yarns as well as in filament yarns. This phenomenon must be solved as a statically indeterminate problem.

11.3   Slippages of fiber ends

Equations of slippage

The single-parametric systems of borderline curves $\sigma_1(\lambda)$ and $\sigma_2(\lambda)$ are illustrated by thin dashed lines in both graphs of Figure 11.3. Further, let us think that the curve A_1XYB_1 shown in Figure 11.3a) represents the function $\sigma(\lambda)$[6] (of a part of 'infinitely' long fiber). Here, parts A_1X and YB_1 are shown by dashed lines. Let us consider that point A_1 has the length coordinate $\lambda = 0$ and the point B_1 has the length coordinate $\lambda = l$.

Now, let us create a staple fiber of length l by cutting the earlier 'infinitely' long fiber at points A_1 and B_1. Naturally, the tensile stresses cannot be a function of the end-points of the fiber. So, the end-point A_1 must 'fall down' to a new position A ($\bullet$, $\sigma(\lambda) = 0$) and the end-point B_1 must 'fall

---

6   In this case, we assume $\sigma(\lambda)$ as a continuous function.

down' to a new position B ($\bullet$, $\sigma(\lambda) = 0$) in Figure 11.3. The other characters of $\sigma(\lambda)$, i.e., the nature of open interval $\lambda \in (0,l)$ remains the same as in case of infinitely short length after cutting. [Thus, the complete curve $\sigma(\lambda)$ is discontinuous at the end-points A and B.] Figure 11.3a and 11.3b shows the behaviour of $\sigma(\lambda)$ from the starting point A to the end-point B by a set of arrows.

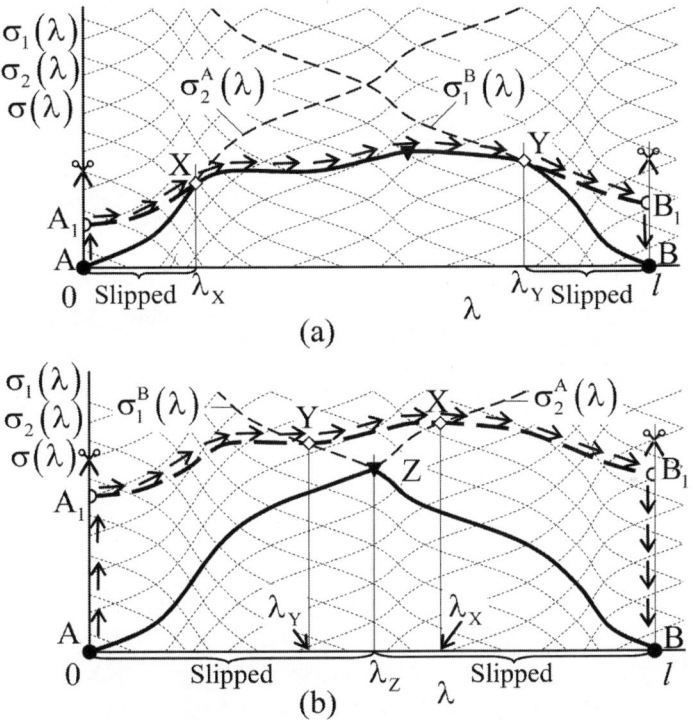

Figure 11.3 Scheme of slippages of fiber ends. Thick curves – tensile stresses $\sigma(\lambda)$: (a) curve for staple fiber with (relatively) small level of tensile stress, (b) curve for staple fiber with (relatively) high level of the tensile stress

Slipped ends of fiber

However, the vertical 'jump' from A to A_1 (see arrows) intersects the system of increasing borderline curves $\sigma_2(0)$ 'from below to above', i.e., against the assumption of static equilibrium[7]. Thus, the original tensile stress $\sigma(\lambda)$

---

7   See the text stated in Section 11.3.

must reduce its values (surrounding A) to the nearest statically balanced position passing through point A, i.e., to the curve $\sigma_2^A(\lambda)$[8]. The following function characterizes this curve in accordance with Equation (11.12):

$$\sigma_2^A(\lambda) = k\left[\int b_\lambda \, d\lambda - \left(\int b_\lambda \, d\lambda\right)_{\lambda=0}\right]. \tag{11.14[9]}$$

The last expression describes the behaviour of a new $\sigma(\lambda)$ curve due to fiber slippage – the thick part AX ($\lambda \in \langle 0, \lambda_X \rangle$) in Figure 11.3a.

On the contrary, the vertical 'jump' from B_1 to B (see arrows) intersects the system of decreasing borderline curves $\sigma_1(l)$ 'from above to below', i.e., against the assumption of static equilibrium. So, the original tensile stress $\sigma(\lambda)$ must reduce its values (surrounding the fiber end B) to the nearest statically balanced position, i.e., curve $\sigma_1^B(\lambda)$ passing through point B. According to Equation (11.11), such a curve is described by the function

$$\sigma_1^B(\lambda) = -k\left[\int b_\lambda \, d\lambda - \left(\int b_\lambda \, d\lambda\right)_{\lambda=l}\right]. \tag{11.15}$$

This equation expresses the behaviour of a new $\sigma(\lambda)$ curve due to fiber slippage – the thick part YB ($\lambda \in \langle \lambda_Y, l \rangle$) in Figure 11.3a.

Non-slipped part of fiber

The fiber part between points X and Y, $\lambda \in (\lambda_X, \lambda_Y)$, does not need to slip because the static equilibrium already exists (from the 'time of infinitely long fiber'), shown in Figure 11.3a. Therefore, the stated fiber part follows the original nature of $\sigma(\lambda)$ without any change. The coordinate λ_X is the root of the following relation (point of intersection):

$$\sigma(\lambda_X) = \sigma_2^A(\lambda_X), \tag{11.16}$$

$$k\left(\int v_\lambda b_\lambda \, d\lambda\right)_{\lambda=\lambda_X} + C = k\left(\int b_\lambda \, d\lambda\right)_{\lambda=\lambda_X} - k\left(\int b_\lambda \, d\lambda\right)_{\lambda=0},$$

---

8   A curve, passing through one definite point, is expressed at this point by a symbol in the superscript. Let us also perceive once more that a steep curve is not possible with the condition of static equilibrium.

9   The second expression stated within the square brackets represents the actual constant C_2/k in Equation (11.12).

$$\left[\int b_\lambda\left(1-v_\lambda\right)\mathrm{d}\lambda\right]_{\lambda=\lambda_X}=\left(\int b_\lambda\,\mathrm{d}\lambda\right)_{\lambda=0}+C/k. \tag{11.17}$$

[Equations (11.13) and (11.14) were used.]

Similarly, the coordinate λ_Y is the root of the following relation (point of intersection):

$$\sigma\left(\lambda_Y\right)=\sigma_1^B\left(\lambda_Y\right), \tag{11.18}$$

$$k\left(\int v_\lambda b_\lambda\,\mathrm{d}\lambda\right)_{\lambda=\lambda_Y}+C=-k\left(\int b_\lambda\,\mathrm{d}\lambda\right)_{\lambda=\lambda_Y}+k\left(\int b_\lambda\,\mathrm{d}\lambda\right)_{\lambda=l},$$

$$\left[\int b_\lambda\left(1-v_\lambda\right)\mathrm{d}\lambda\right]_{\lambda=\lambda_Y}=\left(\int b_\lambda\,\mathrm{d}\lambda\right)_{\lambda=l}-C/k. \tag{11.19}$$

[Equations (11.13) and (11.15) were used.]

Note: If we know the original function $\sigma(\lambda)$ – thick line A_1XYB_1 in Figure 11.3a – then we know functions v_λ, b_λ, k and parameter C.

If we find the roots $\lambda_X<\lambda_Y$, then the new function $\sigma(\lambda)$, $\lambda\in\langle 0,l\rangle$, i.e., thick line AXYB in Figure 11.3a, represents the static equilibrium of the staple fiber.

Note: Let us comment that the mean tensile stress is smaller with the staple fiber than with the earlier 'infinitely' long fiber in consequence of slippage effect near to the fiber ends.

The maximum ($\blacktriangledown$) tensile stress in the staple fiber must lie in the part XY of the function $\sigma(\lambda)$, i.e., in the interval $\lambda\in\langle\lambda_X,\lambda_Y\rangle$. The length of the above-mentioned interval naturally decreases with the increase of yarn tension, i.e., increase of $\sigma(\lambda)$. Such process is continuous until the maximum tensile stress achieves the value of fiber tenacity. Then, the fiber breaks itself.

Fully slipped fiber

A modified situation is arisen when we find $\lambda_X>\lambda_Y$ from Equations (11.17) and (11.19). Then, the function $\sigma(\lambda)$ of tensile stress of the initial 'infinitely' long fiber lies at an enough high position due to Equation (11.13) – the thick dashed curve in Figure 11.3b. We can find the borderline functions by applying the same logic as applied in the previous case. However, the increasing borderline function, passing through the point A – Equation (11.14), and the decreasing borderline function, passing through the point B – Equations (11.15), intersect at point Z (horizontal coordinate λ_Z). This point lies under the curve of original function $\sigma(\lambda)$. The coordinate λ_z is the root of the following relation:

$$\sigma_2^A(\lambda_z) = \sigma_1^B(\lambda_z),\tag{11.20}$$

$$k\left[\left(\int b_\lambda \,d\lambda\right)_{\lambda=\lambda_z} - \left(\int b_\lambda \,d\lambda\right)_{\lambda=0}\right] = -k\left[\left(\int b_\lambda \,d\lambda\right)_{\lambda=\lambda_z} - \left(\int b_\lambda \,d\lambda\right)_{\lambda=l}\right],$$

$$\int_0^{\lambda_z} b_\lambda \,d\lambda = \int_{\lambda_z}^{l} b_\lambda \,d\lambda.\tag{11.21}$$

[Equations (11.14) and (11.15) were used.]

All parts of such a staple fiber are slipped and the values of the original function $\sigma(\lambda)$ are never realized. Thus, the complete function $\sigma(\lambda)$, $\lambda \in \langle 0,l \rangle$, i.e., the thick line AZB in Figure 11.3b, represents a new static equilibrium obtained by the staple fiber.

Note: In addition, the mean tensile stress is smaller with the staple fiber than with the earlier 'infinitely' long fiber. This is in consequence of the slippage effect near to the fiber ends.

The maximum (▼) tensile stress on the staple fiber lies at point Z of the function $\sigma(\lambda)$. If the scheme shown in Figure 11.3b is valid for all fibers in a (twisted) fiber system, i.e., if the borderline frictional shear stress is very small, then the fibers do not break. All the fibers are mutually slipped completely at a very high tension. Such a situation is necessary for the drawn fibrous assembly like roving, etc.

Note: Let us note that the solution of the previous problem does not require to know the stress–strain relation of fiber; this is a case of statically determinate problem.

Simplest theoretical example

The results of calculation are very simple in case where the borderline frictional quantity b_λ is a constant and the relative frictional loading v_λ is equal to zero, i.e., both are independent to the horizontal coordinate λ of the fiber. So

$$b_\lambda = b \text{ is constant}, \quad v_\lambda = 0 \text{ is constant}.\tag{11.22}$$

In addition, $\varphi(\lambda) = \varphi = 0$ in this case – see point 3 in the text written after Equation (11.10).

Note: The note written before Equation (11.11) partly indicates our imagination of the previous assumptions. The constant value of b shows that the highest frictional force, which is able to balance a tensile force in a fiber, is a constant along the complete fiber length (e.g., based on Coulomb's law). It indirectly means that a compression of a fiber by the surrounding fibers is a constant. Moreover, the zero-value of v_λ refers to a fact that the frictional forces are not realized on the fiber surfaces (except the slipped fiber ends).

Then, the following expressions are valid from Equations (11.11), (11.12) and (11.13):

$$\sigma_1(\lambda) = -k \int b \, d\lambda + C_1 = -kb\lambda + C_1, \tag{11.23}$$

$$\sigma_2(\lambda) = k \int b \, d\lambda + C_2 = kb\lambda + C_2, \tag{11.24}$$

$$\sigma(\lambda) = k \int 0 \cdot b \, d\lambda + C = C. \tag{11.25}$$

[The constant value of $\sigma(\lambda)$ is shown in curve II of Figure 11.2.]

Figure 11.4 illustrates straight lines from the last three functions in analogy to Figure 11.2. (The description of the scheme shown in Figure 11.4 is the same as that depicted in Figure 1.2.)

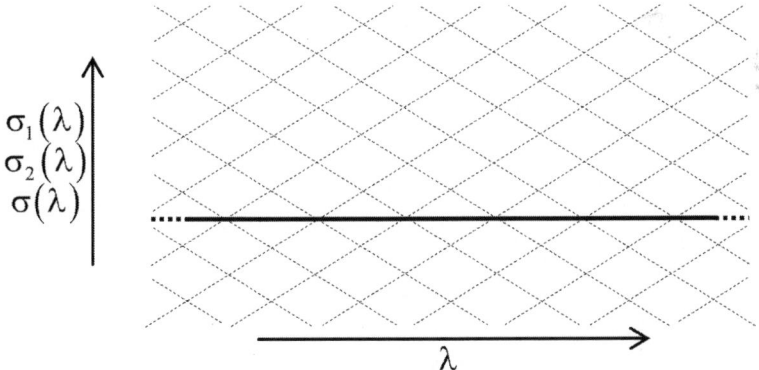

Figure 11.4 Scheme of borderline stresses $\sigma_1(\lambda)$, $\sigma_2(\lambda)$ (thin, dotted) and an example of tensile stress $\sigma(\lambda)$ (thick) for the simplest theoretical example, Equations (11.23), (11.24) and (11.25)

Figure 11.5 represents an analogy to Figure 11.3 in the case of this simplest theoretical example. Using Equations (11.14), (11.15) and (11.22), the borderline increasing function $\sigma_2^A(\lambda)$, passing through the point A, and the borderline decreasing function $\sigma_1^B(\lambda)$, passing through the point B, are

$$\sigma_2^A(\lambda) = k\left[\int b\,d\lambda - \left(\int b\,d\lambda\right)_{\lambda=0}\right] = k\left[b\lambda - b\cdot 0\right] = kb\lambda, \qquad (11.26)$$

$$\sigma_1^B(\lambda) = -k\left[\int b\,d\lambda - \left(\int b\,d\lambda\right)_{\lambda=l}\right] = -k\left(b\lambda - bl\right) = kb\left(l-\lambda\right), \qquad (11.27)$$

respectively. The coordinate λ_X is determined by Equation (11.17) and it is valid to write that

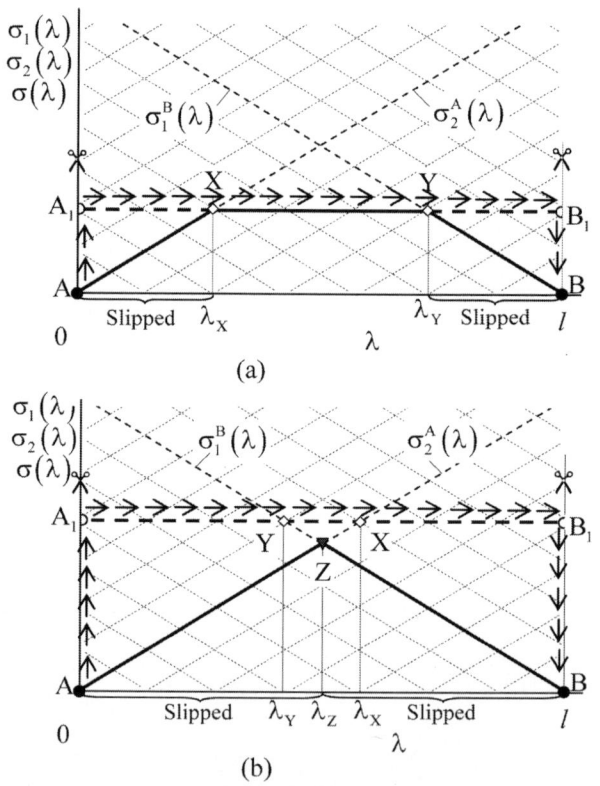

Figure 11.5 Scheme of slippages of fiber ends by the simplest theoretical example. Border lines $\sigma_1(\lambda)$, $\sigma_2(\lambda)$ (thin, dashed), tensile stresses $\sigma(\lambda)$ (thick). (a) (Relatively) small level of the tensile stress. (b) (Relatively) high level of the tensile stress

$$\left[\int b(1-0)\,d\lambda\right]_{\lambda=\lambda_X} = \left(\int b\,d\lambda\right)_{\lambda=0} + C/k,$$

$$b\lambda_X = (b\cdot 0) + C/k, \qquad \lambda_X = C/(bk). \qquad (11.28)$$

Similarly, the coordinate λ_Y is determined by Equation (11.19) and it is valid to write that

$$\left[\int b(1-0)\,d\lambda\right]_{\lambda=\lambda_Y} = \left(\int b\,d\lambda\right)_{\lambda=l} - C/k,$$

$$b\lambda_Y = bl - C/k, \qquad \lambda_Y = l - C/(bk). \tag{11.29}$$

If $\lambda_X < \lambda_Y$, then the following relation is valid according to Equations (11.28) and (11.29):

$$C/(bk) < \left[l - C/(bk)\right], \quad C/(bk) < l/2. \tag{11.30}$$

This case is illustrated in Figure 11.5a. On the contrary, if $\lambda_X > \lambda_Y$, then evidently

$$C/(bk) > l/2, \quad C > bkl/2. \tag{11.31}$$

This relation is depicted in Figure 11.5b, i.e., the case of fully slipped fiber. By using Equation (11.21), the coordinate value λ_Z can be obtained as follows:

$$\int_0^{\lambda_Z} b\,d\lambda = \int_{\lambda_Z}^l b\,d\lambda, \quad b\lambda_Z = bl - b\lambda_Z, \quad \lambda_Z = l/2. \tag{11.32}$$

[This result is also immediately obtained from the symmetry of $\sigma(\lambda)$, shown in Figure 11.5b.] The maximum tensile stress at point Z is now

$$\sigma_2^A(\lambda_Z) = \sigma_1^B(\lambda_Z) = kbl/2. \tag{11.33}$$

[Equations (11.32) and (11.26) or (11.27) were used.]

11.4   Slippages on the 'middle' part of fiber.

A rapid increase or decrease of tensile stress on a short section in the 'middle' part of the fiber (i.e., out of slipped fiber ends) can be originated by tensioning the yarn. Such a statically unstable situation can be solved by considering the slippage of fiber parts.

Rapid continuous increase of $\sigma(\lambda)$

The single-parametric systems of borderline curves $\sigma_1(\lambda)$ and $\sigma_2(\lambda)$ – Equations (11.11) and (11.12) – are illustrated by thin dashed lines in

Figure 11.6. Further, let us think that the thick continuous curve EFPXQGH represents the function $\sigma(\lambda)$ of tensile stress during the first elementary time after adding tension to the yarn – see arrows. (The part of this curve is shown by dashed line from F through PXQ to G.) So, the functions v_λ and b_λ, parameter k, and parameter C are known from Equation (11.13). The curve $\sigma(\lambda)$ increases rapidly in part PQ, so much that it intersects the net-like image of increasing curves $\sigma_2(\lambda)$ 'from bottom to above' (formally $v_\lambda > 1$). It shows static imbalance[10] which must be solved by considering the slippage of fiber parts. In part PQ, it is really valid that $d\sigma(\lambda)/d\lambda \geq d\sigma_2(\lambda)/d\lambda$. By using Equations (11.12), (11.13), we can write

$$kv_\lambda b_\lambda \geq kb_\lambda, \quad v_\lambda \geq 1. \tag{11.34}$$

Thus, finding part of function $\sigma(\lambda)$ where $v_\lambda \geq 1$ – Equation (11.13), we find part PQ including the corresponding horizontal coordinates λ_P and λ_Q.

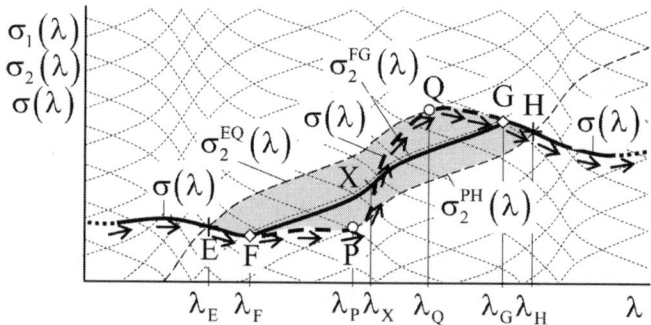

Figure 11.6 A rapid continuous increase of tensile stress $\sigma(\lambda)$ and slippages of fiber elements

The increasing borderline curve $\sigma_2^{PH}(\lambda)$ [11] passing through point P (thicker dashed line PH) satisfies the condition $\sigma_2^{PH}(\lambda_P) = \sigma(\lambda_P)$ (intersection

---

10   Compare this with the text written in front of Section 11.3.

11   Symbol $\sigma_2(\lambda)$ represents the whole net-like image of the curves, based on different values of parameter C_2, according to Equation (11.12). If we select only one curve, passing though given points, then the names of such points are marked by superscript – e.g., PH – in this section.

at point P). Similarly, the increasing border curve $\sigma_2^{EQ}(\lambda)$ passing through the point Q (thicker dashed line EQ) satisfies the condition $\sigma_2^{EQ}(\lambda_Q) = \sigma(\lambda_Q)$ (intersection in the point Q).

The rapid increase of statically unstable part PQ must be changed during slipping up to an increasing borderline curve $\sigma_2(\lambda)$, lying somewhere between $\sigma_2^{PH}(\lambda)$ and $\sigma_2^{EQ}(\lambda)$ – grey area EQHP in Figure 11.6. Each such curve satisfies the requirement of static equilibrium but does not need to satisfy another necessary condition. (It refers to statically indeterminate problem.)

For now, let us assume that the 'right' curve is the curve FG, passing through the point F with the coordinate $\lambda_F \in (\lambda_E, \lambda_P)$, shown in Figure 11.6. This curve satisfies the assumption (intersection):

$$\sigma_2^{FG}(\lambda_F) = \sigma(\lambda_F), \tag{11.35}$$

$$\left(k\int b_\lambda \, d\lambda\right)_{\lambda=\lambda_F} + C_2 = k\left(\int v_\lambda b_\lambda \, d\lambda\right)_{\lambda=\lambda_F} + C, \quad C_2 = k\left[\int b_\lambda (v_\lambda - 1) d\lambda\right]_{\lambda=\lambda_F} + C. \tag{11.36}$$

[Equations (11.12) and (11.13) were used.] Then, the determined borderline curve is

$$\sigma_2^{FG}(\lambda) = k\int b_\lambda \, d\lambda + C_2 = k\int b_\lambda \, d\lambda + k\left[\int b_\lambda (v_\lambda - 1) d\lambda\right]_{\lambda=\lambda_F} + C. \tag{11.37}$$

By applying Equations (11.13) and (11.37), for $\lambda = \lambda_G$, we obtain

$$\sigma_2^{FG}(\lambda_G) = \sigma(\lambda_G), \tag{11.38}$$

$$k\left[\int b_\lambda \, d\lambda\right]_{\lambda=\lambda_G} + k\left[\int b_\lambda (v_\lambda - 1) d\lambda\right]_{\lambda=\lambda_F} + C = k\left[\int v_\lambda b_\lambda \, d\lambda\right]_{\lambda=\lambda_G} + C,$$

$$\left[\int b_\lambda (v_\lambda - 1) d\lambda\right]_{\lambda=\lambda_F} = \left[\int b_\lambda (v_\lambda - 1) d\lambda\right]_{\lambda=\lambda_G}, \tag{11.39}$$

$$0 = \int_{\lambda_F}^{\lambda_G} b_\lambda (v_\lambda - 1) d\lambda, \quad \lambda_G \in (\lambda_Q, \lambda_H). \tag{11.40}$$

The last expression determines the (first) relation between λ_F and λ_G.

Note: By using Equation (11.39) in (11.37), we can also write the function $\sigma_2^{FG}(\lambda)$ as follows:

$$\sigma_2^{FG}(\lambda) = k \int b_\lambda \, d\lambda + C_2 = k \int b_\lambda \, d\lambda + k \left[\int b_\lambda (v_\lambda - 1) \, d\lambda \right]_{\lambda = \lambda_G} + C.$$

$$(11.41)$$

Shifting of fiber elements

The tensile stress–strain function of fiber $\tau = \tau[\varepsilon]$ [12] is a monotonously increasing stress function of fiber strain ε. Let us assume that the same stress–strain relation is valid at all places along the fiber. The inverse function $\varepsilon = \varepsilon[\tau]$ expresses fiber strain as a monotonously increasing function of stress. It is also valid that $\tau = 0$ and $\varepsilon = 0$ at the beginning of fiber straining. So, we use the following expressions:

$$\left. \begin{array}{l} \tau = \tau[\varepsilon], \quad \tau[0] = 0, \\ \varepsilon = \varepsilon[\tau], \quad \varepsilon[0] = 0. \end{array} \right\}$$

$$(11.42)$$

The coordinate λ was introduced as an initial horizontal coordinate along the non-tensioned fiber. However, we observe a higher distance $\Lambda(\lambda)$ along the fiber, when the fiber is elongated by (original) stress $\tau = \sigma(\lambda)$ (before slippage) according to Equation (11.13). It is then valid to write that

$$d\Lambda(\lambda) = d\lambda \left\{ 1 + \varepsilon[\sigma(\lambda)] \right\}.$$

$$(11.43)$$

Let us 'measure' the elongated distance $\Lambda(\lambda)$ from the point F, i.e., from $\lambda = \lambda_F$ in Figure 11.6. It is then valid to write the following expression for each general λ^*:

$$\Lambda(\lambda^*) = \int_{\lambda = \lambda_F}^{\lambda = \lambda^*} d\Lambda(\lambda) = \int_{\lambda_F}^{\lambda^*} \left\{ 1 + \varepsilon[\sigma(\lambda)] \right\} d\lambda, \quad \lambda^* \in \langle \lambda_F, \lambda_G \rangle. \quad (11.44)$$

(We use λ^* in the upper limit so as to differ it from the integration variable λ.)

---

12   Here we use symbol τ for the general tensile stress of fiber. However, a more traditional symbol σ was used for the specific tensile stress of a fiber – see Equations (11.11) to (11.13).

Similarly, we observe another distance $\Lambda_2(\lambda)$ along the fiber, when the fiber is elongated by stress $\tau = \sigma_2^{FG}(\lambda)$ (after slippage) according to Equation (11.37). It is valid to write that

$$d\Lambda_2(\lambda) = d\lambda \left\{ 1 + \varepsilon\left[\sigma_2^{FG}(\lambda)\right]\right\}. \tag{11.45}$$

Then, it is valid to write the following for each general λ^*:

$$\Lambda_2(\lambda^*) = \int_{\lambda=\lambda_F}^{\lambda=\lambda^*} d\Lambda_2(\lambda) = \int_{\lambda_F}^{\lambda^*} \left\{1 + \varepsilon\left[\sigma_2^{FG}(\lambda)\right]\right\} d\lambda, \quad \lambda^* = \langle \lambda_F, \lambda_G \rangle. \tag{11.46}$$

By using Equations (11.44) and (11.46), the slippage of fiber element, lying on the horizontal coordinate λ^*, is then

$$\delta\Lambda_2(\lambda^*) = \Lambda_2(\lambda^*) - \Lambda(\lambda^*) = \int_{\lambda_F}^{\lambda^*} \left\{1 + \varepsilon\left[\sigma_2^{FG}(\lambda)\right]\right\} d\lambda - \int_{\lambda_F}^{\lambda^*} \left\{1 + \varepsilon\left[\sigma(\lambda)\right]\right\} d\lambda,$$

$$\delta\Lambda_2(\lambda^*) = \int_{\lambda_F}^{\lambda^*} \left\{\varepsilon\left[\sigma_2^{FG}(\lambda)\right] - \varepsilon\left[\sigma(\lambda)\right]\right\} d\lambda, \quad \lambda^* \in \langle \lambda_F, \lambda_G \rangle, \tag{11.47}$$

where $\sigma(\lambda)$ and $\sigma_2^{FG}(\lambda)$ are given by Equations (11.13) and (11.37), respectively.

If especially $\lambda^* = \lambda_F$, then $\delta\Lambda_2(\lambda_F) = \int_{\lambda_F}^{\lambda_F} \left\{\varepsilon\left[\sigma_2^{FG}(\lambda)\right] - \varepsilon\left[\sigma(\lambda)\right]\right\} d\lambda = 0$. This corresponds well to Figure 11.6 where point F remains at its original position (position without slippage). Nevertheless, we must obtain the same result also at point G which must stay at its original position (position without slippage) when $\lambda^* = \lambda_G$. Thus, according to Equation (11.47),

$$\delta\Lambda_2(\lambda_G) = \int_{\lambda_F}^{\lambda_G} \left\{\varepsilon\left[\sigma_2^{FG}(\lambda)\right] - \varepsilon\left[\sigma(\lambda)\right]\right\} d\lambda = 0. \tag{11.48}$$

[The stresses in the last expression are given by Equations (11.13) and (11.37).] This Equation (11.48) determines the (second) relation between λ_F and λ_G.

Note: Let us imagine that the curve FXG is limited to the curve EQ, shown in Figure 11.6. Then, $\lambda_F \to \lambda_E$, $\lambda_G \to \lambda_Q$, $\sigma_2^{EQ}(\lambda) \geq \sigma(\lambda)$, here. Hence, $\varepsilon\left[\sigma_2^{FG}(\lambda)\right] \geq \varepsilon\left[\sigma(\lambda)\right]$ in the interval $\langle \lambda_F, \lambda_G \rangle \to \langle \lambda_E, \lambda_Q \rangle$ such that the integral expressed in Equation (11.48) is higher than 0. Further, let us imagine that the curve FXG is limited to the curve PH. Then, $\lambda_F \to \lambda_P$, $\lambda_G \to \lambda_H$, $\sigma_2^{PH}(\lambda) \leq \sigma(\lambda)$, here. Hence, $\varepsilon\left[\sigma_2^{PH}(\lambda)\right] \leq \varepsilon\left[\sigma(\lambda)\right]$ in the interval $\langle \lambda_F, \lambda_G \rangle \to \langle \lambda_P, \lambda_H \rangle$ such that the integral expressed in Equation (11.48) is smaller than 0. It means that the 'right' curve FXG must lie in-between EQ and PH to obtain the zero-value of Equation (11.48).

The right values of coordinates λ_F are λ_G can be determined from the couple of Equations (11.40) and (11.48) by using Equations (11.13) and (11.37).

Note: By applying Equation (11.48) in (11.47), we can express $\delta\Lambda_2(\lambda^*)$ also in an alternative form as follows:

$$\delta\Lambda_2(\lambda^*) = \overbrace{\int_{\lambda_F}^{\lambda_G} \left\{\varepsilon\left[\sigma_2^{FG}(\lambda)\right] - \varepsilon\left[\sigma(\lambda)\right]\right\} d\lambda}^{=0} - \int_{\lambda^*}^{\lambda_G} \left\{\varepsilon\left[\sigma_2^{FG}(\lambda)\right] - \varepsilon\left[\sigma(\lambda)\right]\right\} d\lambda,$$

$$\delta\Lambda_2(\lambda^*) = -\int_{\lambda^*}^{\lambda_G} \left\{\varepsilon\left[\sigma_2^{FG}(\lambda)\right] - \varepsilon\left[\sigma(\lambda)\right]\right\} d\lambda. \tag{11.49}$$

Rapid continuous decrease of $\sigma(\lambda)$

The solution of rapid continuous decrease of the function $\sigma(\lambda)$ of tensile stress just after addition of tension in the yarn is very similar to that of rapid continuous increase of $\sigma(\lambda)$. It is shown that Figure 11.7 is a 'mirror' image of Figure 11.6[13].

---

13   In the following text, a logical description and mathematical deduction can be presented in a similar manner as done in the above-mentioned previous case. (The style of marking symbols remains the same.)

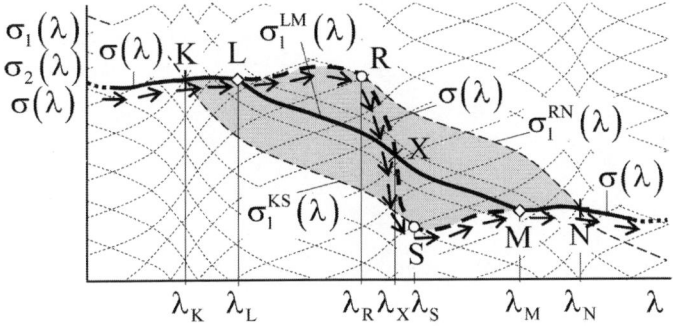

Figure 11.7 A rapid decrease of tensile stress $\sigma(\lambda)$ and slippages of fiber elements

Figure 11.7 illustrates single-parametric systems of borderline curves $\sigma_1(\lambda)$ and $\sigma_2(\lambda)$ [Equations (11.11) and (11.12)] by thin dashed lines, in the same way as in Figures 11.2 and 11.6. The thick curve KLRXSMN represents the function $\sigma(\lambda)$ of tensile stress just after addition of tension in the yarn – see arrows. (The part LRXSM is shown by dashed line.) The functions v_λ, b_λ and parameters k and C are known from Equation (11.13). This curve decreases rapidly in the part RS, intersects the net-like image of the decreasing borderline curves $\sigma_1(\lambda)$ 'from above to below' (formally $v_\lambda < -1$ here) and shows a static imbalance in this part.

The decreasing borderline curve $\sigma_1^{RN}(\lambda)$[14], passing through the point R (thicker dashed line RN) satisfies the condition (intersection at point R) $\sigma_1^{RN}(\lambda_R) = \sigma(\lambda_R)$. Similarly, the decreasing borderline curve $\sigma_1^{KS}(\lambda)$, passing through the point S (thicker dashed line KS), satisfies the condition $\sigma_1^{KS}(\lambda_S) = \sigma(\lambda_S)$ (intersection at point S).

The rapid decrease of unstable part RS must be changed due to slippage up to the decreasing borderline curve $\sigma_1(\lambda)$, lying somewhere between $\sigma_1^{KS}(\lambda)$ and $\sigma_1^{RN}(\lambda)$ – grey area KSNR in Figure 11.7. Each such curve

---

14   Symbol $\sigma_1(\lambda)$ represents the whole net-like image of curves, based on different values of parameter C_1 according to Equation (11.11). If we select only one curve, passing though two points – e.g., R and N, the name of such points will be marked as superscript.

satisfies the requirement of static equilibrium but does not need to satisfy another necessary condition (problem of statically indeterminate).

Now let us assume that the 'right' curve is the curve LM, passing through the point L at the coordinate $\lambda_L \in (\lambda_K, \lambda_R)$. This curve satisfies the condition (intersection):

$$\sigma_1^{LM}(\lambda_L) = \sigma(\lambda_L), \tag{11.50}$$

$$-k\left(\int b_\lambda \, d\lambda\right)_{\lambda=\lambda_L} + C_1 = k\left(\int v_\lambda b_\lambda \, d\lambda\right)_{\lambda=\lambda_L} + C, \quad C_1 = k\left[\int b_\lambda (v_\lambda + 1) \, d\lambda\right]_{\lambda=\lambda_L} + C. \tag{11.51}$$

[Equations (11.11) and (11.13) were used.] Then, the determined borderline curve is

$$\sigma_1^{LM}(\lambda) = -k\int b_\lambda \, d\lambda + C_1 = -k\int b_\lambda \, d\lambda + k\left[\int b_\lambda (v_\lambda + 1) \, d\lambda\right]_{\lambda=\lambda_L} + C. \tag{11.52}$$

By applying Equations (11.13) and (11.52), for $\lambda = \lambda_M$, we obtain

$$\sigma_1^{LM}(\lambda_M) = \sigma(\lambda_M), \tag{11.53}$$

$$-k\left[\int b_\lambda \, d\lambda\right]_{\lambda=\lambda_M} + k\left[\int b_\lambda (v_\lambda + 1) \, d\lambda\right]_{\lambda=\lambda_L} + C = k\left[\int v_\lambda b_\lambda \, d\lambda\right]_{\lambda=\lambda_M} + C,$$

$$\left[\int b_\lambda (v_\lambda + 1) \, d\lambda\right]_{\lambda=\lambda_L} = \left[\int b_\lambda (v_\lambda + 1) \, d\lambda\right]_{\lambda=\lambda_M}, \tag{11.54}$$

$$0 = \int_{\lambda_L}^{\lambda_M} b_\lambda (v_\lambda + 1) \, d\lambda, \quad \lambda_M \in (\lambda_S, \lambda_N). \tag{11.55}$$

The last expression determines the (first) relation between λ_L and λ_M.

Note: Because Equation (11.54) is valid, we can write Equation (11.52) also in the form

$$\sigma_1^{LM}(\lambda) = -k\int b_\lambda \, d\lambda + k\left[\int b_\lambda (v_\lambda + 1) \, d\lambda\right]_{\lambda=\lambda_M} + C. \tag{11.56}$$

Shifting of fiber elements

The slippage of fiber element lying on a horizontal coordinate λ^* can also be obtained by using Equations (11.43) to (11.47)[15] as follows:

$$\delta\Lambda_1\left(\lambda^*\right) = \int_{\lambda_L}^{\lambda^*} \left\{\varepsilon\left[\sigma_1^{LM}\left(\lambda\right)\right] - \varepsilon\left[\sigma\left(\lambda\right)\right]\right\}d\lambda, \quad \lambda^* = \left\langle\lambda_L, \lambda_M\right\rangle. \qquad (11.57)$$

Finally, especially for $\lambda^* = \lambda_M$, it is valid that

$$\delta\Lambda_1\left(\lambda_M\right) = \int_{\lambda_L}^{\lambda_M} \left\{\varepsilon\left[\sigma_1^{LM}\left(\lambda\right)\right] - \varepsilon\left[\sigma\left(\lambda\right)\right]\right\}d\lambda = 0. \qquad (11.58)$$

[This is an expression analogous to Equation (11.48).] The last expression determines stresses by using Equations (11.13) and (11.52). The right values of coordinates λ_M and λ_L can now be determined from the couple of Equations (11.55) and (11.58).

Note: By using Equation (11.58) in (11.57), we obtain $\delta\Lambda_1\left(\lambda^*\right)$ in an alternative form as follows:

$$\delta\Lambda_1\left(\lambda^*\right) = \overbrace{\int_{\lambda_L}^{\lambda_M} \left\{\varepsilon\left[\sigma_1^{LM}\left(\lambda\right)\right] - \varepsilon\left[\sigma\left(\lambda\right)\right]\right\}d\lambda}^{=0} - \int_{\lambda^*}^{\lambda_M} \left\{\varepsilon\left[\sigma_1^{LM}\left(\lambda\right)\right] - \varepsilon\left[\sigma\left(\lambda\right)\right]\right\}d\lambda,$$

$$\delta\Lambda_1\left(\lambda^*\right) = -\int_{\lambda^*}^{\lambda_M} \left\{\varepsilon\left[\sigma_1^{LM}\left(\lambda\right)\right] - \varepsilon\left[\sigma\left(\lambda\right)\right]\right\}d\lambda. \qquad (11.59)$$

Discontinuous increase of $\sigma(\lambda)$

Let us imagine a (black) fiber surrounded by a lot of other (grey) fibers according to the scheme shown in Figure 11.8. The neighbouring light-grey-coloured fibers impart only frictional stress to the black fiber, but the dark-grey-coloured fibers hold the black fiber so tightly that either they do not allow full tension to be developed in the fiber part held in-between or, in

---

15   In contrary to Equations (11.45) to (11.47), we work with $\sigma_1^{LM}\left(\lambda\right)$ in the place of earlier $\sigma_2^{FG}\left(\lambda\right)$.

the contrary, they compel this fiber part to be under higher tension[16]. At the first instant the tensile stress $\sigma(\lambda)$ can rapidly increase ('jump over') or rapidly decrease ('jump down') in a very small place (point) surrounding the tightly held fibers and this causes fiber-to-fiber slippage.

Figure 11.8 Fibers surrounded by fibers in a fibrous assembly

The scheme shown in Figure 11.9 characterizes the rapid increase of $\sigma(\lambda)$ (the case of 'jump over'). The discontinuous curve EFPXQGH represents the function of $\sigma(\lambda)$ just after addition of tension in the yarn. (The part of it is shown by dashed line from F through PXQ to G.) The points P, X, Q have the same horizontal coordinate $\lambda_P = \lambda_X = \lambda_Q$ which we now denote by a common symbol λ_{jump}. [$\lambda = \lambda_{jump}$ determines the point of discontinuity of $\sigma(\lambda)$.]

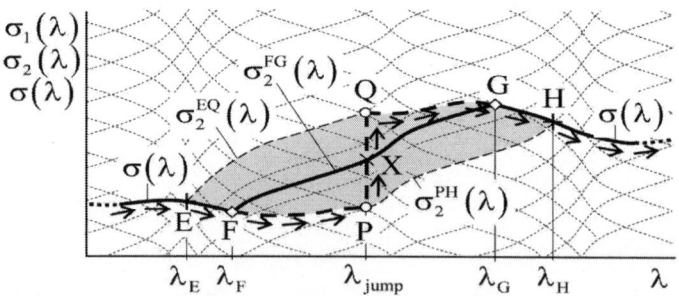

Figure 11.9 A discontinuous increase of tensile stress $\sigma(\lambda)$ and slippages of fiber elements

---

16   The surrounding structure of yarn compensates with this transmission of tensile stress. Such situations can be originated in the initial yarn structure when a fiber part under consideration has many crimps, while other fiber parts are straight.

Equation (11.13) generally determines the function $\sigma(\lambda)$. Now, we assume that the borderline function b_λ remains continuous[17], but the function v_λ of relative frictional loading and the constant C can jump to their values at $\lambda = \lambda_{jump}$. So, Equation (11.13) must be modified as follows:

$$\left.\begin{aligned}
\sigma(\lambda) &= k\int v_\lambda b_\lambda \, d\lambda + C_I, \quad \lambda < \lambda_{jump}, \\
\sigma(\lambda) &= k\int v_\lambda b_\lambda \, d\lambda + C_{II}, \quad \lambda > \lambda_{jump}, \\
\sigma(\lambda) &\ldots \text{not defined}, \quad \lambda = \lambda_{jump}.
\end{aligned}\right\} \tag{11.60}$$

(The values of constants C_I, C_{II} and the function v_λ before and after λ_{jump} are generally different.) The limit values of $\sigma(\lambda)$ at points P and Q are

$$\left.\begin{aligned}
\sigma(\lambda_P) &= \lim_{\lambda \to \lambda_{jump}^-} \sigma(\lambda) = k \lim_{\lambda \to \lambda_{jump}^-} \int v_\lambda b_\lambda \, d\lambda + C_I, \\
\sigma(\lambda_Q) &= \lim_{\lambda \to \lambda_{jump}^+} \sigma(\lambda) = k \lim_{\lambda \to \lambda_{jump}^+} \int v_\lambda b_\lambda \, d\lambda + C_{II}.
\end{aligned}\right\} \tag{11.61}[18]$$

$[\sigma(\lambda_P)$ and $\sigma(\lambda_Q)$ display the 'heights' of points P and Q in Figure (11.9); $\sigma(\lambda_P) < \sigma(\lambda_Q)$.]

The net-like image of borderline functions $\sigma_2(\lambda)$ remains same (b_λ remains continuous) as described in Equation (11.12). The increase of borderline curves $\sigma_2^{PH}(\lambda)$ and $\sigma_2^{EQ}(\lambda)$ (passing through the points P and Q, dashed lines PH and EQ) continue to satisfy the earlier-stated conditions $\sigma_2^{PH}(\lambda_P) = \sigma(\lambda_P)$ and $\sigma_2^{EQ}(\lambda_Q) = \sigma(\lambda_Q)$ (intersections at points P and Q). Equation (11.61) now determines $\sigma(\lambda_P)$ and $\sigma(\lambda_Q)$. Certainly, a particular curve of $\sigma_2(\lambda)$ from the grey area EQHP is the solution to our problem of fiber-to-fiber slippage.

Let us now assume that the 'right' function is the curve FG, which is passing through the point F with coordinate $\lambda_F \in (\lambda_E, \lambda_{jump})$. This is shown in Figure 11.9. This curve satisfies the condition according to Equation

---

17 This function expresses the borderline frictional shear stress along the fiber. This does not express an actual frictional shear along the fiber. Remember the text written before Equation (11.8).

18 Traditionally, the superscript '−' denotes the limit at the left and the superscript '+' indicates the limit at the right.

(11.35), i.e., $\sigma_2^{FG}(\lambda_F) = \sigma(\lambda_F)$. By using Equations (11.12) and (11.60), we can write

$$\left(k\int b_\lambda\,d\lambda\right)_{\lambda=\lambda_F} + C_2 = k\left(\int v_\lambda b_\lambda\,d\lambda\right)_{\lambda=\lambda_F} + C_I, \quad C_2 = k\left[\int b_\lambda(v_\lambda - 1)\,d\lambda\right]_{\lambda=\lambda_F} + C_I.$$

(11.62)

[We used the first expression from Equation (11.60) because $\lambda_F < \lambda_{jump}$.] By substituting C_2 from Equation (11.62) to (11.12), we obtain

$$\sigma_2^{FG}(\lambda) = k\int b_\lambda\,d\lambda + C_2 = k\int b_\lambda\,d\lambda + k\left[\int b_\lambda(v_\lambda - 1)\,d\lambda\right]_{\lambda=\lambda_F} + C_I.$$

(11.63)

The coordinate λ_G of point G is the root of Equation (11.38). Seeing that $\lambda_G > \lambda_{jump}$, it can be stated that the function $\sigma(\lambda_G)$ must follow the second expression from Equation (11.60); $\sigma_2^{FG}(\lambda_G)$ follows Equation (11.63). Thus, the horizontal coordinate λ_G is the root of the following equation:

$$\sigma_2^{FG}(\lambda_G) = \sigma(\lambda_G),$$

$$k\left[\int b_\lambda\,d\lambda\right]_{\lambda=\lambda_G} + k\left[\int b_\lambda(v_\lambda - 1)\,d\lambda\right]_{\lambda=\lambda_F} + C_I = k\left[\int v_\lambda b_\lambda\,d\lambda\right]_{\lambda=\lambda_G} + C_{II},$$

(11.64)

$$0 = C_{II} - C_I + k\int_{\lambda_F}^{\lambda_G} b_\lambda(v_\lambda - 1)\,d\lambda,$$

(11.65)

and/or – if the function v_λ is also discontinuous at point λ_{jump} – we can preferably write

$$0 = C_{II} - C_I + k\int_{\lambda_F}^{\lambda_{jump}} b_\lambda(v_\lambda - 1)\,d\lambda + k\int_{\lambda_{jump}}^{\lambda_G} b_\lambda(v_\lambda - 1)\,d\lambda.$$

(11.66)

[Compare this with Equations (11.38) and (11.40).] Equation (11.66) gives the relation between λ_F and λ_G.

Note: Because Equation (11.64) is valid, we can also rearrange Equation (11.63) in the following form:

$$\sigma_2^{FG}(\lambda) = k\int b_\lambda \, d\lambda + k\left[\int b_\lambda (v_\lambda - 1) d\lambda\right]_{\lambda = \lambda_G} + C_{II}. \qquad (11.67)$$

Equation (11.48) allows us to derive the second relation between λ_F and λ_G. The fiber strain ε depends on a continuous function $\sigma_2^{FG}(\lambda)$ – see Equation (11.63), but also on a discontinuous function $\sigma(\lambda)$ – see Equation (11.60). Thus, we rearrange Equation (11.48) as follows:

$$0 = \int_{\lambda_F}^{\lambda_G} \varepsilon\left[\sigma_2^{FG}(\lambda)\right] d\lambda - \int_{\lambda_F}^{\lambda_{jump}} \varepsilon\left[\sigma\left(\lambda < \lambda_{jump}\right)\right] d\lambda - \int_{\lambda_{jump}}^{\lambda_G} \varepsilon\left[\sigma\left(\lambda > \lambda_{jump}\right)\right] d\lambda.$$

$$(11.68)$$

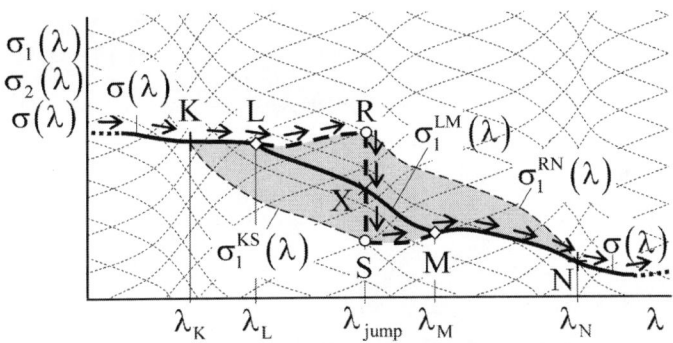

Figure 11.10 A discontinuous decrease of tensile stress $\sigma(\lambda)$ and slippages of fiber elements

Equation (11.68) is the (second) expression that determines the relation between λ_F and λ_G. The right values of coordinates λ_F and λ_G can now be found out from the couple of Equations (11.66) and (11.68).

Discontinuous decrease of $\sigma(\lambda)$

In analogy with Figure 11.9, the scheme shown in Figure 11.10 characterizes a 'mirror', i.e., a 'jump down' of $\sigma(\lambda)$ just after adding tension to the yarn. The function $\sigma(\lambda)$ follows the curve KLRXSMN. Points R, X, S have the same coordinate λ_{jump} which determines the point of discontinuity of the function $\sigma(\lambda)$. This situation can be solved promptly by considering fiber-to-fiber slippage.

As assumed in the previous case, we here assume that the borderline function b_λ remains continuous, but the function v_λ of relative frictional loading and the constant C can jump to their values at $\lambda = \lambda_{\text{jump}}$. So, Equation (11.13) must be modified in analogy with Equation (11.60) as follows:

$$
\left.
\begin{aligned}
&\sigma(\lambda) = k\int v_\lambda b_\lambda \, d\lambda + C_{\text{III}}, \quad \lambda < \lambda_{\text{jump}}, \\
&\sigma(\lambda) = k\int v_\lambda b_\lambda \, d\lambda + C_{\text{IV}}, \quad \lambda > \lambda_{\text{jump}}, \\
&\sigma(\lambda) \ldots \text{not defined}, \quad \lambda = \lambda_{\text{jump}}.
\end{aligned}
\right\}
\tag{11.69}
$$

(Constants C_{III}, C_{IV} and the function v_λ before and after λ_{jump} are generally different.)

The limit values of $\sigma(\lambda)$ at points R and S are

$$
\left.
\begin{aligned}
&\sigma(\lambda_{\text{R}}) = \lim_{\lambda \to \lambda_{\text{jump}}^-} \sigma(\lambda) = k \lim_{\lambda \to \lambda_{\text{jump}}^-} \int v_\lambda b_\lambda \, d\lambda + C_{\text{III}}, \\
&\sigma(\lambda_{\text{S}}) = \lim_{\lambda \to \lambda_{\text{jump}}^+} \sigma(\lambda) = k \lim_{\lambda \to \lambda_{\text{jump}}^+} \int v_\lambda b_\lambda \, d\lambda + C_{\text{IV}}.
\end{aligned}
\right\}
\tag{11.70}
$$

[Compare it with Equation (11.61).]

It stands to reason that a particular curve of the function $\sigma_1(\lambda)$ form those displayed in the grey-coloured area of Figure 11.10 must be the 'right' solution of fiber-to-fiber slippage. Let it be the curve LM, which is passing through the point L; $\lambda_{\text{L}} \in (\lambda_{\text{K}}, \lambda_{\text{jump}})$. This function satisfies the condition according to Equation (11.50), i.e., $\sigma_1^{\text{LM}}(\lambda_{\text{L}}) = \sigma(\lambda_{\text{L}})$ (intersection). By applying Equations (11.11) and (11.69), we can write

$$
-k\left(\int b_\lambda \, d\lambda\right)_{\lambda=\lambda_{\text{L}}} + C_1 = k\left(\int v_\lambda b_\lambda \, d\lambda\right)_{\lambda=\lambda_{\text{L}}} + C_{\text{III}}, \quad C_1 = k\left[\int b_\lambda (v_\lambda + 1)\,d\lambda\right]_{\lambda=\lambda_{\text{L}}} + C_{\text{III}}.
$$

$$
\tag{11.71}
$$

[We used the first expression from Equation (11.69) because $\lambda_L < \lambda_{jump}$.]
By using the value of C_1 in the general Equation (11.11), we obtain

$$\sigma_1^{LM}(\lambda) = -k\int b_\lambda \, d\lambda + C_1 = -k\int b_\lambda \, d\lambda + k\left[\int b_\lambda(v_\lambda + 1)d\lambda\right]_{\lambda=\lambda_L} + C_{III}.$$

(11.72)

By applying Equations (11.69) and (11.72), the horizontal coordinate λ_M of the point M, shown in Figure 11.10, is the root of following expression (intersection):

$$\sigma_1^{LM}(\lambda_M) = \sigma(\lambda_M),$$

$$-k\left[\int b_\lambda \, d\lambda\right]_{\lambda=\lambda_M} + k\left[\int b_\lambda(v_\lambda + 1)d\lambda\right]_{\lambda=\lambda_L} + C_{III} = k\left[\int v_\lambda b_\lambda \, d\lambda\right]_{\lambda=\lambda_M} + C_{IV},$$

(11.73)

$$C_{III} - C_{IV} - k\int_{\lambda_L}^{\lambda_M} b_\lambda(v_\lambda + 1)d\lambda = 0,$$

(11.74)

and/or – if the function v_λ is also discontinuous at point λ_{jump} – we can then preferably write

$$C_{III} - C_{IV} - k\int_{\lambda_L}^{\lambda_{jump}} b_\lambda(v_\lambda + 1)d\lambda - k\int_{\lambda_{jump}}^{\lambda_M} b_\lambda(v_\lambda + 1)d\lambda = 0.$$

(11.75)

Equations (11.74) and (11.75) are the expressions that determine the relation between λ_L and λ_M.

Note: Because Equation (11.73) is valid, we can also rearrange Equation (11.72) in the following form:

$$\sigma_1^{LM}(\lambda) = -k\int b_\lambda \, d\lambda + k\left[\int b_\lambda(v_\lambda + 1)d\lambda\right]_{\lambda=\lambda_M} + C_{IV}.$$

(11.76)

Equation (11.58) expresses the (second) equation between λ_L and λ_M. The fiber strain ε depends on the continuous function of $\sigma_1^{LM}(\lambda)$ – see Equation (11.72), but also on the discontinuous function $\sigma(\lambda)$ – see Equation (11.69). Then, we must rearrange Equation (11.58) as follows:

$$0 = \int_{\lambda_L}^{\lambda_M} \varepsilon \left[\sigma_1^{LM}(\lambda) \right] d\lambda - \int_{\lambda_L}^{\lambda_{jump}} \varepsilon \left[\sigma \left(\lambda < \lambda_{jump} \right) \right] d\lambda - \int_{\lambda_{jump}}^{\lambda_M} \varepsilon \left[\sigma \left(\lambda > \lambda_{jump} \right) \right] d\lambda .$$

(11.77)

[Equations (11.69) and (11.72) determine stresses from the last expression.] Equation (11.77) is the (second) expression which determines the relation between λ_L and λ_M. The right values of coordinates λ_L and λ_M can now be found out from the couple of Equations (11.74) and (11.77).

11.5   Double-sided slippage at 'middle' parts of fiber.

Double-sided slippage for continuous $\sigma(\lambda)$

Once more, let us imagine a (black) fiber surrounded by a lot of other (grey-coloured) fibers according to the scheme shown in Figure 11.8. In contrast to the light-grey-coloured fibers, the couples of dark grey-coloured fibers hold the black-coloured fiber so tightly that either they do not allow full tension to be developed in the fiber part held in-between or, in the contrary, they compel this fiber part to be under higher tension. So, a couple of rapid increase to decrease or rapid decrease to increase of tensile stress $\sigma(\lambda)$ can be origi-nated at the beginning of yarn tensioning.

The possible situations are illustrated by six independent curves on three graphs in Figure 11.11. (The senses and graphical symbols for points, curves and 'grey areas' are identical to those indicated in earlier graphs.[19]) Two quite independent cases of tensile stress $\sigma(\lambda)$ are always displayed on each graph from (a) to (c) – rapid increase to decrease by curves type 1 and rapid de-crease-to-increase by curves type 2. (The original thick curves of tensile stress are shown by dashed line in the part before fiber slippage and by con-tinuous line after slippage.)

Figure 11.11a displays the cases where the parts of rapid increase and rapid decrease of $\sigma(\lambda)$ are so far from each other that the respective 'grey areas' are not mutually covered. We can independently solve such

---

19   For lucidity, the names of points and their λ coordinates are not generally written on all graphs in Figure 11.11. The 'curve type 1' and 'curve type 2' illustrate always two quite independent and possible situations. Therefore, the same marking symbols are used for 'curve type 1' and 'curve type 2' in Figure 11.11c.

configuration for part of rapid increase as well as for part of rapid decrease according to the previous equations related to Figures 11.6 and 11.7.

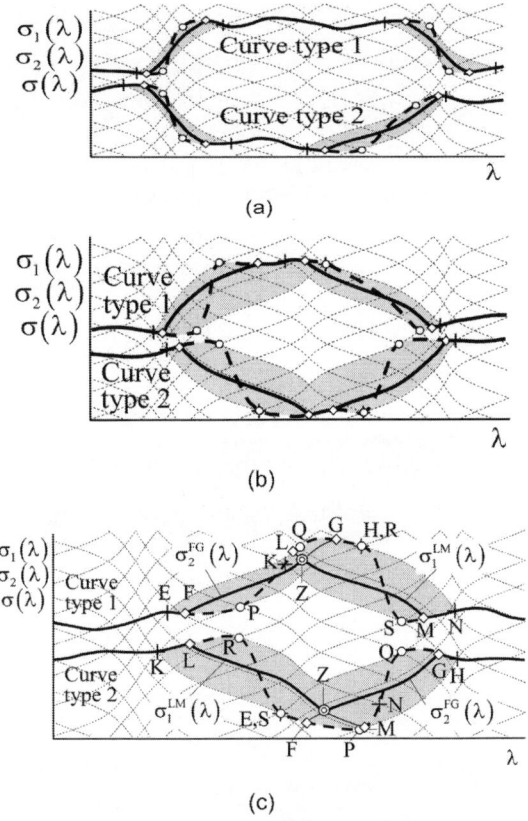

Figure 11.11 Schemes of double-sided slippage by continuous $\sigma(\lambda)$

Figure 11.11b shows the cases where the parts of rapid increase and rapid decrease of $\sigma(\lambda)$ are more closely related to each other, so that the corresponding 'grey areas' are mutually covered to a certain degree. Nevertheless, the curves of slippages ($\Diamond - \Diamond$) are not mutually intersect with each other inside the grey-coloured areas. Also here, we can independently solve such configuration for the part of rapid increase and the part of rapid decrease according to the previous equations related to Figures 11.6 and 11.7.

The two cases shown in Figure 11.11c are more complicated, because the curves of slippages ($\Diamond - \Diamond$) mutually intersected with each other at point Z (inside the grey-coloured areas).

Note: In the previous section, the coordinates λ_F and λ_G of rapid increase were derived by using Equations (11.40) and (11.48), where Equations (11.13) and (11.37) were used. The coordinates λ_L and λ_M of rapid decrease were also derived by using Equations (11.55) and (11.58), where Equations (11.13) and (11.52) were applied. If $\lambda_F < \lambda_M$ and $\lambda_G > \lambda_L$, then the resulting $\sigma(\lambda)$ is displayed in Figure 11.11c. In other cases, the double-sided slippage corresponds to the couple of two 'independent' slippages according to Figure 11.11a and 11.11b.

Let us at first solve the rapid increase-to-decrease case – curve type 1. (The notation of points is fully consistent with the marking of symbols followed in Figures 11.6 and 11.7.) We need to determine the triplet of coordinates $\lambda_F, \lambda_Z, \lambda_M$ of points F, Z and M. The points F and G always lie on the original curve $\sigma(\lambda)$ so that Equation (11.37) of $\sigma_2^{FG}(\lambda)$ is valid here. Likewise, the points L and M always lie on the original curve $\sigma(\lambda)$ so that Equation (11.56) of $\sigma_1^{LM}(\lambda)$ is valid too. Thus, the following expression is valid for the point Z of intersection:

$$\sigma_2^{FG}(\lambda_Z) = \sigma_1^{LM}(\lambda_Z), \quad k\left[\int b_\lambda \, d\lambda\right]_{\lambda=\lambda_Z} + k\left[\int b_\lambda v_\lambda \, d\lambda\right]_{\lambda=\lambda_F} - k\left[\int b_\lambda \, d\lambda\right]_{\lambda=\lambda_F} + C$$

$$= -k\left[\int b_\lambda \, d\lambda\right]_{\lambda=\lambda_Z} + k\left[\int b_\lambda v_\lambda \, d\lambda\right]_{\lambda=\lambda_M} + k\left[\int b_\lambda \, d\lambda\right]_{\lambda=\lambda_M} + C,$$

$$\int_{\lambda_F}^{\lambda_Z} b_\lambda \, d\lambda - \int_{\lambda_Z}^{\lambda_M} b_\lambda \, d\lambda = \int_{\lambda_F}^{\lambda_M} b_\lambda v_\lambda \, d\lambda. \tag{11.78}$$

The above expression binds all three coordinates $\lambda_F, \lambda_Z, \lambda_M$ together.

The function type $\sigma_1(\lambda)$ [e.g., $\sigma_1^{LM}(\lambda)$] characterizes the borderline of fiber slippage in the positive direction of horizontal coordinate λ. On the contrary, the function type $\sigma_2(\lambda)$ [e.g. $\sigma_2^{FG}(\lambda)$] characterizes the borderline of fiber slippage in the negative direction of coordinate λ[20]. Nevertheless, what is happening at the borderline of both parts, i.e., at point Z? The shifting of this point due to fiber slippage shall be equal to zero.

---

20   See the text written after Equation (11.10) in Section 11.2.

Equation (11.47) expresses the shifting $\delta\Lambda_2(\lambda^*)$ of a general fiber element, which is lying at a coordinate λ^* on the curve $\sigma_2^{FG}(\lambda)$. But, the shifting of point Z, i.e., $\lambda^* = \lambda_Z$, must be equal to zero. Thus,

$$\delta\Lambda_2(\lambda_Z) = \int_{\lambda_F}^{\lambda_Z} \left\{\varepsilon\left[\sigma_2^{FG}(\lambda)\right] - \varepsilon\left[\sigma(\lambda)\right]\right\} d\lambda = 0, \qquad (11.79)$$

where $\sigma_2^{FG}(\lambda)$ is given according to Equation (11.37) and $\sigma(\lambda)$ is determined according to Equation (11.13). The last equation represents the relation between λ_F and λ_Z.

Note: Let us note that λ_F appears in Equation (11.37) too.

Equation (11.59) expresses the shifting $\delta\Lambda_1(\lambda^*)$ of a general fiber element, which is lying at a coordinate λ^* on the curve $\sigma_1^{LM}(\lambda)$. Especially, the shifting of point Z, i.e., $\lambda^* = \lambda_Z$, must be equal to zero. Then,

$$\delta\Lambda_1(\lambda_Z) = -\int_{\lambda_Z}^{\lambda_M} \left\{\varepsilon\left[\sigma_1^{LM}(\lambda)\right] - \varepsilon\left[\sigma(\lambda)\right]\right\} d\lambda = 0, \qquad (11.80)$$

where $\sigma_1^{LM}(\lambda)$ is given by Equation (11.56) and $\sigma(\lambda)$ is determined by Equation (11.13). The last equation represents the relation between λ_M and λ_Z.

Note: Let us note that λ_M appears in Equation (11.56) too.

Summarily, the triplet of Equations (11.78), (11.79) and (11.80) determines the coordinates $\lambda_F, \lambda_Z, \lambda_M$. Then, the nature of tensile stress after slippage of fiber is given by Equation (11.37) – from λ_F to λ_Z, and Equation (11.56) – from λ_Z to λ_M. [At $\lambda = \lambda_Z$, the result of both equations is same because Equation (11.78) is valid.]

Further, let us solve the rapid decrease-to-increase case – curve type 2, shown in Figure 11.11c. We need to determine the triplet of coordinates $\lambda_L, \lambda_Z, \lambda_G$ of points L, Z and G. The points L and M always lie on the original curve $\sigma(\lambda)$ so that Equation (11.52) of $\sigma_1^{LM}(\lambda)$ is also now valid. Likewise, the points F and G always lie on the original curve $\sigma(\lambda)$ so that

Equation (11.41) of $\sigma_2^{FG}(\lambda)$ is valid too. Thus, the following expression is valid for point Z of intersection:

$$\sigma_1^{LM}(\lambda_Z) = \sigma_2^{FG}(\lambda_Z), \quad -k\left[\int b_\lambda d\lambda\right]_{\lambda=\lambda_Z} + k\left[\int b_\lambda v_\lambda\, d\lambda\right]_{\lambda=\lambda_L} + k\left[\int b_\lambda\, d\lambda\right]_{\lambda=\lambda_L} + C$$

$$= k\left[\int b_\lambda d\lambda\right]_{\lambda=\lambda_Z} + k\left[\int b_\lambda v_\lambda\, d\lambda\right]_{\lambda=\lambda_G} - k\left[\int b_\lambda\, d\lambda\right]_{\lambda=\lambda_G} + C,$$

$$-\int_{\lambda_L}^{\lambda_Z} b_\lambda\, d\lambda + \int_{\lambda_Z}^{\lambda_G} b_\lambda\, d\lambda = \int_{\lambda_L}^{\lambda_G} b_\lambda v_\lambda\, d\lambda. \tag{11.81}$$

As in the case of curve type 1, the shifting of the point Z due to fiber slippage shall be equal to zero. Equation (11.57) expresses the shifting $\delta\Lambda_1(\lambda^*)$ of a general fiber element, which is lying at a coordinate λ^* on the curve $\sigma_1^{LM}(\lambda)$. Especially, the shifting of point Z, i.e., $\lambda^* = \lambda_Z$, must be zero[21]. Thus,

$$\delta\Lambda_1(\lambda_Z) = \int_{\lambda_L}^{\lambda_Z} \left\{\varepsilon\left[\sigma_1^{LM}(\lambda)\right] - \varepsilon\left[\sigma(\lambda)\right]\right\} d\lambda = 0, \tag{11.82}$$

where $\sigma_1^{LM}(\lambda)$ is given by Equation (11.52) and $\sigma(\lambda)$ is determined by Equation (11.13). The last equation represents the relation between λ_L and λ_Z.

Note: Let us note that λ_L appears in Equation (11.52) too.

Similarly, Equation (11.49) expresses the shifting $\delta\Lambda_2(\lambda^*)$ of a general fiber element, which is lying at a coordinate λ^* on the curve $\sigma_2^{FG}(\lambda)$. Especially, the shifting of point Z, i.e., $\lambda^* = \lambda_Z$, must be zero. Then,

$$\delta\Lambda_2(\lambda_Z) = -\int_{\lambda_Z}^{\lambda_G} \left\{\varepsilon\left[\sigma_2^{FG}(\lambda)\right] - \varepsilon\left[\sigma(\lambda)\right]\right\} d\lambda = 0, \tag{11.83}$$

where $\sigma_2^{FG}(\lambda)$ is given by Equation (11.41) and $\sigma(\lambda)$ is determined by Equation (11.13). The last equation represents the relation between λ_Z and λ_G.

---

21   See the comments made after Equation (11.78).

Note: Let us note that λ_G appears in Equation (11.41) too.

Summarily, the triplet of Equations (11.81), (11.82) and (11.83) determines coordinates $\lambda_L, \lambda_Z, \lambda_G$. Then, the nature of tensile stress after slippage of fiber is given by Equation (11.52) – from λ_L to λ_Z, and Equation (11.41) – from λ_Z to λ_G. [At $\lambda = \lambda_Z$ the result of both equations is same because Equation (11.81) is valid.]

Double-sided slippage for discontinuous $\sigma(\lambda)$

This variant is partly analogical to the previous case. If the increase-to-decrease 'jump' and decrease-to-increase 'jump' are enough far from each other, then such situations are similar to that displayed in Figure 11.11a and 11.11b). Each such case can be solved as a couple of two independent problems displayed in Figures 11.9 and 11.10. The cases of intersection of slippage curves, demonstrated in Figure 11.12, are more complicated.

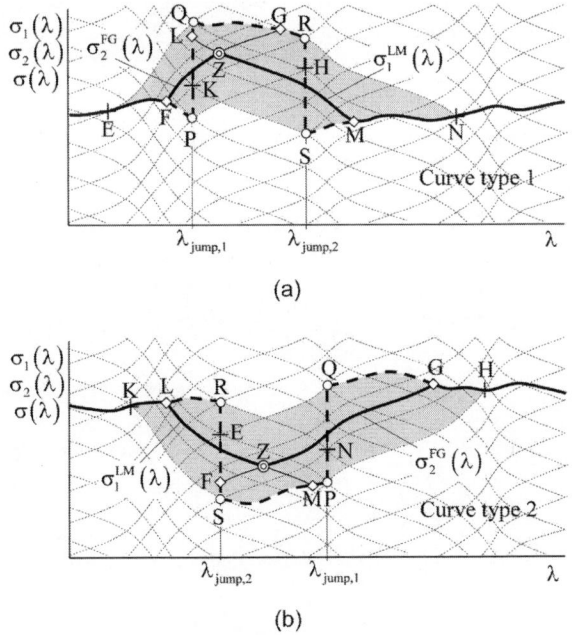

(a)

(b)

Figure 11.12 Schemes of double-sided slippage by discontinuous $\sigma(\lambda)$ – case of intersection of slippage curves

The first scheme a, i.e., curve type 1, illustrates the formation of increase-to-decrease slippage curve. The original tensile stress $\sigma(\lambda)$ (before slippage) is shown by the thick curve EFPQRSMN[22] just after yarn tensioning. (Part FPQRSM is shown by dashed line.) There exist two discontinuities: 'jump over' at the coordinate $\lambda_{\text{jump},1}$ (points P, K, L and Q) and 'jump down' at the coordinate $\lambda_{\text{jump},2}$ (points R, H and S).

The first part EFPQR of the above-mentioned original curve corresponds to the case of discontinuous increase (Figure 11.9), where Equation (11.60) is valid at $\lambda_{\text{jump}} = \lambda_{\text{jump},1}$. Similarly, the part QRSMN corresponds to the case of discontinuous decrease (Figure 11.10), where Equation (11.69) is valid at $\lambda_{\text{jump}} = \lambda_{\text{jump},2}$. Let us observe that the part QR of original tensile stress $\sigma(\lambda)$ is a common part of both cases. Thus, the coefficients C_{II} in Equation (11.60) and C_{III} in Equation (11.69) are identical. We will use a common marking for these coefficients in this case as follows:

$$C_{\text{II}} = C_{\text{III}} = C_{\text{QR}} .\qquad(11.84)$$

So, the original tensile stress is given by expressions stated as follows:

$$\left.\begin{array}{l}
\text{(a)}\ \sigma(\lambda) = k\int v_\lambda b_\lambda\, d\lambda + C_{\text{I}},\quad \lambda < \lambda_{\text{jump},1}, \\[2mm]
\text{(b)}\ \sigma(\lambda) = k\int v_\lambda b_\lambda\, d\lambda + C_{\text{QR}},\quad \lambda \in \left(\lambda_{\text{jump},1}, \lambda_{\text{jump},2}\right), \\[2mm]
\text{(c)}\ \sigma(\lambda) = k\int v_\lambda b_\lambda\, d\lambda + C_{\text{IV}},\quad \lambda > \lambda_{\text{jump},2}, \\[2mm]
\text{(d)}\ \sigma(\lambda)\ldots\text{not defined by } \lambda = \lambda_{\text{jump},1} \text{ and } \lambda = \lambda_{\text{jump},2}.
\end{array}\right\}\qquad(11.85)$$

The limit stresses at points P, Q, R and S can be obtained as follows from Equations (11.61) and (11.70) by applying Equation (11.84):

$$\left.\begin{array}{l}
\sigma(\lambda_{\text{P}}) = k \lim_{\lambda\to\lambda^-_{\text{jump},1}} \int v_\lambda b_\lambda\, d\lambda + C_{\text{I}},\quad \sigma(\lambda_{\text{Q}}) = k \lim_{\lambda\to\lambda^+_{\text{jump},1}} \int v_\lambda b_\lambda\, d\lambda + C_{\text{QR}}, \\[3mm]
\sigma(\lambda_{\text{R}}) = k \lim_{\lambda\to\lambda^-_{\text{jump},2}} \int v_\lambda b_\lambda\, d\lambda + C_{\text{QR}},\quad \sigma(\lambda_{\text{S}}) = k \lim_{\lambda\to\lambda^+_{\text{jump},2}} \int v_\lambda b_\lambda\, d\lambda + C_{\text{IV}}.
\end{array}\right\}\qquad(11.86)$$

---

22   The marking of points correspond to Figures 11.9 and 11.10.

Let us imagine that the borderline curve $\sigma_2^{FG}(\lambda)$, which is passing through point F – Equation (11.63) – and the borderline curve $\sigma_1^{LM}(\lambda)$, which is passing through point M – Equation (11.76) – solve the 'right' characteristic of tensile stress $\sigma(\lambda)$ after fiber-to-fiber slippage shown in Figure 11.12a. Nevertheless, they intersect at each other at point Z whose coordinate is $\lambda_Z \in \left(\lambda_{jump,1}, \lambda_{jump,2}\right)$. So, the relation $\sigma_2^{FG}(\lambda_Z) = \sigma_1^{LM}(\lambda_Z)$ must be valid. [This is also the first expression in Equation (11.78) which displays $\sigma(\lambda)$ as a continuous variant.] By applying Equations (11.63) and (11.76), we obtain the relation as follows:

$$\sigma_2^{FG}(\lambda_Z) = \sigma_1^{LM}(\lambda_Z), \quad k\left[\int b_\lambda \, d\lambda\right]_{\lambda=\lambda_Z} + k\left[\int b_\lambda v_\lambda \, d\lambda\right]_{\lambda=\lambda_F} - k\left[\int b_\lambda \, d\lambda\right]_{\lambda=\lambda_F} + C_I$$

$$= -k\left[\int b_\lambda \, d\lambda\right]_{\lambda=\lambda_Z} + k\left[\int b_\lambda v_\lambda \, d\lambda\right]_{\lambda=\lambda_M} + k\left[\int b_\lambda \, d\lambda\right]_{\lambda=\lambda_M} + C_{IV},$$

$$\int_{\lambda_F}^{\lambda_Z} b_\lambda \, d\lambda - \int_{\lambda_Z}^{\lambda_M} b_\lambda \, d\lambda + \frac{C_I - C_{IV}}{k} = \int_{\lambda_F}^{\lambda_M} b_\lambda v_\lambda \, d\lambda. \tag{11.87}$$

[If $C_I = C_{IV}$, then the last expression is the same as Equation (11.78).] This expression again binds the three coordinates $\lambda_F, \lambda_Z, \lambda_M$ altogether.

We must also apply the same consideration for zero-shifting of fiber element at point Z. This is commented in the text before Equation (11.79). The referred equation must be valid in this present case too. Nevertheless, the function of tensile stress $\sigma(\lambda)$ is given by Equation (11.85). Therefore, Equation (11.79) has the following form now:

$$\delta\Lambda_2(\lambda_Z) = \int_{\lambda_F}^{\lambda_Z} \left\{\varepsilon\left[\sigma_2^{FG}(\lambda)\right] - \varepsilon\left[\sigma(\lambda)\right]\right\} d\lambda = 0,$$

$$\int_{\lambda_F}^{\lambda_Z} \varepsilon\left[\overbrace{\sigma_2^{FG}(\lambda)}^{\substack{\text{According to}\\ \text{Eq. (11.63)}}}\right] d\lambda - \int_{\lambda_F}^{\lambda_{jump,1}} \varepsilon\left[\overbrace{\sigma(\lambda)}^{\substack{\text{According to}\\ \text{Eq. (11.85), a)}}}\right] d\lambda - \int_{\lambda_{jump,1}}^{\lambda_Z} \varepsilon\left[\overbrace{\sigma(\lambda)}^{\substack{\text{According to}\\ \text{Eq. (11.85), b)}}}\right] d\lambda = 0.$$

$$\tag{11.88}$$

This expression binds the two coordinates λ_F, λ_Z together.

Similarly, Equation (11.80), which has the following form in this case, must be valid:

$$\delta\Lambda_1\left(\lambda_Z\right) = -\int\limits_{\lambda_Z}^{\lambda_M}\left\{\varepsilon\left[\sigma_I^{LM}\left(\lambda\right)\right] - \varepsilon\left[\sigma\left(\lambda\right)\right]\right\}d\lambda = 0,$$

$$-\int\limits_{\lambda_Z}^{\lambda_M}\varepsilon\left[\underbrace{\sigma_I^{LM}\left(\lambda\right)}_{\substack{\text{According to}\\ \text{Eq. (11.76)}}}\right]d\lambda + \int\limits_{\lambda_Z}^{\lambda_{jump,2}}\varepsilon\left[\underbrace{\sigma\left(\lambda\right)}_{\substack{\text{According to}\\ \text{Eq. (11.85), b)}}}\right]d\lambda + \int\limits_{\lambda_{jump,2}}^{\lambda_M}\varepsilon\left[\underbrace{\sigma\left(\lambda\right)}_{\substack{\text{According to}\\ \text{Eq. (11.85), c)}}}\right]d\lambda = 0.$$

$$(11.89)$$

This expression binds the two coordinates λ_Z, λ_M together. [Compare the re-arrangement in the last two equations also with a view to Figure 11.12a).]

Summarily, the triplet of Equations (11.87), (11.88) and (11.89) determines the coordinates $\lambda_F, \lambda_Z, \lambda_M$. Equation (11.63) displays the nature of tensile stress after fiber-to-fiber slippage from λ_F to λ_Z and Equation (11.76) exhibits the same from λ_Z to λ_M. [At $\lambda = \lambda_Z$, the result of both equations is same because Equation (11.87) is valid.]

Similarly, the second scheme b, i.e., curve type 2, shown in Figure 11.12 illustrates the formation of decrease-to-increase slippage curve. The original tensile stress $\sigma\left(\lambda\right)$ (before slippage) is shown by the thick curve KLR-SPQGH just after yarn tensioning. (Part LRSPQG is shown by dashed line.) There exhibit two discontinuities – 'jump down' at the coordinate $\lambda_{jump,2}$ and 'jump over' at the coordinate $\lambda_{jump,1}$.

The part KLRSP of the above-mentioned original curve corresponds to the case of discontinuous decrease (Figure 11.10), where Equation (11.69) is valid at $\lambda_{jump} = \lambda_{jump,2}$. The part SPQGH corresponds to the case of discontinuous increase (Figure 11.9), where Equation (11.60) is valid at $\lambda_{jump} = \lambda_{jump,1}$. Because the part SP of original tensile stress $\sigma\left(\lambda\right)$ is a common part, the coefficients C_{IV} and C_I must be identical; hence we now use the following common marking:

$$C_{IV} = C_I = C_{SP}.$$

$$(11.90)$$

So, the original tensile stress is given by the following expressions:

$$(a)\ \sigma(\lambda) = k\int v_\lambda b_\lambda\ d\lambda + C_{III},\quad \lambda < \lambda_{jump,2},$$

$$(b)\ \sigma(\lambda) = k\int v_\lambda b_\lambda\ d\lambda + C_{SP},\quad \lambda \in (\lambda_{jump,2}, \lambda_{jump,1}),$$

$$(c)\ \sigma(\lambda) = k\int v_\lambda b_\lambda\ d\lambda + C_{II},\quad \lambda > \lambda_{jump,1},$$

$$(d)\ \sigma(\lambda)\ldots\text{not defined by } \lambda = \lambda_{jump,2} \text{ and } \lambda = \lambda_{jump,1}.$$

$$(11.91)$$

The limit stresses at points R, S, P and Q follow Equations (11.70) and (11.61) by using Equation (11.84):

$$\sigma(\lambda_R) = k\lim_{\lambda \to \lambda_{jump,2}^-}\int v_\lambda b_\lambda\ d\lambda + C_{III},\quad \sigma(\lambda_S) = k\lim_{\lambda \to \lambda_{jump,2}^+}\int v_\lambda b_\lambda\ d\lambda + C_{SP},$$

$$\sigma(\lambda_P) = k\lim_{\lambda \to \lambda_{jump,1}^-}\int v_\lambda b_\lambda\ d\lambda + C_{SP},\quad \sigma(\lambda_Q) = k\lim_{\lambda \to \lambda_{jump,1}^+}\int v_\lambda b_\lambda\ d\lambda + C_{II}.$$

$$(11.92)$$

Let us think that the curve $\sigma_1^{LM}(\lambda)$, which is passing through the point L – Equation (11.72) – and the curve $\sigma_2^{FG}(\lambda)$, which is passing through the point G – Equation (11.67) – create the 'right' nature of tensile stress $\sigma(\lambda)$ after fiber-to-fiber slippage, shown in Figure 11.12b). Nevertheless, they intersect at each other at point Z, $\lambda_Z \in (\lambda_{jump,2}, \lambda_{jump,1})$. So, it must be valid that $\sigma_1^{LM}(\lambda_Z) = \sigma_2^{FG}(\lambda_Z)$. [See also Equation (11.81) for continuous variant.] By applying Equations (11.72) and (11.67), we obtain the relation as follows:

$$\sigma_1^{LM}(\lambda_Z) = \sigma_2^{FG}(\lambda_Z),\quad -k\left[\int b_\lambda d\lambda\right]_{\lambda=\lambda_Z}$$

$$+k\left[\int b_\lambda v_\lambda d\lambda\right]_{\lambda=\lambda_L} + k\left[\int b_\lambda d\lambda\right]_{\lambda=\lambda_L} + C_{III}$$

$$= k\left[\int b_\lambda d\lambda\right]_{\lambda=\lambda_Z} + k\left[\int b_\lambda v_\lambda d\lambda\right]_{\lambda=\lambda_G} - k\left[\int b_\lambda d\lambda\right]_{\lambda=\lambda_G} + C_{II},$$

$$-\int_{\lambda_L}^{\lambda_Z} b_\lambda d\lambda + \int_{\lambda_Z}^{\lambda_G} b_\lambda d\lambda, +\frac{C_{III}-C_{II}}{k} = \int_{\lambda_L}^{\lambda_G} b_\lambda v_\lambda d\lambda.$$

$$(11.93)$$

[If $C_{III} = C_{II}$, then the last expression becomes same as Equation (11.81).] This expression once again binds the three coordinates $\lambda_L, \lambda_Z, \lambda_G$ altogether.

We must also apply the same general comments for zero-shifting at point Z as mentioned in the text before Equation (11.79). So, Equation (11.82) must be valid[23] too – see Figure 11.12b.

$$\delta\Lambda_1\left(\lambda_Z\right) = \int_{\lambda_L}^{\lambda_Z} \left\{\varepsilon\left[\sigma_1^{LM}\left(\lambda\right)\right] - \varepsilon\left[\sigma(\lambda)\right]\right\} d\lambda = 0,$$

$$\int_{\lambda_L}^{\lambda_Z} \varepsilon\left[\overbrace{\sigma_1^{LM}\left(\lambda\right)}^{\substack{\text{According to}\\\text{Eq. (11.72)}}}\right] d\lambda - \int_{\lambda_L}^{\lambda_{jump,2}} \varepsilon\left[\overbrace{\sigma(\lambda)}^{\substack{\text{According to}\\\text{Eq. (11.91), (a)}}}\right] d\lambda - \int_{\lambda_{jump,2}}^{\lambda_Z} \varepsilon\left[\overbrace{\sigma(\lambda)}^{\substack{\text{According to}\\\text{Eq. (11.91), (b)}}}\right] d\lambda = 0.$$

(11.94)

This expression again binds the two coordinates λ_L, λ_Z together.

Similarly, the following expression, analogical to Equation (11.83), must be valid too:

$$\delta\Lambda_2\left(\lambda_Z\right) = -\int_{\lambda_Z}^{\lambda_G} \left\{\varepsilon\left[\sigma_2^{FG}\left(\lambda\right)\right] - \varepsilon\left[\sigma(\lambda)\right]\right\} d\lambda = 0,$$

$$-\int_{\lambda_Z}^{\lambda_G} \varepsilon\left[\overbrace{\sigma_2^{FG}\left(\lambda\right)}^{\substack{\text{According to}\\\text{Eq. (11.67)}}}\right] d\lambda + \int_{\lambda_Z}^{\lambda_{jump,1}} \varepsilon\left[\overbrace{\sigma(\lambda)}^{\substack{\text{According to}\\\text{Eq. (11.91), (b)}}}\right] d\lambda + \int_{\lambda_{jump,1}}^{\lambda_G} \varepsilon\left[\overbrace{\sigma(\lambda)}^{\substack{\text{According to}\\\text{Eq. (11.91), (c)}}}\right] d\lambda = 0.$$

(11.95)

This expression again binds the two coordinates λ_Z, λ_G together.

Summarily, the triplet of Equations (11.93), (11.94) and (11.95) determines the triplet of coordinates $\lambda_F, \lambda_Z, \lambda_M$. After slippage of the fiber, the nature of tensile stress is given by Equation (11.72) from λ_L to λ_Z and Equation (11.67) from λ_Z to λ_G. [At $\lambda = \lambda_Z$, the result of both equations is the same because Equation (11.93) is valid.]

11.6   Simplest theoretical examples of slippage at 'middle' parts of fiber

It is necessary to know many items of information (inputs) for practical utilization of the derived relations for fiber-to-fiber slippage in a fiber bundle, e.g., yarn. They are:

---

23   Equation (11.82) and also Equation (11.83) were originally derived in the case of continuous tensile stress $\sigma(\lambda)$.

1. The tensile stress $\sigma(\lambda)$ just after tensioning. It contains:

    (a) fiber parameter k according to Equation (11.7) and fiber length l,

    (b) parameter C or alternatively parameters from C_I to C_{IV} [cases of rapid increase and/or rapid decrease of $\sigma(\lambda)$ – Equations (11.13), (11.60), (11.69), (11.85), (11.91)],

    (c) usually continuous function b_λ that expresses together

        (i) fiber tensioning,

        (ii) fiber-to-fiber friction (e.g., Coulomb's law or other laws of friction) and

        (iii) external action of forces on fiber surface, and

    (d) continuous or discontinuous function v_λ characterizing mostly the variability of microstructure around the slipped fiber.

2. The continuously increasing stress–strain relation characterizing the tensile properties of the slipped fiber according to Equation (11.42).

The determination of such a set of inputs depends on the properties of the fibers and on the (mathematical models of) structural behaviour of the surrounding of the slipped fiber.

Based on the discontinuous tensile stress $\sigma(\lambda)$, several simplest examples can be created. They are probably not a solution of a particular real problem; however, they demonstrate the methods of application of the derived expressions.

Example 1 – discontinuous increase of $\sigma(\lambda)$

Let us think that the following relations are valid for Equation (11.60):

$$\left. \begin{array}{l} b_\lambda = b \ldots \text{known constant,} \\[2mm] v_\lambda = 0 \ldots \text{known constant, } \lambda \gtrless \lambda_{jump}, \\[2mm] v_\lambda \ldots \text{not defined, } \lambda = \lambda_{jump}, \\[2mm] k, \lambda_{jump} \text{ and } C_I < C_{II} \ldots \text{known parameters.} \end{array} \right\} \tag{11.96}$$

Note: We can understand that a constant value of b results in a constant effect of friction along a fiber from its surroundings. The zero-value of v_λ can express a non-variable character of microstructure along the fiber (except coordinate λ_{jump}).

Further, let us think that the increasing stress–strain relation of the slipped fiber has the following form – see also Equation (11.42):

$$\tau = E\varepsilon, \quad \varepsilon = \tau/E, \tag{11.97}$$

E...parameter (Young's modulus),

where τ represents general tensile stress of fiber and ε is the corresponding strain of fiber. (The above equation expresses the well-known Hooke's law.)

Equation (11.60) takes the following form by application of Equation (11.96):

$$\left.\begin{aligned}
\sigma(\lambda) &= k\int 0\,b\,d\lambda + C_{\mathrm{I}} = C_{\mathrm{I}}, \quad \lambda < \lambda_{\mathrm{jump}}, \\
\sigma(\lambda) &= k\int 0\,b\,d\lambda + C_{\mathrm{II}}, = C_{\mathrm{II}}, \quad \lambda > \lambda_{\mathrm{jump}}, \\
\sigma(\lambda) &\ldots\text{not defined}, \quad \lambda = \lambda_{\mathrm{jump}}.
\end{aligned}\right\} \tag{11.98}$$

Similarly, it is valid to write for Equation (11.61) as

$$\left.\begin{aligned}
\sigma(\lambda_{\mathrm{P}}) &= k \lim_{\lambda \to \lambda_{\mathrm{jump}}^-} \int 0\,b\,d\lambda + C_{\mathrm{I}} = C_{\mathrm{I}}, \\
\sigma(\lambda_{\mathrm{Q}}) &= k \lim_{\lambda \to \lambda_{\mathrm{jump}}^+} \int 0\,b\,d\lambda + C_{\mathrm{II}} = C_{\mathrm{II}}.
\end{aligned}\right\} \tag{11.99}$$

Figure 11.13 illustrates this special nature of $\sigma(\lambda)$ before slippage by the line EFPXQGH – the part FPXQG is shown by dashed line.

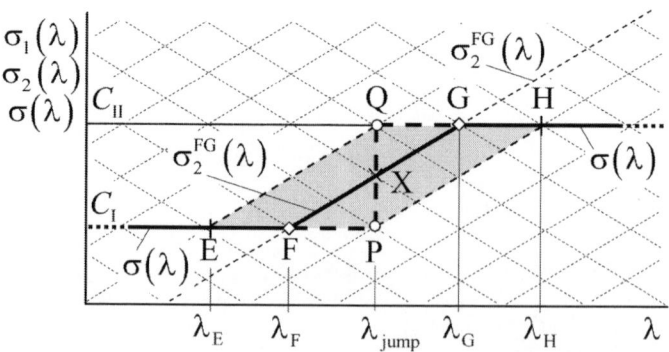

Figure 11.13 Example 1 – discontinuous increase of $\sigma(\lambda)$

The determination of actual fiber-to-fiber slippage requires to apply the present Equations (11.96) to (11.99) to the set of more general Equations

(11.62) to (11.68). Appendix 14 shows the mathematical rearrangement of it.

The borderline curve of tensile stress after slippage, i.e., the line $\sigma_2^{FG}(\lambda)$, is given by Equation (A14.3) in Appendix 14. Then, we obtain

$$\sigma_2^{FG}(\lambda) = \frac{\lambda - \lambda_F}{\lambda_G - \lambda_F}(C_{II} - C_I) + C_I. \tag{11.100}$$

(This is an increasing straight line as shown in Figure 11.13.) Evidently, $\sigma_2^{FG}(\lambda_F) = C_I$ and $\sigma_2^{FG}(\lambda_G) = C_{II}$; these results correspond to Figure 11.13.

The coordinates λ_F and λ_G are determined by Equations (A14.7) and (A14.8) in Appendix 14. Then, we get

$$\lambda_F = \lambda_{jump} - \frac{C_{II} - C_I}{2kb}, \quad \lambda_G = \lambda_{jump} + \frac{C_{II} - C_I}{2kb}. \tag{11.101}$$

(The coordinates λ_F and λ_G are lying symmetrically around the coordinate λ_{jump} in this special case.) The fiber slips in the interval from λ_F to λ_G just after increase of tensile stress discontinuously. The corresponding tensile stresses at points F, G and X are

$$\left. \begin{array}{l} \sigma_2^{FG}(\lambda_F) = C_I, \quad \sigma_2^{FG}(\lambda_G) = C_{II}, \\[2mm] \sigma_2^{FG}(\lambda_X) = \sigma_2^{FG}(\lambda_{jump}) = \dfrac{C_I + C_{II}}{2}. \end{array} \right\} \tag{11.102}$$

[Equations (11.100) and (A14.9) from Appendix 14 were used.]

Figure 11.13 illustrates the tensile stress acting on the fiber after slippage by the thick curve EFXGH.

Example 2 – discontinuous decrease of $\sigma(\lambda)$ [24]

In analogy to the previous case of example 1, let us think that the following relations are valid for Equation (11.69):

---

24   Example 2 is a 'mirror image' of Example 1.

$$b_\lambda = b \ldots \text{known constant,}$$

$$v_\lambda = 0 \ldots \text{known constant, } \lambda \gtrless \lambda_{\text{jump}},$$

$$v_\lambda \ldots \text{not defined, } \lambda = \lambda_{\text{jump}},$$

$$k, \lambda_{\text{jump}} \text{ and } C_{\text{III}} > C_{\text{IV}} \ldots \text{known parameters.}$$

$$\tag{11.103}$$

Let us more think that the increasing stress–strain relation of the slipped fiber has again the increasing linear behaviour according to Equation (11.97) (Hooke's law).

Equations (11.69) and (11.70) take the following forms by application of Equation (11.103):

$$\sigma(\lambda) = k \int 0\, b\, \mathrm{d}\lambda + C_{\text{III}} = C_{\text{III}}, \quad \lambda < \lambda_{\text{jump}},$$

$$\sigma(\lambda) = k \int 0\, b\, \mathrm{d}\lambda + C_{\text{IV}}, = C_{\text{IV}}, \quad \lambda > \lambda_{\text{jump}},$$

$$\sigma(\lambda) \ldots \text{not defined,} \quad \lambda = \lambda_{\text{jump}},$$

$$\tag{11.104}$$

and

$$\sigma(\lambda_{\text{R}}) = k \lim_{\lambda \to \lambda_{\text{jump}}^-} \int 0\, b\, \mathrm{d}\lambda + C_{\text{III}} = C_{\text{III}},$$

$$\sigma(\lambda_{\text{S}}) = k \lim_{\lambda \to \lambda_{\text{jump}}^+} \int 0\, b\, \mathrm{d}\lambda + C_{\text{IV}} = C_{\text{IV}}.$$

$$\tag{11.105}$$

Figure 11.14 illustrates this behaviour of $\sigma(\lambda)$ before slippage along the line KLRXSMN. (The part LRXSM is shown by dashed line.)

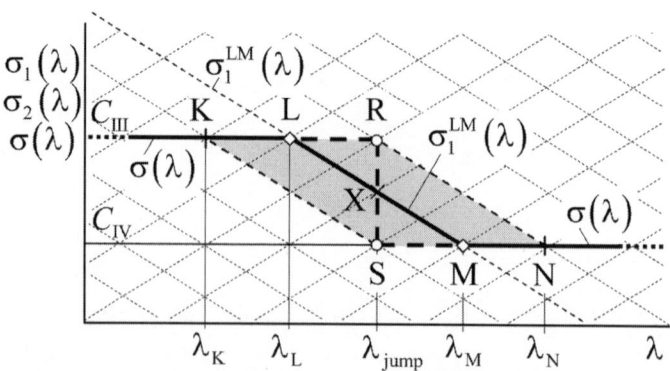

Figure 11.14 Example 2 – discontinuous decrease of $\sigma(\lambda)$

The determination of actual fiber-to-fiber slippage requires to apply Equations (11.97) and (11.103) to (11.105) in the set of more general Equations (11.71) to (11.77). Appendix 14 shows the mathematical rearrangement of it.

The borderline curve of tensile stress after slippage, i.e., the line $\sigma_1^{LM}(\lambda)$, is given by Equation (A14.12) in Appendix 14. Then,

$$\sigma_1^{LM}(\lambda) = -\frac{\lambda - \lambda_L}{\lambda_M - \lambda_L}(C_{III} - C_{IV}) + C_{III} . \tag{11.106}$$

(This represents the decreasing straight line shown in Figure 11.14.) Note that it is valid to write that $\sigma_1^{LM}(\lambda_L) = C_{III}$ and $\sigma_1^{LM}(\lambda_M) = C_{IV}$. This corresponds to Figure 11.14.

The coordinates λ_L and λ_M were derived in Equations (A14.16) and (A14.17) in Appendix 14. Then,

$$\lambda_L = \lambda_{jump} - \frac{C_{III} - C_{IV}}{2kb}, \quad \lambda_M = \lambda_{jump} + \frac{C_{III} - C_{IV}}{2kb} . \tag{11.107}$$

(The coordinates λ_L and λ_M are lying symmetrically around the coordinate λ_{jump} also in this special case.) The fiber slips immediately after origin of discontinuous increase of tensile stress in the interval from λ_L to λ_M.

The corresponding tensile stresses at points L, X and M are

$$\left.\begin{array}{l} \sigma_1^{LM}(\lambda_L) = C_{III}, \quad \sigma_1^{LM}(\lambda_M) = C_{IV}, \\[2mm] \sigma_1^{LM}(\lambda_X) = \sigma_1^{LM}(\lambda_{jump}) = \dfrac{C_{III} + C_{IV}}{2} . \end{array}\right\} \tag{11.108}$$

[Equation (11.106) and Equation (A14.18) from Appendix 14 were used.]

Figure 11.14 illustrates the tensile stress developed on the fiber after slippage by the thick curve KLXMN.

Example 3 – discontinuous increase to decrease of $\sigma(\lambda)$

The double-sided increase-to-decrease 'jumps' are described by Equations (11.84) to (11.89). Nevertheless, the following special relations are valid for probably the simplest case of this problem:

$$b_\lambda = b \dots \text{known constant,}$$

$$v_\lambda = 0 \dots \text{known constant,}\ \lambda \neq \lambda_{\text{jump,1}},\ \lambda \neq \lambda_{\text{jump,2}},$$

$$v_\lambda \dots \text{not defined by } \lambda = \lambda_{\text{jump,1}},\ \lambda = \lambda_{\text{jump,2}},$$

$$k,\ \lambda_{\text{jump,1}} < \lambda_{\text{jump,2}},\ C_{\text{I}} = C_{\text{IV}},\ C_{\text{QR}} > C_{\text{I}}, \dots \text{known parameters.}$$

$$(11.109)$$

Note: The previous assumptions related to Equations (11.96) and (11.103) were used earlier. Furthermore, the equivalency $C_{\text{I}} = C_{\text{IV}}$ is introduced in this simplest case of double-sided slippage.

We also assume the simplest stress–strain relation in fiber, according to Equation (11.97) – Hooke's law with a Young's modulus E.

Equations (11.85) and (11.86) take the special form by applying Equation (11.109):

$$(a)\ \sigma(\lambda) = k \int 0\, b\, d\lambda + C_{\text{I}} = C_{\text{I}},\quad \lambda < \lambda_{\text{jump,1}},$$

$$(b)\ \sigma(\lambda) = k \int 0\, b\, d\lambda + C_{\text{QR}} = C_{\text{QR}},\quad \lambda \in \left(\lambda_{\text{jump,1}}, \lambda_{\text{jump,2}}\right),$$

$$(c)\ \sigma(\lambda) = k \int 0\, b\, d\lambda + C_{\text{I}} = C_{\text{I}},\quad \lambda > \lambda_{\text{jump,2}},$$

$$(d)\ \sigma(\lambda) \dots \text{not defined by } \lambda = \lambda_{\text{jump,1}} \text{ and } \lambda = \lambda_{\text{jump,2}}.$$

$$(11.110)$$

$$\sigma(\lambda_{\text{P}}) = k \lim_{\lambda \to \lambda_{\text{jump,1}}^-} \int 0\, b\, d\lambda + C_{\text{I}} = C_{\text{I}},\quad \sigma(\lambda_{\text{Q}}) = k \lim_{\lambda \to \lambda_{\text{jump,1}}^+} \int 0\, b\, d\lambda + = C_{\text{QR}},$$

$$\sigma(\lambda_{\text{R}}) = k \lim_{\lambda \to \lambda_{\text{jump,2}}^-} \int 0\, b\, d\lambda + C_{\text{QR}} = C_{\text{QR}},\quad \sigma(\lambda_{\text{S}}) = k \lim_{\lambda \to \lambda_{\text{jump,2}}^+} \int 0\, b\, d\lambda + C_{\text{I}} = C_{\text{I}}.$$

$$(11.111)$$

Figure 11.15[25] illustrates this special behaviour of $\sigma(\lambda)$ before slippage by the line FPQRSM (thick dashed line between F and M, and fully drawn line before F and after M).

---

25 The points X_1, L, V, W, G, X_2 and U are shown by smaller fonts for orientation of the graph.

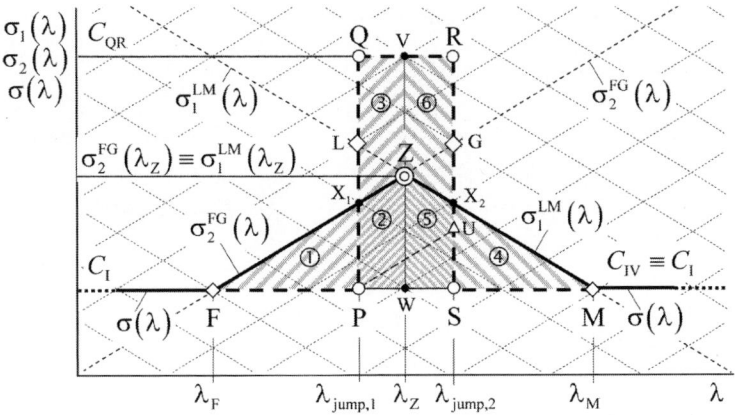

Figure 11.15 Example 3 – discontinuous increase to decrease of $\sigma(\lambda)$

Equations (A14.19) and (A14.20) in Appendix 14 determine the 'right' borderline curves after slippage, i.e., $\sigma_2^{FG}(\lambda)$ and $\sigma_1^{LM}(\lambda)$

$$\sigma_2^{FG}(\lambda) = kb\lambda - kb\lambda_F + C_I,$$ (11.112)

$$\sigma_1^{LM}(\lambda) = -kb\lambda + kb\lambda_M + C_I.$$ (11.113)

The first function is the increasing straight line passing through the point F, $\sigma_2^{FG}(\lambda_F) = C_I$ and the second function is the decreasing straight line passing through the point M, $\sigma_1^{LM}(\lambda_M) = C_I$.

The coordinates $\lambda_Z, \lambda_F, \lambda_M$ were derived in Appendix 14, Equations (A14.32), (A14.33) and (A14.34). Then,

$$\lambda_Z = \frac{\lambda_{jump,1} + \lambda_{jump,2}}{2},$$ (11.114)

$$\lambda_F = \frac{\lambda_{jump,1} + \lambda_{jump,2}}{2} - \sqrt{\frac{(C_{QR} - C_I)}{kb}(\lambda_{jump,2} - \lambda_{jump,1})},$$ (11.115)

$$\lambda_M = \frac{\lambda_{jump,1} + \lambda_{jump,2}}{2} + \sqrt{\frac{(C_{QR} - C_I)}{kb}(\lambda_{jump,2} - \lambda_{jump,1})}.$$ (11.116)

Let us note that λ_Z, according to Equation (11.114), is lying just at the middle between $\lambda_{jump,1}$ and $\lambda_{jump,2}$ in this example. According to the previous triplet of equations, the coordinates λ_F and λ_M are lying symmetrically around the value λ_Z. [See also Equation (A14.21) in Appendix 14.] This is illustrated in Figure 11.15 too.

It is not too easy to imagine clearly a physical (mathematical) logic of creation of λ_F and λ_M immediately from Equations (11.115) and (11.116). Nevertheless, Equations (A14.25) and (A14.30) were derived in Appendix 14, so that it is valid to write

$$\frac{kb}{2}\left(\lambda_Z - \lambda_F\right)^2 = \left(C_{QR} - C_I\right)\left(\lambda_Z - \lambda_{jump,1}\right), \qquad (11.117)$$

$$\frac{kb}{2}\left(\lambda_M - \lambda_Z\right)^2 = \left(C_{QR} - C_I\right)\left(\lambda_{jump,2} - \lambda_Z\right). \qquad (11.118)$$

Let us understand the first equation with a view to the scheme shown in Figure 11.15. It depicts the area of triangle FWZ[26] – i.e., areas ① plus ② – on the left-hand side of Equation (11.117). The right-hand side of the above-mentioned equation expresses the area of rectangle PQVW – i.e., areas ② plus ③. However, the densely shaded area ② is common for both the stated areas, so that the area of triangle FPX_1 must be equal to the area of trapezoid $QVZX_1$. [What is under the slippage line $\sigma_2^{FG}\left(\lambda\right)$ before jump must be over it after jump.] We mentioned in the paragraph after Equation (11.78) that the shifting of points – F and Z – shall be equal to zero due to fiber slippage so that Equation (11.79) is valid. (The derivation of Equation (11.117) really starts from Equation (11.79); this is rearranged first to Equation (11.88), then to Equation (A14.22) in Appendix 14, and finally transferred to this section as Equation (11.117).)

Similarly, we can interpret Equation (11.118). The left-hand side represents the area of triangle MWZ as shown in Figure 11.15, i.e., areas ④ plus ⑤ in Figure 11.15. The right-hand side represents the area of rectangle SRVW, i.e., areas ⑤ plus ⑥. Because the densely shaded area ⑤ is common to both the areas, the area of triangle MSX_2 must be equal to

---

26 Generally, the height h of a right triangle ◺h is $h = b \tan \alpha$. The area of such triangle is $bh/2 = (b^2 \tan \alpha)/2$. The stated tangent of the angle shown in Figure 11.15 – angle ZFW – is the angular coefficient kb of the straight line according to Equation (11.112).

area of trapezoid $RVZX_2$. This interpretation is analogous to the previous argumentation. [The derivation of Equation (11.118) starts from Equation (11.80); this is rearranged first to Equation (11.89), then to Equation (A14.30) in Appendix 14, and finally transferred to this section as Equation (11.118).]

The 'highest' value of tensile stress in the fiber, i.e., the tensile stress $\sigma_2^{FG}(\lambda_Z) = \sigma_1^{LM}(\lambda_Z)$ at point Z, is shown by Equation (A14.35) in Appendix 14. It is valid to write that

$$\sigma_2^{FG}(\lambda_Z) = \sigma_1^{LM}(\lambda_Z) = \sqrt{kb(C_{QR} - C_I)(\lambda_{jump,2} - \lambda_{jump,1})} + C_I.$$

(11.119)

Nevertheless, we also need to check if the solution given in Example 3 corresponds really to the case of 'double-sided slippage' illustrated more generally in Figure 11.12a. It is evident that the calculated tensile stress $\sigma_2^{FG}(\lambda_Z) = \sigma_1^{LM}(\lambda_Z)$ at point Z, i.e., the value according to Equation (11.119), must not exceed the value of $\sigma(\lambda_Z)$ before slippage; this is $\sigma(\lambda_Z) = C_{QR}$ according to Equation (11.110), case b[27], in Example 3. Thus, the following condition must be satisfied:

$$\sqrt{kb(C_{QR} - C_I)(\lambda_{jump,2} - \lambda_{jump,1})} + C_I \le C_{QR}.$$

(11.120)

By rearranging this inequality according to Equation (A14.36) of Appendix 14, we get

$$kb(\lambda_{jump,2} - \lambda_{jump,1}) \le (C_{QR} - C_I) \quad \text{and/or} \quad kb(\lambda_{jump,2} - \lambda_{jump,1}) + C_I \le C_{QR}.$$

(11.121)

It is possible to graphically illustrate the last expression which is valid in Example 3. Let us remind that kb is a common angular coefficient of each line $\sigma_2(\lambda)$, i.e., including the line PU in Figure 11.15. So, the expression $kb(\lambda_{jump,2} - \lambda_{jump,1})$ refers to the length of the segment SU and $(C_{QR} - C_I)$

---

27   The horizontal coordinate λ_Z of point Z must lie in the interval $(\lambda_{jump,1},$ $\lambda_{jump,2})$. Therefore case b is valid.

represents the length of segment SR. In other words, point U must not lie over point R along the vertical line SR.

If the condition according to Equation (11.121) is not valid, then the whole previous calculation of Example 3 is not usable. We must solve such a situation by another way, namely as two independent cases – discontinuous increase and discontinuous decrease – according to Examples 1 and 2, respectively.

Example 4 – discontinuous decrease-increase of $\sigma(\lambda)$

The double-sided decrease-to-increase 'jumps' are described by Equation (11.90) to (11.95). Nevertheless, the following special relations are valid for probably the simplest case of this example:

$$\left.\begin{array}{l} b_\lambda = b\ldots\text{known constant,} \\[2mm] v_\lambda = 0\ldots\text{known constant, } \lambda \ne \lambda_{\text{jump,2}}, \lambda \ne \lambda_{\text{jump,1}}, \\[2mm] v_\lambda \ldots\text{not defined by } \lambda = \lambda_{\text{jump,2}}, \lambda = \lambda_{\text{jump,1}}, \\[2mm] C_{\text{III}} = C_{\text{II}}, C_{\text{SP}} < C_{\text{III}}, \ldots\text{known parameters,} \\[2mm] k, \lambda_{\text{jump,2}} < \lambda_{\text{jump,1}}, \ldots\text{known parameters.} \end{array}\right\} \quad (11.122)$$

Note: These assumptions are related to Equations (11.96) and (11.103). Moreover, the equivalency $C_{\text{III}} = C_{\text{II}}$ is introduced in this simplest case of double-side slippage.

We always assume the validity of Equation (11.97) – Hooke's law with a Young's modulus E.

Equations (11.91) and (11.92) take the following special form by application of Equation (11.109):

$$\left.\begin{array}{l} \text{(a) } \sigma(\lambda) = k\int 0\,b\,d\lambda + C_{\text{III}} = C_{\text{III}}, \quad \lambda < \lambda_{\text{jump,2}}, \\[2mm] \text{(b) } \sigma(\lambda) = k\int 0\,b\,d\lambda + C_{\text{SP}} = C_{\text{SP}}, \quad \lambda \in \left(\lambda_{\text{jump,2}}, \lambda_{\text{jump,1}}\right), \\[2mm] \text{(c) } \sigma(\lambda) = k\int 0\,b\,d\lambda + C_{\text{III}} = C_{\text{III}}, \quad \lambda > \lambda_{\text{jump,1}}, \\[2mm] \text{(d) } \sigma(\lambda)\ldots\text{not defined by } \lambda = \lambda_{\text{jump,2}} \text{ and } \lambda = \lambda_{\text{jump,1}}. \end{array}\right\} \quad (11.123)$$

$$\left.\begin{array}{l} \sigma(\lambda_{\text{R}}) = k \lim_{\lambda\to\lambda_{\text{jump,2}}^-} \int 0\,b\,d\lambda + C_{\text{III}} = C_{\text{III}}, \quad \sigma(\lambda_{\text{S}}) = k \lim_{\lambda\to\lambda_{\text{jump,2}}^+} \int 0\,b\,d\lambda + C_{\text{SP}} = C_{\text{SP}}, \\[3mm] \sigma(\lambda_{\text{P}}) = k \lim_{\lambda\to\lambda_{\text{jump,1}}^-} \int 0\,b\,d\lambda + C_{\text{SP}} = C_{\text{SP}}, \quad \sigma(\lambda_{\text{Q}}) = k \lim_{\lambda\to\lambda_{\text{jump,1}}^+} \int 0\,b\,d\lambda + C_{\text{III}} = C_{\text{III}}. \end{array}\right\}$$

$$(11.124)$$

Figure 11.16 illustrates this special behaviour of $\sigma(\lambda)$ before slippage by the line LRSPQG (thick dashed line between L and G and fully drawn line before L and after G).

Note: Figure 11.16 is in principle a mirror image of Figure 11.15. It means that the derivation of the required relations is quite analogous to those used for previous Example 3.

Equations (A14.37) and (A14.38) in Appendix 14 determine the 'right' borderlines after slippage, i.e., functions $\sigma_1^{LM}(\lambda)$ and $\sigma_2^{FG}(\lambda)$. Thus,

$$\sigma_1^{LM}(\lambda) = -kb\lambda + kb\lambda_L + C_{III}, \tag{11.125}$$

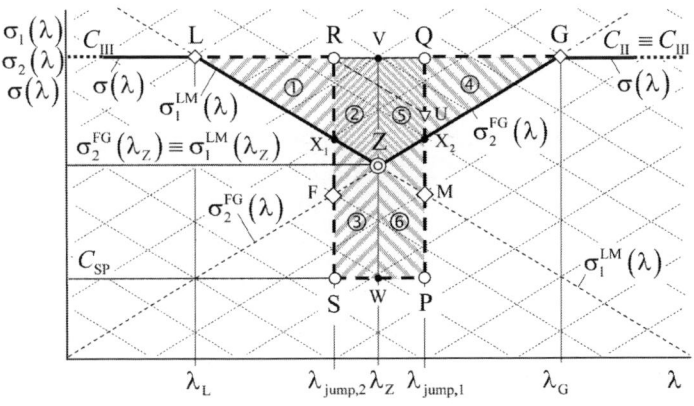

Figure 11.16 Example 4 – discontinuous decrease to increase of $\sigma(\lambda)$

$$\sigma_2^{FG}(\lambda) = kb\lambda - kb\lambda_G + C_{III}. \tag{11.126}$$

The first function is the decreasing straight line passing through the point L $[\sigma_1^{LM}(\lambda_L) = C_{III}]$ and the second function is the increasing straight line passing through the point G $[\sigma_2^{FG}(\lambda_G) = C_{III}]$.

The determined coordinates $\lambda_Z, \lambda_L, \lambda_G$ were derived in Equations (A14.48), (A14.49) and (A14.50), respectively, of Appendix 14. Thus,

$$\lambda_Z = \frac{\lambda_{jump,2} + \lambda_{jump,1}}{2}, \tag{11.127}$$

$$\lambda_L = \frac{\lambda_{jump,2} + \lambda_{jump,1}}{2} - \sqrt{\frac{C_{III} - C_{SP}}{kb}(\lambda_{jump,1} - \lambda_{jump,2})}, \tag{11.128}$$

$$\lambda_G = \frac{\lambda_{jump,2} + \lambda_{jump,1}}{2} + \sqrt{\frac{C_{III} - C_{SP}}{kb}\left(\lambda_{jump,1} - \lambda_{jump,2}\right)}. \tag{11.129}$$

Let us note that λ_Z is lying in the middle of $\lambda_{jump,2}$ and $\lambda_{jump,1}$. Also, the coordinates λ_L and λ_G are lying symmetrically around the value λ_Z in this case. This is shown in Figure 11.16 too.

The logical sense of determination of λ_L and λ_G is better shown in Equations (A14.42) and (A14.46) of Appendix 14. We can also write these equations as follows:

$$\frac{kb}{2}\left(\lambda_Z - \lambda_L\right)^2 = \left(C_{III} - C_{SP}\right)\left(\lambda_Z - \lambda_{jump,2}\right), \tag{11.130}$$

$$\frac{kb}{2}\left(\lambda_G - \lambda_Z\right)^2 = \left(C_{III} - C_{SP}\right)\left(\lambda_{jump,1} - \lambda_Z\right). \tag{11.131}$$

The value kb expresses the tangent of angle VLZ and/or VGZ in Figure 11.16 – see angular coefficients in Equations (11.125) and (11.126). So, the left-hand side of Equation (11.130) represents the area of triangle VLZ[28], i.e., area ① plus area ②. The right-hand side of Equation (11.130) represents the area of rectangle SRVW, i.e., area ② plus area ③. Because Equation (11.130) is valid, the area ① must be same as the area ③. (What is over the line LZ must be also under the line.) Similarly, the left-hand side of Equation (11.131) represents the area of triangle VGZ, i.e., area ④ plus area ⑤. The right-hand side of this equation represents the area of rectangle PQVW, i.e., area ⑤ plus area ⑥. So, the area ④ must be same as the area ⑥. (What is under the line ZG must be also over the line.)

Equation (A14.51) of Appendix 14 shows the value of tensile stress in the fiber at point Z, i.e., the tensile stress $\sigma_1^{LM}\left(\lambda_Z\right) = \sigma_2^{FG}\left(\lambda_Z\right)$. It is valid to write that

$$\sigma_1^{LM}\left(\lambda_Z\right) = \sigma_2^{FG}\left(\lambda_Z\right) = C_{III} - \sqrt{kb\left(C_{III} - C_{SP}\right)\left(\lambda_{jump,1} - \lambda_{jump,2}\right)}. \tag{11.132}$$

---

28   See the footnote given to the text after Equation (11.118).

It is true that we also need to check if the given solution in Example 4 corresponds really to the case of 'double-sided slippage' illustrated more generally in Figure 11.12b. The calculated tensile stress $\sigma_1^{LM}(\lambda_Z) = \sigma_2^{FG}(\lambda_Z)$ at point Z, i.e., the value according to Equation (11.132), must not fall down under the corresponding value $\sigma(\lambda_Z)$ before slippage, i.e., under $\sigma(\lambda_Z) = C_{SP}$ according to Equation (11.123), case b[29]. Thus, the following condition must be satisfied:

$$C_{III} - \sqrt{kb(C_{III} - C_{SP})(\lambda_{jump,1} - \lambda_{jump,2})} \geq C_{SP}. \tag{11.133}$$

Rearranging the previous relation according to Equation (A14.52) of Appendix 14, we can write this condition in the form as follows:

$$(C_{III} - C_{SP}) \geq kb(\lambda_{jump,1} - \lambda_{jump,2}) \quad \text{or,} \quad C_{III} - kb(\lambda_{jump,1} - \lambda_{jump,2}) \geq C_{SP}. \tag{11.134}$$

Let us think about the first inequality in the last expression. The size of its right-hand side shows the length of segment QU in Figure 11.16. The size of its left-hand side represents the length of segment QP. It means that the length QU must not be longer then the length QP.

If the condition according to Equation (11.134) is not valid, then the whole previous calculation of Example 4 is not usable. We must solve such situation as two independent cases – discontinuous decrease and discontinuous increase – according to Examples 2 and 1, respectively.

Note: Let us remark that all expressions that are valid in Example 4 are very analogical to the expressions derived in Example 3. These two cases are really mirror images to each other.

11.7   Closing remarks on fiber-to-fiber slippage in yarns

It has already been mentioned in the introductory text of Sections 11.1 and 11.6 that yarn tensioning and yarn strength are very complex phenomena. This is because each deeper theoretical solution must include the regulations

---

29   The horizontal coordinate λ_z of point Z must lie in the interval $(\lambda_{jump,1}, \lambda_{jump,2})$ so that case b is valid.

of yarn structure (geometry), mechanics of single fibers and fiber interactions inside a yarn, and different random phenomena (variability) mostly related to yarn microstructure and micromechanics. Therefore, the creation of a more general mathematical model of yarn tensioning and yarn strength remains as an open problem to the theoreticians working on yarns.

The fiber-to-fiber slippage is ranked as a very significant but highly 'mysterious' phenomenon, which had not been easy to incorporate into yarn mechanics. In the previous sections of this chapter, our theoretical solution showed more or less a general solution of inherent slippages of fibers, including fibers in a yarn. But it did not say anything about the input conditions that are valid 'around' the slipped fibers in a yarn and also nothing about the variability of fiber properties along the fiber. They are difficult to find it, so that they are perhaps expressed by simplified ideas and simplified mathematical models. The input conditions are concentrated mostly to the triplet of functions b_λ, v_λ and $\varepsilon[\sigma(\lambda)]$.

Borderline function b_λ

This function (together with suitable constants[30]) determines the borderline curves $\sigma_1(\lambda), \sigma_2(\lambda)$ of fiber-to-fiber slippage as it is shown, e.g., by Equations (11.11) and (11.12). Function b_λ determines whether the fiber element will slip or it will be held without slippage by its surrounding fibers. In short, this function above all represents the conditions of fiber-to-fiber friction. It is reasonable and advantageous to interpret this function b_λ as a manifestation of different 'regular' inherent laws. (The different 'local variabilities' are related to function v_λ.)

The forces that induce friction go usually out from the relation between the packing density and the corresponding pressure. (See chapter 5 in the book [1] for this phenomenon.) However, the packing density changes (usually decreases) with the increase of radius inside the yarn so that the pressure also changes (decreases) (see Chapter 8). Moreover, the fiber geometry – i.e., slopes of fibers due to twist, fiber orientation, etc. – also changes with yarn radius and thus influences the above-mentioned relation. If a fiber passes by different radial layers in a yarn (see migration models in Chapter 5 and/or more general models in Chapter 3), then the outer pressures must be different at different places of the fiber. However,

---

30   Constants C_1, C_2 by continual function $\sigma(\lambda)$ or C_I, C_{II}, C_{III}, C_{IV} by discontinuous function $\sigma(\lambda)$.

each fiber lies at only one radius in the idealized case of helical model (see Chapter 4) so that such a solution is theoretically simpler. Naturally, the simplest is the case of ideal helical model, having a constant value of packing density in the whole yarn body but such model is often too far from the real yarn.

The forces around the fiber can evoke frictional forces onto the fiber surface (see Section 11.2). The mutual relation between these two phenomena is known as a law of fiction. Most often, the so-called Coulomb's law – the linear relation between the normal force and the maximum evoked frictional force – is often used[31]. Unfortunately, this law is not too much valid for fiber-to-fiber friction.

Although the model equations in the previous sections are relatively general, they do not distinguish from static and dynamic friction whose differences can also play a significant role in fiber-to-fiber slippage.

Summarily, the determination of the borderline function b_λ of fiber frictional shear stress is markedly difficult. It might be possible that the first step could be started with the helical model of a yarn and Coulomb's model of friction using a 'radial profile' of packing density and its relation to pressure. Such a starting model can then be generalized in pursuant to comparison of theoretical results with several experimental ones.

Relative frictional loading function v_λ

This partial function occurs only in the expression of tensile stress $\sigma(\lambda)$ in fiber (mainly after tensioning of a yarn); see Equation (11.13).

The function v_λ can be interpreted as a variance of tensile stress due to the variability of frictional shear stress along the fiber. Such variability is mostly influenced by the fluctuations in internal microstructure of yarn. The random changes of tensile stress $\sigma(\lambda)$ is originated due to the random nature of fiber-to-fiber contacts, whose density and distribution are decided by the

---

31 Generally, the mutual relation between the forces present around a fiber and the frictional forces evoked thereby is usually interpreted empirically. The exact physical model of it is not yet commonly known. The easiest and oldest variant is known to be the Coulomb's law, i.e., $\Phi = k_f N$, where N is the normal force, k_f is the constant coefficient of fiction (parameter of materials) and Φ is the (maximum) frictional force. Let us note that so-called 'Euler's friction' (friction on a cylindrical surface) is the only one possible application of Coulomb's law.

compression of fibrous material and the orientation of fibers inside the yarn. (See Chapters 3–5 of the book [1].) In general, the fiber-to-fiber contacts generate so small touches that the contact places can be idealized by the contact points. Then the function v_λ can be ideally described by the two examples given in Section 11.3 (slippage of fiber ends) and the four examples in Section 11.6 (slippage at the 'middle' part of fiber).

Another reason for the variable character of the function v_λ lies in the fact that the portions of fibers between their contacts have different levels of crimp. The more crimped portions can provide 'too long' lengths to the neighbouring 'straight and highly tensioned' fiber portions by means of fiber-to-fiber slippage during yarn tensioning. The whole such process is evidently highly variable along the fiber length and among the fibers in the yarn. (See Section 6.4 of the book [1] about the effect of variability of fiber crimp.)

The complexity and variability of all dominant influences might not give too many possibilities to obtain a purely theoretical solution for the function v_λ. Instead of this, we can expect to find a satisfactory phenomenological[32] model for description of function v_λ. In future, this can be developed probably based on the probabilistic theory of random processes.

Strain–stress function $\varepsilon[\tau] = \varepsilon\big[\sigma(\lambda)\big]$ of fiber

The derived equations also work with the stated inverse function of stress–strain relation of fiber – generally Equation (11.42). Usually, we can assume that these relations are monotonously increasing functions and practically independent of the horizontal coordinate λ of fiber. However, different fibers have often different stress–strain functions, even if their shapes are similar. It is then possible to either work with an average function or introduce the regulations of variability in stress–strain relations (see Section 6.4 of the book [1]).

The previous ideas are more or less easily applicable when none of the fibers is broken in the yarn. The situation gets complicated with the breakages of the fibers in the yarn – see the comments in this section.

---

32   The term 'phenomenology' in modern science, especially in physics, indicates a set of knowledge, which is mutually connected to different empirical observations and is consistent with the basic theory but is not derived directly from the theory. This is something in-between pure theory and pure empiricism.

Fiber slippage and stress–strain curve of yarn

The tensile stress of yarn is the 'sum' of partial contributions of stresses from all the fibers at a given strain of the yarn. There are two regions on specific stress–strain curve of yarn – see Figure 9.1 (white and grey parts) in Section 9.1.

The original structural character (white part) of a yarn remains relatively stable and practically all fibers are mechanically stressed in this part, hence we can integrate the mechanical contributions from individual fiber over the whole yarn structure. To a great extent, different fluctuations of tensile stress of individual fibers due to slippage can be 'averaged out' in consequence of the process of integration (except for the case of decrease in tensile stress due to slippage of fiber ends). In this way, some inaccuracies of a mathematical model can be partly corrected by the process of integration.

A solution of the second (grey) region in Figure 9.1 is much more difficult. In this part, a significant number of fibers are broken and the frictional relations change because: (1) the broken fibers are shorter (i.e., the number of slipped fiber ends is increasing), (2) the mutual fiber-to-fiber compression is decreasing so that the fiber-to-fiber slippage become easier to take place and (3) the non-broken fiber portions must take over the tensile forces from the broken fiber parts, so that the stress in the non-broken fiber segments increases so high that an unstable situation appears (a certain 'avalanche' of fiber breakage); this refers to ultimate yarn breakage. (See Chapter 6.4 of the book [1] for a similar consideration in case of parallel fiber bundle.)

A certain processes of fiber-to-fiber slippage can play an important role on the breakages of fibers in a yarn. While the variable tensile stress along the fibers can be 'averaged out' before (significant) fiber breakage, the extraordinary high value of tensile stress, e.g., according to Example 3 in the previous section, can overcome the local fiber tenacity, as a result, the fibers would although the 'average' fiber tension is much lower than the fiber tenacity.

It stands out to reason that the whole such phenomenon is extremely difficult to describe theoretically.

Possibilities of calculation of yarn strength

The above-mentioned complex situation consisting of numerous influences does not give any possibility for creation of a deeper theoretical (mathematical) model in near future. Nevertheless, for the time being, the yarn manufacturers should have an 'instrument' for calculation of yarn strength. With this view, different empirical and/or semi-empirical expressions are derived in this chapter. Hopefully, the spinners of textile industries will be able to make use of this chapter completely.

11.8   Reference

[1] Neckář, B. and Das, D., Theory of Structure and Mechanics of Fibrous Assemblies, Woodhead Publishing India Pvt. Ltd., New Delhi, 2012.

Semi-empirical modelling of yarn strength

12.1 Introduction

Yarn strength is the result of extremely complicated synergy of many partial (material and technological) influences. We partly analyzed some of them in this book as well as in our previous book [1] (see Table 12.1). Nevertheless, a more general solution of adequate theoretical model of yarn strength is still needed[1].

It, however, appears that there are two quantities that play a major role in determining yarn strength. They are yarn fineness T and yarn twist Z. In the following sections, we will primarily work with these two variables to understand their roles in deciding yarn strength.

Note: The readers can read Sections 1.3, 1.4 and 1.5 from Chapter 1 and Sections 8.5 and 8.6 from Chapter 8. It will help them to understand this chapter in a better manner.

12.2 Optimum twisting

General experience and its explanation

Traditionally, we observe that the tenacity of a single staple fiber yarn changes with the increase of yarn twist. We often work with the coefficient of utilization of fiber tenacity $\varphi_\sigma^* = \sigma_Y^*/\sigma_f^*$, as stated in Equation (9.5), where σ_Y^* is tenacity (specific breaking stress) of yarn and σ_f^* is mean tenacity (specific breaking stress) of fibers.

---

1   Needless to say, the research problems mentioned in Table 12.1 are not very comprehensive. It states several themes which the authors of this book feel particularly important. Nevertheless, we think that this table is enough sufficient for illustration of depth and complexity of exact solution of the research problems related to yarn strength.

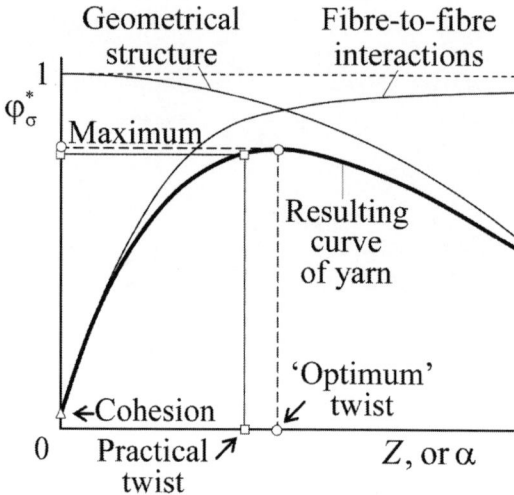

Figure 12.1 Generation of optimum twist

The coefficient of utilization of fiber tenacity is very small for an untwisted bundle of staple fibers. It represents only cohesion among fibers – see the scheme shown in Figure 12.1 ($\triangle$). By increasing twist Z or twist coefficient α, the coefficient of utilization of fiber tenacity φ_σ^* changes its value according to the thick curve shown in Figure 12.1. Initially, we observe that this curve rises to its maximum level (points $\bigcirc$), known as the 'optimum' twist or 'optimum' twist coefficient. However, beyond this value, the curve falls.

Table 12.1 Some phenomena influencing on yarn strength

Spheres of research problems	In this book	In our previous book [1]
I. Role of fibers and fiber bundles		
• Stress-strain relations and breaking characteristics of fibers (fiber bundles)	Sections 9.2, 9.3	
• Mechanics of blended fiber bundles (Hamburger's theory and its generalization)	Sections 9.2, 9.3	Sections 6.2, 6.3
• Variability in stress–strain curves of fibers (fiber bundles)		Section 6.4
• Fiber length, fineness and fiber-to-fiber friction	Chapter 11	

II. Role of yarn structure – macro characteristics

- Yarn fineness and yarn twist   Sections 1.3, 1.4, 8.5, 8.6

- Blend ratio

- Mass and structural unevenness of yarn   Chapter 6

III. Role of yarn structure – micro characteristics

- Yarn as a mechanical continuum   Chapter 8

- Fiber orientation, fiber migration, spinning- Chapters 3, 5
  in coefficient, yarn hairiness, etc.

- Fiber-to-fiber contacts - density and   Chapter 4
  distribution (fiber portions)

- Fiber crimp and crimp of fiber portions   Section 6.4

- Fiber-to-fiber slippage   Chapter 11

IV. Role of other conditions of tensioning and breaking of yarns

- Influence of gauge length   Chapter 10

- Influence of strain rate

Note: The word 'optimum' is traditionally used in English publications, although this term is often logically incorrect. Why? Simply because such twist is not usually optimum. In other languages (Czech, German, Russian, Polish, etc.), this is often termed as critical twist or critical twist coefficient. Mathematically, this can be interpreted as an argument of extreme value. Nevertheless, following English convention, we will use the term optimum (without quotation marks) here.

However, if we apply twists higher than the optimum twist, then we observe a decreasing trend for the coefficient of utilization of fiber tenacity φ_σ^*. This is shown in Figure 12.1. The obvious question is: why is it so?

It is often asked: why does the coefficient of utilization of fiber tenacity first increases and then decreases? There are two dominant influences that are acting together to produce such results.

1. With increase of twist, the fibers are mutually more compressed, their interactions become more intensive, the level of fiber-to-fiber friction increases, the fiber-to-fiber slippages are still smaller and the fiber bundle becomes more 'compact'. In consequence of this occurrence, the partial strength utilization increases monotonously.

Nevertheless, this does never result in making the coefficient of utilization of fiber tenacity equals to one. (This process could be separately imaginable when a non-twisted fiber bundle would be compressed by external forces, analogous to Figure 8.14.) However, the thin curve rises till a value near to one, which represents the effect of fiber-to-fiber interaction on yarn tenacity, as shown in Figure 12.1. (This trend is mainly observed in case of staple fiber yarns.)

2. Also, with the increase of twist, the geometry of fibers in yarn changes as the fibers stand more oblique in relation to yarn axis. It was derived in Chapter 9 that the increase of fiber slope leads to 'poor' mechanical utilization of fibers in yarns; see, for example, the easiest Gegauff's model according to Sections 9.5 and 9.7, including Figures 9.5 and 9.7. (Yarn twist is characterized by the peripheral angle β_D there.) Figure 9.14 explains this effect. The coefficient of utilization of fiber tenacity in case of twisted fiber bundle becomes simply poorer with the increase of twist in consequence of 'poor' yarn geometry. (This trend also pertains to filament yarns.) The decreasing nature of the thin curve, starting from the coefficient of utilization of fiber tenacity is one $\left(\varphi_\sigma^* = 1\right)$, represents the above-mentioned geometrical effect of yarn twisting. This is shown in Figure 12.1.

The common interaction or synergy of both influences finally result into the thick curve, as displayed in Figure 12.1.

Note: We know a method to mathematically describe the influence of geometrical structure in determining yarn tenacity. They are estimated, too idealized, but applicable in principle. On the contrary, it is very complicated to solve the influence of fiber-to-fiber interaction in deciding the tenacity of yarns. This is reported in Chapters 8 and 11. The resulting yarn curves are therefore most often treated on empirical or semi-empirical basis.

Note: Usually, during yarn spinning, we use a smaller (15 %) level of yarn twist than the optimum level of yarn twist. This is because the resulting curve is 'flat' surrounding its maximum value. It means that we can obtain a negligibly smaller value of coefficient of utilization of fiber tenacity by applying a significantly smaller value of yarn twist (see points □ in Figure 12.1). This is very important with a view to yarn price. (However, this consideration is not always valid, e.g., in case of crepe yarns.)

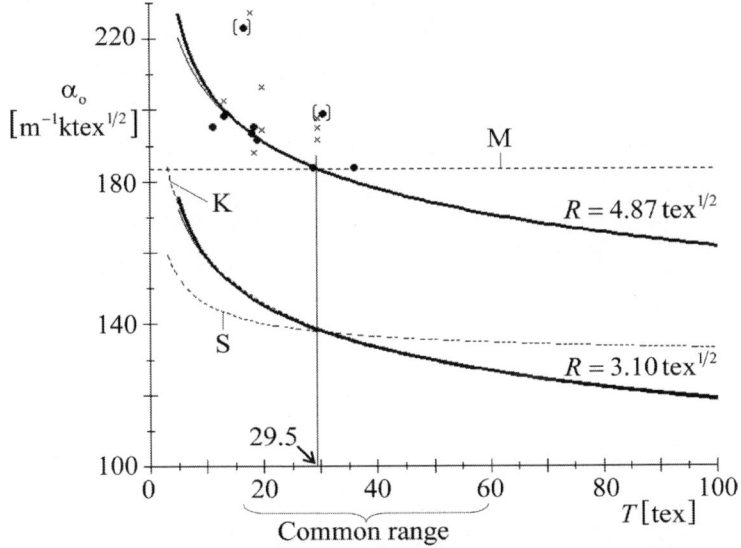

Figure 12.2 Optimum twist coefficient for cotton-carded yarns.
Dashed lines: M – Müller, Equation (12.2); S – Solovev, Equation (12.3); K – Korickij,
Equation (12.4); points: • – Johansen, × – Döttinger; thick continuous lines –
according to 'original method'; thin continuous lines – according to 'approximated
method'

Optimum twist coefficient – a short review

To find out a suitable relation for determination of optimum twist factor is often recognized as a classical old theme in the theory of yarn structure. Müller [2] thought about this coefficient a long ago, sometime in the year of 1880. Many empirical studies were published in German and English literature in the first half of the 20th century. The Russian researchers dealt with the problem of optimum twist factor more intensively in the third quarter of the 20th century. Most of the published work remain empirical or semi-empirical in nature, attempting to evaluate this coefficient for yarns prepared from natural fibers like cotton and wool as well as synthetic fibers such as viscose, polyester, etc. (It appears that the interest of researchers in this theme got declined in the last few decades.) Some of the results related to cotton-carded yarns are illustrated graphically in Figure 12.2.

As mentioned earlier, Müller [2] was one of the early researchers who worked on the problem of optimum yarn twist. His work was based on the second special assumption of Koechlin's concept. This assumption reads as follows: 'The yarns of different finenesses should have the same value of twist intensity'[2] (see Chapter 1, Section 1.4). Then, Koechlin's twist coefficient, according to Equation (1.63), becomes a constant, irrespective of yarn fineness T. Evidently, the angle β_D of the peripheral fibers in the yarn is also a constant, regardless of yarn fineness T, according to Equation (1.52).

If the same idea is used also for optimum twisting, then – for the same fibrous material and same type of spinning technology – the optimum twist coefficient α_o shall be a constant so that the optimum yarn twist can be expressed as stated hereunder

$$Z_o = \frac{\alpha_o}{\sqrt{T}}, \quad \left(Z_{o[m^{-1}]} = \frac{\sqrt{1000}\,\alpha_{o[m^{-1}ktex^{1/2}]}}{\sqrt{T_{[tex]}}} \right), \quad \alpha_o = \text{constant} . \quad (12.1)$$

Müller recommended to use the optimum twist coefficient as

$$\alpha_o = 183\,m^{-1}ktex^{1/2} \quad (12.2)$$

for cotton yarns, based on his experience. (See the horizontal straight line M in Figure 12.2.)

Besides Müller, some researchers published their experimental values of optimum twist coefficient. Figure 12.2 shows such values, reported by Johansen [3] and Döttinger [4], in relation to cotton-carded yarns.

The Russian researchers preferred to create empirical equations for determination of optimum twist coefficient α_o. The popular Solovev's [5] expression for cotton yarn, mentioned also by Zurek [6] and Gusev and Usenko [7], takes the following form:

$$\alpha_{o[m^{-1}ktex^{1/2}]} = \frac{\left(109870 - 673650\sigma^*_{f[N\,tex^{-1}]}\,t_{[tex]}\right)\sigma^*_{f[N\,tex^{-1}]}\,t_{[tex]}}{l_{f[mm]}} + \frac{56.9}{\sqrt{T_{[tex]}}}, \quad (12.3)$$

where $\sigma^*_{f[N\,tex^{-1}]}$ is mean fiber tenacity, $t_{[tex]}$ is fiber fineness and $l_{f[mm]}$ is staple length of fibers. The dashed curve S shown in Figure 12.2 displays an example of this, which is calculated by considering $l_f = 28\,mm$, $\sigma^*_f = 0.28\,N\,tex^{-1}$ and $t = 0.16\,tex$.

---

2   It means for same (or analogical) purpose of end-use.

Korickij [8, 9] presented the following equation, known as his 'second'[3] equation, for optimum twist coefficient α_o :

$$\alpha_{o[m^{-1}ktex^{1/2}]} = 133.35\sqrt{k}\,\frac{t_{[tex]}^{1/4}}{\sqrt{l_{f[mm]}}\,T_{[tex]}^{1/8}},\tag{12.4}$$

where k is a suitable parameter, t is fiber fineness and l_f is staple length of fibers. The dashed curve K shown in Figure 12.2 exhibits an example of this for cotton-carded yarns, calculated by considering $k = 177.27$[4], $l_f = 28\,mm$ and $t = 0.16\,tex$.

Note: Korickij's equation is also presented in his book [11] and in the books of Gusev and Usenko [7] and Sokolov [10].

Optimum twist coefficient – alternative solution

Let us come back to the basic idea of Müller [2], which states that the yarns of different finenesses should have the same value of twist intensity[5] also at the level of their optimum twist. Nevertheless, let us make two differences from Müller's concept:

1. We apply our semi-empirical results from Sections 8.5 and 8.6[6] in the place of Koechlin's theory.

2. We think that the modified angle β', corresponding to the average diameter D' of the peripheral fibers – see Figure 8.16a, is same for all yarn finenesses T at the level of optimum twist.

---

3   According to Sokolov [10], the 'first' Korickij's equation does not correspond to the experimental results very well.

4   Korickij originally recommended the following expression for parameter k: $k = 39/f$ for cotton-carded yarns, where the coefficient of fiber-to-fiber friction f was considered as $f = 0.22$; thus $k = 39/0.22 = 177.27$. This is numerically very close to 180, which is recommended in Reference [8].

5   Let us remind that the geometrical sense of twist intensity κ is the tangent of angle β_D of the peripheral fiber in a yarn – see Equation (1.52) and Figure 1.9. Thus, the same value of twist intensity signifies also the same value of angle β_D .

6   We recommend to remind about Section 8.5 and especially Section 8.6.

According to Equation (8.142), the parameter R must also be the same for all yarn finenesses T at the level of optimum twist. (Fiber material and spinning technology determine the optimum value of parameter $R = R_o$.)

Note: The following relationship is valid from Equation (8.142) for angle β' :

$$\tan \beta' = \sqrt{\frac{4\pi R}{Q\rho}} \quad \left(\tan \beta' = \sqrt{\frac{4\pi R_{[\text{tex}^{1/2}]}}{Q_{[\text{m}^2\text{tex}^{-1/2}]} P_{[\text{kg m}^{-3}]} 10^6}} \right). \tag{12.5}$$

If a constant value $R = R_o$ represents the optimum twist of yarns with different finenesses, then we can find the relationship between yarn count and optimum twist coefficient – $\alpha^*_{o\,\text{Koechlin}}$ and/or $a^*_{o\,\text{Phrix}}$ – in a manner that is described in Section 8.6. [See 'A common method of application' stated after Equation (8.156) – (a) 'original method' and (b) 'approximated method'.]

Example

Let us think about two possible values of optimum twisting parameters, namely $R_o = 3.10\,\text{tex}^{1/2}$ and $R_o = 4.87\,\text{tex}^{1/2}$, for cotton-carded yarns. Then, we can calculate the required parameters, characteristics and relationships by using the methods, described in Section 8.6. Table 12.2 shows the calculated values obtained in the same style as those reported in Table 8.3. By using the values from Table 12.2 in Equation (8.157), we obtain the following expressions for optimum twist coefficient in case of cotton-carded yarns in accordance with the approximated method:

$$\alpha_{o\,\text{Koechlin}[\text{m}^{-1}\text{ktex}^{1/2}]} = \frac{\alpha_{q,o[\text{m}^{-1}\text{tex}^{q_o}]}}{31.623\,T^{q_o-0.5}_{[\text{tex}]}} = \frac{209.59}{T^{0.12215}_{[\text{tex}]}} \quad \text{when} \quad R_o = 3.10\,\text{tex}^{1/2}, \tag{12.6}$$

$$\alpha_{o\,\text{Koechlin}[\text{m}^{-1}\text{ktex}^{1/2}]} = \frac{\alpha_{q,o[\text{m}^{-1}\text{tex}^{q_o}]}}{31.623\,T^{q_o-0.5}_{[\text{tex}]}} = \frac{260.15}{T^{0.10295}_{[\text{tex}]}} \quad \text{when} \quad R_o = 4.87\,\text{tex}^{1/2}. \tag{12.7}$$

Figure 12.2 illustrates the calculated relation obtained by following the original method (numerical solution for each yarn fineness) – continuous thick lines, and the corresponding relation obtained by following the approximated method – continuous thin lines, Equations (12.6) and (12.7). (Both curves are very similar to each other, as it was already mentioned in Section 8.6.)

Table 12.2 Optimum twisting parameters of cotton-carded yarn

Common input parameters:

$t = 0.16\,\text{tex}$, $\rho = 1520\,\text{kg}\,\text{m}^{-3}$ [a], $\mu_m = 0.9$ [a], $g = 1$ [a], $Q = 9.61 \times 10^{-8}\,\text{m}^2\,\text{tex}^{-1/2}$ [a].

Fineness of selected 'middle' yarn $T^* = 29.5\,\text{tex}$

Quantity	Dimension	Equation	Calculated parameters for comparison purpose	
			Solovev, Korickij	Johansen
R_o	$\text{tex}^{1/2}$	Input	3.10	4.87
μ_o^*	–	(8.143)	0.50995[b]	0.57098[b]
A	–	(8.130)	5.0013	6.0861
B	–	(8.130)	7.02930	13.654
V	–	(8.150)	0.03975	0.03975
U	tex^{-V}	(8.150)	0.80975	0.80975
q_o	–	(8.153)	0.62215	0.60295
$\alpha_{q,o}$	$\text{m}^{-1}\text{tex}^q$	(8.156)	6627.8	8226.7
β_o'	deg	(12.5)	27.313	32.914
Values of selected 'middle' yarn $T^* = 29.5\,\text{tex}$				
Z_o^*	m^{-1}	(8.157)	806.11	1069.1
$\alpha^*_{o\,\text{Koechlin}}$	$\text{m}^{-1}\text{ktex}^{1/2}$	(1.66)	138.45	183.63
	$(\text{t.p.i})\,N_e^{-1/2}$ [c]	$\alpha^*_{o\,[\text{m}^{-1}\text{ktex}^{1/2}]}/30.241$	4.5784	6.0722
$a^*_{o\,\text{Phrix}}$	$\text{m}^{-1}\text{ktex}^{2/3}$	(1.68)	76.96	102.1

Superscript * – value for 'middle' yarn with fineness $T^* = 29.5\,\text{tex}$.

Subscript 'o' – quantity describing optimum twisting.
[a]See also Table 8.1.
[b]Found by numerical method.
[c]Commonly used expression for 'dimension' of English twist coefficient (cotton); regular physical dimension is $\left[\text{inch}^{-1}\text{yd}^{-1/2}\,840^{1/2}\text{lb}^{1/2}\right]$.

By comparing the curves shown in Figure 12.2 and based on our previous empirical knowledge, we can state that

1. The earlier results, published by Müller, Johansen and Döttinger, reported relatively high value for optimum twist coefficient, for example, greater than $\alpha_o = 180 \, \mathrm{m^{-1} ktex^{1/2}}$. On the contrary, the newer experience, published by Solovev and Korickij, reported relatively low value for optimum twist coefficient, for example, around $\alpha_o = 140 \, \mathrm{m^{-1} ktex^{1/2}}$. Such different values probably express significant differences in fiber material (e.g., new varieties of cotton with longer staple length) and differences in technological processes of cotton spinning employed in-between the end of 19th century and the second half of 20th century.

2. The constant character of optimum twist coefficient, as recommended by Müller, does not correspond to reality. However, the trend of Johansen's data corresponds to our model satisfactorily at $R_o = 4.87 \, \mathrm{tex^{1/2}}$, except two outlier values (shown in brackets). Döttinger's data did not result in a more remarkable trend.

3. Solovev's curve results in practically same optimum twist coefficient as given by Korickij's equation (near to $\alpha_o = 140 \, \mathrm{m^{-1} ktex^{1/2}}$) in case of 'middle' yarn $T = T^* = 29.5 \, \mathrm{tex}$. Nevertheless, the general trend of Solovev's curve is 'too flat' in relation to Korickij's equation and our Equation (12.6) with $R_o = 3.10 \, \mathrm{tex^{1/2}}$.

4. Korickij's empirical curve is in remarkably good accordance with our results (original method as well as approximated method) with $R_o = 3.10 \, \mathrm{tex^{1/2}}$.

Summarily, we may say that our determination of optimum twist factor [Equation (12.6) using approximated method] expresses the nature of fully empirical Korickij's Equation (12.4) a little deeply in the light of relation to yarn fineness. Equation (12.4) as well as Equation (12.6) describes a good method for determination of optimum twist.

12.3   Yarn strength according to Solovev

Yarn strength is the result of interaction of many partial effects, as shown in Section 12.1. Many empirical studies were conducted to determine the most significant influence on yarn strength. Of them, Solovev's [5] formula is probably the most interesting one. (This was presented and discussed in many articles [6, 7].)

General structure of Solovev's formula

Solovev proposed the following general empirical expression for yarn strength:

$$\sigma_Y^* = \sigma_f^* \, K_n \, K_l \, K_\alpha \, \eta \quad \left(\sigma_{Y[\text{N tex}^{-1}]}^* = \sigma_{f[\text{N tex}^{-1}]}^* \, K_n \, K_l \, K_\alpha \, \eta \right), \tag{12.8}$$

where σ_Y^* and σ_f^* stand for (mean) yarn strength and (mean) fiber strength, respectively, $K_n \leq 1$ is a dimensionless factor characterizing the number of fibers present in the cross-section of yarn, $K_l \leq 1$ is a dimensionless factor related to fiber length, $K_\alpha \leq 1$ is a dimensionless factor of Koechlin's twist coefficient, η is a dimensionless factor characterizing the quality of technological process. Then, the dimensionless coefficient of utilization of fiber tenacity φ_σ^*, defined by Equation (9.5), is

$$\varphi_\sigma^* = \frac{\sigma_Y^*}{\sigma_f^*} = K_n K_l K_\alpha \eta . \tag{12.9}$$

Note: This coefficient must be always smaller than one. In practice, it often lies in an interval ranging from about 0.35 to about 0.6.

Factor K_n – influence of number of fibers

It is generally known that the 'thicker' yarns yield higher values of coefficient of utilization of fiber tenacity than the 'thinner' yarns for a comparable level of yarn twist [12]. The obvious question arises: why is it so?

It is probably because the density of fiber-to-fiber contacts, intensity of fiber-to-fiber friction, mutual pressure among fibers – simply stated, the compactness of fiber material – are smaller in the peripheral layers than in the internal layers of yarn body. It means that the fiber layers near to yarn surface are transporting forces more poorly than those near to yarn core. It is then evident that the mechanical behaviour of fibers cannot be fully utilized in yarn. This effect is more prominent in the case of 'thinner' yarns as compared to the case of 'thicker' yarns.

Solovev had an extremely simplified imagination – a 'black and white' ideal scheme. He assumed that a peripheral layer of thickness δ transfers no axial force to a yarn, whereas the other layers, i.e., internal layers, fully transfer the axial forces to a yarn. Figure 12.3 illustrates this idea. The outer layer at yarn diameter D and thickness δ does not transfer forces while the fibers inside the ring of diameter D', where

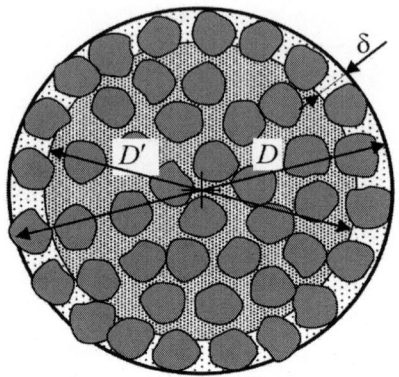

Figure 12.3 Peripheral layer in yarn cross section

$$D' = D - 2\delta , \tag{12.10}$$

fully transfer the forces. Subsequently, it is assumed for further simplification that the local packing density μ is same at all places in the yarn cross section[7].

Then, the total area S_c and the area of fibers S inside the yarn cross section are

$$S_c = \pi D^2 / 4, \quad S = \mu S_c = \mu \pi D^2 / 4 . \tag{12.11}$$

Similarly, the total area S_c' and the area of fibers S' inside the ring of diameter D' are

$$S_c' = \pi D'^2 / 4, \quad S' = \mu S_c' = \mu \pi D'^2 / 4 . \tag{12.12}$$

By using Equations (12.10) and (12.12), the ratio of transferable fiber area divided by the total fiber area in yarn cross section is

$$K_n' = \frac{S'}{S} = \frac{D'^2}{D^2} = \frac{(D - 2\delta)^2}{D^2} = 1 - 4\frac{\delta}{D} + 4\left(\frac{\delta}{D}\right)^2 . \tag{12.13}$$

Our experimental experience reveals that the ratio δ/D is relatively small so that the third quadratic term in Equation (12.13) can be neglected. Then, we can roughly write

---

7   This idealized and oft-used assumption was also considered by ideal helical model in Chapter 4.

$$K'_n \doteq 1 - 4\frac{\delta}{D}. \tag{12.14}$$

Let us consider yarn diameter according to Koechlin's theory, i.e., $D = K\sqrt{T}$, where K is a suitable constant – see Equation (1.67). The thickness of the peripheral layer is not known, but it is possible to assume that this quantity is proportional to the equivalent fiber diameter d – see the image shown in Figure 12.3. Then,

$$\delta = C_\delta d, \quad C_\delta \text{ is constant}. \tag{12.15}$$

We find the following expression from Equation (12.14) by using Equations (1.67), (12.15) and (1.6):

$$K'_n = 1 - 4\frac{C_\delta d}{K\sqrt{T}} = 1 - 4\frac{C_\delta\sqrt{4t/(\pi\rho)}}{K\sqrt{T}} = 1 - 4\frac{C_\delta\sqrt{4/(\pi\rho)}}{K}\sqrt{\frac{t}{T}},$$

$$K'_n = 1 - c_n\sqrt{\frac{t}{T}}, \quad \text{where} \quad c_n = \frac{4C_\delta\sqrt{4/(\pi\rho)}}{K} \text{ is parameter.} \tag{12.16}$$

The last expression creates the basis for the first factor K_n in Equation (12.8). Nevertheless, Solovev probably perceived an influence of yarn unevenness on yarn strength. He, therefore, added a fully empirical quantity, usually written as CH [8], into the previous expression to obtain his final formula for the first factor K_n. This is stated hereunder

$$K_n = K'_n - CH = 1 - CH - c_n\sqrt{\frac{t}{T}}, \quad CH \text{ and } c_n \text{ are parameters}. \tag{12.17}$$

Factor K_l – influence of fiber length

The fiber-to-fiber slippage is one of the partial reasons for the decrease of coefficient of utilization of fiber tenacity in a staple fiber yarn. Especially, the slippages of fiber ends in a yarn body play a significant role. This

---

[8]   Originally, C was a suitable constant and H was yarn unevenness according to the so-called Sommer's method. They were applied in spinning practice before using any mathematical statistics. (H is the first letter of the Russian word 'неровнота' – unevenness.) Presently, the whole quantity CH is interpreted as a 'characteristic of quality of technological process' [6].

phenomenon was deeply analyzed in Section 11.3. The part entitled 'Simplest theoretical example' describes the easiest case of tensile stress acting on a slipping fiber. This image also creates the idea of Solovev's factor K_l.

It is assumed that the maximum possible frictional shear stress, i.e., max-imum frictional shear force related to the areal unit of fiber surface, possesses a constant value[9], $b_\lambda = b$, at all places on the complete fiber length – see Equation (11.22). (Fiber length is denoted by l and the longitudinal coordi-nate along fiber is $\lambda \in (0,l)$.) Then, the local tensile stress along the fiber behaves like a 'trapezoidal' graph AXYB, displayed in Figure 11.5a.

The portion of fiber from $\lambda = \lambda_X$ to $\lambda = \lambda_Y$ does not slip, because the compression due to twist is enough high[10] there. The tensile stress $\sigma(\lambda)$ along this part of fiber is a suitable constant, $\sigma(\lambda) = C$ – see Equation (11.25). The 'fiber end' from $\lambda = 0$ to $\lambda = \lambda_X$ slips so that the tensile stress $\sigma_2^A(\lambda)$ linearly increases according to Equation (11.26), i.e., $\sigma_2^A(\lambda) = kb\lambda$; the con-stant k represents a common fiber parameter according to Equation (11.7). The second 'fiber end' from $\lambda = \lambda_Y$ to $\lambda = l$ also slips and the tensile stress $\sigma_1^B(\lambda)$ linearly decreases according to Equation (11.27), i.e., $\sigma_1^B(\lambda) = kb(l - \lambda)$. It is also easy to derive that $\lambda_X = C/(bk)$ and $\lambda_Y = l - C/(bk)$ – see Equations (11.28) and (11.29). Evidently, $\sigma_2^A(\lambda_X) = \sigma_1^B(\lambda_Y) = \sigma(\lambda) = C$.

The middle (non-slipping) part of the fiber elongates itself together with elongation of the complete yarn body. The maximum possible tension in the fiber corresponds to fiber strength σ_f^*. The scheme shown in Figure 12.4 il-lustrates such a situation by the grey-coloured trapezoid. Then, $C = \sigma_f^*$ so that Equation (11.25) takes the following form:

$$\sigma(\lambda) = \sigma_f^*, \quad \lambda \in (\lambda_X, \lambda_Y). \qquad (12.18)$$

---

9   If the normal pressure and the coefficient of friction are same at all places on fiber surfaces, then, e.g., the known simplest Coulomb's law satisfies the introduced assumption. This was – probably – the basic idea of Solovev too.

10   Thus, we do not think the case displayed in Figure 11.5b, i.e., the slippage of a complete fiber typically happens in roving.

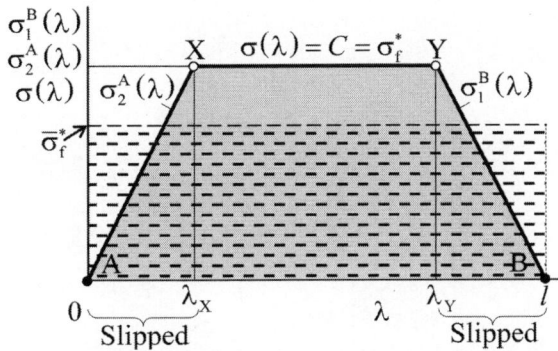

Figure 12.4 Tensile stress and mean tensile stress of the fiber under maximum tension

According to Equations (11.28) and (11.29), the longitudinal coordinates of points X and Y are

$$\lambda_X = \frac{\sigma_f^*}{bk}, \quad \lambda_Y = l - \lambda_X = l - \frac{\sigma_f^*}{bk}. \tag{12.19}[11]$$

Note: Equations (11.26) and (11.27) are valid without any change, because they do not contain the quantity C.

Let us calculate the mean tensile stress $\overline{\sigma}_f^*$ in a fiber when it is under maximum tension. This elementary problem can be understood as a transformation of trapezoidal area AXYB to rectangular area $l \cdot \overline{\sigma}_f^*$, as shown in Figure 12.4 – see the dashed rectangle. Thus, it is valid that

$$\frac{\lambda_X \sigma_f^*}{2} + \left[l - \lambda_X - \left(\overbrace{l - \lambda_Y}^{=\lambda_X} \right) \right] \sigma_f^* + \frac{\left(\overbrace{l - \lambda_Y}^{=\lambda_X} \right) \sigma_f^*}{2} = l \overline{\sigma}_f^*,$$

$$\lambda_X \sigma_f^* + \left(l - 2\lambda_X \right) \sigma_f^* = l \overline{\sigma}_f^*, \quad \sigma_f^* \left(l - \lambda_X \right) = l \overline{\sigma}_f^*,$$

$$\overline{\sigma}_f^* = \sigma_f^* \left(1 - \frac{\lambda_X}{l} \right) = \sigma_f^* \left(1 - \frac{\sigma_f^*}{bkl} \right). \tag{12.20}$$

[Equation (12.19) was used for rearrangement.] The partial strength utilization of fiber due to the slippage of fiber ends is then

---

11   Quantities b and σ_f^* indicate stresses. Physical dimension of parameter k is reciprocal of length – see Equation (11.7).

$$K_l = \frac{\overline{\sigma}_f^*}{\sigma_f^*} = 1 - \frac{c_l}{l}, \; c_l = \frac{\sigma_f^*}{bk} \text{ is fibre parameter.} \qquad (12.21)$$

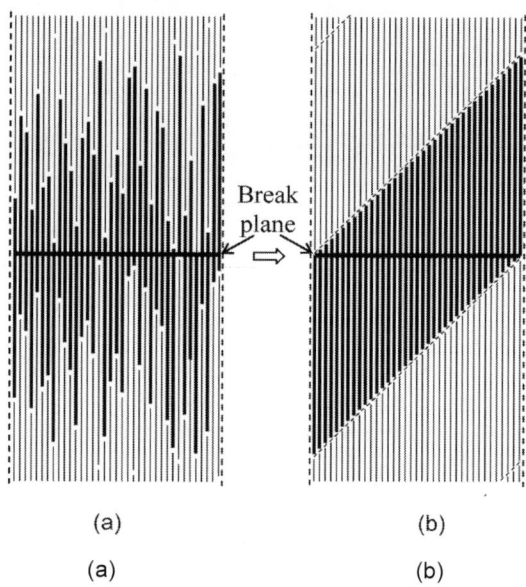

(a)   (b)

(a)   (b)

Figure 12.5 Longitudinal fiber arrangement in a fiber bundle (yarn).
Thick fibers are passing through the plane of breakage of fibers:
(a) original configuration, (b) after rearrangement

The staple fibers are randomly placed along the yarn – see, for example, Chapter 6. If we simplify a yarn by a parallel fiber bundle made up of fibers of same length then such an ideal fiber bundle corresponds to the scheme shown in Figure 12.5a. Fibers, passing by the plane of (future) fiber breakages, are drawn by thick line segments, other fibers are drawn by thin lines.

We can – imaginatively – rearrange the fibers passing by the plane of (future) breakages in compliance with the lengths of their intersected segments. If a plenty of fibers creates the cross section of a bundle (yarn) then the fibers create – more or less – a parallelogram after rearrangement; see Figure 12.5b. This shows that the fibers occur uniformly in all places in the breaking plane of the bundle (yarn). Thus, we can also imagine that each fiber brings mean tensile fiber stress $\overline{\sigma}_f^*$ for the breakage of the bundle (yarn). So, the factor K_l from Equation (12.21) expresses also partial utilization of fiber strength due to slippage of fiber ends.

Factor K_α – influence of twist (twist coefficient)

The influence of yarn twist in determining the strength of yarn is extremely difficult, as it was commented in the first paragraphs of Section 12.2. Solovev's concept derived this influence by the 'distance' of actual (Koechlin's) twist coefficient α from the optimum twist coefficient α_o. This is expressed as follows:

$$\delta_o = \alpha - \alpha_o, \quad \left(\delta_{o\,[m^{-1}\,tex^{1/2}]} = \alpha_{[m^{-1}\,tex^{1/2}]} - \alpha_{o\,[m^{-1}\,tex^{1/2}]}\right), \tag{12.22}$$

and the idea that only this quantity determines the partial influence of yarn twist is under consideration. Solovev and other authors assigned the factor K_α to the quantity δ_o according to Table 12.3.

Table 12.3 Recommended values of factor K_α

$\delta_{o\,[m^{-1}\,tex^{1/2}]}$	K_α * cotton	K_α ** viscose	$\delta_{o\,[m^{-1}\,tex^{1/2}]}$	K_α * cotton	K_α ** viscose
−60	–	0.73	10	0.99	0.98
−50	0.70	0.80	15	0.98	0.97
−40	0.80	0.85	20	0.96	0.95
−30	0.86	0.90	30	0.94	0.94
−25	0.91	0.93	40	0.91	0.91
−20	0.94	0.95	50	0.88	0.87
−15	0.96	0.97	60	0.85	0.82
−10	0.98	0.98	70	0.82	0.78
−5	0.99	0.99	80	0.79	0.73
0	1.00	1.00	90	–	0.68

* Cotton yarns according to Solovev [5].
** Viscose yarns according to Gusev and Usenko [7].

Factor η – quality of technological process

The earlier factors cannot affect all material, technical, and technological influences, as well as (sometimes special) the conditions of testing of yarn strength, which are often specific for given spinning mill and testing laboratory. Therefore, a fully empirical factor η, characterizing the quality of the technological process, is introduced by Equation (12.8). This must be specified according to individual's practical experience. (The initial value is $\eta = 1$.)

Values of parameters

Table 12.4 reports on a set of recommended values of parameters according to the experimental experience of different authors.

Table 12.4 Examples of recommended values

Factor	Equation	Parameter	Cotton carded yarn	Cotton combed yarn	Viscose yarn
K_n	(12.17)	CH	0.17–0.18*,**	0.13–0.15*,**	0.11–0.15**
		c_n		2.65*	2.8**
K_l	(12.21)	$c_{l\,[\text{mm}]}$		5*,**	4.4**
K_α	(12.6)	$\alpha_{o\,[\text{m}^{-1}\text{ktex}^{1/2}]}$		See section 12.2	
	(12.22)	$\delta_{[\text{m}^{-1}\text{ktex}^{1/2}]}$		See Table 12.3	
η		–		0.95–1.1**	

* Zurek [6].
** Gusev and Usenko [7].

Example

Let us think about a cotton-carded yarn of 29.5 tex ($T = 29.5$ tex) fineness. This yarn is prepared from fibers of 30 mm ($l = 30$ mm) staple length and 0.1875 tex fineness ($t = 0.1875$ tex [12]), and 0.28 N tex^{-1} (mean) tenacity ($\sigma_f^* = 0.28$ N tex^{-1}). Let the actual twist factor of this yarn be $\alpha = 120$ m^{-1}ktex$^{1/2}$ and the optimum twist factor be $\alpha_o = 139$ m^{-1} ktex$^{1/2}$. By using the parameters from Table 12.4, we obtain

---

12   The fineness of a medium-staple cotton fiber usually lies from 1.75 to 2 dtex. The quantity used here refers to the average value.

- factor $K_n = 1 - 0.18 - 2.65\sqrt{(0.1875/29.5)} = 0.6087$ – according to Equation (12.17),
- factor $K_l = 1 - 5/30 = 0.8333$ – according to Equation (12.21),
- the quantity $\delta = 120 - 139 = -19\,\text{m}^{-1}\text{ktex}^{1/2}$ and factor $K_\alpha = 0.944$ – according to Table 12.3,
- factor $\eta = 1$ in our case.

Then, according to Equation (12.9), the coefficient of utilization of fiber tenacity is $\varphi_\sigma^* = 0.4789$ and according to Equation (12.8), the yarn tenacity is $\sigma_Y^* = 0.1341\,\text{N tex}^{-1}$.

Note: The readers are requested to consider the calculations for information only. Any real-world application needs to specify and correctly use the parameters in relation to the actual conditions of a given spinning mill.

12.4   References

[1] Neckář, B. and Das, D., Theory of Structure and Mechanics of Fibrous Assemblies, Woodhead Publishing (India) Pvt. Ltd., New Delhi, 2012.

[2] Müller, E., Über die Festigkeit fadenförmiger Fasergebilde in ihrer Abhängingkeit von dem Drahte derselben (About Strength of Yarn Creating Fiber Systems in Relation to Their Twists), Civiling, 1880 (In German).

[3] Johansen, O., Handbuch der Baumwollspinnerei, Rohweberei und Fabrikanlagen, I. Band (Handbook of Cotton Spinning and Weaving Mills. Volume I), B. F. Voigt Publisher, Leipzig, 1930 (In German).

[4] Döttinger, E., Dissertation, Mitteilungen des F.I., Reutlingen 1925 (In German).

[5] Solovev, A.N., Projektirovanie svojstv prjazi v chlopcatobumaznom proizvodstve. (Design of Yarn Properties in Cotton Industry), Gizlegprom Publisher, 1951 (In Russian).

[6] Zurek, W., Struktura liniowych wyrobów włókienniczych. (Structure of Linear Textiles.) Wydawnictwa Naukovo-Techniczne, Warszava, 1989 (In Polish).

[7] Gusev, V.E. and Usenko, V.A., Prjadenie chimiceskogo stapelnogo volokna (Spinning of Chemical Staple Fibers), Legkaya Industrija publisher, Moscow, 1964 (In Russian).

[8] Korickij, K.I., Osnovy proektirovanija svojstv prjazi (Principles of Development of Yarn Properties), Gizlegprom Publisher, Moscow, 1963 (In Russian).

[9] Korickij, K.I., Inzenernoe projektirovanie tekstilnych materialov (Engineering Development of Textile Materials), Legkaya Industrija Publisher, Moscow, 1971 (In Russian).

[10] Sokolov, G.V., Voprosy teorii krucenia voloknistych materialov (Problems of Twisting Theory of Fibrous Materials), Gizlegprom Publisher, Moscow, 1957 (In Russian).

[11] Korickij, K.I., Rascet procnosti niti na rastjazenie (Calculation of Yarn Strength), Proceedings of research works of CNICHBI, Volume XVI, Gizlegprom Publisher, Moscow, 1956 (In Russian).

[12] Uster Statistics, Uster Technologies AG., Accessed on 28 June 2017 from https://www.uster.com/en/service/uster-statistics, 2013.

Appendix 1

Goniometrical solution of Equation (4.39)

Let us rearrange Equation (4.39) as follows:

$$(1-\delta)^2 = \frac{1}{1+x_n^2 \dfrac{4\pi}{\mu\rho} \dfrac{\alpha_i^2}{(1-\delta)^3}}, \quad (1-\delta)^2 + x_n^2 \frac{4\pi}{\mu\rho} \frac{\alpha_i^2}{(1-\delta)} = 1,$$

$$(1-\delta)^3 + x_n^2 \frac{4\pi}{\mu\rho} \alpha_i^2 = (1-\delta), \quad (1-\delta)^3 - (1-\delta) + x_n^2 \frac{4\pi}{\mu\rho} \alpha_i^2 = 0. \quad \text{(A1.1)}$$

Then, by introducing the following quantities

$$y = 1-\delta, \quad p = -1/3, \quad q = x_n^2 \frac{2\pi}{\mu\rho} \alpha_i^2, \quad \text{(A1.2)}$$

we can rewrite Equation (A1.1) as follows:

$$y^3 + 3py + 2q = 0. \quad \text{(A1.3)}$$

As known, this is a 'standard' cubic equation (without a quadratic member). The procedure for solving such equation is demonstrated in many handbooks of mathematics. In this appendix, we used the so-called goniometrical (trigonometric) method [1] for solving this equation. Nevertheless, very similar expressions can also be found in many handbooks, including Reference [2].

By using Equation (A1.2), it is valid to write that

$$p < 0, \quad \text{and} \quad p^3 + q^2 = (-1/3)^3 + \left(x_n^2 \, 2\pi\alpha_i^2/\mu\rho\right)^2 \le 0 \,[1]. \quad \text{(A1.4)}$$

Let us now define two quantities r and φ as

$$r = \sqrt{|p|} = \frac{1}{\sqrt{3}}, \varphi = \arccos\left(\frac{q}{r^3}\right)$$

$$= \arccos\left(x_n^2 \frac{2\pi}{\mu\rho} \alpha_i^2 3\sqrt{3}\right) = \arccos\left(\frac{6\sqrt{3}\pi x_n^2}{\mu} \frac{\alpha_i^2}{\rho}\right), \quad \varphi \in \left(0, \frac{\pi}{2}\right). \quad \text{(A1.5)}$$

---

1   The expressions must also be valid for small twisting, i.e., for $\alpha_i \to 0$.

Under the conditions stated in Equations (A1.4) and (A1.5), the triplet of roots of Equation (A1.3) can be found in accordance with Reference [2] as follows:

$$y_1 = -2r\cos\frac{\varphi}{3} = \frac{-2}{\sqrt{3}}\cos\frac{\arccos\left(\dfrac{6\sqrt{3}\pi x_n^2}{\mu}\dfrac{\alpha_i^2}{\rho}\right)}{3},$$

$$y_2 = 2r\cos\left(\frac{\pi}{3}-\frac{\varphi}{3}\right) = \frac{2}{\sqrt{3}}\cos\left[\frac{\pi}{3}-\frac{1}{3}\arccos\left(\frac{6\sqrt{3}\pi x_n^2}{\mu}\frac{\alpha_i^2}{\rho}\right)\right], \quad (A1.6)$$

$$y_3 = 2r\cos\left(\frac{\pi}{3}+\frac{\varphi}{3}\right) = \frac{2}{\sqrt{3}}\cos\left[\frac{\pi}{3}+\frac{1}{3}\arccos\left(\frac{6\sqrt{3}\pi x_n^2}{\mu}\frac{\alpha_i^2}{\rho}\right)\right].$$

The first root y_1 is negative so that it is not physically real. By applying $y = 1-\delta$, stated in Equation (A1.2), the roots y_2 and y_3 can be expressed together as follows:

$$\delta = 1 - \frac{2}{\sqrt{3}}\cos\left[\frac{\pi}{3}\pm\frac{1}{3}\arccos\left(\frac{6\sqrt{3}\,\pi x_n^2}{\mu}\frac{\alpha_i^2}{\rho}\right)\right]. \quad (A1.7)$$

References

[1] Rektorys, K., Přehled užité matematiky (View of Applied Mathematics), SNTL Publishers, Prague, 1963 (In Czech).

[2] Korn, G. A. and Korn, T. M., Mathematical Handbook, McGraw-Hill, New York, San Francisco, Toronto, London, Sydney, 1968.

Limit of twisting: Hypothesis of zero axial force

The following expression can be written by using Equation (1.54) in the form $\kappa = 2\sqrt{\pi}\,\alpha/\sqrt{\mu\rho}$ and Equation (4.34):

$$x = \sqrt{1+\kappa^2} = \sqrt{1 + \frac{4\pi\alpha^2}{\mu\rho}} = \sqrt{1 + \frac{4\pi\alpha_i^2}{\mu\rho(1-\delta)^3}}\,. \tag{A2.1}$$

By rearrangement, the aforesaid expression can be written as follows:

$$\kappa = \sqrt{x^2-1}, \quad x \geq 1. \tag{A2.2}$$

Then, in accordance with Equation (4.80), it is possible to write the following expression for yarn retraction:

$$\delta = 1 - \frac{\ln x}{x-1}, \quad 0 = 1 - \delta - \frac{\ln x}{x-1} \tag{A2.3}$$

and rearrangement of Equation (4.83) leads to

$$\frac{2\sqrt{\pi}\alpha_i}{\sqrt{\mu\rho}} = \sqrt{x^2-1}\left(\frac{\ln x}{x-1}\right)^{3/2} = \sqrt{(x+1)(x-1)}\frac{(\ln x)^{3/2}}{(x-1)\sqrt{x-1}} = \sqrt{x+1}\frac{(\ln x)^{3/2}}{x-1}\,. \tag{A2.4}$$

It is also possible to interpret the relation $\alpha_i - \delta$ parametrically form Equations (A2.3) and (A2.4) with the help of a parameter $x \in (1,\infty)$.

Values δ, α_i when $x \to 1$ ($\kappa \to 0$)

It is valid to write from Equation (A2.3) that

$$\lim_{x\to 1}\delta = \lim_{x\to 1}\left(1 - \frac{\ln x}{x-1}\right) = 1 - \lim_{x\to 1}\frac{\ln x}{x-1} = 1 - \lim_{x\to 1}\frac{1/x}{1} = 1 - 1 = 0, \quad \lim_{x\to 1}\delta = 0\,. \tag{A2.5}$$

(Here, L'Hospital's rule for evaluation of limits was used.) Further, from Equation (A2.4), we obtain

$$\lim_{x \to 1}\left(\frac{2\sqrt{\pi}\alpha_i}{\sqrt{\mu\rho}}\right) = \lim_{x \to 1}\left[\sqrt{x+1}\frac{(\ln x)^{3/2}}{x-1}\right] = \lim_{x \to 1}\sqrt{x+1}\cdot\lim_{x \to 1}\frac{(\ln x)^{3/2}}{x-1}$$

$$= \sqrt{2}\lim_{x \to 1}\frac{3(\ln x)^{1/2}}{2} = \sqrt{2}\cdot 0 = 0, \quad \lim_{x \to 1}\alpha_i = 0. \tag{A2.6}$$

It is evident that the curve showing the relationship between α_i and δ passes through the point $\alpha_i = 0$, $\delta = 0$.

Values δ, α_i when $x \to \infty$ ($\kappa \to \infty$)

It is valid to write from Equation (A2.3) that

$$\lim_{x \to \infty}\delta = \lim_{x \to \infty}\left(1 - \frac{\ln x}{x-1}\right) = 1 - \lim_{x \to \infty}\frac{\ln x}{x-1} = 1 - \lim_{x \to \infty}\frac{1/x}{1} = 1 - 0 = 1, \quad \lim_{x \to \infty}\delta = 1.$$

$$\tag{A2.7}$$

Further, from Equation (A2.4), we obtain

$$\lim_{x \to \infty}\left(\frac{2\sqrt{\pi}\alpha_i}{\sqrt{\mu\rho}}\right) = \lim_{x \to \infty}\left[\sqrt{x+1}\frac{(\ln x)^{3/2}}{x-1}\right] = \lim_{x \to \infty}\left[\frac{\sqrt{x+1}}{\sqrt{x}}\frac{x}{x-1}\frac{(\ln x)^{3/2}}{\sqrt{x}}\right]$$

$$= 1 \cdot 1 \cdot \lim_{x \to \infty}\left[\frac{(\ln x)^{3/2}}{\sqrt{x}}\right] = \lim_{x \to \infty}\left[\frac{\frac{3}{2}\sqrt{\ln x}\frac{1}{x}}{\frac{1}{2}\frac{1}{\sqrt{x}}}\right] = 3\lim_{x \to \infty}\left[\frac{\sqrt{\ln x}}{\sqrt{x}}\right]$$

$$= 3\lim_{x \to \infty}\left[\frac{\frac{1}{2}\frac{1}{\sqrt{\ln x}}\frac{1}{x}}{\frac{1}{2}\frac{1}{\sqrt{x}}}\right] = 3\lim_{x \to \infty}\left[\frac{1}{\sqrt{\ln x}\sqrt{x}}\right] = 0, \quad \lim_{x \to \infty}\alpha_i = 0. \tag{A2.8}$$

(L'Hospital's rule for evaluation of limits was used in the last two expressions.) Evidently, the curve showing the relationship between α_i and δ passes through the point $\alpha_i = 0$, $\delta = 1$ too.

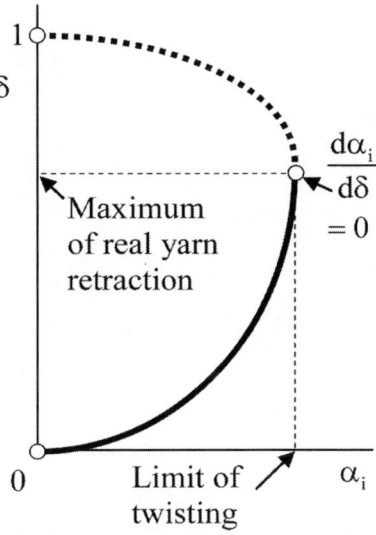

Figure A2.1 Assumed relation $\alpha_i - \delta$

Limit of twisting

Because α_i and δ must not possess negative values, the relationship between α_i and δ is assumed to be what is shown in Figure A2.1. Evidently, there is an extreme, which satisfies $d\alpha_i/d\delta = 0$. This is possible to evaluate by differentiating the second expression in Equation (A2.3)[1] as follows:

$$0 = \left[-1 - \frac{d}{dx}\left(\frac{\ln x}{x-1} \right) \frac{\partial x}{\partial \delta} \right] d\delta + \left[-\frac{d}{dx}\left(\frac{\ln x}{x-1} \right) \frac{\partial x}{\partial \alpha_i} \right] d\alpha_i,$$

$$0 = -\left[1 + \frac{d}{dx}\left(\frac{\ln x}{x-1} \right) \frac{\partial x}{\partial \delta} \right] + \left[-\frac{d}{dx}\left(\frac{\ln x}{x-1} \right) \frac{\partial x}{\partial \alpha_i} \right]\left(\frac{d\alpha_i}{d\delta} \right). \tag{A2.9}$$

However, the expression $d\alpha_i/d\delta = 0$ is valid in the limiting case of twisting. So, the following expression is evident:

---

1   Do not forget that x is not an independent variable. According to Equation (A2.1), it is a function of α_i and δ.

$$0 = 1 + \frac{d}{dx}\left(\frac{\ln x}{x-1}\right)\frac{\partial x}{\partial \delta}. \tag{A2.10}$$

It is valid to write that

$$\frac{d}{dx}\left(\frac{\ln x}{x-1}\right) = \frac{\frac{1}{x}(x-1) - \ln x \cdot 1}{(x-1)^2} = \frac{x-1-x\ln x}{x(x-1)^2}. \tag{A2.11}$$

Further, from Equation (A2.1), we can write the following expression:

$$\frac{\partial x}{\partial \delta} = \frac{\partial}{\partial \delta}\sqrt{1 + \frac{4\pi\alpha_i^2}{\mu\rho(1-\delta)^3}} = \frac{1}{2\sqrt{1 + \frac{4\pi\alpha_i^2}{\mu\rho(1-\delta)^3}}} \cdot \frac{4\pi\alpha_i^2}{\mu\rho} \cdot \frac{-3}{(1-\delta)^4} \cdot (-1)$$

$$= \frac{1}{2\sqrt{1 + \frac{4\pi\alpha_i^2}{\mu\rho(1-\delta)^3}}}\left[\frac{4\pi\alpha_i^2}{\mu\rho(1-\delta)^3}\right]\frac{3}{1-\delta}. \tag{A2.12}$$

Nevertheless, from Equation (A2.1), we can also write the following expression:

$$\frac{4\pi\alpha_i^2}{\mu\rho(1-\delta)^3} = x^2 - 1 = (x+1)(x-1). \tag{A2.13}$$

Now, we rearrange Equation (A2.12) by using Equations (A2.13) and (A2.3) in the form $1-\delta = \ln x/(x-1)$. So, we obtain

$$\frac{\partial x}{\partial \delta} = \frac{1}{2x}(x+1)(x-1)\frac{3(x-1)}{\ln x} = \frac{3}{2}\frac{(x+1)(x-1)^2}{x\ln x}. \tag{A2.14}$$

By substituting Equations (A2.11) and (A2.14) into (A2.10), the following expression is obtained:

$$0 = 1 + \frac{x-1-x\ln x}{x(x-1)^2}\frac{3}{2}\frac{(x+1)(x-1)^2}{x\ln x},$$

$$-\frac{2}{3} = \frac{\left[(x-1) - x\ln x\right](x+1)}{x^2 \ln x}, \quad -\frac{2}{3} = \frac{\left(x^2 - 1\right) - x^2 \ln x - x\ln x}{x^2 \ln x}$$

$$= \frac{\left(x^2 - 1\right) - x\ln x}{x^2 \ln x} - 1,$$

$$\frac{1}{3} = \frac{\left(x^2 - 1\right) - x\ln x}{x^2 \ln x}, \quad \frac{x^2}{3}\ln x = x^2 - 1 - x\ln x, \quad \ln x = \frac{x^2 - 1}{x^2/3 + x}. \quad \text{(A2.15)}$$

The root of Equation (A2.15) characterizes the limiting case of twisting. By using a suitable numerical method, we obtain only one root of Equation (A2.15) as

$$x = 9.5802. \tag{A2.16}$$

By substituting Equation (A2.16) into (A2.1), the following expression is obtained for the limiting case of twisting:

$$\sqrt{1 + \frac{4\pi \alpha_i^2}{\mu\rho(1-\delta)^3}} = 9.5802. \tag{A2.17}$$

$$\text{Integrals } \int \mu_r r \, dr \text{ and } \int \left[\mu_r r \Big/ \sqrt{1+\left(2\pi r Z\right)^2} \right] dr$$

Indefinite integrals

It is valid to write that

$$\int r \, dr = \frac{r^2}{2} + C. \tag{A3.1}[1]$$

$$\int r^2 \, dr = \frac{r^3}{3} + C. \tag{A3.2}$$

$$\int \frac{r \, dr}{\sqrt{1+\left(2\pi r Z\right)^2}} = \int \frac{1}{y} \frac{y \, dy}{\left(2\pi Z\right)^2} = \frac{1}{\left(2\pi Z\right)^2} \int dy = \frac{y}{\left(2\pi Z\right)^2} = \frac{\sqrt{1+\left(2\pi r Z\right)^2}}{\left(2\pi Z\right)^2} + C.$$

Substitution: $y^2 = 1 + \left(2\pi r Z\right)^2$, $2y \, dy = \left(2\pi Z\right)^2 2r \, dr$, $r \, dr = y \, dy \Big/ \left(2\pi Z\right)^2$.

$$\tag{A3.3}$$

$$\int \frac{r^2 \, dr}{\sqrt{1+\left(2\pi r Z\right)^2}} = \int \frac{1}{\sqrt{1+x^2}} \frac{x^2 dx}{\left(2\pi Z\right)^3} = \frac{1}{\left(2\pi Z\right)^3} \int \frac{x^2 dx}{\sqrt{1+x^2}}.$$

Substitution: $x = 2\pi r Z$, $r = \frac{x}{2\pi Z}$, $dr = dx \Big/ \left(2\pi Z\right)$, $\tag{A3.4}$

$$r^2 \, dr = \frac{x^2}{\left(2\pi Z\right)^2} \frac{dx}{2\pi Z} = \frac{x^2 dx}{\left(2\pi Z\right)^3}.$$

The integral mentioned at the right-hand side of the last expression can be solved by using the method of integration by parts, which can be expressed as follows:

---

1   *C* denotes the integral constant in all the expressions wherever it appears in this appendix.

$$\int \frac{x^2 dx}{\sqrt{1+x^2}} = \int \frac{\left(1+x^2\right)-1}{\sqrt{1+x^2}} dx = \int \sqrt{1+x^2}\, dx - \int \frac{dx}{\sqrt{1+x^2}}$$

$$u' = 1, \qquad\qquad u = x,$$
$$v = \sqrt{1+x^2}, \quad v' = x/\sqrt{1+x^2},$$

$$= x\sqrt{1+x^2} - \int \frac{x^2 dx}{\sqrt{1+x^2}} - \int \frac{dx}{\sqrt{1+x^2}}, \qquad (A3.5)$$

$$2\int \frac{x^2 dx}{\sqrt{1+x^2}} = x\sqrt{1+x^2} - \int \frac{dx}{\sqrt{1+x^2}},$$

$$\int \frac{x^2 dx}{\sqrt{1+x^2}} = \frac{x}{2}\sqrt{1+x^2} - \frac{1}{2}\int \frac{dx}{\sqrt{1+x^2}}.$$

Let us now introduce the following substitution for solving the integral stated at the right-hand side of the last expression

$$\sqrt{1+x^2} + x = t, \quad \sqrt{1+x^2} = t - x, \qquad (A3.6)$$

$$1+x^2 = t^2 - 2tx + x^2, \quad 1 = t^2 - 2tx, \quad x = \frac{t^2-1}{2t}. \qquad (A3.7)$$

By using Equations (A3.6) and (A3.7), we obtain

$$\sqrt{1+x^2} = t - \frac{t^2-1}{2t} = \frac{2t^2 - t^2 + 1}{2t} = \frac{t^2+1}{2t}. \qquad (A3.8)$$

Further, the following expression is obtained by differentiating Equation (A3.7):

$$dx = \frac{2t\,2t - \left(t^2-1\right)2}{\left(2t\right)^2} dt = \frac{4t^2 - 2t^2 + 2}{4t^2} dt = \frac{t^2+1}{2t^2} dt. \qquad (A3.9)$$

By using Equations (A3.8), (A3.9) and (A3.6), the following integral can be solved as follows:

$$\int \frac{dx}{\sqrt{1+x^2}} = \int \frac{2t}{t^2+1} \frac{t^2+1}{2t^2} dt = \int \frac{dt}{t} = \ln|t| = \ln\left|\sqrt{1+x^2} + x\right| + C. \quad (A3.10)$$

Now, by substituting Equation (A3.10) into (A3.5), it is found that

$$\int \frac{x^2 dx}{\sqrt{1+x^2}} = \frac{x}{2}\sqrt{1+x^2} - \frac{1}{2}\ln\left|\sqrt{1+x^2} + x\right| + C. \qquad (A3.11)$$

Finally, by substituting Equation (A3.11) into (A3.4) and then by using the substitution of x from Equation (A3.4), we find

$$\int \frac{r^2\, dr}{\sqrt{1+(2\pi r Z)^2}} = \frac{1}{(2\pi Z)^3} \int \frac{x^2 dx}{\sqrt{1+x^2}} = \frac{1}{(2\pi Z)^3}$$

$$\left\{ \frac{x}{2}\sqrt{1+x^2} - \frac{1}{2}\ln\left|\sqrt{1+x^2}+x\right| \right\}.$$

$$\int \frac{r^2\, dr}{\sqrt{1+(2\pi r Z)^2}} = \frac{1}{(2\pi Z)^3}\left\{ \frac{2\pi r Z}{2}\sqrt{1+(2\pi r Z)^2} \right.$$

$$\left. - \frac{1}{2}\ln\left(\sqrt{1+(2\pi r Z)^2}+2\pi r Z\right) \right\} + C. \tag{A3.12}$$

Definite integrals

The (trapezoidal) model of local packing density μ_r, used in Section 4.7, is given by Equation (4.99) as follows:

$$\mu_r = \mu_A \text{ for } r \in \langle 0, r_A \rangle, \quad \mu_r = -\mu_A \frac{r}{R-r_A} + \mu_A \frac{R}{R-r_A} \text{ for } r \in (r_A, R).$$

The following definite integrals are used in the above-mentioned section:

1. By using indefinite integrals as expressed by Equations (A3.1) and (A3.2), we can derive that

$$\int_0^{r_B} \mu_r r\, dr = \int_0^{r_A} \mu_r r\, dr + \int_{r_A}^{r_B} \mu_r r\, dr = \int_0^{r_A} \mu_A r\, dr + \int_{r_A}^{r_B}\left(-\mu_A \frac{r}{R-r_A} + \mu_A \frac{R}{R-r_A} \right) r\, dr$$

$$= \mu_A\left(\frac{r^2}{2}\right)_0^{r_A} - \frac{\mu_A}{R-r_A}\left(\frac{r^3}{3}\right)_{r_A}^{r_B} + \frac{\mu_A R}{R-r_A}\left(\frac{r^2}{2}\right)_{r_A}^{r_B}$$

$$= \frac{\mu_A}{2}\left(r_A^2\right) - \frac{\mu_A}{3(R-r_A)}\left(r_B^3 - r_A^3\right) + \frac{\mu_A R}{2(R-r_A)}\left(r_B^2 - r_A^2\right)$$

$$= \mu_A \frac{3r_A^2(R-r_A) - 2(r_B^3 - r_A^3) + 3R(r_B^2 - r_A^2)}{6(R-r_A)}$$

$$= \mu_A \frac{3Rr_A^2 - 3r_A^3 - 2r_B^3 + 2r_A^3 + 3Rr_B^2 - 3Rr_A^2}{6(R-r_A)} = \mu_A \frac{-3r_A^3 - 2r_B^3 + 2r_A^3 + 3Rr_B^2}{6(R-r_A)},$$

$$\int_0^{r_B} \mu_r r\, dr = \mu_A \frac{-r_A^3 + 3Rr_B^2 - 2r_B^3}{6(R-r_A)}. \tag{A3.13}$$

2. If $r_B \to R$, then Equation (A3.13) is limited to what follows in Equation (A3.14):

$$\int_0^R \mu_r r \, dr = \lim_{r_B \to R} \int_0^{r_B} \mu_r r \, dr = \lim_{r_B \to R} \left[\mu_A \frac{-r_A^3 + 3Rr_B^2 - 2r_B^3}{6(R - r_A)} \right] = \mu_A \frac{-r_A^3 + 3R^3 - 2R^3}{6(R - r_A)}$$

$$= \mu_A \frac{R^3 - r_A^3}{6(R - r_A)} = \mu_A \frac{(R - r_A)(R^2 + Rr_A + r_A^2)}{6(R - r_A)},$$

$$\int_0^R \mu_r r \, dr = \mu_A \frac{R^2 + Rr_A + r_A^2}{6}. \qquad (A3.14)$$

3. By using the indefinite integrals as expressed by Equations (A3.3), (A3.12) and (4.95), it is possible to derive the following expression:

$$\int_0^{r_B} \frac{\mu_r r \, dr}{\sqrt{1 + (2\pi r Z)^2}} = \int_0^{r_A} \frac{\mu_A r \, dr}{\sqrt{1 + (2\pi r Z)^2}} + \int_{r_A}^{r_B} \frac{\left(-\mu_A \dfrac{r}{R - r_A} + \mu_A \dfrac{R}{R - r_A} \right) r \, dr}{\sqrt{1 + (2\pi r Z)^2}}$$

$$= \mu_A \int_0^{r_A} \frac{r \, dr}{\sqrt{1 + (2\pi r Z)^2}} - \frac{\mu_A}{R - r_A} \int_{r_A}^{r_B} \frac{r^2 \, dr}{\sqrt{1 + (2\pi r Z)^2}} + \frac{\mu_A R}{R - r_A} \int_{r_A}^{r_B} \frac{r \, dr}{\sqrt{1 + (2\pi r Z)^2}}$$

$$= \mu_A \left[\frac{\sqrt{1 + (2\pi r Z)^2}}{(2\pi Z)^2} \right]_0^{r_A}$$

$$- \frac{\mu_A}{R - r_A} \left[\frac{1}{(2\pi Z)^3} \left\{ \frac{2\pi r Z}{2} \sqrt{1 + (2\pi r Z)^2} - \frac{1}{2} \ln\left(\sqrt{1 + (2\pi r Z)^2} + 2\pi r Z \right) \right\} \right]_{r_A}^{r_B}$$

$$+ \frac{\mu_A R}{R - r_A} \left[\frac{\sqrt{1 + (2\pi r Z)^2}}{(2\pi Z)^2} \right]_{r_A}^{r_B}$$

$$= \frac{\mu_A}{(2\pi Z)^2} \left[\sqrt{1 + (2\pi r_A Z)^2} - 1 \right]$$

$$- \frac{\mu_A}{2(R - r_A)(2\pi Z)^3} \left[2\pi r_B Z \sqrt{1 + (2\pi r_B Z)^2} - \ln\left(\sqrt{1 + (2\pi r_B Z)^2} + 2\pi r_B Z \right) \right.$$

$$\left. - 2\pi r_A Z \sqrt{1 + (2\pi r_A Z)^2} + \ln\left(\sqrt{1 + (2\pi r_A Z)^2} + 2\pi r_A Z \right) \right]$$

$$+ \frac{\mu_A R}{(R - r_A)(2\pi Z)^2} \left[\sqrt{1 + (2\pi r_B Z)^2} - \sqrt{1 + (2\pi r_A Z)^2} \right]$$

$$= \frac{\mu_A}{(2\pi Z)^2} \frac{R}{R-r_A} \left\{ \frac{\overbrace{+\sqrt{1+(2\pi r_A Z)^2} - \frac{r_A}{R}\sqrt{1+(2\pi r_A Z)^2}}}{R-r_A} \sqrt{1+(2\pi r_A Z)^2} - \frac{\overbrace{-1+\frac{r_A}{R}}}{R-r_A} \right.$$

$$-\frac{r_B}{R} \frac{1}{2} \sqrt{1+(2\pi r_B Z)^2} + \frac{1}{2(2\pi R Z)} \ln\left(\sqrt{1+(2\pi r_B Z)^2} + 2\pi r_B Z\right)$$

$$+\frac{r_A}{R} \frac{1}{2} \sqrt{1+(2\pi r_A Z)^2} - \frac{1}{2(2\pi R Z)} \ln\left(\sqrt{1+(2\pi r_A Z)^2} + 2\pi r_A Z\right)$$

$$\left. +\sqrt{1+(2\pi r_B Z)^2} - \sqrt{1+(2\pi r_A Z)^2} \right\},$$

$$\int_0^{r_B} \frac{\mu_r r \, dr}{\sqrt{1+(2\pi r Z)^2}} = \frac{\mu_A}{(2\pi Z)^2} \frac{R}{R-r_A} \left\{ \left(2-\frac{r_B}{R}\right)\frac{1}{2}\sqrt{1+(2\pi r_B Z)^2} - \frac{r_A}{R}\frac{1}{2}\sqrt{1+(2\pi r_A Z)^2} \right.$$

$$\left. -1+\frac{r_A}{R} + \frac{1}{2(2\pi R Z)} \ln \frac{\sqrt{1+(2\pi r_B Z)^2} + 2\pi r_B Z}{\sqrt{1+(2\pi r_A Z)^2} + 2\pi r_A Z} \right\}.$$

$$\text{(A3.15)}$$

4. If $r_B \rightarrow R$, then Equation (A3.15) is limited to what follows in Equation (A3.16):

$$\int_0^R \frac{\mu_r r \, dr}{\sqrt{1+(2\pi r Z)^2}} = \lim_{r_B \rightarrow R} \int_0^{r_B} \frac{\mu_r r \, dr}{\sqrt{1+(2\pi r Z)^2}}$$

$$= \lim_{r_B \rightarrow R} \left[\frac{\mu_A}{(2\pi Z)^2} \frac{R}{R-r_A} \left\{ \left(2-\frac{r_B}{R}\right)\frac{1}{2}\sqrt{1+(2\pi r_B Z)^2} - \frac{r_A}{R}\frac{1}{2}\sqrt{1+(2\pi r_A Z)^2} \right. \right.$$

$$\left. \left. -1+\frac{r_A}{R} + \frac{1}{2(2\pi R Z)} \ln \frac{\sqrt{1+(2\pi r_B Z)^2} + 2\pi r_B Z}{\sqrt{1+(2\pi r_A Z)^2} + 2\pi r_A Z} \right\} \right]$$

$$= \frac{\mu_A}{(2\pi Z)^2} \frac{R}{R-r_A} \left\{ \left(2-\frac{R}{R}\right)\frac{1}{2}\sqrt{1+(2\pi R Z)^2} - \frac{r_A}{R}\frac{1}{2}\sqrt{1+(2\pi r_A Z)^2} \right.$$

$$\left. -1+\frac{r_A}{R} + \frac{1}{2(2\pi R Z)} \ln \frac{\sqrt{1+(2\pi R Z)^2} + 2\pi R Z}{\sqrt{1+(2\pi r_A Z)^2} + 2\pi r_A Z} \right\},$$

$$\int_0^R \frac{\mu_r r \, dr}{\sqrt{1+\left(2\pi r Z\right)^2}} = \frac{\mu_A}{\left(2\pi Z\right)^2} \frac{R}{R-r_A} \left\{ \frac{1}{2}\sqrt{1+\left(2\pi R Z\right)^2} - \frac{r_A}{R}\frac{1}{2}\sqrt{1+\left(2\pi r_A Z\right)^2} \right.$$

$$\left. -1+\frac{r_A}{R}+\frac{1}{2\left(2\pi R Z\right)}\ln\frac{\sqrt{1+\left(2\pi R Z\right)^2}+2\pi R Z}{\sqrt{1+\left(2\pi r_A Z\right)^2}+2\pi r_A Z} \right\}.$$

$$(A3.16)$$

Rearrangement of definite integrals

The integrals as expressed by Equations (A3.13) to (A3.16) can be rearranged by using the following symbols:

$$\kappa = 2\pi R Z, \quad \rho_A = r_A/R, \quad \rho_B = r_B/R, \quad \xi = \mu/\mu_A.$$

Note that these symbols are introduced in Equations (4.95) to (4.98). It is then valid to write from Equation (A3.13) that

$$\int_0^{r_B} \mu_r r \, dr = \mu_A R^2 \frac{-\dfrac{r_A^3}{R^3}+3\dfrac{R r_B^2}{R^3}-2\dfrac{r_B^3}{R^3}}{6\left(\dfrac{R}{R}-\dfrac{r_A}{R}\right)},$$

$$(A3.17)$$

$$\int_0^{r_B} \mu_r r \, dr = \mu_A R^2 \frac{-\rho_A^3+3\rho_B^2-2\rho_B^3}{6\left(1-\rho_A\right)}.$$

Equation (A3.14) can be rearranged as follows:

$$\int_0^R \mu_r r \, dr = \mu_A R^2 \frac{\dfrac{R^2}{R^2}+\dfrac{R r_A}{R^2}+\dfrac{r_A^2}{R^2}}{6},$$

$$\int_0^R \mu_r r \, dr = \mu_A R^2 \frac{1+\rho_A+\rho_A^2}{6}.$$

$$(A3.18)$$

Equation (A3.15) can be obtained in the following form:

$$\int_0^{r_B} \frac{\mu_r r \, dr}{\sqrt{1+\left(2\pi r Z\right)^2}} = \frac{\mu_A}{\left(2\pi R Z\right)^2} \frac{R^2}{\dfrac{R}{R}-\dfrac{r_A}{R}} \left\{ \left(2-\frac{r_B}{R}\right)\frac{1}{2}\sqrt{1+\left(\frac{r_B}{R}\right)^2\left(2\pi R Z\right)^2} \right.$$

$$\left. -\frac{r_A}{R}\frac{1}{2}\sqrt{1+\left(\frac{r_A}{R}\right)^2\left(2\pi R Z\right)^2} -1+\frac{r_A}{R}+\frac{1}{2\left(2\pi R Z\right)}\ln\frac{\sqrt{1+\left(\frac{r_B}{R}\right)^2\left(2\pi R Z\right)^2}+\frac{r_B}{R}2\pi R Z}{\sqrt{1+\left(\frac{r_A}{R}\right)^2\left(2\pi R Z\right)^2}+\frac{r_A}{R}2\pi R Z} \right\},$$

$$\int_0^{r_B} \frac{\mu_r r \, dr}{\sqrt{1+\left(2\pi r Z\right)^2}} = \frac{\mu_A R^2}{\kappa^2 \left(1-\rho_A\right)} \Bigg\{ \left(1 - \frac{\rho_B}{2}\right)\sqrt{1+\rho_B^2 \kappa^2} - \frac{\rho_A}{2}\sqrt{1+\rho_A^2 \kappa^2}$$

$$-1 + \rho_A + \frac{1}{2\kappa}\ln\frac{\sqrt{1+\rho_B^2 \kappa^2}+\rho_B \kappa}{\sqrt{1+\rho_A^2 \kappa^2}+\rho_A \kappa} \Bigg\} \quad \text{(A3.19)}$$

Finally, Equation (A3.16) can be rearranged as follows:

$$\int_0^R \frac{\mu_r r \, dr}{\sqrt{1+\left(2\pi r Z\right)^2}} = \frac{\mu_A}{\left(2\pi R Z\right)^2} \frac{R^2}{\dfrac{R}{R} - \dfrac{r_A}{R}} \Bigg\{ \frac{1}{2}\sqrt{1+\left(2\pi R Z\right)^2} - \frac{r_A}{R}\frac{1}{2}\sqrt{1+\left(\frac{r_A}{R}\right)^2 \left(2\pi R Z\right)^2}$$

$$-1+\frac{r_A}{R}+\frac{1}{2\left(2\pi R Z\right)}\ln\frac{\sqrt{1+\left(2\pi R Z\right)^2}+2\pi R Z}{\sqrt{1+\left(\dfrac{r_A}{R}\right)^2 \left(2\pi R Z\right)^2}+\dfrac{r_A}{R} 2\pi R Z}\Bigg\},$$

$$\int_0^R \frac{\mu_r r \, dr}{\sqrt{1+\left(2\pi r Z\right)^2}} = \frac{\mu_A R^2}{\kappa^2 \left(1-\rho_A\right)} \Bigg\{ \frac{1}{2}\sqrt{1+\kappa^2} - \frac{\rho_A}{2}\sqrt{1+\rho_A^2 \kappa^2}$$

$$-1 + \rho_A + \frac{1}{2\kappa}\ln\frac{\sqrt{1+\kappa^2}+\kappa}{\sqrt{1+\rho_A^2 \kappa^2}+\rho_A \kappa}\Bigg\}. \quad \text{(A3.20)}$$

Indefinite integrals

The following indefinite integrals are required for solving the expression for radial migration of fibers in yarns (Note that the integral constant is not written in the following expressions.):

$$\int \frac{dy}{\sqrt{y^2-1}} = \int \frac{dz}{z} = \ln|z|, \quad \int \frac{dy}{\sqrt{y^2-1}} = \ln\left|y+\sqrt{y^2-1}\right|.$$

$$\text{Substitution: } y+\sqrt{y^2-1}=z, \quad \left(1+\frac{y}{\sqrt{y^2-1}}\right)dy \qquad (A4.1)$$

$$= \frac{\sqrt{y^2-1}+y}{\sqrt{y^2-1}}\,dt = \frac{z}{\sqrt{y^2-1}}\,dy = dz, \quad dy\Big/\sqrt{y^2-1} = dz/z,$$

and

$$\int\sqrt{y^2-1}\,dy = y\sqrt{y^2-1} - \int \frac{y^2}{\sqrt{y^2-1}}\,dy = y\sqrt{y^2-1} - \int \frac{y^2-1}{\sqrt{y^2-1}}\,dy - \int \frac{dy}{\sqrt{y^2-1}},$$

$$u'=1, \quad u=y, \quad v=\sqrt{y^2-1}, \quad v'=y\Big/\sqrt{y^2-1},$$

$$\int\sqrt{y^2-1}\,dy = y\sqrt{y^2-1} - \int\sqrt{y^2-1}\,dy - \int \frac{dy}{\sqrt{y^2-1}},$$

$$2\int\sqrt{y^2-1}\,dy = y\sqrt{y^2-1} - \int \frac{dy}{\sqrt{y^2-1}}. \qquad (A4.2)$$

By applying Equation (A4.1) in (A4.2), we find

$$\int\sqrt{y^2-1}\,dy = \frac{1}{2}\left(y\sqrt{y^2-1} - \ln\left|y+\sqrt{y^2-1}\right|\right). \qquad (A4.3)$$

We also use the following indefinite integrals:

$$\int \frac{dy}{\sqrt{a^2 + y^2}} = \int \frac{dt}{t} = \ln|t|, \quad \int \frac{dy}{\sqrt{a^2 + y^2}} = \ln\left|y + \sqrt{a^2 + y^2}\right|.$$

Substitution: $y + \sqrt{a^2 + y^2} = t$, $\left(1 + \frac{y}{\sqrt{a^2 + y^2}}\right) dy = dt$, (A4.4)

$$\frac{\sqrt{a^2 + y^2} + y}{\sqrt{a^2 + y^2}} dy = \frac{t}{\sqrt{a^2 + y^2}} dy = dt, \quad \frac{dy}{\sqrt{a^2 + y^2}} = \frac{dt}{t},$$

and

$$\int \sqrt{a^2 + y^2}\, dy = y\sqrt{a^2 + y^2} - \int \frac{y^2 dy}{\sqrt{a^2 + y^2}} dy,$$

$$u' = 1, \quad v = \sqrt{a^2 + y^2}, \quad u = y \quad v' = y/\sqrt{a^2 + y^2},$$

$$\int \sqrt{a^2 + y^2}\, dy = y\sqrt{a^2 + y^2} - \int \frac{a^2 + y^2}{\sqrt{a^2 + y^2}} dy + \int \frac{a^2\, dy}{\sqrt{a^2 + y^2}},$$

$$2\int \sqrt{a^2 + y^2}\, dy = y\sqrt{a^2 + y^2} + a^2 \int \frac{dy}{\sqrt{a^2 + y^2}}.$$ (A4.5)

By applying Equation (A4.4) in (A4.5), we obtain

$$\int \sqrt{a^2 + y^2}\, dy = \frac{1}{2}\left[y\sqrt{a^2 + y^2} + a^2 \ln\left|y + \sqrt{a^2 + y^2}\right| \right].$$ (A4.6)

Further, it is valid to write that

$$\int y\sqrt{a^2 + y^2}\, dy = \int t^2 dt = \frac{t^3}{3} = \frac{1}{3}\left(a^2 + y^2\right)^{3/2}.$$ (A4.7)

Substitution: $a^2 + y^2 = t^2$, $y\, dy = t\, dt$.

We also need the following indefinite integral

$$\int \frac{y\, dy}{\sqrt{a^2 + y^2}} = \int \frac{t\, dt}{t} = t, \quad \int \frac{y\, dy}{\sqrt{a^2 + y^2}} = \sqrt{a^2 + y^2}.$$ (A4.8)

Substitution: $a^2 + y^2 = t^2$, $y\, dy = t\, dt$.

Let us solve the integral

$$\int \frac{y^2 dy}{\sqrt{a^2 + y^2}} = \int \frac{a^2 + y^2}{\sqrt{a^2 + y^2}} dy - \int \frac{a^2}{\sqrt{a^2 + y^2}} dy = \int \sqrt{a^2 + y^2}\, dy - a^2 \int \frac{dy}{\sqrt{a^2 + y^2}}.$$ (A4.9)

By using Equations (A4.4) and (A4.6) in the last expression, we obtain the following solution:

$$\int \frac{y^2 dy}{\sqrt{a^2 + y^2}} = \int \sqrt{a^2 + y^2}\, dy - a^2 \int \frac{dy}{\sqrt{a^2 + y^2}}$$

$$= \frac{1}{2}\left[y\sqrt{a^2 + y^2} + a^2 \ln\left|y + \sqrt{a^2 + y^2}\right| \right] - a^2 \ln\left|y + \sqrt{a^2 + y^2}\right|,$$

$$\int \frac{y^2 dy}{\sqrt{a^2 + y^2}} = \frac{y}{2}\sqrt{a^2 + y^2} - \frac{a^2}{2}\ln\left|y + \sqrt{a^2 + y^2}\right|.$$

$$\text{(A4.10)}$$

Finally, we need to solve the following integral:

$$\int \frac{y^3 dy}{\sqrt{a^2 + y^2}} = \int \frac{\left(t^2 - a^2\right)t}{t} = \int t^2 dt - a^2 \int dt = \frac{t^3}{3} - a^2 t,$$

Substitution: $a^2 + y^2 = t^2$, $\quad y\,dy = t\,dt$,   (A4.11)

$$\int \frac{y^3 dy}{\sqrt{a^2 + y^2}} = \frac{1}{3}\left(a^2 + y^2\right)^{3/2} - a^2 \sqrt{a^2 + y^2}.$$

Some definite integrals

We need the following definite integrals and/or special general integrals for solving the problems of radial migration of fibers in yarns.

According to Equation (A4.3), it is valid to write that

$$\int_1^C \sqrt{t^2 - 1}\, dt = \frac{1}{2}\left[t\sqrt{t^2 - 1} - \ln\left|t + \sqrt{t^2 - 1}\right| \right]_1^C = \frac{1}{2}\left[C\sqrt{C^2 - 1} - \ln\left|C + \sqrt{C^2 - 1}\right| \right].$$

$$\text{(A4.12)}$$

According to Equation (A4.6), we can write that

$$\int_0^1 \sqrt{\frac{1}{4(p/D)^2} + y^2}\, dy = \frac{1}{2}\left[y\sqrt{\frac{1}{4(p/D)^2} + y^2} + \frac{1}{4(p/D)^2}\ln\left|y + \sqrt{\frac{1}{4(p/D)^2} + y^2}\right| \right]_0^1$$

$$= \frac{1}{2}\left[\sqrt{\frac{1}{4(p/D)^2} + 1} + \frac{1}{4(p/D)^2}\ln\left|1 + \sqrt{\frac{1}{4(p/D)^2} + 1}\right| - \ln\left|\frac{1}{2(p/D)}\right| \right]$$

$$= \frac{1}{2}\frac{1}{2(p/D)}\left[\sqrt{1 + 4(p/D)^2} + \frac{1}{2(p/D)}\ln\left|2(p/D) + \sqrt{1 + 4(p/D)^2}\right| \right]. \quad \text{(A4.13)}$$

According to Equation (A4.7), we can get

$$\int_0^1 y\sqrt{y^2(\pi DZ)^2+1}\,dy = \pi DZ\int_0^1 y\sqrt{\frac{1}{(\pi DZ)^2}+y^2}\,dy$$

$$= \pi DZ\left[\frac{1}{3}\left(\frac{1}{(\pi DZ)^2}+y^2\right)^{3/2}\right]_0^1,$$

$$\int_0^1 y\sqrt{y^2(\pi DZ)^2+1}\,dy = \pi DZ\,\frac{1}{3}\left[\left(\frac{1}{(\pi DZ)^2}+1\right)^{3/2}-\frac{1}{(\pi DZ)^3}\right]. \qquad (A4.14)$$

The following indefinite integral is arising from Equation (A4.4):

$$\int\frac{dr}{\sqrt{1+(2\pi rZ)^2}} = \frac{1}{(2\pi Z)}\int\frac{dy}{\sqrt{1+y^2}} = \frac{1}{(2\pi Z)}\ln\left|y+\sqrt{1+y^2}\right|,$$

Substitution: $2\pi rZ = y$, $(2\pi Z)dr = dy$,

$$\int\frac{dr}{\sqrt{1+(2\pi rZ)^2}} = \frac{1}{(2\pi Z)}\ln\left|2\pi rZ+\sqrt{1+(2\pi rZ)^2}\right|. \qquad (A4.15)$$

Further, by using Equation (A4.8), we can solve the following:

$$\int_0^{D/2}\frac{r\,dr}{\sqrt{1+(2\pi rZ)^2}} = \frac{1}{2\pi Z}\int_0^{D/2}\frac{2\pi rZ\,dr}{\sqrt{1+(2\pi rZ)^2}} = \frac{1}{(2\pi Z)^2}\int_0^{\pi DZ}\frac{y\,dy}{\sqrt{1+y^2}},$$

Substitution: $2\pi rZ = y$, $(2\pi Z)dr = dy$,

$$\int_0^{D/2}\frac{r\,dr}{\sqrt{1+(2\pi rZ)^2}} = \frac{1}{(2\pi Z)^2}\left[\sqrt{1+y^2}\right]_0^{\pi DZ} = \frac{1}{(2\pi Z)^2}\left[\sqrt{1+(\pi DZ)^2}-1\right]. \qquad (A4.16)$$

Similarly, by using Equation (A4.8), we get

$$\int_0^{D/2}\frac{r\,dr}{\sqrt{\dfrac{Q^2}{Q^2-1}+(2\pi rZ)^2}} = \frac{1}{2\pi Z}\int_0^{D/2}\frac{2\pi rZ\,dr}{\sqrt{\dfrac{Q^2}{Q^2-1}+(2\pi rZ)^2}} = \frac{1}{(2\pi Z)^2}\int_0^{\pi DZ}\frac{y\,dy}{\sqrt{\dfrac{Q^2}{Q^2-1}+y^2}},$$

Substitution: $2\pi rZ = y$, $(2\pi Z)dr = dy$,

$$= \frac{1}{(2\pi Z)^2}\left[\sqrt{\frac{Q^2}{Q^2-1}+y^2}\right]_0^{\pi DZ} = \frac{1}{(2\pi Z)^2}\left[\sqrt{\frac{Q^2}{Q^2-1}+(\pi DZ)^2}-\sqrt{\frac{Q^2}{Q^2-1}}\right]. \qquad (A4.17)$$

By using Equation (A4.8), we also find that

$$\int_0^{\pi DZ} \frac{x\,dx}{\sqrt{1+x^2}} = \left[\sqrt{1+x^2}\right]_0^{\pi DZ} = \left[\sqrt{1+(\pi DZ)^2}-1\right]. \tag{A4.18}$$

Further, by using Equation (A4.10), it is possible to write the following

$$\int_0^{D/2} \frac{(2\pi rZ)^2\,dr}{\sqrt{\dfrac{Q^2}{Q^2-1}+(2\pi rZ)^2}} = \frac{1}{2\pi Z}\int_0^{\pi DZ} \frac{y^2\,dy}{\sqrt{\dfrac{Q^2}{Q^2-1}+y^2}},$$

Substitution: $2\pi rZ = y$, $(2\pi Z)\,dr = dy$,

$$\int_0^{D/2} \frac{(2\pi rZ)^2\,dr}{\sqrt{\dfrac{Q^2}{Q^2-1}+(2\pi rZ)^2}} = \frac{1}{2\pi Z}\left[\frac{y}{2}\sqrt{\dfrac{Q^2}{Q^2-1}+y^2}\right.$$

$$\left.-\frac{1}{2}\frac{Q^2}{Q^2-1}\ln\left|y+\sqrt{\dfrac{Q^2}{Q^2-1}+y^2}\right|\right]_0^{\pi DZ}$$

$$= \frac{1}{2\pi Z}\frac{1}{2}\left[\pi DZ\sqrt{\dfrac{Q^2}{Q^2-1}+(\pi DZ)^2}\right.$$

$$\left.-\frac{Q^2}{Q^2-1}\ln\left|\pi DZ+\sqrt{\dfrac{Q^2}{Q^2-1}+(\pi DZ)^2}\right|+\frac{Q^2}{Q^2-1}\ln\left|\sqrt{\dfrac{Q^2}{Q^2-1}}\right|\right].$$

$$\tag{A4.19}$$

Let us rearrange the last expression as follows:

$$\int_0^{D/2} \frac{(2\pi rZ)^2\,dr}{\sqrt{\dfrac{Q^2}{Q^2-1}+(2\pi rZ)^2}} = \frac{1}{2\pi Z}\frac{1}{2}\left[\pi DZ\sqrt{\dfrac{Q^2}{Q^2-1}+(\pi DZ)^2}+\frac{Q^2}{Q^2-1}\ln\frac{\sqrt{\dfrac{Q^2}{Q^2-1}}}{\pi DZ+\sqrt{\dfrac{Q^2}{Q^2-1}+(\pi DZ)^2}}\right]$$

$$= \frac{1}{2\pi Z}\frac{1}{2}\sqrt{\dfrac{Q^2}{Q^2-1}}\left[\pi DZ\sqrt{1+\dfrac{Q^2-1}{Q^2}(\pi DZ)^2}\right.$$

$$\left.-\sqrt{\dfrac{Q^2}{Q^2-1}}\ln\left(\pi DZ\sqrt{\dfrac{Q^2-1}{Q^2}}+\sqrt{1+\dfrac{Q^2-1}{Q^2}(\pi DZ)^2}\right)\right],$$

$$\int_0^{D/2} \frac{(2\pi rZ)^2 \, dr}{\sqrt{\dfrac{Q^2}{Q^2-1}+(2\pi rZ)^2}} = \frac{1}{2\pi Z}\frac{1}{2}\sqrt{\frac{Q^2}{Q^2-1}}(\pi DZ)\left[\sqrt{1+\frac{Q^2-1}{Q^2}(\pi DZ)^2}\right.$$

$$\left. -\frac{1}{\sqrt{\dfrac{Q^2-1}{Q^2}(\pi DZ)^2}}\ln\left(\sqrt{\frac{Q^2-1}{Q^2}(\pi DZ)^2}+\sqrt{1+\frac{Q^2-1}{Q^2}(\pi DZ)^2}\right)\right],$$

$$\int_0^{D/2} \frac{(2\pi rZ)^2 \, dr}{\sqrt{\dfrac{Q^2}{Q^2-1}+(2\pi rZ)^2}} = \frac{1}{2\pi Z}\frac{1}{2}\sqrt{\frac{Q^2}{Q^2-1}}(\pi DZ)\left[\sqrt{1+x^2}-\frac{1}{x}\ln\left(x+\sqrt{1+x^2}\right)\right],$$

$$\text{where } x=\sqrt{\frac{Q^2-1}{Q^2}(\pi DZ)^2}.$$

(A4.20)

Also, we need to solve the following definite integral by using Equation (A4.11) as follows:

$$\int_0^{D/2} \frac{(2\pi rZ)^3}{\sqrt{\dfrac{Q^2}{Q^2-1}+(2\pi rZ)^2}} \, dr = \frac{1}{2\pi Z}\int_0^{\pi DZ} \frac{y^3 \, dy}{\sqrt{\dfrac{Q^2}{Q^2-1}+y^2}}$$

Substitution: $2\pi rZ = y$, $(2\pi Z)dr = dy$,

$$= \frac{1}{2\pi Z}\left[\frac{1}{3}\left(\frac{Q^2}{Q^2-1}+y^2\right)^{3/2}-\frac{Q^2}{Q^2-1}\sqrt{\frac{Q^2}{Q^2-1}+y^2}\right]_0^{\pi DZ}$$

$$= \frac{1}{2\pi Z}\left[\frac{1}{3}\left(\frac{Q^2}{Q^2-1}+(\pi DZ)^2\right)^{3/2}\right.$$

$$\left. -\frac{Q^2}{Q^2-1}\sqrt{\frac{Q^2}{Q^2-1}+(\pi DZ)^2}-\frac{1}{3}\left(\frac{Q^2}{Q^2-1}\right)^{3/2}+\frac{Q^2}{Q^2-1}\sqrt{\frac{Q^2}{Q^2-1}}\right]$$

$$= \frac{1}{2\pi Z}\left[\frac{1}{3}\left(\frac{Q^2}{Q^2-1}+(\pi DZ)^2\right)^{3/2}-\frac{Q^2}{Q^2-1}\sqrt{\frac{Q^2}{Q^2-1}+(\pi DZ)^2}+\frac{2}{3}\left(\frac{Q^2}{Q^2-1}\right)^{3/2}\right]$$

$$= \frac{1}{2\pi Z}\left(\frac{Q^2}{Q^2-1}\right)^{3/2}\left[\frac{1}{3}\left(1+\frac{Q^2-1}{Q^2}(\pi DZ)^2\right)^{3/2}-\sqrt{1+\frac{Q^2-1}{Q^2}(\pi DZ)^2}+\frac{2}{3}\right],$$

$$\int_0^{D/2} \frac{(2\pi r Z)^3}{\sqrt{\dfrac{Q^2}{Q^2-1}+(2\pi r Z)^2}}\, dr = \frac{1}{2\pi Z}\left(\frac{Q^2}{Q^2-1}\right)^{3/2}\left[\frac{1}{3}\left(1+x^2\right)^{3/2}-\sqrt{1+x^2}+\frac{2}{3}\right],$$

$$\text{where } x = \sqrt{\frac{Q^2-1}{Q^2}(\pi D Z)^2}.$$

(A4.21)

Statistical properties of variable type y and variable type d

Basic relationships

In statistics, the symbols E, D and v are usually denoted as operators for the mean value, variance and coefficient of variation, respectively. Let us now review some basic relations[1] as follows. The following expression is valid for a constant k:

$$E(k) = k.$$ (A5.1)

The mean value of the product of two independent random variables (or constants) m and x is

$$E(mx) = E(m)E(x).$$ (A5.2)

The variance of the random variable x is

$$D(x) = E\left\{\left[x - E(x)\right]^2\right\} = E\left\{x^2\right\} - E^2(x)$$ (A5.3)

and the variance of the constant is

$$D(k) = E\left\{k^2\right\} - E^2(k) = k^2 - k^2 = 0.$$ (A5.4)

The following expression is valid for the variance of product kx:

$$D(kx) = k^2 D(x).$$ (A5.5)

The coefficient of variation of the variable x is given by the following expression:

$$v^2(x) = D(x)/E^2(x) \quad v(x) = \sqrt{D(x)}/E(x)$$ (A5.6)

---

1   Equations (A5.1) to (A5.7) are known as the fundamental expressions that are usually given in each handbook of mathematical statistics.

and the coefficient of variation of the random variable kx is

$$v(kx) = \text{sgn}(k)v(x). \tag{A5.7}$$

The symbol $\text{sgn}(k)$ denotes the sign of the constant k; $\text{sgn}(k) = 1$ for $k > 0$, $\text{sgn}(k) = -1$ for $k < 0$ and $\text{sgn}(k) = 0$ for $k = 0$.

Random variable y_m

We consider an independent random variable x_i, where $i = 1, 2, \ldots, m$. We also consider that each random variable x_i follows the same probabilistic distribution. Consequently, each random variable has same characteristics of distribution. The common characteristics will be denoted by the symbols without any index – for example, mean value $E(x_i) = E(x)$, mean square $E(x_i^2) = E(x^2)$, variance $D(x_i) = D(x)$, coefficient of variation $v(x_i) = v(x)$, etc.

We define the random variable y_m as follows:

$$\left.\begin{array}{ll} y_m = 0 & \text{for } m = 0, \\ y_m = \displaystyle\sum_{i=1}^{m} x_i & \text{for } m = 1, 2, \ldots, \end{array}\right\} \; m \text{ is given integer quantity}. \tag{A5.8}$$

Statistical characteristics of random variable y_m

The mean value of this random variable can be easily obtained by applying Equation (A5.2) as follows:

$$\left.\begin{array}{ll} E(y_m) = E(0) = 0 & \text{for } m = 0, \\ E(y_m) = E\left(\displaystyle\sum_{i=1}^{m} x_i\right) = m\,E(x) & \text{for } m = 1, 2, \ldots, \\ E(y_m) = m\,E(x) & \text{for } m = 0, 1, \ldots, m. \end{array}\right. \tag{A5.9}$$

The mean square value can be expressed as follows:

$$E(y_m^2) = E(0^2) = 0 \quad \text{for } m = 0 \tag{A5.10}$$

and

$$E\left(y_m^2\right) = E\left[\left(\sum_{i=1}^{m} x_i\right)^2\right] = E\left(x_1^2 + x_1 x_2 + \cdots + x_1 x_m + \right.$$

$$+ x_2 x_1 + x_2^2 + \cdots + x_2 x_m +$$

$$\cdots\cdots\cdots\cdots\cdots$$

$$\left. + x_m x_1 + x_m x_2 + \cdots + x_m^2\right),$$

$$E\left(y_m^2\right) = \sum_{i=1,2,\ldots,m} \overbrace{E\left(x_i^2\right)}^{=E\left(x^2\right)} + \sum_{\substack{i=1,2,\ldots,m \\ j=1,2,\ldots,m \\ i \neq j}} \overbrace{E\left(x_i x_j\right)}^{=E(x)E(x)} \tag{A5.11}$$

$$= mE\left(x^2\right) + \left(m^2 - m\right)E^2\left(x\right) \quad \text{for } m = 1,2,\ldots,$$

$$E\left(y_m^2\right) = m\left[E\left(x^2\right) - E^2\left(x\right)\right] + m^2 E^2\left(x\right) = mD(x) + m^2 E^2\left(x\right) \text{ for } m = 1,2,\ldots \tag{A5.12}$$

We obtain the following expression from Equations (A5.10) and (A5.12):

$$E\left(y_m^2\right) = mE\left(x^2\right) + \left(m^2 - m\right)E^2\left(x\right) = mD(x) + m^2 E^2\left(x\right) \quad \text{for } m = 0,1,\ldots \tag{A5.13}$$

The expression for the variance of random variable y_m is obtained from Equations (A5.9) and (A5.13) as follows:

$$D\left(y_m\right) = E\left(y_m^2\right) - E^2\left(y_m\right) = mE\left(x^2\right) + \left(m^2 - m\right)E^2\left(x\right) - m^2 E^2\left(x\right) =$$

$$= mE\left(x^2\right) - mE^2\left(x\right) = m\left[E\left(x^2\right) - E^2\left(x\right)\right] = mD(x) \quad \text{for } m = 0,1,\ldots. \tag{A5.14}$$

Note: Let us note that according to Equations (A5.9), (A5.13) and (A5.14), the values of $E\left(y_m\right)$, $E\left(y_m^2\right)$ and $D\left(y_m\right)$ depend on a given value of m.

Random variable y

We define a discrete random variable m, which takes the values $0,1,2,\ldots$. The mean value of this random variable is $E(m)$, mean square value is $E\left(m^2\right)$, and variance is $D(m) = E\left(m^2\right) - E^2\left(m\right)$.

Now we define a random variable y by means of a generated function

$$\left.\begin{array}{ll} y = 0 & \text{for } m = 0, \\[2mm] y = \sum_{i=1}^{m} x_i & \text{for } m = 1, 2, \ldots, \end{array}\right\} \quad m \text{ is discrete random variable.} \quad (A5.15)$$

Note: In opposite to the previous random variable y_m, the quantity m is not a parameter but a random variable now.

The previously derived expression for $E(y_m)$ is now considered only as a 'partial' mean value of the subset values y, which originate from the given value of m. The mean value of the random variable y is found from the mean value of these 'partial' mean values generated 'over all m' values, $E(y) = E[E(y_m)]$. By applying Equations (A5.2) and (A5.9), we find

$$E(y) = E[E(y_m)] = E\left[m\overbrace{E(x)}^{=\text{const.}}\right] = E(m) E(x). \quad (A5.16)$$

Accordingly, it is valid for the mean square value of the random variable y that $E(y^2) = E[E(y_m^2)]$. By applying of Equations (A5.2) and (A5.13), we can write

$$\begin{aligned} E(y^2) &= E[E(y_m^2)] = E[mD(x) + m^2 E^2(x)] = E(m)D(x) + E(m^2)E^2(x) \\ &= E(m)D(x) + [D(m) + E^2(m)]E^2(x). \end{aligned}$$

$$(A5.17)$$

By substituting the last two equations into Equation (A5.3), we find the following expression for the variance of the random variable y:

$$\begin{aligned} D(y) &= E(y^2) - E^2(y) \\ &= \{E(m)D(x) + [D(m) + E^2(m)]E^2(x)\} - \{E^2(m)E^2(x)\} \\ &= E(m)D(x) + D(m)E^2(x). \end{aligned}$$

$$(A5.18)$$

Finally, by substituting Equations (A5.16) and (A5.18) into Equation (A5.6), we find the following expression for the square of the coefficient of variation:

$$v^2(y) = \frac{D(y)}{E^2(y)} = \frac{E(m)D(x) + D(m)E^2(x)}{E^2(m)E^2(x)},$$

$$v^2(y) = \frac{1}{E(m)}\left[\frac{D(x)}{E^2(x)} + \frac{D(m)}{E(m)}\right] = \frac{1}{E(m)}\left[v^2(x) + \frac{D(m)}{E(m)}\right]. \qquad \text{(A5.19)}$$

Binomial distribution of random variable m

Especially, if the (discrete) random variable m follows the binomial distribution $B(m)$ then it is valid to write that

$$B(m) = \binom{m_{max}}{m} p^m (1-p)^{m_{max}-m}, \quad m = 0,1,\ldots,m_{max}, \qquad \text{(A5.20)}$$

where m_{max} is the highest possible value of the random variable m and the parameter p has the meaning of probability.

The mean value and variance of this distribution[2] are given by the following expressions:

$$E(m) = pm_{max}, \quad D(m) = pm_{max}(1-p). \qquad \text{(A5.21)}$$

Thus,

$$p = \frac{E(m)}{m_{max}}, \quad \frac{D(m)}{E(m)} = 1 - p = 1 - \frac{E(m)}{m_{max}}. \qquad \text{(A5.22)}$$

By applying Equation (A5.22) into (A5.19), we find the following expression for the square of the coefficient of variation of the random variable y:

$$v^2(y) = \frac{1}{E(m)}\left[v^2(x) + \frac{D(m)}{E(m)}\right] = \frac{1}{E(m)}\left\{v^2(x) + \left[1 - \frac{E(m)}{m_{max}}\right]\right\}. \qquad \text{(A5.23)}$$

---

2   Read any handbook of mathematical statistics for the statistical characteristics of binomial distribution.

Poisson distribution of random variable m

The Poisson distribution expresses the probability $P(m)$ of the discrete random variable m as follows:

$$P(m) = \frac{\left[E(m)\right]^m}{m!} e^{-E(m)}, \quad m = 0, 1, \ldots, \infty. \tag{A5.24}$$

The parameter of this distribution is the mean value $E(m)$ of the discrete random variable m. Another parameter of this distribution, i.e., variance is given by

$$D(m) = E(m), \quad D(m)/E(m) = 1. \tag{A5.25}$$

The coefficient of variation of random variable y **by Poisson distribution of** m

If we apply Equation (A5.25) into (A5.19), then we find the following expression for the coefficient of variation of the random variable y:

$$v^2(y) = \frac{1}{E(m)}\left[v^2(x) + \frac{D(m)}{E(m)}\right] = \frac{1}{E(m)}\left[v^2(x) + 1\right]. \tag{A5.26}$$

Random variable d

If we consider that each random variable x is defined in terms of another random variable d and the constant k by the following equation:

$$x = kd^2, \tag{A5.27}$$

then the mean value of the random variable d is designated by $E(d)$.

Equation (A5.27) can be approximately expressed by the first two elements of Taylor series expanded for the value $E(d)$. The first order derivative of the function expressed in Equation (A5.27) is $dx/dd = 2kd$. Thus

$$x = k E^2(d) + \frac{2k E(d)}{1!}\left[d - E(d)\right] = \left[2k E(d)\right]d - \left[k E^2(d)\right], \tag{A5.28}$$

which is a linear expression as compared to the quadratic expression given by Equation (A5.27).

Note: Equation (A5.28) is acceptable as a linear (tangential) substitution of the quadratic equation (A5.27) only in case that the values of the random variable d are not too different from the mean value $E(d)$.

By considering the validity of Equation (A5.28) and according to the statistical rules[3], it is valid to write that

$$E(x) = \left[2k\,E(d)\right]E(d) - \left[k\,E^2(d)\right] = k\,E^2(d),\qquad\text{(A5.29)}$$

$$D(x) = \left[2k\,E(d)\right]^2 D(d),\qquad\text{(A5.30)}$$

$$v^2(x) = \frac{D(x)}{E^2(x)} = \frac{\left[2k\,E(d)\right]^2 D(d)}{k^2 E^4(d)} = 4\frac{\left[k\,E(d)\right]^2}{k^2 E^2(d)}\overbrace{\left[\frac{D(d)}{E^2(d)}\right]}^{=v^2(d)} = 4\,v^2(d).$$

$$\text{(A5.31)}$$

The last expression establishes a relation between the coefficients of variation of the random variables x and d. This can be used in Equations (A5.19), (A5.23) and (A5.26), and the coefficient of variation of the random variable y can be approximately expressed by the coefficient of variation of the random variable d.

---

3   If v is a random variable and k, q are constants, then the mean value of the random variable $u = kv + q$ is $E(u) = k\,E(v) + q$ and the variance of the random variable is $D(u) = k^2 D(v)$.

Mean value of number of fibers in non-empty bundles

The binomial model considers that the number of fibers in a bundle q_{43} follows the binomial distribution of the following form:

$$B(q_{43}) = \binom{q_{43\,max}}{q_{43}} \left(\frac{\overline{q}_{43}}{q_{43\,max}} \right)^{q_{43}} \left(1 - \frac{\overline{q}_{43}}{q_{43\,max}} \right)^{q_{43\,max} - q_{43}}. \qquad (A6.1)$$

(See also Equations (A5.20) to (A5.22) in Appendix 5.) The symbols signify the mean value $\overline{q}_{43}$ and the maximum number of fibers $q_{43\,max}$ in the bundle. This distribution is valid also for the empty bundles. Each empty bundle possesses zero number of fibers, $q_{43}^{\circ} = 0$, so that the mean number of fibers in all empty bundles is $\overline{q}_{43}^{\circ} = 0$. The non-empty bundles have different number of fibers $q_{43}^{*} > 0$ [1]; and their mean number of fibers is $\overline{q}_{43}^{*} > 0$.

The relative frequency of empty bundles p_{43}° (or also the probability that the bundle is empty) is obtained from Equation (A6.1) as follows:

$$p_{43}^{\circ} = B(q_{43} = 0) = \overbrace{\binom{q_{43\,max}}{0}}^{=1} \overbrace{\left(\frac{\overline{q}_{43}}{q_{43\,max}} \right)^{0}}^{=1} \left(1 - \frac{\overline{q}_{43}}{q_{43\,max}} \right)^{q_{43\,max} - 0} = \left(1 - \frac{\overline{q}_{43}}{q_{43\,max}} \right)^{q_{43\,max}}.$$

$$(A6.2)$$

---

1   The quantities for the empty aggregates are denoted by the right-hand superscript ○ and the same for the non-empty aggregates are denoted by the right-hand superscript *. There is no right-hand superscript used for the quantities related to the aggregates with empty as well as non-empty fibers altogether.

The relative frequency of non-empty bundles p_{43}^{*} (or probability that the bundle is not empty) is

$$p_{43}^{*} = 1 - p_{43}^{\circ} = 1 - \left(1 - \frac{\overline{q}_{43}}{q_{43\,\text{max}}} \right)^{q_{43\,\text{max}}} .$$ (A6.3)

(Naturally, $p_{43}^{\circ} + p_{43}^{*} = 1$.) It is evident that the mean number of fibers $\overline{q}_{43}$ with all (empty and non-empty) bundles is $\overline{q}_{43} = p_{43}^{\circ} \overline{q}_{43}^{\circ} + p_{43}^{*} \overline{q}_{43}^{*}$. By using the previous relationships it is valid to write that

$$\overline{q}_{43} = p_{43}^{\circ} \overline{q}_{43}^{\circ} + p_{43}^{*} \overline{q}_{43}^{*} = \left[\left(1 - \frac{\overline{q}_{43}}{q_{43\,\text{max}}} \right)^{q_{43\,\text{max}}} \right] 0 + \left[1 - \left(1 - \frac{\overline{q}_{43}}{q_{43\,\text{max}}} \right)^{q_{43\,\text{max}}} \right] \overline{q}_{43}^{*},$$

$$\overline{q}_{43}^{*} = \frac{\overline{q}_{43}}{1 - \left(1 - \overline{q}_{43} / q_{43\,\text{max}} \right)^{q_{43\,\text{max}}}} .$$ (A6.4)

Equation (A6.4) expresses the mean number of fibers in a non-empty bundle.

Mean value of number of fibers in non-empty clusters

The binomial model considers that the number of bundles in cluster q_{32} follows the binomial distribution of the following form:

$$B(q_{32}) = \binom{q_{32\,\text{max}}}{q_{32}} \left(\frac{\overline{q}_{32}}{q_{32\,\text{max}}} \right)^{q_{32}} \left(1 - \frac{\overline{q}_{32}}{q_{32\,\text{max}}} \right)^{q_{32\,\text{max}} - q_{32}} .$$ (A6.5)

(See Equations (A5.20) to (A5.22) in Appendix A5 once again.) The symbols represent the mean value $\overline{q}_{32}$ and maximum number $q_{32\,\text{max}}$ of bundles in the cluster.

The probability $\beta(q_{32})$

The probability that the cluster has q_{32} number of bundles and simultaneously each bundle is empty is obtained from the rules of theory of probability as follows:

$$\beta(q_{32}) = \overbrace{B(q_{32})}^{\substack{\text{probability that} \\ \text{cluster contains} \\ q_{32}\ \text{bundles}}} \overbrace{p_{43}^{\circ}}^{\substack{\text{probability that 1-st} \\ \text{bundle is empty}}} \overbrace{p_{43}^{\circ}}^{\substack{\text{probability that 2-nd} \\ \text{bundle is empty}}} \cdots \overbrace{p_{43}^{\circ}}^{\substack{\text{probability that } q_{32}\text{-th} \\ \text{bundle is empty}}} = B(q_{32})(p_{43}^{\circ})^{q_{32}}.$$

(A6.6)

The following expression is obtained by substituting Equation (A6.5) into (A6.6) and then rearranging

$$\beta(q_{32}) = \binom{q_{32\,max}}{q_{32}}\left(\frac{\overline{q}_{32}}{q_{32\,max}}\right)^{q_{32}}\left(1 - \frac{\overline{q}_{32}}{q_{32\,max}}\right)^{q_{32\,max} - q_{32}}(p_{43}^{\circ})^{q_{32}}$$

$$= \binom{q_{32\,max}}{q_{32}}\left(\frac{\overline{q}_{32}}{q_{32\,max}}\right)^{q_{32}}\left(\frac{q_{32\,max} - \overline{q}_{32}}{q_{32\,max}}\right)^{-q_{32}}\left(1 - \frac{\overline{q}_{32}}{q_{32\,max}}\right)^{q_{32\,max}}(p_{43}^{\circ})^{q_{32}}$$

$$= \binom{q_{32\,max}}{q_{32}}\left(\frac{\overline{q}_{32}}{q_{32\,max}}\frac{q_{32\,max}}{q_{32\,max} - \overline{q}_{32}}\right)^{q_{32}}\left(1 - \frac{\overline{q}_{32}}{q_{32\,max}}\right)^{q_{32\,max}}(p_{43}^{\circ})^{q_{32}}$$

$$= \binom{q_{32\,max}}{q_{32}}\left(\frac{\overline{q}_{32}/q_{32\,max}}{1 - \overline{q}_{32}/q_{32\,max}}\right)^{q_{32}}\left(1 - \frac{\overline{q}_{32}}{q_{32\,max}}\right)^{q_{32\,max}}(p_{43}^{\circ})^{q_{32}},$$

$$\beta(q_{32}) = \binom{q_{32\,max}}{q_{32}}\left(\frac{p_{43}^{\circ}\,\overline{q}_{32}/q_{32\,max}}{1 - \overline{q}_{32}/q_{32\,max}}\right)^{q_{32}}\left(1 - \frac{\overline{q}_{32}}{q_{32\,max}}\right)^{q_{32\,max}}.$$

(A6.7)

However, it is valid that

$$\frac{p_{43}^{\circ}\,\overline{q}_{32}/q_{32\,max}}{1 - \overline{q}_{32}/q_{32\,max}} = \frac{p_{43}^{\circ}\,\overline{q}_{32}/q_{32\,max}}{1 - \overline{q}_{32}/q_{32\,max} + p_{43}^{\circ}\,\overline{q}_{32}/q_{32\,max} - p_{43}^{\circ}\,\overline{q}_{32}/q_{32\,max}}$$

$$= \frac{p_{43}^{\circ}\,\overline{q}_{32}/q_{32\,max}}{1 - \overline{q}_{32}/q_{32\,max}\left(1 - p_{43}^{\circ}\right) - p_{43}^{\circ}\,\overline{q}_{32}/q_{32\,max}} = \frac{\dfrac{p_{43}^{\circ}\,\overline{q}_{32}/q_{32\,max}}{1 - \overline{q}_{32}/q_{32\,max}\left(1 - p_{43}^{\circ}\right)}}{1 - \dfrac{p_{43}^{\circ}\,\overline{q}_{32}/q_{32\,max}}{1 - \overline{q}_{32}/q_{32\,max}\left(1 - p_{43}^{\circ}\right)}}$$

$$= \frac{\dfrac{1}{q_{32\,max}}\left[\dfrac{p_{43}^{\circ}\overline{q}_{32}}{1 - \overline{q}_{32}/q_{32\,max}\left(1 - p_{43}^{\circ}\right)}\right]}{1 - \dfrac{1}{q_{32\,max}}\left[\dfrac{p_{43}^{\circ}\overline{q}_{32}}{1 - \overline{q}_{32}/q_{32\,max}\left(1 - p_{43}^{\circ}\right)}\right]} = \frac{a/q_{32\,max}}{1 - a/q_{32\,max}}.$$

(A6.8)

The auxiliary variable a is obtained by rearranging Equations (A6.2) and (A6.4) as follows:

$$a = \frac{p_{43}^{\circ}\overline{q}_{32}}{1 - \frac{\overline{q}_{32}}{q_{32\,max}}\left(1 - p_{43}^{\circ}\right)} = \frac{\left(1 - \overline{q}_{43}/q_{43\,max}\right)^{q_{43\,max}}\overline{q}_{32}}{1 - \frac{\overline{q}_{32}}{q_{32\,max}}\left[1 - \left(1 - \overline{q}_{43}/q_{43\,max}\right)^{q_{43\,max}}\right]} = \frac{\left(1 - \frac{\overline{q}_{43}}{\overline{q}_{43}^{*}}\right)\overline{q}_{32}}{1 - \frac{\overline{q}_{32}}{q_{32\,max}}\frac{\overline{q}_{43}}{\overline{q}_{43}^{*}}}.$$

$$(A6.9)$$

By substituting Equation (A6.8) into (A6.7), the following expression for the probability $\beta(q_{32})$ is obtained:

$$\beta(q_{32}) = \binom{q_{32\,max}}{q_{32}}\left(\frac{a/q_{32\,max}}{1 - a/q_{32\,max}}\right)^{q_{32}}\left(1 - \overline{q}_{32}/q_{32\,max}\right)^{q_{32\,max}}$$

$$= \binom{q_{32\,max}}{q_{32}}\frac{\left(a/q_{32\,max}\right)^{q_{32}}}{\left(1 - a/q_{32\,max}\right)^{q_{32}}}\left(1 - \overline{q}_{32}/q_{32\,max}\right)^{q_{32\,max}}$$

$$= \binom{q_{32\,max}}{q_{32}}\left(\frac{a}{q_{32\,max}}\right)^{q_{32}}\frac{\left(1 - \overline{q}_{32}/q_{32\,max}\right)^{q_{32\,max}}}{\left(1 - a/q_{32\,max}\right)^{q_{32}}} \cdot \frac{\left(1 - a/q_{32\,max}\right)^{q_{32\,max}}}{\left(1 - a/q_{32\,max}\right)^{q_{32\,max}}},$$

$$\beta(q_{32}) = \left\{\binom{q_{32\,max}}{q_{32}}\left(\frac{a}{q_{32\,max}}\right)^{q_{32}}\left(1 - \frac{a}{q_{32\,max}}\right)^{q_{32\,max} - q_{32}}\right\}\frac{\left(1 - \overline{q}_{32}/q_{32\,max}\right)^{q_{32\,max}}}{\left(1 - a/q_{32\,max}\right)^{q_{32\,max}}}.$$

$$(A6.10)$$

Note: The expression written within the curly brackets follows the binomial distribution of the random variable q_{32} with mean value a and maximum value $q_{32\,max}$.

Relative frequency of empty clusters

The relative frequency of empty clusters p_{42}° (or the probability that cluster is empty, without considering the number of empty bundles formed) can be obtained as the summation $\sum_{q_{32}=0}^{q_{32\,max}}\beta(q_{32})$, based on the rules of theory of probability. The following expression is obtained from Equation (A6.10):

$$p_{42}^{\circ} = \sum_{q_{32}=0}^{q_{32\,max}} \beta(q_{32}) = \frac{\left(1 - \overline{q}_{32}/q_{32\,max}\right)^{q_{32\,max}}}{\left(1 - a/q_{32\,max}\right)^{q_{32\,max}}}$$

$$\cdot \overbrace{\sum_{q_{32}=0}^{q_{32\,max}} \left\{ \binom{q_{32\,max}}{q_{32}} \left(\frac{a}{q_{32\,max}}\right)^{q_{32}} \left(1 - \frac{a}{q_{32\,max}}\right)^{q_{32\,max} - q_{32}} \right\}}^{=1}$$

$$= \frac{\left(1 - \overline{q}_{32}/q_{32\,max}\right)^{q_{32\,max}}}{\left(1 - a/q_{32\,max}\right)^{q_{32\,max}}}. \tag{A6.11}$$

Then the relative frequency of the non-empty clusters p_{42}^{*} (or the probability that the cluster is not empty) is

$$p_{42}^{*} = 1 - p_{42}^{\circ} = 1 - \frac{\left(1 - \overline{q}_{32}/q_{32\,max}\right)^{q_{32\,max}}}{\left(1 - a/q_{32\,max}\right)^{q_{32\,max}}}. \tag{A6.12}$$

The mean number of fibers in a non-empty cluster

The empty clusters contain no fibers, i.e., $\overline{q}_{42}^{\circ} = 0$, and the average number of fibers in non-empty clusters is $\overline{q}_{42}^{*}$. The mean number of fibers $\overline{q}_{42}$ with all empty and non-empty clusters is $\overline{q}_{42} = p_{42}^{\circ}\overline{q}_{42}^{\circ} + p_{42}^{*}\overline{q}_{42}^{*}$. By applying the previous relationships, it is valid to write that

$$\overline{q}_{42} = p_{42}^{\circ}\overline{q}_{42}^{\circ} + p_{42}^{*}\overline{q}_{42}^{*} = \left[\frac{\left(1 - \overline{q}_{32}/q_{32\,max}\right)^{q_{32\,max}}}{\left(1 - a/q_{32\,max}\right)^{q_{32\,max}}}\right]0 + \left[1 - \frac{\left(1 - \overline{q}_{32}/q_{32\,max}\right)^{q_{32\,max}}}{\left(1 - a/q_{32\,max}\right)^{q_{32\,max}}}\right]\overline{q}_{42}^{*},$$

$$\overline{q}_{42}^{*} = \frac{\overline{q}_{42}}{1 - \frac{\left(1 - \overline{q}_{32}/q_{32\,max}\right)^{q_{32\,max}}}{\left(1 - a/q_{32\,max}\right)^{q_{32\,max}}}} = \overline{q}_{42}\frac{\left(1 - a/q_{32\,max}\right)^{q_{32\,max}}}{\left(1 - a/q_{32\,max}\right)^{q_{32\,max}} - \left(1 - \overline{q}_{32}/q_{32\,max}\right)^{q_{32\,max}}}.$$

$$\tag{A6.13}$$

The following expressions can be determined by using Equation (A6.9) and then the last expression in Equation (A6.13):

$$1 - \frac{a}{q_{32\,max}} = 1 - \frac{a}{\overline{q}_{32}} \frac{\overline{q}_{32}}{q_{32\,max}} = 1 - \frac{1 - \dfrac{\overline{q}_{43}}{\overline{q}_{43}^{*}}}{1 - \dfrac{\overline{q}_{32}}{q_{32\,max}} \dfrac{\overline{q}_{43}}{\overline{q}_{43}^{*}}} \frac{\overline{q}_{32}}{q_{32\,max}}$$

$$= \frac{1 - \dfrac{\overline{q}_{32}}{q_{32\,max}} \dfrac{\overline{q}_{43}}{\overline{q}_{43}^{*}} - \dfrac{\overline{q}_{32}}{q_{32\,max}} + \dfrac{\overline{q}_{32}}{q_{32\,max}} \dfrac{\overline{q}_{43}}{\overline{q}_{43}^{*}}}{1 - \dfrac{\overline{q}_{32}}{q_{32\,max}} \dfrac{\overline{q}_{43}}{\overline{q}_{43}^{*}}} = \frac{1 - \dfrac{\overline{q}_{32}}{q_{32\,max}}}{1 - \dfrac{\overline{q}_{32}}{q_{32\,max}} \dfrac{\overline{q}_{43}}{\overline{q}_{43}^{*}}},$$

$$1 - \frac{\left(1 - \dfrac{\overline{q}_{32}}{q_{32\,max}}\right)^{q_{32\,max}}}{\left(1 - \dfrac{a}{q_{32\,max}}\right)^{q_{32\,max}}} = 1 - \left(1 - \frac{\overline{q}_{32}}{q_{32\,max}}\right)^{q_{32\,max}} \frac{\left(1 - \dfrac{\overline{q}_{32}}{q_{32\,max}} \dfrac{\overline{q}_{43}}{\overline{q}_{43}^{*}}\right)^{q_{32\,max}}}{\left(1 - \dfrac{\overline{q}_{32}}{q_{32\,max}}\right)^{q_{32\,max}}}$$

$$= 1 - \left(1 - \frac{\overline{q}_{32}}{q_{32\,max}} \frac{\overline{q}_{43}}{\overline{q}_{43}^{*}}\right)^{q_{32\,max}},$$

$$\overline{q}_{42}^{*} = \frac{\overline{q}_{42}}{1 - \dfrac{\left(1 - \overline{q}_{32}/q_{32\,max}\right)^{q_{32\,max}}}{\left(1 - a/q_{32\,max}\right)^{q_{32\,max}}}} = \frac{\overline{q}_{42}}{1 - \left(1 - \dfrac{\overline{q}_{32}}{q_{32\,max}} \dfrac{\overline{q}_{43}}{\overline{q}_{43}^{*}}\right)^{q_{32\,max}}}. \qquad (A6.14)$$

Equation (A6.14) expresses the mean number of fibers in a non-empty cluster.

Limits

When the mean number of fibers in the bundle approaches to 0, the mean number of fibers in the non-empty bundle given by Equation (A6.4) can be expressed with the help of L'Hospital's rule of limit as follows:

$$\lim_{\overline{q}_{43} \to 0} \overline{q}_{43}^{*} = \lim_{\overline{q}_{43} \to 0} \frac{\overline{q}_{43}}{1 - \left(1 - \overline{q}_{43}/q_{43\,max}\right)^{q_{43\,max}}}$$

$$= \lim_{\overline{q}_{43} \to 0} \frac{1}{-q_{43\,max}\left(1 - \overline{q}_{43}/q_{43\,max}\right)^{q_{43\,max}-1}\left(-1/q_{43\,max}\right)}$$

$$= \lim_{\overline{q}_{43} \to 0} \frac{1}{\left(1 - \overline{q}_{43}/q_{43\,max}\right)^{q_{43\,max}-1}} = 1. \qquad (A6.15)$$

Similarly, when the mean number of fibers in the cluster approaches to 0, the mean number of fibers in the non-empty cluster, given by Equation (A6.14), can be expressed by applying the relationship $\bar{q}_{42} = \bar{q}_{43}\bar{q}_{32}$, according to Equations (6.81) and (A6 .15), with the help of L'Hospital's rule of limit as follows:

$$
\lim_{\substack{\bar{q}_{43} \to 0 \\ \bar{q}_{32} \to 0}} \bar{q}_{42}^* = \lim_{\bar{q}_{43} \to 0} \left\{ \lim_{\bar{q}_{32} \to 0} \bar{q}_{42}^* \right\} = \lim_{\bar{q}_{43} \to 0} \left[\lim_{\bar{q}_{32} \to 0} \frac{\bar{q}_{43}\bar{q}_{32}}{1 - \left(1 - \dfrac{\bar{q}_{32}}{q_{32\,max}}\dfrac{\bar{q}_{43}}{\bar{q}_{43}^*}\right)^{q_{32\,max}}} \right]
$$

$$
= \lim_{\bar{q}_{43} \to 0} \left[\bar{q}_{43} \lim_{\bar{q}_{32} \to 0} \frac{\bar{q}_{32}}{1 - \left(1 - \dfrac{\bar{q}_{32}}{q_{32\,max}}\dfrac{\bar{q}_{43}}{\bar{q}_{43}^*}\right)^{q_{32\,max}}} \right]
$$

$$
= \lim_{\bar{q}_{43} \to 0} \left[\bar{q}_{43} \lim_{\bar{q}_{32} \to 0} \frac{1}{-q_{32\,max}\left(1 - \dfrac{\bar{q}_{32}}{q_{32\,max}}\dfrac{\bar{q}_{43}}{\bar{q}_{43}^*}\right)^{q_{32\,max}-1}\left(-\dfrac{1}{q_{32\,max}}\dfrac{\bar{q}_{43}}{\bar{q}_{43}^*}\right)} \right]
$$

$$
= \lim_{\bar{q}_{43} \to 0} \left[\bar{q}_{43} \lim_{\bar{q}_{32} \to 0} \frac{1}{\left(1 - \dfrac{\bar{q}_{32}}{q_{32\,max}}\dfrac{\bar{q}_{43}}{\bar{q}_{43}^*}\right)^{q_{32\,max}-1}\dfrac{\bar{q}_{43}}{\bar{q}_{43}^*}} \right] = \lim_{\bar{q}_{43} \to 0} \left[\bar{q}_{43} \dfrac{1}{\dfrac{\bar{q}_{43}}{\bar{q}_{43}^*}} \right]
$$

$$
= \lim_{\bar{q}_{43} \to 0} \left[\bar{q}_{43}^* \right] = 1. \tag{A6.16}
$$

Each short fiber portion makes an angle ϑ_ζ with the direction of yarn axis (see Figure 7.5). There are two problems which must be solved in the context of equivalent fiber diameter:

1. How is the distribution of angle ϑ_ζ in the hairiness sphere?

2. How is the relation between angle ϑ_ζ and the sectional area of fiber in yarn cross-section, i.e., to what degree the first expression shown in Equation (7.35) is valid?

Both problems can be solved by using a prior hypothesis arising out of our subjective 'feeling' from the microscopic observations of yarns.

Distribution of angle ϑ_ζ

We do not have any direct information about the distribution of angle ϑ_ζ. Intuitively, the 'coat' of hairs on yarns reminds something like a carded 'web' of fibers. Then, we can also observe a trend for the preference of longitudinal direction. This evokes the idea that the distribution of angle ϑ_ζ can be estimated by the probability density function of fiber orientation in carded webs.

Such a probability density function, derived in Reference [1], is introduced by the following expression:

$$f(\vartheta_\zeta) = \frac{1}{\pi} \frac{C}{C^2 - (C^2 - 1)\cos^2 \vartheta_\zeta}, \quad \vartheta_\zeta \in (-\pi/2, \pi/2). \qquad (A7.1)$$

Figure A7.1 illustrates the graphical interpretation of the previous probability density function. The parameter $C \geq 1$ is a 'measure' of intensity of preferential orientation.

Based on our experimental experience reported in Reference [1], the typical value is approximately $C = 1.9$ for carded webs.

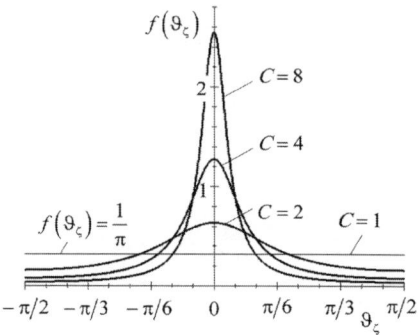

Figure A7.1 Probability density function of orientation of fiber segments according to Equation (A7.1)

Note: Let us note that at $\vartheta_\zeta = \pm\pi/2$ (i.e., $\vartheta_\zeta = \pm90°$), the function $f(\vartheta_\zeta)$ has the minimum but a positive value $1/(\pi C)$; at $C = 1.9$ it is 0.170.

Relations between angle ϑ_ζ and the sectional area of fiber

Figure A7.2 illustrates the sectional areas of differently arched (oblique) cylindrical fibers. The arrows show the directions of the sectioned fiber segments. The angles ϑ_ζ determine the directions of the fiber segments in relation to the vertical direction of the yarn axis.

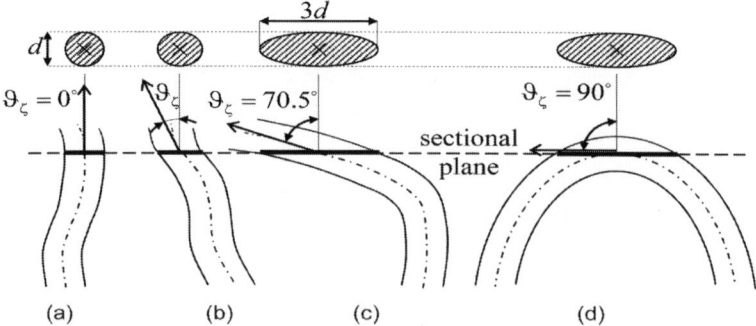

Figure A7.2 Sections of differently arched (oblique) cylindrical fibers

Fiber (a) follows $\vartheta_\zeta = 0$ and its sectional area $s^* = s = \pi d^2/4$ creates a circle having diameter d. fiber (b) follows a general angle ϑ_ζ ($\vartheta_\zeta < 70.5°$) and its sectional area is $s^* = s/\cos\vartheta_\zeta$ – see Equation (1.40). This fiber section looks like an ellipse with the major axis higher than d (but smaller than

$3d$). fiber (c) follows an angle $\vartheta_\zeta = 70.5°$ and its sectional area is $s^* = s/\cos 70.5°$. The major axis of the sectional ellipse is just $3d$ now. In each of these three cases, it is valid that the sectional area of fiber is $s^* = s/\cos \vartheta_\zeta$ (where $s = \pi d^2/4$).

It would show that if the angle ϑ_ζ is limited to $\pi/2$ (90°), then $\cos \vartheta_\zeta$ is limited to zero so that the sectional area of fiber is limited to infinity. But, we usually do not observe such an extremely long fiber section in yarn cross section[1]. Why is it so? The short fiber portion in fiber (d) follows the directional angle $\vartheta_\zeta = 90°$; however, the length of the fiber section (major axis of ellipse) remains $3d$; this is same as fiber (c). In opposite to Figure 1.8 and Equation (1.40) from Chapter 1, it is evident that $s^* \neq s/\cos \vartheta_\zeta$ in the present case. This is because the fiber creates a loop (or a wave) as shown in (d).

To simplify this problem, let us assume that

1. If the angle $\left|\vartheta_\zeta\right| \in \left(0°, 70.5°\right)$, then $s^* = s/\cos \vartheta_\zeta$ is valid.

2. If $\left|\vartheta_\zeta\right| \in \left(70.5°, 90°\right)$, then $s^* \neq s/\cos \vartheta_\zeta$ but

$$s^* = \text{constant} = s/\cos 70.5° = 3s.$$

Note: The simplified 'black-and-white' idea reflected our subjective experience that is gained through experimental observations of yarn cross sections. The long fiber sections occur itself very seldom and they are practically not too long. Nevertheless, the borderline value 70.5° (1.231 radian; $1/\cos 70.5° = 3$) refers to our subjective estimation only.

Mean fiber sectional area

If the expression $s^* = s/\cos \vartheta_\zeta$ would always be valid, then, according to Equation (A7.1), the mean of reciprocal values of cosines, i.e., the value of σ [according to Equations (7.13) and (7.26)] must be

---

1   Some sectional areas of fibers in yarn cross section are very long, for example, the 'bridge fibers' in OE yarns but (1) they lie probably in the inner sphere of the yarn and (2) this phenomenon is specific to OE yarns only.

$$\sigma = \int_{-\pi/2}^{\pi/2} \frac{1}{\cos \vartheta_\zeta} f\left(\vartheta_\zeta\right) d\vartheta_\zeta = \int_{-\pi/2}^{\pi/2} \frac{1}{\cos \vartheta_\zeta} \left[\frac{1}{\pi} \frac{C}{C^2 - \left(C^2 - 1\right)\cos^2 \vartheta_\zeta} \right] d\vartheta_\zeta$$

$$= 2\int_{0}^{\pi/2} \frac{1}{\cos \vartheta_\zeta} \left[\frac{1}{\pi} \frac{C}{C^2 - \left(C^2 - 1\right)\cos^2 \vartheta_\zeta} \right] d\vartheta_\zeta = \infty. \qquad (A7.2)$$

(The function $\cos \vartheta_\zeta$ is known to be an even function.)

However, Equation (7.2) is not right, considering our previous assumptions. Then, we must write that

$$\left. \begin{aligned} \sigma = 2 \int_{0}^{1.231} \frac{1}{\cos \vartheta_\zeta} \left[\frac{1}{\pi} \frac{C}{C^2 - \left(C^2 - 1\right)\cos^2 \vartheta_\zeta} \right] d\vartheta_\zeta \\ + 2 \int_{1.231}^{\pi/2} 3 \left[\frac{1}{\pi} \frac{C}{C^2 - \left(C^2 - 1\right)\cos^2 \vartheta_\zeta} \right] d\vartheta_\zeta. \end{aligned} \right\} \qquad (A7.3)$$

By using the value $C = 1.9$, recommended for the carded webs, we obtain the following result from Equation (A7.3)

$$\sigma = 1.478, \quad \sqrt{\sigma} = 1.216. \qquad (A7.4)$$

(Equation (A7.3) was solved numerically.)

Reference

[1] Neckář, B. and Das, D. Theory of Structure and Mechanics of Fibrous Assemblies, Woodhead Publishing India Pvt. Ltd., New Delhi, 2012.

Appendix 8

Radial functions of packing densities and contraction ratio

The radial function of the contraction ratio η_r determines the change of radial function of packing density from the initial function μ_r to the final function μ_r', followed by a yarn under tension. Clearly, if we know (e.g., experimentally) this change of packing density function, then we can deduce the function of contraction ratio.

Note: It is usually extremely difficult to determine exactly the initial and the final functions of yarn packing density experimentally in the present time, even if it is possible in principle. It might be possible that the future experimental methods will enable us to find it more easily.

Volume of differential layers

Let us imagine a general differential layer[1] of thickness dr, which is situated at a radius r in a non-tensioned (initial) yarn of length ζ – see Figure A8.1. The area of the corresponding differential annulus is $2\pi r\,dr$, and the total volume of this differential layer is

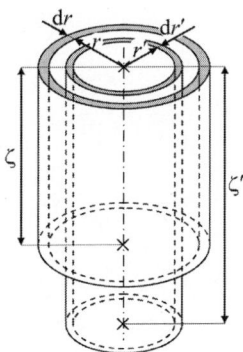

Figure A8.1 General differential layers before and after yarn tension

---

1   See Section 4.1 – Figure 4.3 and the corresponding text – for more details.

$$dV_c = 2\pi r \zeta \, dr \,. \tag{A8.1}$$

The same differential layer changes its position to a new layer of (smaller) radius r' and thickness dr', in a tensioned yarn of length ζ' – see Figure A8.1. The area of the corresponding differential annulus is $2\pi r' dr'$, and the total volume of the differential layer under tension is

$$dV_c' = 2\pi r' \zeta' dr'. \tag{A8.2}$$

We will use the relative quantities, i.e., axial strain of yarn ε_Y, radial strain of differential layer ε_r and contraction ratio η_r as stated follows:

$$\varepsilon_Y = (\zeta' - \zeta)/\zeta, \quad \varepsilon_r = (r' - r)/r, \quad \eta_r = -\varepsilon_r/\varepsilon_Y \,. \tag{A8.3}$$

Note: The same quantities are already determined in Equations (9.29) to (9.31). In place of the earlier used lengths $d\zeta$ and $d\zeta'$, we now use the finite lengths ζ and ζ'.

By using the above-mentioned relations, we can write the expressions as follows:

$$r' = (1 + \varepsilon_r)r = (1 - \eta_r \varepsilon_Y)r, \tag{A8.4}$$

$$\frac{dr'}{dr} = -\frac{d\eta_r}{dr}\varepsilon_Y r + (1 - \eta_r \varepsilon_Y) = 1 - \eta_r \varepsilon_Y - \frac{d\eta_r}{dr}\varepsilon_Y r. \tag{A8.5}$$

Further,

$$dV_c' = 2\pi r' \zeta' dr' = 2\pi \left[(1 - \eta_r \varepsilon_Y)r \right] \left[\zeta(1 + \varepsilon_Y) \right] \left(1 - \eta_r \varepsilon_Y - \frac{d\eta_r}{dr}\varepsilon_Y r \right) dr$$

$$= dV_c (1 - \eta_r \varepsilon_Y)(1 + \varepsilon_Y) \left(1 - \eta_r \varepsilon_Y - \frac{d\eta_r}{dr}\varepsilon_Y r \right),$$

$$\upsilon_c = \frac{dV_c'}{dV_c} = (1 - \eta_r \varepsilon_Y)(1 + \varepsilon_Y) \left(1 - \eta_r \varepsilon_Y - \frac{d\eta_r}{dr}\varepsilon_Y r \right). \tag{A8.6}$$

[Equations (A8.1) to (A8.5) were used.] The last quantity υ_c determines the change in volume of the differential layer as a consequence of straining of yarn.

Fiber volume

Let us think about one non-tensioned fiber of length l and cross-sectional area s – see the left-hand side scheme in Figure A8.2. The equivalent fiber diameter, according to Equation (1.6), is $d = \sqrt{4s/\pi}$. After elongation, the fiber obtains a new (longer) length l' and a new (smaller) cross-sectional area s' – see the right-hand side scheme in Figure A8.2. The new equivalent fiber diameter is $d' = \sqrt{4s'/\pi}$.

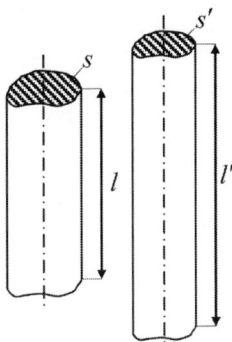

Figure A8.2 Fiber before and after tension

We introduce the fiber strain ε_f, change in equivalent fiber diameter ε_d and fiber contraction ratio $\eta_d(\varepsilon_f)$ as follows:

$$\varepsilon_f = \frac{l'-l}{l} = \frac{l'}{l} - 1, \quad l' = (1+\varepsilon_f)l, \tag{A8.7}^2$$

$$\varepsilon_d = \frac{d'-d}{d} = \frac{d'}{d} - 1, \quad d' = (1+\varepsilon_d)d, \tag{A8.8}$$

$$\eta_d(\varepsilon_f) = \frac{-\varepsilon_d}{\varepsilon_f}, \quad \varepsilon_d = -\varepsilon_f\,\eta_d(\varepsilon_f). \tag{A8.9}^3$$

The volumes of the non-tensioned and tensioned fibers are

$$V_1 = \frac{\pi d^2}{4}l, \quad V_1' = \frac{\pi d'^2}{4}l'. \tag{A8.10}$$

---

2   This quantity was also introduced in Equation (9.32), in place of differential lengths dl, dl', we now use the finite lengths l, l'.

3   The contraction ratio η_d of fiber generally changes its value in relation to fiber strain ε_f; η_d is a function of ε_f, $\eta_d = \eta_d(\varepsilon_f)$. This function can be determined in principle experimentally by progressive straining of the fiber and microscopic measurements of the corresponding fiber diameters.

By using Equations (A8.7) to (A8.10), we obtain

$$V_1' = \frac{\pi d'^2}{4} l' = \frac{\pi}{4}\left[\left(1+\varepsilon_d\right)d\right]^2\left[\left(1+\varepsilon_f\right)l\right]$$

$$= V_1\left(1+\varepsilon_d\right)^2\left(1+\varepsilon_f\right) = V_1\left[1-\varepsilon_f\,\eta_d\left(\varepsilon_f\right)\right]^2\left(1+\varepsilon_f\right),$$

$$\upsilon_f = \frac{V_1'}{V_1} = \left[1-\varepsilon_f\,\eta_d\left(\varepsilon_f\right)\right]^2\left(1+\varepsilon_f\right).\tag{A8.11}$$

The last quantity υ_f determines the change of fiber volume as a consequence of straining of yarn.

Note: If (1) fiber volume is not changed by straining ($V_1 = V_1'$), i.e.,

$$\upsilon_f = 1,\tag{A8.12}$$

and (2) fiber strain is very small ($\varepsilon_f \ll$, i.e., $\varepsilon_f^2 \to 0$ and $\varepsilon_f^3 \to 0$), then the following approximate relation is valid from Equations (A8.11) and (A8.12):

$$1 = \left[1-\varepsilon_f\,\eta_d\left(\varepsilon_f\right)\right]^2\left(1+\varepsilon_f\right) = \left[1 - 2\varepsilon_f\,\eta_d\left(\varepsilon_f\right) + \varepsilon_f^2\,\eta_d^2\left(\varepsilon_f\right)\right]\left(1+\varepsilon_f\right)$$

$$= 1 - 2\varepsilon_f\,\eta_d\left(\varepsilon_f\right) + \overbrace{\varepsilon_f^2\,\eta_d^2\left(\varepsilon_f\right)}^{\to 0} + \varepsilon_f - \overbrace{2\varepsilon_f^2\,\eta_d\left(\varepsilon_f\right)}^{\to 0} + \overbrace{\varepsilon_f^3\,\eta_d^2\left(\varepsilon_f\right)}^{\to 0}$$

$$= 1 - 2\varepsilon_f\,\eta_d\left(\varepsilon_f\right) + \varepsilon_f, \quad \eta_d\left(\varepsilon_f\right) = 0.5.\tag{A8.13}$$

(This is a known result in the theory of technical mechanics.)

Differential equation of contraction ratio η_r

The differential layer in the initial (non-tensioned) yarn – Figure A8.1 – has packing density μ_r which is (generally) a corresponding function of r. The fiber volume inside this differential layer is

$$dV = dV_c\,\mu_r.\tag{A8.14}$$

Similarly, the differential layer in the tensioned yarn has (another) packing density $\mu_{r'}' \equiv \mu'(r')$.

Note: It is possible to measure experimentally the packing densities at different radii r in the initial (non-tensioned) yarn. On the contrary, it can also be (in principle) possible to measure experimentally the packing density

$\mu'_{r'} \equiv \mu'(r')$ [4] in the tensioned yarn but as a function of transformed radii r' only. [Equation (A8.4) is valid between initial radius r and the corresponding radius r' after tension.]

The fiber volume inside the differential layer of the yarn under tension is then

$$dV' = dV'_c \mu'_{r'} = dV'_c \mu'(r'). \tag{A8.15}$$

The ratio of the last two expressions can now be written as

$$\frac{dV'}{dV} = \frac{dV'_c}{dV_c} \frac{\mu'(r')}{\mu_r}. \tag{A8.16}$$

The fibers that create the fibrous material in the given differential layer of the initial yarn have the same helical geometry, i.e., the same angle β at the same radius r – see Figure 4.2 and Equations (4.1) and (4.2) in Chapter 4. Therefore, they also have the same fiber strain ε_f according to Equation (9.37) in the yarn under tension. In this case, we can consider that the ratio of the elemental fiber volumes dV'/dV in the differential layers is same as the ratio $\upsilon_f = V'_1/V_1$ according to Equation (A8.11), derived for one fiber from Figure A8.2. So, we can express the following:

$$\overbrace{\left(\frac{dV'}{dV}\right)}^{=V'_f/V_f=\upsilon_f} = \overbrace{\left(\frac{dV'_c}{dV_c}\right)}^{=\upsilon_c} \frac{\mu'(r')}{\mu_r}, \quad \upsilon_f = \upsilon_c \frac{\mu'(r')}{\mu_r},$$

$$\left[1 - \varepsilon_f \eta_d(\varepsilon_f)\right]^2 (1 + \varepsilon_f) = (1 - \eta_r \varepsilon_Y)(1 + \varepsilon_Y)\left(1 - \eta_r \varepsilon_Y - \frac{d\eta_r}{dr} \varepsilon_Y r\right) \frac{\mu'(r')}{\mu_r},$$

where $r' = (1 - \eta_r \varepsilon_Y) r,$

$$\varepsilon_f = \sqrt{1 + 2\varepsilon_Y\left(\cos^2 \beta - \eta_r \sin^2 \beta\right) + \varepsilon_Y^2\left(\cos^2 \beta + \eta_r^2 \sin^2 \beta\right)} - 1,$$

and $\tan \beta = 2\pi r Z.$

$$\left.\begin{array}{r}\end{array}\right\} \tag{A8.17}$$

[Equations (A8.4), (A8.11), (A8.6), (9.37) and (4.1) were used.]

---

4   Because to state that μ' can be only measured experimentally as a function of radii r', we use the symbol $\mu'(r')$ besides the shorter symbol $\mu'_{r'}$.

If we know (1) yarn twist Z, (2) actual yarn strain ε_Y, (3) function of fiber contraction ratio $\eta_d(\varepsilon_f)$, (4) radial function of packing density μ_r (in the initial, non-tensioned yarn) and (5) radial function of packing densities $\mu'_{r} \equiv \mu'(r')$ (in tensioned yarn), then Equation (A8.17) can be in principle solved as an ordinary differential equation type $f(d\eta_r/dr, \eta_r, r) = 0$.

Boundary condition

Equation (A8.4) is valid for all radii in the yarn, including the radii on yarn surfaces. In the above-mentioned case, $r = D/2$, $r' = D'/2$, where D and D' are yarn diameters of initial (non-tensioned) and tensioned yarns. Thereafter, Equation (A8.4) can be obtained in the following special form:

$$D'/2 = \left(1 - \eta_{r=D/2}\varepsilon_Y\right)D/2, \quad \eta_{r=D/2} = \frac{1 - D'/D}{\varepsilon_Y}. \qquad (A8.18)$$

The diameters D and D' can be known experimentally. Equation (A8.18) creates a boundary condition for the differential equation (A8.17).

Note: The main problem in complete application of the described method for determination of radial function of contraction ratio η_r lies in the fact that the inputs required for this are not possible to determine precisely at the present time. If, in future, we will have a more precise and faster experimental method then the described method can be fully used.

Special cases

Let us assume that the fiber volume is not (significantly) changed (a) in any radius and (b) in any value of yarn strain ε_Y – assumption 1. Then, the change of volume of fiber υ_f, determined according to Equation (A8.11), is equal to 1. [If further small deformations are assumed then Equation (A8.13) is valid.] Then, the differential equation (A8.17) obtains the following special shape:

$$\left.\begin{aligned} &1 = \left(1 - \eta_r\varepsilon_Y\right)\left(1 + \varepsilon_Y\right)\left(1 - \eta_r\varepsilon_Y - \frac{d\eta_r}{dr}\varepsilon_Y r\right)\frac{\mu'(r')}{\mu_r}, \\ &\text{where } r' = \left(1 - \eta_r\varepsilon_Y\right)r. \end{aligned}\right\} \qquad (A8.19)$$

More, if the contraction ratio η_r possesses a constant value for all radii[5] (at a given value of yarn strain ε_Y) – assumption 2 – then the following expressions are valid from Equation (A8.19):

$$\left.\begin{array}{l}
1 = \left(1 - \eta_r \varepsilon_Y\right)^2 \left(1 + \varepsilon_Y\right)\dfrac{\mu'(r')}{\mu_r}, \\[3mm]
\dfrac{\mu'(r')}{\mu_r} = \dfrac{1}{\left(1 - \eta_r \varepsilon_Y\right)^2 \left(1 + \varepsilon_Y\right)} \ldots \text{independent to } r, \\[3mm]
\dfrac{r'}{r} = 1 - \eta_r \varepsilon_Y \ldots \text{independent to } r.
\end{array}\right\} \qquad (A8.20)$$

Further, the following equations are valid in this case according to Equation (A8.18):

$$\eta_r = \eta_{r=D/2} = \frac{1 - D'/D}{\varepsilon_Y}, \quad \left(\eta_r \varepsilon_Y = 1 - \frac{D'}{D}\right). \qquad (A8.21)$$

$$\frac{\mu'(r')}{\mu_r} = \frac{1}{\dfrac{D'^2}{D^2}\left(1 + \varepsilon_Y\right)}. \qquad (A8.22)$$

Note: Equation (A8.21) allows us to determine the corresponding (constant) value of the contraction ratio η_r when we experimentally know yarn diameters D and D' for a given value of yarn strain ε_Y.

Let us further assume that the change of volume of each differential layer υ_c is equal to one (no change) for each yarn strain ε_Y – assumption 3 – and yarn strains ε_Y are small – assumption 4. According to Equation (A8.6), the change of volume of each differential layer is

$$\upsilon_c = \left(1 - \eta_r \varepsilon_Y\right)^2 \left(1 + \varepsilon_Y\right) = \left(1 - 2\eta_r \varepsilon_Y + \eta_r^2 \varepsilon_Y^2\right)^2 \left(1 + \varepsilon_Y\right)$$

$$= 1 - 2\eta_r \varepsilon_Y + \overset{\to 0}{\eta_r^2 \varepsilon_Y^2} + \varepsilon_Y - \overset{\to 0}{2\eta_r \varepsilon_Y^2} + \overset{\to 0}{\eta_r^2 \varepsilon_Y^3}, \qquad (A8.23)$$

$$\upsilon_c \doteq 1 - \varepsilon_Y\left(2\eta_r - 1\right).$$

---

5   We introduce this assumption only for mathematical simplification of the stated problem.

(Assumptions 2 and 4 are used.) Because $\upsilon_c = 1$ in this case, it must be valid

$$1 \doteq 1 - \varepsilon_Y \left(2\eta_r - 1\right), \quad \eta_r = 0.5. \tag{A8.24}$$

[Compare this also with the derivation of Equation (A8.13).] By applying the last two equations in Equation (A8.20), we find

$$\frac{\mu'\left(r'\right)}{\mu_r} \doteq \frac{1}{1 - \varepsilon_Y \left(2\eta_r - 1\right)} = \frac{1}{1 - \varepsilon_Y \left(2 \cdot 0.5 - 1\right)} = 1. \tag{A8.25}$$

The packing density of (each) differential layer is not changed by yarn straining in this special case.

Solution of integral $\int_0^{\beta_D}\left(\cos^2\beta-\eta\sin^2\beta\right)\tan\beta\,d\beta$.

It is valid to write that

$$\int_0^{\beta_D}\left(\cos^2\beta-\eta\sin^2\beta\right)\tan\beta\,d\beta=\int_0^{\beta_D}\cos\beta\sin\beta\,d\beta-\eta\int_0^{\beta_D}\frac{\sin^3\beta}{\cos\beta}\,d\beta.\ \ (A9.1)$$

Let us at first solve the corresponding indefinite integrals. (The integral constant is marked by C.) We can derive

$$\int\cos\beta\sin\beta\,d\beta=\int t\,dt=\frac{t^2}{2}+C=\frac{\sin\beta^2}{2}+C=\frac{1-\cos\beta^2}{2}+C,\ \ (A9.2)$$

Substitution: $\sin\beta=t,\ \cos\beta\,d\beta=dt.$

$$\int\frac{\sin^3\beta}{\cos\beta}\,d\beta=\int\frac{\left(1-\cos^2\beta\right)}{\cos\beta}\sin\beta\,d\beta=\int\frac{\left(1-t^2\right)}{t}(-dt)=-\int\frac{dt}{t}+\int t\,dt$$

Substitution: $\cos\beta=t,\ -\sin\beta\,d\beta=dt,$

$$=-\ln|t|+\frac{t^2}{2}+C=-\ln|\cos\beta|+\frac{\cos^2\beta}{2}+C.\ \ (A9.3)$$

$$\int\cos\beta\sin\beta\,d\beta-\eta\int\frac{\sin^3\beta}{\cos\beta}\,d\beta=\frac{1-\cos\beta^2}{2}-\eta\left[-\ln|\cos\beta|+\frac{\cos^2\beta}{2}\right]+C$$

$$=\frac{1-\cos\beta^2}{2}+\eta\ln|\cos\beta|-\eta\frac{\cos^2\beta}{2}+C$$

$$=\frac{1}{2}\left[\overbrace{1-\eta+\eta}^{\text{added}}-\cos^2\beta+\eta 2\overbrace{\ln|\cos\beta|}^{=\ln\cos^2\beta}-\eta\cos^2\beta\right]+C$$

$$=\frac{1}{2}\left[-\eta+(1+\eta)\left(1-\cos^2\beta\right)+\eta\ln\cos^2\beta\right]+C.\ \ (A9.4)$$

The definite integral, according to Equation (A9.1), is then

$$\int\limits_0^{\beta_D} \left(\cos^2\beta - \eta\sin^2\beta\right)\tan\beta\,d\beta = \int\limits_0^{\beta_D} \cos\beta\sin\beta\,d\beta - \eta\int\limits_0^{\beta_D} \frac{\sin^3\beta}{\cos\beta}\,d\beta$$

$$= \frac{1}{2}\left[-\eta + (1+\eta)\overbrace{\left(1-\cos^2\beta_D\right)}^{=\sin^2\beta_D} + \eta\ln\cos^2\beta_D\right]$$

$$-\frac{1}{2}\left[-\eta + (1+\eta)\overbrace{\left(1-\cos^2 0\right)}^{=0} + \eta\overbrace{\ln\cos^2 0}^{=0}\right]$$

$$= \frac{1}{2}\left[(1+\eta)\sin^2\beta_D + \eta\ln\cos^2\beta_D\right] = \frac{\tan^2\beta_D}{2}\left[(1+\eta)\cos^2\beta_D + \eta\frac{\ln\cos^2\beta_D}{\tan^2\beta_D}\right].$$

$$(A9.5)$$

Oriented angle ψ

In Equations (9.78), (9.80) and (9.81), the following expression is used:

$$\left.\begin{aligned}
& I = \int_0^{\vartheta_u} f(\vartheta) u(\vartheta) d\vartheta, \\
& \text{where } f(\vartheta) = \sigma_f(\varepsilon_f) \cos^2 \vartheta, \\
& u(\vartheta) = \frac{1}{\pi} \frac{C}{C^2 - (C^2 - 1)\cos^2(\vartheta + \beta)} \\
& \qquad + \frac{1}{\pi} \frac{C}{C^2 - (C^2 - 1)\cos^2(\vartheta - \beta)}, \quad C \ge 1, \\
& \varepsilon_f = \varepsilon_Y \left(\cos^2 \vartheta - \eta \sin^2 \vartheta\right), \\
& \vartheta_u = \arcsin\left(1/\sqrt{1+\eta}\right).
\end{aligned}\right\} \qquad (A10.1)$$

[Equations (9.66), (9.73) and (9.74) are also presented in the previous expression.] It is evident that:

- ε_f is a periodic even function of J with a period of $\pi/2$.
- Therefore, $\sigma_f(\varepsilon_f)$ is also a periodic even function of ϑ with a period of $\pi/2$.
- In addition, $f(\vartheta)$ is a periodic even function of ϑ with a period of $\pi/2$.

So we can rearrange the previous integral I as follows:

$$I = \int_0^{\vartheta_u} f(\vartheta) u(\vartheta) d\vartheta$$

$$= \int_0^{\vartheta_u} f(\vartheta) \left[\frac{1}{\pi} \frac{C}{C^2 - (C^2 - 1)\cos^2(\vartheta + \beta)} + \frac{1}{\pi} \frac{C}{C^2 - (C^2 - 1)\cos^2(\vartheta - \beta)} \right] d\vartheta$$

$$= \int_0^{\vartheta_u} f(\vartheta) \frac{1}{\pi} \frac{C}{C^2 - (C^2 - 1)\cos^2(\vartheta + \beta)} d\vartheta$$

$$+ \int_0^{\vartheta_u} f(\vartheta) \frac{1}{\pi} \frac{C}{C^2 - (C^2 - 1)\cos^2(\vartheta - \beta)} d\vartheta$$

Let us rename ϑ to ξ and also ϑ to ψ.

$$= \int_0^{\vartheta_u} f(\xi) \frac{1}{\pi} \frac{C}{C^2 - (C^2 - 1)\cos^2(\xi + \beta)} d\xi$$

$$+ \int_0^{\vartheta_u} f(\psi) \frac{1}{\pi} \frac{C}{C^2 - (C^2 - 1)\cos^2(\psi - \beta)} d\psi$$

$$= - \int_{\vartheta_u}^0 f(\xi) \frac{1}{\pi} \frac{C}{C^2 - (C^2 - 1)\cos^2(\xi + \beta)} d\xi$$

$$+ \int_0^{\vartheta_u} f(\psi) \frac{1}{\pi} \frac{C}{C^2 - (C^2 - 1)\cos^2(\psi - \beta)} d\psi$$

Substitution: $\xi = -\psi$, $d\xi = -d\psi$

$$= \int_{-\vartheta_u}^0 \overbrace{f(-\psi)}^{\text{even function}} \frac{1}{\pi} \frac{C}{C^2 - (C^2 - 1)\underbrace{\cos^2(-\psi + \beta)}_{\text{even function}}} d\psi$$

$$+ \int_0^{\vartheta_u} f(\psi) \frac{1}{\pi} \frac{C}{C^2 - (C^2 - 1)\cos^2(\psi - \beta)} d\psi$$

$$= \int_{-\vartheta_u}^0 f(\psi) \frac{1}{\pi} \frac{C}{C^2 - (C^2 - 1)\cos^2(\psi - \beta)} d\psi$$

$$+ \int_0^{\vartheta_u} f(\psi) \frac{1}{\pi} \frac{C}{C^2 - (C^2 - 1)\cos^2(\psi - \beta)} d\psi,$$

$$\left. \begin{array}{l} I = \int_{-\vartheta_u}^{\vartheta_u} f(\psi) \dfrac{1}{\pi} \dfrac{C}{C^2 - (C^2 - 1)\cos^2(\psi - \beta)} d\psi, \\[4mm] \text{where} \quad f(\psi) = \sigma_f(\varepsilon_f)\cos^2\psi, \\[2mm] \qquad\quad \varepsilon_f = \varepsilon_Y(\cos^2\psi - \eta\sin^2\psi). \end{array} \right\} \qquad (A10.2)$$

Probability density function of ψ

In the last equation, the partial expression

$$g(\psi) = \frac{1}{\pi}\frac{C}{C^2 - (C^2-1)\cos^2(\psi-\beta)}$$

$$= \frac{1}{\pi}\frac{C}{C^2 - (C^2-1)\left[1 - \sin^2(\psi-\beta)\right]}$$

$$= \frac{1}{\pi}\frac{C}{\cos^2(\psi-\beta) + C^2\sin^2(\psi-\beta)}$$

$$= \frac{1}{\pi}\frac{C}{\left[1 + C^2\tan^2(\psi-\beta)\right]\cos^2(\psi-\beta)},$$

$$\psi \in (-\pi/2, \pi/2). \tag{A10.3}$$

represents the probability density function[1] of (oriented) angle ψ with parameters C and β.

Distribution function

The distribution function $G(\psi)$, corresponding to the probability density function $g(\psi)$ according to Equation (A10.3), is defined by the following expression:

$$G(\psi) = \int g(\psi)\,d\psi + Q = \frac{1}{\pi}\int \frac{C}{\left[1 + C^2\tan^2(\psi-\beta)\right]\cos^2(\psi-\beta)}\,d\psi + Q,$$

$$\tag{A10.4}$$

where the integral constant Q follows the assumption $G(-\pi/2) = 0$. Let us remind that $G(\psi)$ is defined in the interval $\psi \in \langle -\pi/2, \pi/2 \rangle$. The corresponding indefinite integral can be derived as follows:

---

1   This probability density function is derived and discussed by Equations (3.44) and (3.45) in the book [1].

$$\frac{1}{\pi}\int \frac{C}{\left[1+C^2\tan^2(\psi-\beta)\right]\cos^2(\psi-\beta)}\,d\psi$$

Substitution: $\psi-\beta=\varphi$, $d\psi=d\varphi$,

$$=\frac{1}{\pi}\int \frac{C}{\left[1+C^2\tan^2\varphi\right]\cos^2\varphi}\,d\varphi=\frac{1}{\pi}\int \frac{C}{\left[1+x^2\right]\cos^2\varphi}\frac{\cos^2\varphi}{C}\,dx$$

Substitution: $x=C\tan\varphi$, $dx=\dfrac{C}{\cos^2\varphi}\,d\varphi$,

$$=\frac{1}{\pi}\int \frac{dx}{\left[1+x^2\right]}=\frac{1}{\pi}\arctan[x]=\frac{1}{\pi}\arctan\left[C\tan(\varphi+n\pi)\right],$$

$$G(\psi)=\frac{1}{\pi}\arctan\left[C\tan(\psi-\beta+n\pi)\right]+Q.$$

$$(A10.5)$$

Note: The function $\tan\varphi$ is a periodic function with a period of π. Therefore, it is valid to write that $\tan\varphi=\tan(\varphi+n\pi)$ for each integer n.

We assume that

(a) The angles β are acute angles, where $\beta\in(0,\pi/2)$.

(b) It is possible to obtain the right-hand side value of tangent (command 'tan') when the angular value lies in the interval $(-\pi/2,\pi/2)$.

Then, the suitable integer value n and the integral constant Q must be determined as follows.

Case 1

Let us think about the distribution function lying in the interval $\psi\in(-\pi/2,-\pi/2+\beta)$. Thus, it is valid that $(\psi-\beta)\in(-\pi/2-\beta,-\pi/2)$. The quantity $\psi-\beta$ takes the smaller values than $-\pi/2$ in this case. However, by selecting $n=1$ in Equation (A10.5), we obtain the following expression:

$$G(\psi)=\frac{1}{\pi}\arctan\left[C\tan(\psi-\beta+\pi)\right]+Q,\qquad (A10.6)$$

where all angles $\psi-\beta+\pi\in(\pi/2-\beta,\pi/2)$ lie in the required interval $(-\pi/2,\pi/2)$.

The integral constant Q follows the requirement $G(-\pi/2) = 0$. By using Equation (A10.5), we find

$$G\left(-\frac{\pi}{2}\right) = 0 = \frac{1}{\pi}\arctan\left[C\tan\left(-\frac{\pi}{2} - \beta + \pi\right)\right] + Q$$

$$= \frac{1}{\pi}\arctan\left[C\tan\left(\frac{\pi}{2} - \beta\right)\right] + Q, \quad Q = -\frac{1}{\pi}\arctan\left[C\tan\left(\frac{\pi}{2} - \beta\right)\right].$$

$$\text{(A10.7)}$$

Then,

$$G(\psi) = \frac{1}{\pi}\arctan\left[C\tan\left(\psi - \beta + \pi\right)\right] - \frac{1}{\pi}\arctan\left[C\tan\left(\frac{\pi}{2} - \beta\right)\right]. \quad \text{(A10.8)}$$

The upper limit of ψ can be obtained as follows:

$$G\left(-\frac{\pi}{2} + \beta\right) = \frac{1}{\pi}\arctan\left[C\tan\left(-\frac{\pi}{2} + \beta - \beta + \pi\right)\right] - \frac{1}{\pi}\arctan\left[C\tan\left(\frac{\pi}{2} - \beta\right)\right]$$

$$= \frac{1}{2} - \frac{1}{\pi}\arctan\left[C\tan\left(\frac{\pi}{2} - \beta\right)\right]. \quad \text{(A10.9)}$$

Case 2

Let us think about the distribution function lying in the interval $\psi \in (-\pi/2 + \beta, \pi/2)$. Thus, it is valid that $(\psi - \beta) \in (-\pi/2, \pi/2 - \beta)$. Here, all the quantities $\psi - \beta$ take the values from required interval $(-\pi/2, \pi/2)$. Therefore, we use integer $n = 0$ in this case and from Equation (A10.5), we obtain

$$G(\psi) = \frac{1}{\pi}\arctan\left[C\tan\left(\psi - \beta\right)\right] + Q. \quad \text{(A10.10)}$$

We determine the integral constant Q from the assumption that the distribution function $G(-\pi/2 + \beta)$ – with the lower limit of the current interval and the upper limit of the previous interval – must be equal to the expression stated in Equation (A10.9). So

$$G\left(-\frac{\pi}{2} + \beta\right) = \frac{1}{2} - \frac{1}{\pi}\arctan\left[C\tan\left(\frac{\pi}{2} - \beta\right)\right] = \frac{1}{\pi}\arctan\left[C\tan\left(-\frac{\pi}{2} + \beta - \beta\right)\right] + Q,$$

$$Q = \frac{1}{2} - \frac{1}{\pi}\arctan\left[C\tan\left(\frac{\pi}{2}-\beta\right)\right] - \frac{1}{\pi}\arctan\left[\overbrace{\underbrace{C\tan\left(-\frac{\pi}{2}\right)}_{=-\infty}}^{=-\pi/2}\right] \quad \text{(A10.11)}$$

$$= 1 - \frac{1}{\pi}\arctan\left[C\tan\left(\frac{\pi}{2}-\beta\right)\right].$$

By substituting Q from the last equation to Equation (A10.10), we get

$$G(\psi) = \frac{1}{\pi}\arctan\left[C\tan(\psi-\beta)\right] - \frac{1}{\pi}\arctan\left[C\tan\left(\frac{\pi}{2}-\beta\right)\right] + 1. \quad \text{(A10.12)}$$

Note: If $\psi = \pi/2$, then $G(\pi/2) = 1$, as expected.

Figure A10.1 illustrates an example of distribution function $G(\psi)$ according to Equation (A10.8) – case 1, and Equation (A10.12) – case 2.

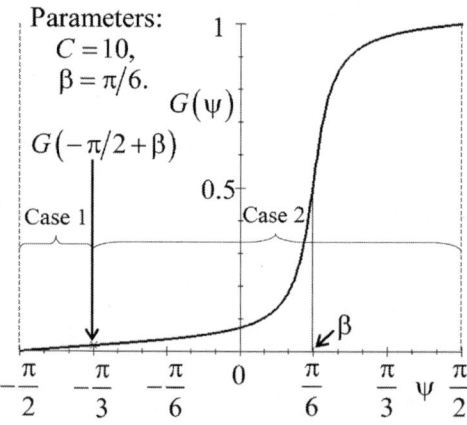

Figure 10.1 Example of distribution function $G(\psi)$

Inverse of function $G(\psi)$

In addition, let us solve for the value y corresponding to a given value of distribution function $G(\psi)$.

1. If $G(\psi) \leq G(-\pi/2+\beta)$ – see Equation (A10.9), then Equation (A10.8) is valid (case 1). Thus,

$$\arctan\left[C\tan(\psi-\beta+\pi)\right]=\pi G(\psi)+\arctan\left[C\tan\left(\frac{\pi}{2}-\beta\right)\right],$$

$$\tan(\psi-\beta+\pi)=\frac{\tan\left\{\pi G(\psi)+\arctan\left[C\tan\left(\frac{\pi}{2}-\beta\right)\right]\right\}}{C},$$

$$\psi=\arctan\frac{\tan\left\{\pi G(\psi)+\arctan\left[C\tan\left(\frac{\pi}{2}-\beta\right)\right]\right\}}{C}+\beta-\pi. \qquad (A10.13)$$

2. If $G(\psi)>G(-\pi/2+\beta)$ – see Equation (A10.9), then Equation (A10.12) is valid (case 2). Thus,

$$\arctan\left[C\tan(\psi-\beta)\right]=\pi\left[G(\psi)-1\right]+\arctan\left[C\tan\left(\frac{\pi}{2}-\beta\right)\right],$$

$$\tan(\psi-\beta)=\frac{\tan\left\{\pi\left[G(\psi)-1\right]+\arctan\left[C\tan\left(\frac{\pi}{2}-\beta\right)\right]\right\}}{C},$$

$$\psi=\arctan\frac{\tan\left\{\pi\left[G(\psi)-1\right]+\arctan\left[C\tan\left(\frac{\pi}{2}-\beta\right)\right]\right\}}{C}+\beta. \qquad (A10.14)$$

Numerical solution of integral I from Equation (A10.2)

The traditional method of numerical integration usually uses a constant step of integration variable ψ. However, it is often not the best idea for solving an integral. It is used to be better to apply variable steps of angle y which correspond to constant steps of distribution function $G(\psi)$. To each value of $G(\psi)$, we can calculate the corresponding value of ψ using Equation (A10.13) – if $G(\psi)\leq G(-\pi/2+\beta)$, or Equation (A10.14) – if $G(\psi)>G(-\pi/2+\beta)$.

Reference

[1] Neckář, B. and Das, D., Theory of Structure and Mechanics of Fibrous Assemblies, Woodhead Publishing India Pvt. Ltd., New Delhi, 2012.

Appendix 11

Normal variables and their distributions and characteristics

Distribution of linearly transformed yarn strength

The linearly transformed yarn strength is defined according to Equation (10.13). If this equation is differentiated, then the following expression is obtained:

$$\left.\begin{array}{l} u = \dfrac{P - \overline{P}_{l_0}}{\sigma_{l_0}}, \quad P = \sigma_{l_0} u + \overline{P}_{l_0}, \quad dP = \sigma_{l_0} du, \\[2em] \left(u_{\min} = \dfrac{P_{\min} - \overline{P}_{l_0}}{\sigma_{l_0}}, \quad u_{\max} = \dfrac{P_{\max} - \overline{P}_{l_0}}{\sigma_{l_0}} \right). \end{array}\right\} \tag{A11.1}$$

[Let us remind that P denotes random yarn strength, measured on an arbitrary chosen gauge length l, which is generally different from l_0. $\overline{P}_{l_0}$ and σ_{l_0} are two constants, i.e., $\overline{P}_{l_0}$ denotes the mean yarn strength and σ_{l_0} indicates the standard deviation of yarn strength. Both of them are related to the (short) gauge length l_0.]

If the random yarn strength at a general gauge length l is P, then the probability density function, according to Equation (10.4), is $f(P,l)$. The corresponding probability density function of linearly transformed random quantity u is then $g(u,l)$. Based on the rules of theory of probability, it must be valid to write that

$$f(P,l)\,dP = g(u,l)\,du. \tag{A11.2}$$

By using expressions from Equations (A11.1) and (A11.2), we obtain

$$g(u,l) = \frac{dP}{du} f(P,l) = \sigma_{l_0} f(P,l). \tag{A11.3}$$

The corresponding distribution function of P is $F(P,l)$ – see, e.g., Equation (10.3). The distribution function $G(u,l)$ of the corresponding linearly transformed random quantity u is found as follows:

$$F(P,l) = \int_{P_{min}}^{P} f(Q,l)\,\mathrm{d}Q = \int_{u_{min}}^{u}\left[g(v,l)\frac{1}{\sigma_{l_0}}\right]\left[\sigma_{l_0}\,\mathrm{d}v\right] = \int_{u_{min}}^{u} g(v,l)\,\mathrm{d}v = G(u,l),$$

Substitution: $Q = \sigma_{l_0} v + \overline{P}_{l_0}$, $\mathrm{d}Q = \sigma_{l_0}\,\mathrm{d}v$,

Lower limit: $v_{lover} = \dfrac{P_{min} - \overline{P}_{l_0}}{\sigma_{l_0}} = u_{min}$,   upper limit: $v_{upper} = \dfrac{P - \overline{P}_{l_0}}{\sigma_{l_0}} = u.$

(A11.4)

[Equation (A11.3) is used for rearrangement. Moreover, this substitution corresponds to Equation (A11.1). Nevertheless, the integration variables u and P are denoted by symbols v and Q, respectively, to avoid any ambiguity among the variables.]

Special case of Gaussian standard distribution by $l = l_0$

If (1) the yarn strengths P are especially related to the (short) gauge length $l = l_0$ and (2) such yarn strength follows normal (Gaussian) distribution, according Equations (10.9) and (10.10) – Peirce's assumption – then the linearly transformed yarn strength u corresponds to the so-called standardized random quantity and its distribution refers to the so-called standard normal distribution. We will denote the standardized probability density function $g(u,l_0)$ by a special symbol $\varphi(u)$ and the standardized distribution function $G(u,l_0)$ by a special symbol $\Phi(u)$. We will use the following expressions:

$$g(u,l_0) \equiv \varphi(u), \quad G(u,l_0) \equiv \Phi(u).$$

(A11.5)

Note: At the same time, the lower limit of standard normal (Gaussian) distribution is $u_{min} = -\infty$ and the upper limit is $u_{max} = \infty$.

It is generally known that the standardized probability density function of normal (Gaussian) distribution is

$$\varphi(u) = g(u,l_0) = \frac{1}{\sqrt{2\pi}}e^{-\frac{u^2}{2}},$$

(A11.6)

and the standardized distribution function is

$$\Phi(u) = G(u, l_0) = \frac{1}{\sqrt{2\pi}} \int_{-\infty}^{u} e^{-\frac{v^2}{2}} \, dv = \frac{1}{\sqrt{\pi}} \int_{-\infty}^{u/\sqrt{2}} e^{-t^2} \, dt . \tag{A11.7}$$

[The last definite integral is the so-called Laplace-Gaussian integral. This must be solved only numerically.]

Distribution of linearly transformed strength

This is derived in Equation (10.4). Let us use Equation (A11.3) in case of general length l and in case of (short) length l_0, i.e., Equations (A11.4) and (A11.5) in Equation (10.4). So we obtain the probability density function of linearly transformed yarn strength as follows:

$$f(P,l) = \frac{l}{l_0}\left[1 - F(P,l_0)\right]^{\frac{l}{l_0}-1} f(P,l_0),$$

$$g(u,l)\frac{1}{\sigma_{l_0}} = \frac{l}{l_0}\left[1 - G(u,l_0)\right]^{\frac{l}{l_0}-1} g(u,l_0)\frac{1}{\sigma_{l_0}},$$

$$g(u,l) = \frac{l}{l_0}\left[1 - \Phi(u)\right]^{\frac{l}{l_0}-1} \varphi(u). \tag{A11.8}$$

The corresponding distribution function of linearly transformed yarn strength follows Equation (10.3). By using Equation (A11.4) for l and l_0 and Equation (A11.5), we can write

$$F(P,l) = 1 - \left[1 - F(P,l_0)\right]^{l/l_0},$$

$$G(u,l) = 1 - \left[1 - G(u,l_0)\right]^{l/l_0} = 1 - \left[1 - \Phi(u)\right]^{l/l_0}. \tag{A11.9}$$

Note: Let us note that we do not require to know the values of parameters $\overline{P}_{l_0}$ and σ_{l_0} for calculation of distribution of linearly transformed yarn strength u.

Statistical characteristics of random variable u

Usually, non-central (general) moments and central moments are used as suitable statistical characteristics of random variable.

A general m-th non-central moment of random variable u, called here $\overline{u^m}$ [1], is defined as follows:

$$\overline{u^m} = \int_{-\infty}^{\infty} u^m g(u,l)\,du,$$

(A11.10)

where m is a natural number and $g(u,l)$ is the probability density function according to Equation (A11.8). We will especially use the symbol $\overline{u}$ of the first non-central moment,

$$\overline{u} = \overline{u^1} = \int_{-\infty}^{\infty} u\,g(u,l)\,du,$$

(A11.11)

because the sense of this is the mean value.

A general central moment of random variable u, called here $\overline{(u-\overline{u})^m}$, is defined as follows:

$$\overline{(u-\overline{u})^m} = \int_{-\infty}^{\infty} (u-\overline{u})^m g(u,l)\,du.$$

(A11.12)

By applying the binomial theorem[2] and Equation (A11.10), we obtain the following expression:

$$\overline{(u-\overline{u})^m} = \int_{-\infty}^{\infty} \sum_{j=0}^{m} \left[(-1)^j \binom{m}{j} u^{m-j}\,\overline{u}^j \right] g(u,l)\,du$$

$$= \sum_{j=0}^{m} \left[(-1)^j \binom{m}{j} \overline{u}^j \int_{-\infty}^{\infty} u^{m-j} g(u,l)\,du \right],$$

$$\overline{(u-\overline{u})^m} = \sum_{j=0}^{m} \left[(-1)^j \binom{m}{j} \overline{u}^j \overline{u^{m-j}} \right].$$

(A11.13)

The following equations can be obtained from the last equation using values $m = 2, 3, 4$.

---

1   The symbol $\overline{u^m}$ represents the mean value of random quantity u^m, while $\overline{u}^m$ denotes m-th power of mean value $\overline{u}$ of random quantity u. Similar symbols are also used in the equations later on.

2   That is, generally $(a \pm b)^m = \sum_{i=0}^{m} \left[(\pm 1)^i \binom{m}{i} a^{m-i} b^i \right]$.

$$\overline{(u-\overline{u})^2} = \sigma_u^2 = \sum_{j=0}^{2}\left[(-1)^j\binom{2}{j}\overline{u}^j\overline{u^{2-j}}\right] = \overline{u^2} - 2\overline{u}^2 + \overline{u}^2 = \overline{u^2} - \overline{u}^2. \quad (A11.14)$$

We use the special symbol σ_u^2 for second central moment $\overline{(u-\overline{u})^2}$; this quantity is the so-called variance and the square root σ_u is known as standard deviation. Further,

$$\overline{(u-\overline{u})^3} = \sum_{j=0}^{3}\left[(-1)^j\binom{3}{j}\overline{u}^j\overline{u^{3-j}}\right] = \overline{u^3} - 3\overline{u^2}\overline{u} + 3\overline{u}^3 - \overline{u}^3 = \overline{u^3} - 3\overline{u^2}\overline{u} + 2\overline{u}^3,$$

$$(A11.15)$$

$$\overline{(u-\overline{u})^4} = \sum_{j=0}^{4}\left[(-1)^j\binom{4}{j}\overline{u}^j\overline{u^{4-j}}\right] = \overline{u^4} - 4\overline{u^3}\overline{u} + 6\overline{u^2}\overline{u}^2 - 4\overline{u}^4 + \overline{u}^4$$

$$= \overline{u^4} - 4\overline{u^3}\overline{u} + 6\overline{u^2}\overline{u}^2 - 3\overline{u}^4. \quad (A11.16)$$

Also, the ratios of non-central and central moments are used for statistical characterization of distribution. It includes namely, coefficient of variation, kurtosis and skewness. (Skewness a is measure of lack of symmetry of the distribution of a dataset, whereas kurtosis e is a measure of whether the data are peaked or flat relative to a normal distribution; a normal distribution is a symmetric distribution where $a = e = 0$.)

By using Equations (A11.11), (A11.14), (A11.15) and (A11.16) step by step, we obtain the following expressions.

- The coefficient of variation of linearly transformed yarn strength u is

$$v_u = \frac{\sigma_u}{\overline{u}} = \frac{\sqrt{\overline{u^2} - \overline{u}^2}}{\overline{u}}. \quad (A11.17)$$

- The skewness of linearly transformed yarn strength u is

$$a = \frac{\overline{(u-\overline{u})^3}}{\sigma_u^3} = \frac{\overline{u^3} - 3\overline{u^2}\overline{u} + 2\overline{u}^3}{\left[\overline{u^2} - \overline{u}^2\right]^{3/2}}. \quad (A11.18)$$

- The kurtosis of linearly transformed yarn strength u is

$$e = \frac{\overline{(u-\overline{u})^4}}{\sigma_u^4} - 3 = \frac{\overline{u^4} - 4\overline{u^3}\overline{u} + 6\overline{u^2}\overline{u}^2 - 3\overline{u}^4}{\left[\overline{u^2} - \overline{u}^2\right]^2} - 3. \quad (A11.19)$$

Note: Let us note that the above-mentioned three characteristics, i.e., v_u, a and e, can be expressed using first of four non-central moments. [These must be calculated from Equation (A11.10) with the help of Equations (A11.6) to (A11.8), using a suitable method of numerical integration.]

Statistical characteristics of random yarn strength P

According to Equation (A11.1)[3], it is valid to write that $P = \sigma_{l_0} u + \overline{P}_{l_0}$. The general m-th non-central moment of yarn strength P is then

$$\overline{P^m} = \int_{-\infty}^{\infty} \left(\sigma_{l_0} u + \overline{P}_{l_0} \right)^m g(u,l)\,\mathrm{d}u = \int_{-\infty}^{\infty} \left\{ \sum_{i=0}^{m} \left[\binom{m}{i} \sigma_{l_0}^{m-i} u^{m-i} \overline{P}_{l_0}^{i} \right] g(u,l) \right\} \mathrm{d}u$$

$$= \sum_{i=0}^{m} \left[\binom{m}{i} \sigma_{l_0}^{m-i} \overline{P}_{l_0}^{i} \int_{-\infty}^{\infty} u^{m-i} g(u,l)\,\mathrm{d}u \right] = \sum_{i=0}^{m} \left[\binom{m}{i} \sigma_{l_0}^{m-i} \overline{P}_{l_0}^{i} \overline{u^{m-i}} \right].$$

(A11.20)

Equations (A11.1) and (A11.10) were used for rearrangement. The Footnotes 1 and 2 are valid analogically.

We will use the symbol $\overline{P}_l$ of the first non-central moment, because the sense of this is the mean value of yarn strength at a general length l:

$$\overline{P}_l = \overline{P^1} = \sum_{i=0}^{1} \left[\binom{1}{i} \sigma_{l_0}^{1-i} \overline{P}_{l_0}^{i} \overline{u^{1-i}} \right] = \sigma_{l_0} \overline{u^1} + \overline{P}_{l_0} = \sigma_{l_0} \overline{u} + \overline{P}_{l_0}.$$ (A11.21)

[The mean of the linearly transformed yarn strength $\overline{u}$ is described by Equation (A11.11) together with Equations (A11.6) to (A11.8).]

The general m-th central moment of yarn strength P is

$$\overline{\left(P - \overline{P}_l \right)^m} = \int_{-\infty}^{\infty} \left\{ \left[\left(\sigma_{l_0} u + \overline{P}_{l_0} \right) - \left(\sigma_{l_0} \overline{u} + \overline{P}_{l_0} \right) \right]^m \right\} g(u,l)\,\mathrm{d}u$$

$$= \sigma_{l_0}^{m} \int_{-\infty}^{\infty} \left(u - \overline{u} \right)^m g(u,l)\,\mathrm{d}u = \sigma_{l_0}^{m} \overline{\left(u - \overline{u} \right)^m}.$$ (A11.22)

Equations (A11.1), (A11.21) and (A11.12) were used for rearrangement.

---

3   Let us remind that P is the random yarn strength at an arbitrary chosen gauge length l. The mean strength $\overline{P}_{l_0}$ and standard deviation σ_{l_0} are related to the short gauge length l_0. u is the linearly transformed yarn strength.

We will use the symbol σ_l^2 for the second central moment, because the meaning of this is the variance of yarn strength.

$$\sigma_l^2 = \overline{\left(P - \overline{P_l}\right)^2} = \sigma_{l_0}^2 \overline{\left(u - \overline{u}\right)^2} = \sigma_{l_0}^2 \left(\overline{u^2} - \overline{u}^2\right) = \sigma_{l_0}^2 \sigma_u^2. \qquad \text{(A11.23)}$$

Equations (A11.1), (A11.21), (A11.22) and (A11.14) were used for rearrangement.

Besides earlier equations and Equations (A11.15) and (A11.16), we ana-logically obtain the expressions for $m = 3, 4$ as follows:

$$\overline{\left(P - \overline{P_l}\right)^3} = \sigma_{l_0}^3 \overline{\left(u - \overline{u}\right)^3} = \sigma_{l_0}^3 \left(\overline{u^3} - 3\overline{u^2}\overline{u} + 2\overline{u}^3\right), \qquad \text{(A11.24)}$$

$$\overline{\left(P - \overline{P_l}\right)^4} = \sigma_{l_0}^4 \overline{\left(u - \overline{u}\right)^4} = \sigma_{l_0}^4 \left(\overline{u^4} - 4\overline{u^3}\overline{u} + 6\overline{u^2}\overline{u}^2 - 3\overline{u}^4\right). \qquad \text{(A11.25)}$$

Now, we can formulate the coefficient of variation v_l of yarn strength P. Using Equations (A11.23), (A11.21) and (A11.17), we obtain

$$v_{l_0} = \frac{\sigma_{l_0}}{\overline{P_{l_0}}}. \qquad \text{(A11.26)}^4$$

We can derive the coefficient of variation v_l according to the following function:

$$v_l = \frac{\sigma_l}{\overline{P_l}} = \frac{\sigma_{l_0} \sigma_u}{\sigma_{l_0} \overline{u} + \overline{P_{l_0}}} = \frac{\dfrac{\sigma_u}{\overline{u}}}{1 + \dfrac{\overline{P_{l_0}}}{\sigma_{l_0} \overline{u}}} = \frac{v_u}{1 + \dfrac{1}{v_{l_0} \overline{u}}}. \qquad \text{(A11.27)}$$

The skewness of probability density function of yarn strength P is

$$a = \frac{\overline{\left(P - \overline{P_l}\right)^3}}{\sigma_l^3} = \frac{\sigma_{l_0}^3 \left(\overline{u^3} - 3\overline{u^2}\overline{u} + 2\overline{u}^3\right)}{\sigma_{l_0}^3 \left(\overline{u^2} - \overline{u}^2\right)^{3/2}} = \frac{\overline{u^3} - 3\overline{u^2}\overline{u} + 2\overline{u}^3}{\left(\overline{u^2} - \overline{u}^2\right)^{3/2}}, \qquad \text{(A11.28)}$$

and the kurtosis of probability density function of yarn strength P is

---

4   The value v_{l_0} represents the coefficient of variation of yarn strength obtained at (short) gauge length l_0.

$$e = \frac{\overline{\left(P - \overline{P}_l\right)^4}}{\sigma_l^4} - 3 = \frac{\sigma_{l_0}^4 \left(\overline{u^4} - 4\overline{u^3\overline{u}} + 6\overline{u^2\overline{u}^2} - 3\overline{u}^4\right)}{\sigma_{l_0}^4 \left[\overline{u^2} - \overline{u}^2\right]^2} - 3$$

$$= \frac{\overline{u^4} - 4\overline{u^3\overline{u}} + 6\overline{u^2\overline{u}^2} - 3\overline{u}^4}{\left[\overline{u^2} - \overline{u}^2\right]^2} - 3. \tag{A11.29}$$

Equations (A11.23) to (A11.25) were used for the derivation of the last two expressions.

Note: By comparing the last two equations with Equations (A11.18) and (A11.19), we observe that the skewness and kurtosis are same for original yarn strength P as well as for the linearly transformed yarn strength u.

Appendix 12
Statistical characteristics of Weibull distribution

Gamma function

We are using a special higher transcendental function, the so-called gamma-function $\Gamma(x)$ as stated below:

$$\left.\begin{array}{l} \Gamma(x) = \int_0^\infty u^{x-1} e^{-u}\, du, \quad u \in \langle 0, \infty \rangle, \\[2mm] \Gamma(n+1) = n!, \quad n \ldots \text{non-negative integer number,} \\[2mm] \Gamma(x+1) = x\Gamma(x), \quad x > 0. \end{array}\right\} \qquad \text{(A12.1)[1]}$$

[See also Equation (10.45).]

Non-central statistical moments of transformed yarn strength u

By using Equations (10.44) and (A12.1), we can formulate the general statistical moments of the transformed yarn strength $u = \left[(P - P_{\min})/q \right]^c$ as follows:

$$\overline{u^x} = \int_0^\infty u^x \psi(u)\, du = \int_0^\infty u^x e^{-u}\, du = \Gamma(x+1), \quad x = 1, 2, \ldots. \qquad \text{(A12.2)[2]}$$

Note: The statistical moments are defined for integer values of x, but the previous equation is valid more generally, i.e., for each (positive) real values of x. (It will be utilized later on.)

---

1 See a mathematical handbook for more information on gamma function.
2 We use same symbols for statistical moments as mentioned in Appendix 11.

Non-central statistical moments of yarn strength P

By using Equations (10.42) and (A12.1) and the binomial theorem[3], we can formulate m-th non-central moment as follows:

$$\overline{P^m} = \int_{P_{min}}^{\infty} P^m f(P,l)\, dP = \int_{P_{min}}^{\infty} P^m \frac{c}{q} \left(\frac{P - P_{min}}{q} \right)^{c-1} \exp\left[-\left(\frac{P - P_{min}}{q} \right)^c \right] dP$$

Substitution: $u = \left(\dfrac{P - P_{min}}{q} \right)^c$, $\quad P = qu^{1/c} + P_{min}$, $\quad dP = \dfrac{q}{c} u^{\frac{1}{c}-1}\, du$,

$$= \int_0^{\infty} \left(qu^{1/c} + P_{min} \right)^m \frac{c}{q} u^{\frac{c-1}{c}} e^{-u} \frac{q}{c} u^{\frac{1}{c}-1}\, du = \int_0^{\infty} \left(qu^{1/c} + P_{min} \right)^m e^{-u}\, du$$

$$= \int_0^{\infty} \left[\sum_{i=0}^{m} \binom{m}{i} q^{m-i} u^{\frac{m-i}{c}} P_{min}^i \right] e^{-u}\, du = \sum_{i=0}^{m} \left[\binom{m}{i} q^{m-i} P_{min}^i \int_0^{\infty} u^{\frac{m-i}{c}} e^{-u}\, du \right]$$

$$= \sum_{i=0}^{m} \left[\binom{m}{i} q^{m-i} P_{min}^i \, \Gamma\left(\frac{m-i}{c} + 1 \right) \right], \quad m = 1, 2, \ldots \qquad (A12.3)$$

We obtain the following four special expressions from the last equation

$$\overline{P^1} = \overline{P_l} = q\Gamma\left(\frac{1}{c} + 1 \right) + P_{min}\Gamma(1) = \frac{q}{c}\Gamma\left(\frac{1}{c} \right) + P_{min}. \qquad (A12.4)$$

Note: This quantity expresses the mean value of P, denoted by $\overline{P_l}$.

$$\overline{P^2} = q^2\Gamma\left(\frac{2}{c} + 1 \right) + 2qP_{min}\Gamma\left(\frac{1}{c} + 1 \right) + P_{min}^2\Gamma(1)$$

$$= \frac{2q^2}{c}\Gamma\left(\frac{2}{c} \right) + \frac{2qP_{min}}{c}\Gamma\left(\frac{1}{c} \right) + P_{min}^2. \qquad (A12.5)$$

Further,

---

3   It is valid to write that $(a \pm b)^m = \displaystyle\sum_{i=0}^{m} \left[(\pm 1)^i \binom{m}{i} a^{m-i} b^i \right]$ for $m = 0, 1, 2, \ldots$. Let us remind that all binomial coefficients $\binom{m}{0} = 1$.

$$\overline{P^3} = q^3\Gamma\left(\frac{3}{c}+1\right) + 3q^2 P_{min}\Gamma\left(\frac{2}{c}+1\right) + 3q P_{min}^2\Gamma\left(\frac{1}{c}+1\right) + P_{min}^3\Gamma(1)$$

$$= \frac{3q^3}{c}\Gamma\left(\frac{3}{c}\right) + \frac{6q^2 P_{min}}{c}\Gamma\left(\frac{2}{c}\right) + \frac{3q P_{min}^2}{c}\Gamma\left(\frac{1}{c}\right) + P_{min}^3, \tag{A12.6}$$

$$\overline{P^4} = q^4\Gamma\left(\frac{4}{c}+1\right) + 4q^3 P_{min}\Gamma\left(\frac{3}{c}+1\right) + 6q^2 P_{min}^2\Gamma\left(\frac{2}{c}+1\right)$$

$$+ 4q P_{min}^3\Gamma\left(\frac{1}{c}+1\right) + P_{min}^4\Gamma(1)$$

$$= \frac{4q^4}{c}\Gamma\left(\frac{4}{c}\right) + \frac{12q^3 P_{min}}{c}\Gamma\left(\frac{3}{c}\right) + \frac{12q^2 P_{min}^2}{c}\Gamma\left(\frac{2}{c}\right) + \frac{4q P_{min}^3}{c}\Gamma\left(\frac{1}{c}\right) + P_{min}^4. \tag{A12.7}$$

Central statistical moments of the yarn strength P

By applying the known operator E of mean value, Equation (10.43), binomial theorem and Equation (A12.2), we can formulate m-th central moment as follows:

$$\overline{\left(P - \overline{P_t}\right)^m} = E\left\{\left[P - \overline{P}\right]^m\right\} = E\left\{\left[\left(qu^{1/c} + P_{min}\right) - \left(q\overline{u^{1/c}} + P_{min}\right)\right]^m\right\}$$

$$= E\left\{\left[qu^{1/c} - q\overline{u^{1/c}}\right]^m\right\} = q^m E\left\{\left[u^{1/c} - \overline{u^{1/c}}\right]^m\right\}$$

$$= q^m E\left\{\sum_{j=0}^{m}\left[(-1)^j\binom{m}{j}u^{(m-j)/c}\left(\overline{u^{1/c}}\right)^j\right]\right\}$$

$$= q^m\sum_{j=0}^{m}\left[(-1)^j\binom{m}{j}\overline{u^{(m-j)/c}}\left(\overline{u^{1/c}}\right)^j\right]$$

$$= q^m\sum_{j=0}^{m}\left[(-1)^j\binom{m}{j}\Gamma\left(\frac{m-j}{c}+1\right)\Gamma^j\left(\frac{1}{c}+1\right)\right]$$

$$= q^m\sum_{j=0}^{m}\left[(-1)^j\binom{m}{j}\Gamma\left(\frac{m-j+c}{c}\right)\Gamma^j\left(\frac{1+c}{c}\right)\right]. \tag{A12.8}$$

Especially for $m = 2,3,4$, we obtain the following equations from general Equation (A12.8):

$$\overline{\left(P - \overline{P}_i\right)^2} = \sigma_i^2 = q^2 \sum_{j=0}^{2}\left[(-1)^j \binom{2}{j}\Gamma\left(\frac{2-j+c}{c}\right)\Gamma^j\left(\frac{1+c}{c}\right)\right]$$

$$= q^2\left\{\Gamma\left(\frac{2+c}{c}\right)\Gamma^0\left(\frac{1+c}{c}\right) - 2\Gamma\left(\frac{1+c}{c}\right)\Gamma^1\left(\frac{1+c}{c}\right) + \Gamma(1)\Gamma^2\left(\frac{1+c}{c}\right)\right\}$$

$$= q^2\left\{\Gamma\left(\frac{2+c}{c}\right) - 2\Gamma^2\left(\frac{1+c}{c}\right) + \Gamma^2\left(\frac{1+c}{c}\right)\right\} = q^2\left\{\frac{2}{c}\Gamma\left(\frac{2}{c}\right) - \frac{1}{c^2}\Gamma^2\left(\frac{1}{c}\right)\right\}.$$

$$(A12.9)$$

Note: The last equation expresses the dispersion of yarn strength, indicated by σ_i^2 in short.

$$\overline{\left(P - \overline{P}_i\right)^3} = q^3 \sum_{j=0}^{3}\left[(-1)^j \binom{3}{j}\Gamma\left(\frac{3-j+c}{c}\right)\Gamma^j\left(\frac{1+c}{c}\right)\right]$$

$$= q^3\left\{\Gamma\left(\frac{3+c}{c}\right)\Gamma^0\left(\frac{1+c}{c}\right) - 3\Gamma\left(\frac{2+c}{c}\right)\Gamma^1\left(\frac{1+c}{c}\right)\right.$$

$$\left. + 3\Gamma\left(\frac{1+c}{c}\right)\Gamma^2\left(\frac{1+c}{c}\right) - \Gamma(1)\Gamma^3\left(\frac{1+c}{c}\right)\right\}$$

$$= q^3\left\{\frac{3}{c}\Gamma\left(\frac{3}{c}\right) - \frac{6}{c^2}\Gamma\left(\frac{2}{c}\right)\Gamma\left(\frac{1}{c}\right) + \frac{2}{c^3}\Gamma^3\left(\frac{1}{c}\right)\right\}, \qquad (A12.10)$$

$$\overline{\left(P - \overline{P}_i\right)^4} = q^4 \sum_{j=0}^{4}\left[(-1)^j \binom{4}{j}\Gamma\left(\frac{4-j+c}{c}\right)\Gamma^j\left(\frac{1+c}{c}\right)\right]$$

$$= q^4\left\{\Gamma\left(\frac{4+c}{c}\right)\Gamma^0\left(\frac{1+c}{c}\right) - 4\Gamma\left(\frac{3+c}{c}\right)\Gamma\left(\frac{1+c}{c}\right)\right.$$

$$\left. + 6\Gamma\left(\frac{2+c}{c}\right)\Gamma^2\left(\frac{1+c}{c}\right) - 4\Gamma\left(\frac{1+c}{c}\right)\Gamma^3\left(\frac{1+c}{c}\right) + \Gamma(1)\Gamma^4\left(\frac{1+c}{c}\right)\right\}$$

$$= q^4\left\{\frac{4}{c}\Gamma\left(\frac{4}{c}\right) - \frac{12}{c^2}\Gamma\left(\frac{3}{c}\right)\Gamma\left(\frac{1}{c}\right) + \frac{12}{c^3}\Gamma\left(\frac{2}{c}\right)\Gamma^2\left(\frac{1}{c}\right) - \frac{3}{c^4}\Gamma^4\left(\frac{1}{c}\right)\right\}.$$

$$(A12.11)$$

We can utilize the best-known statistical characteristics by using the derived equations and the expression $q = Q/l^{1/c}$ according to Equation (10.40).

By using Equation (A12.4), the mean value of yarn strength can be written as

$$\bar{P}_l = \frac{q}{c}\Gamma\left(\frac{1}{c}\right) + P_{\min}, \quad \left(\frac{\bar{P}_l - P_{\min}}{q} = \frac{1}{c}\Gamma\left(\frac{1}{c}\right)\right),$$

$$\bar{P}_l = l^{-\frac{1}{c}}\frac{Q}{c}\Gamma\left(\frac{1}{c}\right) + P_{\min}. \tag{A12.12}$$

The standard deviation of yarn strength is the square root of dispersion according to Equation (12.9). This is shown as follows:

$$\sigma_l = q\sqrt{\frac{2}{c}\Gamma\left(\frac{2}{c}\right) - \frac{1}{c^2}\Gamma^2\left(\frac{1}{c}\right)}, \quad \left(\frac{\sigma_l}{q} = \sqrt{\frac{2}{c}\Gamma\left(\frac{2}{c}\right) - \frac{1}{c^2}\Gamma^2\left(\frac{1}{c}\right)}\right),$$

$$\sigma_l = l^{-\frac{1}{c}}Q\sqrt{\frac{2}{c}\Gamma\left(\frac{2}{c}\right) - \frac{1}{c^2}\Gamma^2\left(\frac{1}{c}\right)}. \tag{A12.13}$$

The coefficient of variation of yarn strength is expressed by the ratio of the last two equations. This is stated as follows:

$$v_l = \frac{\sigma_l}{\bar{P}_l} = \frac{q\sqrt{\frac{2}{c}\Gamma\left(\frac{2}{c}\right) - \frac{1}{c^2}\Gamma^2\left(\frac{1}{c}\right)}}{\frac{q}{c}\Gamma\left(\frac{1}{c}\right) + P_{\min}} = \frac{\sqrt{\frac{2}{c}\Gamma\left(\frac{2}{c}\right) - \frac{1}{c^2}\Gamma^2\left(\frac{1}{c}\right)}}{\frac{1}{c}\Gamma\left(\frac{1}{c}\right) + \frac{P_{\min}}{q}},$$

$$v_l = \frac{\sqrt{\frac{2}{c}\Gamma\left(\frac{2}{c}\right) - \frac{1}{c^2}\Gamma^2\left(\frac{1}{c}\right)}}{\frac{1}{c}\Gamma\left(\frac{1}{c}\right) + \frac{P_{\min}}{Q}l^{1/c}}. \tag{A12.14}$$

Further, we will express the coefficients of skewness a and kurtosis e. The skewness can be defined by the ratio $\overline{(P - \bar{P}_l)^3}/\sigma_l^3$. By using Equations (A12.10) and (A12.13), we obtain

$$a = \frac{\overline{(P - \bar{P}_l)^3}}{\sigma_l^3} = \frac{q^3\left\{\frac{3}{c}\Gamma\left(\frac{3}{c}\right) - \frac{6}{c^2}\Gamma\left(\frac{2}{c}\right)\Gamma\left(\frac{1}{c}\right) + \frac{2}{c^3}\Gamma^3\left(\frac{1}{c}\right)\right\}}{\left[q\sqrt{\frac{2}{c}\Gamma\left(\frac{2}{c}\right) - \frac{1}{c^2}\Gamma^2\left(\frac{1}{c}\right)}\right]^3}$$

$$
= \frac{\dfrac{3}{c}\Gamma\!\left(\dfrac{3}{c}\right) - \dfrac{6}{c^2}\Gamma\!\left(\dfrac{2}{c}\right)\Gamma\!\left(\dfrac{1}{c}\right) + \dfrac{2}{c^3}\Gamma^3\!\left(\dfrac{1}{c}\right)}{\left[\dfrac{2}{c}\Gamma\!\left(\dfrac{2}{c}\right) - \dfrac{1}{c^2}\Gamma^2\!\left(\dfrac{1}{c}\right)\right]^{3/2}}.
\tag{A12.15}
$$

The kurtosis can be defined by the ratio $\overline{\left(P-\overline{P}_l\right)^4}\big/\sigma_l^4 - 3$. By using Equations (A12.11) and (A12.13), we obtain

$$
\begin{aligned}
e &= \frac{\overline{\left(P-\overline{P}_l\right)^4}}{\sigma_l^4} - 3 \\[2mm]
&= \frac{q^4\left\{\dfrac{4}{c}\Gamma\!\left(\dfrac{4}{c}\right) - \dfrac{12}{c^2}\Gamma\!\left(\dfrac{3}{c}\right)\Gamma\!\left(\dfrac{1}{c}\right) + \dfrac{12}{c^3}\Gamma\!\left(\dfrac{2}{c}\right)\Gamma^2\!\left(\dfrac{1}{c}\right) - \dfrac{3}{c^4}\Gamma^4\!\left(\dfrac{1}{c}\right)\right\}}{\left[q\sqrt{\dfrac{2}{c}\Gamma\!\left(\dfrac{2}{c}\right) - \dfrac{1}{c^2}\Gamma^2\!\left(\dfrac{1}{c}\right)}\,\right]^4} - 3 \\[2mm]
&= \frac{\dfrac{4}{c}\Gamma\!\left(\dfrac{4}{c}\right) - \dfrac{12}{c^2}\Gamma\!\left(\dfrac{3}{c}\right)\Gamma\!\left(\dfrac{1}{c}\right) + \dfrac{12}{c^3}\Gamma\!\left(\dfrac{2}{c}\right)\Gamma^2\!\left(\dfrac{1}{c}\right) - \dfrac{3}{c^4}\Gamma^4\!\left(\dfrac{1}{c}\right)}{\left[\dfrac{2}{c}\Gamma\!\left(\dfrac{2}{c}\right) - \dfrac{1}{c^2}\Gamma^2\!\left(\dfrac{1}{c}\right)\right]^2} - 3.
\end{aligned}
\tag{A12.16}
$$

Note: Let us note that the skewness a and the kurtosis e, and also quantity $\left(\overline{P}_l - P_{\min}\right)\big/q$ as well as σ_l/q depend only on one parameter, that is, the exponent c.

Gaussian distribution of random multi-variables

Initial functions

We know the following probability density function of normal (Gaussian) distribution of P_i :

$$f(P_i) = \frac{1}{\sqrt{2\pi}\sigma_{l_0}} \exp\left\{-\frac{\left(P_i - \overline{P_{l_0}}\right)^2}{2\sigma_{l_0}^2}\right\}, \quad \begin{array}{l} P_i \ldots \text{random variable, } P_i \in (-\infty, \infty), \\ \overline{P_{l_0}} \ldots \text{mean value} - \text{parameter,} \\ \sigma_{l_0} \ldots \text{standard deviation} - \text{parameter.} \end{array}$$

$$\text{(A13.1)}$$

Moreover, the Gaussian conditional probability density function of random quantity P_{i+1}, with a condition that the previous quantity is a given value P_i (parameter), is

$$\varphi(P_{i+1} | P_i) = \frac{1}{\sqrt{2\pi}\,\sigma_{l_0}\sqrt{1 - r^2}} \exp\left\{-\frac{\left(P_{i+1} - \left[\overline{P_{l_0}} + r\left(P_i - \overline{P_{l_0}}\right)\right]\right)^2}{2\sigma_{l_0}^2\left(1 - r^2\right)}\right\},$$

$P_{i+1} \ldots$ random variable, $P_{i+1} \in (-\infty, \infty)$,

$P_i, \overline{P_{l_0}}, \sigma_{l_0} \ldots$ previous value, mean value and standard deviation – parameters,

$r = \rho(P_i, P_{i+1}) \ldots$ correlation coefficient between P_i and P_{i+1} – parameter.

$$\text{(A13.2)}$$

The previous pair of equations is consistent with Equations (10.114) and (10.115), as shown in Section 10.5 of this book. They are very well-known equations in the theory of probability. (See a suitable handbook on this topic.)

Conjugate probability density function of pair P_i, P_{i+1}

According to Equations (10.58), (A13.1) and (A13.2), the conjugate probability density function of a pair of strengths P_i, P_{i+1} is

$$f\left(P_i, P_{i+1}\right) = f\left(P_i\right) \varphi\left(P_{i+1} \middle| P_i\right)$$

$$= \frac{1}{\sqrt{2\pi}\sigma_{l_0}} \exp\left\{-\frac{\left(P_i - \bar{P}_{l_0}\right)^2}{2\sigma_{l_0}^2}\right\} \frac{1}{\sqrt{2\pi}\,\sigma_{l_0}\sqrt{1-r^2}} \exp\left\{-\frac{\left(P_{i+1} - \left[\bar{P}_{l_0} + r\left(P_i - \bar{P}_{l_0}\right)\right]\right)^2}{2\sigma_{l_0}^2\left(1-r^2\right)}\right\}$$

$$= \frac{1}{2\pi\sigma_{l_0}^2\sqrt{1-r^2}} \exp\left\{-\frac{\left(P_i - \bar{P}_{l_0}\right)^2}{2\sigma_{l_0}^2} - \frac{\left(P_{i+1} - \left[\bar{P}_{l_0} + r\left(P_i - \bar{P}_{l_0}\right)\right]\right)^2}{2\sigma_{l_0}^2\left(1-r^2\right)}\right\}$$

$$= \frac{1}{2\pi\sigma_{l_0}^2\sqrt{1-r^2}} \exp\left\{-\frac{\left(P_i - \bar{P}_{l_0}\right)^2 - r^2\left(P_i - \bar{P}_{l_0}\right)^2}{2\sigma_{l_0}^2\left(1-r^2\right)} - \frac{\left[\left(P_{i+1} - \bar{P}_{l_0}\right) - r\left(P_i - \bar{P}_{l_0}\right)\right]^2}{2\sigma_{l_0}^2\left(1-r^2\right)}\right\}$$

$$= \frac{1}{2\pi\sigma_{l_0}^2\sqrt{1-r^2}} \exp\left\{-\frac{\left(P_i - \bar{P}_{l_0}\right)^2 - r^2\left(P_i - \bar{P}_{l_0}\right)^2}{2\sigma_{l_0}^2\left(1-r^2\right)}\right.$$

$$\left. -\frac{\left(P_{i+1} - \bar{P}_{l_0}\right)^2 - 2r\left(P_i - \bar{P}_{l_0}\right)\left(P_{i+1} - \bar{P}_{l_0}\right) + r^2\left(P_i - \bar{P}_{l_0}\right)^2}{2\sigma_{l_0}^2\left(1-r^2\right)}\right\},$$

$$f\left(P_i, P_{i+1}\right) = \frac{1}{2\pi\sigma_{l_0}^2\sqrt{1-r^2}} \exp\left\{-\frac{\left(P_i - \bar{P}_{l_0}\right)^2 - 2r\left(P_i - \bar{P}_{l_0}\right)\left(P_{i+1} - \bar{P}_{l_0}\right) + \left(P_{i+1} - \bar{P}_{l_0}\right)^2}{2\sigma_{l_0}^2\left(1-r^2\right)}\right\}.$$

(A13.3)

Note: The last equation is often referred to as a known expression in the handbooks of probability in case of conjugate normal (Gaussian) distribution.

Standardized normal (Gaussian) process

Equation (10.85) determines the probability density function of standardized normal (Gaussian) process U_i. It is valid to write that

$$U_i = \frac{P_i - \bar{P}_{l_0}}{\sigma_{l_0}}, \quad \text{i.e.,} \quad P_i = \sigma_{l_0} U_i + \bar{P}_{l_0}, \quad \text{and} \quad \frac{dP_i}{dU_i} = \sigma_{l_0}. \tag{A13.4}$$

By using Equations (A13.1) and (A13.4), the probability density function of such standardized normal (Gaussian) variable is

$$f(U_i) = f(P_i)\frac{dP_i}{dU_i} = \frac{1}{\sqrt{2\pi}\sigma_{l_0}}\exp\left\{-\frac{U_i^2}{2}\right\}\sigma_{l_0} = \frac{1}{\sqrt{2\pi}}\exp\left\{-\frac{U_i^2}{2}\right\}.\text{(A13.5)}^1$$

Similarly, it is valid for conditional probability density function[2] $\varphi(U_{i+1}|U_i)$ to write the following expression according to Equations (A13.2) and (A13.4):

$$\varphi(U_{i+1}|U_i) = \varphi(P_{i+1}|P_i)\frac{dP_{i+1}}{dU_{i+1}}$$

$$= \frac{1}{\sqrt{2\pi}\,\sigma_{l_0}\sqrt{1-r^2}}\exp\left[-\frac{\left(\overbrace{\sigma_{l_0}U_{i+1}+\overline{P}_{l_0}}^{=P_{i+1}} - \left[\overline{P}_{l_0}+r\left(\overbrace{\sigma_{l_0}U_i+\overline{P}_{l_0}}^{=P_i}-\overline{P}_{l_0}\right)\right]\right)^2}{2\sigma_{l_0}^2\left(1-r^2\right)}\right]\sigma_0$$

$$= \frac{1}{\sqrt{2\pi}\sqrt{1-r^2}}\exp\left\{-\frac{\left(\sigma_{l_0}U_{i+1}-\sigma_{l_0}rU_i\right)^2}{2\sigma_{l_0}^2\left(1-r^2\right)}\right\}$$

$$= \frac{1}{\sqrt{2\pi}\sqrt{1-r^2}}\exp\left\{-\frac{\left(U_{i+1}-rU_i\right)^2}{2\left(1-r^2\right)}\right\}. \qquad\qquad \text{(A13.6)}$$

In analogy with Equation (10.58), we find the conjugate probability density function using Equations (A13.5) and (A13.6) as follows:

$$f(U_i,U_{i+1}) = f(U_i)\varphi(U_{i+1}|U_i)$$

$$= \frac{1}{\sqrt{2\pi}}\exp\left\{-\frac{U_i^2}{2}\right\}\frac{1}{\sqrt{2\pi}\sqrt{1-r^2}}\exp\left\{-\frac{\left(U_{i+1}-rU_i\right)^2}{2\left(1-r^2\right)}\right\}$$

$$= \frac{1}{2\pi\sqrt{1-r^2}}\exp\left\{-\frac{U_i^2-r^2U_i^2}{2\left(1-r^2\right)}-\frac{U_{i+1}^2-2rU_iU_{i+1}+r^2U_i^2}{2\left(1-r^2\right)}\right\},$$

---

1   Quite generally, if a function $y = g(x)$ is valid between two random variables x and y, then the relation $f(y) = f(x)dx/dy$ is valid between their probability density functions. (See a handbook of probability for mathematical derivation.)

2   It is still valid that U_{i+1} is a random variable, but U_i is a (known) value of the parameter.

$$f\left(U_i, U_{i+1}\right) = \frac{1}{2\pi\sqrt{1-r^2}} \exp\left\{-\frac{U_i^2 - 2rU_iU_{i+1} + U_{i+1}^2}{2\left(1-r^2\right)}\right\}. \tag{A13.7}$$

(Also this expression is often presented in the handbooks of probability.)

In analogy to Equation (10.63), we can write the following expression by using Equation (A13.6):

$$\varphi\left(U_{i+k}|U_i\right) = \int\limits_{U_{i+1}=-\infty}^{\infty} \int\limits_{U_{i+2}=-\infty}^{\infty} \cdots \int\limits_{U_{i+k-1}=-\infty}^{\infty} \left[\prod_{j=1}^{k}\varphi\left(U_{i+j}|U_{i+j-1}\right)\right] dU_{i+1}\, dU_{i+2}\dots dU_{i+k-1}$$

$$= \int\limits_{U_{i+1}=-\infty}^{\infty} \int\limits_{U_{i+2}=-\infty}^{\infty} \cdots \int\limits_{U_{i+k-1}=-\infty}^{\infty} \left[\prod_{j=1}^{k}\left(\frac{1}{\sqrt{2\pi}\sqrt{1-r^2}}\exp\left\{-\frac{\left(U_{i+j}-rU_{i+j-1}\right)^2}{2\left(1-r^2\right)}\right\}\right)\right]$$

$$dU_{i+1}\, dU_{i+2}\dots dU_{i+k-1},\ k = 2,3,\dots. \tag{A13.8}$$

Mathematical tools

We need to formulate a mathematical tool for solving the right-hand side of the last equation. Let x, y, z be a triplet of real variables. It is known from mathematical analysis that

$$\int\limits_{-\infty}^{\infty} e^{-a^2x^2}\, dx = 2\int\limits_{0}^{\infty} e^{-a^2x^2}\, dx = \frac{\sqrt{\pi}}{a},\quad a > 0\dots\text{parameter (Integral Laplace-Gauss)}. \tag{A13.9}$$

We also use the following symbol:

$$\mathrm{erfc}\, x = \frac{2}{\sqrt{\pi}}\int\limits_{x}^{\infty} e^{-y^2}\, dy. \tag{A13.10}$$

With reference to Equation (A13.9), it is also valid to write that

$$\left.\begin{aligned}
\mathrm{erfc}\left(-\infty\right) &= \frac{2}{\sqrt{\pi}}\int\limits_{-\infty}^{\infty} e^{-y^2}\, dy = \frac{2}{\sqrt{\pi}}\frac{\sqrt{\pi}}{1} = 2,\\
\mathrm{erfc}\left(\infty\right) &= \frac{2}{\sqrt{\pi}}\lim_{q\to\infty}\int\limits_{q}^{\infty} e^{-y^2}\, dy = 0.
\end{aligned}\right\} \tag{A13.11}$$

We also find the following relations using previous expressions:

$$\int_{\xi}^{\infty} e^{-a^2x^2+bx} \, dx = e^{\frac{b^2}{4a^2}} \int_{\xi}^{\infty} e^{-a^2x^2+bx-\frac{b^2}{4a^2}} \, dx = e^{\frac{b^2}{4a^2}}$$

$$\cdot \int_{a\xi-b/(2a)}^{\infty} e^{-y^2} \frac{dy}{a} = \frac{\sqrt{\pi}}{2a} e^{\frac{b^2}{4a^2}} \operatorname{erfc}\left(a\xi - \frac{b}{2a}\right),$$

$a, b \ldots$ real parameters. Substitution: $y = ax - b/(2a), \quad dy = a\,dx$.

$$\text{(A13.12)}$$

$$\int_{-\infty}^{\infty} e^{-a^2x^2+bx} \, dx = \frac{\sqrt{\pi}}{2a} e^{\frac{b^2}{4a^2}} \operatorname{erfc}(-\infty) = \frac{\sqrt{\pi}}{a} e^{\frac{b^2}{4a^2}}. \qquad \text{(A13.13)}$$

To have a better formalism, let us introduce the following function:

$$p_n(y,x) = \frac{1}{\sqrt{2\pi}\sqrt{1-r^{2n}}} \exp\left[-\frac{(y-r^n x)^2}{2(1-r^{2n})}\right], \quad n=1,2,\ldots. \quad \text{(A13.14)}$$

We will also need the following integral:

$$\int_{-\infty}^{\infty} p_n(y,x)\, p_1(z,y)\, dy$$

$$= \int_{-\infty}^{\infty} \frac{1}{\sqrt{2\pi}\sqrt{1-r^{2n}}} \exp\left[-\frac{(y-r^n x)^2}{2(1-r^{2n})}\right] \frac{1}{\sqrt{2\pi}\sqrt{1-r^2}} \exp\left[-\frac{(z-ry)^2}{2(1-r^2)}\right] dy$$

$$= \frac{1}{2\pi\sqrt{1-r^{2n}}\sqrt{1-r^2}} \int_{-\infty}^{\infty} \exp\left[-\frac{(y-r^n x)^2}{2(1-r^{2n})} - \frac{(z-ry)^2}{2(1-r^2)}\right] dy.$$

$$\text{(A13.15)}$$

It is possible to rearrange the expression mentioned in the square brackets as follows:

$$-\frac{(y-r^n x)^2}{2(1-r^{2n})} - \frac{(z-ry)^2}{2(1-r^2)}$$

$$= -\frac{(1-r^2)(y^2 - 2r^n xy + r^{2n}x^2) + (1-r^{2n})(z^2 - 2ryz + r^2 y^2)}{2(1-r^{2n})(1-r^2)}$$

$$= -\frac{y^2 - 2r^n xy + r^{2n} x^2 - r^2 y^2 + 2r^{n+2} xy - r^{2n+2} x^2}{2\left(1-r^{2n}\right)\left(1-r^2\right)}$$

$$-\frac{z^2 - 2ryz + r^2 y^2 - r^{2n} z^2 + 2r^{2n+1} yz - r^{2n+2} y^2}{2\left(1-r^{2n}\right)\left(1-r^2\right)}$$

$$= -\frac{x^2 r^{2n}\left(1-r^2\right) + z^2\left(1-r^{2n}\right)}{2\left(1-r^{2n}\right)\left(1-r^2\right)} - \frac{y^2\left(1-r^{2n+2}\right) - 2yr\left[r^{n-1}x\left(1-r^2\right) + z\left(1-r^{2n}\right)\right]}{2\left(1-r^{2n}\right)\left(1-r^2\right)}.$$

(A13.16)

If we write

$$a^2 = \frac{1-r^{2n+2}}{2\left(1-r^{2n}\right)\left(1-r^2\right)}, \quad b = \frac{r\left[r^{n-1}x\left(1-r^2\right) + z\left(1-r^{2n}\right)\right]}{\left(1-r^{2n}\right)\left(1-r^2\right)}, \quad \text{(A13.17)}$$

then the expression according to Equation (A13.16) can be rearranged as follows:

$$-\frac{\left(y-r^n x\right)^2}{2\left(1-r^{2n}\right)} - \frac{\left(z-ry\right)^2}{2\left(1-r^2\right)} = -\frac{x^2 r^{2n}\left(1-r^2\right) + z^2\left(1-r^{2n}\right)}{2\left(1-r^{2n}\right)\left(1-r^2\right)} - a^2 y^2 + by.$$

(A13.18)

The integral stated in Equation (A13.15) can be expressed using Equations (A13.13) and (A13.18) as follows:

$$\int_{-\infty}^{\infty} p_n\left(y,x\right) p_1\left(z,y\right) dy$$

$$= \frac{1}{2\pi\sqrt{1-r^{2n}}\sqrt{1-r^2}} \int_{-\infty}^{\infty} \exp\left[-\frac{x^2 r^{2n}\left(1-r^2\right) + z^2\left(1-r^{2n}\right)}{2\left(1-r^{2n}\right)\left(1-r^2\right)} - a^2 y^2 + by\right] dy$$

$$= \frac{1}{2\pi\sqrt{1-r^{2n}}\sqrt{1-r^2}} \exp\left[-\frac{x^2 r^{2n}\left(1-r^2\right) + z^2\left(1-r^{2n}\right)}{2\left(1-r^{2n}\right)\left(1-r^2\right)}\right] \int_{-\infty}^{\infty} e^{-a^2 y^2 + by} dy$$

$$= \frac{1}{2\pi\sqrt{1-r^{2n}}\sqrt{1-r^2}} \exp\left[-\frac{x^2 r^{2n}\left(1-r^2\right) + z^2\left(1-r^{2n}\right)}{2\left(1-r^{2n}\right)\left(1-r^2\right)}\right] \frac{\sqrt{\pi}}{a} e^{\frac{b^2}{4a^2}}.$$

(A13.19)

Nevertheless, the following equation follows Equation (A13.17):

$$\frac{b^2}{4a^2} = \frac{r^2\left[r^{n-1}x\left(1-r^2\right)+z\left(1-r^{2n}\right)\right]^2}{\left(1-r^{2n}\right)^2\left(1-r^2\right)^2} \cdot \frac{2\left(1-r^{2n}\right)\left(1-r^2\right)}{4\left(1-r^{2n+2}\right)}$$

$$= \frac{r^2\left[r^{n-1}x\left(1-r^2\right)+z\left(1-r^{2n}\right)\right]^2}{2\left(1-r^{2n}\right)\left(1-r^2\right)\left(1-r^{2n+2}\right)}$$

$$= \frac{r^2\left[r^{2n-2}x^2\left(1-r^2\right)^2+2r^{n-1}xz\left(1-r^2\right)\left(1-r^{2n}\right)+z^2\left(1-r^{2n}\right)^2\right]}{2\left(1-r^{2n}\right)\left(1-r^2\right)\left(1-r^{2n+2}\right)}$$

$$= \frac{r^{2n}x^2\left(1-r^2\right)^2+2r^{n+1}xz\left(1-r^2\right)\left(1-r^{2n}\right)+r^2z^2\left(1-r^{2n}\right)^2}{2\left(1-r^{2n}\right)\left(1-r^2\right)\left(1-r^{2n+2}\right)}.$$

$$\text{(A13.20)}$$

By substituting Equations (A13.17) and (A13.20) in (A13.19), we obtain the following expression:

$$\int_{-\infty}^{\infty} p_n\left(y,x\right)p_1\left(z,y\right)dy$$

$$= \frac{1}{2\pi\sqrt{1-r^{2n}}\sqrt{1-r^2}}\exp\left[-\frac{x^2r^{2n}\left(1-r^2\right)+z^2\left(1-r^{2n}\right)}{2\left(1-r^{2n}\right)\left(1-r^2\right)}\right]\sqrt{\pi}\cdot\frac{\sqrt{2}\sqrt{1-r^{2n}}\sqrt{1-r^2}}{\sqrt{1-r^{2n+2}}}$$

$$\cdot\exp\left[\frac{r^{2n}x^2\left(1-r^2\right)^2+2r^{n+1}xz\left(1-r^2\right)\left(1-r^{2n}\right)+r^2z^2\left(1-r^{2n}\right)^2}{2\left(1-r^{2n}\right)\left(1-r^2\right)\left(1-r^{2n+2}\right)}\right],$$

$$\int_{-\infty}^{\infty} p_n\left(y,x\right)p_1\left(z,y\right)dy$$

$$= \frac{1}{\sqrt{2\pi}\sqrt{1-r^{2n+2}}}\exp\left[-\frac{x^2r^{2n}\left(1-r^2\right)\left(1-r^{2n+2}\right)+z^2\left(1-r^{2n}\right)\left(1-r^{2n+2}\right)}{2\left(1-r^{2n}\right)\left(1-r^2\right)\left(1-r^{2n+2}\right)}\right.$$

$$\left.-\frac{-r^{2n}x^2\left(1-r^2\right)^2-2r^{n+1}xz\left(1-r^2\right)\left(1-r^{2n}\right)-r^2z^2\left(1-r^{2n}\right)^2}{2\left(1-r^{2n}\right)\left(1-r^2\right)\left(1-r^{2n+2}\right)}\right].$$

$$\text{(A13.21)}$$

The expression stated in the square brackets can be significantly simplified as follows:

$$
-\frac{x^2 r^{2n}\left(1-r^2\right)\left(1-r^{2n+2}\right)+z^2\left(1-r^{2n}\right)\left(1-r^{2n+2}\right)}{2\left(1-r^{2n}\right)\left(1-r^2\right)\left(1-r^{2n+2}\right)}
$$

$$
-\frac{-r^{2n}x^2\left(1-r^2\right)^2-2r^{n+1}xz\left(1-r^2\right)\left(1-r^{2n}\right)-r^2 z^2\left(1-r^{2n}\right)^2}{2\left(1-r^{2n}\right)\left(1-r^2\right)\left(1-r^{2n+2}\right)}
$$

$$
=-\frac{x^2 r^{2n}\left(1-r^2\right)\left(1-r^{2n+2}-1+r^2\right)+z^2\left(1-r^{2n}\right)\left(1-r^{2n+2}-r^2+r^{2n+2}\right)}{2\left(1-r^{2n}\right)\left(1-r^2\right)\left(1-r^{2n+2}\right)}
$$

$$
-\frac{-2r^{n+1}xz\left(1-r^2\right)\left(1-r^{2n}\right)}{2\left(1-r^{2n}\right)\left(1-r^2\right)\left(1-r^{2n+2}\right)}
$$

$$
=-\frac{x^2 r^{2n+2}+z^2-2r^{n+1}xz}{2\left(1-r^{2n+2}\right)}=-\frac{\left(z-r^{n+1}x\right)^2}{2\left(1-r^{2n+2}\right)}=-\frac{\left(z-r^{n+1}x\right)^2}{2\left(1-r^{2(n+1)}\right)}.
$$

$$(A13.22)$$

By using Equation (A13.22) in (A13.21), we find the following expression:

$$
\int_{-\infty}^{\infty} p_n(y,x)\,p_1(z,y)\,\mathrm{d}y = \frac{1}{\sqrt{2\pi}\sqrt{1-r^{2(n+1)}}}\exp\left[-\frac{\left(z-r^{n+1}x\right)^2}{2\left(1-r^{2(n+1)}\right)}\right]. \quad (A13.23)
$$

By comparing Equation (A13.14) with (A13.23), we can also write

$$
\int_{-\infty}^{\infty} p_n(y,x)\,p_1(z,y)\,\mathrm{d}y = p_{n+1}(z,x). \quad (A13.24)
$$

Conditional probability density function $\varphi\left(U_{i+k},U_i\right)$

It is possible to rearrange Equation (A13.8) using Equation (A13.14) and then repeatedly according to Equation (A13.24) in the following manner:

$$\varphi\left(U_{i+k}\middle|U_i\right)=\int_{U_{i+1}=-\infty}^{\infty}\int_{U_{i+2}=-\infty}^{\infty}\cdots\int_{U_{i+k-1}=-\infty}^{\infty}\left[\prod_{j=1}^{k}p_1\left(U_{i+j},U_{i+j-1}\right)\right]dU_{i+1}\,dU_{i+2}\ldots dU_{i+k-1}$$

$$=\int_{U_{i+1}=-\infty}^{\infty}\int_{U_{i+2}=-\infty}^{\infty}\cdots\int_{U_{i+k-1}=-\infty}^{\infty}\left[\prod_{j=3}^{k}p_1\left(U_{i+j},U_{i+j-1}\right)\right]p_1\left(U_{i+1},U_i\right)p_1\left(U_{i+2},U_{i+1}\right)dU_{i+1}\,dU_{i+2}\ldots dU_{i+k-1}$$

$$=\int_{U_{i+2}=-\infty}^{\infty}\cdots\int_{U_{i+k-1}=-\infty}^{\infty}\left[\prod_{j=3}^{k}p_1\left(U_{i+j},U_{i+j-1}\right)\right]\left[\int_{U_{i+1}=-\infty}^{\infty}p_1\left(U_{i+1},U_i\right)p_1\left(U_{i+2},U_{i+1}\right)dU_{i+1}\right]dU_{i+2}\ldots dU_{i+k-1}$$

$$=\int_{U_{i+2}=-\infty}^{\infty}\cdots\int_{U_{i+k-1}=-\infty}^{\infty}\left[\prod_{j=3}^{k}p_1\left(U_{i+j},U_{i+j-1}\right)\right]p_2\left(U_{i+2},U_i\right)dU_{i+2}\ldots dU_{i+k-1}$$

$$=\int_{U_{i+3}=-\infty}^{\infty}\cdots\int_{U_{i+k-1}=-\infty}^{\infty}\left[\prod_{j=4}^{k}p_1\left(U_{i+j},U_{i+j-1}\right)\right]\left[\int_{U_{i+2}=-\infty}^{\infty}p_2\left(U_{i+2},U_i\right)p_1\left(U_{i+3},U_{i+2}\right)dU_{i+2}\right]dU_{i+3}\ldots dU_{i+k-1}$$

$$=\int_{U_{i+3}=-\infty}^{\infty}\cdots\int_{U_{i+k-1}=-\infty}^{\infty}\left[\prod_{j=4}^{k}p_1\left(U_{i+j},U_{i+j-1}\right)\right]p_3\left(U_{i+3},U_i\right)dU_{i+3}\ldots dU_{i+k-1} \qquad (A13.25)$$

$$\vdots$$

By repeating this procedure, we finally obtain

$$\varphi\left(U_{i+k}\middle|U_i\right)=p_k\left(U_{i+k},U_i\right), \qquad (A13.26)$$

and by using Equation (A13.14), we can write the following expression:

$$\varphi\left(U_{i+k}\middle|U_i\right)=\frac{1}{\sqrt{2\pi}\sqrt{1-r^{2k}}}\exp\left[-\frac{\left(U_{i+k}-r^k U_i\right)^2}{2\left(1-r^{2k}\right)}\right]. \qquad (A13.27)$$

Conjugate probability density function $f\left(U_i,U_{i+k}\right)$

In analogy with Equation (10.58) and using Equation (A13.27), we can write the following expression:

$$f\left(U_i,U_{i+k}\right)=f\left(U_i\right)\varphi\left(U_{i+k}\middle|U_i\right)$$

$$=\frac{1}{\sqrt{2\pi}}\exp\left[-\frac{U_i^2}{2}\right]\frac{1}{\sqrt{2\pi}\sqrt{1-r^{2k}}}\exp\left[-\frac{\left(U_{i+k}-r^k U_i\right)^2}{2\left(1-r^{2k}\right)}\right],$$

$$(A13.28)$$

$$f\left(U_i,U_{i+k}\right) = f\left(U_i\right)\varphi\left(U_{i+k}\big|U_i\right)$$

$$= \frac{1}{2\pi\sqrt{1-r^{2k}}}\exp\left[-\frac{U_i^2\left(1-r^{2k}\right)}{2\left(1-r^{2k}\right)} - \frac{\left(U_{i+k}-r^kU_i\right)^2}{2\left(1-r^{2k}\right)}\right]$$

$$= \frac{1}{2\pi\sqrt{1-r^{2k}}}\exp\left[\frac{U_i^2 - r^{2k}U_i^2 + U_{i+k}^2 - 2r^kU_iU_{i+k} + r^{2k}U_i^2}{2\left(1-r^{2k}\right)}\right]$$

$$= \frac{1}{2\pi\sqrt{1-r^{2k}}}\exp\left[\frac{U_i^2 + U_{i+k}^2 - 2r^kU_iU_{i+k}}{2\left(1-r^{2k}\right)}\right]. \tag{A13.29}$$

Statistical characteristics of standardized process

It is generally known that the mean value of standardized normal (Gaussian) process $\bar{U}=0$ and the standard deviation of this process $\sigma_U = 1$ – see Equations (10.86) and (10.88). In analogy with Equation (10.70), the covariance $\mathrm{cov}\left(U_i,U_{i+k}\right)$ (non-standardized correlation function) as well as the (standardized) correlation function $\rho\left(U_i,U_{i+k}\right)$ are

$$\mathrm{cov}\left(U_i,U_{i+k}\right) = \rho\left(U_i,U_{i+k}\right) = \int_{U_i=-\infty}^{\infty}\int_{U_{i+k}=-\infty}^{\infty} U_i U_{i+k} f\left(U_i,U_{i+k}\right)\mathrm{d}U_i\,\mathrm{d}U_{i+k} - \bar{U}^2$$

$$= \int_{U_i=-\infty}^{\infty}\int_{U_{i+k}=-\infty}^{\infty} U_i U_{i+k} f\left(U_i,U_{i+k}\right)\mathrm{d}U_i\,\mathrm{d}U_{i+k}. \tag{A13.30}^3$$

By using Equation (A13.28), we can write the following expression:

---

3   It is valid to write that $\rho\left(U_i,U_{i+k}\right) = \mathrm{cov}\left(U_i,U_{i+k}\right)\big/\sigma_U^2$; however, $\sigma_U = 1$ and simultaneously $\bar{U}=1$.

$$\mathrm{cov}(U_i, U_{i+k}) = \rho(U_i, U_{i+k}) = \int\limits_{U_i=-\infty}^{\infty} \int\limits_{U_{i+k}=-\infty}^{\infty} U_i U_{i+k} \frac{1}{\sqrt{2\pi}} \exp\left[-\frac{U_i^2}{2}\right] \frac{1}{\sqrt{2\pi}\sqrt{1-r^{2k}}}$$

$$\cdot \exp\left[-\frac{\left(U_{i+k}-r^k U_i\right)^2}{2\left(1-r^{2k}\right)}\right] dU_i\, dU_{i+k}$$

$$= \int\limits_{-\infty}^{\infty} U_i \frac{1}{\sqrt{2\pi}} \exp\left[-\frac{U_i^2}{2}\right] \left\{ \int\limits_{-\infty}^{\infty} \frac{U_{i+k}}{\sqrt{2\pi}\sqrt{1-r^{2k}}} \exp\left[-\frac{\left(U_{i+k}-r^k U_i\right)^2}{2\left(1-r^{2k}\right)}\right] dU_{i+k} \right\} dU_i$$

$$\text{Substitution: } V = \left(U_{i+k}-r^k U_i\right)/\sqrt{1-r^{2k}},$$

$$U_{i+k} = V\sqrt{1-r^{2k}} + r^k U_i, \quad dU_{i+k} = dV\sqrt{1-r^{2k}},$$

$$= \int\limits_{-\infty}^{\infty} U_i \frac{1}{\sqrt{2\pi}} \exp\left[-\frac{U_i^2}{2}\right] \left\{ \int\limits_{-\infty}^{\infty} \frac{V\sqrt{1-r^{2k}} + r^k U_i}{\sqrt{2\pi}} \exp\left[-\frac{V^2}{2}\right] dV \right\} dU_i$$

$$= \int\limits_{-\infty}^{\infty} U_i \frac{1}{\sqrt{2\pi}} \exp\left[-\frac{U_i^2}{2}\right] \left\{ \sqrt{1-r^{2k}} \overbrace{\int\limits_{-\infty}^{\infty} V \frac{1}{\sqrt{2\pi}} \exp\left[-\frac{V^2}{2}\right] dV}^{=0} \right.$$

$$\left. + r^k U_i \overbrace{\int\limits_{-\infty}^{\infty} \frac{1}{\sqrt{2\pi}} \exp\left[-\frac{V^2}{2}\right] dV}^{=1} \right\} dU_i$$

$$= r^k \int\limits_{-\infty}^{\infty} U_i^2 \frac{1}{\sqrt{2\pi}} \exp\left[-\frac{U_i^2}{2}\right] dU_i = r^k \overbrace{E\left(U_i^2\right)}^{=1},$$

$$\mathrm{cov}(U_i, U_{i+k}) = \rho(U_i, U_{i+k}) = r^k. \tag{A13.31}[4]$$

Statistical characteristics of general (non-standardized) process

The mean value $\bar{P}_{I_0}$, variance $\sigma_{I_0}^2$ and correlation coefficient r (correlation coefficient between the 'neighboring' quantities) describe a non-standardized SEMG process P_i. The corresponding covariance function and correlation function follow Equations (10.89) and (A13.31) as follows:

---

4   It is valid to write that
$$\sigma_U^2 = E\left(U_i^2\right) - \bar{U}^2, \quad E\left(U_i^2\right) = \sigma_U^2 + \bar{U}^2 = 1 + 0 = 1.$$

$$\text{cov}\left(P_i, P_{i+k}\right) = \sigma_{l_0}^2 \, \text{cov}\left(U_i, U_{i+k}\right) = \sigma_{l_0}^2 \, r^k .$$

(A13.32)

$$\rho\left(P_i, P_{i+k}\right) = \text{cov}\left(U_i, U_{i+k}\right) = r^k .$$

(A13.33)

Strength distribution of the general length by SEMG process

The probability that the general (long) length l of a yarn will be broken by a force P generally describes the distribution function of random variable P, i.e., $G(P,k)$ according to Equation (10.111) stated in Section 10.5. [The value k is a parameter related to the 'long' gauge length l according to Equation (10.108).] Moreover, the corresponding probability density function $g(P,k)$ is related to the distribution function $G(P,k)$, i.e.,

$$g(P,k) = -\frac{\mathrm{d}}{\mathrm{d}P}\left\{ \int_{P_i=P}^{P_{max}} \int_{P_{i+1}=P}^{P_{max}} \cdots \int_{P_{i+k}=P}^{P_{max}} f(P_i)\left[\prod_{j=1}^{k} \varphi\left(P_{i+j}\big|P_{i+j-1}\right) \right] \mathrm{d}P_i \mathrm{d}P_{i+1} \ldots \mathrm{d}P_{i+k} \right\}.$$

(A13.34)

This is generally expressed by Equation (10.112).

In SEMG stochastic process, Equations (A13.1) and (A.13.2) are valid to write, and $P_{min} = -\infty$, $P_{max} = \infty$. By using them in Equation (A13.34), we obtain the following expression:

$$g(P,k) = -\frac{\mathrm{d}}{\mathrm{d}P}\left\{ \int_{P_i=P}^{\infty} \int_{P_{i+1}=P}^{\infty} \cdots \int_{P_{i+k}=P}^{\infty} \frac{1}{\sqrt{2\pi}\sigma_{l_0}} \exp\left(-\frac{\left(P_i - \overline{P}_{l_0}\right)^2}{2\sigma_{l_0}^2} \right) \right.$$

$$\cdot \prod_{j=1}^{k} \left[\frac{1}{\sqrt{2\pi}\,\sigma_{l_0}\sqrt{1-r^2}} \exp\left(-\frac{\left(P_{i+j} - \left[\overline{P}_{l_0} + r\left(P_{i+j-1} - \overline{P}_{l_0}\right)\right]\right)^2}{2\sigma_{l_0}^2\left(1-r^2\right)} \right) \right] \mathrm{d}P_i \, \mathrm{d}P_{i+1} \ldots \mathrm{d}P_{i+k} \bigg\}$$

$$= -\frac{1}{\sqrt{2\pi}\sigma_{l_0}}\left(\frac{1}{\sqrt{2\pi}\,\sigma_{l_0}\sqrt{1-r^2}} \right)^k \frac{\mathrm{d}}{\mathrm{d}P}\left\{ \int_{P_i=P}^{\infty} \int_{P_{i+1}=P}^{\infty} \cdots \int_{P_{i+k}=P}^{\infty} \exp\left(-\frac{\left(P_i - \overline{P}_{l_0}\right)^2}{2\sigma_{l_0}^2} \right) \right.$$

$$\cdot \prod_{j=1}^{k} \left[\exp\left(-\frac{\left(P_{i+j} - \left[\overline{P}_{l_0} + r\left(P_{i+j-1} - \overline{P}_{l_0}\right)\right]\right)^2}{2\sigma_{l_0}^2\left(1-r^2\right)} \right) \right] \mathrm{d}P_i \, \mathrm{d}P_{i+1} \ldots \mathrm{d}P_{i+k} \bigg\}, \qquad k = 1,2,\ldots.$$

(A13.35)

Now, let us define the linearly transformed random quantity, according to Equation (10.134), as follows:

$$p = \frac{P - \overline{P}_{l_0}}{\sigma_{l_0}}, \quad \left(P = \sigma_{l_0} p + \overline{P}_{l_0} \right). \tag{A13.36}$$

[The probability density function of such linearly transformed random quantity p will be denoted by $g(p,k)$.]

Initially, let us solve the following 'helping' integral:

$$I = \int\limits_{P_i = P}^{\infty} \int\limits_{P_{i+1} = P}^{\infty} \cdots \int\limits_{P_{i+k} = P}^{\infty} \exp\left(-\frac{\left(P_i - \overline{P}_{l_0} \right)^2}{2\sigma_{l_0}^2} \right)$$

$$\cdot \prod_{j=1}^{k} \left[\exp\left(-\frac{\left(P_{i+j} - \left[\overline{P}_{l_0} + r\left(P_{i+j-1} - \overline{P}_{l_0} \right) \right] \right)^2}{2\sigma_{l_0}^2 \left(1 - r^2 \right)} \right) \right] dP_i \, dP_{i+1} \ldots dP_{i+k},$$

$$\tag{A13.37}$$

Substitutions: $U_{i+j} = \dfrac{P_{i+j} - \overline{P}_{l_0}}{\sigma_{l_0}}$ (standardized quantities), $P_{i+j} = \sigma_{l_0} U_{i+j} + \overline{P}_{l_0}$,

$$\partial P_{i+j} / \partial U_{i+j} = \sigma_{l_0}; \quad \text{if } j_1 \neq j_2 \text{ then } \partial P_{i+j_1} / \partial U_{i+j_2} = 0,$$

Jakobian determinant:

$$J = \begin{vmatrix} \dfrac{\partial P_i}{\partial U_i} & \dfrac{\partial P_i}{\partial U_{i+1}} & \cdots & \dfrac{\partial P_i}{\partial U_{i+k}} \\ \dfrac{\partial P_{i+1}}{\partial U_i} & \dfrac{\partial P_{i+1}}{\partial U_{i+1}} & \cdots & \dfrac{\partial P_{i+1}}{\partial U_{i+k}} \\ \vdots & \vdots & \ddots & \vdots \\ \dfrac{\partial P_{i+k}}{\partial U_i} & \dfrac{\partial P_{i+k}}{\partial U_{i+1}} & \cdots & \dfrac{\partial P_{i+k}}{\partial U_{i+k}} \end{vmatrix} = \begin{vmatrix} \sigma_{l_0} & 0 & \cdots & 0 \\ 0 & \sigma_{l_0} & \cdots & 0 \\ \vdots & \vdots & \ddots & \vdots \\ 0 & 0 & \cdots & \sigma_{l_0} \end{vmatrix} = \sigma_{l_0}^{k+1},$$

$$I = \int\limits_{U_i = p}^{\infty} \int\limits_{U_{i+1} = p}^{\infty} \cdots \int\limits_{U_{i+k} = p}^{\infty} \exp\left(-\frac{U_i^2}{2} \right)$$

$$\cdot \prod_{j=1}^{k} \left[\exp\left(-\frac{\left(\sigma_{l_0} U_{i+j} - r\sigma_{l_0} U_{i+j-1} \right)^2}{2\sigma_{l_0}^2 \left(1 - r^2 \right)} \right) \right] |J| \, dU_i \, dU_{i+1} \ldots dU_{i+k},$$

$$I = \sigma_{l_0}^{k+1} \int\limits_{U_i=p} \int\limits_{U_{i+1}=p} \cdots \int\limits_{U_{i+k}=p}^{\infty} \exp\left(-\frac{U_i^2}{2}\right)$$

$$\cdot \prod_{j=1}^{k} \left[\exp\left(-\frac{\left(U_{i+j} - rU_{i+j-1}\right)^2}{2\left(1-r^2\right)}\right) \right] dU_i \, dU_{i+1} \ldots dU_{i+k}.$$

(A13.38)

The derivative dI/dP, stated in Equation (A13.35), can also be expressed in the following form:

$$\frac{dI}{dP} = \frac{dI}{dp}\frac{dp}{dP}.$$

(A13.39)

Then, we can write the probability density function $g(P,k)$, expressed by Equation (A13.34), by using Equation (A13.37) as follows:

$$g(P,k) = -\frac{1}{\sqrt{2\pi}\sigma_{l_0}} \left(\frac{1}{\sqrt{2\pi}\,\sigma_{l_0}\sqrt{1-r^2}} \right)^k \frac{dI}{dP}$$

$$= -\frac{1}{\sqrt{2\pi}\sigma_{l_0}} \left(\frac{1}{\sqrt{2\pi}\,\sigma_{l_0}\sqrt{1-r^2}} \right)^k \frac{dI}{dp}\frac{dp}{dP}.$$

(A13.40)

On the basis of theory of probability, the following relation is valid between the probability density functions $g(P,k)$ and $g(p,k)$:

$$g(P,k) = g(p,k)\,dp/dP.$$

(A13.41)

By comparing the right-hand sides of the last two equations, we obtain

$$g(p,k)\frac{dp}{dP} = -\frac{1}{\sqrt{2\pi}\sigma_{l_0}} \left(\frac{1}{\sqrt{2\pi}\,\sigma_{l_0}\sqrt{1-r^2}} \right)^k \frac{dI}{dp}\frac{dp}{dP},$$

$$g(p,k) = -\frac{1}{\sqrt{2\pi}\sigma_{l_0}} \left(\frac{1}{\sqrt{2\pi}\,\sigma_{l_0}\sqrt{1-r^2}} \right)^k \frac{dI}{dp}$$

$$= -\frac{1}{\sqrt{2\pi}\sigma_{l_0}} \left(\frac{1}{\sqrt{2\pi}\,\sigma_{l_0}\sqrt{1-r^2}} \right)^k \frac{d}{dp}\left\{ \sigma_{l_0}^{k+1} \int\limits_{U_i=p} \int\limits_{U_{i+1}=p} \cdots \int\limits_{U_{i+k}=p}^{\infty} \exp\left(-\frac{U_i^2}{2}\right) \right.$$

$$\cdot \prod_{j=1}^{k}\left[\exp\left(-\frac{\left(U_{i+j}-rU_{i+j-1}\right)^{2}}{2\left(1-r^{2}\right)}\right)\right]\mathrm{d}U_{i}\,\mathrm{d}U_{i+1}\ldots\mathrm{d}U_{i+k}\Bigg\},$$

$$g(p,k)=-\frac{1}{\sqrt{2\pi}}\left(\frac{1}{\sqrt{2\pi}\sqrt{1-r^{2}}}\right)^{k}\frac{\mathrm{d}}{\mathrm{d}p}\Bigg\{\int\limits_{U_{i}=p}^{\infty}\int\limits_{U_{i+1}=p}^{\infty}\ldots\int\limits_{U_{i+k}=p}^{\infty}\exp\left(-\frac{U_{i}^{2}}{2}\right)$$

$$\cdot \prod_{j=1}^{k}\left[\exp\left(-\frac{\left(U_{i+j}-rU_{i+j-1}\right)^{2}}{2\left(1-r^{2}\right)}\right)\right]\mathrm{d}U_{i}\,\mathrm{d}U_{i+1}\ldots\mathrm{d}U_{i+k}\Bigg\}.$$

$$(\text{A13.42})$$

Fiber-to-fiber slippage – derivations of examples

Mathematical derivation of results of example 1

The derivation of fiber-to-fiber slippage follows Equations (11.62) to (11.68) under the conditions of Equations (11.96) to (11.99).

The borderline stress $\sigma_2^{FG}(\lambda)$, according to Equations (11.63) and (11.96), is

$$\sigma_2^{FG}(\lambda) = k \int b \, d\lambda + k \left[\int b(0-1) \, d\lambda \right]_{\lambda=\lambda_F} + C_I = kb\lambda - kb\lambda_F + C_I,$$

$$\sigma_2^{FG}(\lambda) = kb(\lambda - \lambda_F) + C_I. \tag{A14.1}$$

(The term kb is the angular coefficient of the straight line – see Figure 11.13.)

By using Equation (11.96) in Equation (11.66), we obtain

$$0 = C_{II} - C_I + k \int_{\lambda_F}^{\lambda_{jump}} b(0-1) \, d\lambda + k \int_{\lambda_{jump}}^{\lambda_G} b(0-1) \, d\lambda$$

$$= C_{II} - C_I - kb(\lambda_{jump} - \lambda_F) - kb(\lambda_G - \lambda_{jump}) = C_{II} - C_I - kb(\lambda_G - \lambda_F),$$

$$kb = \frac{C_{II} - C_I}{\lambda_G - \lambda_F} \quad \text{or} \quad \lambda_G = \frac{C_{II} - C_I}{kb} + \lambda_F. \tag{A14.2}$$

This is the first equation which determines the relation between λ_F and λ_G.

Note: By applying Equation (A14.2) in (A14.1), we can also write

$$\sigma_2^{FG}(\lambda) = \frac{\lambda - \lambda_F}{\lambda_G - \lambda_F}(C_{II} - C_I) + C_I. \tag{A14.3}$$

The first term stated on the right-hand side of Equation (11.68) can be rearranged by using Equations (11.97), (A14.1) and (A14.2) in the following manner:

$$\int_{\lambda_F}^{\lambda_G} \varepsilon\left[\sigma_2^{FG}(\lambda)\right]d\lambda = \int_{\lambda_F}^{\lambda_G} \frac{1}{E}\left[kb(\lambda - \lambda_F) + C_I\right]d\lambda$$

$$= \frac{kb}{E}\frac{\lambda_G^2 - \lambda_F^2}{2} - \frac{kb}{E}\lambda_F(\lambda_G - \lambda_F) + \frac{C_I}{E}(\lambda_G - \lambda_F)$$

$$= \frac{1}{E}\left(\frac{C_{II} - C_I}{\lambda_G - \lambda_F}\right)\frac{(\lambda_G - \lambda_F)(\lambda_G + \lambda_F)}{2}$$

$$- \frac{1}{E}\left(\frac{C_{II} - C_I}{\lambda_G - \lambda_F}\right)\lambda_F(\lambda_G - \lambda_F) + \frac{C_I}{E}(\lambda_G - \lambda_F)$$

$$= \frac{C_{II} - C_I}{E}\left[\frac{\lambda_G}{2} + \frac{\lambda_F}{2} - \lambda_F + \frac{C_I}{C_{II} - C_I}(\lambda_G - \lambda_F)\right]$$

$$= \frac{C_{II} - C_I}{E}\left[\frac{\lambda_G - \lambda_F}{2} + \frac{C_I}{C_{II} - C_I}(\lambda_G - \lambda_F)\right],$$

$$\int_{\lambda_F}^{\lambda_G} \varepsilon\left[\sigma_2^{FG}(\lambda)\right]d\lambda = \frac{C_{II} - C_I}{E}\left[\frac{\lambda_G - \lambda_F}{2} + \frac{C_I}{C_{II} - C_I}\lambda_G - \frac{C_I}{C_{II} - C_I}\lambda_F\right]. \quad (A14.4)$$

The second and third terms stated on the right-hand side of Equation (11.68) can be rearranged by using Equations (11.97) and (11.98) as follows:

$$-\int_{\lambda_F}^{\lambda_{jump}} \varepsilon\left[\sigma(\lambda < \lambda_{jump})\right]d\lambda - \int_{\lambda_{jump}}^{\lambda_G} \varepsilon\left[\sigma(\lambda > \lambda_{jump})\right]d\lambda =$$

$$= -\int_{\lambda_F}^{\lambda_{jump}} \frac{1}{E}C_I d\lambda - \int_{\lambda_{jump}}^{\lambda_G} \frac{1}{E}C_{II} d\lambda = -\frac{1}{E}\left[C_I(\lambda_{jump} - \lambda_F) + C_{II}(\lambda_G - \lambda_{jump})\right]$$

$$= -\frac{1}{E}\left[-C_I\lambda_F + C_{II}\lambda_G - \lambda_{jump}(C_{II} - C_I)\right]$$

$$= -\frac{C_{II} - C_I}{E}\left[-\frac{C_I}{C_{II} - C_I}\lambda_F + \frac{C_{II}}{C_{II} - C_I}\lambda_G - \lambda_{jump}\right]. \quad (A14.5)$$

By applying Equations (A14.4) and (A14.5) in (11.68), we obtain

$$0 = \frac{C_{II} - C_I}{E}\left[\frac{\lambda_G - \lambda_F}{2} + \frac{C_I}{C_{II} - C_I}\lambda_G - \frac{C_I}{C_{II} - C_I}\lambda_F\right]$$

$$- \frac{C_{II} - C_I}{E}\left[-\frac{C_I}{C_{II} - C_I}\lambda_F + \frac{C_{II}}{C_{II} - C_I}\lambda_G - \lambda_{jump}\right],$$

$$0 = \frac{\lambda_G - \lambda_F}{2} + \frac{C_I}{C_{II} - C_I}\lambda_G - \frac{C_I}{C_{II} - C_I}\lambda_F + \frac{C_I}{C_{II} - C_I}\lambda_F - \frac{C_{II}}{C_{II} - C_I}\lambda_G + \lambda_{jump},$$

$$0 = \frac{\lambda_G - \lambda_F}{2} - \lambda_G + \lambda_{jump}, \qquad \lambda_{jump} = \lambda_G - \frac{\lambda_G - \lambda_F}{2} = \frac{\lambda_F + \lambda_G}{2}.$$

$$(A14.6)$$

This is the second equation which determines the relation between λ_F and λ_G.

By substituting Equation (A14.2) in (A14.6), we find the coordinate λ_F as follows:

$$\lambda_{jump} = \frac{\lambda_F}{2} + \frac{\dfrac{C_{II} - C_I}{kb} + \lambda_F}{2} = \lambda_F + \frac{C_{II} - C_I}{2kb}, \qquad \lambda_F = \lambda_{jump} - \frac{C_{II} - C_I}{2kb}. \quad (A14.7)$$

The coordinate λ_G is obtained from Equations (A14.6) and (A14.7) as

$$2\lambda_{jump} = \lambda_F + \lambda_G = \left(\lambda_{jump} - \frac{C_{II} - C_I}{2kb}\right) + \lambda_G, \qquad \lambda_G = \lambda_{jump} + \frac{C_{II} - C_I}{2kb}.$$

$$(A14.8)$$

Note: The tensile stress after slippage at point X, i.e., when $\lambda = \lambda_{jump}$, follows Equation (A14.1) by using (A14.7) as

$$\sigma_2^{FG}\left(\lambda_{jump}\right) = kb\left(\lambda_{jump} - \lambda_F\right) + C_I$$

$$= kb\frac{C_{II} - C_I}{2kb} + C_I = \frac{C_{II} - C_I}{2} + C_I = \frac{C_I + C_{II}}{2}. \qquad (A14.9)$$

Mathematical derivation of results of example 2

The derivation of fiber-to-fiber slippage follows Equations (11.71) to (11.77) under the conditions of Equations (11.97) and (11.103) to (11.105).

The borderline stress $\sigma_1^{LM}(\lambda)$ is derived from Equation (11.72) by using (11.103) as follows:

$$\sigma_1^{LM}(\lambda) = -k\int b\,d\lambda + k\left[\int b(0+1)\,d\lambda\right]_{\lambda=\lambda_L} + C_{III} = -kb\lambda + kb\lambda_L + C_{III},$$

$$\sigma_1^{LM}(\lambda) = -kb(\lambda - \lambda_L) + C_{III}.$$

$$(A14.10)$$

(The term $-kb$ is the angular coefficient of this straight line – see Figure 11.14.)

By using Equation (11.103) in (11.75), we find

$$C_{III} - C_{IV} - k\int_{\lambda_L}^{\lambda_{jump}} b(0+1)\,d\lambda - k\int_{\lambda_{jump}}^{\lambda_M} b(0+1)\,d\lambda = 0,$$

$$C_{III} - C_{IV} - kb(\lambda_{jump} - \lambda_L) - kb(\lambda_M - \lambda_{jump}) = C_{III} - C_{IV} - kb(\lambda_M - \lambda_L) = 0,$$

$$kb = \frac{C_{III} - C_{IV}}{\lambda_M - \lambda_L}, \quad \text{or} \quad \lambda_M = \frac{C_{III} - C_{IV}}{kb} + \lambda_L. \qquad (A14.11)$$

This result represents the first expression that determines the relation between λ_L and λ_M.

Note: By applying Equation (A14.11) in (A14.10), we can also write

$$\sigma_1^{LM}(\lambda) = -\frac{\lambda - \lambda_L}{\lambda_M - \lambda_L}(C_{III} - C_{IV}) + C_{III}. \qquad (A14.12)$$

The first term on the right-hand side of Equation (11.77) can be rearranged by using Equations (11.97), (A14.10) and (A14.11). So, we obtain

$$\int_{\lambda_L}^{\lambda_M} \varepsilon\left[\sigma_1^{LM}(\lambda)\right]d\lambda = \int_{\lambda_L}^{\lambda_M} \frac{1}{E}\left[-kb(\lambda - \lambda_L) + C_{III}\right]d\lambda =$$

$$= -\frac{kb}{E}\frac{(\lambda_M^2 - \lambda_L^2)}{2} + \frac{kb}{E}\lambda_L(\lambda_M - \lambda_L) + \frac{C_{III}}{E}(\lambda_M - \lambda_L)$$

$$= -\frac{1}{E}\left(\frac{C_{III} - C_{IV}}{\lambda_M - \lambda_L}\right)\frac{(\lambda_M + \lambda_L)(\lambda_M - \lambda_L)}{2}$$

$$+ \frac{1}{E}\left(\frac{C_{III} - C_{IV}}{\lambda_M - \lambda_L}\right)\lambda_L(\lambda_M - \lambda_L) + \frac{C_{III}}{E}(\lambda_M - \lambda_L)$$

$$= -\frac{C_{\mathrm{III}} - C_{\mathrm{IV}}}{E} \frac{\lambda_{\mathrm{M}} + \lambda_{\mathrm{L}}}{2} + \frac{C_{\mathrm{III}} - C_{\mathrm{IV}}}{E} \lambda_{\mathrm{L}} + \frac{C_{\mathrm{III}}}{E} (\lambda_{\mathrm{M}} - \lambda_{\mathrm{L}})$$

$$= \frac{C_{\mathrm{III}} - C_{\mathrm{IV}}}{E} \left[-\frac{\lambda_{\mathrm{M}}}{2} - \frac{\lambda_{\mathrm{L}}}{2} + \lambda_{\mathrm{L}} + \frac{C_{\mathrm{III}}}{C_{\mathrm{III}} - C_{\mathrm{IV}}} \lambda_{\mathrm{M}} - \frac{C_{\mathrm{III}}}{C_{\mathrm{III}} - C_{\mathrm{IV}}} \lambda_{\mathrm{L}} \right],$$

$$\int_{\lambda_{\mathrm{L}}}^{\lambda_{\mathrm{M}}} \varepsilon \left[\sigma_{1}^{\mathrm{LM}} (\lambda) \right] \mathrm{d}\lambda = \frac{C_{\mathrm{III}} - C_{\mathrm{IV}}}{E} \left[-\frac{\lambda_{\mathrm{M}} - \lambda_{\mathrm{L}}}{2} + \frac{C_{\mathrm{III}}}{C_{\mathrm{III}} - C_{\mathrm{IV}}} \lambda_{\mathrm{M}} - \frac{C_{\mathrm{III}}}{C_{\mathrm{III}} - C_{\mathrm{IV}}} \lambda_{\mathrm{L}} \right].$$

$$(A14.13)$$

The second and third terms stated on the right-hand side of Equation (11.77) can be rearranged by using Equations (11.97) and (11.104) in the following manner:

$$-\int_{\lambda_{\mathrm{L}}}^{\lambda_{\mathrm{jump}}} \varepsilon \left[\sigma \left(\lambda < \lambda_{\mathrm{jump}} \right) \right] \mathrm{d}\lambda - \int_{\lambda_{\mathrm{jump}}}^{\lambda_{\mathrm{M}}} \varepsilon \left[\sigma \left(\lambda > \lambda_{\mathrm{jump}} \right) \right] \mathrm{d}\lambda$$

$$= -\int_{\lambda_{\mathrm{L}}}^{\lambda_{\mathrm{jump}}} \frac{1}{E} C_{\mathrm{III}} \mathrm{d}\lambda - \int_{\lambda_{\mathrm{jump}}}^{\lambda_{\mathrm{M}}} \frac{1}{E} C_{\mathrm{IV}} \, \mathrm{d}\lambda = -\frac{1}{E} \left[C_{\mathrm{III}} \left(\lambda_{\mathrm{jump}} - \lambda_{\mathrm{L}} \right) + C_{\mathrm{IV}} \left(\lambda_{\mathrm{M}} - \lambda_{\mathrm{jump}} \right) \right]$$

$$= -\frac{1}{E} \left[-C_{\mathrm{III}} \lambda_{\mathrm{L}} + C_{\mathrm{IV}} \lambda_{\mathrm{M}} + \lambda_{\mathrm{jump}} \left(C_{\mathrm{III}} - C_{\mathrm{IV}} \right) \right]$$

$$= -\frac{C_{\mathrm{III}} - C_{\mathrm{IV}}}{E} \left[-\frac{C_{\mathrm{III}}}{C_{\mathrm{III}} - C_{\mathrm{IV}}} \lambda_{\mathrm{L}} + \frac{C_{\mathrm{IV}}}{C_{\mathrm{III}} - C_{\mathrm{IV}}} \lambda_{\mathrm{M}} + \lambda_{\mathrm{jump}} \right]. \qquad (A14.14)$$

By using Equations (A14.13) and (A14.14) in (11.77), we find

$$0 = \frac{C_{\mathrm{III}} - C_{\mathrm{IV}}}{E} \left[-\frac{\lambda_{\mathrm{M}} - \lambda_{\mathrm{L}}}{2} + \frac{C_{\mathrm{III}}}{C_{\mathrm{III}} - C_{\mathrm{IV}}} \lambda_{\mathrm{M}} - \frac{C_{\mathrm{III}}}{C_{\mathrm{III}} - C_{\mathrm{IV}}} \lambda_{\mathrm{L}} \right]$$

$$- \frac{C_{\mathrm{III}} - C_{\mathrm{IV}}}{E} \left[-\frac{C_{\mathrm{III}}}{C_{\mathrm{III}} - C_{\mathrm{IV}}} \lambda_{\mathrm{L}} + \frac{C_{\mathrm{IV}}}{C_{\mathrm{III}} - C_{\mathrm{IV}}} \lambda_{\mathrm{M}} + \lambda_{\mathrm{jump}} \right],$$

$$0 = -\frac{\lambda_{\mathrm{M}} - \lambda_{\mathrm{L}}}{2} + \frac{C_{\mathrm{III}}}{C_{\mathrm{III}} - C_{\mathrm{IV}}} \lambda_{\mathrm{M}} - \frac{C_{\mathrm{III}}}{C_{\mathrm{III}} - C_{\mathrm{IV}}} \lambda_{\mathrm{L}} +$$

$$\frac{C_{\mathrm{III}}}{C_{\mathrm{III}} - C_{\mathrm{IV}}} \lambda_{\mathrm{L}} - \frac{C_{\mathrm{IV}}}{C_{\mathrm{III}} - C_{\mathrm{IV}}} \lambda_{\mathrm{M}} - \lambda_{\mathrm{jump}},$$

$$= -\frac{\lambda_{\mathrm{M}} - \lambda_{\mathrm{L}}}{2} + \lambda_{\mathrm{M}} - \lambda_{\mathrm{jump}}, \quad \lambda_{\mathrm{jump}} = \lambda_{\mathrm{M}} - \frac{\lambda_{\mathrm{M}} - \lambda_{\mathrm{L}}}{2} = \frac{\lambda_{\mathrm{L}} + \lambda_{\mathrm{M}}}{2}. \quad (A14.15)$$

This is the second equation that determines the relation between λ_L and λ_M.

By using Equations (A14.11) to (A14.15), we find the coordinate λ_L as

$$\lambda_{jump} = \frac{\lambda_L}{2} + \frac{\dfrac{C_{III} - C_{IV}}{kb} + \lambda_L}{2} = \lambda_L + \frac{C_{III} - C_{IV}}{2kb}, \quad \lambda_L = \lambda_{jump} - \frac{C_{III} - C_{IV}}{2kb}.$$

(A14.16)

By using the last expression in Equation (A14.15), we obtain the coordinate λ_M in the following manner:

$$2\lambda_{jump} = \lambda_L + \lambda_M = \lambda_{jump} - \frac{C_{III} - C_{IV}}{2kb} + \lambda_M, \quad \lambda_M = \lambda_{jump} + \frac{C_{III} - C_{IV}}{2kb}.$$

(A14.17)

Note: The tensile stress after slippage at point X, i.e., when $\lambda = \lambda_{jump}$, follows Equation (A14.10) by using (A14.16) as follows:

$$\sigma_1^{LM}\left(\lambda_{jump}\right) = -kb\left(\lambda_{jump} - \lambda_L\right) + C_{III} = -kb\frac{C_{III} - C_{IV}}{2kb} + C_{III} = \frac{C_{III} + C_{IV}}{2}.$$

(A14.18)

Mathematical derivation of results of example 3

The double-side increase–decrease 'jumps' is described by Equations (11.84) to (11.89) under the conditions of Equations (11.97) and (11.109) to (11.111).

The borderline curve $\sigma_2^{FG}(\lambda)$ – generally Equation (11.63) – and the borderline curve $\sigma_1^{LM}(\lambda)$ – generally Equation (11.76) – are solved for the 'correct' behaviour of tensile stress $\sigma(\lambda)$ after fiber-to-fiber slippage. By using Equation (11.109) in the above-mentioned general expressions, we obtain

$$\sigma_2^{FG}(\lambda) = k\int b\,d\lambda + k\left[\int b(0-1)\,d\lambda\right]_{\lambda=\lambda_F} + C_I = kb\lambda - kb\lambda_F + C_I. \quad \text{(A14.19)}$$

$$\sigma_1^{LM}(\lambda) = -k\int b\,d\lambda + k\left[\int b(0+1)\,d\lambda\right]_{\lambda=\lambda_M} + C_I = -kb\lambda + kb\lambda_M + C_I.$$

(A14.20)

(The terms kb and $-kb$ are the angular coefficients of the previous straight lines see Figure 11.15.)

Equation (11.87) with (11.109) determines the relation among $\lambda_F, \lambda_Z, \lambda_M$ – from intersection $\sigma_2^{FG}(\lambda_Z) = \sigma_1^{LM}(\lambda_Z)$ – which is shown as follows:

$$\int_{\lambda_F}^{\lambda_Z} b\,d\lambda - \int_{\lambda_Z}^{\lambda_M} b\,d\lambda + \frac{C_I - C_I}{k} = \int_{\lambda_F}^{\lambda_M} b\,0\,d\lambda, \quad b(\lambda_Z - \lambda_F) - b(\lambda_M - \lambda_Z) = 0,$$

$$\lambda_Z - \lambda_F = \lambda_M - \lambda_Z, \quad \lambda_Z = \frac{\lambda_M + \lambda_F}{2}. \tag{A14.21}$$

Equation (11.88) has the following form by using Equation (A14.19) in place of Equation (11.63) and by using Equation (11.110) in place of Equation (11.85).

$$\int_{\lambda_F}^{\lambda_Z} \varepsilon\left[\overbrace{\sigma_2^{FG}(\lambda)}^{\substack{\text{According to}\\\text{Eq. (A14.19)}}}\right]d\lambda - \int_{\lambda_F}^{\lambda_{jump,1}} \varepsilon\left[\overbrace{\sigma(\lambda)}^{\substack{\text{According to}\\\text{Eq. (11.110), (a)}}}\right]d\lambda - \int_{\lambda_{jump,1}}^{\lambda_Z} \varepsilon\left[\overbrace{\sigma(\lambda)}^{\substack{\text{According to}\\\text{Eq. (11.110), (b)}}}\right]d\lambda = 0.$$

$$\tag{A14.22}$$

Also, Equation (11.97) – Hooke's law – must be used now. The first term in the last equation can be expressed as follows:

$$\int_{\lambda_F}^{\lambda_Z} \varepsilon\left[\overbrace{\sigma_2^{FG}(\lambda)}^{\substack{\text{According to}\\\text{Eq. (A14.19)}}}\right]d\lambda = \int_{\lambda_F}^{\lambda_Z} \frac{1}{E}[kb\lambda - kb\lambda_F + C_I]\,d\lambda$$

$$= \frac{kb}{E}\frac{\lambda_Z^2 - \lambda_F^2}{2} - \frac{kb}{E}\lambda_F(\lambda_Z - \lambda_F) + \frac{C_I}{E}(\lambda_Z - \lambda_F)$$

$$= \frac{kb}{E}(\lambda_Z - \lambda_F)\left[\frac{\lambda_Z + \lambda_F}{2} - \lambda_F\right] + \frac{C_I}{E}(\lambda_Z - \lambda_F)$$

$$= \frac{kb}{2E}(\lambda_Z - \lambda_F)^2 + \frac{C_I}{E}\lambda_Z - \frac{C_I}{E}\lambda_F. \tag{A14.23}$$

The second and the third terms mentioned in Equation (A14.22) can be stated in the following manner:

$$-\int_{\lambda_F}^{\lambda_{jump,1}} \varepsilon\left[\overbrace{\sigma(\lambda)}^{\substack{\text{According to}\\\text{Eq. (11.110), (a)}}}\right]d\lambda - \int_{\lambda_{jump,1}}^{\lambda_Z} \varepsilon\left[\overbrace{\sigma(\lambda)}^{\substack{\text{According to}\\\text{Eq. (11.110), (b)}}}\right]d\lambda = -\int_{\lambda_F}^{\lambda_{jump,1}} \frac{1}{E}C_I\,d\lambda - \int_{\lambda_{jump,1}}^{\lambda_Z} \frac{1}{E}C_{QR}\,d\lambda$$

$$= -\frac{C_{\mathrm{I}}}{E} \lambda_{\mathrm{jump,1}} + \frac{C_{\mathrm{I}}}{E} \lambda_{\mathrm{F}} - \frac{C_{\mathrm{QR}}}{E} \lambda_{\mathrm{Z}} + \frac{C_{\mathrm{QR}}}{E} \lambda_{\mathrm{jump,1}}. \tag{A14.24}$$

By substituting last two expressions in Equation (A14.22), we obtain

$$\frac{kb}{2E}(\lambda_{\mathrm{Z}} - \lambda_{\mathrm{F}})^2 + \frac{C_{\mathrm{I}}}{E}\lambda_{\mathrm{Z}} - \frac{C_{\mathrm{I}}}{E}\lambda_{\mathrm{F}} - \frac{C_{\mathrm{I}}}{E}\lambda_{\mathrm{jump,1}} + \frac{C_{\mathrm{I}}}{E}\lambda_{\mathrm{F}} - \frac{C_{\mathrm{QR}}}{E}\lambda_{\mathrm{Z}} + \frac{C_{\mathrm{QR}}}{E}\lambda_{\mathrm{jump,1}} = 0,$$

$$\frac{kb}{2}(\lambda_{\mathrm{Z}} - \lambda_{\mathrm{F}})^2 + C_{\mathrm{I}}\lambda_{\mathrm{Z}} - C_{\mathrm{I}}\lambda_{\mathrm{jump,1}} - C_{\mathrm{QR}}\lambda_{\mathrm{Z}} + C_{\mathrm{QR}}\lambda_{\mathrm{jump,1}} = 0,$$

$$\frac{kb}{2}(\lambda_{\mathrm{Z}} - \lambda_{\mathrm{F}})^2 - \left(C_{\mathrm{QR}} - C_{\mathrm{I}}\right)(\lambda_{\mathrm{Z}} - \lambda_{\mathrm{jump,1}}) = 0, \tag{A14.25}$$

$$\frac{kb}{2}(\lambda_{\mathrm{Z}} - \lambda_{\mathrm{F}}) - \left(C_{\mathrm{QR}} - C_{\mathrm{I}}\right)\frac{\lambda_{\mathrm{Z}} - \lambda_{\mathrm{jump,1}}}{\lambda_{\mathrm{Z}} - \lambda_{\mathrm{F}}} = 0. \tag{A14.26}$$

This is the relation between λ_{F} and λ_{Z}.

Equation (11.89) has the following form by using (A14.20) in place of (11.76) and by using Equation (11.110) in place of (11.85):

$$-\int_{\lambda_{\mathrm{Z}}}^{\lambda_{\mathrm{M}}} \varepsilon \left[\overbrace{\sigma_1^{\mathrm{LM}}(\lambda)}^{\substack{\text{According to}\\ \text{Eq. (A14.20)}}} \right] \mathrm{d}\lambda + \int_{\lambda_{\mathrm{Z}}}^{\lambda_{\mathrm{jump,2}}} \varepsilon \left[\overbrace{\sigma(\lambda)}^{\substack{\text{According to}\\ \text{Eq. (11.110), (b)}}} \right] \mathrm{d}\lambda + \int_{\lambda_{\mathrm{jump,2}}}^{\lambda_{\mathrm{M}}} \varepsilon \left[\overbrace{\sigma(\lambda)}^{\substack{\text{According to}\\ \text{Eq. (11.110), (c)}}} \right] \mathrm{d}\lambda = 0.$$

$$\tag{A14.27}$$

Also, Equation (11.97) – Hooke's law – is still valid now. The first term in the last equation is

$$-\int_{\lambda_{\mathrm{Z}}}^{\lambda_{\mathrm{M}}} \varepsilon \left[\overbrace{\sigma_1^{\mathrm{LM}}(\lambda)}^{\substack{\text{According to}\\ \text{Eq. (A14.20)}}} \right] \mathrm{d}\lambda = -\int_{\lambda_{\mathrm{Z}}}^{\lambda_{\mathrm{M}}} \frac{1}{E}[-kb\lambda + kb\lambda_{\mathrm{M}} + C_{\mathrm{I}}]\mathrm{d}\lambda$$

$$= \frac{kb}{E}\frac{\lambda_{\mathrm{M}}^2 - \lambda_{\mathrm{Z}}^2}{2} - \frac{kb}{E}\lambda_{\mathrm{M}}(\lambda_{\mathrm{M}} - \lambda_{\mathrm{Z}}) - \frac{C_{\mathrm{I}}}{E}(\lambda_{\mathrm{M}} - \lambda_{\mathrm{Z}})$$

$$= \frac{kb}{E}(\lambda_{\mathrm{M}} - \lambda_{\mathrm{Z}})\left[\frac{\lambda_{\mathrm{M}} + \lambda_{\mathrm{Z}}}{2} - \lambda_{\mathrm{M}}\right] - \frac{C_{\mathrm{I}}}{E}(\lambda_{\mathrm{M}} - \lambda_{\mathrm{Z}})$$

$$= \frac{kb}{E}(\lambda_{\mathrm{M}} - \lambda_{\mathrm{Z}})\left[-\frac{\lambda_{\mathrm{M}} - \lambda_{\mathrm{Z}}}{2}\right] - \frac{C_{\mathrm{I}}}{E}(\lambda_{\mathrm{M}} - \lambda_{\mathrm{Z}})$$

$$= -\frac{kb}{2E}(\lambda_{\mathrm{M}} - \lambda_{\mathrm{Z}})^2 - \frac{C_{\mathrm{I}}}{E}\lambda_{\mathrm{M}} + \frac{C_{\mathrm{I}}}{E}\lambda_{\mathrm{Z}}. \tag{A14.28}$$

The second and the third terms stated in Equation (A14.27) can be expressed as follows:

$$\int_{\lambda_Z}^{\lambda_{\text{jump,2}}} \varepsilon \left[\overbrace{\sigma(\lambda)}^{\substack{\text{According to} \\ \text{Eq. (11.110), (b)}}} \right] d\lambda + \int_{\lambda_{\text{jump,2}}}^{\lambda_M} \varepsilon \left[\overbrace{\sigma(\lambda)}^{\substack{\text{According to} \\ \text{Eq. (11.110), (c)}}} \right] d\lambda = \int_{\lambda_Z}^{\lambda_{\text{jump,2}}} \frac{1}{E} C_{\text{QR}} \, d\lambda + \int_{\lambda_{\text{jump,2}}}^{\lambda_M} \frac{1}{E} C_{\text{I}} \, d\lambda$$

$$= \frac{C_{\text{QR}}}{E} \lambda_{\text{jump,2}} - \frac{C_{\text{QR}}}{E} \lambda_Z + \frac{C_{\text{I}}}{E} \lambda_M - \frac{C_{\text{I}}}{E} \lambda_{\text{jump,2}}. \tag{A14.29}$$

By substituting the last two expressions in Equation (A14.27), we obtain

$$-\frac{kb}{2E}(\lambda_M - \lambda_Z)^2 - \frac{C_{\text{I}}}{E} \lambda_M + \frac{C_{\text{I}}}{E} \lambda_Z + \frac{C_{\text{QR}}}{E} \lambda_{\text{jump,2}}$$

$$-\frac{C_{\text{QR}}}{E} \lambda_Z + \frac{C_{\text{I}}}{E} \lambda_M - \frac{C_{\text{I}}}{E} \lambda_{\text{jump,2}} = 0,$$

$$-\frac{kb}{2}(\lambda_M - \lambda_Z)^2 + C_{\text{I}}\lambda_Z + C_{\text{QR}}\lambda_{\text{jump,2}} - C_{\text{QR}}\lambda_Z - C_{\text{I}}\lambda_{\text{jump,2}} = 0, \tag{A14.30}$$

$$-\frac{kb}{2}(\lambda_M - \lambda_Z)^2 + (C_{\text{QR}} - C_{\text{I}})(\lambda_{\text{jump,2}} - \lambda_Z) = 0,$$

$$\frac{kb}{2}(\lambda_M - \lambda_Z) - (C_{\text{QR}} - C_{\text{I}}) \frac{\lambda_{\text{jump,2}} - \lambda_Z}{\lambda_M - \lambda_Z} = 0. \tag{A14.31}$$

This is the relation between λ_Z and λ_M.

Both expressions stated on the left-hand sides of Equations (A14.26) and (A14.31) are equal to zero so that the following expression is obtained by using Equation (A14.21):

$$\frac{kb}{2}(\lambda_Z - \lambda_F) - (C_{\text{QR}} - C_{\text{I}}) \frac{\lambda_Z - \lambda_{\text{jump,1}}}{\lambda_Z - \lambda_F} = \frac{kb}{2}(\lambda_M - \lambda_Z) - (C_{\text{QR}} - C_{\text{I}}) \frac{\lambda_{\text{jump,2}} - \lambda_Z}{\lambda_M - \lambda_Z},$$

$$\frac{kb}{2}(\lambda_M - \lambda_Z) - (C_{\text{QR}} - C_{\text{I}}) \frac{\lambda_Z - \lambda_{\text{jump,1}}}{\lambda_M - \lambda_Z} = \frac{kb}{2}(\lambda_M - \lambda_Z) - (C_{\text{QR}} - C_{\text{I}}) \frac{\lambda_{\text{jump,2}} - \lambda_Z}{\lambda_M - \lambda_Z},$$

$$\lambda_Z - \lambda_{\text{jump,1}} = \lambda_{\text{jump,2}} - \lambda_Z, \qquad \lambda_Z = \frac{\lambda_{\text{jump,1}} + \lambda_{\text{jump,2}}}{2}.$$

$$\tag{A14.32}$$

Rearranging Equation (A14.25) by using the last equation, we can write

$$\frac{kb}{2}(\lambda_Z - \lambda_F)^2 = (C_{\text{QR}} - C_{\text{I}})(\lambda_Z - \lambda_{\text{jump,1}}),$$

$$\lambda_F = \lambda_Z - \sqrt{\frac{2}{kb}(C_{QR} - C_I)(\lambda_Z - \lambda_{jump,1})}$$

$$= \frac{\lambda_{jump,1} + \lambda_{jump,2}}{2} - \sqrt{\frac{2}{kb}(C_{QR} - C_I)\left(\frac{\lambda_{jump,1} + \lambda_{jump,2}}{2} - \lambda_{jump,1}\right)},$$

$$\lambda_F = \frac{\lambda_{jump,1} + \lambda_{jump,2}}{2} - \sqrt{\frac{(C_{QR} - C_I)}{kb}(\lambda_{jump,2} - \lambda_{jump,1})}. \tag{A14.33}$$

Further, we express the coordinate λ_F from Equation (A14.21) and rearrange it by using Equations (A14.32) and (A14.33) in the following manner:

$$\lambda_M = 2\lambda_Z - \lambda_F$$

$$= 2\frac{\lambda_{jump,1} + \lambda_{jump,2}}{2} - \left[\frac{\lambda_{jump,1} + \lambda_{jump,2}}{2} - \sqrt{\frac{(C_{QR} - C_I)}{kb}(\lambda_{jump,2} - \lambda_{jump,1})}\right],$$

$$\lambda_M = \frac{\lambda_{jump,1} + \lambda_{jump,2}}{2} + \sqrt{\frac{(C_{QR} - C_I)}{kb}(\lambda_{jump,2} - \lambda_{jump,1})}. \tag{A14.34}$$

The tensile stress after fiber-to-fiber slippage at point Z of intersection is given according to Equation (A14.19) and/or (A14.20) when $\lambda = \lambda_Z$. We apply Equations (A14.21) and (A14.33) in Equation (A14.19) and obtain

$$\sigma_2^{FG}(\lambda_Z) = kb\lambda_Z - kb\lambda_F + C_I$$

$$= kb\left[\frac{\lambda_{jump,1} + \lambda_{jump,2}}{2} - \frac{\lambda_{jump,1} + \lambda_{jump,2}}{2} + \sqrt{\frac{(C_{QR} - C_I)}{kb}(\lambda_{jump,2} - \lambda_{jump,1})}\right] + C_I,$$

$$\sigma_2^{FG}(\lambda_Z) = kb\sqrt{\frac{(C_{QR} - C_I)}{kb}(\lambda_{jump,2} - \lambda_{jump,1})} + C_I \tag{A14.35}$$

$$= \sqrt{kb(C_{QR} - C_I)(\lambda_{jump,2} - \lambda_{jump,1})} + C_I.$$

The inequality $\sigma_2^{FG}(\lambda_Z) \leq C_{QR}$ can be rearranged by using the last equation in the following manner:

$$\sigma_2^{FG}\left(\lambda_Z\right)=\sqrt{kb\left(C_{QR}-C_I\right)\left(\lambda_{jump,2}-\lambda_{jump,1}\right)}+C_I\le C_{QR},$$

$$\sqrt{kb\left(C_{QR}-C_I\right)\left(\lambda_{jump,2}-\lambda_{jump,1}\right)}\le\left(C_{QR}-C_I\right),$$

$$\sqrt{kb\left(\lambda_{jump,2}-\lambda_{jump,1}\right)}\le\sqrt{C_{QR}-C_I},$$

$$kb\left(\lambda_{jump,2}-\lambda_{jump,1}\right)\le\left(C_{QR}-C_I\right)\ \text{and/or}\ kb\left(\lambda_{jump,2}-\lambda_{jump,1}\right)+C_I\le C_{QR}.$$

(A14.36)

Mathematical derivation of results of example 4

The double-side decrease–increase 'jumps' is described by Equations (11.90) to (11.95) under the conditions of Equations (11.97) and (11.122) to (11.124).

Note: The present derivations of all the required equations are so much similar to the derivations shown in the previous example 3 that we can use a shorter style of interpretation.

By applying Equation (11.122) in Equations (11.72) and (11.67), we obtain

$$\sigma_1^{LM}\left(\lambda\right)=-k\int b\,d\lambda+k\left[\int b\left(0+1\right)d\lambda\right]_{\lambda=\lambda_L}+C_{III}=-kb\lambda+kb\lambda_L+C_{III}.$$

(A14.37)

$$\sigma_2^{FG}\left(\lambda\right)=k\int b\,d\lambda+k\left[\int b\left(0-1\right)d\lambda\right]_{\lambda=\lambda_G}+C_{III}=kb\lambda-kb\lambda_G+C_{III}.$$

(A14.38)[1]

It is valid that $\sigma_1^{LM}\left(\lambda_Z\right)=\sigma_2^{FG}\left(\lambda_Z\right)$ at the point Z of intersection. Using Equation (11.122) in (11.93), we find

$$-\int_{\lambda_L}^{\lambda_Z}b\,d\lambda+\int_{\lambda_Z}^{\lambda_G}b\,d\lambda,+\frac{C_{III}-C_{III}}{k}=\int_{\lambda_L}^{\lambda_G}b\,0\,d\lambda,\ \ -b\left(\lambda_Z-\lambda_L\right)+b\left(\lambda_G-\lambda_Z\right)=0,$$

$$\lambda_G-\lambda_Z=\lambda_Z-\lambda_L,\ \text{and/or}\ \lambda_Z=\frac{\lambda_L+\lambda_G}{2}.$$

(A14.39)

---

1   We prefer the behaviour (shape) of Equation (11.67) to that of Equation (11.63) for this derivation.

Equation (11.94) takes the following form by using Equation (A14.37) in place of (11.72) and by using Equation (11.123) in place of (11.91):

$$
\int_{\lambda_L}^{\lambda_Z} \varepsilon \left[\overbrace{\sigma_1^{LM}(\lambda)}^{\substack{\text{According to} \\ \text{Eq. (A14.37)}}} \right] d\lambda - \int_{\lambda_L}^{\lambda_{jump,2}} \varepsilon \left[\overbrace{\sigma(\lambda)}^{\substack{\text{According to} \\ \text{Eq. (11.123), (a)}}} \right] d\lambda - \int_{\lambda_{jump,2}}^{\lambda_Z} \varepsilon \left[\overbrace{\sigma(\lambda)}^{\substack{\text{According to} \\ \text{Eq. (11.123), (b)}}} \right] d\lambda = 0 \ .
$$

$$(A14.40)$$

Let us note that Equation (11.97) – Hooke's law with Young modulus E – is still valid.

The integrals shown on the left-hand side of the previous equation are

$$
\int_{\lambda_L}^{\lambda_Z} \varepsilon \left[\overbrace{\sigma_1^{LM}(\lambda)}^{\substack{\text{According to} \\ \text{Eq. (A14.37)}}} \right] d\lambda = \int_{\lambda_L}^{\lambda_Z} \frac{1}{E} \left[-kb\lambda + kb\lambda_L + C_{III} \right] d\lambda
$$

$$
= -\frac{kb}{E} \frac{\lambda_Z^2 - \lambda_L^2}{2} + \frac{kb}{E} \lambda_L (\lambda_Z - \lambda_L) + \frac{C_{III}}{E} (\lambda_Z - \lambda_L)
$$

$$
= -\frac{kb}{E} (\lambda_Z - \lambda_L) \left[\frac{\lambda_Z + \lambda_L}{2} - \lambda_L \right] + \frac{C_{III}}{E} (\lambda_Z - \lambda_L)
$$

$$
= -\frac{kb}{2E} (\lambda_Z - \lambda_L)^2 + \frac{C_{III}}{E} \lambda_Z - \frac{C_{III}}{E} \lambda_L,
$$

$$
-\int_{\lambda_L}^{\lambda_{jump,2}} \varepsilon \left[\overbrace{\sigma(\lambda)}^{\substack{\text{According to} \\ \text{Eq. (11.123), (a)}}} \right] d\lambda - \int_{\lambda_{jump,2}}^{\lambda_Z} \varepsilon \left[\overbrace{\sigma(\lambda)}^{\substack{\text{According to} \\ \text{Eq. (11.123), (b)}}} \right] d\lambda = -\int_{\lambda_L}^{\lambda_{jump,2}} \frac{1}{E} C_{III} \, d\lambda - \int_{\lambda_{jump,2}}^{\lambda_Z} \frac{1}{E} C_{SP} \, d\lambda
$$

$$
= -\frac{C_{III}}{E} \lambda_{jump,2} + \frac{C_{III}}{E} \lambda_L - \frac{C_{SP}}{E} \lambda_Z + \frac{C_{SP}}{E} \lambda_{jump,2} .
$$

$$(A14.41)$$

By applying Equations (A14.41) in (A14.40), we find

$$
-\frac{kb}{2E} (\lambda_Z - \lambda_L)^2 + \frac{C_{III}}{E} \lambda_Z - \frac{C_{III}}{E} \lambda_L - \frac{C_{III}}{E} \lambda_{jump,2}
$$

$$
+ \frac{C_{III}}{E} \lambda_L - \frac{C_{SP}}{E} \lambda_Z + \frac{C_{SP}}{E} \lambda_{jump,2} = 0,
$$

$$
-\frac{kb}{2} (\lambda_Z - \lambda_L)^2 + (C_{III} - C_{SP})(\lambda_Z - \lambda_{jump,2}) = 0, \qquad (A14.42)
$$

$$
-\frac{kb}{2} (\lambda_Z - \lambda_L) + (C_{III} - C_{SP}) \frac{\lambda_Z - \lambda_{jump,2}}{\lambda_Z - \lambda_L} = 0 \qquad (A14.43)
$$

Equation (11.95) takes the following form by using Equation (A14.38) in place of (11.67) and by using Equation (11.123) in place of (11.91):

$$-\int_{\lambda_Z}^{\lambda_G}\varepsilon\left[\overbrace{\sigma_2^{FG}(\lambda)}^{\substack{\text{According to}\\ \text{Eq. (A14.38)}}}\right]d\lambda+\int_{\lambda_Z}^{\lambda\text{-jump},1}\varepsilon\left[\overbrace{\sigma(\lambda)}^{\substack{\text{According to}\\ \text{Eq. (11.123), (b)}}}\right]d\lambda+\int_{\lambda\text{-jump},1}^{\lambda_G}\varepsilon\left[\overbrace{\sigma(\lambda)}^{\substack{\text{According to}\\ \text{Eq. (11.123), (c)}}}\right]d\lambda=0\,.$$

$$(A14.44)$$

Let us note that Equation (11.97) – Hooke's law with Young modulus E – is still valid.

The integrals mentioned on the left-hand side of the last expression are

$$-\int_{\lambda_Z}^{\lambda_G}\varepsilon\left[\overbrace{\sigma_2^{FG}(\lambda)}^{\substack{\text{According to}\\ \text{Eq. (A14.38)}}}\right]d\lambda=-\int_{\lambda_Z}^{\lambda_G}\frac{1}{E}\left[kb\lambda-kb\lambda_G+C_{III}\right]d\lambda$$

$$=-\frac{kb}{E}\frac{\lambda_G^2-\lambda_Z^2}{2}+\frac{kb}{E}\lambda_G\left(\lambda_G-\lambda_Z\right)-\frac{C_{III}}{E}\left(\lambda_G-\lambda_Z\right)$$

$$=-\frac{kb}{E}\left(\lambda_G-\lambda_Z\right)\left[\frac{\lambda_G+\lambda_Z}{2}-\lambda_G\right]-\frac{C_{III}}{E}\left(\lambda_G-\lambda_Z\right)$$

$$=\frac{kb}{2E}\left(\lambda_G-\lambda_Z\right)^2-\frac{C_{III}}{E}\lambda_G+\frac{C_{III}}{E}\lambda_Z,$$

$$\int_{\lambda_Z}^{\lambda\text{-jump},1}\varepsilon\left[\overbrace{\sigma(\lambda)}^{\substack{\text{According to}\\ \text{Eq. (11.123), (b)}}}\right]d\lambda+\int_{\lambda\text{-jump},1}^{\lambda_G}\varepsilon\left[\overbrace{\sigma(\lambda)}^{\substack{\text{According to}\\ \text{Eq. (11.123), (c)}}}\right]d\lambda=\int_{\lambda_Z}^{\lambda\text{-jump},1}\frac{C_{SP}}{E}d\lambda+\int_{\lambda\text{-jump},1}^{\lambda_G}\frac{C_{III}}{E}d\lambda$$

$$=\frac{C_{SP}}{E}\lambda_{\text{jump},1}-\frac{C_{SP}}{E}\lambda_Z+\frac{C_{III}}{E}\lambda_G-\frac{C_{III}}{E}\lambda_{\text{jump},1}.$$

$$(A14.45)$$

By applying the integrals from Equations (A14.45) to (A14.44), we find

$$\frac{kb}{2E}\left(\lambda_G-\lambda_Z\right)^2-\frac{C_{III}}{E}\lambda_G+\frac{C_{III}}{E}\lambda_Z+\frac{C_{SP}}{E}\lambda_{\text{jump},1}$$

$$-\frac{C_{SP}}{E}\lambda_Z+\frac{C_{III}}{E}\lambda_G-\frac{C_{III}}{E}\lambda_{\text{jump},1}=0,$$

$$\frac{kb}{2}\left(\lambda_G-\lambda_Z\right)^2-\left(C_{III}-C_{SP}\right)\left(\lambda_{\text{jump},1}-\lambda_Z\right)=0. \qquad (A14.46)$$

$$\frac{kb}{2}\left(\lambda_G-\lambda_Z\right)-\left(C_{III}-C_{SP}\right)\frac{\lambda_{\text{jump},1}-\lambda_Z}{\lambda_G-\lambda_Z}=0\,. \qquad (A14.47)$$

Summation of Equations (A14.43) and (A14.47) by using (A14.39) leads to

$$-\frac{kb}{2}\underbrace{\left(\lambda_{\mathrm{Z}}-\lambda_{\mathrm{L}}\right)}_{=\lambda_{\mathrm{G}}-\lambda_{\mathrm{Z}}}+\left(C_{\mathrm{III}}-C_{\mathrm{SP}}\right)\frac{\lambda_{\mathrm{Z}}-\lambda_{\mathrm{jump},2}}{\underbrace{\lambda_{\mathrm{Z}}-\lambda_{\mathrm{L}}}_{=\lambda_{\mathrm{G}}-\lambda_{\mathrm{Z}}}}+$$

$$\frac{kb}{2}\left(\lambda_{\mathrm{G}}-\lambda_{\mathrm{Z}}\right)-\left(C_{\mathrm{III}}-C_{\mathrm{SP}}\right)\frac{\lambda_{\mathrm{jump},1}-\lambda_{\mathrm{Z}}}{\lambda_{\mathrm{G}}-\lambda_{\mathrm{Z}}}=0,$$

$$\lambda_{\mathrm{Z}}-\lambda_{\mathrm{jump},2}=\lambda_{\mathrm{jump},1}-\lambda_{\mathrm{Z}},\quad\lambda_{\mathrm{Z}}=\frac{\lambda_{\mathrm{jump},2}+\lambda_{\mathrm{jump},1}}{2}. \tag{A14.48}$$

Rearranging Equation (A14.42) by using the last expression, we find

$$\frac{kb}{2}\left(\lambda_{\mathrm{Z}}-\lambda_{\mathrm{L}}\right)^{2}=\left(C_{\mathrm{III}}-C_{\mathrm{SP}}\right)\left(\lambda_{\mathrm{Z}}-\lambda_{\mathrm{jump},2}\right),$$

$$\lambda_{\mathrm{L}}=\lambda_{\mathrm{Z}}-\sqrt{\frac{2}{kb}\left(C_{\mathrm{III}}-C_{\mathrm{SP}}\right)\left(\lambda_{\mathrm{Z}}-\lambda_{\mathrm{jump},2}\right)},$$

$$\lambda_{\mathrm{L}}=\frac{\lambda_{\mathrm{jump},2}+\lambda_{\mathrm{jump},1}}{2}-\sqrt{\frac{2}{kb}\left(C_{\mathrm{III}}-C_{\mathrm{SP}}\right)\left(\frac{\lambda_{\mathrm{jump},2}+\lambda_{\mathrm{jump},1}}{2}-\lambda_{\mathrm{jump},2}\right)},$$

$$\lambda_{\mathrm{L}}=\frac{\lambda_{\mathrm{jump},2}+\lambda_{\mathrm{jump},1}}{2}-\sqrt{\frac{C_{\mathrm{III}}-C_{\mathrm{SP}}}{kb}\left(\lambda_{\mathrm{jump},1}-\lambda_{\mathrm{jump},2}\right)}. \tag{A14.49}$$

Further, rearranging Equation (A14.46) by using (A14.48), we get

$$\frac{kb}{2}\left(\lambda_{\mathrm{G}}-\lambda_{\mathrm{Z}}\right)^{2}=\left(C_{\mathrm{III}}-C_{\mathrm{SP}}\right)\left(\lambda_{\mathrm{jump},1}-\lambda_{\mathrm{Z}}\right),$$

$$\lambda_{\mathrm{G}}=\lambda_{\mathrm{Z}}+\sqrt{\frac{2}{kb}\left(C_{\mathrm{III}}-C_{\mathrm{SP}}\right)\left(\lambda_{\mathrm{jump},1}-\lambda_{\mathrm{Z}}\right)},$$

$$\lambda_{\mathrm{G}}=\frac{\lambda_{\mathrm{jump},2}+\lambda_{\mathrm{jump},1}}{2}+\sqrt{\frac{2}{kb}\left(C_{\mathrm{III}}-C_{\mathrm{SP}}\right)\left(\lambda_{\mathrm{jump},1}-\frac{\lambda_{\mathrm{jump},2}+\lambda_{\mathrm{jump},1}}{2}\right)},$$

$$\lambda_{\mathrm{G}}=\frac{\lambda_{\mathrm{jump},2}+\lambda_{\mathrm{jump},1}}{2}+\sqrt{\frac{C_{\mathrm{III}}-C_{\mathrm{SP}}}{kb}\left(\lambda_{\mathrm{jump},1}-\lambda_{\mathrm{jump},2}\right)}. \tag{A14.50}$$

The tensile stress after fiber-to-fiber slippage at point Z of intersection is given according to Equation (A14.37) and/or (A14.38) when $\lambda=\lambda_{\mathrm{Z}}$. By applying Equations (A14.48) and (A14.49) to (A14.37), we obtain

$$\sigma_1^{LM}\left(\lambda_Z\right) = -kb\lambda_Z + kb\lambda_L + C_{III} = kb\left[-\lambda_Z + \lambda_L\right] + C_{III}$$

$$= kb\left[-\frac{\lambda_{jump,2} + \lambda_{jump,1}}{2} + \frac{\lambda_{jump,2} + \lambda_{jump,1}}{2} - \sqrt{\frac{C_{III} - C_{SP}}{kb}}\left(\lambda_{jump,1} - \lambda_{jump,2}\right)\right] + C_{III}$$

$$= -kb\sqrt{\frac{C_{III} - C_{SP}}{kb}}\left(\lambda_{jump,1} - \lambda_{jump,2}\right) + C_{III}$$

$$= C_{III} - \sqrt{kb\left(C_{III} - C_{SP}\right)\left(\lambda_{jump,1} - \lambda_{jump,2}\right)}. \qquad \text{(A14.51)}$$

Finally, the inequality $\sigma_1^{LM}\left(\lambda_Z\right) \geq C_{SP}$ can be rearranged as follows:

$$C_{III} - \sqrt{kb\left(C_{III} - C_{SP}\right)\left(\lambda_{jump,1} - \lambda_{jump,2}\right)} \geq C_{SP},$$

$$\left(C_{III} - C_{SP}\right) \geq \sqrt{kb\left(C_{III} - C_{SP}\right)\left(\lambda_{jump,1} - \lambda_{jump,2}\right)},$$

$$\sqrt{C_{III} - C_{SP}} \geq \sqrt{kb\left(\lambda_{jump,1} - \lambda_{jump,2}\right)},$$

$$\left(C_{III} - C_{SP}\right) \geq kb\left(\lambda_{jump,1} - \lambda_{jump,2}\right), \quad C_{III} - kb\left(\lambda_{jump,1} - \lambda_{jump,2}\right) \geq C_{SP}.$$

$$\text{(A14.52)}$$

Index